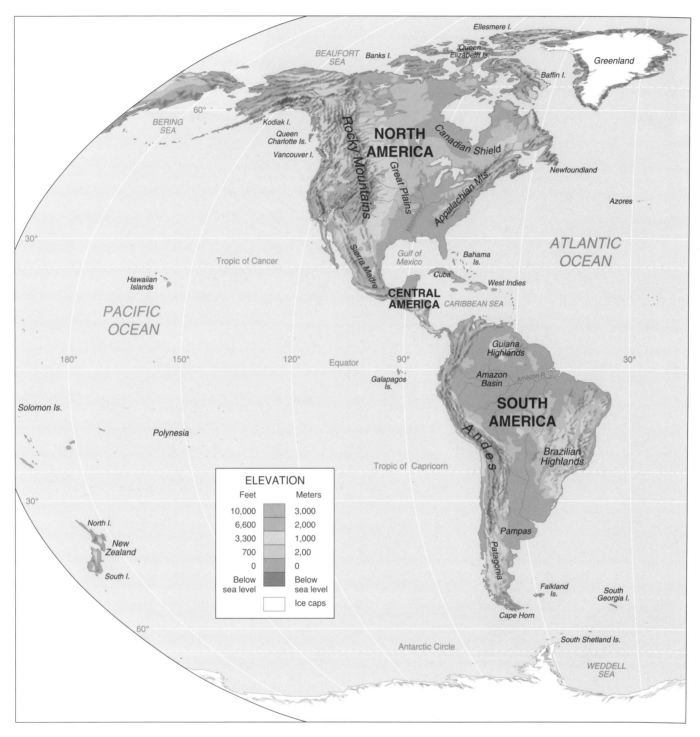

Compare this map of Earth's topography with Earth's population distribution shown on the map inside the back cover. Humans tend to cluster on flat areas, often near navigable waters.

INTRODUCTION TO GEOGRAPHY

INTRODUCTION TO
GEOGRAPHY

PEOPLE, PLACES & ENVIRONMENT

FIFTH EDITION

Carl T. Dahlman
Miami University, Oxford, Ohio

William H. Renwick
Miami University, Oxford, Ohio

Edward F. Bergman
*Lehman College of the City
University of New York*

Prentice Hall

Boston Columbus Indianapolis New York San Francisco Upper Saddle River Amsterdam
Cape Town Dubai London Madrid Milan Munich Paris Montréal Toronto Delhi
Mexico City São Paulo Sydney Hong Kong Seoul Singapore Taipei Tokyo

Geography Editor: Christian Botting
Editorial Director: Frank Ruggirello
VP/Executive Director, Development:
    Carol Trueheart
Senior Development Editor: Melissa Parkin
Marketing Manager: Maureen McLaughlin
Editorial Project Manager: Anton Yakovlev
Editorial Assistant: Christina Ferraro
Marketing Assistant: Nicola Houston
Managing Editor, Geosciences and Chemistry: Gina M. Cheselka
Senior Project Manager, Science: Beth Sweeten
Design Director: Mark Stuart Ong
Interior Designer : Gary Hespenheide
Cover Designer: Riezebos Holzbaur

Senior Technical Art Specialist: Connie Long
Art Studio: Spatial Graphics
Senior Manufacturing and Operations Manager:
    Nick Sklitsis
Operations Supervisor: Maura Zaldivar
Senior Media Producer: Angela Bernhardt
Media Producer: Ziki Diekel
Senior Media Production Supervisor: Liz Winer
Associate Media Project Manager: David Chavez
Photo Research Manager: Elaine Soares
Photo Researcher: Maureen Raymond
Composition/Full Service: PreMediaGlobal
Production Editor, Full Service: Karpagam Jagadeesan
Cover Credit: Corbis Images/Bruno Morandi

Library of Congress Cataloging-in-Publication Data

Dahlman, Carl (Carl T.)
    Introduction to geography : people, places & environment / Carl T. Dahlman
Edward F. Bergman, William H. Renwick. — 5th ed.
        p. cm.
    Rev. ed. of: Introduction to geography / Edward F. Bergman.
    Includes bibliographical references and index.
    ISBN-13: 978-0-321-69531-4 (alk. paper)
    ISBN-10: 0-321-69531-3 (alk. paper)
 1. Geography.   I. Bergman, Edward F.   II. Renwick, William H.
III. Bergman, Edward F. Introduction to geography.   IV. Title.
    G128.D35 2011
    910—dc22

                                                    2010021707

Printed in the United States
10 9 8 7 6 5 4 3 2 1

ISBN: 0-321-69531-3 / 978-0-321-69531-4 (Student Edition)
ISBN: 0-321-69585-2 / 978-0-321-69585-7 (Books á la Carte)

Prentice Hall
is an imprint of

www.pearsonhighered.com

CONTENTS

4 Biogeochemical Cycles and the Biosphere 121

5 Population and Migration 155

13 Global Challenges and the Scale of Response 515

Competing interest

Protecting the Global Environment 517

PREFACE

Colleges and universities recognize the need for educated citizens to be geographically literate and sensitive to the diversity of human culture and the processes that link us in a global society. *Introduction to Geography: People, Places & Environment* provides the groundwork for such literacy and sensitivity. This is an introduction to the conditions and interactions about which a working knowledge is essential to successfully negotiate the world of the twenty-first century. Primarily, this is a book about the ways that humans cope with and change the world around them. We also seek to provide readers with frameworks for evaluating the qualities and consequences of the relationships between different places and peoples. This further entails understanding the effects that global linkages have on local places and populations, as well as the degree to which some places influence the rest of the world. We are keen to promote an integrated view of geography, one that emphasizes the many interrelationships between the breadth of human activities and an environment that ranges from the tropical wilderness to the thoroughly engineered city.

In a globally connected world, distant places that were once culturally, economically, and physically very different from each other now interact with an intensity that was formerly unknown. A generation ago, one could go through daily life in his or her own country without needing to know much more than the political alignments of countries beyond those borders, with little regard for differences in language, religion, or agricultural systems. A relatively small number of specialists bothered to learn the languages of our political allies or adversaries, and few companies had trading relationships with distant counterparts. Americans were aware of certain distant places, and some fought wars in places around the globe. For most, however, knowledge of one's local region was sufficient for thoughtful conduct of everyday life.

Today, we expect educated individuals to have significant understanding of both the diversity of environments and cultures around the world and of the processes that connect them. We recognize that in heating our homes or powering our vehicles, we are participating directly in global energy markets and emitting pollutants that travel around the globe with likely long-term effects on climate. Global trade and finance have made far-flung places more dependent on one another. We interact daily in both large cities and small towns with individuals born and raised in distant places and who retain close linkages to those places. Events in their homelands have repercussions in our towns and vice versa.

New to the Fifth Edition

The fifth edition of *Introduction to Geography: People, Places & Environment* represents a thorough revision based on a number of substantive changes:

- This edition introduces a new author—Carl Dahlman—who has extensive experience teaching geography and conducting research in a number of different world regions.
- New coverage of the rapidly changing debate on global warming, including the IPCC's Fourth Assessment Report and the 2009 Copenhagen Climate Conference (Chapters 2 and 4).
- Discussion of recent earthquakes in China, Haiti, and Chile, including the relationships among levels of economic development and susceptibility to natural hazards (Chapter 3).
- Expanded treatment of the global carbon cycle and related connections among the biosphere, atmosphere, and human resource use (Chapters 4 and 9).
- Coverage of the expanding globalization of grain production trade, including soybean production in Brazil and its relation to global energy prices (Chapter 8).
- New coverage on food insecurity, famine, and hunger as it relates to global food markets (Chapter 8).
- Coverage of the 2008–2009 financial crisis as it pertains to urban geography (Chapter 10) and the world economy (Chapter 12).
- A new concluding chapter on the Geography of Contemporary Problems. Chapter 13 focuses on four major global problems: managing environmental change, global security and human rights, regional trade and cooperation, and human development. This chapter utilizes the concepts introduced in this book to provide a geographical understanding of these issues. It also details how human society has tried to respond to these problems through cooperation at different scales. These challenges are very real topics of current political debate; some will face humankind throughout your lifetime.
- Complete updating of statistics (tables, graphs, and so on) on climate, energy, natural resources, population, and economics.
- In addition, the online Multimedia and Assessment for *Introduction to Geography: People, Places & Environment*, found at www.mygeoscienceplace.com, now includes a fully integrated eBook, numerous geography video clips, MapMaster™ interactive maps, activities, assessments, flash cards, RSS feeds, and additional resources such as further readings and film, music, and popular culture references.

Three Themes in This Textbook

This textbook emphasizes three themes throughout the study of geography. First, geography examines the interrelationships between humans and their natural environment; second, many basic principles of human geography can be studied and demonstrated in your own hometown; and third, geography is dynamic.

Geography Explores the Interrelationships Between Humans and the Environment

The study of Earth's climates, soils, vegetation, and physical features is called *physical geography*, and physical geography sets the stage upon which humans act out their lives. A great deal of human effort is spent wresting a living from the environment, adjusting to it, or altering it.

The first few chapters of this book, therefore, offer an overview of Earth's physical environment. The discussion emphasizes processes operating in the landscape, such as atmospheric circulation, landform change, and vegetation growth. An understanding of these processes is necessary to comprehend the mechanisms whereby humans transform Earth's environments. The theme of human–environmental interaction then weaves through the book. People interpret and evaluate possibilities in their environment, and they alter natural conditions. Any alteration in one of Earth's natural, physical, and ecological systems, however, will trigger changes throughout the entire system. Because these systems are tremendously complex, many changes may be unexpected or even harmful. For example, people "tame" rivers and build on their flood plains, but that construction can increase damage during a flood. People redistribute plants and animals around Earth, but relocated species can cause unexpected ecological damage. Chapter 2 notes how cities change local climates; these changes affect human health. Pollution threatens the air we breathe, the water we drink, and the soil we till. Chapter 13 emphasizes that environmental protection is one issue that today challenges all humankind, requiring international collaboration for a positive goal.

Geography Is Not Restricted to the Study of Exotic Lands and Peoples

The basic principles of geography can be studied in your hometown and even on campus. How do local temperatures and rainfall vary throughout the year? What processes shaped the landforms, and how are these processes changing the land today? What natural hazards affect people in your area? Where have new arrivals in your community come from, and why did they move? Where are local food crops and manufactured goods sold? Are the boundaries of city council districts manipulated to give one group an unfair advantage in elections? How many religions are represented by houses of worship in your town, and are these centers of identifiable residential communities? Can you map the rents on commercial properties in your town, and how do these values reflect accessibility or, perhaps, perception of which neighborhoods are the most elegant?

Study of geography introduces a great range of careers, even though the professionals at work in many fields may not call themselves *geographers*. People studying or "doing" geography include planners designing new suburbs, scientists working to reduce water pollution, transportation consultants routing new highways, advertisers targeting "junk mail" to zip codes where residents have specific income levels, diplomats negotiating treaties to regulate international fishing, and still more various occupations.

Modern Geography Is Dynamic

It is important to know the current distributions of landforms, people, languages, religions, cities, and economic activities, but none of these patterns is static. Earth's surface is constantly changing, sometimes slowly as when erosion wears down a mountain chain over millions of years, and sometimes spectacularly and rapidly as when a new volcano explodes and builds on Earth's surface or a flood prompts a river to change its course. Transportation and communication ties among peoples and regions have multiplied, so social, political, and economic forces constantly redistribute human activities. Maps of these activities reveal only temporary balances between forces for change and forces for stability. What happens *at* places depends more and more on what happens *among* places, and we can understand maps of economic or cultural activity only if we understand the patterns of movement that create them. Modern geography explores the forces at work behind the maps.

Each day, the news media reports events in which what happens is directly related to where it happens, and these events trigger changes in geography: A volcano erupts in Mexico; a bountiful harvest in Argentina reduces food prices and thus improves the diet available to Africans; Canadian scientists synthesize a substitute for a mineral previously imported; a new government in Africa redirects international alliances, economic links, and migration streams. American movies and music diffuse U.S. language and culture around the world, while Americans themselves adopt new foods and words, such as sushi. Developing countries join the already-developed world in sprouting new industrial sources of air pollution, poisoning their own citizens, and changing the chemical composition of Earth's atmosphere sufficiently to change Earth's climate. Protestant Christianity wins converts throughout Latin America; women are accepted into the priesthood, and governments open family planning clinics. Meanwhile, in some African and Asian countries, Islamic fundamentalists win political power and curb women's rights. These events remap world cultural, political, and economic landscapes. Today's dynamic geography doesn't just exist; it *happens*. In every topic covered in this text, it is our goal not only to *describe* distributions and locations but to *explain* the distributions and locations.

Contemporary Issues in Geography

What you learn in your reading and in your geography classroom can help you better understand current events and form your own opinions on important questions of the day. Each chapter of this book provides background material for understanding the news. A number of topics in the book demonstrate this benefit, but here we mention just two: environmental protection and development.

Each one of Earth's billions of people aspires to a high level of material welfare, yet today many people live in conditions of deprivation. The world distribution of wealth and welfare does not coincide with the world distribution of raw material resources. If the possession of raw materials were the key to wealth, then the Republic of Congo and Mexico would count among the richest countries in the world, and Japan and Switzerland would count among the poorest. In fact, the Republic of Congo and Mexico are poor, and Japan and Switzerland are rich. An understanding of the resolution of this paradox is essential to understanding the world today.

This book goes beyond merely describing where there is wealth and where there is not. It weaves together a number of threads of understanding, making clear the *why* of the *where*. Relevant economic principles are individually introduced in the text where appropriate, including adding value to raw materials, sectoral evolution, locational determinants for manufacturing, various nations' economic policies, and the patterns of world trade. New considerations arise virtually every day, such as the discovery of new resources, new technologies, and new governments with new policies. These changes redistribute advantages. Chapter 12 suggests how and why the balance among the factors that support economic development is continuously shifting, and Chapter 13 highlights how these geographic shifts redistribute global sources of pollution. New industries in developing countries generate pollution. Furthermore, those new industries allow rising standards of living, and that in turn allows increased consumption of consumer goods and increased production of waste products.

Economic development is only one aspect of development. Human development includes adequate nutrition, education, political liberty, and the opportunity to live in a healthy, unpolluted environment. Chapter 5 analyzes the geography of health, and Chapter 8 explains the distribution of world food production and trade. Chapter 9 discusses problems and challenges in the consumption of natural resources and the treatment of wastes and disposed items. Chapter 13 includes a new section that discusses human development and the major initiatives under way to raise people out of poverty. In each case, the text suggests what factors might redistribute these activities in the future.

Maps, Cartograms, and GIS

A great variety of maps illustrate this book. Some of them include relief shading. You will probably be familiar with traditional maps that illustrate distributions as mosaic patterns of color. On other maps, called *flow maps*, arrows and lines represent movements of people or of goods, and on these the widths of the arrows and lines often convey the quantity of each flow—the numbers of passengers flying major airline routes across the United States, for example (Figure 1-15).

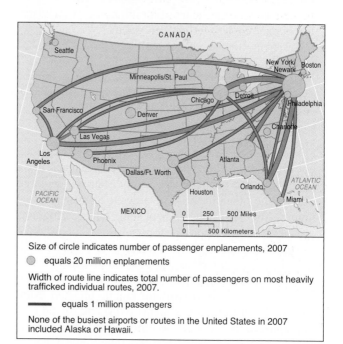

Size of circle indicates number of passenger enplanements, 2007

○ equals 20 million enplanements

Width of route line indicates total number of passengers on most heavily trafficked individual routes, 2007.

▬▬▬ equals 1 million passengers

None of the busiest airports or routes in the United States in 2007 included Alaska or Hawaii.

Figure 1-15 Major air-passenger routes in the United States. This map is a good indicator of the urban hierarchy of the United States, although it is a bit distorted by airlines' routing systems. New York is certainly the principal focus of national life. Los Angeles, Dallas, Atlanta, and Chicago are functional "capitals" of individual regions of the country. Minneapolis/St. Paul and Denver have large numbers of total enplanements but few or no individual routes with more than 2 million passengers. These cities serve as hubs of many local regional routes with small numbers of fliers on each.

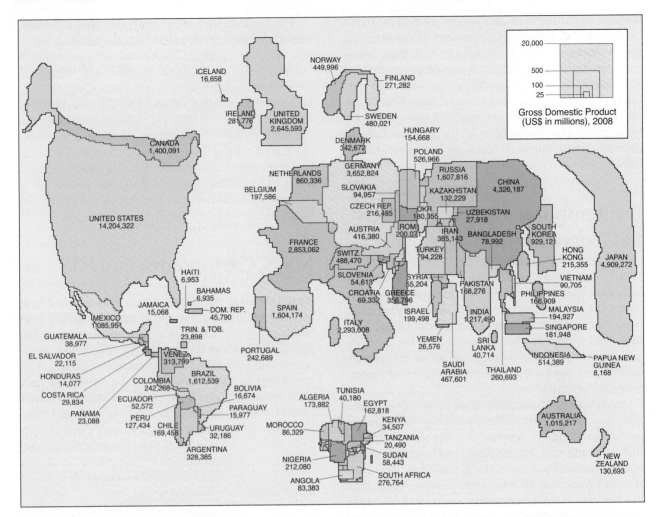

Figure 12-5 The relative sizes of nations' gross domestic products. The size of the countries on this cartogram corresponds to their total gross domestic product. The United States, Europe, and Japan grow considerably beyond the sizes of their relative territories. Africa, in contrast, shrinks.

Several cartograms have also been specially designed for this book. A *cartogram* is a visual device much like a map, but on a cartogram physical distance is replaced with some other measure in order to convey a visual impression of the magnitude of something. Cartograms of such things as countries' populations, for example (Figure 5-2), or of countries' economies (Figure 12-5), visually convey the relative population or wealth of different countries better than standard land area maps do.

Today maps are drawn and rapidly updated using geographic information systems (GIS) and digital cartographic tools. GIS have become essential and ubiquitous on geographers' desks, and their importance is clear in this book. An expanded discussion of GIS technology has been added in Chapter 1, and we have illustrated how GIS are used in geographic problem solving throughout the book.

A variety of other visual devices with which you are probably familiar are included as well: tables, bar graphs, and pie graphs. In each case, the captions have been carefully written to help you read these sophisticated images. Each contains a great deal of information.

Special care has been taken in preparing the captions for all illustrations. Each caption guides your eye through or into the figure. Many captions include questions to get you thinking about the information in the image. Many ask you to compare it with information shown on another image in the book in order to develop hypotheses of explanation.

A Word About Numbers

This book contains many numbers—measurements of populations, economic conditions, production of various commodities, world trade, and more. These measures come from a variety of sources—private organizations, national governments, international organizations—and they are the best available. Such numbers, however, must always be read with two considerations in mind: reliability and date.

The compilation of measures is a tremendously difficult task. For example, the United States is the world's richest country, with many highly skilled government workers, yet the government admits that the national census is probably inaccurate by a factor of 5 to 7 percent. What degree of accuracy, then, can we expect from a poor country that does not have a sophisticated bureaucracy? It is said that the government of India does not have even an accurate count of the number of cities in India. We do not want to promote cynicism about the value or reliability of statistics, but an educated person does exercise judgment about the probable exactitude of any figure.

The second caution is that the measures themselves change. The population of the United States changes every day, yet the United States officially counts its residents only once every 10 years—although periodic estimates are more frequently published. It takes a long time to gather and compile statistics, so the measures may seem out of date by the time they are published. This is especially true of international comparative statistics. For example, each year the United Nations Conference on Trade and Development (UNCTAD) publishes a handbook of statistics of world trade, but the book appears three or four years after its date, and many statistics recorded were measured years before the date of the volume. The figures used to make world maps, as in this text, are sometimes older than figures given for many individual countries because advanced countries are able to compile numbers faster than poorer countries, and then international organizations need time to compile all of the national figures. Furthermore, governments sometimes change the way they measure things. For example, for many years governments counted and published a statistic called gross national product (GNP), but today that statistic has been replaced by a slightly different measure called the gross national income (GNI). The meaning of GNI is explained in Chapter 12.

The statistics in this textbook are as up to date as could possibly be obtained from the most reliable sources as of early 2010. The text notes the direction in which many of these measures are changing, and in many cases we have dared to predict their future direction. The U.S. population will probably continue to rise, and the percentage of the national labor force working in manufacturing will probably continue to fall. We encourage you to go to the library or to search the World Wide Web to update those measures to today.

The Teaching and Learning Package

For Instructors

Instructor Resource Center on DVD (978-0-321-67542-2 • 0-321-67542-8): Everything instructors need where they want it. The Pearson Prentice Hall Instructor Resource Center helps make instructors more effective by saving them time and effort. All digital resources can be found in one well-organized, easy to-access place. The IRC on DVD includes:

- All textbook images as JPEGs, PDFs, and Power-Point™ presentations
- Pre-authored Lecture Outline PowerPoint™ presentations, which outline the concepts of each chapter with embedded art, and can be customized to fit instructors' lecture requirements
- CRS "Clicker" Questions in PowerPoint™ format, which correlate to the revised U.S. National Geography Standards and Bloom's Taxonomy
- The TestGen software, Test Bank questions, and answers for both MACs and PCs
- Electronic files of the *Instructor Resource Manual* and *Test Bank*
- Numerous geography video clips

This Instructor Resource content is also available completely online via the Instructor Resources section of www.mygeoscienceplace.com and www.pearsonhighered.com/irc.

Instructor Resource Manual Download (978-0-321-69586-4 • 0-321-69586-0): The Instructor Resource Manual is intended as a resource for both new and experienced instructors. It includes a variety of lecture outlines, additional source materials, teaching tips, advice about how to integrate visual supplements (including the Web-based resources), and various other ideas for the classroom. www.pearsonhighered.com/irc

TestGen® Download (978-0-321-67543-9 • 0-321-67543-6): TestGen® is a computerized test generator that lets instructors view and edit Test Bank questions, transfer questions to tests, and print the test in a variety of customized formats. This Test Bank includes approximately 1,000 multiple-choice, true/false, and short answer/essay questions. Questions correlate to the revised U.S. National Geography Standards and Bloom's Taxonomy to help instructors better map the assessments against both broad and specific teaching and learning objectives. The Test Bank is also available in Microsoft Word® and is importable into Blackboard and WebCT. www.pearsonhighered.com/irc

Television for the Environment *Earth Report* Geography Videos on DVD (978-0-321-66298-9 • 0-321-66298-9): This three-DVD set is designed to help students visualize how human decisions and behavior have affected the environment, and how individuals are taking steps toward recovery. With topics ranging from the poor land management promoting the devastation of river systems in Central America to the struggles for electricity in China and Africa, these 13 videos from Television for the Environment's global *Earth Report* series recognize the efforts of individuals around the world to unite and protect the planet.

Television for the Environment *Life* World Regional Geography Videos on DVD (978-0-13-159348-0 • 0-13-159348-X): From Television for the Environment's global *Life* series, this two-DVD set brings globalization and the developing world to the attention of any world regional geography course. These 10 full-length video programs highlight matters such as the growing number of homeless children in Russia, the lives of immigrants living in the United States trying to aid family members still living in their native countries, and the European conflict between commercial interests and environmental concerns.

Television for the Environment *Life* Human Geography Videos on DVD (978-0-13-241656-6 • 0-13-241656-5): This three-DVD set is designed to enhance any human geography course. These DVDs include 14 full-length video programs from Television for the Environment's global *Life* series, covering a wide array of issues affecting people and places in the contemporary world, including the serious health risks of pregnant women in Bangladesh, the social inequalities of the "untouchables" in the Hindu caste system, and Ghana's struggle to compete in a global market.

Aspiring Academics: A Resource Book for Graduate Students and Early Career Faculty (978-0-13-604891-6 • 0-13-604891-9): Drawing on several years of research, this set of essays is designed to help graduate students and early career faculty start their careers in geography and related social and environmental sciences. This teaching aid stresses the interdependence of teaching, research, and service—and the importance of achieving a healthy balance in professional and personal life—in faculty work and does not view it as a collection of unrelated tasks. Each chapter provides accessible, forward-looking advice on topics that often cause the most stress in the first years of a college or university appointment.

Teaching College Geography: A Practical Guide for Graduate Students and Early Career Faculty (978-0-13-605447-4 • 0-13-605447-1): This book provides a starting point for becoming an effective geography teacher from the very first day of class. It is divided in two parts. The first set of chapters addresses "nuts-and-bolts" teaching issues in the context of the new technologies, student demographics, and institutional expectations that are the hallmarks of higher education in the twenty-first century. The second part explores other important issues: effective teaching in the field; supporting critical thinking with GIS and mapping technologies; engaging learners in large geography classes; and promoting awareness of international perspectives and geographic issues.

Online Resources

Premium Website A dedicated Premium Website with eText offers a variety of resources for students and professors, including an eText version of the textbook with linked/integrated multimedia, MapMaster™ interactive maps with activities, geography videos with assessments, geoscience animations with assessments, flashcards, RSS feeds, weblinks, annotated resources for further exploration, and Class Manager & GradeTracker Gradebook functionality for instructors. www.mygeoscienceplace.com

AAG Community Portal for Aspiring Academics & Teaching College Geography This website is intended to support community-based professional development in geography and related disciplines. Here you will find activities providing extended treatment of the topics covered in both books. The activities can be used in workshops, graduate seminars, brown bags, and mentoring programs offered on campus or within an academic department. You can also use the discussion boards and contributions tool to share advice and materials with others. www.pearsonhighered.com/aag/

Course Management Systems Pearson Prentice Hall offers content specific to *Introduction to Geography: People, Places & Environment*, fifth edition, within the BlackBoard and CourseCompass course management system platforms. Each of these platforms lets the instructor easily post his or her syllabus, communicate with students online or offline, administer quizzes, and record student results and track their progress. www.pearsonhighered.com/elearning/

For Students

Mapping Workbook (978-0-321-75099-0 • 0-321-75099-3): This workbook, which can be used in conjunction with the fifth edition text or an atlas, features political and physical base maps of every global region, printed in black and white. These maps, along with the names of the regional key locations and physical features, are the basis for identification exercises. Conceptual exercises are included to further examine students' comprehension of the key points presented in the main text chapters. These exercises review both the physical and human environments of the regions.

Goode's World Atlas, 22nd Edition (978-0-321-65200-3 • 0-321-65200-2): *Goode's World Atlas* has been the world's premiere educational atlas since 1923, and for good reason. It features over 250 pages of maps, from definitive physical and political maps to important thematic maps that illustrate the spatial aspects of many important topics. The 22nd edition includes 160 pages of new, digitally produced reference maps, as well as new thematic maps on global climate change, sea level rise, CO_2 emissions, polar ice fluctuations, deforestation, extreme weather events, infectious diseases, water resources, and energy production.

Encounter World Regional Geography Workbook & Website by Jess C. Porter (978-0-321-68175-1 • 0-321-68175-4): *Encounter World Regional Geography* uses online geobrowser technology to explore world regions through the themes of Environment, Population, Culture, Geopolitics, and Economy & Development. Each Exploration consists of a worksheet, online quizzes, and a corresponding Google Earth™ KMZ file, available for download from www.mygeoscienceplace.com.

Encounter Geosystems: Interactive Explorations of Earth Using Google Earth™ Workbook and Website by Charlie Thomsen and Robert Christopherson (0-321-63699-6 • 978-0-321-63699-7): *Encounter Geosystems* provides rich, interactive explorations of physical geography concepts through Google Earth™ explorations. All chapter explorations are available in print format as well as online quizzes, accommodating different classroom needs. All worksheets are accompanied by corresponding Google Earth™ media files, available for download from www.mygeoscienceplace.com.

Encounter Earth: Interactive Geoscience Explorations Workbook & Website by Steve Kluge (978-0-321-58129-7 • 0-321-58129-6): *Encounter Earth* uses online geobrowser technology to explore introductory geoscience topics. Each Exploration consists of a worksheet, online quizzes, and a corresponding Google Earth™ KMZ file, available for download from www.mygeoscienceplace.com.

Dire Predictions by Michael Mann and Lee R. Kump (978-0-13-604435-2 • 0-13-604435-2): Periodic reports from the Intergovernmental Panel on Climate Change (IPCC) evaluate the risk of climate change brought on by humans. But the sheer volume of scientific data remains inscrutable to the general public, particularly to those who may still question the validity of climate change. In just over 200 pages, this practical text presents and expands upon the essential findings of the 4th Assessment Report in a visually stunning and undeniably powerful way to the lay reader. Scientific findings that provide validity to the implications of climate change are presented in clearcut graphic elements, striking images, and understandable analogies.

Acknowledgments

Countless colleagues, librarians, and generous individuals both in government and in the private sector helped with information for this text. We wish to thank in a special way Professor James M. Rubenstein of Miami University, Ohio, for his many contributions to our thinking. Special thanks for colleagues who aided in deliberations with new material, especially: Edward Carr, University of South Carolina; Marcia Castro, Harvard University; Kirstin Dow, University of South Carolina; Owen Dwyer, Indiana University-Purdue University Indianapolis; Brent McCusker, University of West Virginia.

We also wish to thank our scholarly colleagues who provided thoughtful suggestions for improving the book over the years. These include:

Gillian Acheson, Southern Illinois University, Edwardsville
Tanya Allison, Montgomery College
Holly R. Barcus, Morehead State University
Lee Berman, Southern Connecticut State University
Daniel Block, Chicago State University
Paul L. Butt, University of Central Arkansas
Jim Byrum, University of South Carolina
Edward Carr, University of South Carolina
Joseph M. Cirrincione, University of Maryland
Bruce Davis, Eastern Kentucky University
Bryce Decker, University of Hawaii, Manoa
Stanford Demars, Rhode Island College
Leslie Dienes, University of Kansas
Gary Fowler, University of Illinois, Chicago
Michael Fox, Carleton University
Roberto Garza, San Antonio College
Jennifer Gebelein, Florida International University
Brooks Green, University of Central Arkansas
Mark Guizlo, Lakeland Community College
Rene J. Hardy, Shoreline College
James Harris, Metropolitan State College of Denver
Erick Howenstine, Northeastern Illinois University
James C. Hughes, Slippery Rock University
Mark Jones, University of Connecticut
Tulasi R. Joshi, Fairmont State College
Walter Jung, Central Oklahoma University
Angelina Kendra, Central Connecticut State University
Rob Kent, University of Akron
Lori Krebs, Salem State College
Miriam K. Lo, Mankato State University
José Lopez, Minnesota State University
Ruben A. Mazariegos, University of Texas, Pan American
Ian A. McKay, Wilfrid Laurier University
Roger Miller, Black Hills State University
Holly Porter-Morgan, GIS Laboratory, New York Botanical Garden
G.L. "Jerry" Reynolds, University of Central Arkansas
Scott C. Robinson, University of Nebraska, Omaha
Lallie Scott, Northeast Oklahoma State University
Henry Sirotin, Hunter College of the CUNY
Christa Smith, Clemson University
James N. Snaden, Central Connecticut State University
David M. Solzman, University of Illinois, Chicago
Robert C. Stinson, Macomb Community College
Christopher J. Sutton, Northwestern State University of Louisiana
Melissa Tollinger, East Carolina University
James Tyner, Kent State University
Thomas B. Walter, Hunter College of the CUNY

Gerald R. Webster, University of Alabama

Kathy Williams, Bronx Community College

John Wright, New Mexico State University

Charles T. Ziehr, Northeastern State University

We owe a debt of gratitude to many people. At Pearson, Christian Botting, Acquisitions Editor, managed the project from its beginning stages through the journey to production. Melissa Parkin lent a keen eye to every detail during the editing and production process; this is a better book thanks to her. James Harris and Holly Porter-Morgan contributed important comments and suggestions; we thank them for their careful work. Thanks to Christina Ferraro for managing the review process. Maureen Raymond, photo researcher, did an exceptional job of finding excellent imagery; Ziki Dekel produced and managed the website for the book; and Beth Sweeten, Kelly Keeler, and Karpagam Jagadeesan provided invaluable assistance during production. We have enjoyed working with all of these people, and we thank them. Contemporary geography is a wide field that covers many topics and, quite literally, the entire world. We have strived to present our field in its diversity by selecting carefully from the work of our peers and others. We welcome suggestions and ideas for how to improve our efforts in service to the teaching of our discipline.

Carl T. Dahlman
William H. Renwick
Edward F. Bergman

Carl T. Dahlman earned degrees in sociology, music, and urban affairs before receiving his Ph.D. in geography from the University of Kentucky in 2001. He is Associate Professor of Geography at Miami University, where his teaching focuses on political geography, migration and mobility, and globalization. His current research includes the role of European integration in the geopolitics of Southeastern Europe. He enjoys photography and hunting for fossils with his son.

William H. Renwick earned a B.A. from Rhode Island College in 1973 and a Ph.D. in geography from Clark University in 1979. He has taught at the University of California, Los Angeles, and Rutgers University and is currently Associate Professor of Geography at Miami University. A physical geographer with interests in geomorphology and environmental issues, his research focuses on impacts of land-use change on rivers and lakes, particularly in agricultural landscapes in the Midwest. When time permits, he studies these environments from the seat of a wooden canoe.

Edward F. Bergman was born in Wisconsin and received a B.A. from the University of Wisconsin. He earned an M.A. and a Ph.D. from the University of Washington. He taught at the City University of New York and widely in Europe, South America, and South Africa. Now retired as a professor emeritus, he still travels and occasionally lectures, and advises museums on the writing of labels for exhibits.

Connect to the geography all around us

What happens *in* places depends increasingly on what happens *among* places. To understand mapped patterns, we need to recognize the movement that creates and continuously rearranges them. By examining what happens in one set of geographic processes—and how those processes affect others—***Introduction to Geography: People, Places & Environment,* Fifth Edition** presents the major tools, techniques, and methodological approaches of geography.

Chapter Opener

An anecdote and related photo begin each chapter, detailing real-world questions and problems related to the chapter topics and themes.

The soaring arched dome of Hagia Sofia ("Holy Wisdom" in Istanbul) remains a masterpiece of world architecture.

7

The Geography of Languages and Religions

Since its construction in the sixth century, the building known as Hagia Sophia has witnessed enormous cultural change. It was built during the reign of Justinian the Great, ruler of the Byzantine Empire, during the early centuries of Christianity. Marble and other materials, weighing as much as 70 tons, were brought from all over the Empire to Istanbul to construct what would be for several centuries the largest church in the world. The freestanding arched dome measures 31 meters (102 feet) across. The building withstood numerous earthquakes and fires, each time ingeniously repaired to strengthen its delicate form. A perhaps more stunning change came in 1453 when the Ottoman Turkish Empire captured Istanbul. Their leader, Mehmed the Conqueror, ordered the renovation of Hagia Sophia to repair the damage incurred during the decline of Byzantium. He also converted the church into a mosque, removing the specifically Christian elements, such as the altar, and adding Islamic ones, including a minaret (a tower for the call to prayer). The Ottoman Empire made numerous changes to the building during their almost five centuries in Istanbul. Some of the mosaics showing Christian and Byzantine figures were covered over with geometric tiling and Koranic inscriptions. A religious school and other buildings were added to the grounds. The Ottomans later restored many of the Christian mosaics. When the Ottoman Empire came to an end in 1924, the first president of Turkey, Mustafa Kemal Ataturk, ordered that the building no longer be used as a mosque but as a museum. Today, fifteen centuries of art, architecture, and history from two major religions can be found under the magnificent dome of Hagia Sophia.

A Look Ahead

Defining Languages and Language Regions

The great variety of languages spoken today testifies to the relative isolation of groups in the past. Any language's distribution illustrates the dispersal pattern of its original speakers or their cultural impact on others.

The Development and Diffusion of Languages

If groups of people who speak a common language disperse, then each breakaway group will develop its own language but retain elements of the common ancestral language. Languages that are thus related make up a language family. The geography of written language also records the origins and dispersal of peoples and cultures. Toponymy, the study of place names, records natural features or the origins or values of a place's inhabitants.

Linguistic Differentiation in the Modern World

The use of some languages is diffusing around the world, gaining greater power and influence in world affairs, or winning new adherents to their ideas. Other languages are disappearing.

The Teachings, Origin, and Diffusion of the World's Major Religions

Each religion originated in one place and spread out from there, but it is difficult to identify precise reasons why any religion diffuses. Some religious groups try to convert others to their religious beliefs, but others do not.

The Political and Social Impact of the Geography of Religion

Religion is a major influence on of human behavior, so the geography of people holding and following a set of teachings affects politics, economics, agriculture, and diet, as well as many other aspects of human life.

249

A Look Ahead

Each chapter begins with a brief outline of the main points of the chapter.

CONNECTIONS

Geography, Geographic Information Systems, and the Global Carbon Budget

Concern about the accumulation of carbon dioxide in the atmosphere is mounting, and countries are moving toward implementation of an international treaty to control carbon dioxide emissions. Therefore, the need for accurate information on carbon dioxide release and uptake is increasing. One of the most important parts of the puzzle is to compile detailed knowledge of the amounts of carbon released and taken up by the biosphere. Geography and geographic information systems (GISs) are critical to providing that information.

Many different flows must be estimated in calculating the carbon dioxide budget of the atmosphere, hydrosphere, biosphere, and lithosphere. Some of these are easier to estimate than others. We know fairly closely, for example, how much carbon is emitted by fossil-fuel combustion because we know approximately how much coal, oil, and natural gas each country burns. Another large number in the budget—the amount that goes from the atmosphere to the oceans each year—can be calculated based on knowledge of the physical processes that control the exchange of carbon dioxide between water and air. On the other hand, estimating the amount of carbon released by deforestation or taken up by forest growth is much more difficult, because it is highly variable from place to place. Other key processes, too, such as the accumulation and decay of organic matter in the bottoms of reservoirs, the release of carbon dioxide from decaying peat bogs, or the buildup and loss of organic matter from agricultural soils, are very poorly known. These processes, however, are the very ones that can be managed directly by humans, so they must be accounted for in any international agreement.

Geography and GISs provide the tools for extrapolating from a few isolated measurements to estimate processes spread over large areas, so geographers are playing a big role in studies of the carbon budget. For example, we can directly measure rates of carbon uptake by photosynthesis or release through deforestation for small areas like a few square meters or a few hectares. Such measurements involve tasks like collecting, weighing, and analyzing vegetation, or carefully sampling and analyzing carbon dioxide in air samples. We could not possibly do this field research for every hectare of land in the United States, but we can do this at many individual sites representing different land-cover types—coniferous forest, deciduous forest, lakes, wetlands, farmlands, and so forth. Then, using a GIS, we can make detailed maps of land cover (often based on satellite imagery, as in the figure) showing the spatial extent of each different kind of surface, and we can assign rates of carbon exchange for each of these types of surface based on our field measurements. The GIS then does the counting for us to estimate total carbon movements over large areas. Not only is a GIS useful in that accounting process, but, in combination with satellite imagery, it can help monitor and account for changing land-cover characteristics over time (see, for example, Figure 1-24). Geography and geographers will continue to play a central role in this work.

(a)

(b)

Disappearing arctic lakes. A gre… gions, especially low-lying coastal pl… as these of Siberia in 1973 (above) a… disappearance (arrows) associated w… Drainage of these lakes exposes sedi… carbon, and may accelerate release … the atmosphere. (http://earthobservat… NewImages/images.php3?img-id-16986)

Connections Essays

Connections essays help students make the link between geographical concepts in different chapters that, together, promote a fuller understanding of an issue or event.

GLOBAL AND LOCAL

The Diffusion of "News"

Western media dominate the gathering and dissemination of news to such an extent that most people in non-Western lands learn about the affairs of other non-Western lands—and even about their own national affairs—through Western media.

Alternative sources of global news are appearing. The year 2005 saw the launch of Telesur, a television network aimed to provide an alternative to U.S.-based news and analysis for Latin America. Telesur is based in Caracas, Venezuela, and financed by Venezuela, Argentina, Cuba, Bolivia, Ecuador, and Nicaragua. The U.S. government has criticized the network's news coverage as biased.

Al-Jazeera, a Qatar-based television news service, is expanding its global coverage. When founded in 2001, al-Jazeera provided only Arab-language broadcasts, but in 2006, it launched an English-language news service. British journalist David Frost accepted a position with al-Jazeera, arguing, "We in the West have been broadcasting our views to the non-Western world for many years. It is only fair that these non-Western areas should have the chance to return the compliment." Then-U.S. Secretary of Defense Donald Rumsfeld criticized the network as offering an "anti-American worldview," but then-Secretary of State Condoleezza Rice chose to engage the network and appeared as a guest. Al-Jazeera's independence startles even many Muslim governments. It criticizes them, for example, for their lack of democracy and subjection of women. These topics are avoided in the state-owned media monopolies in most Muslim states.

The competitors to Al-Jazeera include the more moderate Al-Arabiya, based in the United Arab Emirates and backed by Saudi Arabia. When President Barack Obama granted Al-Arabiya the first interview after his inauguration, many viewed his choice as trying to promote healthy competition among news agencies. America's own government-subsidized Voice of America (VOA) has won tens of millions of listeners in 45 languages. Its charter states that the service should be "a reliable and authoritative source of news" and that it should be "accurate, objective and comprehensive," but some American critics demand that it presents exclusively a pro-American point of view.

Television remains an important medium for diffusing news in non-Western countries, where television viewership is often widespread. The growing use of the Internet in these countries has also increased demand for online news content… pers and new agencies have quickly developed web… so, too, have the new television agencies like Al-Jaz… sites offer content in the local language, but many… ers also read English-language news sources. Interes… competition for online readers has caused Western… tailor their news sites to a more international audien…

News presenters on Al-Jazeera dress in Wester… and women appear without veils. These news… broadcast from the network's Kuala Lumpur broadc…

Global and Local Essays

More than any other discipline, geography explains the connections between global forces and local places. *Global and Local* essays examine in detail how particular places respond to global impacts.

RAPID CHANGE

Who Killed the Record Store?

The cultural landscape can change quickly. Not long ago, the record store was a basic feature of any U.S. town. It was more than just a place to buy music. Going to the record store was an important cultural practice, especially for U.S. youth. Going to the record store alone, without parents, marked a coming of age for many young people. It was an exposure to new music, played in the store by a usually knowledgeable clerk who tried to play something good but relatively unknown. Staff and other shoppers might share suggestions for music, providing descriptions about not only the music but cultural cues about who listens to the music and why. A bulletin board listed local concerts, advertised used musical gear for sale, and sought new members for garage bands.

Going down to the record store meant wearing the right clothes, affecting the appropriate attitude, using the right language, and, above all, picking the right music to define oneself and one's peer group. As with youth today, there were different ways of being "cool," but each had its own set of "rules" and social expectations. In short, each musical style had is its own subculture. T-shirts, buttons, patches, jewelry, and other visible markings of one's identity were sold alongside the music.

Ironically, the anxiety over being authentic (not a "poser"!) was also a part of becoming a modern consumer. Sorting out one's identity through musical genres, clothing, and friends was a way to learn about one's own tastes, preferences, and lifestyle in relation to others. Even if your peer group and taste preferences changed, the experience was a valuable lesson in the social importance of melding behavioral traits to particular peer subcultures while also being in mainstream society.

But record stores, as described here, are fast disappearing. Small independent record stores struggle to stay open, but most cannot. Four in five national chain stores, such as Tower Records, have closed since 1991. Yet music remains a lynchpin of American popular and youth culture. So who killed the record store? Some blame mass marketers, like Best Buy, and discount marketers, like Target, who certainly took sales from record stores. Though record stores have been an important place in the cultural landscape, they are a business built around a rapidly changing technology. The same cultural preferences for selection, convenience, and portability that increased sales and diversified musical genres have more recently shifted demand toward digital downloads. Instead of paying for whole albums, consumers can buy just a few tracks and carry them everywhere. Virtual stores, such as iTunes, eliminate the need for physical stores. But the shift to digital didn't stop there. Digital technology makes it easy to "share" songs, and an estimated 40 billion tracks were illegally downloaded in 2008. That's 95 percent of all music downloads. Not all those tracks would translate to legal sales, but "sharing" does explain why fewer and fewer young people go to music stores anymore. The few independent record stores that remain open cater to a "graying" clientele. These trends have enormous economic implications for the music industry as well as perhaps profound repercussions on the role of music in popular culture.

The last of the independent record stores?

Rapid Change Essays

These essays emphasize the issues that arise as local places contend with environmental, economic, cultural, and political changes that occur at unprecedented speed.

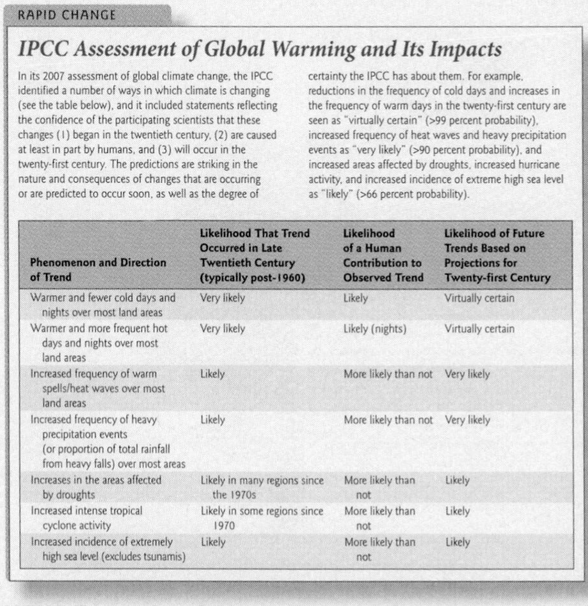

RAPID CHANGE

IPCC Assessment of Global Warming and Its Impacts

In its 2007 assessment of global climate change, the IPCC identified a number of ways in which climate is changing (see the table below), and it included statements reflecting the confidence of the participating scientists that these changes (1) began in the twentieth century, (2) are caused at least in part by humans, and (3) will occur in the twenty-first century. The predictions are striking in the nature and consequences of changes that are occurring or are predicted to occur soon, as well as the degree of certainty the IPCC has about them. For example, reductions in the frequency of cold days and increases in the frequency of warm days in the twenty-first century are seen as "virtually certain" (>99 percent probability), increased frequency of heat waves and heavy precipitation events as "very likely" (>90 percent probability), and increased areas affected by droughts, increased hurricane activity, and increased incidence of extreme high sea level as "likely" (>66 percent probability).

Phenomenon and Direction of Trend	Likelihood That Trend Occurred in Late Twentieth Century (typically post-1960)	Likelihood of a Human Contribution to Observed Trend	Likelihood of Future Trends Based on Projections for Twenty-first Century
Warmer and fewer cold days and nights over most land areas	Very likely	Likely	Virtually certain
Warmer and more frequent hot days and nights over most land areas	Very likely	Likely (nights)	Virtually certain
Increased frequency of warm spells/heat waves over most land areas	Likely	More likely than not	Very likely
Increased frequency of heavy precipitation events (or proportion of total rainfall from heavy falls) over most areas	Likely	More likely than not	Very likely
Increases in the areas affected by droughts	Likely in many regions since the 1970s	More likely than not	Likely
Increased intense tropical cyclone activity	Likely in some regions since 1970	More likely than not	Likely
Increased incidence of extremely high sea level (excludes tsunamis)	Likely	More likely than not	Likely

NEW! Extensive Climate Change Coverage

The rapidly changing debate on global warming is discussed, including the IPCC's Fourth Assessment Report and the 2009 Copenhagen Climate Conference (Chapters 2 and 4). Expanded treatment of the global carbon cycle and related connections among the biosphere, atmosphere, and human resource use are provided in Chapters 4 and 9.

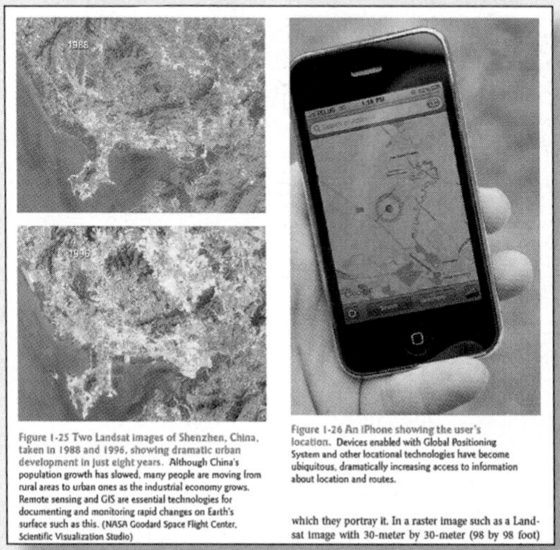

Figure 1-25 Two Landsat images of Shenzhen, China, taken in 1988 and 1996, showing dramatic urban development in just eight years. Although China's population growth has slowed, many people are moving from rural areas to urban ones as the industrial economy grows. Remote sensing and GIS are essential technologies for documenting and monitoring rapid changes on Earth's surface such as this. (NASA Goddard Space Flight Center, Scientific Visualization Studio)

Figure 1-26 An iPhone showing the user's location. Devices enabled with Global Positioning System and other locational technologies have become ubiquitous, dramatically increasing access to information about location and routes.

which they portray it. In a raster image such as a Landsat image with 30-meter by 30-meter (98 by 98 foot)

NEW! Modern Geographic Tools

Today, maps are drawn and updated rapidly using geographic information systems (GIS) and digital cartographic tools. GIS has become essential and ubiquitous on geographers' desks, and their importance is clear in this book. An expanded discussion of GIS technology has been added in Chapter 1, and the use of GIS in geographic problem-solving is illustrated throughout.

Explore unparalleled multimedia resources

PEARSON
mygeoscience™ place

www.mygeoscienceplace.com

This unique portal contains the book's premium website, which offers a variety of multimedia resources for students and instructors.

Tailored to *Introduction to Geography: People, Places, & Environment*, Fifth Edition

Features include:

- **eText version of the book** with linked/integrated media
- **NEW!** Assignable *MapMaster*™ Interactive Maps with assessments
- **NEW! Interactive Geoscience Animations** with assessments
- **NEW! Geography Videos** with assessments
- **NEW! Flashcards**
- **NEW! RSS Feeds**
- **Annotated Resources** for further exploration
- **Quizzes** and **Problems** mapped closely to the book to aid assessment goals
- **Web links**
- **Class Manager** and **GradeTracker Gradebook** for instructors

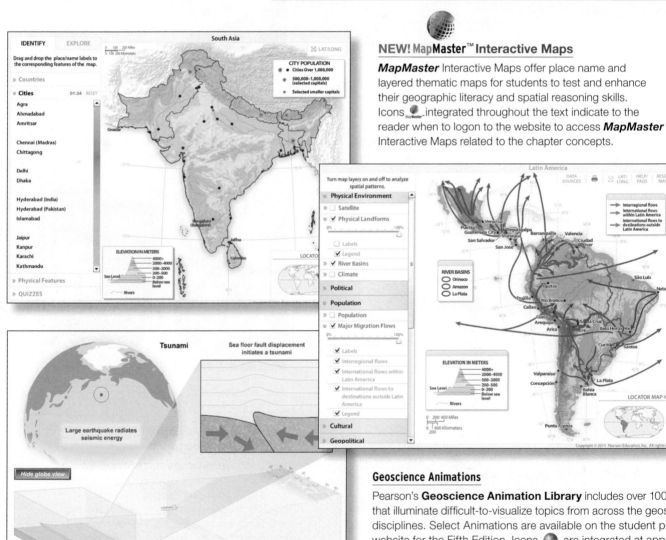

NEW! MapMaster™ Interactive Maps

MapMaster Interactive Maps offer place name and layered thematic maps for students to test and enhance their geographic literacy and spatial reasoning skills. Icons integrated throughout the text indicate to the reader when to logon to the website to access *MapMaster* Interactive Maps related to the chapter concepts.

Geoscience Animations

Pearson's **Geoscience Animation Library** includes over 100 animations that illuminate difficult-to-visualize topics from across the geoscience disciplines. Select Animations are available on the student premium website for the Fifth Edition. Icons are integrated at appropriate points throughout the text to indicate when students should logon to the website to view an Animation corresponding to the chapter material.

Geography Videos

These videos from **Television for the Environment's** global *Earth Report* and *Life* series cover a wide array of issues affecting people and places today, such as:

- the serious health risks of pregnant women in Bangladesh
- Ghana's struggle to compete in the global market
- poor land management promoting the devastation of river systems in Central America
- the struggles for electricity in China and Africa

...and more.

Select Videos are available on the student premium website. Icons 🌀 are integrated at appropriate points throughout the text to indicate when students should logon to the website to view a video related to the chapter concepts.

Learn geography concepts with Google Earth™

Pearson's *Encounter Series* provides rich, interactive explorations of geoscience concepts through **Google Earth** activities. All chapter explorations are available in the print workbooks as well as via online quizzes and downloadable PDFs, accommodating different classroom needs. All explorations include corresponding Google Earth KMZ media files, available for download at **www.mygeoscienceplace.com**

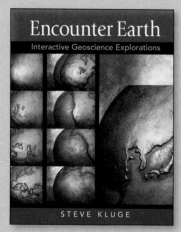

Encounter Earth: Interactive Geoscience Explorations
by Steve Kluge
© 2009 • Paper • 72 pages
978-0-321-58129-7 • 0-321-58129-6

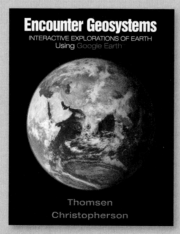

Encounter Geosystems: Interactive Explorations of Earth Using Google Earth
by Charles E. Thomsen and Robert W. Christopherson
© 2010 • Paper • 100 pages
978-0-321-63699-7 • 0-321-63699-6

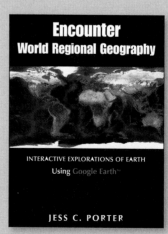

NEW! *Encounter World Regional Geography: Interactive Explorations of Earth Using Google Earth*
by Jess C. Porter
© 2011 • Paper • 176 pages
978-0-321-68175-1 • 0-321-68175-4

About Our Sustainability Initiatives

This book is carefully crafted to minimize environmental impact. The materials used to manufacture this book originated from sources committed to responsible forestry practices. The paper is Forest Stewardship Council (FSC) certified.

The printing, binding, cover, and paper come from facilities that minimize waste, energy consumption, and the use of harmful chemicals.

Pearson closes the loop by recycling every out-of-date text returned to our warehouse. We pulp the books, and the pulp is used to produce items such as paper coffee cups and shopping bags. In addition, Pearson aims to become the first climate neutral educational publishing company.

The future holds great promise for reducing our impact on Earth's environment, and Pearson is proud to be leading the way. We strive to publish the best books with the most up-to-date and accurate content, and to do so in ways that minimize our impact on Earth.

Mixed Sources
Product group from well-managed forests, controlled sources and recycled wood or fiber
www.fsc.org   Cert no. SW-COC-002985
©1996 Forest Stewardship Council

Prentice Hall
is an imprint of

U.S. National Geography Standards

In 1994 the U.S. Congress adopted Goals 2000: The Educate America Act (Public Law 103-227). This act listed geography among the fundamental subjects of a national curriculum. Geographical understanding, wrote Congress, is essential to achieve "productive and responsible citizenship in the global economy." Several academic and scholarly geographical organizations collaboratively produced an agreed-upon core of geographic material and ideas, which was published as *Geography for Life: The National Geography Standards*. These 18 standards specify the geographical subject matter and skills that U.S. students should master.

The goals demonstrate the degree to which geographic knowledge is essential for both understanding and effectively managing environmental and human relations in the twenty-first century. They were established in the hope that all persons educated in the public school system become geographically knowledgeable. These goals were not reached by 2000, but they remain a national target. In this book, we go beyond these standards in the treatment of both subject matter and thinking skills, but we provide here the outline of the goals in order to demonstrate the great breadth of the field.

The geographically informed person knows and understands:

The World in Spatial Terms

1. How to use maps and other geographic representations, geospatial technologies, and spatial thinking to understand and communicate information.
2. How to use mental maps to organize information about people, places, and environments in a spatial context.
3. How to analyze the spatial organization of people, places, and environments on Earth's surface.

Places and Regions

4. The physical and human characteristics of places.
5. That people create regions to interpret Earth's complexity.
6. How culture and experience influence people's perceptions of places and regions.

Physical Systems

7. The physical processes that shape the patterns of Earth's surface.
8. The characteristics and spatial distribution of ecosystems and biomes on Earth's surface.

Human Systems

9. The characteristics, distribution, and migration of human populations on Earth's surface.
10. The characteristics, distribution, and complexity of Earth's cultural mosaics.
11. The patterns and networks of economic interdependence on Earth's surface.
12. The processes, patterns, and functions of human settlement.
13. How the forces of cooperation and conflict among people influence the division and control of Earth's surface.

Environment and Society

14. How human actions modify the physical environment.
15. How physical systems affect human systems.
16. The changes that occur in the meaning, use, distribution, and importance of resources.

The Uses of Geography

17. How to apply geography to interpret the past.
18. How to apply geography to interpret the present and plan for the future.

Maintenance workers at a solar powerplant in Wuhan city, China. While China is still heavily dependent on fossil fuels, it has also become a major producer and consumer of renewable energy technology.

1

Introduction to Geography

*S*hine Science and Technology Co., Ltd., manufactures solar panels in Jaingsu Province, China. The company manufactures photovoltaic (PV) cells, solar modules, solar panels, polycrystalline solar modules, monocrystalline solar modules, and solar power systems that are marketed around the world. China is positioning itself in two important ways: (1) to generate large amounts of energy to meet its own needs, using solar technology, and (2) to be one of the leading, if not the top, suppliers of solar panels in the world. New plants that will build solar panels are under construction in many parts of China, as are facilities that will use those panels.

Silicon is one of the raw materials used in PV cells, and our experience in the past few decades with another silicon-based technology—computer chips—suggests that economies of scale can build, rapidly leading to dramatic decreases in prices and increases in performance. This is good news because solar-generated electricity is still not fully competitive with conventional technologies such as coal or nuclear power, but modest reductions in the cost of solar power could lead to rapid growth in this relatively clean technology.

Rapid changes in the global distribution of wealth, industrial production, and natural resource use are underway, with profound implications for all aspects of human society and physical environments. Geography's integrative perspective on these changes helps us to understand their far-reaching effects.

What Is Geography?

Many readers of this textbook may not have studied geography since grade school, when *geography* may have meant simply memorizing place names. Knowing place names is a tool for studying geography, just as counting is a tool for studying mathematics and reading is a tool for studying literature. In geography, as in any other field of study, we begin by gathering some basic information. In the case of geography, this is the *where*. Knowing where places are, however, is not all there is to geography.

Once we know the names and locations of environmental features, people, and activities, we can proceed to the challenging questions of significance: *Why* are people and activities located where they are? How do the features and activities at any one place *interact* to make that place unique, and what are the *relationships* among different places? What factors or forces *cause* these distributions of things and activities? How and why are these distributions *changing*? And how, perhaps, can we *alter* those distributions for greater human convenience, profit, or well-being?

Geography is the study of the interaction of all physical and human phenomena at individual places and of how interactions among places form patterns and organize space. **Physical geography** studies the characteristics of the physical environment. When geography concentrates on topics such as climate, soil, and vegetation, it is a natural science. **Human geography** studies human groups and their activities, such as language, industry, and the building of cities; human geography is a social science. **Cultural geography**, a subfield of human geography, focuses specifically on the role of human cultures. Physical and human geographers share both their approach and a great deal of information, and their analyses of the landscape always weave their understanding together. Thus, geography bridges the physical sciences and the social sciences. **Cartography** (mapmaking), **remote sensing** (mapping Earth from satellites and aircraft), and **geographic information systems (GIS)** provide tools that help both physical and human geographers store, display, and analyze geographic data.

Geographers investigate the processes underlying all observed distributions and patterns. Technology has accelerated the pace of change, and knowledge of forces of change is critical to understanding and managing global problems that might arise in the future.

Geography offers a way of thinking about problems, and geographers are particularly well equipped to understand interactions among different forces affecting a place. For example, to understand hunger in Africa, geographers examine the relationships among climate, soils, agricultural practices, population growth, food prices, environmental degradation, international economic forces, and political unrest. The interrelationships among factors affecting places help us understand why humans behave as they do. No other scientific discipline takes this integrative approach.

MapMaster™
Place Names

The Development of Geography

Geography in the classical Western world   Every language has a word for the study that we call "geography," but the Greek word that we use, "Γεωγραφία" which means "Earth description," was the title of a book by the Greek scholar Eratosthenes (c. 276–194 B.C.). He was the director of the library in Alexandria, Egypt. The library was the greatest center of learning in the Mediterranean world for hundreds of years. No copies of Eratosthenes's book have survived, but we know from other authors that Eratosthenes accepted the idea that Earth is round and that he even calculated its circumference with amazing precision. Eratosthenes also drew detailed maps of the world as it was known to him. Hipparchus (180–125 B.C.), a Greek scholar and later director of the library, was the first person to draw a grid of imaginary lines on Earth's surface to locate places precisely.

After the great age of Greek civilization gave way to the Roman era, the Roman Empire produced a number of geography scholars and compilations of geographic learning. These culminated in *Geography* by Ptolemy, a Greek scholar who worked at the library in Alexandria between A.D. 127 and 150. Subsequently, in the long period between the fall of the Roman Empire (A.D. 476) and the European Renaissance of the fifteenth century, Western civilization accumulated little additional geographic knowledge. Much knowledge was actually lost in Western Europe, but preserved by Arab scholars.

Geography in the non-European world   Outside Europe, geography made considerable advances. The expansion of the religion of Islam (discussed in Chapter 7) fostered pilgrimages to holy sites and forged contacts among Muslims from Spain to India and beyond. The Arabs who first practiced Islam were also involved in extensive trade networks that required practical knowledge of geography. Muslim scholars, including al-Edrisi (1099?–1154), ibn-Battuta (1304?–1378), and ibn-Khaldun (1332–1406), produced impressive texts describing and analyzing both physical environments and also the customs and lifestyles of different peoples.

China also developed an extensive geographic literature. The oldest-known work, *The Tribute of Yu*, dates from the fifth century B.C. and describes the physical geography and natural resources of the various provinces of the Chinese empire. It interprets world geography as a nest of concentric squares of territory. The innermost zone is China's imperial domain. The imperial domain is surrounded by the empire, and beyond the empire lies an outermost zone inhabited by "barbarians." The world's oldest map that shows distances marked with numbers was engraved on copper plate about 325 B.C. In A.D. 267, Phei Hsiu (often called the father of Chinese cartography) made an elaborate map of China's empire. In the following centuries, Chinese Buddhists occasionally wrote accounts of travel to India to visit places sacred to the history of Buddhism, which had spread into China. Chinese maritime trade extended throughout Southeast Asia and the Indian Ocean to East Africa, but the Chinese eventually withdrew from more extensive exploration. Some scholars argue that the Chinese may even have reached North America, but the evidence for this is insubstantial since few Chinese descriptive geography texts of any Chinese explorations survive.

The Japanese and the Koreans also engaged in East Asian trade. In fact, to our knowledge, the greatest world map produced before European exploration was made in Korea in 1402 (Figure 1-1). This map, the Kangnido, drew on the combined knowledge of Korea, Japan, and China, including Islamic sources that were known to the Chinese. As a result, this map includes not only East Asia but also India, the countries of the Islamic world, Africa, and even Europe. It demonstrates a far more extensive knowledge of the world than the Europeans had at the time.

The peoples of Africa and of the Western Hemisphere may have had access to geographic accounts of the parts of the world known to them, but either these works have not survived or modern scholars have not yet fully inventoried and studied them.

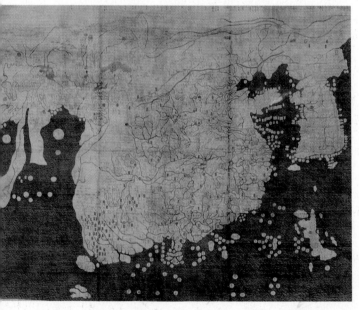

Figure 1-1 The Kangnido (detail). When Columbus embarked on his voyage, he knew nothing of Korea, yet the world's finest map was then hanging in the royal palace there. The Kangnido included not only China and Japan but showed India, the countries of the Middle East, and Europe itself. Spain was easy to recognize on this map, which even showed Columbus's native Genoa along the shores of the Mediterranean. The image of Europe was not exact, but no map in Europe had an image of Asia that was as good as this map's image of Europe.

The revival of European geography Beginning in the fifteenth century, Europe's world exploration and conquest introduced so much new geographic knowledge that Europeans were challenged to devise new methods of cataloging, or organizing, it all.

One attempt at organizing the information gathered by the European voyages of discovery is the book *General Geography* (1650) by Bernhard Varen (*Varenius* in Latin; 1622–1650), a German who taught at the University of Leiden, in Holland. Varen differentiated two approaches to geography: *special geography* and *general geography*. Special geography describes and analyzes places in terms of categories such as local population, customs, politics, economy, and religion. Today, we call this approach **regional geography.** Regional geographers usually begin regional studies with a description of the local physical environment—the climate, landforms, soils, and other physical attributes—and then proceed with an inventory much like Varen's, weaving these factors together as they interrelate. Throughout this book we make frequent references to regions such as Europe or Latin America.

What Varen labeled general geography examines topics of universal application or occurrence. We might study, for example, the geography of climate, water, vegetation, or minerals. This notion corresponds to what we now refer to as **topical geography,** or **systematic geography,** and it is the basic approach taken for the outline of this textbook. Individual geographers concentrate on topics as diverse as the geography of soils (pedology), of life forms (biogeography), of politics (political geography), of economic activities (economic geography), and of cities (urban geography). The Association of American Geographers (AAG), a professional scholarly organization, currently recognizes 55 topical specialties (Table 1-1). Regional and topical geography are complementary. A regional study covers all topics in the region under review, whereas a topical study notes how a particular topic varies across regions.

Varen's book was a standard reference for over 100 years. Even Sir Isaac Newton (1642–1727) edited two editions in Latin, one of which was studied by students at Harvard in the eighteenth century.

The human–environment tradition We cannot fully understand the physical environment without understanding humankind's role in altering it, nor can we fully understand human life without understanding the physical environment in which we live.

The German explorer and naturalist Alexander von Humboldt (1769–1859) wove together virtually all knowledge of Earth sciences and anthropology in his great multivolume book, *Cosmos*. Von Humboldt's brilliant interrelating of the phenomena of physical nature and the world of humankind had an enormous intellectual impact in the United States (Figure 1-2). George Perkins Marsh (1801–1882), an American scholar and diplomat, expanded on von Humboldt's

TABLE 1-1 Topical Specialties Recognized by the Association of American Geographers

Agricultural Geography	Geographic Information Systems	Planning, Regional
Applied Geography	Geographic Theory	Planning, Urban
Arid Regions	Geomorphology	Political Geography
Biogeography	Global Change	Population Geography
Cartography	Hazards	Quantitative Methods
Climatology	History of Cartography/Historical Cartography	Recreational Geography
Cultural Ecology	Historical Geography	Regional Geography
Cultural Geography	History of Geography	Remote Sensing
Developmental Studies	Land Use and Conversion	Resource Geography
Earth Science	Librarianship, Geographical	Rural Geography
Economic Development	Location Theory	Social Geography
Economic Geography	Marine Resources	Soils Geography
Educational Geography	Marketing Geography	Teaching Techniques
Energy	Medical Geography	Transportation and Communication
Environmental Perception	Military Geography	University/Other Administration
Environmental Science	Mountain Environments	Urban Geography
Environmental Studies (Conservation)	Oceanography	Water Resources
Field Methods	Physical Geography	
Gender	Planning, Environmental	

Figure 1-2 Alexander von Humboldt. This portrait bust of the explorer and naturalist Alexander von Humboldt was the first statue of a non-U.S. citizen to be erected in New York City's Central Park. It was installed in 1869, marking the 100th anniversary of von Humboldt's birth. Today, he gazes across the street at the American Museum of Natural History, where statues of the explorers Lewis and Clark gaze back.

theme of the interconnections between humankind and the physical environment. While Marsh served as U.S. ambassador to several Mediterranean countries, he was struck by humankind's destruction of an environment that ancient authors had described as lush and rich. Marsh's book *Man and Nature, or Physical Geography as Modified by Human Action* (1864) was one of the earliest key works in what would become today's environmental movement.

Why study geography?   Who is a geographer today? And why should anyone study geography? Today, almost all of us are geographers, whether or not we call ourselves that. Geographers include people who plan for future schools; develop real estate; manage parklands, forests, or water supplies; use maps to plan journeys or decide on a neighborhood in which to live; farmers who choose which crop to plant on the basis of world markets; and journalists, diplomats, and executives of multinational corporations.

The information that any citizen needs in order to make an informed decision about an important question of the day is largely geographic. And as world events daily force rapid change upon us, the geography of our world changes too. Almost any topic of the day's news will bring about redistributions—new geographies—of people and activities. Studying geography will not always help you predict the future, but it will help you understand the forces behind these changes. Therefore, when changes come, you may be less surprised. Sometimes you will be able to see changes coming, make changes work for you, or even make the changes yourself.

Geography is one of the most diverse fields of academic study, and it offers an introduction to a unique breadth of careers. A great variety of professions in-

Figure 1-3 A geographer at work. Kristy York, NRCS Soil Conservationist in Audubon County, Iowa, uses a Global Positioning System (GPS) and a personal digital assistant (PDA) to record natural resource data in the field for the National Resources Inventory.

volve studying or "doing" geography, even though the persons at work may not call themselves geographers (Figure 1-3). As you read through the topics in this text, which range from newspaper distribution to agricultural supplies, from urban planning to the delineation of voting districts, and from multinational marketing of consumer goods to international treaties on environmental preservation, think of the many ways you might develop a career analyzing or managing any of these activities.

Geographers' studies of *where, what, when, why,* and *why there* necessarily prompt thoughts about *where will* to predict future patterns, and of *how ought.* Geographers' expertise can inform public debates on major issues of the day—not just to help us understand the world, but to help improve it.

Contemporary Approaches in Geography

All geographers collect and analyze geographic information, but they focus on different topics and employ different analytical approaches. Most contemporary geographers employ three analytical methods:

- Area analysis, which integrates the geographic features of an area or a place.
- Spatial analysis or locational analysis, which emphasizes interactions among places.

- Geographic systems analysis, which emphasizes the understanding of physical and human systems and the interactions among them.

Let us look briefly at each of these methods.

Area Analysis

Geographers have a long tradition of surveying, describing, and compiling geographic data about places. Each place in the world occupies a unique location and possesses a unique combination of human behavior and environmental processes that give it a special character. In compiling and analyzing Earth's places, geographers rely on three basic ideas: site, situation, and the concept of region.

Site   Geographers often speak of specific sites and site characteristics. **Site** describes the exact location of a place and can be described either in terms of latitude and longitude (discussed below) or in terms of the characteristics of the place. Each place has a unique combination of physical and human characteristics. Travelers do not describe the latitude and longitude of a vacation spot they have visited. Instead, they report on the interesting people and customs, different foods, and varied landscapes that made their trip memorable. They may, however, also report that they spoke the same language, used the same credit cards, and saw the same movies as at home. This mix of unique features and common features is what geographers mean when they talk about place.

Relative location and situation   The location of a place relative to other places is called its **situation**, and knowledge of a place's situation helps us understand how it interacts with the rest of the world. **Relative location** is location in reference to another place. Relative location may describe accessibility, which is indicated by such terms as *nearer* and *farther, easier* or *more difficult to reach, between,* and *on the way* or *out of the way.* Accessibility can be a resource as valuable as mineral deposits or fertile soil.

Relative location is almost constantly changing, as changes in transportation routes and technologies, as well as developments in communication continuously redistribute accessibility. The opening of a new highway, for example, stimulates the development of new housing and new roadside services, such as shopping malls. At the same time, however, the new highway can reroute traffic away from older shopping malls, and cut them off from customers.

Great cities rise and countries often prosper if they are advantageously situated within national or international patterns of trade, but isolated areas usually lose their wealth and stagnate. One of history's greatest redefinitions of relative location occurred when Europeans learned how to sail around Africa in order to reach Asia.

Much of the commerce that had previously moved by long and difficult caravan journeys by land across Asia and Africa was drawn to the seacoasts. New seacoast cities sprang up, such as Bombay (today Mumbai) in India, Rangoon in Burma (today known as Yangon in Myanmar), and Hong Kong on the coast of China. Major cities in the interior of the Eurasian and African continents were suddenly "on the back road." Some of these, such as Kashgar, a city in western China visited by the explorer Marco Polo in 1275, and Samarkand, a city thousands of years old in today's Uzbekistan, have never recovered their earlier importance (Figure 1-4).

Changes in transportation technology also redistribute accessibility. In the 1950s, the Suez Canal was the principal route of oil shipments to Western Europe from the Middle East. New oil tankers, however, grew so large that they were incapable of squeezing through the canal, so they had to sail around Africa. This rerouting placed Cape Town, South Africa, unaccustomed to such traffic, "between" the Persian Gulf and Rotterdam, Europe's chief port for importing oil. Cape Town's ship supply and repair facilities won new business.

Each day's news reports the opening of a new transportation or communication facility that will redistribute activities. In 2006, new pipelines opened that carry oil and natural gas from the Caspian Sea area through Azerbaijan, Georgia, and Turkey, delivering gas to markets in Turkey and oil to the port of Cehyan on the Mediterranean (Figure 1-5). These pipelines run through earthquake-prone and politically volatile regions, and this complicates their operation. They bind Turkey, and therefore Europe, to the Caucasus region both politically and economically for decades to come.

A place's relative location can also change if the territorial scale of an activity's organization changes—that is, the extent of territory within which that activity occurs. For example, when trade barriers between countries fall, activities redistribute themselves. Significant redistributions are occurring throughout North America as the economies of the United States, Canada, and Mexico merge through the trade pacts discussed in Chapter 13. Activities in Europe are readjusting from a national to the supranational scale of organization represented by the European Union.

Improvements in transportation and communication links have "shrunk" Earth, so today many activities have expanded their scale of organization to cover the whole globe, which is a process called **globalization.** An advertisement for one U.S. bank shows a satellite photograph of Earth drifting in space. The headline reads, "We do business in only one place." Economic globalization has far outpaced cultural or political integration, especially since 9/11 triggered new national security concerns around the world. Tensions among the different economic, cultural, and political organizing activities generate much of today's international news.

Conversely, when a large territory is broken up into smaller individually organized territories, activities are redistributed and places' accessibility reduced. Until 1991, the U.S.S.R. organized 15 percent of Earth's land surface under one highly unified government before it broke up into 15 independent countries (see Figure 13-23). The

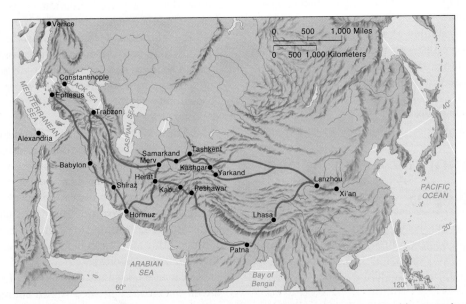

Figure 1-4 Internal Asian trade routes in 1400. Trabzon, Kashgar, and Merv were once great cities on major trade routes, but when Europeans opened sea routes around Africa and Asia, these and many other cities, like the ones shown here, declined. Few of these cities ever recovered their former economic strength.

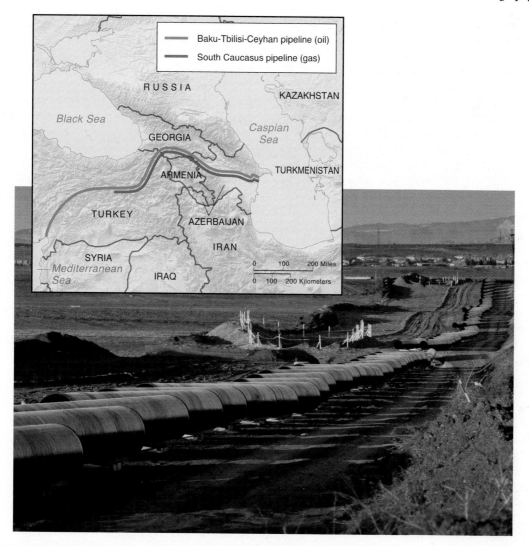

Figure 1-5 An oil pipeline under construction connecting Baku, Azerbaijan, on the Caspian Sea coast, with Ceyhan, Turkey, on the Mediterranean Sea. The pipeline strengthens economic linkages between the Caspian Sea region and Europe.

political change required reorganizing the scale of all political, economic, and cultural activities throughout the vast territory.

The concept of a region   For purposes of geographic study, we divide Earth into **regions,** which are areas defined by one or more distinctive characteristic or feature, such as climate, soil type, language, or economic activity. Geographers define regions and describe and analyze similarities and differences among them.

Probably the most familiar of all maps of regions is the map of states (Figure 1-6). People have divided Earth's land surface into almost 200 countries, ranging from Russia, occupying one-ninth of Earth's land area, to microstates such as Andorra, Monaco, and Singapore. This regionalization of Earth influences almost all other human activities. Geographers study such patterns of how people organize their societies and occupy land.

Of all the features that can be found in any landscape, a geographer chooses only a few specific criteria and then maps those criteria to define a region. The criteria chosen may be either physical or cultural.

Geographers distinguish three kinds of regions. A **formal region** is one that exhibits essential uniformity in one or more physical or cultural features, such as a country or a mountain range. A **functional region,** by contrast, is one defined by interactions among places, such as trade or communication. The city of Chicago, for example, is a formal region, whose government covers the legal limits of the city's incorporation. The many commuters and shoppers who circulate daily throughout the city and its suburbs, however, would more readily identify a region larger than Chicago; they would identify the functional region of the Chicago metropolitan area, which includes parts of Illinois, Indiana, and even Wisconsin. The desire of people in northwest Indiana

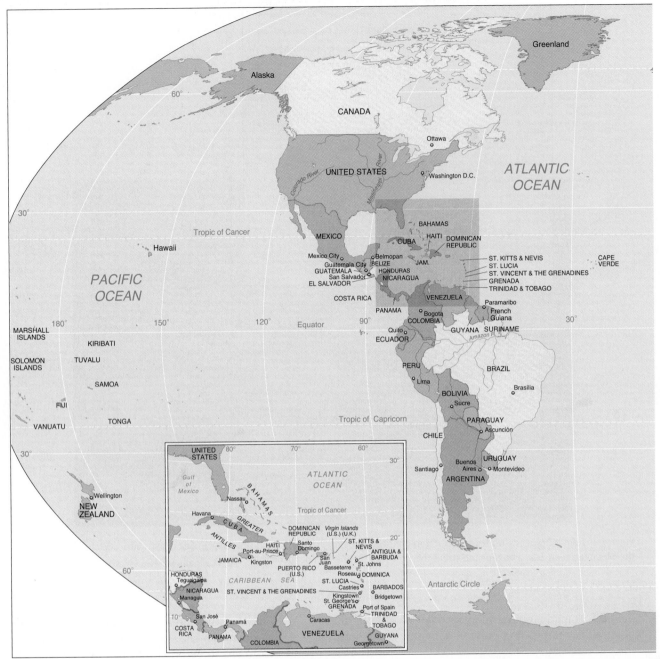

Figure 1-6 Political map of the world. The world includes almost 200 sovereign states. The number has increased markedly in recent decades, and many of the names of states shown here may be unfamiliar.

MapMaster
Place Names

to be in the same time zone as Chicago confirms that northwest Indiana is part of the Chicago functional region (Figure 1-7).

The third type of region is a **vernacular region.** *Vernacular* means "everyday language," and vernacular regions are defined by widespread popular perception of their existence by people within or outside them (Figure 1-8).

The choice of criteria to define a region follows from the kinds of questions a geographer asks about a set of places. Therefore, a region is a concept, an abstract idea. The geographer's concept of a region is similar to the historian's concept of an "era" or "period." Time, like

space, is continuous, but historians divide time into units of analysis called periods to suit their purposes. Art historians define "styles," sociologists define "societies," and many other scholarly fields define their conceptual units of analysis.

Sometimes the features geographers choose to demarcate regions are clear in the landscape. For example, a region defined as a certain forest could be mapped distinctly. Individual countries are usually clearly defined on the landscape by national borders.

In other cases, however, the regions that geographers define are less distinct on the landscape. For example, Chapter 2 maps regions of different climates. On Earth,

those climatic regions are not sharply defined, but rather they merge into one another imperceptibly. On our map, however, the precise location of the line between one climatic region and its neighbor is determined by the geographer's definition and trained discrimination.

Regions defined by cultural phenomena often merge or overlap as well. For example, people who live between two cities may listen to radio stations located in both cities. If, however, a geographer wants to delineate sharply the market regions of two radio stations broadcasting from the two cities, he or she might choose to draw a line dividing the area where most households listen to the station broadcasting from City A from the area where most households listen to the station broadcasting from City B (Figure 1-9).

Sometimes people confuse regions based on natural phenomena with regions based on cultural criteria. For example, many people refer to Africa as a region. Africa is a continent, but the physical environment is not homogeneous across that vast continent, nor do its many peoples share one culture or historical experience. Historically, the Sahara Desert divided peoples more effectively than the Mediterranean Sea did.

The countries of the world also can be grouped into regions according to different criteria. Grouping them by continent is fairly easy, as is grouping by membership in certain international organizations such as NATO (North Atlantic Treaty Organization) or OPEC (Organization of Petroleum Exporting Countries). At other times, it is useful to categorize them by broadly similar characteristics,

Figure 1-7 Metropolitan Chicago, a functional region. The metropolitan region of Chicago includes parts of Wisconsin and Indiana, as well as northeastern Illinois. Commuter trains link Chicago with surrounding communities. While most of Indiana is in the Eastern Time Zone, several counties in northwestern Indiana are in the Central Time Zone to facilitate their links with Chicago, which is on Central Time.

such as large countries, democratic countries, industrial countries, rich countries, and poor countries.

There are enormous differences in levels of wealth among the almost 200 countries of the world today, and this distinction is a particularly important one for geographers. Wealth affects virtually every aspect of a country's human environment, from food availability and disease to pollution and political power. It is useful to make generalizations about geographic patterns of wealth and power, yet by doing so we necessarily obscure important differences between countries. It is important that you pay close attention to the ways such regions are defined.

In this book, we will sometimes refer to *rich* and *poor* countries, but these should be interpreted as referring to the overall material and economic wellbeing of countries relative to one another. In addition, not everyone in a rich country has a lot of money and not everyone in the world's poorest countries is poor. In Chapter 13, we will describe the challenge of development, which means improving people's material well-being, especially in poorer countries. We will sometimes describe countries where people enjoy advantages such as better health, higher incomes, and effective education systems as *developed* or *more developed*; countries trying to reach these goals are described as *developing* or *underdeveloped*. The patterns of wealthier versus impoverished countries are sometimes described geographically as the *core* and *periphery* or the *Global North* and *Global South*.

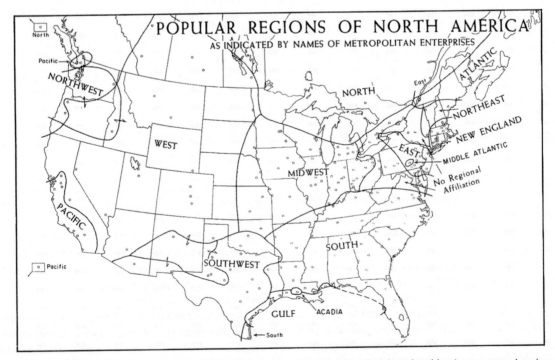

Figure 1-8 Vernacular regions of North America. The regions on this map reflect how local business enterprises in metropolitan areas (the small circles) define themselves in telephone directories. For example, within the region labeled "the South" on this map, businesses most probably identify themselves by that regional designation, such as the Southern Telephone Company. Thus, the regions drawn here are vernacular, defined by popular local everyday language.

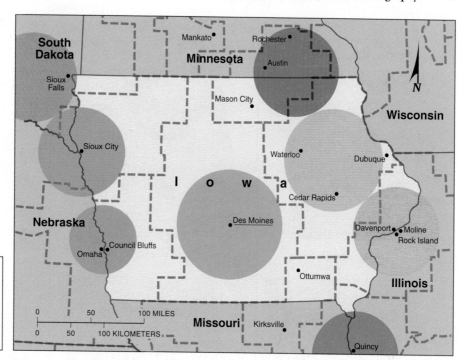

Figure 1-9 Iowa's broadcast television viewing areas. The colored circles on this map are known as areas of dominant influence (ADIs). The circles are broadcast areas within which at least 50 percent of viewers are watching the NBC affiliate station for that market region. Cable television is changing this geography of viewership.

These terms are useful generalizations for describing our world, but they are never fully accurate and should be interpreted with care.

Spatial Analysis

A second approach to geographic inquiry, called spatial analysis or locational analysis, looks for patterns in the distribution of human actions and environmental processes and in movements across Earth's surface.

Distribution The **distribution** of a phenomenon means its position, placement, or arrangement throughout space. Geographers recognize distributions and use three terms to define the distribution of any phenomenon under study: density, concentration, and pattern (Figure 1-10).

Density **Density** describes the frequency of occurrence of a phenomenon in relation to geographic area. Density is usually expressed as a number per square kilometer or square mile. Examples include road density (the number of kilometers of roads per square kilometer) and population density (the number of people per square kilometer).

Many statistics are reported using a political unit as their measure, such as by state or country. In 2000, the population of Russia was 147 million, and the population of Belgium was 10 million, so Russia had almost 15 times as many people as Belgium. Russia, however, is much larger than Belgium, so the population density of Russia was only 9 persons per square kilometer

(23 per square mile), whereas that of Belgium was 333 per square kilometer (862 per square mile)—37 times that of Russia.

Concentration **Concentration** refers to the distribution of a phenomenon within a given area. If all the occurrences are found in close proximity, the distribution would be described as concentrated, but if they are scattered far from each other, the distribution would be described as dispersed. For example, in many parts of the world farmers live in villages and travel to their fields in the countryside. The population of such an agricultural landscape is concentrated in villages. In North America, on the other hand, most farmers live in isolated farmhouses located on the land they farm, so the population is dispersed.

Pattern **Pattern** refers to the geometrical arrangement of objects within an area. For example, in most modern cities, streets are arranged in a rectangular grid pattern, whereas in older cities the street layout is more irregular. In areas where rock structures exert strong controls on stream erosion, we sometimes see streams with many right-angle bends; this pattern is called a trellis pattern. Conversely, in areas without structural control, streams and their tributaries form branching or treelike patterns, called dendritic patterns.

Movement

Distance Interaction among people and places occurs across distance, but there are several ways to measure distance. **Distance** can be measured absolutely, in

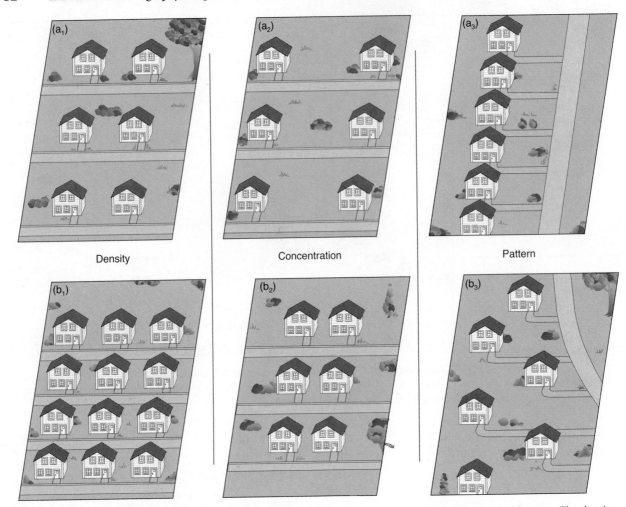

Figure 1-10 Density, concentration, and pattern. Assume that the area represented in all figures is 1 hectare. The density is six houses per hectare in a_1, whereas in b_1 it is 12 houses per hectare. The concentration of the houses might be described as dispersed in a_2 and clustered in b_2. The pattern of the distribution of houses is linear in a_3 but irregular in b_3.

terms of miles or kilometers, but it can also be measured in other ways. It can, for instance, be measured in time. Most people use time as a distance measurement in their daily lives without realizing it: "The store is about 20 minutes from here." Sometimes it can be quicker to reach a place that is far away than to reach another place that is, in absolute distance, closer. This may be the result of rugged **topography** (surface relief) in one direction, direct or winding routes, or different methods of transportation (Figure 1-11).

Geographers also frequently measure distances in terms of the cost required to overcome the distance. Transportation over land is usually more expensive than transportation over water, so two seaports that are far apart in absolute distance may be close in *cost distance*. They may even be closer to each other in cost distance than either is to cities inland from itself. That explains why many political units grew up around the shores of bodies of water. Classical Greece, for example, consisted of cities surrounding the Aegean Sea, and the Roman Empire developed around the Mediterranean Sea.

No matter how we choose to measure distance, we must make some effort to overcome distance when we want to move or transport items. We call that effort or cost the **friction of distance.** The friction of distance usually limits interactions across great distances.

Distance decay   The presence or impact of any phenomenon may diminish away from its origin, just as the volume of a sound diminishes the further it gets from its source. This phenomenon is called **distance decay.** As we travel away from any city, for example, the percentage of the local population that reads that city's newspapers, tunes in to its radio and television stations, or relies on its other services shrinks. For example, we might try to define and map the functional regions of the cities of Los Angeles and San Diego. To the immediate south of Los Angeles, people probably subscribe to Los Angeles newspapers, watch Los Angeles television stations, and rely on service from the Los Angeles airport. As we move southward toward San Diego, the influence of Los Angeles diminishes and yields to that of San Diego.

Physicists have a precise mathematical formula to express the gravitational force that one object exerts on another, and geographers have tried to find analogous mathematical equations to describe patterns of human

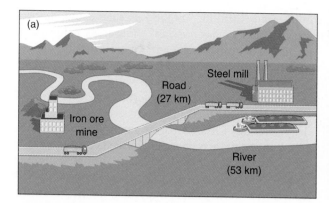

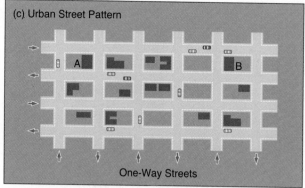

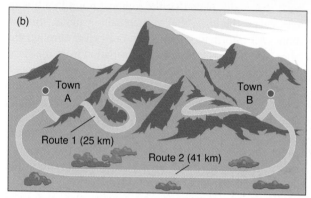

Figure 1-11 Measuring distance. When distance is measured by time or by cost, it may seem to be warped when compared to distance measured by absolute units "as the crow flies." For example, drawings (a) and (b) illustrate how the least expensive or fastest distance between two points may not be the shortest. Transportation by water is almost always cheaper than transportation over land, so heavy iron ore will be moved by water if possible. Traveling on a level road is often cheaper and faster than choosing a shorter route crossing rugged land. Drawings (c) and (d) illustrate how the time or cost distance from any point A to a point B may not be the same as from point B back to point A. Most drivers are at one time or another stymied by a situation such as that in drawing (c).

activities. For example, a geographer might try to write a mathematical equation to define what percentage of households at varying distances from downtown Los Angeles subscribes to Los Angeles newspapers. Perhaps circulation drops 10 percent for every 8 kilometers (5 miles) away from the city center. That equation could then serve as a model to compare the circulation of newspapers in metropolitan Los Angeles to the circulation of newspapers in other metropolitan areas. A **model** is an idealized, simplified representation of reality that can be used as a standard to compare individual cases in the real world. Comparing the equation model of the distribution of newspapers in metropolitan Los Angeles with patterns in New York, St. Louis, and other metropolitan regions might help us understand how shopping patterns, local professional sports loyalties, or even involvement in metropolitan politics vary from one metropolitan region to another. In some metropolitan regions, the downtown newspapers might circulate farther out into the suburbs, building a greater sense of metropolitan identity.

Diffusion **Diffusion** is the process of an item or feature spreading through time. Any innovation—the use of a tool, a new clothing fashion, or the development of a new technology—originates at a place called a

hearth. Tracing diffusion of innovation suggests how peoples and cultures interact and influence one another. The simplest image of diffusion suggests concentric waves spreading out from a stone dropped into a pool. Paths of cultural diffusion, however, are seldom so simple. Geographers define three basic paths of diffusion: relocation diffusion, contiguous or contagious diffusion, and hierarchical diffusion.

Relocation diffusion occurs between widely separated points. A nomadic tribe, for example, might wander until it finds a physical environment similar to the one that it left, and then the tribe might settle down. The ancestors of today's Hungarians wandered all the way from Central Asia to the plains of today's Hungary. The ideas and practices they brought with them were spread by relocation diffusion.

Contiguous, or contagious, diffusion occurs from one place to a neighboring place through direct contact. Contagious disease provides an obvious example of how a virus can spread between people and communities that are contiguous or next to each other. Contiguous diffusion that occurred in the past is often revealed by the dispersion of artistic styles. For example, in the fourth century B.C., Alexander the Great's conquests carried Hellenistic Greek art far into Asia. Central Asian statues of Buddha

Figure 1-12 Siddhartha Buddha. This richly dressed young man is the Indian prince Siddhartha before he became the Enlightened One (Buddha) and gave up his worldly wealth. The statue exhibits a mix of styles: the curly hair, stocky physique, relaxed stance, and deeply carved drapery are Greek characteristics, but the facial expression is typical of Indian mysticism. The nimbus of light around the prince's head originated with Iranian statues of sun gods. Where could all these cultural influences have come together for this extraordinary figure to have been carved? The answer is the region of Peshawar in today's Pakistan, which was at the crossroads of Eurasian travel routes, where the Khyber Pass opens from Central Asia into the plains of the Indian subcontinent. This statue dates from the late second or early third century A.D.

sculpted during the centuries immediately following these conquests resemble Greek gods (Figure 1-12).

Sometimes diffusion does not occur contiguously across space, but downward or upward in a hierarchy of organization. When such **hierarchical diffusion** is mapped, it shows up as a network of spots, rather than as an inkblot, spreading across a map. The Roman Catholic Church illustrates both an organizational hierarchy and a geographic hierarchy (Figure 1-13). Each parish priest is answerable to his bishop, who from his cathedral church

(*cathedra* is Latin for "throne") presides over a diocese, which includes many parishes. The bishops, in turn, look up to the Vatican. An announcement from the Vatican diffuses to the cathedrals, to the dioceses and finally down to the parishes around the world.

As a general rule, more information travels up and down a hierarchy than across any level of the hierarchy. Therefore, news from a parish about an innovation in community service or in a church ceremony might reach the cathedral and even the Vatican before it reaches neighboring parishes. This principle of diffusion up and down in a hierarchy is very important, and we will return to it repeatedly. Note that the topmost point in a hierarchy, called the *apex*, may not actually generate innovations. All information, however, does clear through the apex, and most places receive information from the apex.

Advanced societies usually exhibit well-developed hierarchies of cities, so we can study how the hierarchy evolved through time. Once a hierarchy is developed, the diffusion of any specific phenomenon might be traced from the few biggest cities down to the many more smaller cities and then further down to the even more numerous small towns and villages until the landscape is covered (Figure 1-14).

The map of America's most heavily traveled domestic airline routes reflects the national hierarchy of cities (Figure 1-15). To fly from one small city to another small city, a passenger may have to fly first to a nearby big city (which is a movement up the hierarchy), then from that big city to another big city near the ultimate destination (across the hierarchy), and finally from that big city to the destination small city (down the hierarchy again).

When scientists began to study the path of the diffusion of AIDS throughout sub-Saharan Africa, they first assumed that it had diffused contiguously, as is common in poor, underdeveloped regions, where most contact is directly from one village to the next contiguous village. It was discovered, however, that AIDS was being spread across Africa in a hierarchical fashion. Epidemiologists (individuals who study the spread of disease) wondered how this could be occurring. They found that it was being spread by long-distance truck drivers who travel long distances between large towns and cities and often hire prostitutes in each town along the way. The knowledge of this method of diffusion in Africa caused world health officials to check truck drivers in India, another poor region suffering from soaring rates of infection, and to educate them about the necessity for safe-sex techniques. Thus, understanding of the path of diffusion of AIDS in Africa may have saved millions of lives in India and elsewhere.

Many phenomena diffuse hierarchically around the world today. The spread of a disease or of a cultural innovation such as a clothing fashion or a hit song can often be traced among the world's principal metropolises before it reaches down into the smaller cities in

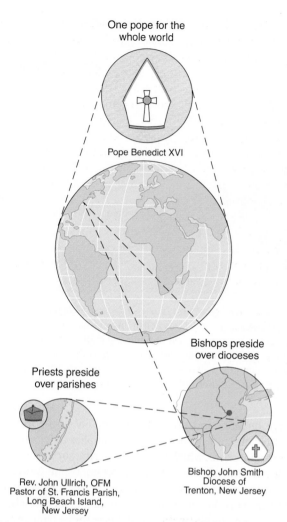

Figure 1-13 Territorial and organizational hierarchy. The Roman Catholic Church illustrates both a territorial and organizational hierarchy. The pope presides over the entire Church, but the Church is subdivided into dioceses, each presided over by a bishop, and the dioceses are subdivided into parishes.

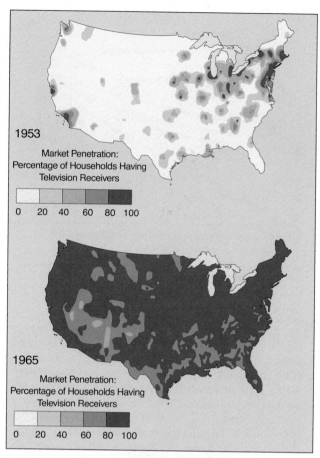

Figure 1-14 Market penetration by television. These maps show how the percentage of U.S. households that had television sets increased from 1953 to 1965. The spots on the first map show that only large cities had television in 1953. The second map shows that by 1965, the innovation had diffused to cover the contiguous 48 states. This occurred before cable television or satellite dishes, so viewers had to receive transmissions from metropolitan centers. How would the pattern of diffusion differ with today's technology?

each country. This path of diffusion reveals the world's interconnectedness.

Barriers to diffusion There are many barriers to cultural diffusion. Oceans and deserts, for example, increase the cost distance or time distance between places. Other topographical features, such as mountains and valleys, historically have blocked human communication and contributed to cultural isolation. In Asia, the high Himalayan mountains, and the steep valleys of the Tongtian, the Mekong, Nu, and Brahmaputra rivers make overland travel between China and the Indian subcontinent extremely difficult. Therefore, these two cultures and great concentrations of population are distinctly isolated from one another.

Other barriers to cultural diffusion include political boundaries and even the boundaries between two culture realms. Hostile misunderstanding, distrust, and competition between two cultural groups can hinder any communication and exchange between them.

Geographic Systems Analysis

A third approach in geographic inquiry views Earth as a set of interrelated environmental and human systems. A **system** is an interdependent group of items that interact in a regular way to form a unified whole. Models of systems help geographers see how factors are interrelated—how one activity, force, or event affects others. Individual environmental systems include the climatic processes that produce precipitation, the hydrologic processes that determine what happens to the rain when it reaches Earth's surface, and the characteristics of the river valleys that receive the water runoff. These individual physical systems are themselves interrelated. For example, an increase in precipitation affects an area's soils, landforms, and vegetation. Geography also emphasizes interactions between humans and the environment. Human activities are influenced by environmental conditions, but humans in turn modify the physical environment in which they live.

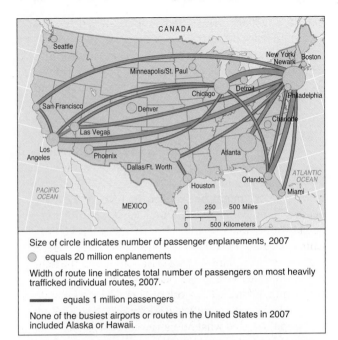

Size of circle indicates number of passenger enplanements, 2007

⬤ equals 20 million enplanements

Width of route line indicates total number of passengers on most heavily trafficked individual routes, 2007.

━━━ equals 1 million passengers

None of the busiest airports or routes in the United States in 2007 included Alaska or Hawaii.

Figure 1-15 Major air-passenger routes in the United States. This map is a good indicator of the urban hierarchy of the United States, although it is a bit distorted by airlines' routing systems. New York is certainly the principal focus of national life. Los Angeles, Dallas, Atlanta, and Chicago are functional "capitals" of individual regions of the country. Minneapolis/St. Paul and Denver have large numbers of total enplanements but few or no individual routes with more than 2 million passengers. These cities serve as hubs of many local regional routes with small numbers of fliers on each.

In the following sections we will examine some of the ways in which geographers have studied the interactions among different aspects of the human and physical environment.

Earth's physical systems   Geographers study natural processes in terms of four systems: the atmosphere (air), the hydrosphere (water), the lithosphere (Earth's solid rocks), and the biosphere, which encompasses all of Earth's living organisms (Figure 1-16).

The **atmosphere** is a thin layer of gases surrounding Earth to an altitude of less than 480 kilometers (300 miles). Pure, dry air in the lower atmosphere contains about 78 percent nitrogen and 21 percent oxygen by volume. It also includes about 0.9 percent argon (an inert gas) and as of 2010 is about 0.039 percent carbon dioxide (a crucial percentage, as you will see in upcoming chapters). Air is a mass of gas molecules held to Earth by gravity creating pressure. Variations in air pressure from one place to another cause winds to blow, as well as create storms and control precipitation patterns.

The **hydrosphere** is the water realm of Earth's surface. Water can exist as a vapor, liquid, or solid, such as the oceans, surface waters on land (lakes, streams, rivers), groundwater in soil and rock, water vapor in the atmosphere, and ice in glaciers. Over 97 percent of the world's water is in the oceans in liquid form. The oceans sustain a

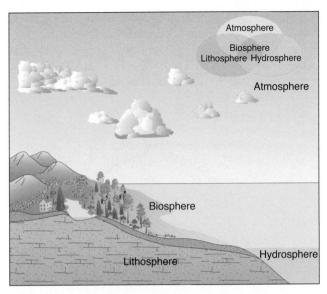

Figure 1-16 Earth's physical systems. Geographers regard natural processes as consisting of four open systems, including three composed of nonliving matter (the atmosphere, hydrosphere, and lithosphere), plus the biosphere, which comprises all of Earth's living organisms. The four systems are interrelated and overlap with each other.

large quantity and variety of marine life in the form of both plants and animals. Seawater supplies water vapor to the atmosphere, which returns to Earth's surface as rainfall and snowfall. These are the most important sources of fresh water, which is essential for the survival of plants and animals. Water changes temperature very slowly, so oceans also moderate seasonal extremes of temperature over much of Earth's surface. Oceans also provide humans with food and a surface for transportation.

The **lithosphere** is the solid Earth, composed of rocks and sediments overlying them. Earth's core is a dense, metallic sphere about 3,500 kilometers (2,200 miles) in radius. Surrounding the core is a mantle about 2,900 kilometers (1,800 miles) thick. A thin, brittle outer shell, the crust is 8 to 40 kilometers (5 to 25 miles) thick. The lithosphere consists of Earth's crust and a portion of upper mantle directly below the crust, extending down to about 70 kilometers (45 miles). Powerful forces deep within Earth bend and break the crust to form mountain chains and shape the crust to form continents and ocean basins. The shape of Earth's crust influences climate. If the surface of Earth were completely smooth, then temperature, winds, and precipitation would form orderly bands at each latitude.

The **biosphere** consists of all living organisms on Earth. The atmosphere, lithosphere, and hydrosphere function together to create the environment of the biosphere, which extends from the depths of the oceans through the lower layers of atmosphere. On the land surface, the biosphere includes giant redwood trees, which can extend up to 110 meters (360 feet), as well as the microorganisms that live many meters down in the soil, in deep caves, or in rock fractures.

These four spheres of the natural environment interact in many ways. Plants and animals live on the surface of the lithosphere, where they obtain food and shelter. The hydrosphere provides water to drink and physical support for aquatic life. Most life forms depend on breathing air, and birds and people also rely on air for transportation. All life forms depend on inputs of solar energy.

Humans also interact with each of these four spheres. We waste away and die if we are without water. We pant if oxygen levels are reduced in the atmosphere, and we cough if the atmosphere contains pollutants. We need heat, but excessive heat or cold is dangerous. We rely on a stable lithosphere for building materials and fuel for energy. We derive our food from the rest of the biosphere.

Plants and animals interact with each other through exchange of matter, energy, and stimuli. An ecological system, or **ecosystem**, is a group of organisms and the nonliving physical and chemical environment with which they interact. The scientific study of ecosystems is called **ecology.** Ecologists study the interrelationships among life forms and their environments in particular ecosystems, as well as among various ecosystems in the biosphere.

Human–Environmental Interaction

The physical environment affects human activities in many ways. Human societies must adapt to local climatic conditions, vegetation, water resources, and other attributes of a local environment as they learn to exploit the local resources. Human societies are not, however, passive. As any human society adapts, it actively and deliberately alters its surrounding natural conditions. The vegetation that covers much of Earth, for example, is cropland or pastureland that is maintained for human benefit. Thus, interaction between the environment and humankind is reciprocal: The environment affects human life and cultures, and humans alter and transform the environment. In the world today, there are no inhabited environments that have not undergone fundamental transformation by human activities, and many uninhabited places have been significantly altered as well.

Human culture and cultural landscapes

Geography is one of the social sciences that contributes to our understanding of human **culture.** The word culture is often used to mean only fine paintings or symphonic music, but to social scientists it means everything about the way people live: what sort of clothes they wear (if any); how they gather or raise food; whether they recognize marriages and, if they do, whom (and how many spouses) they think it proper to marry as well as how they celebrate marriage ceremonies; what sorts of shelters they build for themselves; which languages they speak; which religions they practice; and whether they keep any animals for pets. Human cultures vary significantly in all these aspects. Of all the ways in which the behavior and beliefs of individuals differ, a social scientist chooses a few specific criteria and then defines that bundle of attributes of shared behavior or belief as a culture.

People modify their local landscape in the process of making a living, housing themselves, and carrying on their lives. We say that they transform a **natural landscape,** one without evidence of human activity, into a **cultural landscape,** one that reveals the many ways people modify their local environment. Aspects of cultural landscapes include the treatment of the natural environment as well as houses and other parts of the built environment.

The evidence of human activity is so ubiquitous on Earth that we might be justified in saying that today nearly all of Earth is a cultural landscape. Travelers quickly notice that the cultural landscape is varied. Rural China does not look like rural France or rural Nigeria. One factor in explaining these variations is that the natural environments in these places were different even before humankind set to work on them. The "raw material" that human societies transformed into cultural landscapes affected what humans would do in each.

Another factor in explaining variations in cultural landscapes, however, is that each cultural group creates a distinct cultural landscape, and the better we can learn to interpret a landscape, the more it will reveal to us about the culture that produced it. In the past, human societies developed in greater isolation from one another than today, and the extraordinary diversity of human cultures and cultural landscapes testifies to human ingenuity. Different peoples who live in very similar environments but are isolated from one another developed astonishingly different lifestyles. Conversely, some aspects of cultures that have developed in different physical environments are startlingly similar. No direct cause-and-effect relationship between a physical environment and any aspect of culture can be assumed.

Describing Earth

Earth is nearly spherical, with a diameter of about 12,735 kilometers (8,000 miles) and a circumference of about 40,000 kilometers (25,000 miles). Its highest point is almost 8,850 meters (29,000 feet) above sea level, and the deepest spot in the oceans is about 10,920 meters (35,800 feet) below sea level. Earth's shape varies slightly from true sphericity due to the effect of centrifugal force. Any rotating body tends to bulge in the "middle" and flatten at the "ends"; thus, the equatorial diameter of Earth is slightly greater (43 kilometers, or 27 miles) than its polar diameter.

The Geographic Grid

Earth rotates continuously about an axis that penetrates Earth's surface at the North Pole and the South Pole. An imaginary plane is perpendicular to the axis of rotation and passes through Earth halfway between the poles. It is called the *plane of the equator,* and intersects Earth's surface at Earth's imaginary midline, or "waist," called the **equator**. We can use the North Pole, the South Pole, and the equator as reference points for defining positions on Earth's surface.

Latitude is the angular distance measured north and south of the equator. We can project a line from any given point on Earth's surface to the center of Earth (Figure 1-17a). The angle between this line and the equatorial plane is the latitude of that point. Latitude is expressed in degrees, minutes, and seconds. There are 360 degrees (°) in a circle, 60 minutes (′) in a degree, and 60 seconds (″) in a minute. Latitude varies from 0° at the equator to 90° at the North and South poles.

Lines connecting all points of the same latitude are called **parallels** because they do not intersect. Parallels are imaginary lines. There can be an infinite number of them—one for each degree of latitude, or for every minute, or for any fraction of a second (Figure 1-17b).

Longitude is the angular distance measured east and west on Earth's surface. Longitude is also measured in degrees, minutes, and seconds. It is represented by imaginary lines, called **meridians**, extending from pole to pole and crossing all parallels at right angles (Figure 1-17c).

Meridians are not parallel to one another. They are farthest apart at the equator, become increasingly close northward and southward, and converge at the poles.

The **prime meridian**, from which longitude is measured, was chosen by an international conference in 1884. It is the meridian passing through the Royal Observatory in Greenwich, England, just east of London. Longitude is measured both east and west of the prime meridian to a maximum of 180° in each direction. Thus, the location of any spot on Earth's surface can be described with great precision by reference to its latitude and longitude. For instance, if we say that the dome of the U.S. Capitol in Washington is located at 38°53′23″ north latitude and 77°00′33″ west longitude, we have described its position within about 20 steps.

The same international conference that agreed on the prime meridian also agreed to divide the world into 24 standard time zones (Figure 1-18). When people first began keeping track of time, they defined local time in terms of when the sun rose and set, with noon defined as the time at which the sun is highest in the sky. This method of timing depends on longitude. The standardization of time from one place to another

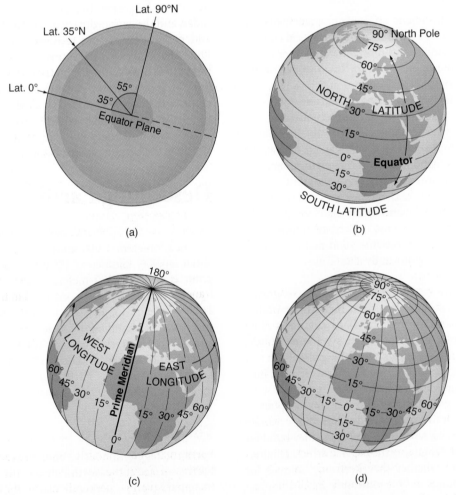

(a)

(b)

(c)

(d)

Figure 1-17 The geographic grid. These drawings show the development of the geographic grid: (a) the measure of latitude, (b) parallels of latitude, (c) meridians of longitude, (d) the complete grid system.

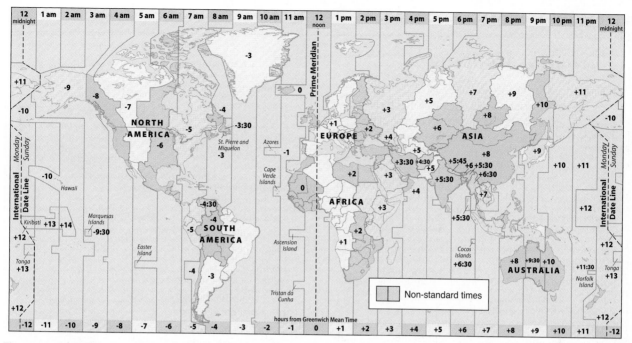

Figure 1-18 World time zones. Each of Earth's 24 time zones represents an average of 15° of longitude, or one hour of time. The number in each time zone on this map represents the number of hours to be added to the local time to find Greenwich Mean Time. The colored stripes indicate the standard time zones; gray areas have a fractional deviation. These irregularities in the zones are dictated by political and economic factors.

became necessary when travel between places was fast enough that differences in time from one place to another became important. In the United States, standardization became important because of the development of the transcontinental rail network.

When we standardize time across large east-west distances, the usual link between the position of the sun in the sky and time can be stretched considerably. Relative to Earth, the sun travels 360° in 24 hours, or 15° per hour, so time zones that standardize time within each 15°-wide slice of Earth's surface offer a solution. The time at the prime meridian, or 0° longitude, is designated **Greenwich Mean Time (GMT)**. Earth rotates eastward, so a clock advances one hour from GMT for each 15° traveled east from the prime meridian. For each 15° west of the prime meridian, a clock moves back to one hour earlier than GMT.

The meridian for 75° west longitude runs through the eastern United States, so time in the eastern United States is five hours earlier than GMT. (The 75° difference in longitude ÷ by 15° per hour = 5 hours.) When the time is 11 A.M. GMT, the time in the eastern United States is five hours earlier, or 6 A.M. The 48 conterminous U.S. states and the Canadian provinces share four standard time zones, known as Eastern, Central, Mountain, and Pacific. Most of Alaska is in the Alaska Time Zone, which is nine hours earlier than GMT. Canada has an Atlantic Time Zone, which is four hours earlier than GMT, and a Newfoundland Time Zone, which is three and a half hours earlier than GMT.

In Figure 1-18, you can see that the **International Date Line** mostly follows 180° longitude, although it

deviates in several places to avoid dividing land areas. When travelers cross the International Date Line heading east (toward North America), the calendar moves back one day. When travelers cross it going west toward Asia, the calendar moves ahead one day.

Today's world time-zone map does not consist of parallel vertical stripes, but rather it bends for the convenience of individual countries to reduce the number of situations in which a given region is divided among multiple time zones. In many cases, it can be much more convenient to be in one time zone than another. In the United States, for example, the Eastern Time Zone is much wider from east to west than longitude alone would dictate: more than 20° instead of the normal 15°. This is because most people in the eastern United States would rather be on the same time as the financial and political centers on the East Coast than an hour different. For instance, most of the state of Indiana is in the Eastern Time Zone. As seen in Figure 1-7, the northwestern part of the state, near Chicago, is on Central Time because Chicago is on Central time, and northwest Indiana residents find it more convenient to be on the same time as Chicago.

Communicating Geographic Information: Maps

One of geographers' most characteristic tools is a map. **Maps** are two-dimensional (flat) representations of some portion of Earth's surface. They are necessarily smaller than the area portrayed, so they cannot show all the things present on the actual Earth surface. Maps show only selected information according to the particular

Thematic Mapping

Thematic maps can be used to convey information about many different geographic topics. This book uses them extensively and students should understand some of their basic characteristics to better interpret the information each one presents. There are two primary considerations for each map type. First, the geography of the data relates to how information is collected and stored. Data are collected according to different kinds of spaces: point features, such as for a town on a national map; line features, such as along a path or road; and areal features, such as a country. Second, map symbols are selected to convey differences in the data. Some symbols display quantitative differences, such as rainfall totals or rates of poverty. Other symbols show different types of things, called nominal or categorical data, for example, agricultural areas or the location of different types of power plants.

MapMaster
Layered Thematic Maps

- Isoline maps connect places of equal value to show variation across an area. The patterns of the lines convey shapes of surfaces. A topographic map is an isoline map where the lines show equal elevation. Figure 2-24 is an isobar map, a kind of isoline map, where lines connect points of equal barometric pressure.

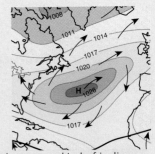

- When we want to identify a characteristic of a place that has qualitative rather than quantitative meaning, such as the name of a place, we use nominal areal symbol maps in which color or shading designates differences between areas. A map showing countries or U.S. states is of this type. Figure 4-14 shows different soil regions.

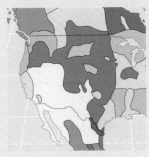

- Dot-distribution maps use points to show the location or distribution of a feature. Some display quantities by assigning a value to each point. The visual impression in such maps is of some areas with dense concentrations of dots and other areas with very few dots, conveying the spatial distribution of the phenomenon. For example, Figure 3-5b shows the location of earthquake epicenters. Figure 8-4 displays the approximate source of every 100,000 metric tons of potatoes and rice grown around the world.

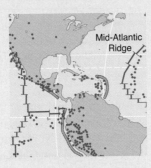

Mid-Atlantic Ridge

- Graduated symbol maps use different sizes of symbols to show differences in quantity across locations. A large symbol shows the location of large quantities, while small symbols show locations of smaller quantities. For example, Figure 10-1 uses dot size to show the location and proportional quantity of major urban populations.

Beijing
Seoul
Tokyo
Tianjin
Wuhan
Osaka/Kobe
hongqing
Shanghai
Guangzhou
Dongguan
Shenzhen
Hong Kong
Bangkok
Manila
Ho Chi Minh City

- Choropleth maps use color or shading to show quantities at different areas, usually with darker shades indicating larger quantities. For example, Figure 5-8 uses several shades of the same color to represent rates of population growth. Choropleth maps often lump different values into a smaller number of ranges or classes; each class is then assigned a different shade.

- Flow maps use different line thicknesses to show different quantities moving along a path. Figure 5-25 displays the movement of slaves out of Africa.

NORTH AMERICA
British North America
Spanish America
British Caribbean
French Caribbean
Danish Caribbean
Dutch Caribbean
PACIFIC OCEAN
SOUTH AMERICA
Brazil

- Sometimes we find it useful to convey an idea by deliberately distorting features on the Earth's surface to indicate some characteristic of those features. Cartograms display different quantities by intentionally distorting features, usually areas. Figure 5-2 distorts the land size of different countries in proportion with each country's population.

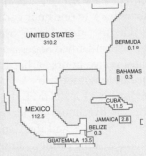

UNITED STATES
310.2
BERMUDA
0.1
BAHAMAS
0.3
MEXICO
112.5
CUBA
11.5
JAMAICA 2.8
BELIZE
0.3
GUATEMALA 13.5

purpose of the map. They are often used to show things that may not be visible on the landscape, but are nonetheless important—political boundaries, for example. Maps are simplifications of Earth—models—designed to store and display geographic data.

Maps take a great variety of forms. Some are just simple drawings intended to show relative location or a path from one place to another. Perhaps the most common map with which you are familiar is a road map like the ones shown on Internet map sites or mobile GPS units, showing the locations of towns and cities and the automobile routes that connect them. **Topographic maps** are more specialized and show elevation and terrain features, especially fixed objects such as roads, buildings, and powerlines. Airplane pilots use specialized maps that also show the elevation of the ground surface and tall objects; landmarks visible from airplanes such as rivers, lakes, urbanized areas, and large highways; areas of the sky in which planes are required to follow certain rules; and the locations of towers and navigational beacons, and landing fields. Photos taken from airplanes are maps, as are images collected from satellites that are often displayed on your TV screen as part of a weather forecast.

Cartography is both a highly technical and a somewhat artistic pursuit, combining the tools of mathematics and engineering with those of graphic design. Maps should be accurate, portraying matter as it really exists rather than as distorted, improperly located, or mislabeled information. They should be visually easy to use, prominently displaying the material a user needs without clutter from unnecessary information. This is why road maps, for example, usually do not show mountains and hills except in the simplest ways. To do so would add many extra lines to a map that is already filled with lines representing roads. Maps that display specific types of information are called **thematic maps**, and cartographers must carefully choose the proper technique for conveying different kinds of data.

In converting geographic data from their original form on Earth's surface to a simplified form on a map, many decisions must be made about how this information is to be represented. No matter how we draw a map, we cannot possibly make it show the world exactly as it is in all its detail, nor would we want to. Scale and projection are two fundamental properties of maps that determine how information is portrayed.

Scale Maps are smaller than the areas they represent. To know how large an area or how great a distance is represented by a given area or distance on a map, we must know its scale. The **scale** of a map is a quantitative statement of the relative sizes of an object on the map and in reality. Map scales are typically expressed in one of three ways: a written statement (1 inch = 1 mile), a representative fraction (1:63,360, meaning that one

unit of length on the map corresponds to 63,360 units on the ground), or a graphic scale in which a bar on the map is labeled with ground distances.

Common map scales vary widely. The maps in this textbook showing the entire world, including the oceans, on one page have a horizontal distance of roughly 25,000 miles (1,584,000,000 inches) on a printed area 6 inches wide, for a scale of 6/1,584,000,000, or about 1:264,000,000. A typical scale used for a road map of a state in the United States might be 1:1,000,000 to 1:2,000,000. A street map of a city might have a scale of 1:10,000, while a surveyor's map of a residential lot might be drawn at 1:100. Maps like the world maps in this book show the land in a very small space, so we call them **small-scale maps**. A map such as a street map that shows a given area in a large space is called a **large-scale map**. A map with a large number in the denominator of a fraction expressing scale (for example, 1/2,000,000) is a small-scale map, whereas a map with a small number in the denominator (for example, 1/100) is a large-scale map (Figure 1-19). A trick for remembering which is small and which is large is that 1/100 is a *larger* number than 1/1000.

The amount of detailed information that can be shown about a place is a function of map scale. On a small-scale map there is not enough room to show a lot of detail. If we showed all the numbered highways of the United States on a map the size of a page in this book, the map would be virtually all highway and nothing else. On a large-scale map we have room for more detail about a place, but we cannot show the relations between that place and another that is far away. Figure 1-20 shows Seattle, Washington, viewed at a range of map scales. When we zoom out from a larger scale to a smaller scale, not only do we show less detail, but some of the information actually changes. For example, wiggles in roads or rivers disappear, making the lines straighter than they actually are. Thus, the scale of a map affects its accuracy.

Projection A second fundamental characteristic of maps that governs the way we show information is **projection**, the transferral of locations on Earth's surface to locations on a flat map. For practical reasons, most maps are flat. Flat maps can be printed on paper and folded, or displayed on a video monitor. However, every cartographer faces a problem when drawing our spherical Earth on a flat piece of paper. Some distortion unavoidably results. Cartographers have invented hundreds of projections, but none is completely free of distortion.

So how do we make a flat model of a spherical Earth? One way might be to draw a map on a spherical rubber balloon and then cut the balloon and stretch it in such a way that it could be glued on a page. Or we could draw our map on a transparent globe, place a light bulb inside it, and project the map onto a screen. For many projections the "screen" is not flat, but we can imagine it wrapped around the globe and then flattened to make

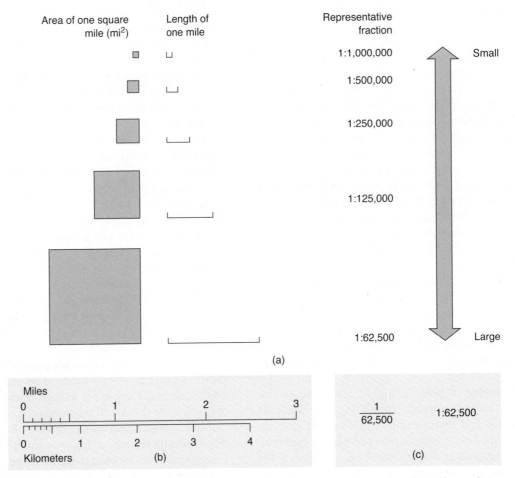

Figure 1-19 Map scales. Drawing (a) shows that a small-scale map can portray a large part of Earth's surface, whereas a large-scale map can show only a small part of the surface. Drawing (b) is a graphic scale, and (c) shows fractional scales.

the maps we use (Figure 1-21). This is, in effect, what is done, except that the conversion of the geometry of the sphere to the geometry of the screen is achieved mathematically rather than optically. Regardless of how we make a flat map, the resulting features on the map are different from their equivalents on Earth's surface. The map cannot be an exact scale model of Earth in the same way a globe can. The process of projecting the globe onto the map inevitably introduces distortion.

Several kinds of distortion occur when an image of a round Earth is fit to a flat surface, and as much as we would like to eliminate this distortion, there is no way to do so. When we are mapping relatively small areas of Earth's surface, the difference between a plane and the surface of a sphere is insignificant for most purposes. For large areas, however, and especially for world maps, the distortions are severe. All projections must distort either size (and distance), shape (and orientation), or both.

A distortion of size and distance means that the scale in one part of a map is different from that in another part—objects in one area appear larger or smaller than they actually are in relation to objects in another part of the map. It is possible to distort size but preserve the shapes of objects, effectively blowing them up or shrinking them. Maps that distort size but preserve shape are called **conformal maps**.

A distortion of shape and orientation means that in moving from a round surface to a flat one, the shapes of objects are stretched more in some parts than in others. A change in shape inevitably means that orientation changes too, such as causing a line that on Earth runs northeast-southwest to instead appear to run east northeast–west southwest. It is possible, however, to make such changes while preserving the size of objects relative to each other. Maps that preserve size but distort shape are called **equal-area maps**, or equivalent projections.

One of the most common conformal map projections in use is the Mercator projection (Figure 1-22). It is named for its Flemish inventor, Gerardus Mercator, who developed it in 1569 for the purpose of navigation. It serves that purpose well, because it preserves shape and, thus, orientation. On the Mercator projection, lines of latitude and longitude form a perfect rectangular grid. North is always the same direction on the map, and a straight line can be traced by following a constant compass bearing. The rectangular shape of the Mercator map adds another attraction. It fits conveniently on a printed page or on a wall. This, more than its utility

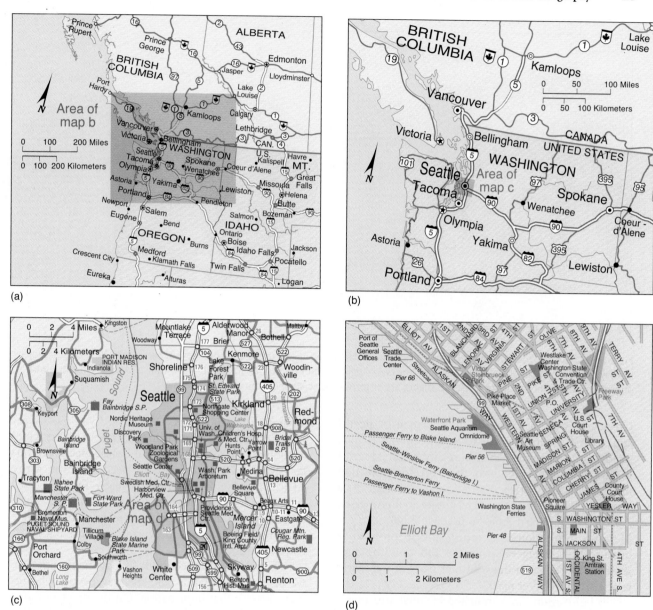

Figure 1-20 Seattle, Washington, at different scales. These four maps show the city of Seattle, Washington, at a variety of scales. Only major landscape features, cities, and highways are shown in Figure 1-20a. When we zoom in on this map, the amount of information displayed changes, with more detail shown at larger scales. In Figure 1-20b, the names of small towns are shown, as are secondary roads. In Figure 1-20c, some details of the lakefront shoreline emerge, as do the names of a few major streets. The names of all the individual streets are shown in Figure 1-20d.

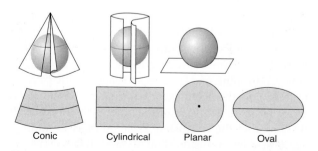

Figure 1-21 General classes of map projections.

in navigation, contributes to the continued widespread use of this projection.

The problem with the Mercator projection is its distortion of size. The distance around the globe at latitude 60° north is half the distance around the globe following the equator, and yet these two distances are the same on a Mercator map. If north–south distances are exaggerated in the same way, as they must be to preserve shape, then the area of a region at the latitude of the Arctic Circle is shown four times its actual size relative to a region at the equator. The distortion increases poleward. This is why on a Mercator map

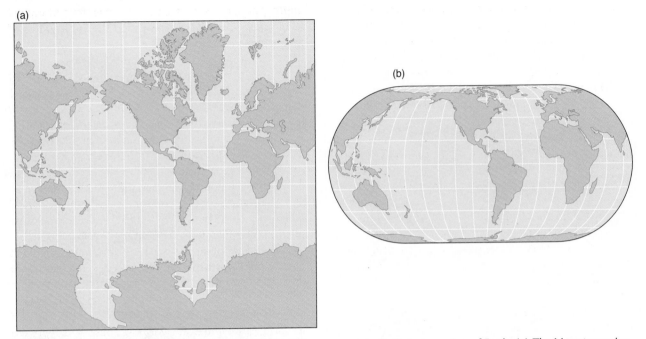

Figure 1-22 Conformal and equal-area projections. Flat maps necessarily distort our view of Earth. (a) The Mercator projection, which is a conformal projection, displays the correct shapes but exaggerates sizes in high-latitude areas. (b) An equal-area projection can portray accurate sizes, but shapes are distorted, especially near the edges of the map.

Greenland (which is centered a little north of 70° north) appears to be similar in size to South America (which straddles the equator), when in reality South America is nine times larger. Similarly, the Mercator projection makes North America, Europe, and northern Asia look much larger than they are relative to South America, Africa, and southern Asia. Some people object to the Mercator projection and other similar projections because they exaggerate the size and thus the suggested importance of the wealthier parts of the world relative to the poorer parts.

Size distortions are eliminated in equal-area maps, of which Figure 1-22b is an example. Equal-area maps of the world are much better than Mercator maps for most purposes, because they communicate more accurately the sizes of countries and continents; when comparing one country with another, size is usually more important than shape. Shape distortions become important, however, because they alter distances. Also, because the polar regions must be drawn much smaller than equatorial regions, there is usually considerable distortion of direction (especially north–south) near the edges of the map.

In addition to shape and size distortions, all flat maps must divide Earth's surface at some point. The surface of a globe is continuous—we can go all the way around without leaving the surface—but a flat map must have edges. If we follow the convention of placing north at the top of the map, then the usual location to split the Earth's image is in the middle of the Pacific Ocean, which causes the least disruptions of land areas. This is also convenient because the prime meridian can be placed at the center of the map with relatively little

separation of land areas. Of course if you live in, say, New Zealand, you might think this layout makes it look as though islands in the central Pacific Ocean are very far away and that one would need to travel across Africa and South America to reach them. A New Zealander might prefer a map centered on 180° longitude. Maps produced in the United States, for example, often have their edges in central Asia, placing North America at the center.

The choice of projection is fundamental to the way we display information and the way people who use maps receive that information. On the one hand, projections can be as dispassionate as a mathematical equation that describes the alteration of a spherical surface into a planar one. However, maps communicate information about places, and the form of the map inevitably influences how and what the map says about a place. Those who use such maps therefore often have very strong opinions about what kinds of distortions are introduced in the projection process.

Insets and other conventions   Other conventions of map projections or printing can lead to misunderstanding. For example, to save space, many maps of the United States tuck a separate map of Alaska into the corner below the U.S. Southwest. This is called an *inset* map. This common practice leads many U.S. schoolchildren to think of Alaska as an island off the coast of California. If the mapmaker also prints the inset map at a smaller scale than that of the other states, as used to be common, the confusion about Alaska's location is aggravated by confusion about its size. In fact, Alaska is nearly one-fifth the size of the lower 48 states combined.

In January 1990, in order to counter this misrepresentation and the resulting confusion in some students' minds about exactly where Alaska is, the Alaska state legislature passed a resolution "respectfully requesting that all major United States magazines, newspapers, textbook publishers, and map publishers . . . place Alaska in its correct geographical position . . . in the northwest corner."

Several world maps in this book cut out great expanses of the world's oceans so that we can map human activities on the land in greater detail. This shifts the relative positions of the continents and tucks Australia and New Zealand south of Asia, to the west of where they really are. Clear divisions on the maps mark the lines along which the map has been cut, and each portion has markings of latitude and longitude for proper positioning. The first of these maps is Figure 2-32. If you want to recheck the continents' relative positions shown more accurately, you can always refer to a globe.

It is conventional to put north at the top of a map, but top and bottom are meaningless in space. In fact, any conventional view conceals a great deal of information simply because it is conventional; we usually do not notice or consider something we have seen often before.

The inability to understand mapmakers' conventions troubled Mark Twain's fictional character Huckleberry Finn when he and Tom Sawyer were blown east in a balloon. Huck insisted that they were not over Indiana yet because the color of the land below had not changed from the green of Illinois to the pink of Indiana, as shown on his school atlas. "What's a map for?" he demanded. "Ain't it to learn you facts?" Our answer is yes, but a map works for you only if you understand its projection and the explanation of its colors and symbols as defined in its key, usually at the bottom of the map.

Geographic Information Technology

Computers have transformed many aspects of social, intellectual, and business life in the last few decades; it is only logical that their impact has been highly significant in geography. Until the 1970s, most maps were being drawn with ink pens and rulers, but now they are composed on computers and printed by machine. Location and land-use information used to be collected by optical surveys and visual inspection of aerial photographs, but now it is collected using global positioning systems (GPSs) and digital satellite imagery as well as by conventional technologies. Countless hours of work with paper, pencil, and calculator were necessary to perform routine statistical analyses, but now such work is performed in seconds on a computer. Geographic information systems (GISs) integrate all of these tasks by using computers to assemble automatically and rapidly collected digital data, analyze such data together with other information that may be on hand, and generate complex and specialized maps for many purposes. GIS technology has revolutionized geographic research and expanded the applications of that work to everyday life.

Automated cartography   Anyone who has worked in traditional pen-and-ink technical drawing can attest that such work is tedious and, for most, not particularly exciting. A map is a highly complex technical drawing, and map production using manual techniques is very expensive. When maps had to be updated because of changes in the landscape—new roads, expanded urban areas, changed boundaries, and place names—the costs were substantial. For this reason, when *computer-aided design (CAD)* technology emerged, it was quickly applied to mapmaking to develop sophisticated, specialized digital cartography systems. These systems make it possible to draw new features on maps and edit them on a video monitor before any paper map is created, in much the same way that word processors make it possible for you to compose a term paper on screen and edit it, correcting typing mistakes and making the needed refinements before you print a paper document. Given the complexity of mapmaking and the specialized skills it requires, this is obviously an important improvement for the cartographer.

Remote-sensing satellites   Another important dimension of GIS is its ability to make use of digital satellite imagery. The acquisition of data about Earth's surface from a satellite orbiting the planet or from high-flying aircraft is known as remote sensing. Geographic applications of remote sensing include mapping of vegetation and other surface cover; gathering data for large unpopulated areas, such as measuring the extent of the winter ice cover on oceans; and monitoring changes such as weather patterns and deforestation.

Landsat satellites, the first of which was launched by the United States in 1972, were the first satellites intended primarily for mapping characteristics of Earth's surface. Passive sensors in the Landsat satellites measure the amount of radiation emanating from Earth's surface in particular wavelengths. Sensors primarily measure the amount of radiation of the various colors of visible light, although some measure infrared (heat) energy. In a few cases, active sensors like radar send radiation to Earth and measure the radiation that is reflected back to the satellite. The most recent Landsat satellite to be launched was *Landsat 7*, in April 1999; the next is planned for 2012.

Remote-sensing satellites scan Earth's surface much as a television camera scans an image in the thin lines you can see on a television screen. The sensor is moved first across the landscape in a line; it is then moved slightly to scan other lines in succession. At any moment, a sensor is recording the amount of energy from only one place, an area called a picture element, or pixel. A map created by remote sensing is essentially a grid containing many rows of pixels.

The smallest feature on Earth's surface that can be detected by a sensor is determined by the size of the pixel. This is called the resolution of the scanner. Early Landsat sensors had a pixel size of 59 meters by 59 meters (194 feet by 194 feet), compared to 30 meters by 30 meters (98 feet by 98 feet) on later Landsat versions or 10 meters by 10 meters (33 feet by 33 feet) and even 1.5 meters by 1.5 meters on the *IKONOS* satellite. Even smaller pixel sizes are available on military and intelligence satellites; these data are not, however, available to civilians. The importance of a smaller pixel size is that smaller objects can be detected (Figure 1-23).

Weather satellites take a broader view than Landsat satellites. They have very large pixels, covering several kilometers on a side, which enables them to map rapidly a large area such as a continent. Weather forecasters need data about a large area very quickly, because weather systems change rapidly.

Dozens of satellites orbit Earth and send back information about its atmosphere, oceans, and land surface, so the amount of information we have available is enormous. Some early GISs were developed to make use of satellite data, and, because of the growth of powerful desktop microcomputers in the 1980s and 1990s, the use and capabilities of GISs increased exponentially.

Figure 1-23 A satellite image of Marathon Key, Florida, showing both the above-water parts of the island and the fringing coral reefs. Images such as this are used worldwide to monitor changes in coral reefs.

Satellites and planes equipped with specialized sensing devices can even read history as it is recorded below the surface. Information available includes changes in plant growth or drainage patterns brought about by human activity and densely compacted soils and erosion associated with agriculture. Archaeologists have also used this kind of information to find long-lost cities.

Human activities cause new and continuing changes in the landscape, and remote sensing is critical to monitoring and evaluating these impacts. For example, climate change and sediment accumulation can affect water resource availability (Figure 1-24). Other important uses of satellite imagery include such diverse applications as estimating nutrient levels in farmland soils, quantifying loss of wetlands in coastal areas, forecasting stream flow based on snow cover, and documenting urban growth (Figure 1-25).

National governments also use satellite images to gather information about other countries or potential enemies. Although the United States and a few other countries have dedicated spy satellites, they are not all-seeing. In fact, most of the images gathered by U.S. satellites have never been analyzed, meaning most information must come from other sources. Governments are not the only ones interested in satellite images; corporate espionage uses commercially available sources to observe their competitors' operations.

Global positioning systems   Determining the location of a place on Earth's surface has become much easier with the development of **global positioning systems (GPSs)**. A GPS is a navigational tool originally developed by the U.S. government for military use, but is now available for civilian purposes worldwide. The system consists of a fleet of satellites that orbit Earth, broadcasting digital codes. A portable receiver "listens" to these signals. By measuring very small differences in arrival times of the signals, it determines its own location. Inexpensive receivers are capable of determining locations to within a few meters, whereas more elaborate systems have accuracies of less than one meter (Figure 1-26). The limitations of the system are relatively few—an antenna must be outside and unobstructed by trees and the satellite signals may be withdrawn from civilian use at the discretion of the government, so that during wartime an enemy cannot use them. On May 1, 2000, however, President Clinton signed an Executive Order allowing civilian GPS devices to have the same degree of accuracy as military devices.

GPS is revolutionizing many business operations, especially those related to transportation and mapping. Surveyors, particularly those working in remote or rural areas, determine their locations by GPS instead of using optical devices that are useful only for distances of a mile or two at best, assuming a clear line of sight.

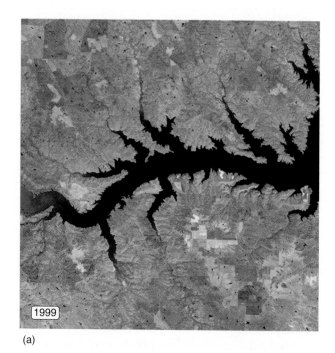

(a)

(b)

Figure 1-24 Effects of drought and sedimentation on the Oahe Reservoir, South Dakota. The Oahe Reservoir, built in 1958, impounds the Missouri River and provides flood control, hydroelectric power, recreational opportunities, and water for irrigation. The first of these two images (a) was taken in 1999. Dark areas are deep water, and blue is shallow. Accumulated sediment has reduced water depths in the upstream (left-hand) portion of the reservoir shown. The second image (b) was taken in 2004, after a prolonged drought. The reduction in water depth and the extent of the reservoir are easily visible.

Figure 1-25 Two Landsat images of Shenzhen, China, taken in 1988 and 1996, showing dramatic urban development in just eight years. Although China's population growth has slowed, many people are moving from rural areas to urban ones as the industrial economy grows. Remote sensing and GIS are essential technologies for documenting and monitoring rapid changes on Earth's surface such as this.

Scientists and land managers use GPS to determine their locations when making environmental measurements. Airplanes and ships now navigate by GPS instead of relying on older, ground-based radio systems.

As GPSs became more widely used, their price dropped rapidly, which led to a further expansion of applications. Many new automobiles now offer GPS-based mapping systems that continually track the vehicle's location and display a moving digital map on the console. Some of these devices incorporate address-locating and route information, so that the driver can enter an address as a destination and the device will calculate an appropriate route from the present location to the destination, then tell the driver when and where to make turns. Hikers carry GPSs to avoid getting lost, runners carry them to monitor their speed and distance, and boaters carry them to return to their favorite fishing holes or to navigate home in a fog.

Figure 1-26 An iPhone showing the user's location.
Devices such as smartphones enabled with Global Positioning System and other locational technologies have become ubiquitous, dramatically increasing access to information about location and routes.

GIS: A Type of Database Software

A GIS is a special form of database software in which spatial information is an important part of the database. A simple database consists of lists of items—people, commodities, transactions, etc. Each item in the list is associated with multiple bits of information. In a database of land parcels maintained by a city clerk's office, for example, one might expect to find information on the owner (name, address, etc.), the value of the parcel, what kinds of structures exist on the parcel, and so forth. In a GIS, one of the attributes of an item in the database is either its spatial characteristics (boundary information, in the case of a land parcel) or something else (such as an address or name of the district in which it is located) that can be used to locate the item in space.

By using such a database, we can create layers of information, each representing a different characteristic, or attribute, of a place. A single layer can be displayed by itself or combined with other layers to show relations among different kinds of information (Figure 1-27). A GIS allows the user to create rapidly many different specialized maps from a single database.

Types of geographic data in a GIS Geographic data are of many types, and each different type has certain characteristics that determine how the information can be analyzed, interpreted, and displayed. One fundamental distinction is between raster and vector data (Figure 1-28). Raster data are arranged in a rectangular grid of cells, which are all the same size. A computer display is a raster device. For example, many computer monitors display information in a grid that is 768 rows high by 1024 columns wide, containing 786,432 grid cells. Each of these cells contains specific information—in this case, the color that is to be displayed on the monitor. The pixels in a remotely sensed image usually contain information about the amount of light or energy sensed as coming from a particular place on Earth's surface.

Vector data are based on points with X and Y coordinates that specify location. Vector data are of three basic types: points, lines, and polygons. A point is defined by a single pair of X and Y coordinates. A straight line can be defined by just two points. More complex lines, such as would represent roads or rivers, are defined by a series of points, each defining a straight-line segment of the line. Regions or areas are represented with polygons, which are simply lines that begin and end at the same point. Only four points are needed to define the border of a rectangular region (Colorado, for example), but many more are needed to define the outline of most geographic regions.

Raster and vector data differ in the kind of information they can portray and the spatial accuracy with which they portray it. In a raster image such as a Landsat image with 30-meter by 30-meter (98 by 98 foot) pixels, every place within the pixel is assumed to have the same value for a given characteristic. If the pixel happens to be located in the middle of a lake, this assumption isn't very important—every place within it is water. If, however, the pixel is along the shore of that lake, then only a part of the pixel is water and a part is land. In a satellite image, the value that will be assigned to that pixel will probably fall somewhere between the appropriate value for water and that for land, thus representing a surface cover type that does not, in fact, exist at that location. A vector line, on the other hand, could be created that would show exactly where the shoreline falls so as to accurately portray this feature. As long as pixels are small in relation to the objects being portrayed, raster formats are very useful for storing and manipulating information about complex maps such as a satellite image of a vegetated landscape, with many variations in the colors of vegetation and shapes of vegetation patches. Vectors, on the other hand, are

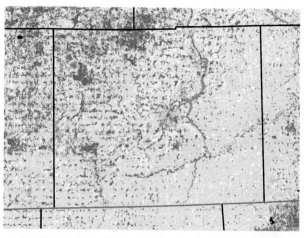

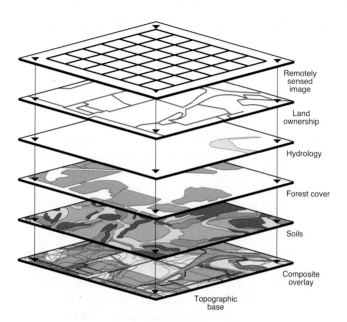

Figure 1-27 Layers of a GIS. A geographic information system involves storing information about a location in layers. Each layer represents a different piece of human or environmental information. The layers can be viewed individually or in combination.

Figure 1-29 A land use/land cover map of Lenawee County, Michigan, showing both raster and vector data. The raster data are the U.S. Geological Survey's National Land Cover Dataset, which shows land use/land cover interpreted from 30-meter (98-foot) resolution Landsat imagery. Agricultural land is in yellow, urban land is in red, forest is in green, water is blue, and wetlands are in purple. The county (black) and state (red) boundaries are vector data, from the U.S. National Atlas.

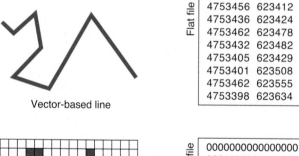

	Flat file
Vector-based line	4753456 623412
	4753436 623424
	4753462 623478
	4753432 623482
	4753405 623429
	4753401 623508
	4753462 623555
	4753398 623634

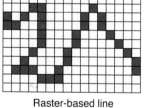

Raster-based line

Flat file
```
0000000000000000
0001100000100000
1010100001010000
1100100001010000
0000100010001000
0000100100000100
0001000100000010
0010000100000001
0111001000000001
0000111000000000
0000000000000000
```

Figure 1-28 Raster and vector data types. Raster data are composed of grids of cells that are coded to indicate a characteristic of a given area, while vector data consist of lines defined by X and Y coordinates of points (vertices) along the line. Remotely sensed images and photographs are typically raster images, whereas maps showing regional boundaries (such as political maps) are usually stored as vector data.

very efficient for storing information about regions, such as political units, or lines, such as roads. Modern GIS devices are able to analyze and display both raster and vector data (including points, lines, and regions) simultaneously (Figure 1-29).

Acquiring digital geographic information A few decades ago, all maps were in paper (analog) form, but today nearly all mapping is done digitally using GIS. One of the biggest tasks in the first few decades of GIS technology has been to create the digital geographic database from which maps are made. Creation of this digital resource is done primarily in two ways. One of these is converting existing maps from paper to digital form. To a large extent this has been accomplished for the wealthier parts of the world; however, digital data are much less available for developing countries. The other way of creating a digital database is through use of remote-sensing satellites, which create digital data directly rather than going through a paper form first.

One way of digitizing paper maps involves manual tracing of lines using a specialized table, following the lines with a device somewhat like a computer mouse. As you might imagine, this is very tedious and time-consuming work. Today, most conversion of maps is accomplished using scanners coupled with software that is able to "trace" features on the scanned image digitally and convert them to vectors (lines). While this technology has improved dramatically, it still requires considerable expertise as well as expensive hardware and software. This helps explain why creation of a digital database in developing countries is lagging well behind the effort in the United States.

Today, a very diverse database is readily available in the United States, much of it freely available on the Internet. For example, the 1:24,000 topographic maps are available in two forms. One of these is simply a scanned image of the map called a digital raster graphic (DRG). This version can be easily displayed on a computer using common image-viewing software. The other version, called a digital line graph (DLG), consists of most of the lines from those maps—boundaries, streams, contours, roads, etc.—in vector form. These DLGs are the form that is more often used in GIS analyses. Digital versions of older paper maps are also available at scales of 1:100,000, 1:250,000, and 1:2,000,000.

In addition to these basic map resources, many specialized maps are now available in digital form. For example, the U.S. Geological Survey makes available online land use/land cover maps derived from satellite imagery, maps of hydrologic features and watersheds, and digital elevation models that are used to portray the landscape in three dimensions (Figure 1-30). The U.S. Fish and Wildlife Service distributes digital versions of its National Wetlands Inventory, and the National Oceanographic and Atmospheric Administration's marine navigational charts are available in digital form. The U.S. Census Bureau produces digital maps that show all roads and streets along with the numbering of addresses on them. These data allow you to type an address into a computer and immediately get a map showing the location of the address; such data are also essential for the emergency response systems used by police and other emergency services. Many local governments have put property maps online so that anyone can search these public records via the Internet to determine who owns a particular piece of property.

Application of GIS to spatial analysis   As discussed earlier, spatial analysis includes topics such as distribution, density, concentration, pattern, and movement. GIS facilitates quantitative measurement of these geographic properties. Calculation of population density simply involves combining (or overlaying) population data with a map of regions such as counties.

Using a GIS, we can determine the total number of people living within the boundary of the region and divide that by the area of the region to determine density. Similarly, a GIS can be used to count the total number of people living within a given distance of some specified location, as part of a measurement of concentration. Changes in the distribution of people, businesses, or forest clearances from one time period to another, as represented in two different layers of a GIS, tell us about movements of people and economic activities over time.

A recent study of the distribution of small ponds illustrates the role of GIS in modern geographic research. An inventory was made of small lakes across the conterminous (48) United States to determine how many lakes there are and their locations. The analysis began with a satellite-based land-cover map produced by the U.S. Geological Survey. Because it was derived from satellite images, it is a raster map, where individual pixels are identified as "row crops," "deciduous forest," "high-density urban," etc. Rivers, lakes, and ocean bays are identified as "water" without further distinction.

The first task was to eliminate from the map all the areas that were clearly not lakes. This was done using map layers that included rivers and the ocean shore. Anything within 1 kilometer (0.6 miles) of a river or 5 kilometers (3 miles) of the shore was blanked out, so these water areas would not be counted as lakes. Next, lines were drawn around each remaining group of water pixels that touched each other, creating a vector layer in which each lake is a single entity. These lakes were then counted; the total is nearly 2.6 million! Figure 1-31 is a map showing their distribution.

The analysis revealed that a very large number of lakes are located in the eastern Great Plains, in areas

Figure 1-30 A three-dimensional perspective view of Mt. Everest. This Google Earth image was made by "draping" an aerial photograph on a digital elevation model. Similar technology is used in many visualization applications, from television weather forecasts to flight simulation in movies.

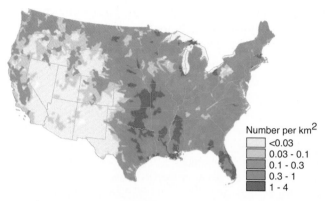

Figure 1-31 The distribution of small lakes in the conterminous United States. The greatest number of lakes are found in the eastern Great Plains, where the vast majority are farm ponds built to store water for farming operations. Large numbers of lakes are also found in areas where they occur naturally, such as the Great Lakes states and Florida. Even in those places, many of these lakes were built by humans.

where natural lakes are rare. In fact, Texas has more lakes than any other state—about 10 percent of the total. Minnesota, which claims "10,000 lakes" on its license plates, in fact ranks seventh according to this study, behind Texas, Florida, Oklahoma, Kansas, Missouri, and Mississippi. The good news for Minnesotans is that they are being modest: The actual number of lakes there is more like 100,000. While the study does not distinguish between natural and artificial lakes, the great concentrations of lakes in areas where natural lakes are rare confirms that the vast majority of these features are of human origin. Most are small ponds built on farms in the latter half of the twentieth century, many of them as part of government programs. They are most numerous in the eastern Great Plains because this is an area that has enough rainfall to fill the ponds in the spring and early summer, yet also has a dry season in which this stored water is particularly important.

In addition to the basic task of counting lakes, this study demonstrates the value of GIS for monitoring environmental conditions in a fast-changing world. Each of these farm ponds is so small that, by themselves, they seem insignificant. Taken together, however, they constitute an enormous modification to the environment. In some areas of the Great Plains, for example, lakes number more than 3 per square kilometer (1 for every 80 acres or so). They store water for cattle, but much lake water that would otherwise soak into the ground is lost to evaporation, principally during the summer. In some areas the amount of water that is lost to evaporation from farm ponds is enough to reduce the average annual flow in streams by 10 percent—an important amount in a water-stressed area.

This example illustrates many of the approaches of geography we have discussed in this chapter. It demonstrates that answering "where" questions remains an important part of geography. Earth's surface is

constantly changing, so even though virtually all of Earth's land surface has been explored and mapped to some level, the maps must continually be redrawn. It illustrates important locational concepts such as distribution and concentration. Spatial interactions are seen in ponds' effects on downstream water availability, the transport of sediment and other substances by streams, and even the migration and distribution of wildlife. The lakes also exemplify the myriad ways in which human activity—in this case especially culture, food production, and government policy—and the physical environment interact.

Finally, it is important to note that this study was carried out using data that are freely available on the Internet, and the data were analyzed on ordinary desktop computers. Anyone with the interest, a modest amount of training in GIS, and the appropriate software could have done this study. It would not, however, have been possible with the technology existing as recently as the mid-1990s. Some of the data were not available or were available only to a select few, much expensive computer time would have to have been devoted to the job, and some of the software needed had not yet been written. Recent advances in GIS technology have opened up many such opportunities for geographers to explore the changing face of Earth's surface.

Integration of Information Technologies

In the 1960s, the 911 call system was developed in the United States to ease reporting of emergencies to police and fire departments and to speed their response to such reports. Similar systems were developed elsewhere. Since the 1980s, these systems have made use of digital telephone-call routing technology that transmits the telephone number of the caller to the receiving telephone (caller ID). Prior to the advent of mobile telephones, telephone numbers were all associated with physical addresses, and it was a relatively simple matter to have a computer linked to the phone system so it could read the number of the incoming phone call, look it up in a list of phone numbers and addresses, and display the address on a dispatcher's computer monitor. The dispatcher could then relay the location by radio to the emergency responder without the caller having to say where he or she was. The system has evolved in many ways since then, including the display of a map on the computer monitor showing the caller's location, and the automatic transmission of the information to a computer in a police car. The growth of cellular telephone use is a problem because the phone is not associated with a fixed location, but the call is routed through a particular antenna so the approximate location can be determined automatically. Integration of GPS with mobile phones can solve this problem.

As computer technology has expanded throughout our society, information increasingly has been stored in

Online Mapping

When the first GIS systems became available for general use in the 1970s, they ran on large computers known as "mainframes." Their analytical and data-management capacity was tiny by today's standards, but so were the amounts of spatial data available in digital form. In the 1980s, desktop computers became powerful enough to handle the large amounts of data used in GIS work, and the numbers of people using GIS grew rapidly. Data limitations continued, however, and for many purposes it was necessary to digitize the data needed for a project, rather than downloading it from an online source. In the late 1990s, the combination of large numbers of Internet users and the availability of comprehensive digital map databases made online mapping possible.

One of the first online mapping programs to be widely used was MapQuest. It performed address matching according to user input and displayed a map of the place specified by the user. The databases available in the late 1990s had little detail and numerous errors, so the locations mapped by these early programs were often only approximate. In the first decade of the twenty-first century, the capability of these address-mapping systems increased dramatically to include providing directions to get from one place to another, showing a choice of conventional road map, satellite/aerial imagery or topography, offering a choice of route types ("avoid highways" or "avoid tolls," for example), and linking to data about places such as street-level photographs or names and phone numbers of businesses. We still experience the limitations of the data underlying these systems from time to time, such as when the route from A to B is not as direct or fast as one that we know from experience to be superior. But Internet-based GIS systems have become very sophisticated and today are much more accurate than they were just a few years ago.

TerraServer is another early map display system, developed in part by the U.S. Geological Survey. It displays images of topographic maps and aerial photos of a place, according to user specifications. Its capabilities are limited, however, and other commercial services have been developed that are faster and offer greater functionality and are used much more widely today.

Modern online mapping systems provide GIS analytical capabilities to users anywhere. The actual GIS software and hardware is located on a server somewhere, but controlled by users around the world. Some of these systems have relatively limited analytical capabilities and focus instead on information storage and display, including offering users the opportunity to add data to the system. For example, Google Earth gives users the opportunity to add ground-level photos to the database. It has thus become something of a "wiki" of geographic information, complete with all the problems associated with only limited quality control. The display capabilities of systems like Google Earth are highly sophisticated, offering dimensional views, rapid zoom-in and zoom-out, and "fly-by" visuals.

Some online map servers allow the user to carry out highly complex analyses. For example, anyone working with water-related issues needs to be able to know the extent of the drainage basin of a stream. We will discuss this in much more detail in Chapters 3 and 4, but it is essentially the geographic area from which a stream derives its water. Drawing the boundaries of drainage basins requires highly detailed analysis of surface topography, and the process is inherently error prone. The U.S. Geological Survey has developed an online mapping system called StreamStats (you can search for it online), which will draw the boundaries of the drainage basin for any stream according to user specifications (see figure). Then, based on the characteristics of the drainage basin, the software will estimate the amount of water flowing in the stream, the sizes of floods, and other important information.

Online mapping is still young, and many new and useful applications will no doubt appear in the coming years. Geographers will play a central role in this process and will of course be among the users of the systems that they develop.

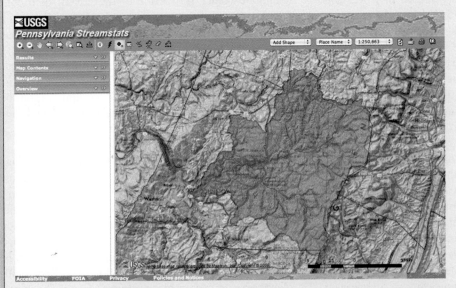

StreamStats map of the drainage basin of the Little Conemaugh River, Johnstown, Pennsylvania.

digital rather than paper form. We still produce paper documents and communicate with each other using paper, but even the information that now is printed on paper is first entered in a computer, and remains stored there long after the paper form has gone to the landfill. How many documents that you have produced on your personal computer are still there, occupying storage space? Financial records, real estate ownership, newspaper articles, credit card transactions, medical records, automobile maintenance histories, product shipments—all are recorded in digital databases.

Sophisticated software systems called relational databases have been developed to manage this information explosion. Different categories of information may be organized in different ways, and each type of information has particular requirements regarding the way it is stored in a computer in order to maximize the efficiency of data storage and retrieval. Relational database software makes it possible to link different databases to each other. For example, an instructor in a geography course keeps records of the grades in a student's class, and these grades are submitted as a list to a central computer at the end of the course. The list has each student's name, probably a unique identification number, and the grade. The information on that list and others submitted by other instructors needs to be reorganized so that, for example, a student can get a report of all the grades earned in all the classes she/he has taken. Relational database software does that reorganizing.

Although many databases are held confidential so that information they contain cannot be easily linked with other databases, the integration of data across different aspects of society is increasing rapidly, and this integration allows powerful new tools for information users. For example, many retail businesses, such as gas stations and supermarkets, offer special discount cards for repeat customers. When a customer uses such a card the details of the transaction (time, location, what was purchased, etc.) are recorded in a database. This information is also associated with the name and address of the customer, often along with other demographic information such as age, gender, and, potentially, many other aspects of the customer's life. Credit card transaction records do this also.

A street address is a spatial location system. It identifies anything associated with that address with a specific place on Earth's surface. The combination of detailed databases with spatial information can be a powerful tool for many applications. Geographic information systems thus make it possible to map virtually anything about which information is available. Suppose, for example, a retailer that sells outdoor recreational equipment wants to open up a new store. Where should that store be located so as to maximize sales? The answer to that question could come from a GIS analysis of the distribution of the sorts of people who would buy recreational equipment. Perhaps these people tend to be young, educated, and relatively affluent.

Where do these people live? Some of this information may come from Census Bureau data. Additional information may come from databases maintained by private corporations and sold expressly for this purpose. This could be combined with road maps and traffic information to give an indication of the driving time from a customer's residence to a potential store location. These sorts of analyses have become routine, and most large retail corporations and many smaller consulting firms employ geographers with GIS training to analyze data in support of business decisions.

Geographic information systems have become prominent on the Internet as well. Wherever map data have become available in digital form, these data have been integrated in online mapping systems that allow the user to view a map of any particular place, at any particular scale. Perhaps the best known of these systems today is the one produced by Google, which is available in both a basic online version and various versions with greater capabilities using both locally installed software and online databases. These systems integrate polygon, line, and point data for boundaries, roads, and cities with satellite or aerial photographic images in raster form. The software displays the data in a scale-dependent fashion, increasing the amount of detail that is shown as scale increases.

The availability of digital maps, remotely sensed images, and other data that can be integrated with GIS is expanding rapidly around the world. But large disparities in information availability remain between rich and poor parts of the world, and in some countries (most notably China) controls on access to information reduce the utility of these technologies to ordinary citizens.

Looking Forward

The twenty-first century is presenting new and difficult challenges to citizens, businesses, and governments. Critical issues include:

- Global-scale environmental change, including global warming, soil degradation, and stressed water resources
- Rates of population growth in some regions that will strain cultural systems, deplete scarce resources, and produce environmental changes
- Continuing disparities in wealth and the threat of conflict between rich and poor countries and within individual countries
- Systems of food production and distribution that do not meet the nutritional needs of many people in developing countries

Confronting problems of this magnitude will require a great deal of creativity and hard work from all segments of society, including geographers. Geography offers several unique perspectives for understanding contemporary issues.

First, geography emphasizes interdisciplinary approaches to contemporary issues. Geographers are trained partly as natural scientists and partly as social scientists, so they have a wide breadth of understanding that makes geography a unique discipline. Geographers work well in interdisciplinary teams analyzing major problems that involve combinations of human and physical factors. They can communicate with specialists in a variety of natural and social sciences, so geographers are often asked to lead group efforts.

Second, geographers have a global perspective. A century ago, a famine or war in a distant part of the world had little consequence outside the immediate region, but today, local events have worldwide impact. New information technology and transportation systems have created global markets for many commodities. What happens in one place has ramifications in other areas and often throughout the world. Concern about an endangered owl in the U.S. Pacific Northwest that

lives only in old-growth forests closed certain easily accessed forests to harvesting. This caused lumber prices to rise not only throughout the United States but in Canada and in East Asia as well. The 1986 accident at Chernobyl, a nuclear power plant in Ukraine (then part of the Soviet Union), spread radioactive fallout around the world, heightened fears of nuclear power, and thus contributed to curtailment of nuclear power development worldwide.

Third, geography has a strong component of applied research. Since the earliest days of the discipline, geographers have worked on problems of practical significance, such as finding new routes for commerce and evaluating the natural resources of sparsely inhabited lands. Geographers emphasize studying everyday phenomena about places. They are aware of conditions in the world around them and are accustomed to applying their knowledge to them. GIS has become central to this endeavor, and is likely to remain so.

Chapter Review

Summary

Geography is the study of the interaction of all physical and human phenomena at individual places and of how interactions among places form patterns and organize space. Physical geography studies the characteristics of the physical environment. Human geography studies human groups and activities, such as languages, industries, and cities.

Geographers divide Earth into regions, which are areas defined by one or more distinctive characteristic or feature, such as climate, soil type, language, or economic activity. Area analysis is an approach that looks at these features of Earth's surface in a regional context.

Geographers study the distribution of objects across Earth's surface and processes by which human and environmental phenomena move from one place to another. Three properties of spatial distribution include density, concentration, and geometric pattern. Movements of matter, people, and information result in interactions among places along networks through diffusion processes.

The place where a culture originates is called the culture hearth. Various aspects of cultures may spread out and be adopted by other peoples in a process called cultural diffusion. The impact or frequency of any cultural attribute may diminish away from its hearth area; this phenomenon is called distance decay. Some aspects of culture may also

develop variations as people who carry that culture wander outward and away from one another. Tracing a course of diffusion may teach us a great deal about how peoples and cultures interact. Diffusion may be by relocation, or it may be contiguous or hierarchical. Barriers to cultural diffusion may be topographic, political, or even cultural. Diffusion does not explain the distribution of all cultural phenomena. Sometimes the same phenomenon occurs spontaneously and independently at two or more places.

Earth's physical environment results from a process of interaction among four systems: the atmosphere, hydrosphere, lithosphere, and biosphere. Cultural systems include customary beliefs, social forms, and material traits. Some geographers regard the distribution of human activities as the result of underlying regularities in cultural systems, while others emphasize the diversity and uniqueness of human behavior in explaining cultural systems. The physical environment influences human actions, although humans can adjust and choose a course of action from many alternatives.

Geographers utilize maps to depict the location of places and to interpret underlying patterns. Remote sensing from satellites and GIS devices are two recently developed tools that help geographers more fully analyze patterns on Earth's surface. GISs allow rapid integration of many different types of geographic data as well as automated map production. Rapid expansion of information technology and the integration of diverse databases with GIS technology are leading to new applications of GPS and GIS in many aspects of our lives, as we will see throughout but this book.

Key Terms

atmosphere p. 16
biosphere p. 16
cartography p. 2

concentration p. 11
conformal map p. 22
contiguous, or contagious,
  diffusion p. 13
cultural geography p. 2

cultural landscape p. 17
culture p. 17
density p. 11
diffusion p. 13
distance p. 11

Questions for Review and Discussion

1. What is geography? Describe differences among physical, human, and regional geography.

2. What contributions did ancient and medieval geographers make to the development of geographic thought?

3. What are the three ways to indicate scale on a map?

4. What are the four types of distortions that can result from map projections?

5. How do remote sensing and GIS devices contribute to geographic study?

6. What is the difference between a meridian (or longitude) and a parallel (or latitude)?

7. What is the difference between formal regions, functional regions, and vernacular regions? Give examples of each of the three types of regions.

8. What is the difference between density and concentration?

9. What do geographers mean by the concept of spatial interaction?

10. What is the difference between relocation diffusion and contiguous diffusion?

11. What are Earth's four physical systems?

Thinking Geographically

1. Cartography is not simply a technical exercise in penmanship and coloring, nor are decisions confined to scale and projection. Mapping is a politically sensitive undertaking. If you were a resident of New Zealand, what changes might you suggest to the world map projections used in this book? Look at how maps in this book distinguish the borders of India with China and Pakistan and of Israel with its neighbors. Can you identify other logical ways to draw these boundaries?

2. Imagine that a transportation device (perhaps the one in *Star Trek*) would enable all humans to travel instantaneously to any location on Earth. What impact might that invention have on spatial interaction? What if such travel were expensive, so only the rich could afford it?

3. When earthquakes, hurricanes, or other so-called "natural" disasters strike, humans tend to blame nature and see themselves as innocent victims of a harsh and cruel natural world. To what extent do environmental hazards stem from unpredictable nature, and to what extent do they originate from human activity? Should victims blame nature, other people, or themselves for the disaster? Why?

4. Geographic approaches, such as area analysis, spatial analysis, and systems analysis, are supposed to help explain contemporary issues. Find a story in your newspaper that can be explained through application of geographic concepts.

Log in to www.mygeoscienceplace.com for videos, animations, **MapMaster**™ interactive maps, RSS feeds, case studies, and self-study quizzes to enhance your study of Introduction to Geography.

MapMaster™

Winter travel on an ice road near Cambridge Bay, Nunavut, Canada.

2

Weather and Climate

Climates are changing throughout much of the world, with the most rapid changes occurring in high latitudes. In the Northern Hemisphere, where there is significant habitable land area at high latitudes, these changes are affecting the way people live. Paradoxically, some northern communities are more isolated in the summer than in the winter because the cheapest way to travel and in some cases the only way to carry heavy, bulky goods is over frozen terrain in the winter, on "ice roads." Many areas are inaccessible by roads in the summer because permanent roads do not exist. For example, much of the heavy equipment needed for energy and mineral development in northern areas can be carried to those areas only in winter. Warming climate can thus increase isolation by shortening the season in which inexpensive land travel is possible.

The most dramatic weather events in the news are those that cause major damage either in the short term, such as tornadoes, or over longer time periods, such as droughts. Floods, mostly caused by heavy rains but sometimes augmented by high sea levels during hurricanes, cause more damage than any other natural hazard. In addition to such extremes, day-to-day weather influences culture and social customs from clothing to agriculture. This chapter explores the mechanisms that cause **weather**—the day-to-day variations in temperature, precipitation, and so forth—and **climate,** the statistical summary of weather over time. As we will see later in the chapter, while we have generally thought of the statistical temperature average

A Look Ahead

Energy and Weather

Solar energy drives the circulation of Earth's atmosphere. Energy inputs to the atmosphere vary with latitude and season, and with elevation in the atmosphere. Atmospheric circulation redistributes energy and, in the process, generates weather and storms.

Precipitation

Precipitation may be created when air rises, causing water vapor to condense in the atmosphere. Air may be forced to rise because it is heated at Earth's surface, when passing over mountains, or in storm systems.

Circulation Patterns

Variations in atmospheric pressure from one place to another cause wind. Air blows from regions of high pressure to regions of low pressure. Pressure differences help explain prevailing wind directions and seasonal differences in winds.

Climate

Earth's climates can be grouped into 11 major types, within which there is considerable variability. The distribution of climate types reflects general global and regional circulation patterns.

Climate Change

Climates have changed significantly over the past few million years, and large variations have occurred even within the past 20,000 years. Many different factors contribute to climatic change, including human activities that are raising the level of carbon dioxide in the atmosphere.

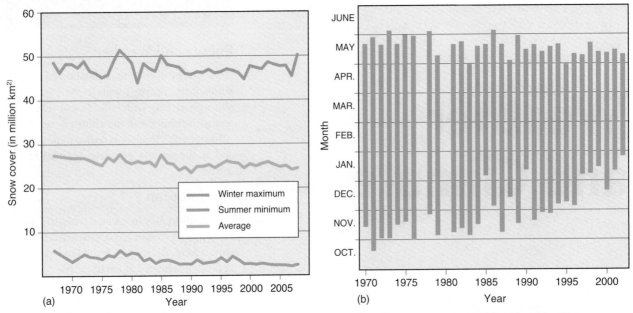

Figure 2-1 Changing Arctic climates. (a) The maximum extent of winter snow cover in the Northern Hemisphere has generally remained constant, but areas that formerly had snow all year in at least some years are now without snow cover in the summer. The snow season is getting shorter as well, which is reflected in a decrease in the length of the period in which tundra roads are without snow cover, shown as green bars in graph (b). The road snow cover data can be affected by factors other than climate, such as safety rules for winter travel, so we have to be careful how we interpret such data. Nonetheless, it is clear that Arctic climates are warming and these changes are affecting the way people live.

as relatively stable in human timescales, we now know that climate is changing (Figure 2-1).

Weather conditions—such as storms, snowfall, clear skies, and warmth or cold—are caused by radiant energy received from the sun. This solar energy is redistributed from tropical latitudes to the polar regions, from lower levels to upper levels of Earth's atmosphere, and eventually back into space. The movement of energy creates air movements that carry warm air from the tropics to high latitudes, and cold, wintry air from polar regions toward warmer climates. Air carries moisture from the oceans over the land and returns drier air to ocean areas.

In the process of air and water moving from one place to another, weather is created. Weather varies from day to day because atmospheric circulation is constantly changing. But varied as it is, circulation tends to follow certain patterns. A zone of rising air is usually found near the equator, causing frequent heavy rains. Over northern Africa, in the region of the Sahara, the air usually descends, causing clear skies and aridity. In the midlatitudes, the air tends to move from west to east most of the time and less often in other directions. In southern and eastern Asia, the wind blows from the land to the ocean in winter and from the ocean to the land in summer, causing intense monsoon rains from India to southern China. All these events are linked together in a dynamic atmospheric circulation.

These weather patterns are one of the fundamental features of Earth's surface that make places different from each other. They make agriculture possible in some places and not in others; they provide abundant

water for hydroelectricity in some places and sunshine on beaches elsewhere. More than any other physical attribute of Earth, weather and climate regulate natural systems and constrain the ways in which humans use their environments. Yet even though Earth's circulation cannot be controlled deliberately, it can be modified by humans. Many people believe that droughts such as the prolonged drought affecting the southwestern United States in recent years become more common as a result of human effects on the atmosphere. In this chapter we will explore the processes that govern weather, the distribution of climates that results from those processes, and some of the ways in which human activities are modifying weather and climate.

Energy and Weather

The movement of energy in the atmosphere can be likened to that in an automobile engine, which uses energy from gasoline to make a vehicle move. An engine releases energy inside cylinders through the burning of gasoline. This concentrated energy drives the vehicle and is dissipated mostly through friction.

The fuel that drives the atmosphere is **solar energy**. Some solar energy directly heats Earth's atmosphere, but most heat the Earth's surface; which in turn, heats the air, creating pressure differences that generates wind and ocean currents. These movements carry heat and moisture from one part of the planet to another before the energy is dissipated into space. In an automobile, motion takes the form of pistons moving back and forth, shafts and gears turning, and wheels spinning. In the atmosphere, air is in motion from one part of the globe to another, sometimes in steady flows and sometimes in intense spinning or vertical movements that produce storms.

The Sun is a large thermonuclear reactor, in which hydrogen atoms combine to produce helium. Most important to us is a by-product of this reaction: the release of vast amounts of energy. The Sun radiates this energy into space in all directions, and Earth intercepts a tiny fraction of it. Yet this small, steady energy flow to Earth is sufficient to power the circulation of the atmosphere and oceans and to support all life on Earth.

As Earth revolves along its elliptical orbit around the Sun each year, it receives this energy across a void averaging about 150 million kilometers (93 million miles). Variations of about 3 percent seasonally in the distance between Earth and the Sun cause small variations in the amount of energy reaching Earth, but these are minor in comparison to the effects of latitude and the tilt of Earth's axis.

Incoming Solar Radiation

The amount of solar energy intercepted by a particular area of Earth, or **insolation,** depends on two factors: the intensity of solar radiation, or the amount arriving per unit of time, and the number of hours during the day that the solar radiation is striking.

Intensity of solar radiation   The intensity of solar radiation at a particular place on Earth depends largely on the angle at which the Sun's rays hit that place. Daily and seasonal differences in intensity are caused by variations in the **angle of incidence,** which is the angle at which solar radiation strikes a particular place at any point in time. The angle of incidence at a particular place on Earth varies during a day and among seasons.

To illustrate the angle of incidence, consider what happens when a 1-square-meter beam of sunlight strikes a spherical object such as Earth (Figure 2-2). If this beam strikes Earth's surface perpendicularly (that is, at a right angle of 90°), the surface it touches also will have an area of 1 square meter. At an angle of less than 90° (that is, an oblique angle), the area illuminated by that beam will be greater than 1 square meter. The intensity of the radiation is reduced because the beam's energy is spread over the larger area.

As the angle is reduced even further, energy is distributed over a still larger area, and the intensity of radiation becomes less and less. Compared to the level of radiation at a 90° angle, the intensity of energy is about 86 percent at an angle of 60° and 50 percent at an angle of 30°. Thus, the intensity of radiation per unit of surface area is affected greatly by the angle of the Sun's rays, that is, the position of the Sun in the sky. When the Sun is overhead, sunlight is more intense, and the level of energy received is more concentrated than when the Sun is low in the sky.

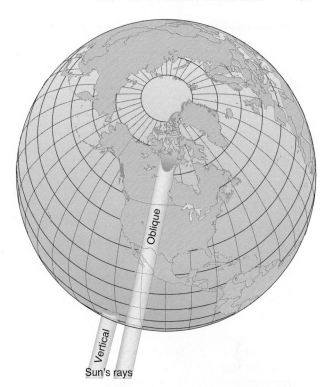

Figure 2-2 The angle of incidence. Radiation is most concentrated when the Sun's rays are perpendicular or vertical to Earth's surface, as occurs when the Sun is directly overhead. As shown, a place receiving more perpendicular insolation has this energy concentrated over a smaller area, making the area warmer. A place receiving more oblique (that is, less perpendicular) insolation has it spread over a larger area, making the area less warm.

On a daily basis, the Sun is most intense at noon, when the Sun is highest in the sky, and least intense at sunrise and sunset. We are more likely to get sunburned around noon than early in the morning or late in the afternoon. Why? Because insolation varies at different times of day.

The angle at which the Sun's energy strikes Earth's surface varies with the season as well as within a day. Throughout the year, the area of Earth's surface where the Sun is overhead keeps shifting due to Earth's tilt and continual revolution around the Sun (Figure 2-3).

The amount of insolation a place receives in a particular season depends on its latitude (Figure 2-4). At noon on the Northern Hemisphere's **vernal (spring) equinox** (March 20 or 21) and **autumnal equinox** (September 22 or 23), the perpendicular rays of the Sun strike the equator, and the Sun is directly overhead at the equator (Figure 2-4a). At places along the **Tropic of Cancer** (23.5° north latitude) and the **Tropic of Capricorn** (23.5° south latitude), the intensity of solar radiation is reduced to 92 percent of the level at the equator. At places along latitude 45° north and south, the intensity of radiation is 71 percent of the level at the equator, and at places along latitude 60° north and south, the intensity is only 50 percent of the level at the

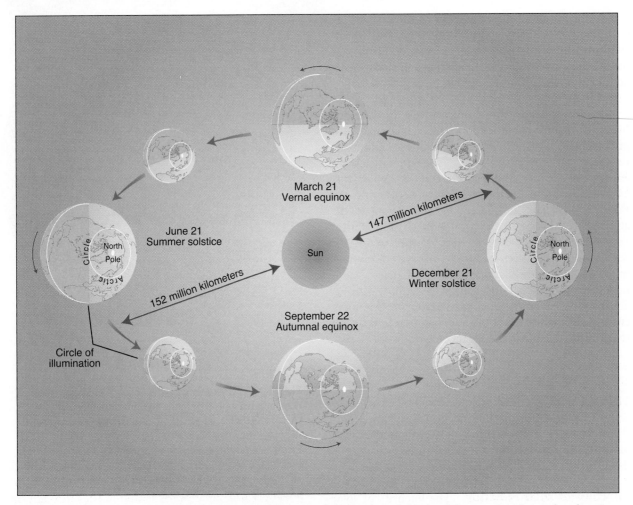

Figure 2-3 Earth's orbit around the Sun. Earth revolves around the Sun in an orbit that follows an imaginary plane known as the plane of the ecliptic. Earth's path around the Sun is shaped like an ellipse. As it revolves around the Sun, Earth rotates around the axis of rotation, which is the line passing through the North and South poles perpendicular to the equator. The line passing through the North and South poles is tilted at an angle of 66.5° away from the plane of the ecliptic, or 23.5° away from a line perpendicular to the plane of the ecliptic. This tilt is constant throughout the year, no matter where Earth is in its orbit. The tilt causes the Northern Hemisphere to be more exposed to the Sun's radiation from March to September (Northern Hemisphere spring and summer), while the Southern Hemisphere receives more radiation from September to March (Southern Hemisphere spring and summer).

Animation
Earth-Sun
Relations

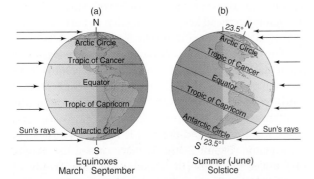

Figure 2-4 Variation in angle of solar radiation. The angle at which solar radiation strikes a particular latitude varies by season. These globes show how Earth receives solar radiation (insolation) on the first day of spring and fall (the equinoxes) (a) and on the first day of Northern Hemisphere summer (b).

equator. At the North and South poles on the equinoxes, the Sun is seen on the horizon and horizontal surfaces do not receive any direct radiation.

From the vernal equinox until the autumnal equinox (from late March until late September), the Sun is directly overhead at places along some latitudes in the Northern Hemisphere. The northernmost latitude receiving direct insolation is the Tropic of Cancer (23.5° north latitude), where the Sun is directly overhead at noon on the **summer solstice** (June 20 or 21). At the September equinox, the Sun is again directly overhead at noon at the equator. From then until the March equinox, places in the southern latitudes receive direct insolation, and the Sun is directly overhead of places along the Tropic of Capricorn (23.5° south latitude) at noon on the Northern Hemisphere's **winter solstice** (December 21 or 22).

Day length The total amount of heat that a particular place on Earth receives in a day is determined by the number of hours during which the Sun's energy strikes the place, as well as the intensity with which it strikes. Places on the equator always receive 12 hours of sunlight and 12 hours of night. But in higher latitudes, the amount of daylight varies considerably with the seasons. For example, Winnipeg, Manitoba, Canada, at 50° north latitude, receives about $16\frac{3}{4}$ hours of daylight at the summer solstice but only about $7\frac{1}{4}$ hours of daylight at the winter solstice. In Winnipeg, the Sun is 63.5° above the horizon at noon on the summer solstice, compared to only 16.5° above the horizon at noon on the winter solstice. In a 24-hour day at the summer solstice, Winnipeg receives nearly six times as much solar radiation, measured at the top of the atmosphere, as it does at the winter solstice.

Variations in the length of day from place to place result from the 23.5° tilt of Earth's axis away from a perpendicular relation to the Sun. Figure 2-4b illustrates the illumination of the globe at the Northern Hemisphere summer solstice in June. Because the North Pole is located at Earth's axis of rotation, it maintains the same position with respect to the Sun throughout the 24-hour rotation of Earth at the solstice. At the summer solstice, if you stand at the North Pole, you can watch the Sun travel in a circle around you at a constant elevation of 23.5° above the horizon. On the day of the solstice, the North Pole is in full sunlight for the entire 24 hours, a phenomenon called the Midnight Sun (Figure 2-5).

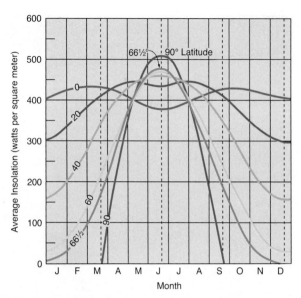

Figure 2-6 Solar radiation at different latitudes by season. Because Earth's axis of rotation is tilted, seasonal variations in the amount of incoming solar radiation are less at places near the equator than at places in higher latitudes. Places at higher latitudes receive somewhat more solar radiation than do places at lower latitudes during the summer, but they receive much less during the winter.

Spatial and seasonal variations in radiation inputs

The amount of solar radiation reaching the top of the atmosphere at a particular latitude varies through the year because of seasonal changes in the angle of incidence, day length, and distance from the Sun. Tropical areas at low latitudes generally are warm throughout the year because they receive large amounts of insolation in every season, whereas places at high latitudes have strong seasonal contrasts in the amount of sunshine and level of temperature (Figure 2-6). The tilt of Earth's axis makes life possible on a seasonal basis at relatively high latitudes; without that tilt, there would not be a warm season at mid and high latitudes, and life would probably be much more concentrated near the equator.

The effects of these variations in energy inputs are clearly visible in world temperature patterns in January and July (Figure 2-7). In general, the highest temperatures occur in low latitudes in both maps. This is because the Sun is highest in the sky in these areas, and therefore the intensity of solar radiation is highest. High solar elevation angles occur throughout the year, and this makes the tropics (the area between the Tropic of Cancer and the Tropic of Capricorn) consistently warm.

High temperatures also occur in high latitudes such as in North America, Europe, and Asia in July. This is because at high latitudes the Sun is high in the sky in summer, and although it is not as high as in the tropics, the day lengths are much greater, so the total amount of

Figure 2-5 Midnight sun. On the day of the summer solstice, poleward of 66.5°, the Sun shines the entire 24-hour day, and the Sun appears to travel above the horizon without setting.

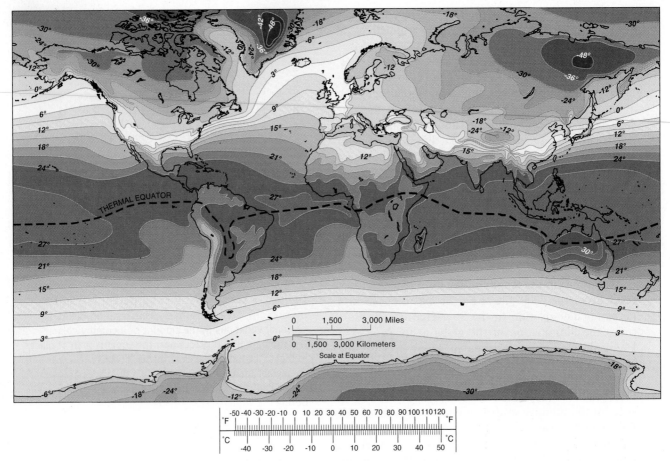

(a) January

Figure 2-7 Global temperatures for January (convertible to Fahrenheit by means of the scale). In the Southern Hemisphere, the warmest areas are over land, especially Australia. The coldest areas are in continental areas in the Northern Hemisphere, especially Greenland and Siberia. In the Northern Hemisphere, the oceans are warmer than land areas at similar latitudes.

sunshine arriving at Earth's surface in a 24-hour day in midsummer is actually greater at high latitudes than in the tropics. The opposite effect is seen in winter, when high latitudes experience both low Sun angles and short days, causing temperatures to be much lower. Seasonal contrasts in temperature are greatest in high latitudes because seasonal contrasts in incoming solar radiation are greatest there.

Storage of Heat in Land and Water

When heat is absorbed by an object, its temperature rises, and when heat is released, an object cools. The ability of an object to store heat depends on what it is made of. Some materials can absorb or release a large amount of heat with only small changes in temperatures, while others heat and cool quickly with only small inputs and releases of energy. Water can absorb and release much larger quantities of heat for a given temperature change than can land. The main reason for this is that a water body such as an ocean can be stirred by the wind and thus carry heat down below its

surface to depths of 10 meters (33 feet) or more seasonally. In contrast, land surfaces warm and cool to only about 2 meters (6 feet) depth seasonally. Oceans can store a season's heat in a much larger volume of matter than can land areas, so they do not heat up as much in summer nor do they cool down as much in winter.

The effect on temperature of water's ability to store heat is clearly visible in Figure 2-7. Large landmasses in high latitudes, such as North America and Asia, have very large temperature differences between July and January. Average July temperatures in Siberia reach above 10°C (50°F), whereas average January temperatures reach −40°C (−40°F). The difference between July and January temperatures exceeds 50°C (90°F) in eastern Siberia. In contrast, the January and July temperatures in the Aleutian Islands of the north Pacific, at about the same latitude as Siberia, are about 15°C (59°F) in July and 5°C (41°F) in January, a range of only about 10°C (18°F). The moderate climates of areas near the ocean compared with the climatic extremes of midcontinent regions are a result of this heat storage in water.

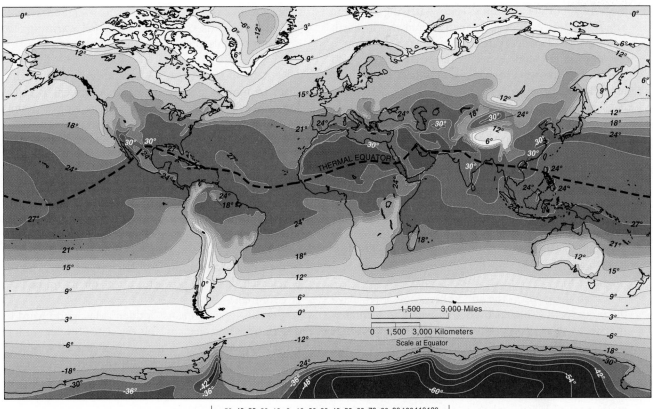

(b) July

Figure 2-7 Global temperatures for July. The warmest areas in the Northern Hemisphere are over land, especially central Asia and North America. The greatest temperature differences between January and July are over land areas, especially the large landmasses of the Northern Hemisphere.

Heat Transfer Between the Atmosphere and Earth

Have you ever noticed that on a winter day, you feel colder sitting inside near a window or exterior walls than near the building's interior walls, even though air temperatures at both places are about the same? Have you ever thought about how moisture on your skin helps cool your body? And why does wind blowing on your skin cool it faster than still air? These effects on the temperature of your skin result from processes of heat transfer. Similar processes of heat transfer operating in the atmosphere as well as on your skin are responsible for the movement of vast amounts of energy from place to place on Earth.

Radiation The most important process of heat transfer in the environment is **radiation.** Energy transmitted by electromagnetic waves, including radio, television, light, and heat, is radiation, or radiant energy. You feel heat radiating from a fire. Heat travels at the speed of light from the fire to your skin, which senses it. Radiation

can travel through the vacuum of space and through materials, although materials may restrict radiant energy flow. Radiant energy allows us to see the Sun's light energy, feel its heat energy, and listen to its radio energy using electronic machines.

Radiant energy waves have different lengths. The **wavelength** is the distance between successive waves, like waves on a pond. Wavelength affects the behavior of the energy when it strikes matter; some waves are reflected, and some are absorbed.

The Sun's energy comes to Earth as radiant energy, for this is the only way energy can travel through the vacuum of space. Two ranges of wavelengths—called **shortwave energy** and **longwave energy**—are most important for understanding how solar energy affects the atmosphere. Most insolation is shortwave, with wavelengths between 0.2 and 5 microns (a micron, or micrometer, is one-millionth of a meter). Wavelengths visible to the human eye account for a small but significant portion of this shortwave energy, from about 0.4 to 0.7 microns. In contrast, most of the energy reradiated by Earth is longwave, in wavelengths between 5 and 30 microns.

It is important that most energy arriving from the Sun is shortwave, while most energy reradiated by Earth is longwave. As energy from the Sun passes through the atmosphere, some wavelengths are absorbed, which warms the atmosphere, while others pass through or are reflected, either to be absorbed elsewhere or to travel back into space. The atmosphere is relatively transparent to incoming shortwave radiation, so this shortwave energy from the Sun easily passes through the atmosphere to reach Earth's surface. When the surface reradiates longwave energy, however, much of it is blocked and absorbed by the atmosphere. The blockage of outgoing longwave energy causes Earth's atmosphere to heat. This heating of the atmosphere is called the **greenhouse effect** because of its similarity to the way glass allows solar energy to enter a greenhouse but limits the loss of heat, causing the temperature inside to rise (Figure 2-8).

Of all the gases in the atmosphere, only a few have the property of being relatively transparent to incoming shortwave solar energy but absorbing outgoing longwave radiation. Gases with these properties are called **greenhouse gases,** and they are critical to heat exchange in the atmosphere. Among the most important ones are water vapor, **carbon dioxide (CO_2), ozone (O_3),** and **methane (CH_4).** Although these gases together constitute a small fraction of 1 percent of the atmosphere, they are the most important from the standpoint of atmospheric heating. Water vapor contributes the most to atmospheric heating. Human activities are increasing the amount of

some greenhouse gases in the atmosphere, and this is believed to be the chief cause of **global warming.** We will return to this topic later in the chapter.

Latent heat exchange   A second important mechanism of moving energy in the environment is called **latent heat exchange.** It transfers tremendous amounts of energy from low latitudes to high ones, and it is also the mechanism most influential in causing precipitation.

We can distinguish between two types of heat—sensible and latent. **Sensible heat** is detectable by your sense of touch. It is heat you can feel, from sunshine or a hot pan, and you can measure it with a thermometer. The atmosphere, oceans, rocks, and soil all have sensible heat, for you can feel their relative warmth or coldness. **Latent heat** is "in storage" in water and water vapor. You cannot feel latent heat, but when it is released, it has a powerful effect on its immediate environment.

Latent—which means "hidden"—is a good word to describe the heat that controls the state of water, for it is invisible stored heat. When ice melts, it must absorb heat energy from its surroundings. This is why ice melting in your hand feels so cold—it is absorbing heat from your hand. The heat becomes stored in the meltwater as latent heat.

Latent heat also is stored in water vapor (Figure 2-9). If you ever have had a finger scalded by steam, you know the startling amount of latent heat that was stored in the vapor and conducted from the condensing water into

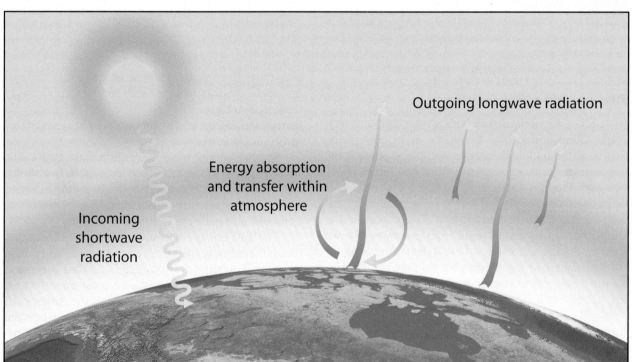

Outgoing longwave radiation

Energy absorption and transfer within atmosphere

Incoming shortwave radiation

Figure 2-8 Greenhouse effect. The atmosphere is transparent to a substantial portion of incoming shortwave radiation from the sun but absorbs much of the outgoing longwave radiation from the earth. This warms the atmosphere. Absorbed heat is circulated within the atmosphere before being reradiated to space.

Animation
Atmospheric
Energy Balance

Figure 2-9 Mist rising off the warm surface of a lake into cold air. The evaporation of water from the lake removes sensible heat from the lake, cooling the liquid water and transferring latent heat to the air.

your finger! Latent heat exchange is what happens when water changes state (Figure 2-10).

Here is an analogy that describes the concept of latent heat: Consider a glass containing a mixture of ice and water on a hot summer afternoon. Because the ice water is much cooler than the air and other objects around it, heat energy is absorbed into the water. The heat begins to melt the ice, but as long as the water contains some ice, its overall temperature remains at 0°C (32°F). The water is absorbing enough heat to convert some ice to liquid, but the temperature does not change. The heat the water absorbs to melt the ice is latent heat, because it cannot be directly sensed as temperature.

When water is converted from liquid to vapor, additional heat is required—in fact, much more is needed than is required to melt ice. The latent heat of vaporization—the process that turns liquid to vapor—is about 540 calories per gram, compared to about 80 calories per gram for the latent heat of melting. A calorie is the amount of energy required to raise the temperature of 1 gram of water 1°C.

In the ice–water example, a portion of the heat used to melt the ice came from sensible heat—heat you can sense—expressed as the temperature of the environment around the glass. The remainder came from the latent heat released by water vapor condensing to liquid on the surface of the glass. Thus, the melting of the ice water involved three transfers of heat:

- Latent heat was converted to sensible heat by condensation of water vapor on the glass.
- Sensible heat was transferred directly from the warm air to the cool glass.
- Sensible heat was converted to latent heat in the melting of the ice.

Figure 2-10 Dew formation. Longwave radiation from the ground to the atmosphere during the night cools air near the ground. When the relative humidity reaches 100 percent, condensation occurs, and water droplets accumulate on solid surfaces such as on these blades of grass.

Now let us transfer this knowledge of latent heat to the much larger scale of the atmosphere. The amount of energy involved in latent heat transfers in the atmosphere is vast, especially for major weather systems and hurricanes. Transfer of heat from the land and water surface to the atmosphere by latent heat exchange amounts to about 40 percent of the solar energy absorbed by the surface. Hurricanes gather strength by drawing in warm, humid air and converting the latent heat it contains into sensible heat, which drives the motion of the storm.

Insolation supplies shortwave energy to warm the seawater. As it evaporates, countless molecules of water vapor—each with its own latent heat—hover above the water, available to join the storm as it passes over. As the hurricane accumulates this water vapor, it is also gathering an enormous amount of latent heat energy. The vapor rises in the rotating storm, cools, and condenses. Latent heat is released as sensible heat, a process roughly analogous to adding gasoline to a forest fire. The amounts of energy involved are very large—the rate of energy released in a typical hurricane, for example, has been estimated as roughly half the total of global electrical generating capacity.

Heat Exchange and Atmospheric Circulation

Liquid water changes to vapor primarily at Earth's surface. This water vapor is carried aloft by rising air. In the atmosphere, vapor is converted back to liquid (clouds and rain) or solid (ice clouds and snow) through the formation of precipitation, which we will discuss later in the chapter. This movement of water back and forth between the surface and the atmosphere acts like a great conveyor belt for energy, picking up heat at the surface, carrying it aloft, and releasing it in the atmosphere. The conveyor is driven by convection.

Convection is movement in any fluid, caused when part of the fluid (whether gas or liquid) is heated. The heated portion expands and becomes less dense. Therefore, it rises up through the cooler portion. Convection causes the turbulence you see in boiling water and in turbulent clouds overhead. As air is warmed, it expands and becomes less dense. In becoming less dense—that is, weighing less per unit of volume—warm air rises above cooler, denser air, just as a hot-air balloon rises through the cooler air surrounding it.

Think of an island on a sunny day (Figure 2-11). Solar energy passes down through the atmosphere and is absorbed by the sandy surface. As the surface warms, it reradiates longwave energy, some of which is absorbed by the air just above the ground. The water surface warms less than the land does because some of the radiation is used to evaporate water, and some penetrates the water to warm lower depths of the sea. The air therefore grows much warmer over the island than over the water. This warmer air over the island expands, grows less dense, and begins to rise. As the

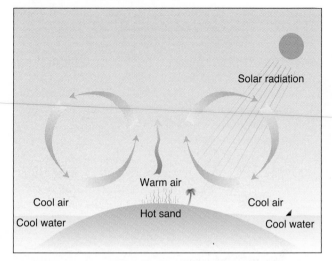

Figure 2-11 Convection in the atmosphere. Convection occurs when air is warmed from beneath, expands, and rises. Cooler air descends to replace it. A common example is where land (such as an island, the Florida peninsula, or Cape Cod) is surrounded by water. Convection is strongest in the daytime, when the surface is warmed by the Sun. Because of a stronger convection process, the wind tends to be gustier in the daytime than at night.

warm air rises, cooler air descends to take its place. The cooler air is then warmed by the surface, and the convection process continues.

Convection also causes horizontal movements of air. Large horizontal transfers of air—and the latent heat contained in water vapor in the air—are major components of Earth's energy system. Horizontal transfer such as this is called **advection.** Heat is advected from tropical areas toward the poles when warm winds blow poleward. Heat in ocean currents moving toward the polar regions is another form of advection.

CONNECTIONS

Climates in Urban Areas

Climates in major urban areas usually differ markedly from those of surrounding rural landscapes. Localized climate characteristics caused by topographic and land-surface characteristics are called **microclimates.** Typically urban microclimates are warmer, both in winter and in summer. The area of warm microclimates associated with an urban area is called the **urban heat island.** Several factors contribute to these differences. First, cities lack large expanses of vegetation and soil that absorb rainfall and later allow it to evaporate. Evaporation cools the land surface, and without this evaporation, the surface remains warmer. Second, the walls of

buildings form vertical surfaces that help to trap solar energy during the daytime. Concrete, asphalt, and similar building materials store heat energy during daylight and reradiate it at night. At night these warm surfaces shield the ground between buildings from exposure to a large expanse of relatively cool sky and thus keep it warmer than it would otherwise be. Third, air pollutants, common in cities, may absorb heat and keep the air warmer. Finally, heat released from energy use, such as in heating buildings, adds to the heat available in the city (although this amount is very small in comparison to the amount of energy derived from the Sun).

Let us look at a small-scale example. During one day's time, warm air rising over the land allows cooler air from the sea to advect over the land. In other words, wind in some coastal areas blows from the sea toward and over the land. On a large scale, during summertime over central Asia, the season's heating causes winds to blow onto the land from over oceans to the south and east. During winter, the land area cools more than the sea, so cold air sinks over the land, warmer air rises over the ocean, and the winds tend to blow from central Asia toward the warmer Pacific Ocean. Such a circulation pattern is called a monsoon; we will discuss this later in the chapter.

Precipitation

People rely on precipitation to be "normal"—amounts that are typical, expected, not unusual. Normal rains and snowmelt are necessary for consistent agriculture, to feed Earth's 6.8 billion humans. People who live on lowlands along rivers rely on "normal" precipitation to keep the river within its banks and out of their basements. All plants and animals are adapted to a "normal" amount of moisture for their environment. However, "normal" does not always happen.

Much of the U.S. Midwest suffered a severe drought in 1988 that heavily damaged crops and lowered the level of the Mississippi River so much that barges became stranded. In 1990 and 1993, the same regions of the Midwest experienced record wet years, with floods on the Mississippi, Missouri, and other rivers. From 1998 to 2002 and again from 2005–2007 the southeastern United States suffered severe droughts.

Such extremes of precipitation are rare during a typical human lifetime, but they actually occur quite frequently, viewed over thousands or millions of years. And at any given time, while most places are experiencing "normal" precipitation, someplace on Earth is experiencing unusual precipitation. Because variations in the amount of rainfall sooner or later affect us all, unusually wet or dry periods make headlines.

Condensation

Precipitation is both a part of energy flows in the atmosphere and a consequence of them. Recall the "conveyor belt" of rising air that carries water vapor, with its latent heat, high enough to become cooled, condense, and cause precipitation. Precipitation is part of the flow of energy, releasing heat in the atmosphere and returning the liquid water back to Earth where it can absorb more heat. To understand this, let us look at the process of **condensation,** the conversion of water from vapor to liquid state.

Air contains water in gaseous or vapor form. Air may hold very little water vapor, as in dry desert air, or it may be filled with water vapor, as in a steamy jungle. Air's ability to hold water-vapor molecules is limited, and this limit varies according to air temperature. We measure the water-vapor content of air by the pressure that the water molecules exert. The maximum water vapor that air can hold is called the **saturation vapor pressure.** Precipitation begins when the pressure of moisture in the air exceeds the saturation amount.

Relative humidity tells us how wet air is. **Relative humidity** is the actual water content of the air, expressed as a percentage of how much water the air could hold at a given temperature. For example, if a 30°C (86°F) sample of air contains half the water vapor that it could hold at that temperature, its relative humidity is 50 percent. But if cooled to 22°C (71°F), that same air sample with the same amount of water vapor would be three-fourths saturated, at 75 percent relative humidity. And if cooled to 15°C (60°F), that same air sample would be saturated, at 100 percent relative humidity, fog would form, and wooden doors would stick. However, if saturated air at 0°C (32°F) were heated to 22°C (71°F), its relative humidity would drop to less than 30 percent. This is why indoor air in winter is so dry: We start with air that might have a high relative humidity outdoors and then warm it substantially indoors. Relative humidity tends to fluctuate daily with changing air temperatures because the amount of water vapor in the air holds fairly steady while temperature rises and falls. Relative humidity typically is lower in the warm afternoon and higher at nighttime.

When air cools, its relative humidity rises, and if the air is cooled enough, it can become saturated. If cooled still further, it becomes supersaturated and contains more water than it can hold in a vapor state. Condensation results from supersaturated air.

Condensation in the atmosphere produces clouds. Clouds form when water-vapor molecules condense around tiny particles of dust, sea salt, pollen, and other particles. The clouds contain small droplets of liquid water or particles of ice—depending on temperature—that are too light to fall. But if these water droplets or ice particles continue to grow as more moisture from the air condenses on them, they eventually become too heavy to be supported by air currents, and they fall to the surface as rain or snow.

Causes of Precipitation

The movement of air causes precipitation in three ways:

- Convection—in which air warmer than its surroundings rises, expands, and cools by this expansion
- Orographic uplift—in which wind forces air up and over mountains
- Frontal uplift—in which air is forced up a boundary (front) between cold and warm air masses

Convectional precipitation On a warm, humid summer day, the sky is clear in the morning, and the Sun is bright. The Sun warms the ground quickly, and the air temperature rises. Most of the warming of the air takes place close to the ground, because the humid air is a good absorber of longwave radiation, which is being reradiated from the ground. As the air near the ground warms, it expands, becomes less dense, and rises through the surrounding cooler air above, like a hot air balloon. Convection is in progress, conveying humid air higher into the atmosphere.

Because air is a gas, it is compressed by the weight of overlying air. When it rises, air has less weight above it, and the lower pressure allows the air to expand. Compressing a gas causes an increase in temperature, whereas expanding it causes a decrease (for example, a spray can gets cold in your hand when you use it). The decrease in temperature that results from expansion of rising air is called **adiabatic cooling;** the word *adiabatic* means "without heat being involved."

As air rises, it rapidly cools adiabatically (by expansion) at a rate of about 10°C for every 1000 meters (5.5°F per 1000 feet) elevation. When the air reaches several hundred meters above the ground, it has cooled to the point where it is saturated, and clouds begin to form (Figure 2-12). Rising warm air may mix with the surrounding cooler air, slowing the convection. If the

convection were driven only by heating at the ground surface, it would probably not be strong enough to cause water droplets to grow big enough to fall as precipitation.

As soon as condensation begins, latent heat—another important source of energy—is released. Latent heat further warms the rising column of air and makes it less dense than the surrounding cool air. The process is self-reinforcing, as the cloud grows rapidly, with strong vertical motion and rapid condensation. The result is often gusty winds and intense rain: a thunderstorm. Thunderstorms are a common example of intense convectional storms (Figure 2-13).

Convectional storms are responsible for a large portion of the world's precipitation (Figure 2-14). In tropical climates, where strong insolation makes temperatures high, all that is needed for intense daily convectional storms is a source of humidity. In midlatitude climates, such storms occur mostly in the summer because higher temperatures allow the air to hold more moisture, and this means more latent heat can be released, causing strong convection. These storms are especially common in situations in which some other factor is present to favor uplift, the trigger that starts the self-reinforcing growth of the storm. Mountains and fronts, especially cold fronts, often provide the trigger.

Orographic precipitation Precipitation sometimes occurs when the horizontal winds move air against mountain ranges, forcing air to rise as it passes over the mountains. This is called **orographic precipitation.** As the air rises, it cools adiabatically (by expansion), the cooling causes condensation, and precipitation results. After air has moved up the windward side of a mountain and over the top, it then descends on the leeward side. As it does so, its relative humidity drops significantly

Animation
Atmospheric
Stability

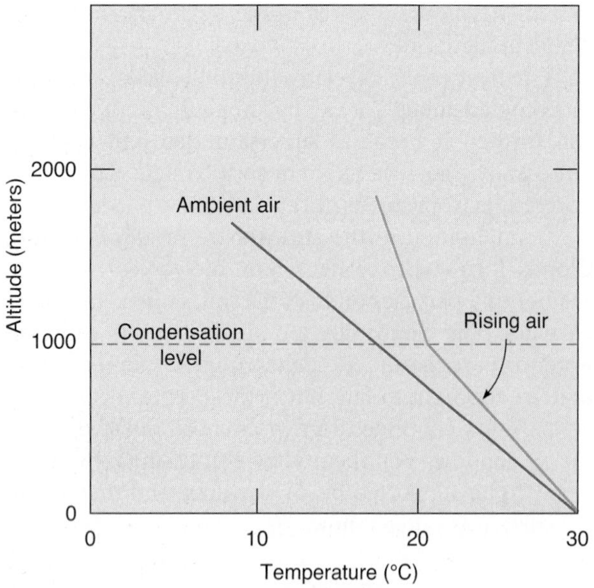

Figure 2-12 Adiabatic cooling. As air rises, it cools at a rate of about 1°C per 100 meters (1.8°F per 330 feet). The temperature of the rising parcel of air relative to the temperature of the air around that parcel determines its behavior. If the parcel is warmer than surrounding air, it tends to rise and therefore is unstable. If it is the same temperature or cooler, it resists movement and therefore is stable. In this example, the air parcel is warmer, and thus unstable. Condensation in the air above 1,000 meters (3,300 feet) elevation causes the air to be warmed by the release of latent (or stored) heat, warming the air. This further contributes to the instability of the air. Intense thunderstorms can form under such conditions.

Figure 2-13 A convectional storm in Australia. Such storms are driven by the release of latent energy in the condensation of water.

Figure 2-15 The Sierra Nevada mountains in California, seen from the east. Because of orographic effects, the mean annual precipitation at the crest of the Sierras exceeds 100 centimeters (40 inches). But in the foreground, Owens Valley is in a rain shadow and has mean annual precipitation of less than 30 centimeters (12 inches).

Figure 2-14 Clouds revealing convection over the Florida peninsula. The view is to the south. The land is warmer than the adjacent Atlantic Ocean and Gulf of Mexico, and this warmth stimulates convection and cloud formation. Note the absence of clouds over Lake Okeechobee, the large lake in the center of Florida, because it is cooler, like the ocean.

at the same time as its temperature increases. The leeward side of a mountain range is often much drier than the rainy windward side. A dry region on the leeward side of a mountain range is called a rain shadow. Some of the world's major deserts are arid because they are situated on the leeward side of a mountain range (Figure 2-15).

The western United States provides an excellent example of orographic effects on rainfall (Figure 2-16). Moist air from the Pacific Ocean, the region's major source of moisture, travels from west to east, producing precipitation as it rises first over coastal mountain ranges in Washington, Oregon, and California and then the Sierra Nevada and Cascade ranges. Some of the highest average U.S. rainfall totals occur in the Sierras and Cascades; but east of the mountain ranges lies the rain shadow, causing dry conditions in eastern Washington, Oregon, Nevada, and Utah. The region is arid because it is cut off from Pacific moisture by the Sierras and Cascades and since the atmosphere circulates mainly in a west to east direction at these latitudes moisture from the Atlantic does not reach the region.

Orographic precipitation also occurs in the Rocky Mountains, but the amount is less than in the Sierras

because the Rockies are more isolated from moisture sources. In the eastern United States and Canada, the Gulf of Mexico and Atlantic Ocean supply ample moisture for precipitation east of the Mississippi Valley. The Appalachians receive more rain than adjacent lowlands because of orographic effects, but the eastern side of the Appalachians does not experience a rain shadow because the region has moisture sources on both sides: the Gulf of Mexico to the southwest and the Atlantic Ocean to the east.

Frontal precipitation Frontal precipitation forms along a **front,** which is a boundary between two air masses. An air mass is a large region of air—hundreds or thousands of square kilometers—with relatively uniform characteristics of temperature and humidity. An air mass acquires these characteristics from the land or water over which it forms.

In North America, air masses that form over central Canada tend to be cool (because of Canada's relatively high latitude) and dry (because of the region's isolation from moisture sources). Air of this type is called *continental polar air.* In contrast, air masses that form over tropical water, such as over the Gulf of Mexico, tend to be warm and moist and are called *maritime tropical air masses.*

When a cool air mass, such as continental polar air from Canada, meets a warm air mass, such as maritime tropical air from the Gulf of Mexico, a boundary, or front, may form between them. Because cool air is relatively dense, it tends to move under less dense warm air, while warm air tends to rise over cool air (Figure 2-17). Fronts are regions characterized by ascending air, cloudiness, windiness, and precipitation.

(a)   (b)

Figure 2-16 Topography and precipitation in Oregon. (a) A shaded relief map showing the north-south-trending Coast Ranges and Cascades separated by the Willamette Valley in western Oregon and plateaus of eastern Oregon. Green colors indicate low elevations, pale tones intermediate, and brown high. (b) Mean annual precipitation. Purple shades indicate high precipitation, green intermediate, and tan low. Note that precipitation is highest in the Coast Ranges and Cascades, somewhat lower in the Willamette Valley, and quite low in eastern Oregon. The high rainfall in the mountains is due to orographic lifting of air as it moves eastward off the Pacific Ocean. By the time this air reaches the plateau east of the Cascade Range, most of the moisture has been removed, and rainfall there is a fraction of that received along the coast.

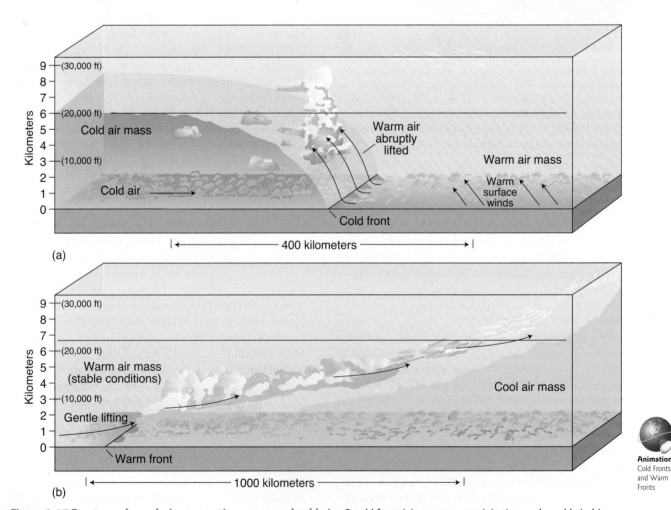

Figure 2-17 Fronts are boundaries separating warm and cold air. A cold front (a) generates precipitation as the cold air drives under the warm air like a wedge, lifting the warm air. This rapid vertical motion of air produces the tall, puffy cumulus clouds characteristic of a cold front. In a warm front (b), warm air rides up over cold air, causing precipitation to fall from broad, flat layers of stratus clouds.

As air masses migrate across Earth's surface, the fronts move with them. When a cold air mass advances against a warmer one, the boundary is called a **cold front.** As the cold, denser air advances, it wedges beneath the warm air, forcing it to rise. This generates clouds, usually resulting in precipitation (Figure 2-17a). A cold front can move quite fast and generate intense thunderstorms if the warm air is sufficiently moist.

The passage of a well-developed cold front in central and eastern North America includes very distinctive weather. Before the front arrives, the air is warm and moist, with the wind typically from the south or southwest. As the front arrives, intense precipitation falls, and in the summer, thunderstorms usually form. As the front passes, the wind shifts to the west or northwest, and the temperature falls quickly. Then the sky clears, and cool dry air arrives, with blue skies and bright sunshine.

When, on the other hand, a warm air mass advances against cooler air, the boundary is a **warm front.** The warm air rides up over the cool air as though the cool air were a gentle ramp (Figure 2-17b). Precipitation along a warm front is less intense than along a cold front, and a warm front often passes without significant precipitation. In winter, a warm front sometimes causes freezing rain, as rain formed when the warm air aloft falls through colder air below freezes as it reaches the ground. Air masses and the front between them may become stalled and not move, creating a stationary front. Note the cold and warm fronts depicted on the weather map (Figure 2-18).

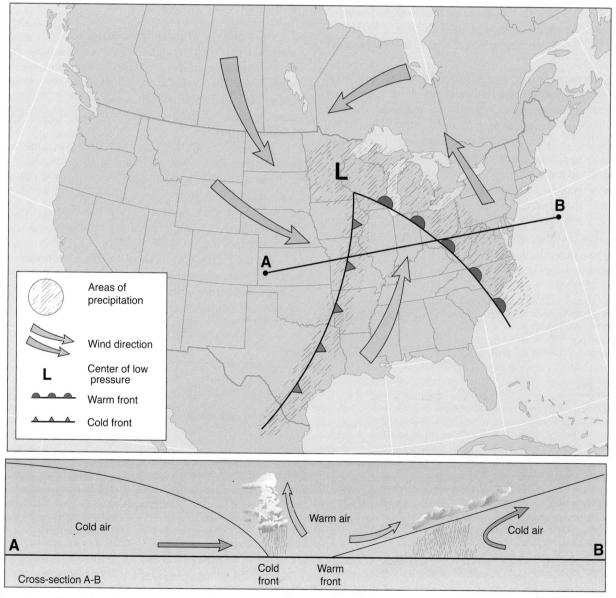

Figure 2-18 Patterns of winds, clouds, and precipitation around an idealized large midlatitude storm. The map view shows the position of the fronts at the ground surface. The cross section shows the locations of the fronts aloft.

Because a cold front normally involves rapid vertical motion of warm and relatively humid air, the clouds that form along it are typically cumulus clouds, which are tall, puffy clouds with billowy tops. In warm fronts, the warm air rises gradually over the cold air, and the stable layering of warm air over cold produces broad, flat layers of clouds known as stratus clouds.

Circulation Patterns

Air is made of molecules, so air has mass. Earth's gravity attracts this mass of air to the surface, giving it weight. We think of air as a lightweight substance, but the thickness of air over your head has considerable weight, pressing on you and Earth's surface at an average of about 1 kilogram per square centimeter (14.7 pounds per square inch). This means that directly above each square centimeter of Earth is a column of air over 480 kilometers (300 miles) tall that weighs on average about 1 kilogram (2.2 pounds) (Figure 2-19).

Atmospheric pressure varies with altitude because the higher you go, the less air exists above you. Thus, atmospheric pressure is greater at sea level than in "mile-high" Denver or atop Mount Everest (8.8 kilometers, or 5.5 miles, high). Because sea level is a surface that is at about the same height worldwide, scientists use the average atmospheric pressure at sea level as a world standard. Atmospheric pressure is measured with a *barometer*. The average atmospheric pressure at sea level read on a barometer is 1,013.2 millibars (a force that supports a column of mercury 29.92 inches tall in a mercury barometer).

Pressure and Winds

At any location, atmospheric pressure varies with conditions. For example, the air above the island in Figure 2-11 is warmer than the air around it, and therefore it is less dense. Because warm air is less dense, it weighs less, and the atmospheric pressure over the island is lower than the pressure over the cool sea. As the warm, lighter air over the island rises, higher pressure over the cool sea forces air horizontally toward the island to replace the rising warm air. The difference in pressure between the two places is the *pressure gradient*.

Differences in pressure produce wind. In Figure 2-20, if pressure at the surface is high, air descends, and winds blow away (diverge) from the area of high pressure. But if pressure at the surface is low, air ascends vertically, and horizontal winds converge toward the area of low pressure.

Coriolis effect   If Earth did not rotate, winds would simply blow in a straight line from areas of high pressure to areas of low pressure. However, on our real spinning planet, winds follow an indirect, curving path. This deflection of wind (and any other object moving above Earth's rotating surface) is called the **Coriolis effect.**

To understand the Coriolis effect, imagine two people throwing a ball back and forth on the back of a flatbed truck (Figure 2-21a). Both people and the ball are moving at the same speed and in the same direction, so they do not change position relative to one another. The ball game is the same as if the truck were not moving.

On the merry-go-round, when a ball is thrown in a straight line, the turning platform moves the other person away from where the ball arrives (Figure 2-21b). As the two people throw the ball back and forth, the ball

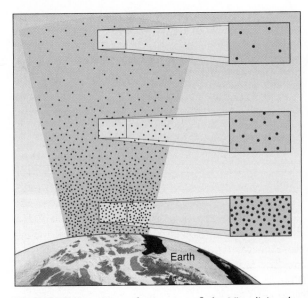

Figure 2-19 Density and pressure of air. Visualizing the density of air (greater near Earth's surface) and atmospheric pressure.

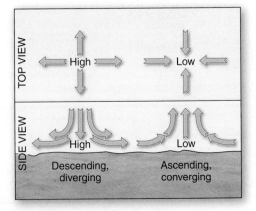

Figure 2-20 Wind movement is partly determined by pressure. Air descends in regions of high pressure. Air ascends in regions of low pressure. Wind blows from areas of high pressure to areas of low pressure.

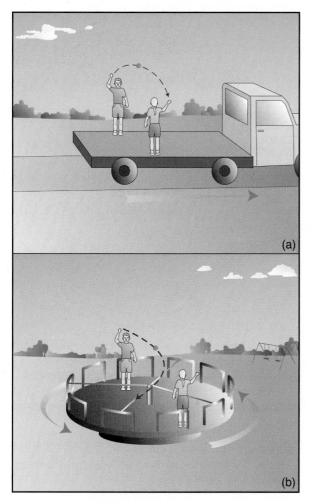

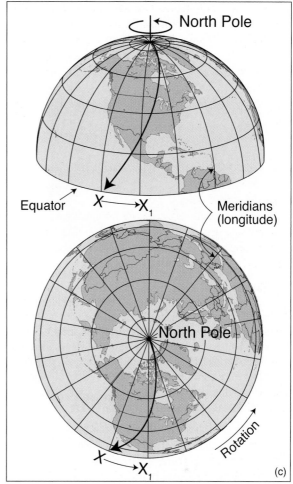

Animation
Coriolis Effect

Figure 2-21 Coriolis effect. (a) When two people on a moving flatbed truck throw a ball to each other, the people and the ball are moving at the same speed and in the same direction, so they do not change position relative to one another. (b) When two people on a merry-go-round throw a ball to each other, the ball follows a straight-line path, but the people have moved. To the people, the ball appears to be following a curve. This apparent curving of the ball demonstrates the Coriolis effect. (c) If an airplane flies in a straight line from the North Pole to the equator, it will actually land west of its intended destination because Earth rotates eastward during the flight.

follows its expected straight-line path, but the receiver has been moved by the merry-go-round. To the players, though, it appears that the ball has followed a curving path. Similarly, as Earth rotates beneath moving air, the air appears to move in a curve. This deflection of motion is only apparent to those on the spinning surface. Viewed from above, the path would be straight.

Near the equator, Earth is like the flatbed truck in Figure 2-21a; the Coriolis effect is zero at the equator. But toward the poles, rotating Earth is more like a spinning disk, and the Coriolis effect is strongest in the polar regions. This is shown in Figure 2-21c, where a plane flies from the North Pole in a straight line toward the equator, but arrives far to the west of its intended destination because Earth rotated eastward during the hours that the plane was in flight.

The net result of the Coriolis effect is a deflection to the right (from the standpoint of a moving object) in the Northern Hemisphere. In the Southern Hemisphere the deflection is to the left. The effect is significant only when considerable distance is involved: A pilot must consider it in setting the flight path of a several-hour flight across the ocean, but not for a model airplane across a backyard. The Coriolis effect also helps weather forecasters understand the behavior of storms.

Figure 2-22 shows the result when the Coriolis effect is added to the vertical air movements in high- and low-pressure areas. It puts a spin on the winds, adding a spiral motion to create cells that rotate clockwise and counterclockwise. In the Northern Hemisphere, high-pressure cells are rotating clockwise, and low-pressure cells are rotating counterclockwise (Figure 2-22a). The opposite patterns hold in the Southern Hemisphere (Figure 2-22b). To summarize, differences in air pressure create winds, and Earth's spin curves their flow.

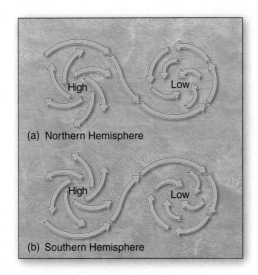

Figure 2-22 Wind movement with Coriolis effect added. (a) The Coriolis effect, caused by Earth's rotation, deflects wind to the right of its expected path in the Northern Hemisphere. This causes spiraling circulation around high- and low-pressure centers, in the directions shown. (b) The same pattern reversed for the Southern Hemisphere. Compare with Figure 2-19.

Animation
Cyclones and
Anticyclones

Global Atmospheric Circulation

Understanding global circulation patterns is very important. These patterns control our weather and climate, and affect agriculture. At the global scale, atmospheric circulation operates like the convection system over the warm island we described previously. Beginning at the equator and moving toward the poles, we can identify four zones of circulation patterns:

- Intertropical convergence zone
- Subtropical high-pressure zones
- Midlatitude low-pressure zones
- Polar high-pressure zones

Intertropical convergence zone In the tropics, dependable year-round inputs of solar energy heat the air, expanding it and creating low pressure. As a result, convectional rising of air occurs daily above Earth's equator. This forms the **intertropical convergence zone (ITCZ),** so called because it is a zone between the Tropics of Cancer and Capricorn where surface winds converge (Figure 2-23).

As air rises in the ITCZ, convectional precipitation occurs, usually as afternoon thunderstorms. Air converges toward the equator at the surface, replacing the rising air. The Coriolis effect deflects this moving air to the right in the Northern Hemisphere to form the Northeast Trade Winds and to the left in the Southern Hemisphere to form the Southeast Trade Winds. The **trade winds** are so called because they were very important to sailing ships that were involved in commerce. For example, the northeast trades in the Atlantic carried European ships to the Caribbean. Aloft, air circulates away from the ITCZ both

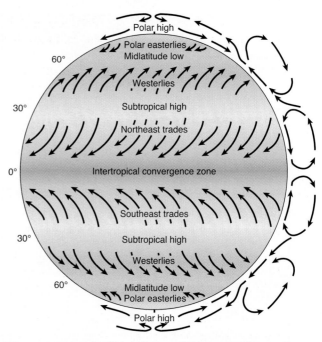

Figure 2-23 General circulation of the atmosphere. This simplified illustration of average atmospheric circulation patterns shows the major circulation zones by latitude: intertropical convergence zone (ITCZ), subtropical high-pressure (STH) zones, midlatitude low-pressure zones, and polar high zones.

northward and southward, toward subtropical latitudes. This air has lost most of its moisture in the daily rainfalls, and it is now warm and dry.

Subtropical high-pressure zones The warm, dry air that spreads poleward from the ITCZ descends at about 30° north and south latitudes. This creates zones of high pressure that are especially strong over the oceans. These **subtropical high-pressure (STH) zones** are areas of dry air, bright sunshine, and little precipitation.

This descending dry air associated with subtropical high-pressure cells creates an arid climate because up-lift is needed for condensation, cloud formation, and precipitation. As a result, most of the world's major desert regions are on land in this zone, that strattles 30° north and south latitudes. Further, most are on the western edges of continents, including deserts in northern Mexico and southwestern United States, the Atacama in Peru and Chile, the Sahara in northern Africa, the Namib in southern Africa, and the Australian desert. The eastern sides of these continents are under the same high-pressure cells, but they tend to be more humid because cyclonic and convection circulation brings in warm, humid tropical air.

On the poleward sides of the subtropical high-pressure cells, circulation is toward the poles. But these winds are deflected by the Coriolis effect, so prevailing winds are from the southwest in the Northern Hemisphere and from the northwest in the Southern Hemisphere. The prevailing westerly wind makes an eastbound flight from Los Angeles to New York normally about an hour shorter than a westbound flight.

Midlatitude low-pressure zones Poleward of the subtropical high-pressure zones are the **midlatitude low-pressure zones.** These lower-pressure areas experience convergence of warm air blowing from subtropical latitudes and cold air blowing from polar regions. The warm and cold air masses collide in swirling low-pressure cells that move along the boundary between the two air masses, which is known as the **polar front.** Winds in these regions generally blow from west to east and are therefore called *westerlies.* Winds are named for the direction from which they blow.

Polar high-pressure zones In the polar regions, the intense cold caused by meager insolation creates dense air and high pressure. In these **polar high-pressure zones,** the air is so cold that it contains very little moisture, and convection and precipitation are limited. The snow and ice around the poles comes from snow accumulating over many thousands of years.

Seasonal Variations in Global Circulation

Global circulation patterns vary with the seasons. Let us now illustrate typical conditions during January and July (Figure 2-24).

Global circulation in January During January, Earth's average atmospheric pressure and winds are arranged into broad zones according to latitude (Figure 2-24a). Conditions in the Northern and Southern hemispheres roughly mirror each other. The ITCZ is generally at about 5° to 10° south latitude.

High-pressure cells predominate in the subtropical regions just north and south of the ITCZ, especially over the subtropical oceans. The Southern Hemisphere, where it is summer in January, displays the clearest tendency for high pressure over the oceans and low pressure over land. The Northern Hemisphere has a high-pressure region in eastern Asia, as well as in the eastern Pacific and Atlantic oceans.

In the midlatitudes, less continuous low-pressure regions in the Northern Hemisphere appear most clearly over oceans. The midlatitude low-pressure zone is much more consistent in the Southern Hemisphere, because of the absence of land between 40° and 70°. A monthly average map obscures the presence of storms, but it is useful to show the general condition of the atmosphere during Northern Hemisphere winter.

Global circulation in July By July, January's major pressure and wind zones have moved northward. In some cases, they have moved from being over land to water areas, or vice versa (Figure 2-24b). The ITCZ, which was south of the equator in January, is almost entirely in the Northern Hemisphere during July, as far north as 30° north latitude in southern Asia. The subtropical highs are still present over the oceans, although they are strengthened in the Northern Hemisphere and weakened in the Southern.

The most notable change in the subtropics and midlatitudes is the replacement of high pressure with low pressure over Asia. In fact, this region has the world's highest average pressures during January and the world's lowest average pressure during July. As a result, wind directions in the vicinity are reversed seasonally. This seasonal reversal of pressure and wind produces a **monsoon circulation,** in which winter winds from the Asian interior produce extremely dry winters in most of south and east Asia, while summer winds blowing inland from the Indian and Pacific oceans result in wet summers. These wet summers are the well-known monsoon season of heavy rains in southern Asia.

Seasonal changes also occur in the midlatitudes. The extreme low-pressure regions over the northern oceans in January are replaced in July with generally weak low pressure and inconsistent winds, although storms with strong winds may periodically pass through the regions. In the Southern Hemisphere winter (July), the low-pressure system is especially deep and consistent around the globe because of the absence of large landmasses to break it up as in the Northern Hemisphere. The south polar high is similarly strong in July.

Remember that with the shift in seasons between January and July, solar radiation also changes. During July in the Northern Hemisphere, solar radiation inputs are highest north of the equator, while in January, insolation is greater south of the equator. Circulation patterns also shift seasonally, bringing strong seasonal contrasts in weather as wind direction and zones of frequent storms move north and south.

Ocean Circulation Patterns

When wind blows over the ocean, it exerts a frictional drag on the sea surface, creating waves and currents (Figure 2-25). Continuing wind adds energy to the waves, building them into larger waves that may travel thousands of kilometers until they break onto a distant shore. The continuing drag of prevailing winds also causes broad currents in the ocean's surface layers. Contributing to these currents are differences in seawater temperature and salinity. These give water different densities, promoting movement of currents from areas of greater density to those of less density. This is similar to the way a high-pressure area in the atmosphere causes wind currents to blow toward a low-pressure area. Both of these currents—oceanic and atmospheric—are important in redistributing heat around Earth, a role played by the Gulf Stream, for example (Figure 2-26).

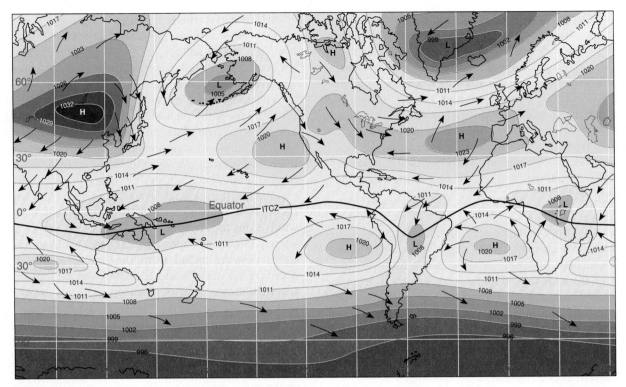

(a) January

Figure 2-24 Worldwide average atmospheric pressures (in millibars) and winds in a Northern Hemisphere winter in January and summer in July. (a) In January, the intertropical convergence zone (ITCZ) is generally south of the equator, especially over land areas. In the Northern Hemisphere, the subtropical high-pressure cells are weak or absent, and a large and intense high-pressure cell dominates interior Asia, with deep low-pressure areas over the North Pacific and North Atlantic.

Animation
Global Wind
Patterns

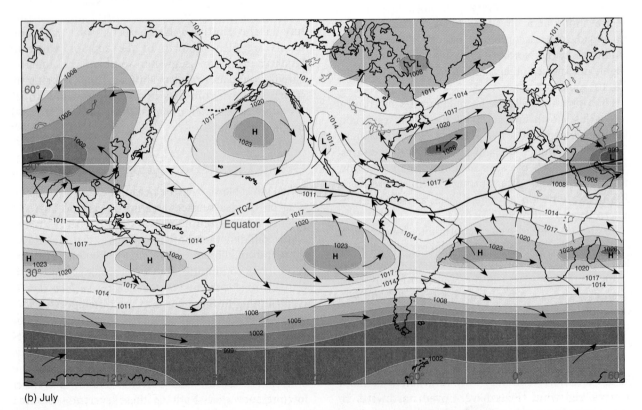

(b) July

Figure 2-24 (b) In July, the ITCZ has shifted to the north, and low pressure has replaced high pressure over interior Asia. Weak low pressure also is seen over North America, and the subtropical highs are prominent. In the Southern Hemisphere, the midlatitude low-pressure belt along latitudes 40° to 60° south is very strong.

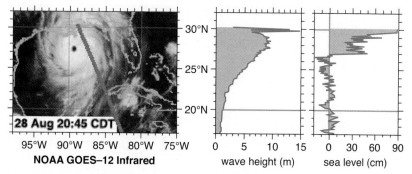

NOAA GOES–12 Infrared

wave height (m)   sea level (cm)

Figure 2-25 Wave height and sea level in the vicinity of Hurricane Katrina. As the storm approached the coast, a satellite recorded sea level and wave heights. The graph at right shows wave heights in meters and sea level variations in centimeters. As the storm was approaching the coast, sea level near the coast was already elevated by about 1 meter (3 feet), with waves 15 meters (49 feet) high. Note that the profile of sea level does not pass directly through the center of the storm; sea level would have been higher there. As the storm approached the shore, the combination of elevated sea level and wind driving the water landward caused sea level at the shore to rise by 3 to 4 meters (10 to 13 feet), with wave crests rising much higher.

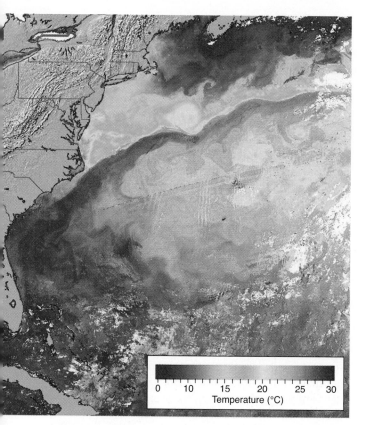

Figure 2-26 The Gulf Stream. In this composite of satellite images of the North American East Coast from late May 1996, red shades indicate warm water, and blue indicates cool water. The strong flow of warm water from southwest to northeast is plainly visible, including swirling eddies in the flow where the warm current moves past cooler water.

Gyres are prominent features of oceanic circulation. These wind-driven circular flows mirror the movement of prevailing winds (Figure 2-27); you can see this by comparing Figure 2-27 with Figure 2-24. Gyres form beneath tropical high-pressure cells. The Gulf Stream forms the western limb of the gyre in the North Atlantic.

Where ocean currents circulate warm water from low equatorial latitudes to higher latitudes, they are carrying heat poleward by advection. Such flows are balanced by cool currents traveling equatorward, most notably along the west coasts of subtropical land areas. Such cold currents cool the lower portions of the atmosphere above them, reducing convection. With less convection occurring, adjacent landmasses may be very dry. Some of the driest areas on Earth, most notably the Atacama Desert of Peru and Chile, owe their aridity partly to this effect.

Storms: Regional-Scale Circulation Patterns

Variations in circulation such as El Niño (also known as ENSO, for El Niño–Southern Oscillation) and the Asian monsoons are examples of recurring changes in the atmosphere. These range in scale from global phenomena such as ENSO, to a continent-size land-water convection pattern such as the Asian monsoon, to regional-scale systems such as hurricanes, down to the scale of a local thunderstorm.

Storms are especially important because they are areas of concentrated convection that bring precipitation and sometimes strong winds. On a regional scale, large storms can affect areas hundreds to thousands of kilometers across. Such storms, which are called **cyclones,** are large low-pressure areas in which winds converge in a counterclockwise swirl in the Northern Hemisphere and clockwise in the Southern Hemisphere. We will look at two types of cyclones: tropical cyclones (hurricanes or typhoons) and midlatitude cyclones.

Tropical cyclones Tropical cyclones are intense, rotating convectional systems that develop over warm ocean areas in the tropics and subtropics, primarily during the warm season. When such storms have wind velocities exceeding 119 kilometers per hour (74 miles

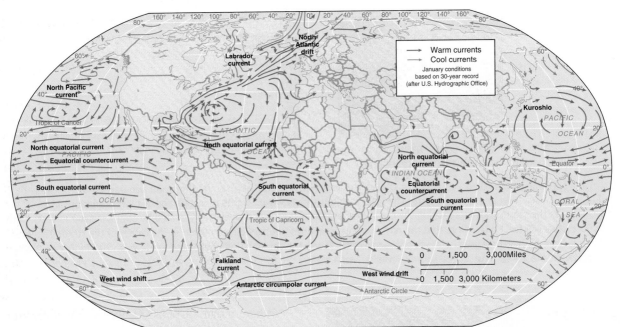

Animation
Ocean
Circulation

Figure 2-27 Oceanic circulation, showing major currents. Note how the ocean circulation is clockwise in the Northern Hemisphere and counterclockwise in the Southern Hemisphere, driven mainly by circulation around the subtropical high-pressure zones.

per hour), we call them **hurricanes** in North America, **typhoons** in the north western Pacific, and *cyclones* in the Indian Ocean and northwestern Australia.

These storms typically develop in the eastern portion of an ocean within the trade-wind belt. They begin as areas of low pressure (rising air) and converging winds, drawing in warm, moisture-laden air. This humid tropical air contains a great deal of energy, both as sensible heat and latent heat in the water vapor, but especially latent heat. As this energy is drawn into the developing

GLOBAL AND LOCAL

El Niño/La Niña

Animation
El Niño
and La Niña

The close linkage between oceanic and atmospheric circulation is demonstrated by a phenomenon called **El Niño.** The term is Spanish for "the (male) child," a reference to the Christ child, because the phenomenon occurs around Christmastime. El Niño is a circulation change in the eastern tropical Pacific Ocean that occurs every few years. In this change, the usual cool flow from South America westward is slowed and sometimes reversed, replaced by a warm-water flow from the central Pacific eastward. The counterpart of El Niño, in which especially cool waters are found in this region, is known as **La Niña** ("the female child"). The change in ocean currents is linked to a change in atmospheric circulation patterns in the Pacific Ocean known as the Southern Oscillation; the two phenomena together are referred to as El Niño–Southern Oscillation (ENSO).

Peruvian fishermen named El Niño, and they knew its effect: The typical (La Niña) circulation causes deep ocean water to rise to the surface off the coast of Peru, delivering nutrients that support fish populations. The reversed flow of water contributed to the collapse of the Peruvian anchovy

industry in the 1970s. ENSO events are far reaching because a modification of circulation in one part of the globe may cause circulation patterns to change elsewhere in North and South America and the Pacific region. For example, El Niño events are linked to flooding in the U.S. Southwest, droughts in Australia, and reduced rainfall in India. La Niña events often bring wet weather in south and Southeast Asia, and dry conditions in the southern United States.

In 1997, a particularly strong El Niño event developed, with ocean temperatures in the eastern Pacific about 5°C (9°F) above non-El Niño levels. This led to greatly increased convection in the eastern Pacific and rains in normally dry coastal Peru. At the same time, the western equatorial Pacific was cooler than normal, causing drought in Southeast Asia. Forest fires raged out of control in Indonesia, causing severe air-pollution problems in Kuala Lumpur, Malaysia, and intense storms pounded coastal California. By 2000, the pattern had reversed, and La Niña conditions dominated. This, along with warm water in the western Pacific, was a major contributor to the 1998–2002 drought in the southern United States.

storm, condensation and the resulting release of latent heat intensifies the convection. The center of low pressure grows more intense, and as wind speed grows, the Coriolis effect causes a spiraling circulation to develop, counterclockwise in the Northern Hemisphere and clockwise in the Southern Hemisphere.

These tropical storms move with the general circulation over the subtropical Atlantic, Pacific, and Indian Oceans, from east to west. As they travel across the warm ocean surface, they often intensify. By the time they reach the western portion of the ocean, their atmospheric pressures may be as much as 10 percent lower than average sea level pressure, with winds exceeding 150 kilometers per hour (93 miles per hour).

Hurricanes thrive on warm, moist air, so they are most intense over ocean areas during the warm season. They lose intensity over land because they lose their source of energy. Furthermore, the smooth ocean surface favors development of high winds. In contrast, the hills and trees of land areas slow the wind by friction.

Hurricanes' greatest threat to humans is therefore in tropical and subtropical coastal areas, on the eastern margins of the continents and in southern Asia, where monsoon circulation draws air northward from the Indian Ocean. When a hurricane strikes land, the combination of intense wind and extremely low pressure causes a **storm surge,** an area of elevated sea level in the center of the storm that may be several meters high. The surge carries large waves crashing inland onto low-lying coastal areas, with devastating results. In fact, the majority of hurricane deaths and damage result from storm surges. Adding to this hazard, tornadoes commonly are spawned as a hurricane comes ashore.

Among the coastlines most vulnerable to tropical cyclones is that of the Bay of Bengal, including Bangladesh and West Bengal in eastern India. The land in this area is low (elevation less than 2 meters, or 6 feet), consisting mostly of the Ganges and Brahmaputra River Deltas, and it is densely populated. Bangladesh has millions of people and very limited resources for coping with periodic storms. Although satellite-based warning systems help evacuate people in advance of approaching storms and many cyclone shelters have been built, transportation is difficult, and people often are unable to reach high ground. In recent decades, several severe storms have struck the area, including one in 1991 that killed more than 138,000 people.

Animation
Midlatitude
Cyclones

Midlatitude cyclones In midlatitude areas another type of storm dominates the weather: the midlatitude cyclone. The midlatitude cyclone has little in common with the tropical cyclone (hurricane), for it occurs in cooler, less humid latitudes that cannot provide the immense volume of heat and water vapor needed to generate a hurricane. **Midlatitude cyclones** are centers of low pressure that develop along the polar front. They move from west to east along that front, following

the general circulation in the midlatitudes. Midlatitude cyclones are usually much less intense than hurricanes, but they are much more common.

In a midlatitude cyclone, air is drawn toward the center of low pressure from both the warm and the cold sides of the polar front. Where warm air is drawn toward cold, typically on the eastern side of the storm, a warm front develops. On the western side, the spiraling motion causes cold air to drive under the warm air, forming a cold front. As the center of low pressure moves eastward, these fronts move with it, bringing precipitation to areas over which they pass. The passage of a front at the surface usually is marked by a significant change of temperature, precipitation, and shifting winds. The repeated passage of such storms creates highly variable weather conditions, with alternating cold and warm air as the polar front moves back and forth across the land. Occasionally, especially in North America, strong cold fronts associated with midlatitude cyclones produce tornadoes.

Tornadoes are an extreme form of weather created when energy conditions in the atmosphere are such that extremely intense convection occurs. They generate the highest surface wind velocities known to occur on Earth, and the destruction they cause can be near total, though fortunately very localized.

Tornadoes are intense columns of rising air, usually associated with thunderstorms. As the air rises, it creates a partial vacuum (an area of very low pressure) that draws air in toward it. As this air is drawn in, it creates a swirling vortex, much like the swirling motion of water as it drains out of a sink. As the air swirls in, it gains tremendous speed, approaching 160 kilometers per hour (100 miles per hour) in moderate tornadoes and as much as 480 kilometers per hour (300 miles per hour) in extreme storms. This vortex usually moves horizontally, sometimes touching the ground and sometimes rising above it, leaving an erratic and usually very narrow path of destruction (Figure 2-28).

Most tornadoes occur in one of two circulation patterns: in association with intense thunderstorms embedded in hurricanes and along especially strong cold fronts in the midlatitudes. The south-central United States has the greatest frequency of tornadoes of any place in the world. In the United States, tornadoes occur mostly in spring and early summer, when contrasts between cold, dry continental air and moist, humid air from the Gulf of Mexico are greatest. Such conditions create cells of rapidly rising air, including very intense thunderstorms and sometimes tornadoes.

A Snapshot of Global Circulation and Weather

The global and regional circulation patterns just described are visible in satellite images of Earth, such as the one shown in Figure 2-29. These patterns may not be obvious

Figure 2-28 Path of a tornado in Collin County, TX.
Seventeen homes were destroyed and three people were killed
by this tornado in May 2006.

Figure 2-29 Weather patterns on Earth. Full-disk image
of Earth, September 12, 2009. A Snapshot of Global Circulation and Weather" for a description of the patterns shown in
this image.

at first glance, but a trained eye can identify wind directions, fronts, and precipitation patterns even in a satellite
image. Here they are shown with appropriate weather-
map symbols on the image. The image shown in Figure 2-
29 shows most of the Western Hemisphere in late
summer. In this view of Earth, from about 35,000 kilometers (22,000 miles) out in space, we see a little less than half
the globe. The South and North Poles are not visible, but
the view extends to almost 80° north and south.

The intertropical convergence is visible as a broken
line of cloudy areas extending across the North Pacific
at about 10° north, continuing east across the Caribbean
Sea. The ITCZ consists mainly of convective storms,
some of which are scattered and some of which are concentrated in areas of low pressure. On the northern

margins of the ITCZ, we see one tropical cyclone in the
early stages of formation in the tropical Atlantic. This is
Tropical Storm Fred.

In the midlatitudes we see sweeping curves of clouds,
many of which show the locations of fronts. In August, it
is winter in the Southern Hemisphere, and two particularly well-developed storms are visible there. A third can
be seen in the North Atlantic, approaching Europe.

The atmosphere is continually in motion, and on
any given day, the atmosphere contains many storms,
some intense and some mild. This circulation occurs in
the vertical dimension as well as horizontally; areas of
rising air are cloudy and may produce precipitation,
while areas of sinking air are cloud free. The locations of
storms and high-pressure cells change from day to day,
producing weather variability from one day to the next.

Climate

The atmospheric circulation patterns we see on any
given day are the result of energy distribution patterns
occurring at that time, and these change rapidly from
hour to hour and day to day. Geographers also study
temperature, precipitation, and other atmospheric
phenomena that persist for much longer periods than
a thunderstorm or a heat wave.

Climate is the summary of weather conditions over over 30 or more years- a place's weather pattern over time. The vegetation, natural resources, and human activities that characterize a particular region of Earth are heavily influenced by its climate. At first glance, describing the climate of a place may appear to be a straightforward task of calculating annual averages of temperature and precipitation. In reality, climate is extremely difficult to describe. Air temperature and precipitation change not only day to day but year to year, and other factors such as windiness, humidity, and occurrence of storms must also be considered. We will discuss this in detail later in the chapter.

Climate changes over time. Earth has alternated between warming periods and cooling periods. Glaciers have covered large areas during cool periods, and they have melted during warm periods. The human population has survived by adapting to the alternating periods of warming and cooling. Ironically, human activity now may be causing climatic changes that will fundamentally alter the character of the planet.

People are affected by various elements of climate. The number of hours of sunshine in a day and the degree of cloudiness are important to anyone who is active outdoors—gardeners, farmers, hunters, builders, vacationers at the beach—and residents of homes heated by solar energy. Amounts of rainfall and snowfall make great differences in how we build our homes and roads from place to place. The windiness of a place is important in designing structures and harnessing wind power. Pollutants in the air can have both short-term and long-term effects on human and animal health. Counts of pollen and mold spores are important to those who suffer allergies.

Air Temperature

To understand a region's distinctive climate, the two most important measures are air temperature and precipitation. In everyday speech, we refer to "hot or cold climates" and "wet or dry climates." The most obvious differences in air temperature on Earth are those between the tropics and the poles and those between winter and summer. These variations are caused by latitudinal and seasonal variations in solar energy inputs.

Air temperature also varies with elevation. This variation occurs because the atmosphere is warmed by longwave radiation from Earth's surface and because of the adiabatic cooling of rising air described earlier. A very general rule of thumb is that temperature drops about 6.4°C per 1,000 meters (3.5°F per 1,000 feet). This is why mountain regions are cooler than adjacent lowlands (Figure 2-30). Air temperature also varies considerably within a few meters of the ground surface. In a desert, the ground surface temperature can reach 70°C (160°F) during the daytime, while air that lies

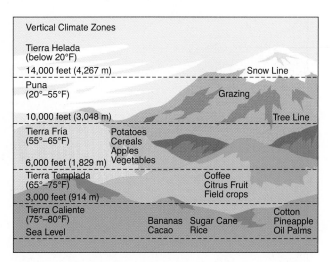

Figure 2-30 Vertical climate zones. Throughout much of Latin America, elevation defines these vertical climate zones. Distinct crops characterize each zone.

2 meters (6 feet) above the ground rarely exceeds 55°C (130° F). At night, the desert ground may be significantly cooler than the air above.

To minimize the effects of the ground surface on temperature records, climatologists usually measure air temperature at a height of about 1.4 meters (55 inches) inside a ventilated shelter that shades the thermometer and other instruments from direct sunlight. Climatologists at these weather stations usually report high and low temperatures for each day and calculate the average daily temperature as the average of the two readings.

Temperature patterns also show the effects of topography, proximity to the ocean, and moisture availability (Figure 2-31).

In both summer and winter, the higher elevations of the Rockies and Appalachians are cooler than adjacent lowlands. At very high elevations, temperatures are so low that year-round snow cover is possible. Permanent snow occurs at an elevation above approximately 4,000 meters (13,100 feet) in the southern Rockies and as low as 1,000 meters (3,300 feet) in the Northern Rockies and the mountains of Alaska.

The smallest seasonal differences in temperatures in North America are found in areas near the ocean, especially along the Pacific Coast. Westerly winds bring air from ocean areas to coastal regions in the west, while on the Atlantic Coast the airflow is normally from the continent, so seasonal temperature variations are greater. The effects of moisture availability and sunshine on temperature are visible in the deserts of southwestern United States and northwestern Mexico, where generally clear skies allow ample sunshine, but because of lack of precipitation water is not available to dissipate that heat through evaporation. The highest recorded temperatures in North America occur in this region.

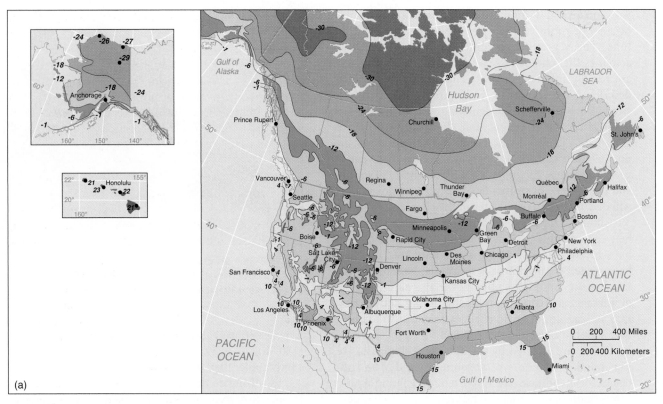

Figure 2-31(a) Average January temperatures. Average temperatures (°C) in North America during January, based on data collected from 1931 to 1960. In January, the lowest temperatures occur in the interior of the continent and in high-altitude areas. Higher temperatures occur in the south and along the West Coast. (Alaska and Hawaii not to scale.)

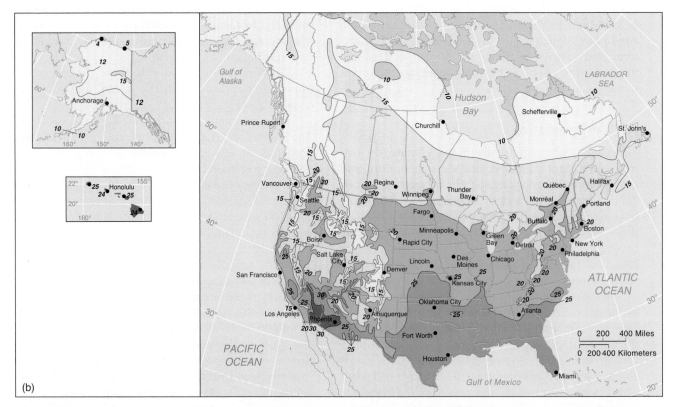

Figure 2-31(b) Average July temperatures. Average temperatures (°C) in North America during July, based on data collected from 1931 to 1960. In July, the highest temperatures occur in the interior of the continent, especially in the desert southwest. Temperature differences between northern and southern portions of the United States are less than in January, and storm systems traveling along the polar front are less intense as a result. Be careful when trying to interpolate between temperature lines, because large changes may occur over short distances, depending on altitude, slope, soil type, vegetation, and urban development. (Alaska and Hawaii not to scale.)

Precipitation

The amount of precipitation is extremely variable between places and over time. A thunderstorm can unleash heavy rain in one place and a light sprinkling just a short distance away. Worldwide, precipitation generally ranges from virtually none in some desert areas to more than 300 centimeters (120 inches) in tropical areas. A few tropical mountain areas have recorded more than 10 meters (33 feet) of rainfall per year.

Precipitation usually is measured by collecting rain or snow in a cylindrical container that is marked in millimeters or hundredths of an inch. Snowfall normally is recorded as an amount of liquid water rather than the depth of snow accumulated on the ground, because the amount of water in snow varies widely with the snow's texture. Weather stations typically report daily precipitation, although the instruments are capable of recording the amount for other periods of time, such as a minute or an hour.

Figure 2-32 shows worldwide average precipitation. Note how the levels correspond to global circulation patterns: the rainy ITCZ around the equator, the midlatitude high-pressure areas with their dry air, the mountain rain shadow areas, and the dry polar regions.

Average rainfall amounts tell us much about the climate of a region, but there is also much they do not tell. The timing and reliability of rainfall are equally important. In midlatitude areas, precipitation frequently takes the form of gentle rainfall, whereas in tropical regions, it may come in torrential and destructive downpours.

In northern Nigeria, for example, 90 percent of all rain falls in storms of more than 25 millimeters (1 inch) per hour. That is, by way of comparison, half the average monthly rainfall of London, England. In Ghana, cloudbursts regularly occur at a rate of 200 millimeters (8 inches) per hour, four times London's monthly total. In Java, an Indonesian island, a quarter of the annual rainfall comes in downpours of 60 millimeters (2.4 inches) per hour, more than Berlin, Germany, gets in an average month.

In sudden, intense storms, a greater portion of the rainfall runs off the land rather than soaking in, sometimes increasing soil erosion. Studies in the West African nation of Burkina Faso found that in one year, nearly 90 percent of all erosion took place in just six hours.

The regularity of rainfall is also important. In Western Europe and eastern North America, rainfall does not vary much from month to month. In its rainiest month, New York receives only one and a half times as much rain as in its driest month, London's rainiest month is two times greater, and Berlin's is two and a half times greater. Delhi receives about the same total annual rainfall as London, but Delhi receives only 10 millimeters (0.4 inch) of rain in November and more than 175 millimeters (7 inches) in both July and August. Zungeru, in central Nigeria, gets 54 times as much rain in the wettest month as in the driest.

Soil can absorb and hold only a limited amount of water for crops to use. Think of this as a bank account deposited in the wet season and drawn on

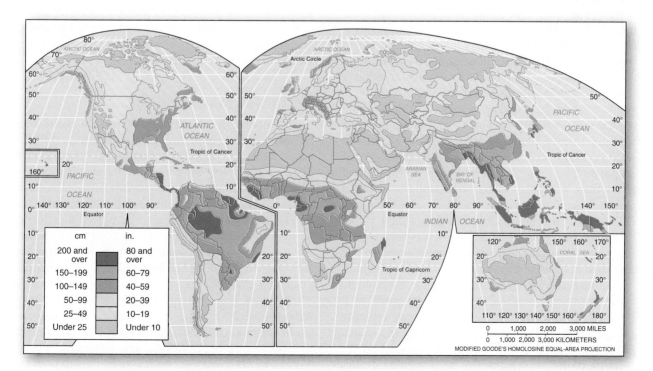

Figure 2-32 Average annual precipitation worldwide. Generally, climates with higher temperatures require greater rainfall to maintain vegetation. This small-scale map is highly generalized, masking local variability.

in the dry season. Although many tropical countries get what seems to be adequate annual rainfall, they get it in a lump sum, like a huge win at a casino. Most is squandered, and the little that is left in the soil bank is used up within a few months. For example, in Agra, India, where the Taj Mahal is located, rainfall exceeds the current needs of vegetation only in July and August. As the rains decrease in September, the water stored in the soil is used up in just three weeks, leaving nine months of the year in which water supplies are inadequate.

Human societies can adjust to variations in rainfall if these variations are regular or predictable. Irrigation can help even out the supply of water over the growing season. In the Sahel, south of the Sahara, farmers grow fast-maturing crops such as sorghum and millet that shoot from seed to ripened ear in three or four months.

It is much more difficult, however, to cope with rains that come irregularly from year to year. In Europe and wetter portions of North America, the amount of annual rainfall varies by less than 15 percent per year on average. In many of the tropical and subtropical regions, on the other hand, it fluctuates from 15 to 20 percent, and in the semiarid and arid lands, it fluctuates by up to 40 percent or more. In the lean years, crops fail. If the rains are late, the growing season is cut short, and yields are greatly diminished. If imported food is not available, famine can follow.

Water availability   Evaluating water availability involves much more than just measuring rainfall. Plants demand water and thus play a key role in what happens to the precipitation. Plants take up water through their roots and evaporate it through their leaves, releasing it into the atmosphere as water vapor, a process called **transpiration.** Therefore, most climate classification systems consider precipitation in relation to what a region's vegetation needs. Plants consume very large amounts of water, totaling about two-thirds of all precipitation that falls on Earth's land areas. A single tree may remove 100 or more liters of water from the soil in a single day. The amount of water that could evaporate from damp soil or be transpired by plants if all of their needed moisture were available is called *potential evapotranspiration* (POTET; see Chapter 4). POTET indicates the moisture demand that plants make on their environment. Evapotranspiration is a fundamental part of the hydrologic cycle, which will be discussed in detail in Chapter 4.

Because heat energy must be available to evaporate water, warmer climates make possible a greater POTET. In tropical climates, where temperatures typically exceed 20°C (68°F) throughout the year, annual POTET can exceed 150 centimeters (60 inches). In contrast, in midlatitude climates, annual POTET totals 50 to 90 centimeters (20 to 35 inches), and annual POTET is only a few centimeters in polar regions.

In distinguishing between arid and humid climates, measurement of potential evapotranspiration is important. Parts of tropical East Africa and central Minnesota both receive the same average annual rainfall of about 80 centimeters (30 inches). Yet trees are sparse in East Africa and abundant in Minnesota. The reason is that East Africa is hotter and therefore has a much greater POTET than central Minnesota. East Africa's POTET exceeds its annual precipitation, so it cannot support as many trees.

Classifying Climate

Classifying climate types is useful for many purposes. Knowing which places have certain climates allows analysis and planning by climatologists, geographers, geologists, social scientists, government, and industry. Describing and naming specific climate types gives us vocabulary to use in describing places and communicating such information to others.

Most climate classifications combine information about temperature and precipitation to take into account the effects of temperature on water availability for vegetation. One such classification was devised in 1918 by German climatologist Wladimir Köppen (Table 2-1 and Figure 2-33; see also Appendix II). Köppen used the distribution of plants to help him draw boundaries between climate regions. He identified five basic climate types and subdivided each to reveal important distinctions. A vegetation-based approach has the advantage of defining climates in terms that have more meaning for other environmental features rather than just selecting an arbitrary temperature or precipitation level as the boundary between two climates. Climatologists have modified Köppen's classification many times but have retained its essential elements. It remains the most widely used classification system.

Earth's Climate Regions

Take a moment to become acquainted with Figure 2-33, because the discussion throughout this section will refer to it. Climate zones form horizontal bands around Earth, roughly following lines of latitude. Tropical (A) climates predominate in the very low latitudes, whereas polar (E) climates predominate in the higher latitudes. These horizontal bands show the powerful influence of temperature on climate. As explained earlier, higher latitudes receive less insolation than do lower latitudes. Observe the striking correspondence between this map and the patterns shown in the world maps of temperature, precipitation, and atmospheric circulation (Figures 2-7, 2-32, and 2-24, respectively). In particular, clear correspondence exists between dry (B) climates and subtropical high-pressure cells. Dry climates occur on the western sides of continents in the latitudes between 20° and 35°, in both the Northern and Southern Hemispheres.

TABLE 2-1 Major Climate Types and Their Köppen Equivalents

Climate Type	Climate Characteristics
Tropical: Climates that are warm all year	
Humid tropical	
Af	Tropical, constantly warm and humid, with no dry season
Am	Tropical, constantly warm and humid, but with a short dry season
Seasonally humid tropical	
Aw	Tropical, constantly warm and humid, but with a pronounced dry low-sun season and wet high-sun season
Dry climates	
Desert	
BWh	Hot desert climate
BWk	Cool desert climate
Semiarid	
BSh	Hot semiarid (steppe) climate
BSk	Cool semiarid (steppe) climate
Midlatitude climates with warm summers and cool winters	
Humid subtropical	
Cfa	Humid, warm subtropical climate, with hot summers and no dry season
Cw	Humid, warm subtropical climate, with hot summers and dry winters
Marine west coast	
Cfb	Marine west coast climate, with warm summers and no dry season
Cfc	Marine west coast climate, with cool summers and no dry season
Mediterranean	
Cs	Mediterranean climate, with dry, warm summers and cool, wet winters
Midlatitude climates with warm summers and cold winters	
Humid continental	
Dfa	Humid continental climate, with hot summers, cold winters, and no dry season
Dwa	Humid continental climate, with hot summers and dry, cold winters
Dfb	Humid continental climate, with warm summers, cold winters, and no dry season
Dwb	Humid continental climate, with warm summers and dry, cold winters
Subarctic	
Dfc	Moist subarctic climate, with cool summers, very cold winters, and no dry season
Dwc	Moist subarctic climate, with cool summers and very cold, dry winters
Dfd	Moist subarctic climate, with cool summers, frigid winters, and no dry season
Dwd	Moist subarctic climate, with cool summers and frigid, dry winters
Polar climates	
Tundra	
ET	Tundra climate, with very cool, short summers and frigid winters
Icecap and ice sheets	
EF	Ice cap climate, with temperatures consistently below freezing

Note: In mountainous areas, large differences in climate occur over short distances, causing detailed climatic patterns that cannot be shown on a world map. These areas are mapped as H, or highlands.
Source: Courtesy Institute for International Economics.

You also can see the influence on climate caused by mountain ranges. The North American Rockies, the Andes of South America, and the Tibetan Plateau in Asia are mapped as H, or highland, climates. These areas are generally colder than lowland areas because of their high altitude, but they are also areas of great spatial variability in climate. East of the Rockies and Andes are rain shadow deserts. Africa's Sahara is a **desert climate,** or B climate, because it is dominated by the subtropical high-pressure zone of the eastern Atlantic. This desert extends into interior Asia because that region is isolated from oceanic moisture sources, both by mountains and distance from the sea.

In the following descriptions, we will focus on the core characteristics of climates, which are briefly summarized in Table 2-1 and described in detail in Appendix II. Graphs called *oleographs* depict the monthly average temperature and precipitation for a typical site in each climate zone. You should recognize, however, that enormous variability occurs within each climatic region and that broad transitions, not sharp boundaries, exist between them. In high mountain regions, the high altitude causes cold conditions most of the year. Most mountain regions have much variation in elevation, with both snowcapped peaks and vegetated valleys.

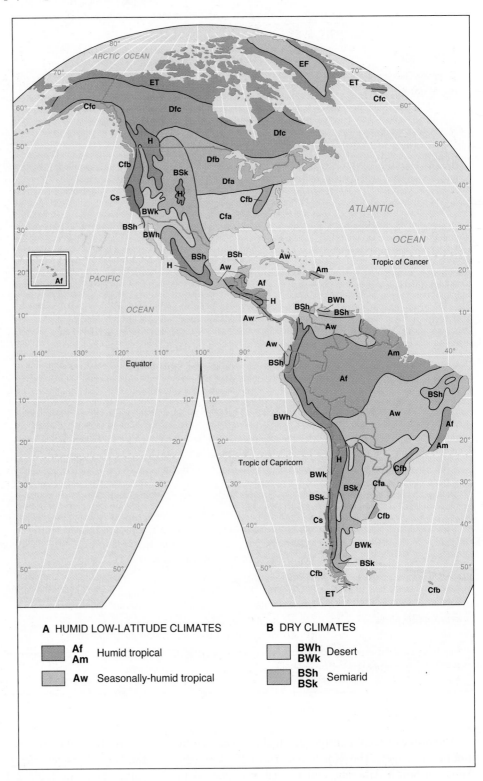

Figure 2-33 World climates. This world map of climate patterns is largely based on the Köppen system of climate classification.

Humid Low-Latitude Tropical Climates (A)

Humid tropical climates (Af, Am)   **Humid tropical climates** (Af and Am climates) lie mostly within 10° north and south of the equator but can extend to 20° north and south (Figure 2-33). These areas are under the rainy ITCZ and include the world's tropical rain forests—its steamy jungles. Tropical climates share important characteristics: Throughout the year, they

are warm and humid. Because tropical areas are warm throughout the year, the variation in temperature in a single day is greater than the difference between the warmest and coolest months of the year. (This is the opposite of what most midlatitude residents experience. There, temperature differences between winter and summer are far greater than between day and night.) In the tropics the air can hold a large amount of moisture

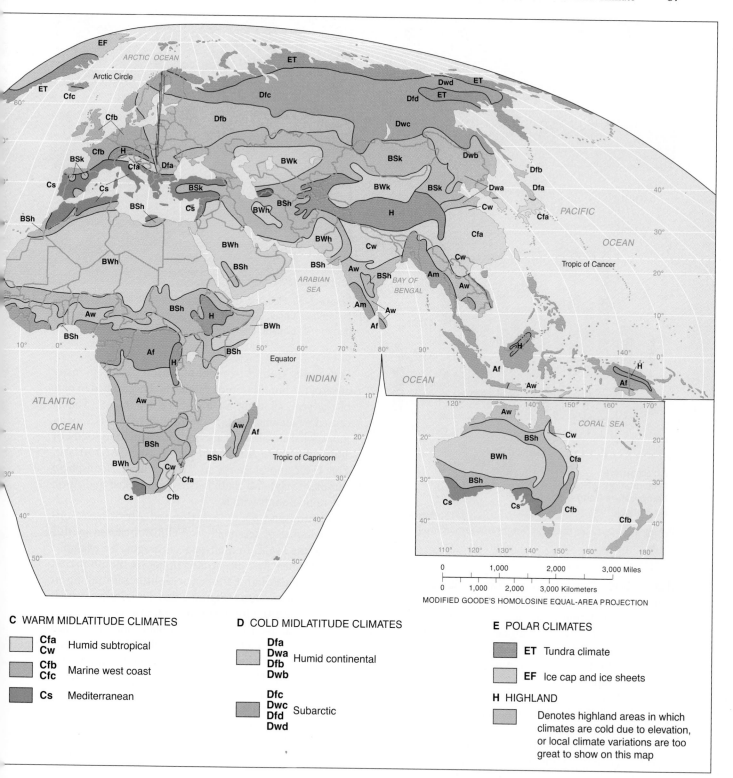

C WARM MIDLATITUDE CLIMATES

Cfa **Cw**	Humid subtropical
Cfb **Cfc**	Marine west coast
Cs	Mediterranean

D COLD MIDLATITUDE CLIMATES

Dfa **Dwa** **Dfb** **Dwb**	Humid continental
Dfc **Dwc** **Dfd** **Dwd**	Subarctic

E POLAR CLIMATES

ET	Tundra climate
EF	Ice cap and ice sheets

H HIGHLAND

Denotes highland areas in which climates are cold due to elevation, or local climate variations are too great to show on this map

MODIFIED GOODE'S HOMOLOSINE EQUAL-AREA PROJECTION

because of the high temperatures, and precipitation usually is heavy.

High temperatures and ample rainfall mean that tropical climates support abundant plant growth and productive agriculture; if water supply and soil fertility are adequate, two or three harvests per year are possible. The high temperatures mean that a great deal of energy is available to evaporate water, so POTET is very high.

Because precipitation is greater than evapotranspiration during most of the year, the tropics generally are humid. But many tropical areas experience distinct wet and dry seasons, as will be discussed below.

The most extensive areas of humid tropical climate are in the Amazon River Basin of South America, equatorial Africa, and the islands of Southeast Asia. The small country of Singapore typifies a humid tropical climate

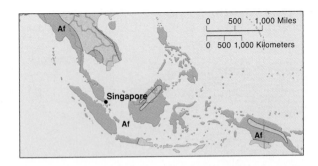

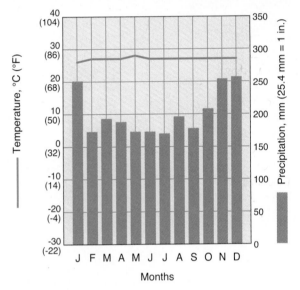

Af: Singapore
Precipitation: 241.3 cm
Temperature Range: 2°C
Lat/long: 1°10'N, 103°51'E

Figure 2-34 Locator map for Singapore and climograph.

Figure 2-35 Flooding at a market in Manila Philippines, caused by intense rainfall. Such rainfall is characteristic of the tropical humid (Af) climate, often contributing to flooding problems.

(Figure 2-34). Temperatures are warm throughout the year, averaging between 26°C and 28°C (79° and 82°F). Annual precipitation totals about 240 centimeters (94 inches), and every month brings at least 150 millimeters (6 inches). Annual precipitation substantially exceeds evapotranspiration, so the area is very humid and features lush plant growth.

Humid tropical climates are influenced by the ITCZ most of the year, so rainfall in these areas primarily is convectional (Figure 2-35). Atmospheric instability results from intense daily insolation, and the high moisture content of the atmosphere favors development of severe convectional storms. Rain may fall in the afternoon or evening nearly every day, with mornings generally clear. The annual totals of rainfall exceed 200 centimeters (80 inches) in many areas, and POTET is generally 120 to 170 centimeters (48 to 68 inches) per year. In oceanic areas of this climate, tropical cyclones are generated, forming hurricanes or typhoons. Southeastern Asia, the Pacific islands of Oceania, and the Caribbean are especially prone to these tropical storms. Some humid tropical climates experience a brief dry season. These are designated "Am."

Seasonally humid tropical climates (Aw) In many areas of the humid tropics, rainfall is concentrated in part of the year, allowing for a distinct dry season. This is caused by seasonal shifts in the location of the ITCZ or by monsoonal circulation patterns (for example, in southern and southeastern Asia and to a lesser extent West Africa). The ITCZ moves north in the Northern Hemisphere summer (May) and south between November and April, and rainfall shifts with it.

Central and South America and Africa have substantial areas of seasonally wet climates caused by this shift in the ITCZ. In Managua, Nicaragua, for example, rainfall is heavily concentrated in the six-month period from May through October, with little rain falling in the other six months (Figure 2-36). Hurricanes, which occur during the high-Sun period, also contribute to high rainfall averages. Annual rainfall totals about 120 centimeters (48 inches), roughly half that of Singapore. Average monthly temperatures vary between 25° and 29°C (77° and 84°F). The seasonal contrasts are caused by movement of the ITCZ. The lack of moisture in the dry season contributes to strong seasonality of vegetation growth (Figure 2-37).

In southern and southeastern Asia, precipitation is also seasonal because of the monsoonal circulation pattern. In the summer, the Asian landmass heats and develops an extensive low-pressure region that draws air in from all directions. The low-pressure cell draws over the land a steady flow of moist air from the Indian and Pacific oceans, and the rainfall is intense. Monsoonal rains are critical to agriculture in Asia. The timing of this circulation change varies from year to year, and if the rain arrives too early or too late, crop yields may be severely affected.

Some of the highest average rainfall totals in the world (5–10 meters, or 16–33 feet, or even more) occur

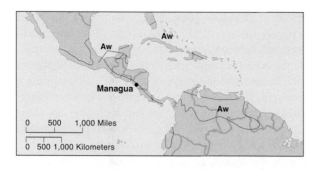

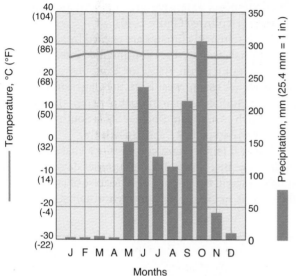

Aw: Managua
Precipitation: 120.4 cm
Temperature Range: 2°C
Lat/long: 1°0'N, 86°20'W

Figure 2-36 Locator map for Managua and climograph.

Figure 2-37 Savanna vegetation in eastern Africa. This type of vegetation is characteristic of seasonally humid tropical climates. In the background, Mt. Kilimanjaro, Tanzania, at 5,895 meters (19,340 feet) elevation, has an ice cap climate even though it is located just 3° south of the equator. Global warming has contributed to rapid shrinkage of the ice cover on Mt. Kilimanjaro; the ice may be gone by 2015.

on the southern slopes of the Himalayas, where mountains cause orographic lifting. In winter, however, intense high pressure develops over central Asia, and the wind in southern and eastern Asia blows from the north and northwest. The air warms as it descends from the Tibetan Plateau, and extremely dry conditions result to the south of the Himalayas.

Dry Climates (BW and BS) Dry climates cover 35 percent of Earth's land area, although fewer than 15 percent of the world's inhabitants live in such areas. Drylands are generally located in bands immediately to the north and south of the low-latitude humid climates. The most extensive region of dry lands extends from North Africa to Central Asia; the African portion is known as the Sahara (the Arabic word for "desert"). Sand dunes cover only a small portion of the Sahara and other drylands; more common is a cover of stone or gravelly soil.

Dry climates are distinguished by a formula that relates average annual temperature to average annual rainfall. Higher rainfall totals are necessary to classify a place as humid (A, C, or D) under warm conditions than under cold conditions. Dry climates can be subdivided into warm types (BWh, BSh) and cool types (BWk, BSk).

Desert climates (BWh, BWk) The most extensive regions of warm, dry climate are found in the subtropics on the western sides of continents. It is interesting to study this in Figure 2-33. The deserts (BWh) of northwestern Mexico and the Atacama Desert of coastal Peru border the Pacific; the Sahara and Namib/Kalahari deserts of northern and southern Africa border the North and South Atlantic; and the Australian Desert borders the Indian Ocean. These BWh climate regions result from air descending in the subtropical high-pressure zones, over the eastern parts of the oceans (in other words, off the west coasts of the continents).

Cold ocean currents can also increase the aridity of these areas. Ocean currents generally circulate clockwise in the Northern Hemisphere and counterclockwise in the Southern Hemisphere. Thus, on the eastern side of the ocean, the currents flow from high latitude to low latitude, bringing cooler water toward the tropics. The cold ocean surface cools the lower layers of the atmosphere, inhibiting evaporation and convection and therefore reducing precipitation.

In warm desert climates like that of Cairo, Egypt, high temperatures exceeding 40°C (104°F) are common (Figure 2-38). The high temperatures combined with low rainfall produce the greatest differences between annual precipitation and POTET. Annual POTET exceeds 150 centimeters (60 inches) in many BW regions; but as precipitation totals less than 20 centimeters (8 inches), the demand greatly exceeds the supply. In these conditions, only plants with specialized water-collecting and water-holding characteristics can survive (Figure 2-39).

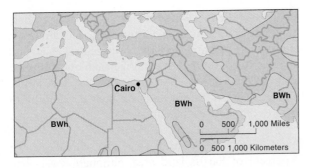

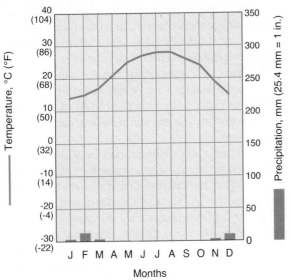

BWh: Cairo
Precipitation: 3.0 cm
Temperature Range: 14°C
Lat/long: 30°1'N, 31°14'E

Figure 2-38 Locator map for Cairo and climograph.

Figure 2-39 A desert environment in Morocco.
Potential evapotranspiration in this warm desert is
much greater than precipitation, and only vegetation
adapted to severe moisture stress can survive.

Semiarid climates (BSh, BSk)   Not all of the world's
drylands are barren of vegetation; in many areas, sufficient rain falls to support extensive seasonal or perennial
plant cover. A climate with enough moisture to support
such vegetation is semiarid, and typically regions of
semiarid climate lie in transitional areas between deserts
and more humid regions. Semiarid climates are designated BS in the Köppen scheme. Annual precipitation is
only 20 to 50 centimeters (8 to 20 inches). Most of the
world's grasslands, or **steppes,** are in this climate type,
which supports extensive grazing activities but only
limited agriculture (Figure 2-40).

Some semiarid climates are caused by rain shadow
effects that isolate the continental interiors from
moisture sources. In Asia, for example, a mountain
belt stretches east to west from the Caucasus to China,
blocking Indian Ocean moisture from the interior.
The North American Rockies and South American
Andes similarly block Pacific moisture.

Seasonal temperature contrasts are characteristic of
midlatitude BSk climates. Long summer days and bright
sunshine warm the ground, and because of lack of water,
this heat is not dissipated through evaporation. Unlike

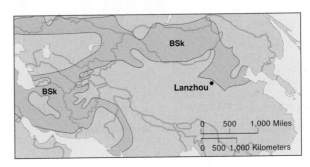

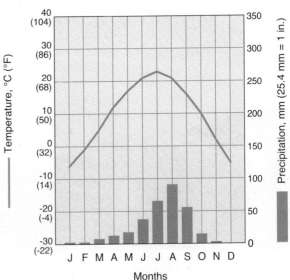

BSk: Lanzhou
Precipitation: 31.2 cm
Temperature Range: 29°C
Lat/long: 36°4'N, 103°44'E

**Figure 2-40 Locator map for Lanzhou
and climograph.**

Figure 2-41 Caucasus Mountains. Semiarid climates have enough moisture for significant plant growth at least part of the year, but moisture shortages are common. Grasses, such as those in this area in the Caucasus Mountains of Georgia, southwestern Asia, are particularly well adapted to semiarid climates.

tropical deserts, however, midlatitude dry climates are cool part of the year, so that plants require less moisture to survive (Figure 2-41). In some areas, such as interior basins of the western United States or western China, agriculture is possible with water imported from wetter mountain areas.

Warm Midlatitude Climates (C) In the midlatitudes, seasonal variations of insolation profoundly influence temperature. In summer, these latitudes receive more radiation per day than equatorial areas. In winter, however, insolation is cut to half or even a third of summer levels. Therefore, the midlatitudes experience a distinct cool season, commonly known as a winter. In subtropical locations, this winter season may be a month or two in which frost can occur. Winter causes life to become restricted to organisms that can tolerate at least occasional freezing conditions.

Lower temperatures mean that annual precipitation and evapotranspiration are generally less in midlatitude climates than in the tropics. Like the tropics, however, midlatitude climates include a wide range of precipitation patterns, including continuous humidity and distinctly seasonal rainfall. Precipitation in the midlatitudes is heavily influenced by the polar front—the boundary between warm tropical air and cold polar air along which midlatitude cyclones form and travel. The significance of these storm systems is that daily weather tends to be more variable in the midlatitudes than in the tropics. This climatic zone experiences rainy spells separated by dry spells, and rainfall can occur at any time of day rather than concentrated in the afternoon and evening.

In many midlatitude environments—in the central United States, for example—seasonal precipitation patterns also are influenced by temperature. Because

warmer air holds more moisture, greater rainfall occurs in summer than in winter. In the southeastern United States, midlatitude cyclones are less frequent in summer than in winter. But summer rainfall is more intense because more water is present in the air, so winter and summer receive similar amounts of rainfall.

Humid subtropical climates (Cfa, Cw) **Humid subtropical climates** occur in latitudes between about 25° and 40° on the eastern sides of continents and between about 35° and 50° on the western sides. These climates are relatively warm most of the year but have at least occasional freezing temperatures during the winter. Frost occurs, so many plants adapted to tropical humid climates cannot survive, and most humid subtropical climates have deciduous species that lose their leaves in autumn and become dormant in winter. Eastern China, the southeastern United States, and parts of Brazil and Argentina are the largest areas of humid subtropical climates.

The subtropical high-pressure zones that create deserts on the western margins of continents (the eastern margins of oceans) create humid conditions on the eastern sides of continents. The circulation around these high-pressure cells brings warm, moist air from low latitudes to the eastern margins of the continents (clockwise in the Northern Hemisphere and counterclockwise in the Southern Hemisphere). This moisture supplies energy to midlatitude cyclones that generate plentiful rainfall in these areas.

Precipitation may occur all year or be concentrated in summer. Summer rainfall may result from monsoonal effects, as in southeastern China, or from greater moisture in the warmer air, as in the southeastern United States (Figure 2-42). Cool winter temperatures bring occasional

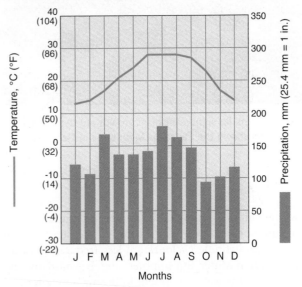

Cfa: New Orleans
Precipitation: 161.5 cm
Temperature Range: 15°C
Lat/long: 30°0'N, 90°5'W

Figure 2-42 Locator map for New Orleans and climograph.

Figure 2-43 A mixed agricultural and forested landscape near Asheville, North Carolina. Humid subtropical climates support a natural cover of forests, but many of the world's most productive agricultural regions are in this climate.

frost in low latitudes, and snowfall occurs poleward, but the ground is not frozen or snow-covered for long. The warm, humid conditions of Cfa climates support a long growing season with forest vegetation under natural conditions and high crop yields in agricultural regions, sometimes with two harvests in a year (Figure 2-43).

In parts of southern and southeastern Asia, the monsoon circulation causes a distinct winter dry spell, producing a subtropical winter-dry climate (Cw). The reduced precipitation falls at a time of year when POTET is low, and it does not have as much of an impact on agriculture or water resources as it would if it occurred in the summer.

Marine west coast climates (Cfb, Cfc)   On continental west coasts between about 35° and 65° are mild climates with small temperature variations and plentiful moisture year-round. Major regions include western North America between northern California and coastal Alaska, southern Chile, northwestern Europe, southeastern Australia, and New Zealand. These areas are cooler than humid subtropical climates (Cfa), especially

in summer. They are subdivided into cooler types (Cfc) and warmer types (Cfb).

Marine west coast climates are moderated by ocean temperatures. Typical summer temperatures are 15° to 25°C (60° to 77°F), rather than the 25°C to 35°C (77° to 95°F) inland at these latitudes. Frost-tolerant plants stay green all year in the mild winters of −5° to 15°C (23° to 59°F), even if snow falls. Maritime influence on temperature is so great that winter temperatures normally associated with subtropical latitudes are found as far poleward as 55° (Figure 2-44). Palm trees survive in coastal areas of southwest England, at the same latitude as Winnipeg, Manitoba, in south-central Canada. Kodiak, Alaska, in a maritime location at latitude 57°N, has the same January average temperature as Richmond, Virginia, which is under continental climate influence at latitude 37°N. Clearly a location on the west coast in midlatitudes significantly raises winter temperatures relative to east coast or interior locations.

The North Atlantic and the northeastern and southern Pacific are known for stormy weather, and these storms bring plenty of moisture to the adjacent land areas (Figure 2-45). Midlatitude cyclones are frequent, especially in winter. The low temperatures mean that air contains less moisture than at lower latitudes, so drizzle is common. Annual rainfall is about 70 to 120 centimeters (28 to 47 inches). Where mountains are near the coast, as in western North America, Chile, and Norway, orographic effects contribute to high rainfall, 2 to 3 meters (6 to 9 feet) per year. The current highest average annual rainfall in the mainland United States is in a marine climate—the west coast of the Olympic Peninsula in Washington, where annual rainfall totals 3 to 4 meters (10 to 13 feet).

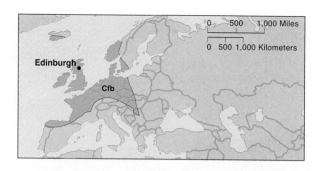

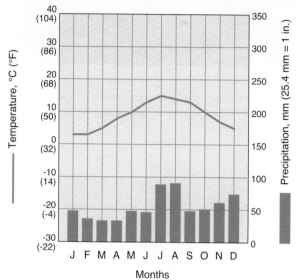

Cfb: Edinburgh
Precipitation: 65.8 cm
Temperature Range: 12°C
Lat/long: 55°57'N, 3°12'W

Figure 2-44 Locator map for Edinburgh and climograph.

Figure 2-45 The coast of British Columbia. This region has mild winters despite its relatively high latitude (49° north).

Mediterranean climates (Cs) On the western margins of continents, another distinctively seasonal climate occurs. The dry-summer **Mediterranean climate** envelops the Mediterranean region of Europe and parts of northern Africa, lending this climate type its familiar name. A similar climate occurs in extreme southern Africa, southwestern Australia, California, and central

coastal Chile. Precipitation is seasonal and caused by northward and southward movement of the subtropical high-pressure zones. In summer, these zones move poleward and bring aridity; in winter, they move toward the equator and are replaced by more frequent storms of the midlatitude low-pressure zone.

In Mediterranean climates, cool, rainy winters bring occasional frost or snow, with frequent midlatitude cyclones. Warm, dry summers have temperatures of 20° to 35°C (68° to 95°F). In Rome, Italy, for example, little or no rain may fall between May and October (Figure 2-46).

Mediterranean climates have significant water-availability problems. Precipitation occurs in winter, when human, animal, and vegetative demand for water is low. However, very little rainfall occurs in the summer, when POTET is highest. Plants with roots that can reach deeper soil moisture stay green during the summer, when shallow-rooted grasses dry out. Because water is scarce in the summer, many plants grow mostly in the winter and spring, even though there is less sunlight then (Figure 2-47). Agriculture in Mediterranean climate regions requires storage of winter precipitation in reservoirs for irrigation in summer.

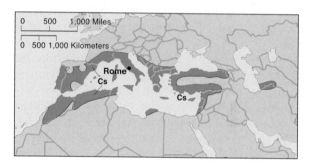

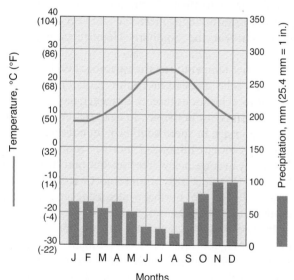

Cs: Rome
Precipitation: 71.4 cm
Temperature Range: 16°C
Lat/long: 41°54'N, 12°30'E

Figure 2-46 Locator map for Rome and climograph.

(a)

(b)

Figure 2-47 Variation in water availability. Seasonal variations in water availability change the appearance of the landscape in Mediterranean climate regions, such as coastal California. The wet season (left) is typically winter and spring; the dry season (right) is summer and autumn.

Cold Midlatitude Climates (D)

Humid continental climates (Dfa, Dwa, Dfb, Dwb)

Like Mediterranean climates, **humid continental climates** have strong seasonal contrasts. In the latter case, however, the contrasts are of temperature rather than moisture availability. Continental climates are so named because they are remote from the ocean and therefore deprived of the sea's input of moisture and moderating influence on temperature. These climates occur between about 35° and 60° latitude in the interior and eastern portions of Northern Hemisphere continents. Many eastern coastal areas at high latitudes also have strong seasonal contrasts because the prevailing westerly winds extend this humid continental climate to them. D climates occur exclusively in the Northern Hemisphere because very little landmass exists in the Southern Hemisphere at these latitudes.

Summer temperatures are warm because long days and high solar angles cause more radiation in a day than tropical locations receive. In winter, days are short and the Sun is low in the sky, so temperatures often fall below freezing. In more southerly humid continental areas like Chicago, Illinois (42° north), January temperatures may average below freezing, but rain still occurs in winter (Figure 2-48). In more northerly places like Moscow, Russia (56° north), winter brings many weeks of subfreezing temperatures.

In most humid continental climates, precipitation occurs throughout the year from frequent midlatitude cyclones. Typical annual rainfalls are 60 to 150 centimeters (24 to 60 inches), usually ample for plants, with an annual evapotranspiration of 50 to 90 centimeters (20 to 35 inches). In winter, plants need little water, but in summer they demand every drop that falls and sometimes more (Figure 2-49). During summer dry spells, farm fields may require supplemental irrigation.

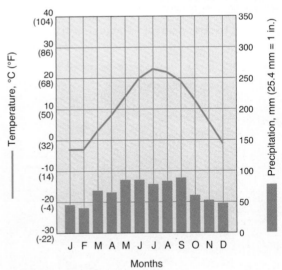

Dfa: Chicago
Precipitation: 80.8 cm
Temperature Range: 26°C
Lat/long: 41°45'N, 87°40'W

Figure 2-48 Locator map for Chicago and climograph.

Subarctic climates (Dfc, Dwc, Dfd, Dwd)

At the northern edge of the humid continental climates, winter temperatures are cold enough and the growing season short enough that agriculture is generally not possible (Figure 2-50). The only plants that survive in these areas

Figure 2-49 Birch woodland. Humid continental climates have strong seasonal contrasts, as seen in these two views of a birch woodland in winter (left) and summer (right).

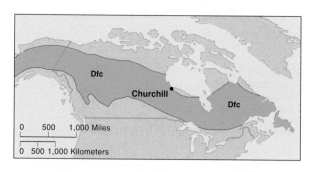

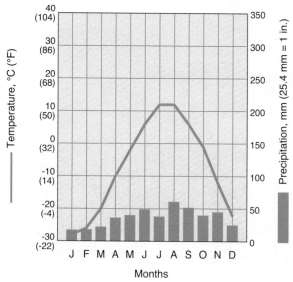

Dfc: Churchill
Precipitation: 44.3 cm
Temperature Range: 40°C
Lat/long: 58°45'N, 94°5'W

Figure 2-50 Locator map for Churchill and climograph.

are those able to weather extreme cold. No abrupt temperature or precipitation boundary exists between the humid continental climate and the **subarctic climate;** the difference literally is one of degrees. Conifers dominate; the broad expanse of spruce, fir, and larch covering much of northern Canada and Siberia is called *boreal* (meaning "northern") *forest*, lending the name boreal forest climate to the subarctic climate (Figure 2-51). Like the humid continental climate, this climate is essentially limited to the Northern Hemisphere, because so little landmass exists at these latitudes in the Southern Hemisphere.

Extremely cold −20° to −10°C (−4° to +14°F) subarctic winters feature snow cover and months of temperatures well below freezing. As days lengthen in spring, temperatures rise rapidly, and 15- to 20-hour summer

Figure 2-51 Winter near Sheregesh, Siberia, Russia.

days bring mild temperatures approaching those of much warmer climates. The growing season is short, generally June to September, in a cool summer (10° to 20°C, 50° to 68°F), with warm spells above 30°C (86°F). Temperatures warm enough for plant growth occur as early as April, but frosts may continue as late as June. Similarly, the first frost is often in September, even though temperatures to 20°C (68°F) occur into early October.

Annual rainfall is low, generally 20 to 50 centimeters (8 to 20 inches), but ample for plants because virtually no evapotranspiration occurs for six months of the year. Winter precipitation usually is locked up as snow and ice until the spring melt releases this water to plants. Summer water shortages are rare, even though evapotranspiration may exceed precipitation during part of the summer.

Polar Climates (E)

High-latitude climates are characterized by two important features: low average temperatures and extreme seasonal variability. These climates occur at and around the poles and are differentiated by temperature alone, because temperature determines whether water is present in liquid form. These climates result from scant insolation, varying from zero during winter at the poles to among the highest daily totals on Earth for a brief period at midsummer.

Tundra climates (ET)

Tundra temperatures are low all the year, rising above freezing only for a brief, cool summer (Figure 2-52). The annual average temperature is below freezing. Winters below −20°C (−4°F) are common, typically staying below freezing for three to five months. Midsummer daytime highs rarely exceed 10° to 15°C (50° to 59°F), even though the Sun can shine 20 to 24 hours a day. The soil 1.5 to 2 meters (5 to 6.5 feet) below the surface is insulated from seasonal temperature changes, so at this depth soil temperatures are close to the annual average. The condition of permanently frozen ground is known as **permafrost.**

Annual precipitation rarely exceeds 30 centimeters (12 inches). Most falls as snow, covering the ground for months except where the ground is bared by the wind. Although annual rainfall is slight, water is plentiful because evapotranspiration rates are low. Most of the year water is present as ice rather than liquid. In the brief summer when the ground thaws, the water may not be able to drain from the soil because of permafrost below the surface; saturated soils are therefore common.

At its northern limit, the boreal forest gives way to a treeless landscape in which vegetation survives the long winter by staying close to the ground, where it is protected from winter's extremes (Figure 2-53). This low-lying hardy vegetation, called *tundra*, gives the **tundra climate** its name. While the vegetation is hardy, this landscape is also highly vulnerable to disruption associated with human activities or climatic warming.

Ice cap climates (EF)

Near the poles, and high in some mountains at lower latitudes, are climates in which even the warmest month averages below freezing

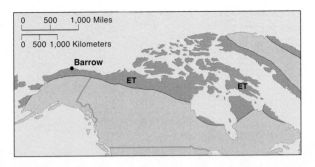

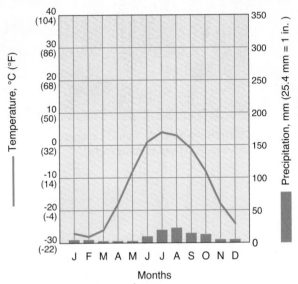

ET: Barrow
Precipitation: 11.0 cm
Temperature Range: 32°C
Lat/long: 71°28'N, 15°60'W

Figure 2-52 Locator map for Barrow and climograph.

Figure 2-53 Arctic tundra on Ellesmere Island, Canada. Flowers bloom in the brief summer.

(Figure 2-54). Virtually all of Antarctica and most of the Arctic have this **ice cap climate.** Where there is land, permafrost is extensive and thick beneath a year-round cover of snow or ice. The extremely low

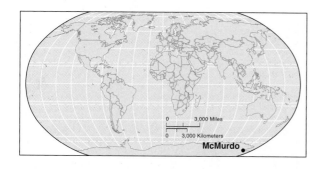

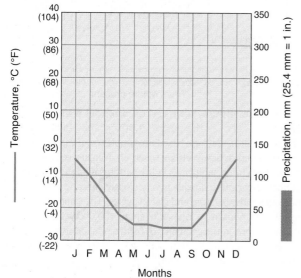

EF: McMurdo
Precipitation: 0 cm
Temperature Range: 21°C
Lat/long: 77°0'S, 170°0'E

Figure 2-54 Locator map for McMurdo and climograph.

Figure 2-55 The edge of the Antarctic ice cap. In ice cap climates, temperatures are only rarely above freezing, and glaciers form from the accumulation of snow. Virtually all of Antarctica has an ice cap climate, with little glacial melting. The glaciers flow to the sea, forming icebergs.

temperatures prevent the air from holding much water. Annual precipitation usually is less than 10 centimeters (4 inches), falling as snow and compacting into a thick mass of glacial ice that flows to the sea rather than melting in place (Figure 2-55). Due to the low temperatures very little precipitation falls in Antarctica and in some places desert-like with areas that are free from snow or ice cover.

Climate Change

So far, we have seen that solar energy (mediated by latitude, land, and water contrasts) and atmospheric processes, such as convection and precipitation, drive global and regional circulation patterns. These circulation patterns, in turn, explain the spatial distribution of climates across Earth's surface. It follows that if any of the factors that determine circulation patterns change, climates will also change. The world climate map described in the previous section is based on averages over periods of a few decades in the mid-twentieth century. However, the world climate map has not always looked like this. Considerable change has taken place over time and is likely to continue in the future.

Humans have long been able to modify environments, sometimes radically, by building settlements, clearing forests for agriculture, domesticating animals, manufacturing goods, damming rivers, extracting resources, and spreading waste throughout the air and seas and across the land. Earth's land surface is profoundly different today from what it was only 200 years ago. For example, the forests of the eastern United States were almost completely removed between 1600 and 1900, and an entire ecological community—the tall-grass prairie of the Great Plains—has been replaced by agricultural crops since 1800. Only recently have we come to realize that humankind, in addition to modifying Earth's surface, has set in motion global-scale climatic changes. Debate continues about the nature of these changes.

To interpret the significance of future climate change, we must understand patterns of past climate changes. These can give us valuable clues in determining whether recent changes are similar to those of the past or whether they are unique because of human actions.

Climatic Change over Geologic Time

Viewed over geologic time, which is about 4.6 billion years, the climate of the past 2 million years has been quite exceptional. This period, which includes our present time, is known to geologists as the **Quaternary Period.** Earth has experienced more climatic variability during the past 2 million years than in most of the previous 200 million years. Periods of similar variability probably occurred prior to 200 million years ago, but the record of the first 4 billion years of Earth's history is not very clear. In any case, it is clear that the Quaternary glaciations are an exceptional episode in recent Earth history.

Climate during the Quaternary Period has included intervals in which global average temperature was as much as 10°C (18°F) cooler than the present and warm intervals in which the climate was as warm as or warmer than today. In the past 1 million years, there have been about 10 of these cold intervals, occurring fairly regularly about once per 100,000 years (Figure 2-56). The portion of the Quaternary Period in which these glaciations occurred is called the **Pleistocene Epoch.** In everyday language, we describe these times of significantly lower global mean temperatures as the ice ages. During the cooler periods, great continental ice sheets, like those covering much of Antarctica and Greenland today, extended over much of North America, northern Europe, and northern Asia. The changes in temperature associated with the appearance and disappearance of continent-sized glaciers are the most striking and easy-to-measure aspects of these climate changes, but all aspects of climate, from precipitation to wind and seasonal weather patterns, were most certainly altered. Periods of glaciation were also periods of lowered sea level because water was taken out of the sea and stored on land in the form of glacial ice. Even areas not covered by ice were much different in the past than they are today. Areas that today are covered with deciduous forest were frozen tundra, and lakes existed in places that now are deserts. Subtropical climates extended well into the midlatitudes.

Earth's climate shifted between warm and cold between 10 and 30 times during the Pleistocene Epoch; the exact number of cold periods is not known. The last cold period reached its maximum, as indicated by the extent of glacial ice, only about 18,000 years ago—within the period of the human archaeological record. At that time, global average temperatures were about 5°C (9°F) cooler than at present. The melting of the ice back to its present extent was completed only about 9,000 years ago, or about 7000 B.C. We are now within a period of relatively warm climate.

Within the past 1,000 years, climate has varied, though less dramatically than in the distant past. Recent climate changes have been important in European history. For example, between A.D. 800 and 1000, seafaring pirates and adventurers from present-day Scandinavia, called the Vikings, extensively explored the North Atlantic Ocean and established settlements in Greenland and North America. More hospitable coastal environments

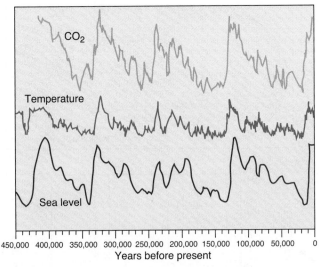

Figure 2-56 Atmospheric carbon dioxide, temperature, and sea level over the past 450,000 years. Carbon dioxide and temperature data are based on samples from the Vostok ice core, Antarctica. Sea levels are based on deep-sea sediment cores. The strong similarities in the patterns demonstrate that atmospheric carbon dioxide is very closely linked to climate, particularly temperature and glacial cycles. The timing of the major glaciations at an interval of about 100,000 years suggests that variations in the geometry of Earth's orbit around the Sun are key, although interactions among the biosphere, oceans, and atmosphere are also important.

and a lack of sea ice during those centuries as a result of especially warm temperatures likely aided the Vikings' exploration.

Cooling occurred in the Middle Ages, beginning about A.D. 1200. The period from about 1500 to 1750, when temperatures were especially cool, is known as the **Little Ice Age.** Glaciers advanced in Europe, North America, and Asia. Since the early 1800s and especially after 1900, climates have warmed significantly. Most of Earth's glaciers have been shrinking since the early 1800s. Short cooling intervals occurred in the 1800s, including one in 1884–1887 caused by the eruption of the volcanic island of Krakatau in Indonesia. Krakatau spewed so much ash and sulfur dioxide into the atmosphere that it prevented some of the shortwave insolation from reaching Earth's surface. The 1930s and 1940s were relatively warm, and cooling occurred from about 1945 to 1970. Global mean temperatures have risen dramatically since 1975.

These climatic fluctuations have occurred well within the period of human occupation of Earth. As noted, glacial melting from the most recent ice age was particularly rapid about 15,000 to 10,000 years ago. Archaeologists have found that agriculture and cities probably developed about 10,000 to 8,000 years ago. Many scholars of early human culture believe climate change was fundamental to the development of civilizations, influencing the availability of water for crops and perhaps driving major migrations.

The period of glacial melting and warming between 15,000 and 5,000 years ago coincided with expansion of human settlement in northern Europe and North America. Settlement extended gradually northward on the European mainland and became well established throughout much of the British Isles by 2000 B.C. Meltwater from the glaciers raised sea level, flooding low-lying areas and separating land areas. Water filled the English Channel, isolating the British Isles from Europe, and the land bridge that once connected North America and Asia went underwater, halting migration from Asia into North America. Rising sea level also led to rapid filling of the Black Sea about 7,500 years ago.

Modern human inventions such as heating, housing, transportation, and industry have made us less dependent on day-to-day weather conditions. But we remain strongly influenced by the dynamics of the atmosphere, both short term (weather) and long term (climate, the average of weather over 30 years or more). Many believe that as a result of human actions, climate will change much more rapidly in the future.

Possible Causes of Climatic Variation

Why does climate change over time? Many potential causes exist, and it is difficult to say which are important and which we can ignore. Understanding the causes of climatic change is of critical importance, because if we learn that climate is affected by human activity, such as burning fossil fuels (coal, oil, and natural gas), then we have the opportunity to take action to limit such effects. On the other hand, if climatic change were entirely governed by natural processes, then there would be no need for concern about potential human impacts on the atmosphere. We will summarize three of the main possible causes of climatic change: astronomical factors, geologic processes, and human modification of Earth's surface and atmosphere.

Astronomical hypotheses Inputs of solar energy drive atmospheric circulation and thus climate. Changes in insolation could alter climate. Humans know little about variations in the amount of solar radiation reaching Earth, because it is difficult to observe the Sun through the filter of the atmosphere. Recent studies based on satellite observations of the Sun are beginning to suggest that variations in solar output could be responsible for some of the climatic variations observed in the past few hundred years, causing some of the warming of the twentieth century, but probably a small portion in comparison to other factors.

In the long run—over tens of thousands of years—we know that the geometry of Earth's revolution around the Sun fluctuates (Figure 2-57). Earth spins like a top, but a slightly unbalanced one, and because of this imbalance, it wobbles slightly and very slowly as it spins. The wobble causes the tilt of Earth's axis to vary between 22° and 24° instead of staying constant at around 23½°. Other variations in the geometry of Earth's orbit around the Sun also occur, with periods of tens of thousands of years. It appears that the 100,000-year timing of major cold and warm periods over the past 1 to 2 million years corresponds to these variations in Earth–Sun geometry, although the magnitude of temperature response in the atmosphere appears to be much larger than would be expected from relatively subtle variations in insolation, indicating that other factors are also at work. These variations in Earth–Sun geometry also do not explain shorter-term climatic change.

One short-term astronomical factor that may cause variation in the Sun's radiation output is sunspots. Sunspots are relatively cool regions on the surface of the Sun that vary in number and appear and disappear over a cycle of 11 years. Sunspot numbers also vary over longer periods. Sunspots affect the output of solar energy. Some climate variations appear to have periods close to the 11-year sunspot cycle. The sunspot cycle also affects concentrations of ozone in Earth's upper atmosphere.

Geologic hypotheses Geologic factors may cause short-term and long-term climatic change. One long-timescale mechanism is plate tectonics. This will be discussed further in Chapter 3, but briefly, it is the movement of great crustal plates that make up Earth's surface at rates of up to a few centimeters per year. Over the course of millions of years, continents might move great distances, opening up oceans or

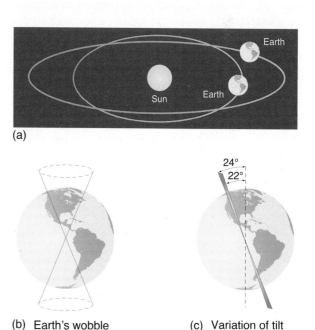

(a)

(b) Earth's wobble   (c) Variation of tilt

Figure 2-57 Variations in Earth's orbital geometry. Climate may vary with long-term variations in the geometry of Earth's orbit around the Sun. These variations include (a) the annual variation of distance between Earth and the Sun, (b) the orientation of the Earth's axis, and (c) the degree of tilt of that axis relative to our orbit around the Sun.

draining them. In addition to affecting ocean shapes and currents, the locations of continents and oceans directly affect many features of atmospheric circulation, including the ITCZ, the monsoon circulation of Asia, midlatitude cyclones, and the stormy zone around Antarctica.

Movements of continental plates have also caused the formation of major mountain ranges, such as the Himalayas, Andes, and Rocky Mountains. Continental movements are ponderously slow, so they cannot explain variations within the Quaternary Period. But they might help explain why no major glaciations occurred for more than 200 million years before then.

Volcanic eruptions can influence climate for a few years by injecting large amounts of dust and gases—especially sulfur dioxide—into the upper atmosphere. These gases reduce the amount of solar radiation filtering through the atmosphere to Earth. Past periods of more frequent volcanic eruptions may have lowered temperatures at other times.

Human causes   Processes such as sunspot cycles and plate tectonics are beyond human control. However, humans are active participants in the changing climate. Two important ways that humans influence climate are altering the atmosphere and removing vegetation.

The carbon dioxide content of the atmosphere has increased dramatically since the start of the Industrial Revolution in the late eighteenth century (Figure 2-58). The elevated CO_2 levels cause warming because CO_2 is a greenhouse gas. Analyses of air trapped in Antarctic ice reveal that the concentration of carbon dioxide was higher during past warm periods and lower during glacial periods (Figure 2-56). Elevated CO_2 concentrations in the atmosphere have been identified as the principal

cause of global warming, discussed below. Other atmospheric pollutants that contribute to global warming include methane (derived from fossil fuel operations, cattle, rice cultivation, landfills, and other sources) and halocarbons (used in refrigeration and many industrial processes). Some atmospheric pollutants, notably sulfur oxides (largely from coal combustion), actually cool the atmosphere.

Altering vegetation cover can affect climate in several ways. One is by changing land surface cover, with barren surfaces typically absorbing less solar energy than vegetated surfaces. This effect is, however, overwhelmed by the fact that vegetation facilitates cooling by evapotranspiration, so that vegetated surfaces tend to be cooler as well as effective in recycling moisture back to the atmosphere. Vegetation also plays a major role in atmospheric CO_2: Deforestation is a source of CO_2 to the atmosphere, while regrowth of forests can remove it from the atmosphere. Interactions between the biosphere and atmospheric carbon dioxide are complex; we will explore these in more detail in Chapter 4.

Global Warming

During the twentieth century, Earth's temperature increased slightly less than 1°C (1.8°F) (Figure 2-59). Throughout the 1990s, the body of scientific evidence linking this temperature rise to emissions of CO_2 accumulated rapidly, and today few scientists doubt that global warming is occurring and that increased CO_2 is the principal cause. The consensus is that unless output of CO_2 slows dramatically, world average temperature could rise by a few degrees Celsius or more in the next century.

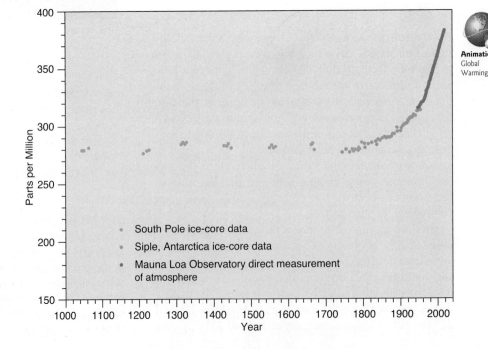

Animation
Global Warming

Figure 2-58 Concentration of carbon dioxide in the atmosphere. Carbon dioxide concentrations are now about one-third higher than they were two centuries ago, as a result of burning fossil fuels in factories, cars, and homes.

- South Pole ice-core data
- Siple, Antarctica ice-core data
- Mauna Loa Observatory direct measurement of atmosphere

Parts per Million
Year

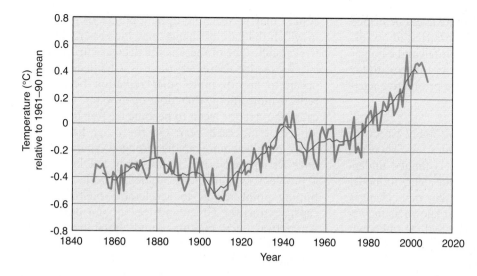

Figure 2-59 World surface temperature history of the past 148 years, based on records of measured temperatures. Temperatures are shown relative to the average for 1961–1990. The trend shows three distinct periods of warming: 1855–1880, 1910–1945, and 1975–2005. The total warming over the period of record is about 1°C (1.8°F).

Evidence of global warming Most of our understanding of global warming is based on computer models, because of course we have no observations of future climates. The models that we use to predict future global warming are similar to the models that we use to predict the weather two or three days in the future. We put in a computer program as much as we know about the present state of the atmosphere and land/ocean surface, along with equations that represent how radiant energy modifies temperatures, how temperature and pressure affect precipitation, wind, etc. The computer then simulates future atmospheric circulation.

How can we know that the computer models used to predict future climate are correct, when we have no observations of the future that we can use to check them? Climate change has occurred naturally for millions of years, so how can we be sure that the warming we have observed is caused by humans?

The most convincing answers to these questions come from the use of models to "predict" historic climates. If we start with the climate of, say, 1910 and add to it what we know about historic changes in solar activity and volcanic eruptions, we can predict the temperatures of the past 100 years. The predictions don't match very well with the observations, particularly since about 1950, when observed temperatures are much higher than the predicted ones. However, when we also add to the model human impacts on the atmosphere, such as emissions of carbon dioxide, methane, and other pollutants, the models do a very good job of predicting the temperatures of the past 100 years. The results demonstrate both that the models can predict temperatures with a reasonable degree of accuracy and that humans are causing global warming.

The most significant summary of our understanding of global warming comes from the work of the Intergovernmental Panel on Climate Change (IPCC), a group of scientists appointed by the governments of more than 130 countries around the world. While the process of summarizing and reporting information at the IPCC is not totally free of political influence, and some scientists have criticized the IPCC's objectivity, its reports are generally regarded as credible syntheses of a very complex body of science. The IPCC's predictions are striking in the breadth and magnitude of consequences foreseen, as well as the degree of certainty with which the predictions are made.

Consequences of global warming One significant result of global warming is a rise in sea level. Worldwide sea level has risen about 17 centimeters (7 inches) in the past 100 years, and the rate of sea level rise has increased in the last two decades. The current rate of sea level rise is about 3 mm (1/8 inch) per year. About half of this rise comes from net glacial melt, and the other half comes from expansion of water in the oceans as it warms. While current sea level rise is itself significant for those living in coastal zones, an even greater concern is the dramatic sea level rise that would result from accelerated melt of the major ice caps in Greenland and Antarctica. This could raise sea level by perhaps 1 to 5 meters (3 to 16 feet) over the next century or so. Danger to coastal residents would not be from constant inundation because sea level would rise very gradually over years, allowing people time to relocate, raise structures, or build dikes. Rather, the danger would be from occasional severe storms that would cause sudden flooding farther inland, such as what happened in Hurricane Katrina. The Dutch have shown that well-built dikes can hold back the sea, but poorer countries cannot afford such protection.

Another possibility is that climate change could reduce water supply in some regions. Semiarid regions and densely populated subhumid areas are particularly vulnerable because they depend on river flow for irrigation, drinking water, and waste removal. The kind of prolonged drought that has plagued much of the western United States in recent years is thought to be more likely in the future, and there is evidence that it is caused by global warming. However, in other areas, warming could bring greater precipitation, and in combination with warmer temperatures, agriculture could be helped rather than harmed. The Canadian prairies, for example, could benefit from increased rainfall and a longer growing season, and northern areas that are today too cool for farming could become productive agricultural regions.

Tipping points and uncertainties   Many aspects of expected global warming are highly uncertain. There are three types of major uncertainties: (1) limited understanding of the climate system and our ability to model it accurately; (2) unforeseen and possibly rapid changes in conditions that affect climate (also known as *tipping points* or, more properly, *thresholds*); and (3) unknown future conditions of factors controlling climate, such as the rate of carbon dioxide emissions.

While the accuracy of the computer models used to predict future climate have improved dramatically in recent years, they still have limits. For example, the 10-day forecasts shown on some Internet weather sites aren't very accurate 6 to 10 days in the future. This is in part because we don't have highly detailed information on the present conditions of temperature, pressure, and so on for every place and altitude and in part because the rules we use to estimate atmospheric flows may not be completely representative of the complexities of the environment. We face similar problems with long-term climate projections. For them, the problem is a little simpler because we are predicting averages rather than specific weather events. But averages are made up of events: Mean annual precipitation can be significantly affected by just a few large storms, for example. Our models are getting better, but still have important limits. For example, while the models of global climate we have at present are reasonably consistent in predicting future temperature changes, they aren't as consistent for precipitation, in part because of the importance of storms and storm tracks.

Our models also suffer from incomplete understanding of some important atmospheric processes. For example, earlier in this chapter, we noted that water vapor is a greenhouse gas that traps more outgoing radiation than any other gas in the atmosphere. Water is naturally present in the atmosphere, and so it is often not included in the list of human-derived substances that may alter climate. However, human-caused global warming can alter the water content of the atmosphere. We believe that atmospheric water content is more likely to increase than decrease. But if it does increase, would the added water trap more outgoing radiation and thus increase warming, or would it instead lead to increased cloud cover that reflects incoming energy? The answer to this question depends on a detailed understanding of atmospheric circulation and processes of cloud formation. We still have much to learn about this important problem.

In recent years, concern has grown that with ongoing climate change, certain parts of the Earth–atmosphere system may be reaching tipping points, or conditions in which the pace of change may increase rapidly and irreversibly. One of these is the decrease in Arctic sea ice. As sea ice decreases, there is a tendency for the ocean to absorb more solar radiation because ice reflects more energy than does open water. This could cause the Arctic Ocean to become ice free for a much longer period each year. On the other hand, open water loses heat more rapidly through evaporation than does ice-covered water, so this may not be as great a problem as some have argued. Another possible tipping point relates to accelerated melting of ice caps, such as the Greenland ice cap and portions of Antarctica. Such melting could destabilize the ice so that it would flow rapidly into the sea, causing a substantial increase in sea level in a short time period. While we are relatively certain that such mechanisms of instability exist, most of the ones being considered are not well understood, and it is impossible to know whether the rapid changes envisioned are likely or not, and if they are likely, whether they will happen soon or far in the future.

Perhaps the greatest uncertainty regarding global warming is the one humans control: the rate of increase in atmospheric carbon dioxide concentrations. The IPCC has developed a series of scenarios that describe possible future levels of economic activity and reliance on fossil fuels.

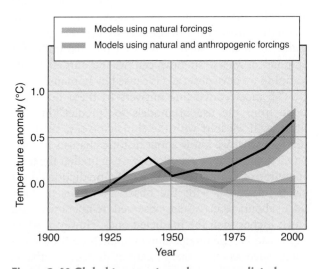

Figure 2-60 Global temperature changes predicted by models and observed. The black line represents observations, and the shaded areas show the range of predictions and thus give an indication of uncertainties. Pink shaded areas are predictions based on models that include human impacts such as carbon dioxide, while blue shaded areas are predictions based only on natural factors such as solar radiation and volcanic eruptions. The fact that the models are relatively good at predicting past climates when human impacts are included shows both that the models are reliable predictors of future climates and that observed global warming is largely caused by humans.

Dealing with global warming Three basic approaches are available to manage global warming and its impacts: (1) develop means to adapt to climate change, (2) reduce carbon emissions in an attempt to slow it, and/or (3) actively take measures to counteract the effects of our carbon emissions.

People already adjust to changing weather, commodity prices, and technology from year to year, so why shouldn't we be able to adjust to climate change, too? The United States refused to join the Kyoto Protocol, an international treaty that would limit CO_2 emissions (see Chapter 13); the United Sates argued that a strategy of adapting to global warming would be less costly and more effective than attempting to reduce it. Adapting to global warming could include measures such as building seawalls to protect low-lying coastal areas or relocating communities in these zones, or supplying people with air conditioning.

Reducing carbon dioxide concentrations will be difficult because we depend on fossil fuels in our daily lives, and energy producers employ many people and earn billions of dollars each year. Significantly reducing fossil-fuel use is possible only if we consume less energy or shift to alternative energy sources. There are other reasons to reduce fossil fuel use, of course, such as to reduce urban air pollution and the price of oil (see Chapter 13). Measures to reduce fossil fuel use are promoted on these grounds as well as on their effects on carbon emissions. We still face the challenge of rapid growth in electricity demand and a limited range of options for large-scale electric production. An alternative to reducing fossil-fuel use is to try to trap and store (*sequester*) CO_2. Several

---

RAPID CHANGE

IPCC Assessment of Global Warming and Its Impacts

In its 2007 assessment of global climate change, the IPCC identified a number of ways in which climate is changing (see the table below), and it included statements reflecting the confidence of the participating scientists that these changes (1) began in the twentieth century, (2) are caused at least in part by humans, and (3) will occur in the twenty-first century. The predictions are striking in the nature and consequences of changes that are occurring or are predicted to occur soon, as well as the degree of certainty the IPCC has about them. For example, reductions in the frequency of cold days and increases in the frequency of warm days in the twenty-first century are seen as "virtually certain" (>99 percent probability), increased frequency of heat waves and heavy precipitation events as "very likely" (>90 percent probability), and increased areas affected by droughts, increased hurricane activity, and increased incidence of extreme high sea level as "likely" (>66 percent probability).

Phenomenon and Direction of Trend	Likelihood That Trend Occurred in Late Twentieth Century (typically post-1960)	Likelihood of a Human Contribution to Observed Trend	Likelihood of Future Trends Based on Projections for Twenty-first Century
Warmer and fewer cold days and nights over most land areas	Very likely	Likely	Virtually certain
Warmer and more frequent hot days and nights over most land areas	Very likely	Likely (nights)	Virtually certain
Increased frequency of warm spells/heat waves over most land areas	Likely	More likely than not	Very likely
Increased frequency of heavy precipitation events (or proportion of total rainfall from heavy falls) over most areas	Likely	More likely than not	Very likely
Increases in the areas affected by droughts	Likely in many regions since the 1970s	More likely than not	Likely
Increased intense tropical cyclone activity	Likely in some regions since 1970	More likely than not	Likely
Increased incidence of extremely high sea level (excludes tsunamis)	Likely	More likely than not	Likely

alternatives are being discussed, from storing CO_2 in the deep ocean or underground to increasing photosynthesis by fertilizing plants in the ocean. Any of these alternatives would be expensive, and there are concerns about whether the CO_2 would leak back into the atmosphere. Would people prefer to cut back on their use of coal-generated electricity or spend money on new ways to produce electricity? Either alternative is expensive and inconvenient, and possibly ineffective.

Another approach being considered is "geoengineering," in which active attempts are made to regulate the global carbon cycle or directly alter climate. This is highly controversial. Some argue that we are unlikely to be able to reduce our carbon emissions sufficiently to slow or stop global warming, and so deliberate intervention in the climate system is our only hope of preventing further damage to the atmosphere. Others are concerned that such interventions are very likely to have unanticipated and unintended negative consequences and that the outcomes could be worse than the outcomes if we did nothing.

Chapter Review

Summary

Solar radiation received by places at a particular latitude varies with season and by day length. The annual total radiation is highest and varies least over the seasons in low latitudes, while day length and angle of incidence cause greater seasonal variations at high latitudes. Energy arrives from the Sun as shortwave energy and is reradiated by Earth as longwave radiation that heats the atmosphere. Insolation absorbed by Earth's surface and atmosphere is eventually returned to space. Human activities may be leading to global warming, especially through emission into the atmosphere of greenhouse gases such as carbon dioxide.

Precipitation forms when humid air rises and is cooled, and the water vapor in it condenses. Three processes create most precipitation: convection, orographic uplift over mountain ranges, and interaction between cold and warm air masses along fronts. Convectional precipitation dominates the tropics, while the main form of precipitation in the midlatitudes is frontal.

Wind blows from areas of high pressure to areas of low pressure, influenced by the Coriolis effect, which is a result of the rotational force of Earth. Global circulation patterns include bands of low pressure in the tropics, high pressure-feeding trade winds in the subtropics, low pressure in the midlatitudes, and high pressure at the poles.

Climate is weather and its seasonal variations averaged over time. The most important variables in defining climate are temperature and precipitation. The Köppen scheme classifies climate into five main categories: A, B, C, D, and E. Humid tropical (A) climates occur in the tropical low-pressure zone and are dominated by the intertropical convergence. Dry climates (B) predominate in the subtropics, generally on the western sides of continents, and in continental areas isolated from moisture sources. Warm midlatitude climates (C) occur in subtropical areas on the eastern sides of continents and on west coasts at higher latitudes. Cool midlatitude climates (D) occur in continental areas, mostly in the Northern Hemisphere. Polar climates (E) occur at high latitudes. A sixth category (H) is mapped in mountain areas.

Earth's climate has varied greatly over the past 2 million years. Average global temperatures have been both higher and much lower than today. Glaciers covered much of Earth's surface during cold periods, the last of which ended about 12,000 years ago. Several possible causes for these variations have been proposed, including changes in the geometry of Earth's orbit around the Sun, geologic factors such as plate tectonics and volcanic eruptions, and changes in the composition of the atmosphere, some human caused. Human-induced global warming, caused primarily by burning fossil fuels that release carbon dioxide into the atmosphere, is under way, demonstrating the key role humans play in Earth's physical processes.

Key Terms

adiabatic cooling p. 48
advection p. 46
angle of incidence p. 39
autumnal equinox p. 39
carbon dioxide (CO_2) p. 44
climate p. 37
cold front p. 51
condensation p. 47
convection p. 46
Coriolis effect p. 52

cyclone p. 57
desert climate p. 65
El Niño p. 58
front p. 49
global warming p. 44
greenhouse effect p. 44
greenhouse gases p. 44
gyre p. 57
humid continental climate p. 74
humid subtropical climate p. 71
humid tropical climate p. 66
hurricane p. 58

ice cap climate p. 76
insolation p. 39
intertropical convergence zone (ITCZ) p. 54
La Niña p. 58
latent heat p. 44
latent heat exchange p. 44
Little Ice Age p. 78
longwave energy p. 43
marine west coast climate p. 72
Mediterranean climate p. 73
methane (CH_4) p. 44

Questions for Review and Discussion

1. When do minimum and maximum daily amounts of solar radiation arriving at the top of the atmosphere occur at your latitude? How do angle of incidence and day length affect these seasonal differences?

2. What processes are responsible for the major energy exchanges between Earth's surface and the atmosphere?

3. Why, in terms of energy availability and convectional processes, do hurricanes occur (a) mostly over oceans and (b) in late summer and autumn?

4. How do climatic conditions in coastal areas differ from those in continental interiors? Why do these differences occur?

5. What are the major features of a midlatitude cyclone? What sequence of weather would one expect to observe as a midlatitude cyclone passes?

6. In what way does temperature affect the definition of a climate as humid or arid? Why?

7. For each of the 11 climate types described in this chapter, describe how the characteristics of the climate relate to the typical circulation patterns in the atmosphere.

8. Describe major variations of global average temperature that have occurred during (a) the past 1,000 years, (b) the past 10,000 years, and (c) the past 100,000 years.

9. What are the major causes of global warming? What are some of its likely effects on human activities?

Thinking Geographically

1. For a two-week period, keep a daily journal of the weather, including factors such as air temperature, wind direction, cloudiness, and precipitation. For the same period, clip the weather map from a daily newspaper. Then compare the two records.

2. In what ways do topography and/or land cover affect the weather where you live?

3. The following is a good group research activity. From your library or an online source, such as the National Climatic Data Center (**http://www.ncdc.noaa.gov**), obtain a list of daily record-high and -low temperatures for your location, since record-keeping began, if possible. Enter the data into a spreadsheet or another program that will perform sorting, and sort the records by year. What percentage of record-high temperatures occur in the second half of the period of record? What percentage of record-low temperatures occur in the second half of the period of record? Can you spot any other trends?

4. How might an increase in average annual temperature of 5°C (9°F) affect your day-to-day life?

5. Compare the map of climates with the world population map (Figure 2-33) and rear endpaper). In what climate regions are the greatest concentration of people found? Why?

Log in to www.mygeoscienceplace.com for videos, animations, **MapMaster**™ interactive maps, RSS feeds, case studies, and self-study quizzes to enhance your study of Weather and Climate.

MapMaster™

Collapsed buildings in Port-au-Prince, Haiti, following the earthquake of January 2010.

3

Landforms

The earthquake of January 12, 2010, in Haiti was the most devastating natural disaster ever to strike that nation. It had a magnitude of 7.0 on the Richter scale and an epicenter just 25 km (16 miles) from the capital city of Port-au-Prince, which had a population of about 3 million. The quake killed about 230,000 thousand people, most of whom were trapped in buildings that collapsed. Less than two months later, an earthquake with Richter magnitude 8.8 struck central Chile, killing more than 200 people and causing enormous damage in many cities. In 2008, an earthquake in Sichuan, China, with Richter magnitude 7.9 killed about 80,000, nearly leveled entire cities, and triggered huge landslides that dammed rivers and created lakes; one of the larger dams created threatened to fail and release a catastrophic flood. The powerful 9.0 magnitude earthquake created a tsunami that struck Indonesia and much of the Indian Ocean shoreline in December 2004 killing more than 150,000. These disasters illustrate that with increasing population and population densities, especially in poorer countries that are less able to protect people from these hazards and their consequences, the dynamics of Earth's surface remain a critical dimension of human–environment interactions.

Some changes on Earth's surface take place gradually, at rates of a few millimeters or less per year. Others, like the earthquakes described above, are sudden and dramatic (Figure 3-1). Whether gradual or catastrophic, changes in the Earth occur everywhere. In this chapter, we will learn about the causes and rates of such movements and how they shape mountains, hills, and valleys. We will look at Earth's rocks, soil, and surface landforms, which comprise the lithosphere, the outermost part of one of the four "spheres." (Look back at Figure 1-17.) **Geomorphology** is the study of landforms and the processes that create them.

Earthquakes and the gradual sinking of the land are newsworthy mostly because of their effects on humans. They also are notable because most of the time Earth's lithosphere appears to be fixed and unchanging relative to the other three spheres—the atmosphere, the biosphere, and the hydrosphere—which change rapidly. The atmosphere changes daily—hot to cold, wet to dry, windy to still. We can see seasonal changes in the biosphere—plants and animals appear, grow, reproduce,

and die. The hydrosphere is visibly dynamic—rivers flood and later dry up. But the lithosphere seems static. Landforms like plains, hills, and valleys do not change noticeably in a lifetime; soil is soil; rock outcrops seem permanent as monuments.

Landforms, in their remarkable variety, did not just appear at some distant time as we see them today. Landforms constantly change, although often imperceptibly slowly. Like the atmosphere, the lithosphere is driven by continual transfers of energy and matter, and interacts with the other three spheres. Some of the energy that modifies the lithosphere comes from deep within Earth. Heat contained in Earth's core influences movement on the planet's surface.

Occasionally, Earth's surface changes quickly. This reminds us that the lithosphere is quite dynamic—earthquakes shake the land, volcanoes spew forth hot gases and molten rock, and ocean waves pound the shore. Humans get in the way of these dramatic changes in landforms when they build settlements in places where earthquakes, volcanoes, and hurricanes

Figure 3-1 Earthquake damage in Sichuan, China. The upper image shows devastation in Beichuan, China, just after the May 2008 earthquake. The river stopped flowing after the earthquake because a landslide dammed the river upstream from Beichuan, creating a temporary lake. In June 2008 the water was released from the lake in a controlled flood, shown in the lower image.

are likely to occur. Because of the rarity of these events, people discount their importance and thus become vulnerable when they do strike.

Our actions have profoundly changed the lithosphere, just as they have changed the other spheres. Agricultural practices are depleting soil fertility. In semiarid areas, overgrazing by animals has produced desert-like conditions. In many agricultural areas, soil is eroding at rates 10 to 100 times greater than natural rates of erosion. Some estimates of human-caused earth moving suggest that more earth is moved by humans than by all natural processes combined.

Geographers studying the shape of Earth's surface—its topography—recognize that it includes many features that seem to have distinctive characteristics see (Look at the map inside the front cover). Elements of Earth's surface that have such identifiable form—its mountains, valleys, hills, and depressions—are called **landforms.** They are built through a combination of endogenic (internal) and exogenic (external) processes. **Endogenic processes** are forces that cause movements beneath or at Earth's surface, such as mountain building and earthquakes. These internal mechanisms move portions of Earth's surface horizontally and vertically, raising some parts and lowering others. Even as these internal forces are building Earth's features, these features are simultaneously attacked by **exogenic processes,** which are forces of erosion, such as running water, wind, and chemical actions that occur at the Earth's surface.

Endogenic and exogenic forces continually move and shape Earth's crust (Figure 3-2). Endogenic processes form rocks and move them to produce mountain ranges, ocean basins, and other topographic features. As these rocks become exposed, exogenic activities go to work. They erode materials, move them down hillslopes, and deposit them in lakes, oceans, and other low-lying areas. We will examine all these processes.

Plate Tectonics

If you were to view Earth from a great distance, as we sometimes view other planets through telescopes, the most distinctive topographic features would be the enormous mountain ranges arranged in linear patterns that extend for thousands of kilometers. Four especially prominent mountain ranges of Earth's land surface include the Rockies of North America, which continue through Central America to the Andes of South America; the Himalayas, extending across Asia; and the north–south system of mountains in eastern Africa. Large as these highly visible mountain ranges are, none rank as the world's longest. That title belongs to a mountain system beneath the oceans, the interconnecting mid-ocean ridges that are more than 64,000 kilometers (40,000 miles) long.

For millennia, people believed that Earth's continents and oceans were fixed in place for all time and

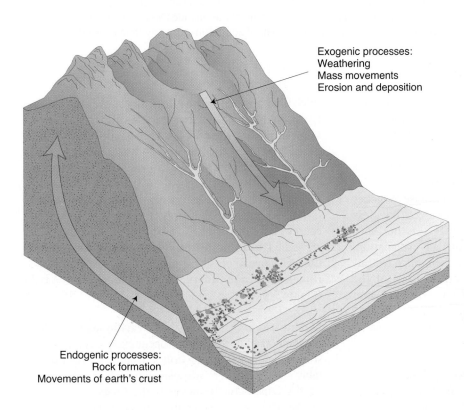

Exogenic processes:
Weathering
Mass movements
Erosion and deposition

Endogenic processes:
Rock formation
Movements of earth's crust

Figure 3-2 Endogenic and exogenic processes. The landforms around you are the product of interaction among endogenic and exogenic processes. Endogenic processes involve movement of Earth's crust. Exogenic processes wear down rocks once they reach the surface.

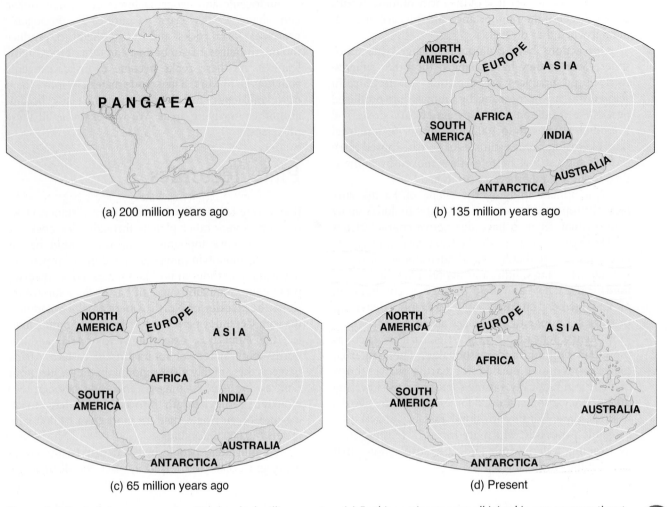

(a) 200 million years ago

(b) 135 million years ago

(c) 65 million years ago

(d) Present

Figure 3-3 Past plate movements. Two hundred million years ago (a) Earth's continents were all joined in one supercontinent known as Pangaea. This continent gradually broke apart, beginning with the opening of the Atlantic Ocean from south to north and the movement of India away from Antarctica toward Asia, and the movement of South America away from Africa (b and c). The layout of the world of today is shown in (d).

Animation
Convergent Margins: India-Asia Collision

that Earth was only a few thousand years old. This notion of a "fixed Earth" was challenged early in the twentieth century by German earth scientist Alfred Wegener. He argued that Earth's land areas once had been joined in a single "supercontinent," now known as Pangaea, and that over thousands of years the continents had moved apart (Figure 3-3). Because Wegener could not explain why the continents moved, his ideas were rejected at the time. But researchers in the 1960s vindicated Wegener by working out an explanation, called **plate tectonics theory.** Since then, plate tectonics theory has provided us with explanations of the origins of Earth's great mountain chains, volcanoes, and many other important phenomena.

Earth's Moving Crust

Earth resembles an egg with a cracked shell. Earth's crust is thin and rigid, averaging 45 kilometers (28 miles) in thickness. Beneath this rigid crust, the rock is like a very

thick fluid and is slowly deformed by movements within Earth. While far from the free-flowing substances we know as liquids, the rock just beneath the crust, known as the **mantle,** is fluid enough to move slowly along in convection currents, driven by heat within Earth's core. These currents are analogous to winds in the atmosphere, which carry heat away from Earth's surface. Geologists believe that this motion of the mantle causes pieces of Earth's rigid crust to move; those pieces are called **tectonic plates.** This is the plate tectonics theory. Movement of the plates causes earthquakes to rumble, volcanoes to erupt, and mountains to be built (Figure 3-4).

Earthquakes Thousands of **earthquakes**—sudden movements of Earth's crust—occur every day. Figure 3-5 shows major earthquake zones around the world, notably clustered where two plates meet. The place where Earth's crust actually moves is the **focus of an earthquake.** The focus is generally near the surface but can be

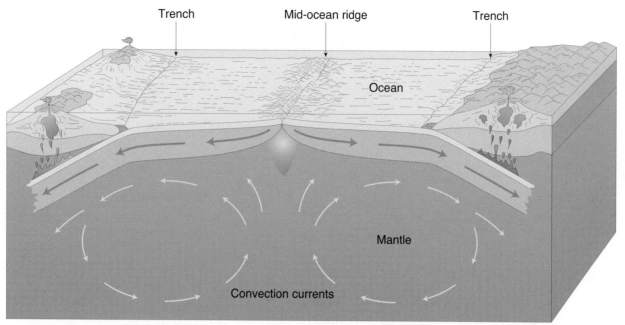

Trench   Mid-ocean ridge   Trench

Ocean

Mantle

Convection currents

Animation
Divergent
Boundary
Formation

Figure 3-4 Mechanisms of plate movement. Plate tectonics theory explains the occurrence of moving continents, mid-ocean ridges, deep-sea trenches, and earthquakes. At the center of the diagram, rock magma melted in Earth's interior rises along mid-ocean ridges, emerging to chill into new crust. Along this ridge, plates of oceanic crust are spreading apart. Where crustal plates collide (at left and right), crust is forced downward (subducted) and recycled (melted) back into the interior. This generates new magma, which migrates toward the surface, sometimes emerging as a volcano. This plate tectonic process is believed to be driven by convection currents in Earth's mantle, shown by arrows.

as deep as 600 kilometers (372 miles) below Earth's surface. The point on the surface directly above the focus is the **epicenter** (Figure 3-6). The tremendous energy released at the focus travels worldwide in all directions and at various speeds through different layers of rock.

Animation
Seismographs

Most earthquakes are too small for people to feel, and they are detectable only with a **seismograph,** which is a device that records the quake's **seismic waves, or vibrations.** Earthquake intensity is measured on a 0-to-9 logarithmic scale developed by Charles F. Richter in 1935. Earthquakes with a magnitude of 3 to 4 on the Richter scale are minor; magnitude 5 to 6 quakes can break windows and topple weak buildings; and magnitude 7 to 8 quakes are devastating killers if they affect populated areas.

Several factors in addition to an earthquake's intensity determine the damage it causes. Generally, damage is greater at places closer to the epicenter and at places built on ground that is subject to landsliding or collapse. Earthquake damage is also greater where buildings are not designed to absorb the shaking. To appreciate this, let us contrast two major earthquakes: one in California and one in Armenia. In 1989, an earthquake struck the San Francisco Bay area. The Bay Bridge buckled, freeways crumbled, and 67 people died. Some 100,000 buildings were damaged or destroyed, many of them in the Marina District, a neighborhood built on unstable fill in the San Francisco Bay. The death and destruction was limited, however, considering the quake's high magnitude of 7.1 on the

Richter scale. In contrast, in 1988, an earthquake in northwestern Armenia killed nearly 55,000 people, injured 15,000, and left at least 400,000 homeless. Although registering 6.9 on the Richter scale—lower than the California quake—and occurring in a city much smaller than San Francisco, this earthquake trapped many people inside collapsing buildings not designed to withstand earthquakes. People in less economically developed societies cannot finance the cost of earthquake-proofing their structures, as is routinely done with new buildings in quake-prone communities of wealthy societies like the United States.

Earthquake prediction is unreliable. Potential movement zones are closely monitored and computer models attempt to replicate conditions along plate boundaries. But predicting an earthquake is like trying to forecast the weather several months from now. It just cannot be done accurately with current technology.

Volcanoes Like earthquakes, volcanoes are clustered along boundaries between tectonic plates, as Figure 3-5 shows. Movement within Earth and between the plates generates **magma** (molten rock). Being less dense than the surrounding rock, magma migrates toward the surface. Some reaches the surface and erupts, and is then called **lava.** A **volcano** is the surface vent where lava emerges. The magma may flow over the surface, forming a plain of volcanic rock, or it may build up to form a mountain. The chemistry of the magma/lava determines its texture and therefore the type of landform it builds.

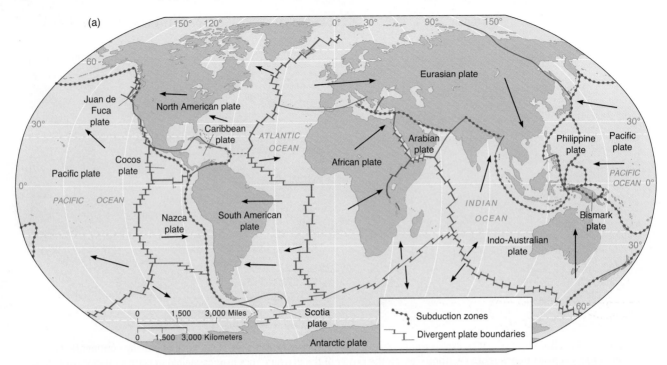

Figure 3-5 Earth's tectonic plates. (a) The major plates of Earth's crust move relative to one another, generally at rates of a few centimeters per year.

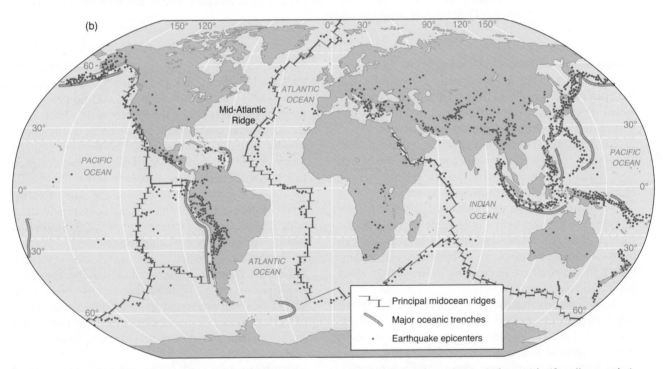

Figure 3-5 (b) These motions cause earthquakes that are concentrated along plate boundaries. Ridges with rift valleys at their centers are formed where plates are moving away from each other, generally in ocean areas. Mountain ranges are created where plates converge, sometimes with deep-ocean trenches on the seaward side of the convergence area.

The chemistry of magma/lava also determines whether the eruption is violent or relatively gradual.

Shield volcanoes erupt runny lava that cools to form a rock called *basalt*. They are called **shield volcanoes** because of their shape (Figure 3-7). Each of the Hawaiian Islands is a large shield volcano, although the only currently active ones are Mauna Loa and Kilauea, on the island of Hawaii (the "Big Island"). These generally sedate volcanoes make news on the rare occasions when they grow more active, and flows of lava threaten settlements. The mid-ocean ridges are formed of similar basaltic lava.

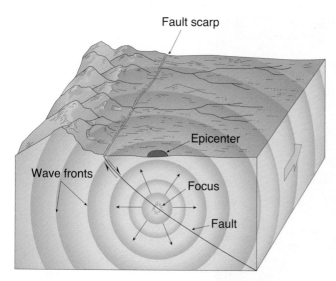

Figure 3-6 Earthquake focus and epicenter. The focus of an earthquake is where Earth's crust actually moves. The point on the surface directly above is the epicenter. Displacement of the surface creates a fault scarp.

Explosive volcanoes that cause death and destruction are more likely to be **composite cone volcanoes** (Figure 3-8). Composite cones are made up of a mixture of lava and ash. Their magma is thick and gassy, and it may erupt explosively through a vent. The eruption sends ash, glassy cinders (called *pyroclasts*), and clouds of sulfurous gas high into the atmosphere. It may also pour lethal gas clouds that are heavier than air and dangerous mudflows down the volcano's slopes. Repeated eruptions build a cone-shaped mountain, made up of a mixture of lava and ash layers.

Eruptions of composite cone volcanoes have killed tens of thousands of people at a time, but such disasters are much less frequent than severe earthquakes. One of the greatest volcanic explosions in recorded history was the 1883 eruption of the island of Krakatau in present-day Indonesia. Two-thirds of the island was destroyed, and the event killed about 36,000 people, most of whom died in a flood triggered by the eruption. Ash discharged into the atmosphere by Krakatau significantly blocked sunlight

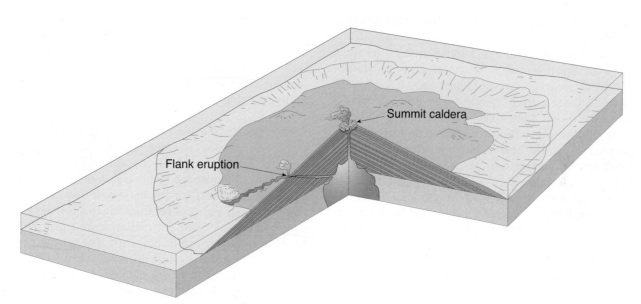

Figure 3-7 Shield volcano. Gentle shield volcanoes, exemplified by the Hawaiian Islands, are the largest volcanoes on Earth. Lava may erupt on the flank of the volcano or from the craterlike caldera at the top of the mountain.

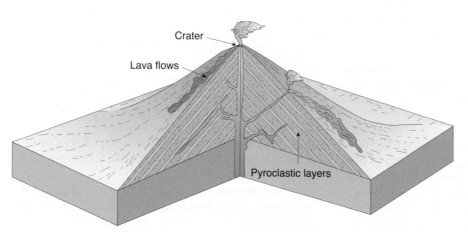

Figure 3-8 Composite cone volcano. Composite cone volcanoes, composed of alternating lava and ash, or pyroclastic layers, are relatively explosive.

Figure 3-9 Eruption of undersea Volcano This eruption occurred near Tonga in the South Pacific in March 2009.

and caused noticeable cooling of Earth's climate for a couple of years.

Thousands of volcanoes stand dormant (inactive, but with the potential to erupt) around the world. Many volcanic eruptions occur under water (Figure 3-9). About 600 are actively spewing lava, ash, and gas—some daily—but they rarely cause damage. Others, however, have not erupted in hundreds of years, so people have settled nearby. In some areas earthquake watch centers provide warnings of volcanic eruptions. But when warnings are not available, the danger can be great. The 1985 eruption of Nevado del Ruiz, in Colombia, triggered giant mudslides that buried most of a town, killing 23,000 people. In general, predicting volcanic eruptions is more accurate than predicting earthquakes, because volcanoes give many warnings before erupting.

Types of Boundaries Between Plates

Three types of boundaries form between moving plates of Earth's crust. The type of boundary depends on whether the plates are spreading apart, pushing into each other, or grinding past each other.

Divergent plate boundaries   A boundary where plates are spreading apart is a **divergent plate boundary.** People seldom are aware of plates spreading apart, first because it happens at a rate of only a few centimeters per year (2.54 centimeters = 1 inch) and, second, because it happens mostly deep in mid-ocean. The Mid-Atlantic ridge is a well-known example. Divergent plate boundaries occur on land, too. The rift valleys of East Africa are an example. These valleys are hundreds of meters deep and extend thousands of kilometers from Mozambique in the south to the Red Sea in the north.

Where two plates are diverging on the seafloor, a phenomenon called **seafloor spreading,** lava continually erupts. Rapidly chilled by seawater, it solidifies to form a new seafloor crust. This is how much of the seafloor crust forms, by very slowly spreading from mid-ocean ridges.

Convergent plate boundaries   A boundary where plates push together is a **convergent plate boundary.** Material from one plate—the subducting plate—is slowly forced downward by the collision, back into the mantle. Because seafloor crust is denser than continental crust, when a plate of continental crust collides with a plate of oceanic crust, the denser oceanic plate sinks beneath the lighter continental crust. The oceanic plate is carried into Earth's mantle, where some of it is remelted. This magma then migrates toward the surface, causing volcanic eruptions at sites above the plunging plate. This occurs, for example, to the south and southwest of Indonesia where the Eurasian and Indo-Australian plates converge.

Transform plate boundaries   A boundary where the plates neither converge nor diverge, but grind past each other, is a **transform plate boundary.** California's San Andreas Fault is an example. Along this fault, the Pacific Plate is moving northwest relative to the North American Plate. The boundary between these plates is not a smooth one, and ridges and mountains are built as the two plates grind against one another. The plates bind for long periods and then abruptly slip, causing the earthquakes that frequently strike California.

Vertical movements of Earth's crust   Parts of the crust move vertically as well as horizontally. As two plates collide, material may be forced downward into Earth's interior or upward to form mountains. Over millions of years, vertical movements along plate boundaries produce mountain ranges thousands of meters high.

Vertical movement of crust also occurs because the crust "floats" on the underlying mantle, much like a boat floating in water. If material is added to the crust, it sinks, and if material is removed, it rises. Deposition of sediment or accumulation of ice in glaciers can cause the crust to sink. These vertical movements caused by loading or unloading the crust are called **isostatic adjustments**. Removal of material by erosion or melting of glaciers allows the crust to adjust isostatically, or "rebound." In some places, crust that was buried under continental ice sheets has risen vertically over 100 meters (330 feet) in the past 15,000 years because of glacial melting.

Rock Formation

Although by human standards Earth's surface moves very slowly—by at most a few centimeters per year—this movement produces Earth's great diversity of rocks. As Earth's crust moves, its materials are eroded and deposited, heated and cooled, buried and exposed.

Types of rocks   Rocks can be grouped into three basic categories that reflect how they form.

Igneous rocks are formed when molten crustal material cools and solidifies. The name derives from the Greek word for *fire*, which is the same root as for the English word *ignite*. Examples of igneous rocks are basalt, which is common in volcanic areas, including much of the ocean floor, and granite, which is common in continental areas.

Sedimentary rocks result when rocks eroded from higher elevations (mountains, hills, plains) accumulate at lower elevations (like swamps and ocean bottoms). When subjected to high pressure and the presence of cementing materials to bind their grains together, rocks like sandstone, shale, conglomerate, and limestone are formed.

Metamorphic rocks are created when rocks are exposed to great pressure and heat, altering them into more compact, crystalline rocks. In Greek, the name means "to change form." Examples include marble (which metamorphosed from limestone) and slate (which metamorphosed from shale).

Minerals   Minerals are natural substances that comprise rocks. Each type of mineral has specific chemical and crystalline properties. Earth's rocks are diverse in part because the crust contains thousands of minerals. The density of rocks depends on the kinds of materials they contain. Denser rocks are dominated by compounds of silicon, magnesium, and iron minerals; they are called **sima** (for *si*licon-*ma*gnesium). Less dense rocks are dominated by compounds of silicon and aluminum minerals; they are called **sial** (for *si*licon-*al*uminum).

Denser sima rocks make up much of the oceanic crust. Less dense sial rocks make up much of the continental crust. The lower density and greater thickness of sial rocks cause the continents to have higher surface elevations than the oceanic crust, just as a less dense dry log will float higher in water than a denser wet one.

The formation and distribution of many minerals is caused by the movements of Earth's crust. Vast areas of the continental crust, known as **shields**, have not been significantly eroded or changed for millions of years (Figure 3-10).

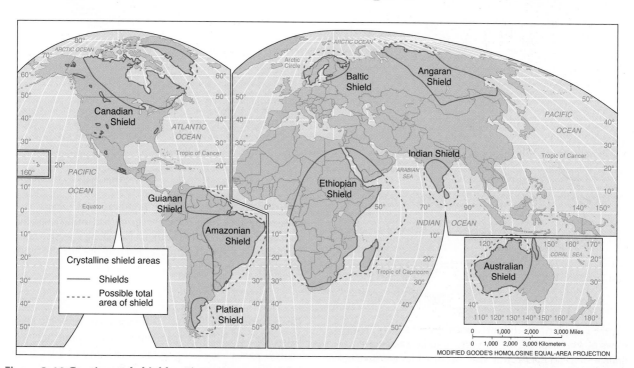

Figure 3-10 Continental shields. The ancient cores of the continents are exposed in several areas of Earth's surface. The rocks exposed in these areas are typically rich in mineral ores.

Shield areas often contain rich concentrations of minerals, such as metal ores and fossil fuels. Shields are located in the core of large continents such as Africa, Asia, and North America. Many of the world's mining districts exist where these continental shields are exposed at the surface.

Stress on rocks   Crustal movements along plate boundaries exert tremendous stress on rocks. Despite their rigidity, rocks bend and fold. When stressed far enough, they fracture along cracks called **faults**. The fractured pieces may then be transported to new locations. Fracturing takes place in different ways, depending on the type of boundary. Near a divergent plate boundary, rocks break apart because they are stretched; the resulting fracture is called a *normal fault* (Figures 3-11 and 3-12). Near a convergent plate boundary, rocks fracture because they are compressed; such fractures are called *reverse faults* (because they are the opposite of normal faults), or *thrust faults* if there is large horizontal movement (Figure 3-13). Alternatively, the crust may rumple like a rug, creating folds. The Appalachian Mountains and the Himalayas are examples of mountain ranges created by faulting and folding that happened along convergent plate boundaries. Faulting also occurs along transform boundaries. Rock movement near a transform boundary is mostly parallel to a plate boundary rather than perpendicular, as in the other two types of boundaries.

Figure 3-12 The Wasatch Fault. The Wasatch Mountains in Utah rise abruptly from the Salt Lake valley along the Wasatch Fault, an active normal fault just east of Salt Lake City.

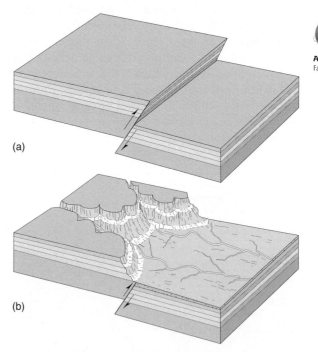

Animation
Faults

(a)

(b)

Figure 3-13 A reverse fault. (a) Rocks break when stressed by forces that compress them. (b) Erosion alters the surface of rocks that have been uplifted along a reverse fault. Later erosion shapes the surface, and may obscure the actual orientation of the fault.

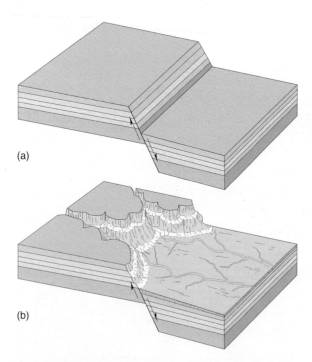

(a)

(b)

Figure 3-11 A normal fault. The diagrams show cutaway views of what happens beneath Earth's surface along a normal fault. (a) Rocks break apart because they are stretched. (b) Erosion alters the block that has been uplifted along a normal fault.

Rocks and landforms   Differences in geologic structures and rock types from one place to another are a critical part of the geographic variability of Earth's surface. These geologic features influence the surface in three different ways. (1) Movement of the crust such as

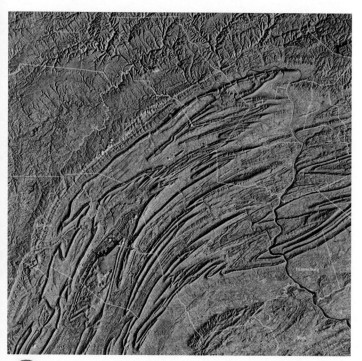

Animation
Folds

Animation
Erosion of Deformed Sedimentary Rock

Figure 3-14 Satellite image of the central Appalachians in Pennsylvania. The winding ridges are formed by resistant layers in folded sedimentary rocks, exposed by erosion of weaker layers that now occur in valleys.

that along faults creates landforms like those illustrated in Figures 3-11, 3-12, and 3-13. (2) Variations in the resistance of rocks to exogenic processes cause weak rocks to be removed more rapidly while more resistant rocks remain in place. This creates **structural landforms**, in which the shape of the land reflects the underlying rock structures. The ridges and valleys of the central Appalachians are formed in this way (Figure 3-14). The ridges are formed by resistant layers, often sandstone and conglomerate, whereas the valleys are underlain by weaker materials such as shale. (3) Differences in the minerals contained in rocks affect the characteristics of soils that form on them. Soils overlying limestone, for example, tend to be rich in calcium, a major component of limestone. The following section describes the exogenic processes that help to expose these geologic influences on the landscape.

Slopes and Streams

Plate tectonics theory helps us to understand the causes of mountain-building motions in Earth's crust. But landforms created through endogenic processes are attacked by exogenic processes as they are being formed. The wearing down of Earth's crust through exogenic processes reshapes Earth's crust into new landforms. Exogenic processes shape Earth's surface in two principal steps: Rocks are first broken down into smaller

pieces through weathering, then they are carried by gravity down the slopes of hills or are transported by water, wind, or ice from one place to another.

Weathering

Weathering is the process of breaking rocks into pieces ranging in size from boulders to pebbles, sand grains, and silt down to microscopic clay particles and dissolved solids. Without weathering, the force of gravity and the agents of water, wind, and ice would have nothing to move. Rocks begin to break down the moment they are exposed to the weather at Earth's surface. They are attacked by water, oxygen, carbon dioxide, and temperature fluctuations. Weathering is the first step in the formation of soil, which will be discussed in more detail in Chapter 4. Weathering takes place in two ways: as chemical weathering and as mechanical weathering.

Chemical weathering   Rocks may be broken down as a result of **chemical weathering**, which is a change in the elements that compose rocks when they are exposed to air and water (Figure 3-15). Chemical weathering occurs faster in places with warm temperatures and abundant water, such as in humid tropical environments. Acids released by decaying vegetation also chemically weather rocks. Some of the dissolved products of chemical weathering are carried away by water seeping through soil and rocks, which is a process called *leaching*. The water eventually may carry these materials to rivers and then to the sea. This is the source of the salinity (dissolved salt) of the oceans.

Figure 3-15 Weathering. Granite exposed along the coastline of Nova Scotia reveals the effects of weathering on rocks. Water has penetrated the rock and weakened the bonds between grains, allowing the rock to be easily broken open. Freezing and thawing along with salt spray from the sea have contributed to the weathering.

Animation
Physical Weathering

One example of chemical weathering is oxidation. Iron is a common element in rocks, and it combines with oxygen in the air to form iron oxide, or rust. Iron oxide has very different properties from the original iron: It is physically weaker and more easily eroded. You can see the effects of oxidation on iron or steel surfaces exposed to the weather; the rusty oxide easily flakes away.

Another example of chemical weathering is the decomposition of calcium carbonate. It is a major component of limestone and other sedimentary rocks. Calcium carbonate dissolves in water, separating into ions of calcium and carbonate that are carried by streams into the sea. In some areas of limestone bedrock, underground water may remove large quantities of rock via chemical weathering beneath the surface. Water flowing underground dissolves passageways and even carves out large caverns in the limestone. If the caverns collapse, they create depressions called sinkholes at the surface. Such underground erosion produces a distinctive form of topography called **karst,** named after a region in Croatia with this type of landscape. Karst topography is found in many parts of the world, including the Caribbean (especially Puerto Rico, Cuba, and Jamaica), several parts of the southeastern United States (especially Florida, Kentucky, and Missouri), and southeastern China.

Mechanical weathering Rocks are also broken down by physical force. This process is called **mechanical weathering**. Rocks expand and contract with frequent changes in temperature, and this action causes them to break apart. Highway potholes in the northern United States and Canada illustrate these processes. Rainwater seeps into roadway cracks and freezes into ice crystals when the temperature turns colder. The water, which expands about 9 percent when frozen, pushes apart the pavement and opens up the potholes in a phenomenon called frost-wedging. Plant roots growing in cracks between rocks also contribute to mechanical weathering; you probably have observed sidewalks that have been heaved by tree roots.

Mechanical and chemical weathering work together to break down rocks. Often, mechanical forces open cracks, and water seeps in to weather the rock chemically.

Moving Weathered Material

Once rocks are weathered, they may be carried from one place to another. Material most commonly moves downhill by gravity. This happens in two ways: by mass movement or by surface erosion. In **mass movement**, rocks roll, slide, or freefall downhill under the steady pull of gravity. In **surface erosion**, water—which flows downhill because of gravity—carries solid rock particles with it. Surface erosion may also result when wind or ice carries material from one place to another.

Material moves faster down steeper hills than down gentler ones. The steepness of a hill is measured through its slope, which is the difference in the elevation between two points (known as the *relief*, or *rise*) divided by the horizontal distance between the two points (known as the *run*). The greater the rise and shorter the run between the two points, the faster the movement of materials down the hill. Wherever slopes occur, gravity is available to move material. Even the gentlest slope provides the potential energy necessary to move at least some material downward, either through mass movement or surface erosion. Erosion is usually much more rapid, however, on steep slopes of land than on gentle slopes.

Mass movement The most common form of mass movement is **soil creep**. As the name suggests, creep is a very slow, gradual movement of material down the slope of a hill (Figure 3-16). A tiny movement can cause creep—a rodent digging, a worm burrowing, an insect pushing aside soil. Creep occurs near the surface, in the top 1 to 3 meters (3 to 10 feet) of soil.

More dangerous and dramatic mass movements, such as rock slides and mudflows, can occur on steep slopes, especially during wet conditions. Steep slopes are prone to rock slides because the force of gravity pushing

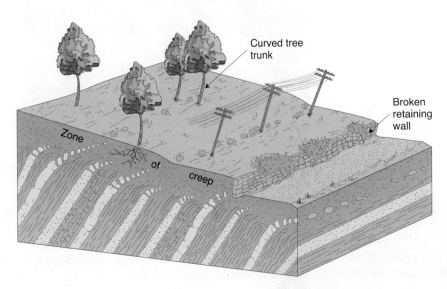

Figure 3-16 Soil creep. Soil creep is the most common form of mass movement. Tilted trees and other objects are often indicators of creep.

Figure 3-17 Landslide. This landslide occurred on a steep slope on Santa Cruz Island, California, following heavy rain. The heavy rain weakened the rocks, triggering the landslide. Small scars near the top of the landslide reveal fractured surfaces in the rock. Once the weathered rock began to slide downhill, it was jostled and broken apart by the motion and turned to mud. By the time it reached the bottom of the slope (the foreground in the photo), the landslide had become a mudflow.

down on the rocks is likely to exceed the strength of the rocks. Landslides on steep slopes can follow intense rains, because material with a high water content is heavier, weaker, and less able to resist the force of gravity. The sliding material may break down into fluid mud, which flows downhill (Figure 3-17). Houses built on very steep slopes, such as along the West Coast of North America, risk damage from landslides and mudflows.

Surface erosion The most common form of surface erosion is caused by rainfall. Intense rain sometimes falls faster than soil can absorb it. Water that cannot infiltrate, or soak into the ground, must run off the surface. As it runs off the surface, water picks up soil particles and carries them down the slope. With enough runoff, water can carve channels into the landforms.

The smallest channels eroded by the flow of water—only a few centimeters deep—are called *rills*. Rills are so small that soil creep or a farmer's plow can obliterate them. If channels gather enough water, however, they become larger and permanent carriers of water. As these stream channels deepen, they gather water and eroded soil from adjacent slopes. When the streams gather enough water, they form ravines, valleys, and canyons.

Surface erosion by water is relatively slow in most natural environments because the ground is covered by grass and trees. But on large parts of Earth's surface, humans have removed vegetation by clearing forests and plowing fields. Once the vegetative ground cover is removed, slow surface erosion can suddenly become severe erosion (Figure 3-18). Ground where vegetation has been removed can suffer more surface erosion in a few months than it experienced during the previous several thousand years. The eroded soil contributes to water pollution downstream, and the remaining soil may be less productive for agriculture.

Stream drainage Streams collect water from two sources: groundwater and overland flow (Figure 3-19). When rain falls on the land surface, most of it soaks into the soil and accumulates as **groundwater**. Groundwater migrates slowly through the soil and underlying rocks. During dry periods, most of the water flowing into

Figure 3-18 Soil erosion in Brazil. Increasing demand for food and energy products—in this case sugar cane likely used for ethanol production—is leading to increases in land under cultivation and thus exposed to erosion.

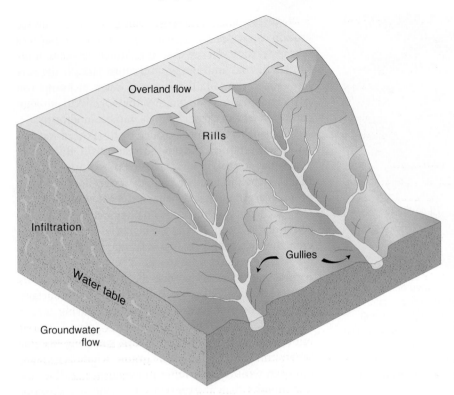

Figure 3-19 Sources of stream-flow. Surface water (overland flow) and groundwater (water below the water table) are the two sources of water for streams. Under the pull of gravity, rainwater infiltrates (soaks into) the soil. The water in the ground slowly migrates through soil and rock. The surface flow follows the lowest path available, sometimes forming small channels called gullies, and eventually running into streams.

streams is supplied not by rainfall but by groundwater. If rain falls intensely, the soil may not be able to absorb it as fast as it falls. Water then may run directly to streams through **overland flow**.

A stream drains groundwater and overland flow from an area called its **drainage basin**. Drainage basins may be as small as a farm field or as large as a major portion of a continent. (Smaller basins are nested within

CONNECTIONS

Wealth and Natural Hazards

When we hear about natural hazards such as wildfire, earthquake, landslides, and coastal erosion, one place that often appears in the headlines is Los Angeles, which seems to have more than the typical share of such calamities. Dramatic scenes of houses toppled by landslides are splashed across the TV screens, and often the homes are very expensive ones. Some of the most hazardous neighborhoods in Los Angeles are also among the most desired. The hills overlooking Hollywood and Malibu and the Palos Verdes Peninsula, with its spectacular ocean views, are examples. The steep slopes scattered on mountainsides around Los Angeles offer isolation from city traffic and spectacular views—at least on clear days. But the steep slopes are also the places most vulnerable to important hazards such as wildfire and landsliding caused by winter rains. In contrast, the poorest neighborhoods in Los Angeles are generally flat and far removed from such hazards. South-central Los Angeles is an example. It does seem that in Los Angeles the rich people have chosen to live in the more hazardous environments. This may cause an increase in the property damages that occur when disasters strike, because luxury homes cost more than lower-income housing.

Los Angeles is just one city, and a unique one at that. But what about other cities? In some Latin American cities, the wealthier neighborhoods tend to be closer to the city center rather than on the outskirts. In the case of coastal cities like Rio de Janeiro, Brazil, the wealthy tend to live on lowlands, while many poor people live in shantytowns on steep hillsides. When landslides happen there, the poor may be more vulnerable than the rich. The 2010 earthquake in Haiti seems to have affected everyone in the area, regardless of wealth or class, but also demonstrates that a country with limited economic resources is much more vulnerable to the after-effects of a disaster than is one with greater resources and effective means to deliver them to those in need.

Because every place is unique, it is difficult to make generalizations about the relation between wealth and vulnerability to hazards. Most people don't think about such hazards when they choose a place to live; other considerations are usually much more important. When we hear about disasters triggered by earthquakes, floods, or other natural hazards, however, we can make one generalization. The outcomes in poor countries are often very different from outcomes of similar events in rich countries.

larger basins.) The greater the area of its drainage basin, the more water a stream must carry. Rivers in dry areas may be exceptions because they may lose water to evaporation. A basin with plentiful runoff from groundwater and overland flow might carve a complex network of many channels to remove the water and sediment. Small rills deliver water and material to larger streams, which join others to form still larger rivers, which flow to the sea.

The volume of water that a stream carries per unit of time is its **discharge**. Discharge ranges from a few cubic centimeters per second in rills to over 200,000 cubic meters (7 million cubic feet) per second at the mouth of the Amazon River, the world's largest. Discharge of any stream usually increases after storms and decreases during dry spells.

Drainage density is the combined length of all of the stream channels in a basin, divided by the area of the drainage basin. A basin that has soil capable of absorbing and storing most of the rain will usually have a low drainage density. Landscapes with soils that cannot absorb rainfall very rapidly are more easily eroded to form channels and tend to have higher drainage densities.

Streams shape their channels by alternately eroding and depositing material on their beds and banks. The turbulent, swirling motion of the water particles form the channel. The water transports these particles, along with loose sediment in the channel and minerals dissolved in the water. This movement of material in

a stream is called **sediment transport**. The amount of sediment a stream carries increases as discharge increases, and larger streams typically carry more sediment than do smaller ones. Transport also increases greatly after a heavy rain. During periods of lower flow, less sediment is transported downstream, possibly causing a channel to partially fill with sediment.

A stream is also responsible for shaping its **floodplain,** which is a nearly level surface at the bottom of the valley through which the stream is flowing (Figure 3-20). The surface of the floodplain is formed from deposits of sediment where the stream periodically floods. Flowing water tends to **meander**, or change direction from side to side, which contributes to widening the channel. The channel continually shifts from side to side as the stream erodes material from one side of the channel, where the current is swifter, and deposits it on the other side, where the slower current has less energy (Figures 3-21 and 3-22).

By continually eroding and depositing material in channels and floodplains, streams tend toward a stable condition, known as their **grade**. A graded stream transports exactly as much sediment as it has collected (Figure 3-23). Streams rarely operate at a condition of grade for long, because daily changes in weather and disturbances from erosion and human activities continually upset the balance. As the stream's stable condition is upset and the transport of sediment increases or decreases, the shape of the stream channel may change.

Figure 3-20 Floodplain of the La Lima. This image shows the floodplain of a stream passing through La Lima, Honduras. The map was constructed using LiDAR (light detection and ranging), a new technology that allows us to generate highly accurate topographic maps. The light blue areas are expected to be inundated by up to 1 meter (3.3 feet) of water in a flood that would occur on average once in 50 years; the medium blue areas would be inundated by 1 to 2 meters (3.3 to 6.6 feet) of water, and the dark blue areas more than 2 meters (more than 6.6 feet).

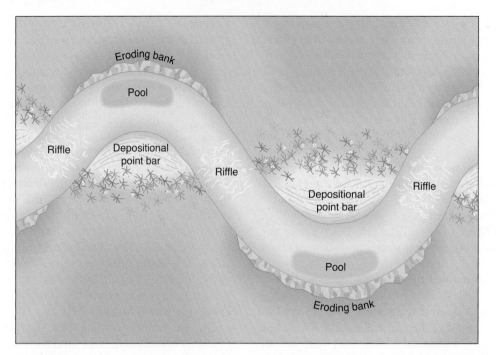

Animation
Meandering
Streams

Figure 3-21 Features of a meandering channel. Erosion occurs primarily on the outsides of bends while deposition occurs on the insides. Depth usually alternates between deep portions or pools and shallows or riffles.

An especially heavy flow following a storm may cause the channel to shift. When increased erosion upstream generates more sediment than a stream can carry, the excess is deposited in the channel or on the floodplain (Figure 3-24).

The deposition of excess sediment slowly raises the elevation of a stream, which in turn reduces the difference in elevation between places upstream and downstream, and also reduces the stream's slope. Lowering the slope reduces the amount of sediment arriving from upstream.

Most sediment carried by a stream does not move to its final resting place in a single step. Typically, sediment becomes temporarily stored in a floodplain, then eroded, transported, and deposited a second time, then a third time, and so on for many times along its journey. Eventually, most sediment reaches the sea (Figure 3-25). Where a river enters the sea, the water velocity drops abruptly, and the sediment may form a large area of deposited sediment called a **delta**.

Rainfall also is an important erosional agent in dry areas, even though this is rare. Because deserts lack

Figure 3-22 The floodplain of the Mississippi River near Herculaneum, Missouri, downstream from St. Louis. Most of the floodplain is on the east side of the river in this area and is clearly visible in the pattern of agricultural fields taking advantage of the rich alluvial soil. Most of the floodplain visible in this image was inundated in the 1993 flood.

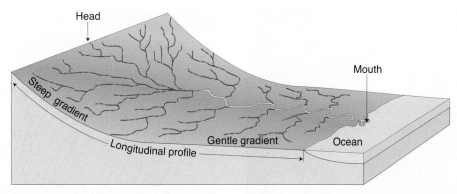

Figure 3-23 An idealized profile of a stream and its numerous tributaries. Note the concave-upward shape of the profile, with steeper gradient toward the heads of streams and gentler gradient downstream. As the stream flows downhill, it gathers more and more water, and its erosive power is greatly increased. The gentler gradient downstream slows the flow, balancing the erosive power of the stream with the amount of sediment it must carry.

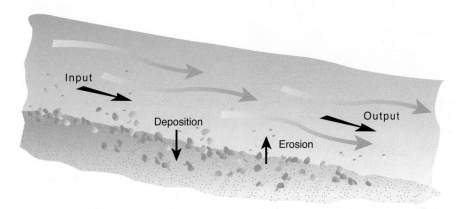

Animation
Sediment
Transport
by Streams

Figure 3-24 Stream erosion. Streams erode and deposit material in response to variations in streamflow and its ability to transport sediment. Sediment flows in and out of a portion of the channel. If more sediment enters than leaves, the excess is deposited, and the channel shrinks or the floodplain accumulates sediment. On the other hand, if more sediment leaves than enters this portion of the channel, the channel erodes and enlarges.

vegetation, the occasional rainfalls cause rapid erosion. Unvegetated soils are much more prone to erosion than are vegetated ones—look at any construction site after a heavy storm to see small channels and muddy runoff. Thus, a dry landscape often is covered with gullies and dry stream channels, even though only a few centimeters of rain may fall per year (Figure 3-26).

A prominent landform in many deserts is the **alluvial fan**. Alluvial fans are broad, gently sloping deposits formed from sand and gravel where a fast-moving stream emerges from a narrow canyon onto a broad valley floor (Figure 3-27). When the stream emerges, the water quickly spreads out and loses velocity. Given the limited amount and frequency of rainfall, a stream in a desert is unlikely to carry sediment very far. Instead, sediment that the stream can no longer carry is deposited in an alluvial fan. Occasional floods crossing the alluvial fan can disturb the pattern and encourage wind erosion. Large depositional areas in desert valleys form important sources of wind-eroded sediment.

Increased erosion from human activity In addition to shaping landforms over millions of years, erosional

forces are important in human time scales because human modification of Earth's surface is usually much more rapid than that occurring over geologic time. Human activities, such as deforestation, agriculture, and urban development, sharply increase the amount of sediment being eroded into streams.

People clear forests because they want to use trees for fuel, lumber, and paper, or they want to use the land for another purpose, especially agriculture and urban development. Erosion has increased as a result of the elimination of the vegetation cover for both reasons, but it is particularly severe where agriculture has replaced forest.

For much of Earth's agricultural land, erosion of soil into streams is a major loss. To meet the needs of a growing population, food production expands in two principal ways—by opening up new land for agriculture and by using existing farmland more intensively. But both strategies can result in erosion of the rich soil necessary for productive agriculture.

In places with rapid population growth, new lands may be opened for agriculture that may not be suited for intensive farming. For example, farmers in East

Figure 3-25 A MODIS view of the mouth of the Yangtze River, China. The Yangtze enters from the west (left). The Yangtze carries an enormous sediment load, and construction of the Three Gorges Dam will reduce this flow of sediment to the sea, with unknown consequences for the coastal region.

Africa are clearing, tilling, and planting extremely steep mountain slopes. The rate of surface lowering by erosion in such areas can be as much as a few millimeters per year. Within a few years, the rich topsoil may be completely eroded in such areas. The consequences

Figure 3-26 Erosion by water in the desert. Erosion by the San Juan River has exposed folded sedimentary rocks in southeastern Utah.

Figure 3-27 Alluvial fan. Death Valley, California

of this erosion for food production will be discussed in Chapter 8.

Opening up new land for agriculture was a major contributor to increased erosion in the United States during the eighteenth and nineteenth centuries. More than 2 million square kilometers (800,000 square miles) of forest were cleared and replaced with plowed fields and pastures in the eastern and midwestern United States. This deforestation probably increased the rate of soil erosion by 10 to 100 times.

Erosion of agricultural land is further increasing as farmers use existing fields more intensively. The soil on about one-fifth of U.S. cropland is being lost faster than it can be replaced by natural soil formation. Also, to obtain higher yields, farmers plant profitable crops like corn, soybeans, and wheat every year, instead of periodically planting cover crops such as clover or alfalfa that restore nutrients to the soil. By failing to restore nutrients to the soil, farmers can exhaust its productivity.

Soil erosion lessened in much of the eastern United States during the twentieth century because farmland was abandoned. Its soil was depleted of nutrients, and more productive land was available in the Midwest. Much of the former farmland in the eastern United States has returned to forest, so erosion is less severe there today than it was in the past. When the prairies west of the Mississippi were brought under the plow in the late 1800s, erosion became a problem there, reaching crisis proportions in the 1920s. Since the 1930s, government-sponsored soil conservation measures have significantly reduced erosion on U.S. farms, but the problem remains serious in some areas.

A heavy rainfall or rapid snowmelt may dump an overload of eroded sediment into a stream, which may be deposited on the floodplain downstream. Such sediment can bury nutrient-filled soils already in the floodplain, as can roads and buildings constructed on the floodplain. Flooding may increase if a sediment-choked stream overflows its banks.

Even after the rate of soil erosion into a stream declines or the sediment supply from upland is reduced by dams, a large quantity of sediment from the valley bottom may continue to be sent downstream. Many U.S. rivers have high sediment loads in part because they are currently excavating sediment from their floodplains that was deposited during a past time of severe agricultural erosion.

Urban development also increases erosion. Vegetation is cleared from land for building new houses, factories, and shops. Erosion rates during the construction phase often are hundreds of times greater than for undisturbed land. Evidence of this erosion is usually clear right after a storm—small channels that deepen downslope and mud washed onto adjoining areas, for example. If construction proceeds slowly, the land may be subject to high rates of erosion for several years.

Once land has been developed, surfaces once covered with vegetation are covered instead with roofs, streets, and parking lots. Because rain does not soak into these nonporous surfaces, it must run off into stream channels. Sewers collect storm water in cities and often discharge it into a stream at a rapid rate. If the sewers are unable to handle the flow during a heavy rainfall, low-lying areas in the city may flood. To handle increased discharge, streams in urban areas may enlarge their channels through erosion of their banks and damage adjacent property.

Ice, Wind, and Waves

In some parts of the world, running water is a less powerful erosional agent than other forces. In some cold places, for example, the land is covered with ice that flows downhill, grinding away at rocks. The wind can cause surface erosion in places with sparse vegetation, especially in the world's extensive deserts. Wind is a much less powerful erosional agent than running water because wind can carry only smaller particles up to the size of sand, and it can carry these only short distances. In coastal areas, a combination of water and wind can cause surface erosion. Waves, driven by winds, pound the shore with turbulent water. Even the strongest rocks and vegetation are eventually broken down by the continual rushing forth and back of the waves.

Glaciers

Greenland, the South Pole, and many high mountain areas are currently covered with thick layers of moving ice, called **glaciers**. **Alpine glaciers** form wherever snow accumulates year after year without melting, such as on the peaks of individual mountains.

Continental glaciers over 3 kilometers (almost 2 miles) thick cover vast areas of Greenland and Antarctica.

Glaciers are rivers of ice that flow from places where snow accumulates yearly to warmer places where the ice melts. Water enters the head of the glacier as snow (Figure 3-28). The glacier flows downhill until the ice eventually melts and leaves the glacier at its terminus as meltwater or icebergs.

Glaciers flow very slowly, usually at rates of a few meters to a few hundred meters per year. Glaciers may change size from one year to the next, depending on variations in weather and climate. They grow when they receive more snowfall at their head, and they shrink if warm temperatures increase the melting at the terminus. For the past 200 years, most of the glaciers of the world have been shrinking as a result of climatic warming (Figure 3-29). Glaciers can also act erratically. For many years, a glacier may move only a few meters per year, then for a few years it may suddenly surge forward at several hundred meters (hundreds to thousands of feet) per year, before slowing again.

Figure 3-28 Alaskan glacier. In this view of the Chugach Mountains, the darker-toned ice in the foreground is older, covered with sediment left behind as the glacier melts. The bright white surface in the distance is snow less than a year old.

New Orleans: Rising Sea Level, Hurricanes, and Coastal Vulnerability

On August 29, 2005, one of America's greatest cities was hit by a hurricane that broke the levees that had protected the city from the Mississippi River. Hurricane Katrina killed 1,800 people, caused $81 billion of damage, and destroyed of 350,000 homes and 30 oil platforms. The city population dropped by half; by late 2009, it had recovered to about 70 percent of its pre-storm level, but it may never recover completely. This tragedy resulted from diverse factors of both human and natural origin.

Despite its location in the flat, tectonically stable interior of North America, the lower Mississippi River is a remarkably dynamic environment. This tendency for rapid change, aggravated by human activity, causes much of the land to be low in elevation and reduces the protection normally provided by coastal wetlands, two factors that contribute to flood hazards in that region. New Orleans lies in the midst of this unstable environment, and because it is a major city, the options for adapting to instability are limited. In 2005, we saw the consequences of that unfortunate combination of instability and human activity.

The Mississippi River drains nearly 3 million square kilometers (1.25 million square miles), which is nearly one-third of the United States. Its vast floodplain, 100 kilometers (60 miles) wide in some parts, is the resting place of sediment eroded from the Rocky Mountains in the west, the Appalachians in the east, and most of the land between. Ultimately, sediment is carried to the mouth of the river, where it accumulates in the Mississippi Delta. Deposition of sediment each year extends the delta further out into the Gulf of Mexico. At the same time, sediment in the delta settles and is compacted under its own weight. Isostatic depression of the crust also causes the delta to sink slowly over time. This means that while global sea level is rising, in areas where the land is sinking the effect is even greater. In the New Orleans area, the sea is rising relative to the land at an average rate of about 10 millimeters (0.4 inch) per year. The site of the city was established in 1718, and in the three centuries since then, much of the city has sunk considerably. This is why so much of the city is 2 to 3 meters (7 to 10 feet) below sea level today and thus is vulnerable to flooding. The rate at which the land is sinking is highly variable from place to place, and in some areas, the land has sunk rapidly. For example, some of the flood walls in New Orleans were originally built in the 1960s to a height of 4.57 meters (15 feet) above mean sea level, but by 2005, they had sunk to a little more than 3.66 meters (12 feet) above sea level.

The subsidence problem is compounded by the fact that over the past several decades, thousands of dams have been built on streams throughout the Mississippi basin (see Figure 1-30). These dams trap sediment that would otherwise be delivered to the Mississippi River and ultimately to the delta. Therefore, the supply of sediment that, under natural conditions, would more than adequately replace the land lost to subsidence is no longer available. Thousands of acres of coastal Louisiana wetlands are lost each year because the land is sinking and not being replaced by new deltaic sedimentation. Because wetland vegetation can slow down water flow, the loss of wetlands allows storm surges to penetrate much further inland than they would otherwise, which, again, contributes to the flood hazard in that region.

For these reasons, geographers, urban planners, engineers, and others have long been concerned about the vulnerability of New Orleans to a major hurricane. But despite many warnings, little was done to reduce this vulnerability. After all, what

---

A glacier is like a conveyor belt because it picks up sediment from areas of erosion and drops it in depositional areas. As ice accumulates and begins to flow, the glacier picks up more material, and where the glacier melts, sediment is deposited. The accumulation area is the erosion site and the melt area is the deposition site. Glacial deposits play an especially important role in shaping landforms in places with rapid melting because large quantities of material are dropped in these places, forming **moraines** (Figure 3-30). A **terminal moraine** is a ridge of material dumped at the end of the glacier.

Meltwater leaving a glacier deposits some of the debris close to the glacier in a broad, gently sloping plain, known as an **outwash plain**. The outwash plain contains a thick layer of rocks deposited close to the glacier in a layer of sand and gravel that can exceed a thickness of 100 meters (330 feet). The finer silt and clay materials are usually carried much farther and may be deposited in lakes, seas, or distant valleys.

Impact of Past Glaciations

Only 20,000 years ago—a very short time in Earth's 4.6-billion-year history—glaciers covered much of North America, Europe, and northern Asia (Figure 3-31).

These glaciers shaped landforms as they advanced. When Earth's atmosphere warmed, these rivers of ice melted back toward the poles or upslope to cold areas, and left behind debris that shapes the landforms we see today in many regions. As they advanced and retreated, alpine glaciers created distinctive landforms. In an unglaciated mountainous area, a stream may carve a V-shaped valley (Figure 3-32a).

can one do? It simply isn't possible to move a city and its inhabitants, especially one with a history as rich as that of New Orleans, out of a dangerous place. The alternative is to build defenses, which the engineers did, but until a major storm strikes, it is hard to find the political will to invest billions of dollars to prevent a disaster that few can imagine.

Katrina occurred near the middle of what was the most active hurricane season on record, with 28 named storms. The storm developed quite close to the North American mainland and moved across southern Florida as a category 1 (wind speed of 119 km/h (75 mph or more)) storm on August 25. When it moved over water again, it strengthened over a thick layer of warm water in the Gulf of Mexico. It reached category 5 (winds over 1,249 km/h (155 mph)), but weakened to category 3 (winds of 178–209 km/h (111–130 mph)) as it approached the coast of Louisiana and Mississippi.

Warnings were issued in advance of the storm, and residents from the central Mississippi Delta to Navarre Beach, Florida, were evacuated. Many people did leave New Orleans, but many did not. Some did not have transportation available, but many stayed because they didn't understand the magnitude of what might happen.

The eye of the storm crossed the coast near the tip of the Mississippi Delta and continued northward, passing about 50 kilometers (30 miles) east of New Orleans. The strongest winds on the east side of the storm hit in the vicinity of Biloxi, Mississippi—not far from where a category 5 storm—Camille—had struck 37 years earlier. At Biloxi, the storm surge was 8 meters (26 feet) above mean sea level, while in the New Orleans area it was about 3 meters (10 feet). On Lake Pontchartrain to the north of New Orleans, waves were driven by winds from the north (the center of the storm was to the east, so the counterclockwise circulation would cause winds to blow from the north in this area), and wave heights there reached 2 meters (7 feet). Flood walls failed at several locations in New Orleans, allowing the sea to flow in. In most cases, the walls failed because the pressure of high water on the seaward side caused the wall to slide landward, rather than by water actually washing over the walls. In a matter of hours, 80 percent of the city was flooded with water up to 6 meters (20 feet) deep, and because the flooded area was below sea level, the water would not drain. Thousands of people were forced to flee to upper stories, rooftops, or attics. In the days that followed, some were able to make their own way to higher ground, others were rescued by boats or helicopters, and still others remained trapped in their attics. It took about four weeks to repair levees and pumping stations and to pump the water out.

Now, years later, the flood walls have been repaired, and basic services such as electricity and water have been restored to much of the city. But the citizens of New Orleans still wrestle with the question of whether it is wise to continue to occupy an environment that is inherently unstable and exposed, by virtue of its elevation and location, to the very real prospect of future disasters.

Long-term changes in the environment—subsidence, sea level rise associated with global warming, and loss of wetlands caused by lack of sediment delivery to the delta—will only increase the vulnerability of New Orleans in the future. Following the 1993 Mississippi River flood in Iowa, Missouri, and Illinois, large areas of the floodplain were abandoned, and entire towns were moved to higher ground. But New Orleans is a major city, not a small town, and its distinctive character is inseparable from the site on which it is built. To abandon the site is to lose forever one of our national treasures, yet to rebuild without enormous new investments in improvements of the flood-control system is to face increasing risk of another disaster.

As an alpine glacier flows through a V-shaped valley, it scours away the rock and rounds the bottom into a U-shape (Figure 3-32b). When the ice melts, the U-shaped valley remains, surrounded by knife-edged ridges (Figure 3-32c).

The advance and retreat of continental glaciers have influenced the distribution of many human activities. The following are examples of how glaciation has affected three important natural resources in North America: soils, drinking-water supplies, and transport routes.

Soils The advance and retreat of glaciers in southern Ontario, Canada, and in the U.S. Midwest broke up bedrock, ground it into sand and silt, and left a great layer of debris covering the land. This rock contains important nutrients such as calcium and magnesium that are released by chemical weathering. The soils that have resulted are highly fertile and help make this region one of the most productive agricultural areas in the world. Other factors also contribute to the productivity of this region, and we will see in Chapter 8 how soil and climate interact with human input to determine regions' agricultural productivity.

Water supplies Retreating glaciers left sand and gravel deposits capable of yielding a large supply of high-quality groundwater (Figure 3-33). Long Island, New York, which is more than 160 km (100 miles) long and about 30 km (18 miles) wide, is the top of a terminal moraine deposited about 20,000 years ago, marking the southern extent of the most recent glaciation. Much of Long Island consists of outwash deposits of sand and gravel that absorb precipitation readily. This groundwater is the primary source of water for 2.5 million inhabitants of Long Island.

Figure 3-29 Retreat of the Muir glacier. The Muir glacier, like many glaciers around the world, has retreated dramatically in the last 2–3 hundred years.

With rapid population growth in recent years, however, demand for water has exceeded the rate at which the groundwater is being recharged. As more of Long Island is paved or built over, precipitation is blocked from entering the ground. As a result, the level of

groundwater has lowered, and saltwater from the Atlantic Ocean has seeped in to replace the depleted fresh groundwater in some places. The quality of the groundwater has also suffered from pollution by landfills, industry, and residential septic systems. Similar problems are occurring in other urbanized areas of the northeastern United States and Europe that depend on groundwater left behind by retreating glaciers.

Transportation routes Before the last continental ice sheet formed tens of thousands of years ago, the Missouri River flowed northward into Hudson Bay in Canada instead of southward to the Gulf of Mexico. The Ohio River did not exist, but another river to the north carried water from the western slope of the Appalachians to the Mississippi. As the glaciers melted and began to recede, these rivers were blocked, and their water could not take the more northerly routes to the sea. Instead, these rivers, plus the meltwater from the glaciers, were forced to flow along the edge of the glacier until they emptied into the Mississippi River. The lower Missouri and the Ohio rivers were thus created as meltwater channels following the edge of the glaciers. When the glaciers melted, they remained in these new positions. The receding glaciers also carved the Great Lakes, which explorers eventually used to reach the interior of the continent.

European exploration of the North American continent during the seventeenth and eighteenth centuries

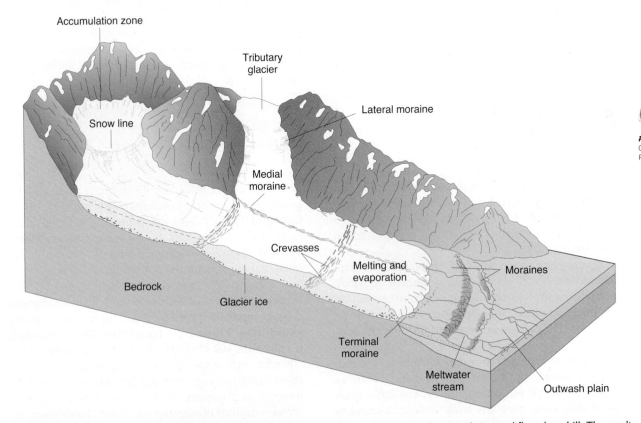

Animation
Glacial
Processes

Figure 3-30 Landforms of alpine glaciation. Alpine glaciers are fed by snow in a zone of accumulation and flow downhill. They melt at lower elevations, leaving debris to accumulate at the end (terminal moraine), side (lateral moraine), or middle (medial moraine) of the glacier or be carried away in meltwater streams.

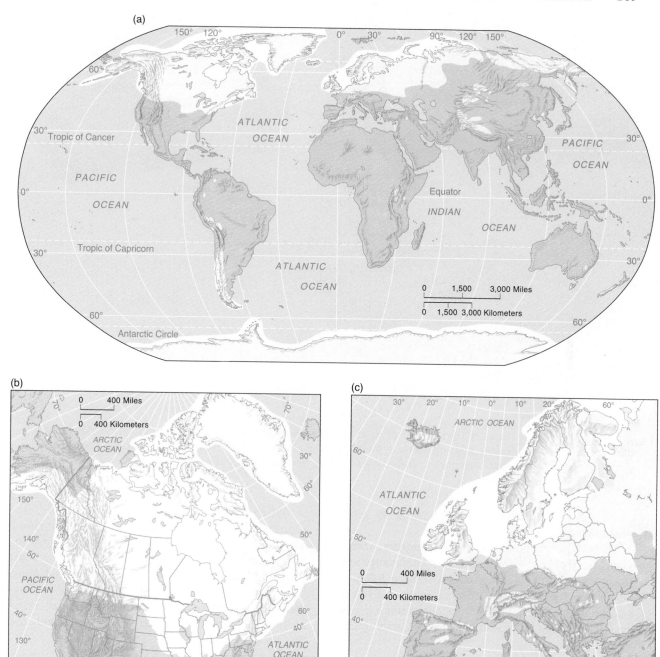

Figure 3-31 Extent of continental glaciation during the last ice age. Much of the Northern Hemisphere was ice-covered 20,000 years ago. These maps show the extent of glaciation (a) worldwide, (b) North America, (c) Europe.

and the eventual settlement of the North American interior were heavily influenced by water transport. Exploration and settlement followed the St. Lawrence River to the Great Lakes and the major inland rivers feeding into the Mississippi, such as the Ohio, Missouri, and Illinois. Travel from the East Coast to the interior was difficult, because no convenient direct water route existed from the Atlantic to the Great Lakes and the Mississippi River system. After the United States gained its independence from Britain at the end of the eighteenth century, many canals were built in **meltwater channels** left behind by receding glaciers (Figure 3-34).

The Champlain Canal follows a meltwater channel to connect the Hudson River (which flows into the Atlantic at New York City) with Lake Champlain (which empties into the St. Lawrence River). The Erie Canal follows a meltwater channel for part of its route across New York State to connect the Mohawk River with Lake Ontario. The Illinois Barge Canal connects the Mississippi River with Lake Michigan at Chicago by following a meltwater spillway occupied by the Illinois River.

Effects of Wind on Landforms

Wind is an important shaper of Earth's landforms, especially in dry regions. Wind erosion is also significant wherever the soil is not well covered with vegetation. This includes deserts, farmland, and coastal areas

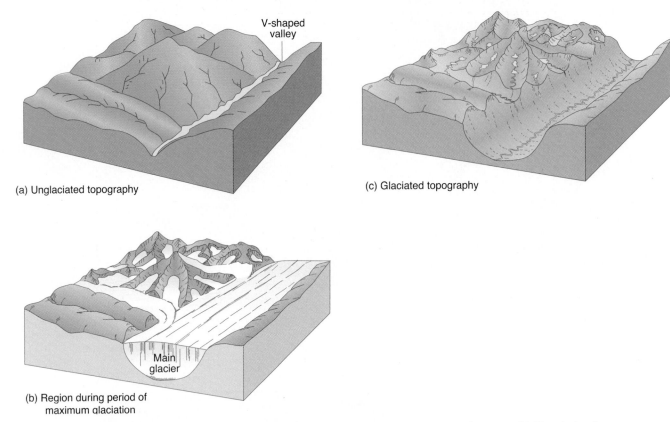

(a) Unglaciated topography

(c) Glaciated topography

(b) Region during period of maximum glaciation

Figure 3-32 Typical features of glaciated landscapes. (a) Stream-eroded terrain prior to glaciation. (b) The glaciers have their way with the land, gouging, scraping, scouring, bulldozing, and plucking rock and soil. (c) After the climate warms and the glacier melts, distinctive landforms remain: U-shaped glacial valleys, new lakes and streams, and sharp-edged mountain ridges.

where beaches are kept free of vegetation by waves washing the shore.

Wind is not capable of moving large particles as big as gravel, but it can carry great amounts of fine-grained sediment such as sand. Where wind velocities are lower, or where topography encourages deposition, the sand accumulates in **dunes**. These accumulations of shifting sand are difficult places for vegetation to become established.

When the fine sand is eroded from the soil surface, larger rocks, pebbles, and gravel are left behind. The landforms in the desert can form a hard, armored surface, called **desert pavement**. The most popular image of a desert landform is a sand dune, but about 90 percent of Earth's desert areas are covered with desert pavement rather than sand. Wind erosion is minimized once desert pavement has

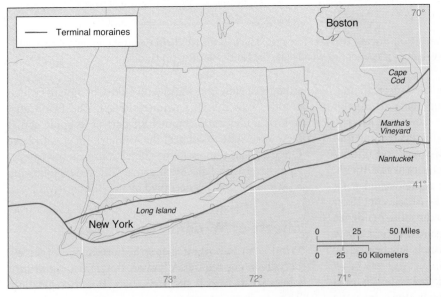

Animation
Sediment
Transport
by Wind

Figure 3-33 Glacial deposits in Long Island, New York. Long Island is covered by glacial deposits. The island is actually the top of a system of terminal moraines that also includes Martha's Vineyard, Nantucket, and Cape Cod, Massachusetts. The sandy soils of Long Island absorb water readily, and groundwater resources once were abundant. Today, however, these resources are threatened by overuse, intrusion of saltwater from the ocean, and pollution.

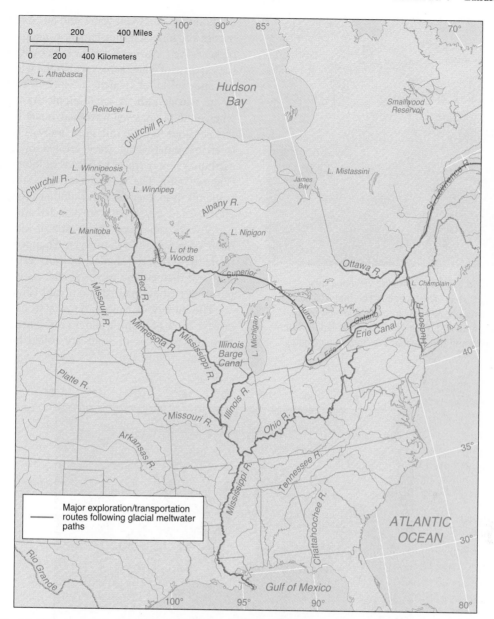

Figure 3-34 Glacial effects on North American drainage. Glacial meltwater channels profoundly altered the drainage of North America and created paths that were exploited by early explorers and later in the construction of canals. The Ohio and Missouri rivers were created by meltwater flowing around the margins of the continental ice sheet (compare this map with Figure 3-31b). Some of these lakes and rivers are connected by canals, often constructed along glacial meltwater channels.

formed, because the remaining rocks are too heavy to be moved by it.

Wind also can affect landforms in humid regions, because very fine-grained sediment can be carried some distance by the wind before being deposited (Figure 3-35). Thick layers of windblown silt, called **loess**, blanket many areas, including central China and the Mississippi River valley of the United States. Originally, this material was carried by meltwater from northern glaciers to nearby valley bottoms. Wind then carried the sediment to adjoining areas, forming loess, a fine agricultural soil. Running water subsequently has carved deep gullies and ravines in these loess deposits.

Windblown sand is common in coastal zones as well as in dry climates because of the lack of vegetation

in both locations. Along coasts, however, vegetation is usually scarce because of the pounding of waves, rather than because of the climate.

Coastal Erosion

A coast is an especially active area of erosion because an enormous amount of energy is concentrated on the shorelines from pounding waves. Land may be lost to erosion, or gained through deposition, at rates up to several meters per year.

Waves   Winds blow across the sea surface, transferring their energy to the water by generating waves. Waves are a form of energy, traveling horizontally along

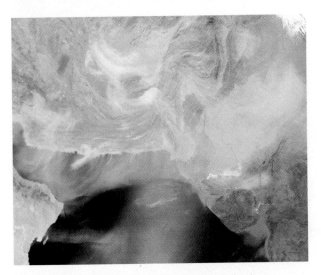

Figure 3-35 Wind erosion. Satellite view of a dust storm in southwest Asia. This image, taken May 2, 2003, shows dust blowing on the Arabian peninsula (left), Iran, Pakistan, and Afghanistan. Plumes of sediment-laden water are also visible entering the Arabian Sea from India (right).

the boundary between water and air. As the wind blows harder and longer, it transfers more energy and generates bigger waves. Waves also grow larger with an increase in the expanse of water across which the wind blows.

The speed at which a wave travels is affected by its **wavelength**. A small ripple travels very slowly, whereas an ocean swell wave may travel 10 to 50 kilometers per hour (6 to 30 miles per hour). A **tsunami**, which is an extremely long wave created by an underwater earthquake, may travel hundreds of kilometers per hour (Figure 3-36). The earthquake that caused devastating tsunamis in southeastern and southern Asia in December 2004 created waves that were detected around the globe.

A wave can travel thousands of kilometers across deep ocean water relatively unchanged, but when it nears the shore, the shallow bottom restricts water motion and distorts the wave's shape. As it slows, the top part of it rushes forward and breaks (Figure 3-37).

The energy of a wave is released as a tremendous erosive force of rushing water on the beach. Beach pebbles and granules of sand are rolled back and forth,

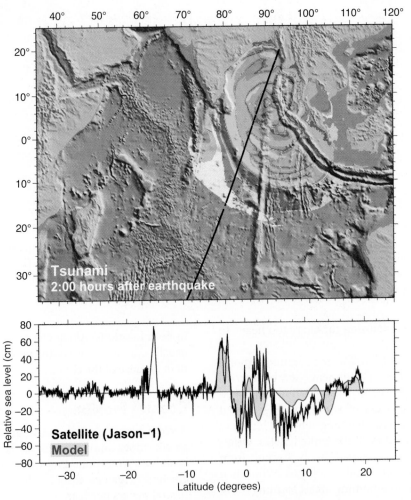

Animation
Tsunami

Figure 3-36 Tsunami. Earthquakes beneath the sea cause *tsunami* (the Japanese word for "harbor wave"). A seafloor fault slips, violently displacing the water above, creating a very low wave at the surface that moves at a very high speed. In this image, the waves that devastated the eastern Indian Ocean in December 2004 were detected by a satellite called Jason-1 using radar. The graph shows the observed sea level change caused by the wave (in black), and waves predicted by computer models (blue). The propagation of waves away from the earthquake epicenter is clearly visible.

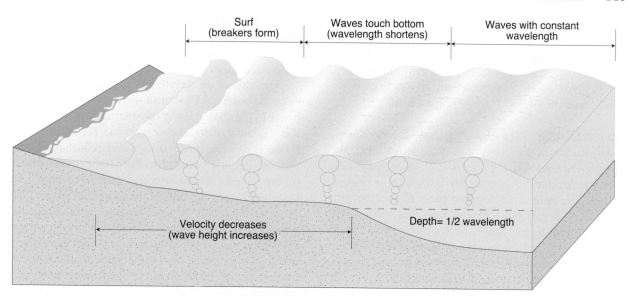

Figure 3-37 Breaking waves. As waves reach shallow water, they change shape, with long, low waves becoming short and tall, eventually breaking and releasing their energy on a beach.

constantly being ground ever finer. The finest particles are carried into deep water and settle on the seafloor. Their larger sand- and gravel-size particles are left behind on the shore to form a **beach**, which is a surface on which waves constantly break and move sand up and down.

A beach reflects the characteristics of the waves that form it. A coastal area pounded by larger waves will likely form a beach of very coarse material, because the waves carry the finer sand offshore and deposit it in quieter waters. The shape and size of the beach can vary if storms hit during only some but not all seasons.

Longshore current   When you watch waves break on a beach, the most obvious motion of the water is perpendicular to the shoreline: The waves move up the beach and then recede. When the waves break, their energy gives a push to the water in a direction parallel to the shore, and the repeated breaking of many waves generates a **longshore current** traveling parallel to the shore (Figure 3-38). The longshore current is like a river, carrying sediment through **longshore transport** from areas where it is eroded by waves and depositing it where breaking waves lose the energy to carry it— usually in deep water. Longshore currents can carry enormous amounts of sediment great distances.

Like rivers, landforms along shorelines are shaped by the balance between sediment arriving on a portion of the shore and then being removed from it. Distinctive

Animation
Beach
Drift and
Longshore
Currents

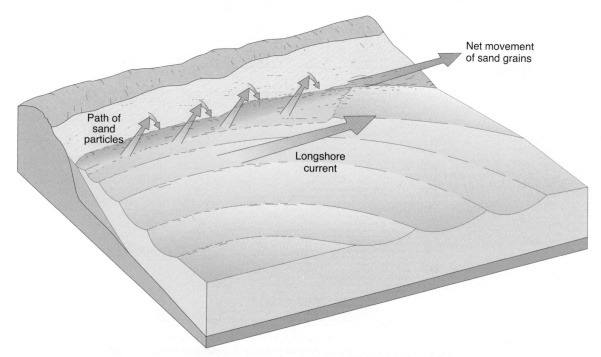

Figure 3-38 Longshore current. The longshore current is driven by waves breaking at an angle to the coastline. The current flows parallel to the shore. Breaking waves stir beach sand, causing it to be carried along with the current.

landforms develop, such as the beaches, spits, and barrier islands, as illustrated in Figure 3-39. If more sediment is removed than arrives, the coast is eroded—which is the condition of most of the world's shorelines. But in some areas, more sediment arrives than is removed, and the land area grows (Figure 3-40).

Sea level change   The edge of the land—the shoreline—is defined by the elevation of the sea, or **sea level**. We think of sea level as being fixed, but on most shorelines, it is not constant; it continually rises or falls relative to the adjacent land. In the short term, the sea can rise or fall several meters because of tides and storms. But two long-term factors can cause sea level to rise or fall: climate change and movements of Earth's crust.

Over the past few hundred years, sea level as a whole has risen on Earth, at about 2 millimeters (.08 inch) per year along coasts that are otherwise stable. Sea level has risen because the volume of seawater has increased. The increase results from the melting of glaciers, due to the overall warming trend since the eighteenth century, as well as from warming of the sea itself. This change in sea level is small compared to the sea level rise of about 85 meters (280 feet) that occurred at the end of the most recent ice age.

Sea level changes are significant at two different time scales. In the short term—a few decades—the direction of sea level change affects how a shoreline erodes. If sea level rises, the water offshore becomes deeper and waves break closer to the land, causing more erosion. If sea level falls, the shallower water causes waves to break further offshore, dissipating their energy and reducing shoreline erosion. Because many shorelines have gentle slopes, a minor increase in sea level can translate into a much larger landward migration of the shoreline (Figure 3-41).

Over thousands of years, large sea level changes can reshape shorelines. During continental glaciation, the sea was substantially lower because more of the world's water was frozen in glacial ice on the land. With a lower sea level, rivers in coastal areas cut deep valleys as they approached the sea. When the glaciers melted, sea level rose worldwide, about 15,000 to 20,000 years ago, drowning the river valleys and creating large bays such as the lower St. Lawrence River, Chesapeake Bay, and Delaware Bay.

Sea level has fallen rather than risen relative to the adjacent land in some places. This has left inland soils,

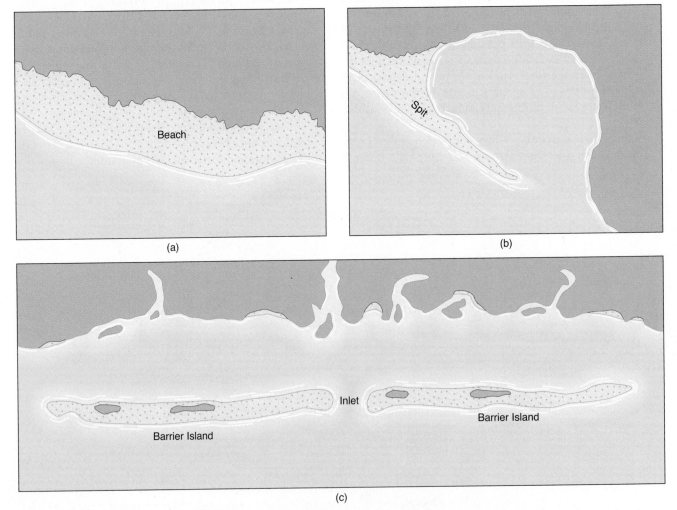

(a)

(b)

(c)

Figure 3-39 Coastal deposits. Deposition along coastlines creates beaches in small embayments (a), spits across the mouths of some bays (b), and barrier islands on gently sloping coastlines (c).

Figure 3-40 Port-Au-Prince, Haiti. This view of the coastline of Haiti's capital city, taken in 2006, shows densely populated areas close to sea level. Much of the land visible here is within 1 or 2 meters (3-6 feet) of sea level. In the center of the image is a recently built delta composed of sediment and solid waste, on which tens of thousands of people have built homes. Damage to structures in this delta in the 2010 earthquake was less dramatic than elsewhere because the buildings are typically only one story tall. Nonetheless, if a major hurricane or tsunami were to strike this coast, the area would be devastated.

established most of the world's densest populations near the sea. The shoreline is an attractive place to build houses and recreational facilities, so coastal areas worldwide have become focal points for settlement and investment.

A drawback of coastal living is erosion from shifting shorelines. In wealthy societies such as the United States, the typical response is to try to control coastal processes by modifying the coastline. These structures achieve the desired erosion control and stabilize the shoreline—but only temporarily. For example, people build groins perpendicular to the shore to slow longshore transport of sediment (Figure 3-43). A groin constructed in one location along the shore to interrupt the movement of sand will cause less sand to be deposited on a beach farther along the shore, worsening the erosion problem there. People also build sea walls parallel to the shore, but the constant pounding of waves removes sand from around the sea walls and ultimately undermines them.

The shoreline is one of the most dynamic features of Earth's surface. On an eroding barrier island of the eastern United States, erosion over the past few decades could average between 10 centimeters and 2 meters (4 inches to 7 feet) per year. At that rate, someone who takes out a 30-year mortgage to buy a house that is 30 meters (100 feet) from the beach may see waves lapping at its edge by the time the last payment is made. If a hurricane hits, the house may be washed away even sooner.

Homebuilders may not know about the threat of erosion when they build new houses on an attractive beach. A coastal housing development is usually well established before its residents discover the dynamic nature of the coast. By then it is too late to abandon the substantial investment, and more money is spent to protect the structures already built. Most of the erosion and damage to these communities takes place during major storms such as hurricanes. After such a disaster, sympathy for the victims induces the government to help rebuild communities at considerable public expense. Over the long run, though, buying up the properties and relocating the families would probably be cheaper than repeatedly rebuilding shorefront communities.

animals, and vegetation that are typical of beaches. For example, much of the U.S. West Coast has been tectonically uplifted in the past few million years as the North American and Pacific plates have ground together. As a result, the region lacks the many deep river mouths of the U.S. East Coast, but it does have former shorelines that are now above sea level, forming **marine terraces** (Figure 3-42).

Human impact on coastal processes The sea has been important to humans for thousands of years for trade, communication, and food. This long tradition has

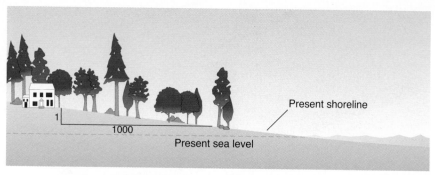

Figure 3-41 Sea level and coastal erosion. Sea level rise and shoreline position are directly related. On gently sloping shorelines, such as coastal plains, relatively low rates of sea level rise can translate into significant rates of shoreline movement. For example, if the land rises only 1 meter for each kilometer inland (a ratio of 1:1,000), a 1-millimeter (.04-inch)-per-year rate of sea level rise could mean horizontal shoreline displacement of 1 meter (3 feet) per year. Such rates of shoreline erosion usually result in major problems in developed coastal areas.

Figure 3-42 Marine terrace. The broad green area in this photo of California's Big Sur coast is a marine terrace, a former beach raised above the sea by tectonic uplift. Such terraces are common on the West Coast of the United States.

The Dynamic Earth

Rates of Landform Change

We can now see how past geologic events have shaped the world we live in and how, occasionally, dramatic Earth movements occur. Most landforms change so slowly that they have become metaphors for permanence, such as "everlasting hills" and "rock of ages." Rates of horizontal movement of continents relative to one another are typically millimeters per year to centimeters per year. Vertical movements of the land are somewhat slower—generally less than a few millimeters per year. Rates of landscape lowering by weathering and erosion are even slower. The surface of a mountain may erode at 5 to 50 millimeters (0.2 to 2 inches) per century and a gentle slope at only 0.1 to 1 millimeter (.004 to .04 inch) per century.

Change may be somewhat faster where processes are strong and materials are weak. For example, massive granite usually erodes very slowly, but a steep slope of fractured shale may erode rapidly. Similarly, waves erode a shoreline of glacial sand and gravel much faster than one of solid rock. Much of the Massachusetts shoreline, which consists of glacial deposits, is eroding horizontally at a rate of 0.5 to 2 meters (2 to 7 feet) per year. But in Maine, resistant rocky coasts are eroding at less than 10 millimeters (0.4 inch) a year.

One way to assess the significance of a rate of change is to determine how long it takes to create a landform that reflects the prevailing environmental conditions. For example, how long does it take a river to create a channel of suitable size to carry its runoff? How long does it take waves to shape a coastline into a smooth beach? How long does it take volcanic rock to develop into mature soil? The time required to shape the land varies greatly, depending on the processes and the materials involved (Figure 3-44).

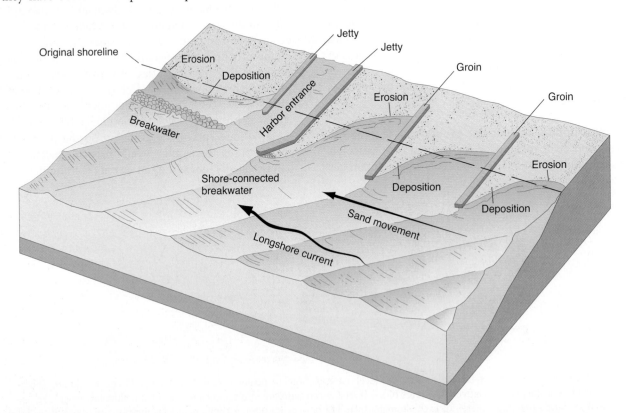

Animation
Coastal Stabilization Structures

Figure 3-43 Human modification of coastal processes. Attempts to modify the original shoreline only change the pattern and cause problems elsewhere. In this example groins, jetties, and breakwaters cause local deposition of sand in some places, but also erosion in others.

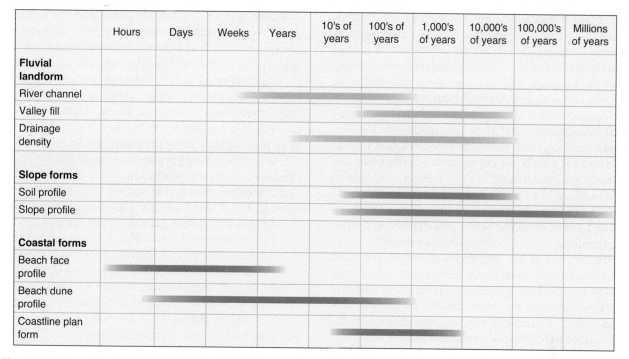

	Hours	Days	Weeks	Years	10's of years	100's of years	1,000's of years	10,000's of years	100,000's of years	Millions of years
Fluvial landform										
River channel										
Valley fill										
Drainage density										
Slope forms										
Soil profile										
Slope profile										
Coastal forms										
Beach face profile										
Beach dune profile										
Coastline plan form										

Figure 3-44 Length of time to form selected landscape features. The length of time needed to create a landform is short where processes are strong and materials weak. But if large amounts of material must be moved, landforms can take very long periods to be shaped by weak processes.

In virtually every landscape, features that take a long time to develop can retain the marks of past events, such as climate and sea level changes from tens of thousands of years ago. These same landscapes also contain more dynamic features that respond to the weather or changes in land use, and so they reflect the environmental conditions of the present or very recent past.

Another way to view rates of change is to see their effects on human activity in particularly sensitive areas, such as along coastlines. In the Mississippi Delta, land is sinking relative to the sea at about 10 millimeters (0.4 inch) per year. The city of New Orleans was founded in 1718 at a site close to sea level. If the land sinks at 10 millimeters per year for 200 years, the total subsidence is 2 meters (7 feet), a significant amount for a low-lying city. This subsidence also complicates management of floods in the lower Mississippi Valley.

Environmental managers must understand natural rates of landform change so they can predict and manage human impacts on the land. To maintain naturally productive soils, for example, we cannot allow soil to erode severely and then expect it to recover in just a few years. On the other hand, if we pave an area and thus alter the amount of water entering a stream, we should not be surprised that the stream quickly floods and causes increased erosion.

Environmental Hazards

Some environmental processes, such as volcanic eruptions, earthquakes, landslides, tornadoes, and hurricanes, occur so rapidly and violently that they cause widespread death and destruction. From a human perspective, these changes are momentous, because at any given location they are rare in human time scales. From a geologic perspective, however, they are routine.

In some areas, rapid change is frequent enough for people to recognize the threat and avoid it. But a coastal lowland may experience a severe hurricane only once in several decades, so people feel they can safely build homes there. Because many environmental hazards are once-in-a-lifetime events, people often rebuild after a disaster rather than move to a safer area. They believe that another similar event is unlikely in the near future, making the risk too small to justify giving up a livelihood or a home. Rebuilding often is accompanied by public pressure to build new structures to protect the community, such as a levee or sea wall.

Unfortunately, structures such as sea walls on the shoreline or levees along rivers cannot protect people and property against the most extreme events in a given environment. And in many cases, these structures worsen the problem. For example, levees confine floods within a narrow channel instead of letting the water spread out across a floodplain, thus increasing the water height during a flood. Protective structures also worsen the problem by giving people a false sense of security, leading them to make greater investments, thus increasing their vulnerability. When the next flood, storm, or wildfire strikes, the damage is even greater than from the last.

With changing climates, the frequency or magnitude of hazards such as floods may increase. This is highly likely with respect to coastal flooding because of sea level rise, but it is also predicted to result from more frequent or more intense storms. If this occurs, will we recognize that an environment that formerly was relatively safe is now more dangerous, and take appropriate action? In some places, where the resources needed to take such actions are available, this is happening already. But in others, especially poor countries, it probably will not and the vulnerability of populations to unprecedented disasters will rise.

As discussed in Chapter 1, humans adapt to the natural environment, and they also modify it. Geographers studying these problems recognize that natural disasters are usually not purely natural. Rather, they are caused by a combination of natural events and human vulnerability. In a flood, a house is destroyed for a simple reason: Someone chose to build where a flood can occur. In most cases, the best approach is not to control the threat but instead to change human behavior.

In the wake of the 1993 floods on the Missouri and Mississippi rivers, some levees were neither repaired nor rebuilt. Instead, flood-prone land was bought or leased by the government so that future floods can inundate this land instead of damaging valuable properties again. Allowing land to be inundated instead of keeping out floodwaters has multiple benefits: Damage is reduced to structures in the path of the flood, and downstream communities are protected. Approaches to hazard management that focus on avoiding hazards, rather than confronting them, are usually much cheaper in the long run.

Earth's surface includes some features that have remained much the same for millions of years and others that have changed dramatically in the past few hundred years. Where landforms are stable and unchanging, humans usually do not have significant impacts on the land. Human activity can destabilize landforms, however, creating danger for people living in that environment. A destabilized drainage basin might now have more frequent floods that erode riverbanks. An unstable coastline might now experience severe erosion. A deforested hillslope now might be prone to catastrophic landsliding.

As Earth's population grows and more land is developed for human use, people must learn to live with instability rather than try to control it. When Earth is moving, either in the sudden jolts of an earthquake or through gradual erosion, human structures are likely to fail. Future urban growth and land development must be sensitive to Earth's dynamic nature by avoiding hazardous areas. Floodplains and other low-lying areas should be left undeveloped or used for parks. Construction of housing on landslide-prone slopes should be discouraged, and soils that are easily eroded should be protected with conservation measures.

Geographers' training and experience in both the natural and social sciences place them in a unique position to help solve problems of human occupancy of hazardous environments. Geographers have been leaders in identifying natural hazards and developing strategies for coping with them. Such skills are increasingly important on our densely populated and dynamic Earth.

Chapter Review

Summary

Major landforms of the world are created by a combination of endogenic (internal) and exogenic (external) landforming processes. Endogenic mechanisms are forces that cause movement beneath Earth's surface, raising some portions and lowering others. The most significant of these are movements associated with plate tectonics. This motion can create earthquakes, volcanoes, and mountains, depending on whether the boundaries between plates are convergent, divergent, or transform.

Exogenic processes are forces of erosion that wear down Earth's crust and reshape it into new landforms. Rocks are first broken down into smaller pieces through chemical and mechanical weathering. Then they are carried by gravity down the slopes of hills or by wind from one place to another.

Streams collect water from groundwater and overland flow and transport the water down to the sea. Streams are conveyor belts for sediment from hillsides, and they play a major role in shaping landforms.

In many mountain and poleward environments, glaciers flow across the land, eroding rock and depositing it at places having higher temperatures that melt the ice. Wind plays a major role in moving material in deserts, and the absence of vegetation allows even infrequent rainfall to shape the land. Along coastlines, waves caused by wind blowing across the ocean surface cause intensive erosion and rapidly change landforms.

Most change on Earth's surface is slow in human terms, usually taking thousands of years to significantly reshape the land. But in some areas, change may be dramatic. Geographers have learned to identify areas that are subject to rapid or sudden change and to understand better how natural processes occur and how they affect human settlements. In general, we have learned that in dynamic environments it is better to live with nature and avoid hazards than to attempt to control them.

Key Terms

alluvial fan p. 103
alpine glacier p. 105
beach p. 113
chemical weathering p. 97

composite cone volcano p. 93
continental glacier p. 105
convergent plate
   boundary p. 94
delta p. 102
desert pavement p. 110

discharge p. 101
divergent plate boundary p. 94
drainage basin p. 100
drainage density p. 101
dune p. 110
earthquake p. 90

Questions for Review and Discussion

1. What are endogenic and exogenic processes? Make a list of mechanisms that affect the shape of Earth's surface and classify each as endogenic or exogenic.

2. What three types of relative motion occur at plate boundaries? What are examples of boundaries where each of these three types of motion is occurring?

3. What is isostatic adjustment, and what are its causes?

4. What environments favor rapid rates of mass movements?

5. What is a drainage basin? What measurable characteristics of drainage basins are important to understanding river processes?

6. What landforms provide evidence of sea level rise? Sea level fall?

7. How has soil erosion from agricultural lands modified sediment transport, erosion, and deposition in rivers?

8. What are some examples of landforms whose shape can be understood in terms of the processes acting on them? How long does it take for these landforms to develop?

Thinking Geographically

1. What were the environmental conditions where you live (or go to school) at the time of the last glacial maximum (20,000 years ago)? Were glaciers present? If not, how was the climate different? How was the vegetation different? What is the evidence that reveals these differences?

2. In your library, obtain a copy of a soil survey for your area. Soil surveys in the United States are published by the Natural Resource Conservation Service and typically are available at most libraries. Read the sections pertaining to the development of soils in the area. Select two different sites that exemplify different kinds of landforms or deposits and visit those sites, comparing what you see with what is written in the soil survey.

3. Look at a topographic map of an area that includes a river. Can you identify the meanders of the channel? Measure the width of the river. Now measure a distance along the channel that is long enough to include several bends of the river. Count the number of right-hand bends in the measured portion of the river and divide that number into the distance. The result is the meander wavelength, normally 10 to 15 times the river width. Try these measurements on other maps to see how consistent the relation is.

4. Mount St. Helens in Washington State erupted violently in 1980. The Loma Prieta earthquake in California occurred in 1989. How far apart are these two points on a map? Was there a possible connection between these two events? List reasons why there might be a connection and why there might not.

MapMaster™

The Hexi Corridor, China. The green area in the upper center is an irrigated valley in an otherwise arid environment. Runoff and glacial melt from the mountains to the south provides the water to support that irrigation.

4

Biogeochemical Cycles and the Biosphere

*R*ising water tables are causing flooding in the Hexi Corridor, an ancient trade route through Gansu Province in western China. The corridor is part of the Silk Road, a collection of trade routes that have connected China with India, the Middle East, and the Mediterranean for millennia. The climate of the Hexi Corridor is arid, so crops must be irrigated using water from springs that emerge from the base of the Qilian Mountains to the south. These springs have been heavily exploited in recent decades, and water was becoming scarce. But since 2003, water scarcity has been replaced by chronic flooding.

The reason for the flooding appears to be accelerated melt in glaciers high in the Qilian Mountains. Climates in the area have warmed substantially in recent years, and glaciers are shrinking rapidly. Similar observations of shrinking glaciers and associated increased flooding have been made in other areas of central and southern Asia. If glacier melt is the cause of increased flooding, the phenomenon will be short-lived, because at some point the glaciers will melt away. When that happens, glacial volume will be diminished to the point that streams will receive water from current or recent precipitation only, and flow will decline. These patterns illustrate the intimate and yet complex connections among climate change, the hydrologic cycle, and resource availability.

Traditionally, studies of diverse topics such as vegetation, weather, human population, and industrial activity have been carried out in isolation from each other. But scientists have increasingly recognized the intimate interaction among all Earth's systems, including natural and human-dominated phenomena. The linkages among Earth's systems are particularly important in a time of significant change in climate. Climate drives the hydrologic cycle, vegetation growth, and soil formation, and therefore profoundly influences natural ecosystems and human resources.

In this chapter, we examine biogeochemical cycles and the biosphere, including the hydrologic, or water, cycle. Water moves among the hydrosphere, biosphere, lithosphere, and atmosphere. It plays a critical role in regulating environmental processes, determining spatial patterns of vegetation and soil on Earth's surface, and linking all of Earth's subsystems. The biosphere is the thin layer of living things on Earth's surface, of which we are an inseparable part.

The biosphere is intimately connected to Earth, ocean, and air. Plants depend on solar energy for **photosynthesis,** which is the formation of food in green plants using light energy. Plants also depend on atmospheric circulation for temperature regulation, carbon supply, and water supply that make growth possible. At the same time, plants and animals absorb and release oxygen and carbon dioxide to the atmosphere and, in so doing, powerfully influence atmospheric composition and climate. Plants and animals help convert rocks to soil, the dynamic medium that stores water and nutrients and supports life, and they also regulate nutrient supplies in the soil. Because of these interactions among climate, soil, living things, and rocks, patterns of plant and animal life correspond closely with patterns of climate, topography, and rock types.

The flow of water among the lithosphere, biosphere, and atmosphere is an example of a biogeochemical cycle. **Biogeochemical cycles** are recycling processes that supply essential substances such as carbon, nitrogen, and other nutrients to the biosphere. Like movement of water, biogeochemical cycles also connect Earth's subsystems.

In previous chapters, we described environmental changes that occurred over the past few million years. These changes have been driven mostly by natural climatic variability. But during the last 10,000 years (since the end of the Pleistocene Epoch), and especially the last 1,000 years, the human population has expanded so rapidly that we now profoundly influence global environmental patterns as much as the most dramatic climatic variations did in the Pleistocene Epoch. Just a few thousand years ago, the human species had merely a minor environmental impact. Today, however, our large numbers, our enormous consumption of natural resources, and our widespread construction projects and other environmental modifications significantly influence global biospheric processes such as energy

exchange, food webs, and movements of materials on the surface of Earth. As a consequence, we can understand Earth's environments today only by recognizing how humans modify and regulate them.

Biogeochemical Cycles

Processes operating in the biosphere—the growth and decay of plants and animals and, indeed, all life processes—depend on exchanges of energy and matter. Earth receives a constant supply of energy in the form of light from the Sun. Matter, including the essential substances water and carbon, is available in the atmosphere, hydrosphere, and lithosphere, as well as in the biosphere itself.

The law of conservation of energy states that energy cannot be created or destroyed under ordinary conditions, but it may be changed from one form to another. Similarly, the law of conservation of matter states that matter may be changed from one form to another under ordinary conditions but cannot be created or destroyed. "Ordinary conditions" describe any situation except nuclear reactions, in which matter is converted to energy.

Although energy and matter are not created or destroyed, they are constantly being transformed. Biogeochemical cycles are the pathways by which energy and matter are transformed and recycled in Earth systems (Figure 4-1). The dynamic earth processes that we have discussed so far (plate tectonics, mountain building, erosion) all involve the exchange of energy and matter. Processes operating on the living organisms of the biosphere also depend on exchanges of energy and matter.

This section focuses on one of the two most important biogeochemical cycles, the hydrologic cycle. The next section examines another important cycle, the carbon cycle. Water is essential for all life to survive on Earth. Carbon is the most important element in biological processes on Earth.

The Hydrologic Cycle

Water is central to every part of the biosphere. Some of the distinctive features of water that make it so important include the following:

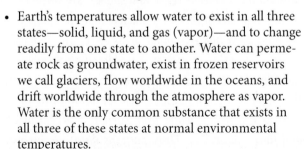

Animation
Earth's Water and the Hydrologic Cycle

- Earth's temperatures allow water to exist in all three states—solid, liquid, and gas (vapor)—and to change readily from one state to another. Water can permeate rock as groundwater, exist in frozen reservoirs we call glaciers, flow worldwide in the oceans, and drift worldwide through the atmosphere as vapor. Water is the only common substance that exists in all three of these states at normal environmental temperatures.
- Relatively large amounts of heat energy are involved in water's changes among solid, liquid, and gaseous states.

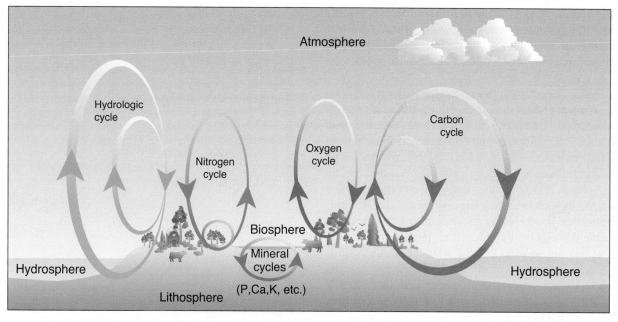

Figure 4-1 Biogeochemical cycles. These cycles transfer matter among the atmosphere, biosphere, hydrosphere, and lithosphere. The cycles represented here are shown in greatly simplified form. Important minerals cycling in the environment include phosphorus (P), calcium (Ca), and potassium (K), among others.

- Water is an excellent solvent, readily dissolving many substances, making them more mobile and available for chemical reactions with whatever water contacts.
- Most living things are primarily water (roughly 70 percent in the case of humans).
- As we saw in Chapter 2, geographic variation in water availability is the key to understanding many environmental variations at the global scale.

Water cycles through the atmosphere, lithosphere, and hydrosphere by means of evaporation, condensation, precipitation, and runoff. Evaporation converts liquid water in lakes and oceans into vapor, returning it into the atmosphere. Water falls from the atmosphere to the ground and ocean through condensation and precipitation. **Runoff** carries water from the land to the sea. This flow is the **hydrologic cycle.**

Water is stored in the atmosphere, hydrosphere, and lithosphere in solid, gaseous, and liquid forms (Figure 4-2). The salty oceans are by far the largest reservoirs of water—about 97.3 percent of all water on Earth. A little more than 2 percent is stored in glacial ice. Most of the remaining 0.6 percent is available as freshwater, with groundwater being the major storage. Only about 0.008 percent of all water is in rivers and lakes. Although the atmosphere is powerful in controlling weather and climate, it contains a mere 0.001 percent of all of the world's water.

The total quantities of water present in various forms change little from year to year for Earth as a whole. But over hundreds or thousands of years, climate substantially alters the size of glaciers and, thus, the volume of

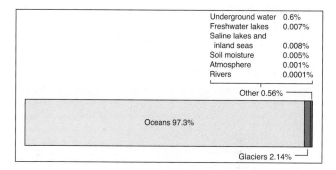

Figure 4-2 Quantities of water in storage. The oceans hold the most water; glaciers hold the next largest amount; and the remainder is in streams, lakes, groundwater, and the atmosphere.

water in the oceans. As explained in Chapter 3, about 20,000 years ago, the amount of water stored in glacial ice was much larger, and as a result, sea level was substantially lower than today.

Water Budgets

A **water budget** is an accounting of all the inflows and outflows of water in a given system over a specified time period. Water budgets are essential tools for water-resource managers because they determine the water available for human use. For example, river water is used for domestic and commercial supplies, and as cooling water for power plants and industries, hydro-electric power generation, irrigation, fish and wildlife habitats, navigation, recreation, and waste removal.

Knowing the amount available for these purposes is important, yet in many areas we have only limited information on streamflows. Measuring precipitation and determining evapotranspiration allows us to predict average river flows (Figure 4-3).

Water budgets help us understand why river flows are so much greater in humid regions than arid regions. For example, Canada's Mackenzie River has a drainage area of 1.8 million square kilometers (0.7 million square miles), most of which is well watered. The Mackenzie has an average flow of 9,600 cubic meters per second (339,000 cubic feet per second). The Nile River in Egypt has a much larger drainage area of 3 million square kilometers (1.15 million square miles) but discharges only 950 cubic meters per second (33,500 cubic feet per second) because so much of the basin is arid, with both low precipitation and high evaporation. The Mackenzie's annual flow is equivalent to a depth of about 17 centimeters (6.7 inches) of water spread over the entire basin area; the Nile's annual flow is equivalent to an average depth of only 1 centimeter (0.4 inch).

Estimates of the global average rates of evaporation, condensation, precipitation, and runoff are shown in the global water budget (Figure 4-4). Each process varies geographically with the amount of water available. Excess water evaporated into the atmosphere over the oceans is carried by wind over land areas, where it

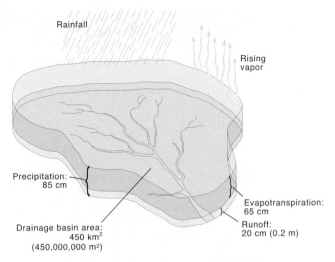

Volume of runoff: 0.2 m x 450,000,000 m²
= 90,000,000 m³ per year
= 2.85 m³ per second (average)

Figure 4-3 Calculating streamflow. Average streamflow can be calculated from the water budget for a drainage basin, to show the close connection between water resources and climate. The average annual evapotranspiration is subtracted from the annual precipitation to determine the average annual runoff, expressed as a depth of water on the land. This value, when multiplied by the drainage basin area (in comparable units), gives the volume of runoff carried by the stream in a year.

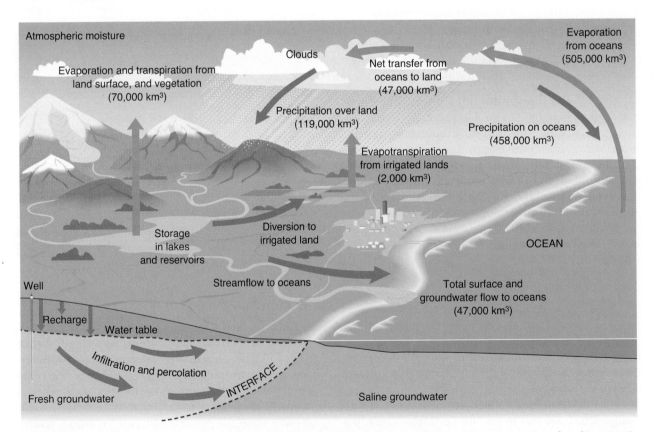

Figure 4-4 The hydrologic cycle: Flows of water among land, sea, and air. The quantities of water transferred in an entire year are shown. Red arrows are flows of water in vapor form; blue arrows are liquid water.

condenses into clouds and falls as precipitation. About two-thirds of the water that falls on land areas evaporates there, and the remaining one-third drains into rivers, which return this excess to the sea as runoff (Figure 4-5). A significant portion of river flow (an average of about 5 percent worldwide) is returned to the atmosphere via plant-water use on irrigated lands rather than flowing to the ocean as liquid water.

The water budget shown in Figure 4-4 does not show the variability of water from place to place and season to season. Evaporation from the sea depends on the insolation available to heat water. Precipitation and evaporation occur in both land and ocean areas, but not in equal amounts; more evaporation than precipitation occurs over the oceans.

Evapotranspiration and local water budgets

When we focus on local water budgets, we recognize the important role of plants in removing water from the soil. In well-vegetated areas, much of the conversion of water from liquid to vapor takes place in the leaves of plants. This water is replaced by water drawn from the soil through plant roots. We call this plant mechanism transpiration, and when combined with evaporation we call the process **evapotranspiration (ET).**

ET occurs whenever water is available for plants to transpire and the air is not saturated with humidity. But ET rates vary tremendously over time and from place

to place. Energy is necessary to evaporate water. When water vaporizes from a liquid to a gas, energy is absorbed (it becomes latent heat in the water vapor), which cools the plant leaves. Thus, the rate of evapotranspiration depends mostly on energy availability. ET occurs fastest under warm conditions and it virtually halts below freezing. Atmospheric humidity and wind speed also are significant: ET is faster on dry, windy days than on humid, calm days. Evapotranspiration is low in winter and high in summer, reflecting active transpiration by plants.

In the winter, when water use is low, precipitation exceeds evapotranspiration. The surplus water either runs off or is stored as groundwater. In springtime, plants develop leaves, and water use rises rapidly, so that by early summer, evapotranspiration exceeds precipitation. Initially, this excess water demand is met by water stored in the soil, but by late summer, soil moisture is exhausted and water use is limited. At this time, shallow-rooted grasses may wither, while deep-rooted shrubs and trees remain green. In autumn, evapotranspiration falls with the temperature, and once again precipitation exceeds it. At this time, excess precipitation restores soil moisture to higher levels, and runoff increases again.

In warm weather, conditions may favor high ET, but the soil may be too dry to supply this demand. Therefore, we distinguish between potential ET and actual ET. **Potential evapotranspiration (POTET)** is the amount

Figure 4-5 The mouth of the Amazon. The Amazon is the largest river on Earth, with an annual flow that represents about 18 percent of all the river flow to the oceans of the world. In this image, we see the sediment-laden water of the Amazon entering blue Atlantic water. Convective clouds are scattered over the land areas but noticeably absent over water areas.

of water that would be evaporated if it were available. **Actual evapotranspiration (ACTET)** is the amount that actually is evaporated under existing conditions. If water is plentiful, then ACTET equals POTET. But if water is in short supply, ACTET is less than POTET.

ACTET never exceeds POTET. Comparing POTET with precipitation is a useful way to describe water availability in various climates. For example, if precipitation always is greater than POTET, plants have plenty of water, and the climate is quite humid. If precipitation is less than POTET most of the time, then plants never get as much water as they need, the natural vegetation is adapted to dryness, and the climate is arid. Farmers using irrigation to supply water to plants can calculate the difference between POTET and precipitation and thereby know how much water they must apply to crops. POTET can be calculated from mean monthly temperature data.

A local water budget compares precipitation and evapotranspiration. For a humid midlatitude site such as Chicago, Illinois, a graph of a water budget reveals that precipitation occurs fairly consistently throughout the year (Figure 4-6).

Water budgets vary tremendously from one climate to another (Figure 4-7). In some humid climates, precipitation almost always meets demand, and stored soil moisture provides the small additional amount of water needed during brief dry spells. In semiarid and arid climates, the demand for water considerably exceeds precipitation during the year, so soil moisture is never fully replenished and plants must withstand severe moisture deficits.

Geography
Video
Back From the
Brink?

Soil's role in the water budget Most of Earth's land area is covered with a layer of rock debris and organic matter known as *soil*. Soil is critical in the water budget, for it stores water and makes it available for evapotranspiration; soil is a water "bank." How effectively soil does its job depends on how fast it can absorb precipitation, how much water it can store in the root zone, and the ability of local vegetation to transpire water into the atmosphere.

Soil's **infiltration capacity** determines whether rainfall infiltrates (soaks into) the soil or runs off. Infiltration is good and most precipitation will soak in if dense vegetation exists at ground level, a layer of organic litter covers the soil, and the soil is kept porous by worms and insects. If the soil is bare or compacted, infiltration is poor, less water is stored, and runoff is greater.

Soil texture—the size of mineral particles in the soil—affects how much water soil can hold. Sandy soils drain quickly and hold little water (pour water on sand and watch how quickly it soaks in). Soils with high contents of finer particles—silt and clay—drain slowly (pour water on modeling clay and see how quickly it runs off). Deep-rooted plants (several meters, or 10 feet or more) have a longer opportunity to absorb water as it drains through the soil, but shallow-rooted plants (half a meter, or 18 inches or less) must rely on moisture contained in the upper soil. Thus, shallow-rooted grasses and herbs generally use less water (and potentially allow more runoff) than deep-rooted trees. Soil and vegetation management is critical for maintaining adequate water resources for agriculture and other human needs.

Vegetation and the Hydrologic Cycle

Forests demonstrate the close association between vegetation and the water budget. Forests occur where ample moisture is available for most of the year. Trees demand large volumes of water: A single large tree can transpire 1,000 liters (35 cubic feet) per day in warm weather (about twice the volume of a typical refrigerator). Trees gather water through extensive and usually deep root systems and transpire through leaves that often are tens of meters above the ground surface. Trees use so much water that they are the dominant medium for returning rainwater to the atmosphere in large forested regions like eastern North America and the Amazon basin.

Different types of forests have water-management implications. In the southeastern United States, areas formerly covered with slow-growing deciduous trees (that is, trees that shed their leaves in autumn) have

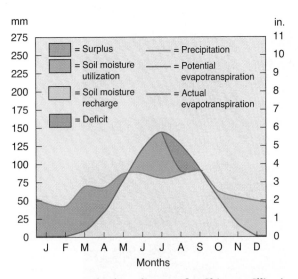

Figure 4-6 Water budget diagram for Chicago, Illinois, a typical humid midlatitude site. Precipitation is shown as a blue line, and potential evapotranspiration is shown in red. In the cool months precipitation is greater than potential evapotranspiration, and the excess becomes runoff (dark blue shaded area). In the warm season potential evapotranspiration is greater than precipitation. Part of this excess demand is made up by soil-moisture use (green shading), but not all. The difference between potential evapotranspiration and actual evapotranspiration is a deficit (brown shading). The soil water that was used in early summer is recharged in autumn (light blue shading). The orange area at the bottom of the graph is water that falls as precipitation and is part of actual evapotranspiration.

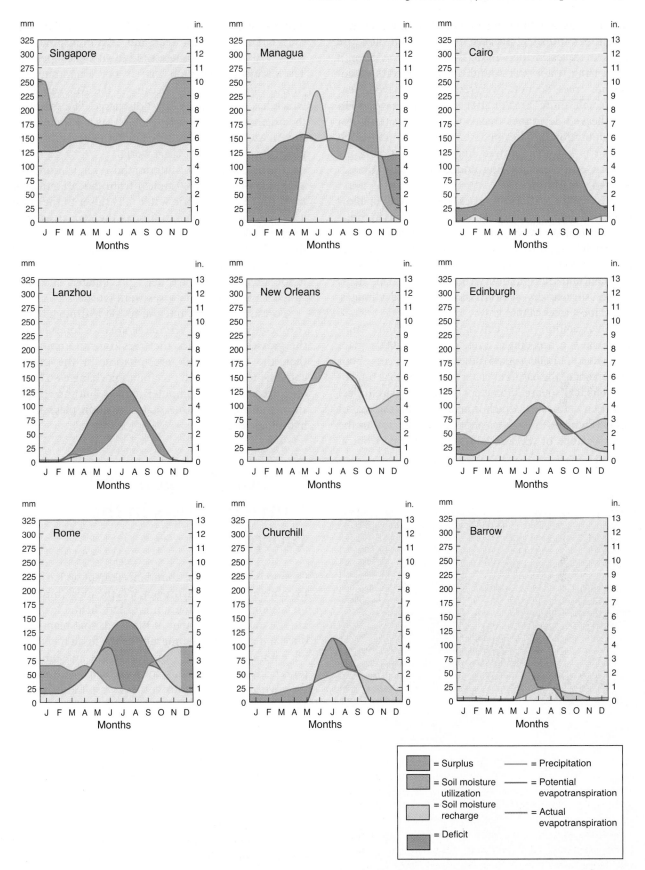

Figure 4-7 Water budget diagrams for selected stations. Compare with the climatic data and descriptions on pages 64 to 77 in Chapter 2.

been replanted with faster-growing evergreens to feed the timber industry. Because evergreens do not lose all their leaves in the winter, they keep growing and transpiring water year-round, whereas forests of deciduous trees do not. Evergreen needle-leaved trees also catch more precipitation on their leaves than broadleaved deciduous trees do, and this water is lost to evaporation. Studies in an experimental watershed in North Carolina showed that this measurably increased evapotranspiration and correspondingly decreased streamflow.

In contrast to trees, grasses are relatively shallow rooted, so they experience significant variations in moisture availability. When soil water is plentiful, grasses grow quickly and transpire at rates similar to trees. But during periods of limited soil moisture, grasses become dormant and transpiration virtually ceases. This helps grasses survive dry seasons, as in semiarid climates, where most trees cannot grow.

Deforestation—clear-cutting of forest—in the Amazon basin is proceeding at a rate of about 30,000 square kilometers (11,600 square miles) per year, or about 0.6 percent of the basin's area per year. By 2000, a total of over 650,000 square kilometers (250,000 square miles) had been deforested, which is an area about the size of Texas (Figure 4-8). Deforestation may be critical to the water balance of the Amazon region. In this tropical rain forest, rainfall averages 200 to 300 centimeters (80 to 120 inches) per year, and evapotranspiration is about 110 to 120 centimeters (44 to 48 inches) per year. The excess precipitation that is not evapotranspired runs off to the Atlantic Ocean via the Amazon River, which has the world's greatest discharge. The Atlantic is the original source of water that falls as precipitation on the Amazon basin. This recycling presents a fine example of the hydrologic cycle at work.

The Amazon basin's interior, however, is more than 2,000 kilometers (1,200 miles) from the Atlantic, and most of the atmospheric water falls closer to the coast. This precipitated water infiltrates the soil, trees transpire it back into the air, and easterly winds carry it further inland. This cycle repeats, moving water step by step westward. Thus, rainfall in the Amazon basin's interior already may have been precipitated and transpired several times in its westward journey.

Forests clearly are integral to the hydrologic cycle. If we continue to cut very large areas of forests and if the grasses that replace the trees transpire much less than the trees would, precipitation in the interior could be reduced. Such precipitation changes so far are only predictions in computer models, but they help to show the important role that vegetation plays in the hydrologic cycle.

Figure 4-8 Fires, roads, and deforestation in the Amazon. This satellite image of eastern Bolivia in October 2007 shows fires burning (red dots) and the resulting plumes of smoke.

Carbon, Oxygen, and Nutrient Flows in the Biosphere

Carbon is not the most abundant element on Earth, but in combination with hydrogen, it is the most important for sustaining life. Compounds of carbon and hydrogen are the major component of the foods that plants produce and animals consume, and of the fossil fuels (coal, oil, natural gas) that are our most important sources of power. Living things constantly exchange carbon with the environment by photosynthesis, respiration, eating, and disposal of waste.

Exchanges of carbon among the biosphere, atmosphere, oceans, and rocks are collectively called the carbon cycle (Figure 4-9). An oxygen cycle is inextricably linked with the carbon cycle, and we will consider them together.

The Carbon and Oxygen Cycles

In the **carbon cycle,** carbon in the form of atmospheric carbon dioxide (CO_2) is incorporated into carbohydrates in plant tissues through photosynthesis. Animals consume plants and use the plant carbohydrates for their own life processes. Through respiration, which occurs in almost all living organisms, carbon is returned to the

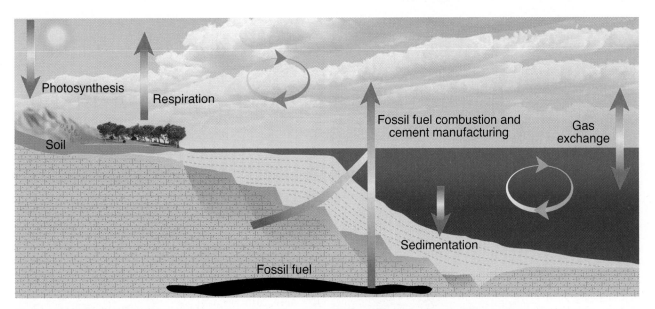

Figure 4-9 The carbon cycle. Carbon is taken from the atmosphere and stored in biomass through photosynthesis. It is returned to the atmosphere through respiration. Combustion of fossil fuels and manufacture of cement release vast quantities of carbon to the atmosphere from long-term storage in rocks. Large quantities are also exchanged between the biosphere and atmosphere via photosynthesis and respiration, and between the atmosphere and oceans via gas exchange.

atmosphere (as CO_2). Carbon dioxide is also added to the atmosphere by the combustion of fossil fuels. Thus, these three processes—photosynthesis, respiration, and combustion—cycle carbon and oxygen back and forth between living things and the environment.

Photosynthesis can be shown by the following equation:

$$\text{carbon dioxide} + \text{water} + \text{energy} \rightarrow \text{carbohydrates} + \text{oxygen}$$

For this reaction, land plants obtain carbon dioxide from the air, water from the soil, and energy from solar radiation. They store carbohydrates in tissue for later use and release oxygen to the atmosphere. Plants are the source of atmospheric oxygen, without which animals could not exist.

Respiration involves the opposite reaction to photosynthesis:

$$\text{carbohydrates} + \text{oxygen} \rightarrow \text{carbon dioxide} + \text{water} + \text{energy (heat)}$$

In respiration, carbohydrates are broken down when they combine with atmospheric oxygen to CO_2 and water. Energy is released in the process. Some of this energy is lost as heat and some is stored in chemical compounds for later use in other life processes.

The lithosphere is a major storehouse of carbon. Through geologic time, carbon enters the lithosphere slowly through rock formation, principally from oceanic sediments such as limestone but also through creation of fossil fuels such as coal.

The spatial and temporal distribution of photosynthesis is determined mostly by climate (Figure 4-10).

Seasonal cycles of solar radiation are reflected directly in the CO_2 content of the atmosphere. The air we breathe has been monitored for carbon dioxide concentration since the late 1950s at the Mauna Loa observatory in Hawaii (Figure 4-11):

- CO_2 varies annually with the seasons by a few parts per million. The greatest CO_2 concentrations in the Northern Hemisphere are during spring, because during winter, photosynthesis is reduced while respiration continues, returning CO_2 to the atmosphere.
- The lowest CO_2 concentrations are during autumn. CO_2 concentration drops during the summer because plants are actively photosynthesizing, removing CO_2 from the atmosphere and storing it in the biosphere.
- The Mauna Loa data also clearly reveal the steady increase in atmospheric CO_2 levels that results from fossil-fuel consumption, and also from the manufacture of cement from limestone.

The extraction and combustion of coal, oil, and natural gas since the Industrial Revolution began in the late 1700s is a relatively new process affecting the carbon cycle. Rates of carbon released into the atmosphere through this process have increased steadily over the past 200 years, and they are expected to continue to increase at least well into the twenty-first century. The long-term carbon dioxide increase that is so evident in Figures 2-56 and 2-58 illustrates this increase. Uncertainty about future fossil-fuel use makes it difficult to predict future atmospheric carbon dioxide concentrations.

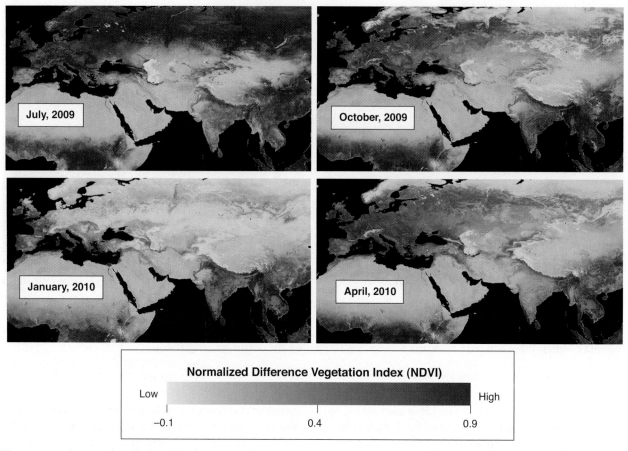

Figure 4-10 Seasonal variations in plant growth. Seasonal maps of Eurasia and northern Africa, showing vegetation growth. In these images, derived from satellite data, green tones are growing vegetation; tan indicates absence of vegetation growth.

The Global Carbon Budget

Exchanges of carbon between the atmosphere and the biosphere and lithosphere are evident in the trend shown in Figure 4-11. Large amounts of carbon are also exchanged between the oceans and the atmosphere, as well as within various parts of the biosphere. Table 4-1 provides a summary of the major flows, along with an indication of the uncertainties surrounding our knowledge. The values shown in the table are average annual values, so they do not include the seasonal exchanges shown in Figure 4-11.

Over time, the total amount of carbon in the atmosphere increases. In the 1980s, the amount of the increase was about 3.3 gigatons (Gt) of carbon per year (a gigaton is a billion metric tons, or 10^{15} grams), and by the period 2000–2005 it was about 4.1 Gt per year. This increase is caused primarily by fossil fuel combustion and cement manufacture, which totaled about 7.3 Gt per year in 2000–2005—much more than the rate of accumulation in the atmosphere.

The amount of carbon accumulating in the atmosphere is less than our emissions because carbon is

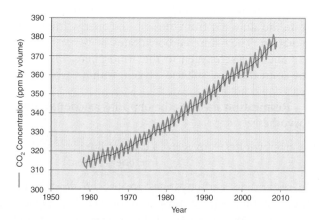

Figure 4-11 Atmospheric carbon dioxide concentrations measured at Mauna Loa, Hawaii. Mauna Loa, located at 19.5° north latitude in the mid-Pacific Ocean, provides data that reflect general Northern Hemisphere conditions. The annual increase and decrease of carbon dioxide concentrations corresponds to seasonal patterns of plant growth and decay, with atmospheric concentrations decreasing in the Northern Hemisphere summer and increasing during the winter.

TABLE 4-1	The Global Carbon Budget		
	1980s	**1990s**	**2000–2005**
Atmospheric increase	3.3 ± 0.1	3.2 ± 0.1	4.1 ± 0.1
Fossil carbon dioxide emissions	5.4 ± 0.3	6.4 ± 0.4	7.2 ± 0.3
Net ocean-to-atmosphere flux	−1.8 ± 0.8	−2.2 ± 0.4	−2.2 ± 0.5
Net land-to-atmosphere flux	−0.3 ± 0.9	−1.0 ± 0.6	−0.9 ± 0.6

Values are in gigatons (Gt; billions of metric tons) per year.
Source: Data from IPCC.

taken up by the land and oceans. The largest share of this uptake, about 2.2 Gt per year, occurs in the oceans. Ocean uptake is driven by the relative amounts of CO_2 in the atmosphere and oceans. Carbonated beverages illustrate this process. When they are made, CO_2 is added to the beverage either through the biological processes of fermentation under pressure or through direct injection of gaseous CO_2 into the beverage. When a pressurized carbonated beverage is opened, the gas is released back to the atmosphere in bubbles that rise to the surface. Humans have added large amounts of CO_2 to the atmosphere in recent decades, raising the CO_2 pressure in the atmosphere. This causes the net flow to be from the atmosphere to the oceans. The CO_2 will not bubble out of the oceans, however, unless we somehow lower the CO_2 concentration of the atmosphere sometime in the future.

A second reason that accumulation of carbon in the atmosphere is less than our fossil emissions is that there is a substantial net flow from the atmosphere to the biosphere. Carbon constitutes a large part of **biomass**— living plant and animal matter in the biosphere. When forests are cut down, most of the carbon in the trees is released to the atmosphere, through decomposition or perhaps fire. But when forests grow back, there is a net flow of carbon from the atmosphere to the biosphere, reducing atmospheric CO_2 and thus limiting global warming. In many parts of the world, especially the humid tropics, deforestation and urban development are reducing carbon storage in the biosphere and sending the carbon to the atmosphere (Figure 4-12). But in other areas, such as the eastern United States, forest area is actually increasing, and young forests are growing, storing carbon in the process. Similarly, when undeveloped lands are first converted to agriculture there is usually a loss of organic matter and thus carbon from the soil, but later that carbon can be replaced if the soils are allowed to recover. At present, the available evidence indicates that the net balance of all these land-use-related exchanges is a transfer of carbon from the atmosphere to the biosphere, but as you can see by the ranges of estimates shown in Table 4-1 there is considerable uncertainty. Later in this chapter, we will discuss processes operating in the biosphere in more detail.

It is clear that the carbon cycle is affected by many different aspects of human activity, including agricultural management practices, production and recycling of paper and other materials made from biomass, our choices with regard to energy production and conservation, and the way we manage water resources. In subsequent sections of this book, we will see examples of the ways in which resource use interacts with the carbon cycle.

Soil

The interface between the lithosphere and the biosphere/atmosphere is the soil. **Soil** is a dynamic, porous layer of mineral and organic matter in which plants grow.

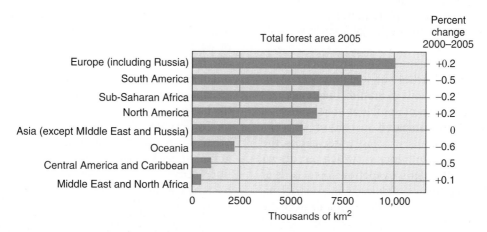

Figure 4-12 Forest extent and deforestation rates by region. The most extensive forests in the world are in Siberia and the Americas, but major forest areas also exist in Africa and southern Asia. Deforestation is most rapid in the tropics; forest area is increasing in Europe and North America.

Soil is the uppermost part of the lithosphere, and at the same time is a key portion of the biosphere. Soil is a storage site for water, carbon, and plant nutrients.

Soil properties are attributable to five major factors:

1. *Climate* regulates both water movements and biological activity.
2. **Parent material** is the mineral matter from which soil is formed.
3. *Biological activity*—plants and animals—moves minerals and adds organic matter to the soil.
4. *Topography* affects water movement and erosion rates.
5. All these factors work over *time*, typically requiring many thousands of years to create a mature soil.

Soil Formation

The first step in soil formation is weathering, or the breakdown of rock into smaller particles and new chemical forms. Chapter 3 explained the two ways that rocks weather: mechanical weathering (such as ice expansion or tree roots growing in the cracks between rocks) and chemical weathering (such as dissolving limestone in water in the soil). The parent material from which soil is formed is important because it influences soil's chemical and physical characteristics, especially in young soils.

Water plays an important role in rock weathering and soil formation. A water budget indicates how much water moves through the soil in a place and measures the significance of water in local weathering. For example, if mean annual precipitation is 100 centimeters (39 inches) and evapotranspiration is 70 centimeters (27 inches), then 30 centimeters (12 inches) of water would seep down through the soil in an average year. This water would be a powerful weathering agent. However, if an area has 70 centimeters of precipitation and the atmosphere is capable of evaporating 150 centimeters (59 inches) of water, virtually all water is evapotranspired by plants, so little would be left to percolate down through the soil to help weather the rocks.

In a very humid climate, much water passes through the soil and leaches out soluble minerals on its way. Because of this, soils in humid climates generally have lower amounts of soluble minerals such as sodium and calcium compared to soils in dry climates. However, in semiarid areas, water enters the soil and picks up soluble minerals, which are drawn toward the surface as water is evapotranspired; soils of semiarid and arid climates often have a layer rich in calcium and other relatively soluble minerals near the surface.

Plant and animal activities are also critical to soil formation. Plants produce organic matter that accumulates on the soil surface, and animals redistribute this organic matter through the soil. Plants and animals also play a role in weathering processes. Topography also affects the amount of water present in the soil, largely through controlling drainage and erosion. Steeply sloping areas generally have better drainage than flat or low-lying areas, and they are often more eroded. Finally, soil formation is a slow process that takes place very gradually over thousands of years. Soils that have only been forming for a few hundred or even a few thousand years have very different characteristics from those that have been modified by chemical and biological processes for tens of thousands of years.

Soil Horizons

Soil is a complex medium, containing six principal components:

1. *Rocks* and *rock particles*, which constitute the greatest portion of the soil and which may weather, releasing nutrients needed for plant growth.
2. *Organic matter*, which is composed of dead and decaying plant and animal remains that holds water, supports soil organisms, and supplies nutrients.
3. *Dissolved substances*, including phosphorus, potassium, calcium, and other nutrients needed for plant growth.
4. *Organisms*, including animals such as insects and worms and many microorganisms, including bacteria, and fungi.
5. *Water from rainfall*, which is necessary for plant growth and which helps to distribute other substances through the soil.
6. *Air*, which shares soil pore spaces with water and which is necessary for respiration by plant roots and soil organisms.

These substances are not uniformly distributed in soils but found in layers called **soil horizons** (Figure 4-13). Soil horizons are formed through the vertical movement of water, minerals, and organic matter in the soil and also by variations in biological and chemical activity at different depths. In this way, they are different from layers in sediments, which are deposited in sequence from the bottom up.

For example, litter—leaves, twigs, dead insects, and other organic matter—accumulates to form a horizon at the surface known as the *O* (organic) horizon. As this litter decays, organisms such as insects, worms, and bacteria consume it and carry it underground, where it helps form the *A* horizon. Waste from these burrowing animals as well as their dead carcasses add more organic matter to the *A* horizon. In many soils, the *A* horizon contains much of the nutrients that support plant life.

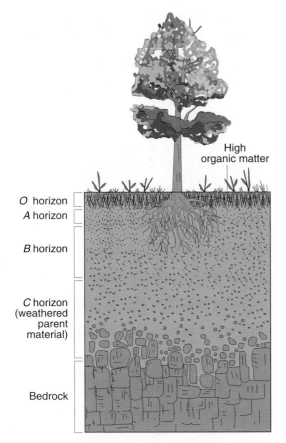

High
organic matter

O horizon
A horizon

B horizon

C horizon
(weathered
parent
material)

Bedrock

Figure 4-13 A soil profile. The sequence from the bottom up shows the typical steps of soil formation. Solid rock at the bottom weathers into large pieces, which weather further into ever smaller particles. At the top is the organic layer, which contains organic matter from decayed plants and animals. Designations like *A* are given to horizons in well-developed soils. Poorly developed soils lack some of these layers.

Water may erode materials from the soil surface, or carry some substances down from the *A* horizon to the *B* horizon. Organisms also play an important role in moving materials between the *A* and *B* horizons. Clay minerals formed from chemical weathering often accumulate in the *B* horizon, and in dry regions soluble minerals such as calcium accumulate in the *B* horizon. The *C* horizon, beneath the *B* horizon, contains weathered parent materials that have not been altered as completely by soil-forming processes as materials above it have.

Thousands of Soils

Soils are divided into 11 broad clusters, called **soil orders** (Figure 4-14 and Figure 4-15). These 11 orders are divided into 47 suborders, about 230 great groups, 1,200 subgroups, 6,000 families, and thousands of soil series. In other words, Earth's dynamic lithosphere, hydrosphere, atmosphere, and biosphere have worked as a team to cre-

ate an enormous diversity of soil. There is nothing simple about "ordinary dirt"!

Farmers and land-use planners use detailed maps developed by the U.S. Department of Agriculture's Natural Resources Conservation Service that show the great local variability of soils. Often, two or three different great groups of soils occur within a few kilometers of each other, and these groups are subdivided into subgroups, families, and series of more specific characteristics. Because of this great local variability, generalizations about soils must be made with caution.

Climate, Vegetation, Soil, and the Landscape

On a global scale, distinctive environmental regions derive from intimate connections among climate, vegetation, and soil:

- *Humid tropical* and *subtropical soils* such as oxisols and ultisols are highly weathered. They have lost soluble minerals due to heavy precipitation, which may deplete soils of essential nutrients, so fertilization is necessary to increase crop yields. Tropical soils are usually oxidized and red in color.
- *Arid region soils* such as aridisols are high in soluble minerals because little leaching occurs with the scant rainfall. These soils are low in organic matter, but they can be very productive if sufficiently irrigated.
- *Midlatitude humid soils* such as alfisols and spodosols are moderately leached where they are dominated by deciduous forest vegetation and heavily leached where they are developed beneath the acidic litter of coniferous forests.
- *Midlatitude subhumid soils* such as mollisols are fertile and are the resource base for many important grain-producing regions.

These climatic soil regions are interrupted by the major mountain chains of the world. Poorly developed soils such as inceptisols predominate because erosion removes soil as rapidly as it is formed.

In Figure 4-16 we zoom in on a portion of the world maps of Figure 4-14 to examine vegetation and soil at a larger map scale. A close relation among climate, vegetation, and soil also exists at the scale of local ecosystems. You can see that vegetation and soils are far more diverse than is revealed in the large-scale map. This semiarid area of northwestern Colorado is in a sandstone and shale upland, incised with narrow valleys. It is an open woodland dominated by pinyon pine and Utah juniper trees and a shrub-grass community that is mostly sagebrush (Figure 4-17). Within a few square kilometers, you may see forest of one or more types, disturbed areas, shrublands, and

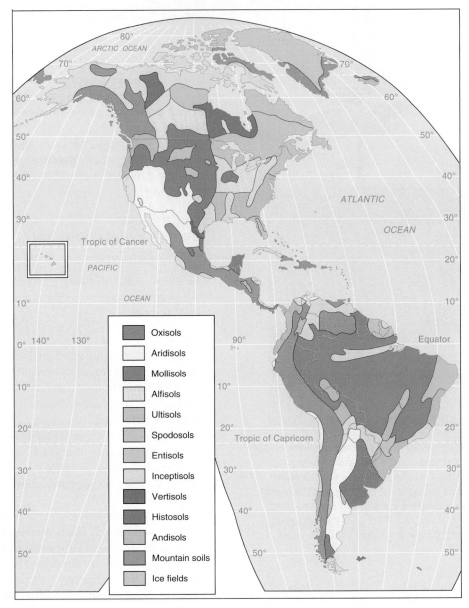

Figure 4-14 World soils.
Oxisol—a deeply weathered, heavily oxidized soil of the humid tropics, red in color.
Aridisol—dry desert soil with limited organic matter, limited chemical weathering, and accumulations of soluble minerals.
Mollisol—grassland soil of subhumid–semiarid land, dark with plentiful organic material. Mollisols are an important agricultural soil in the U.S. Midwest.
Alfisol—a moderately leached soil of humid subtropical and midlatitude forests, with clayish *B* horizon.
Ultisol—red- to yellow-colored, well-weathered soil with clayish *B* horizon typically found in warm, humid climates.
Spodosol—acidic soil formed under coniferous forest, with a light-colored, sandy *A* horizon and red-brown *B* horizon in which iron accumulates.

grasslands. The caption with Figure 4-16 describes the soil–vegetation relations here.

Soil Problems

In modern commercial agriculture and in many traditional systems, nutrients are added artificially to the soil. These nutrients may be delivered in the form of animal manure, as manufactured inorganic fertilizers, or by plowing a soil-enriching crop into the soil. The more intensively nutrients are removed in crop harvests, the more important such additions become. Unfortunately, increasingly intensive agricultural activity has meant that in many areas, nutrients, especially organic matter, are not being replaced as rapidly as they are withdrawn. In such cases, soil fertility declines over

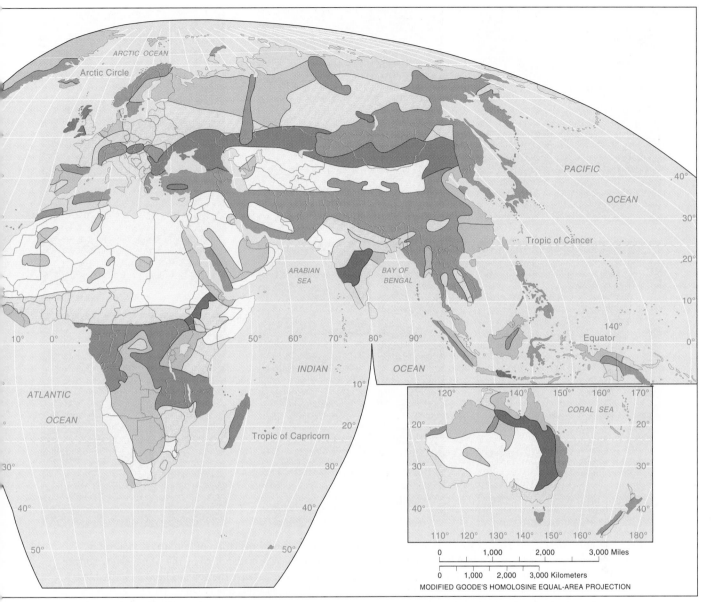

Entisol—poorly developed soil, usually resulting from low weathering rates, with characteristics dominated by parent material.
Inceptisol—weakly developed soils often found in mountainous areas where erosion limits soil development.
Vertisol—soils with high content of clay minerals that swell and shrink on wetting and drying. Deep cracks open up in the dry season.
Histosol—organic soils found in peat bogs and other areas of organic-matter accumulation.
Andisol—young volcanic soils developing on ash and lava deposits.
Mountain soils and ice fields are also mapped. Areas mapped as mountain soils have too much local variability to show individual soil types at this map scale. Ice fields have no soil.

decades, so the problem is less apparent than if it occurred over just a few years. Variations in yield from year to year caused by weather, insects, plant diseases, and changing technology mask the effects of long-term soil degradation. Nevertheless, a long-term decline in soil quality is in progress in many parts of the world.

To this problem, add erosion accelerated by cultivation. Plowing a field breaks up the soil-holding system of plant roots and exposes the soil to attack by rain and wind. Erosion removes the uppermost part of the soil, which is usually the most fertile part. This removes both nutrients and the ability to store them. In many agricultural regions, topsoil loss has ranged from several centimeters (1 to 2 inches) to tens of centimeters (4 inches to a foot or more). Often, the depth of erosion caused by long-term exposure of the soil surface amounts to a significant part of the entire *A* horizon.

In semiarid areas, the process can degrade the soil so that only plants adapted to lower soil moisture

Figure 4-15 Six soil orders. (a) This oxisol in Rwanda has a reddish color caused by oxidation of iron. (b) A mollisol in Nebraska has a dark, organic-rich and fertile *A* horizon. (c) This spodosol in New York is identified by the white, sandy surface horizon. (d) An inceptisol in Massachusetts shows only weak horizon development because the soil has been subjected to erosion. (e) Areas of volcanic activity often have andisols, with distinct ash layers, like this one in Japan. (f) This entisol in Connecticut has not had much time to develop, and so does not have distinct horizons. Compare with soil descriptions on pages 134–135.

and nutrient availability, as in deserts, can survive. **Desertification,** a process by which semiarid vegetation and soil become more desert-like as a result of human use, has occurred in semiarid lands around the world. Sometimes, desertification may be irreversible, at least over durations of a few decades. In such cases, grassland and other relatively productive vegetation types can be replaced with only sparse grasses—shrubs forming clumps of vegetation separated by bare ground (Figure 4-18). Soil organic matter is reduced, and when rain falls, it is more likely to run off the surface, increasing erosion and decreasing the soil's ability to support vegetation. As a result, many fewer animals can be supported by the available forage. Even though the climate today is not fundamentally altered, the land appears more desert-like than it did before excessive exploitation by humans.

Soil Fertility: Natural and Synthetic

Soil fertility refers to the ability of a soil to support plant growth by making nutrients available. Under a given set of conditions, more fertile soils produce higher crop yields than less fertile soils. Fertility, however, is a difficult thing to measure, because other factors that affect yields are also very important, such as the way the soil is plowed and the way crops are planted. Also, some soils may be poorly suited to one crop while another would thrive. Potatoes are grown profitably in northeastern Maine and Long Island, because of the sandy soils there; these soils would not be especially productive for corn.

Plants need sunlight, carbon dioxide, oxygen, and water. But they also require nutrients for growth,

Figure 4-16 A soil map for part of Rio Blanco County, Colorado. This portion of a detailed soil map shows the close correspondence between landforms, soils, and vegetation. Sagebrush prefers deep, well-drained soils (pale green); pinyon–juniper woodland prefers thin, rocky soils (dark green). On gently sloping uplands, moderately thick soils have developed from sandstone bedrock; vegetation is dense-to-open woodland of pinyon pine and juniper, with scattered sagebrush and grasses (yellow); small patches of drier, windblown material show a distinctive fine-textured soil with sagebrush (pink).

Figure 4-17 The landscape of Rio Blanco County, Colorado. This view is taken from the lower center of the map in Figure 4-16, looking north (toward the top of the page) The vegetation in the foreground is a mix of grasses and sagebrush; the darker woodlands in the distance are mainly pinyon pine and juniper.

137

Geography, Geographic Information Systems, and the Global Carbon Budget

Concern about the accumulation of carbon dioxide in the atmosphere is mounting, and countries are moving toward implementation of an international treaty to control carbon dioxide emissions. Therefore, the need for accurate information on carbon dioxide release and uptake is increasing. One of the most important parts of the puzzle is to compile detailed knowledge of the amounts of carbon released and taken up by the biosphere. Geography and geographic information systems (GISs) are critical to providing that information.

Many different flows must be estimated in calculating the carbon dioxide budget of the atmosphere, hydrosphere, biosphere, and lithosphere. Some of these are easier to estimate than others. We know fairly closely, for example, how much carbon is emitted by fossil-fuel combustion because we know approximately how much coal, oil, and natural gas each country burns. Another large number in the budget—the amount that goes from the atmosphere to the oceans each year—can be calculated based on knowledge of the physical processes that control the exchange of carbon dioxide between water and air. On the other hand, estimating the amount of carbon released by deforestation or taken up by forest growth is much more difficult, because it is highly variable from place to place. Other key processes, too, such as the accumulation and decay of organic matter in the bottoms of reservoirs, the release of carbon dioxide from decaying peat bogs, or the buildup and loss of organic matter from agricultural soils, are very poorly known. These processes, however, are the very ones that can be managed directly by humans, so they must be accounted for in any international agreement.

Geography and GISs provide the tools for extrapolating from a few isolated measurements to estimate processes spread over large areas, so geographers are playing a big role in studies of the carbon budget. For example, we can directly measure rates of carbon uptake by photosynthesis or release through deforestation for small areas like a few square meters or a few hectares. Such measurements involve tasks like collecting, weighing, and analyzing vegetation, or carefully sampling and analyzing carbon dioxide in air samples. We could not possibly do this field research for every hectare of land in the United States, but we can do this at many individual sites representing different land-cover types—coniferous forest, deciduous forest, lakes, wetlands, farmlands, and so forth. Then, using a GIS, we can make detailed maps of land cover (often based on satellite imagery, as in the figure) showing the spatial extent of each different kind of surface, and we can assign rates of carbon exchange for each of these types of surface based on our field measurements. The GIS then does the counting for us to estimate total carbon movements over large areas. Not only is a GIS useful in that accounting process, but, in combination with satellite imagery, it can help monitor and account for changing land-cover characteristics over time (see, for example, Figure 1-24). Geography and geographers will continue to play a central role in this work.

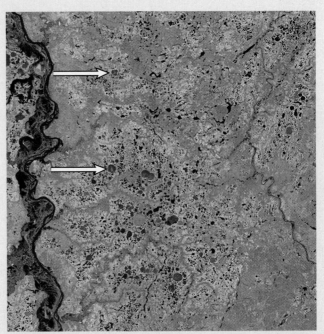

(a)

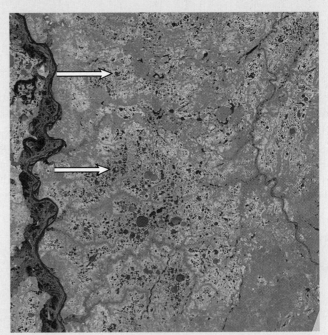

(b)

Disappearing arctic lakes. A great many lakes dot arctic regions, especially low-lying coastal plains. Satellite images such as these of Siberia in 1973 (above) and 2002 (below) show lake disappearance (arrows) associated with melting permafrost. Drainage of these lakes exposes sediments rich in organic carbon, and may accelerate release of this stored carbon to the atmosphere.

Figure 4-18 Desertification. Cattle grazing on sparse grass cover near Chaco Canyon, New Mexico. Only 300 years ago, large herds of native grazers and newly introduced cattle could be supported on extensive grasslands. But today, only a few animals can be supported in this semiarid environment.

including nitrogen, phosphorus, potassium, calcium, magnesium, sulfur, and others. In general, fertile soils are those that store and release sufficient quantities of nutrients and water for optimal plant growth. Nutrients are present in many different forms—dissolved in soil water, as part of organic matter, attached to the surfaces of mineral grains, and in the crystalline structure of minerals. Some forms are more available to plants than others. Organic matter, both living and dead, stores nutrients and makes them available to plants. In most soils, depletion of organic matter reduces fertility, but the addition of manure to the soil helps replace lost organic matter. Most available nutrients occur in the uppermost soil horizons: the O, A, and upper B horizons. Cultivating the soil and harvesting agricultural crops accelerates the breakdown of organic matter and losses of nutrients stored in the soil. Sustaining soil fertility requires maintaining soil nutrients and maintaining the soil's ability to store and release them to plants.

Farmers long have known that measures must be taken to restore soil fertility. In traditional agricultural systems, this has meant a *fallow period*, which is a season when a field is allowed to rest and nothing is harvested from it. During this rest period, the natural processes of plant growth, organic matter accumulation, mineral weathering, and biological activity can restore fertility. However, Earth's burgeoning population demands more agricultural products, so fallow periods have been reduced or eliminated. Many areas that once were fallowed are now cultivated every year, thereby reducing or eliminating the opportunity to restore natural soil fertility.

The low cost, availability, and easy application of inorganic fertilizers has led to replacing organic (natural) fertilizers with these manufactured versions. Manufactured inorganic fertilizers contain valuable nutrients, but only organic matter can help maintain the ability of the soil to store those nutrients. In the short run (a single crop season), nutrients can be added to soil to supply plants' needs, but in the long run, the ability of the soil to support plant growth is diminished.

Farmers and resource managers around the world increasingly worry about the impact of soil degradation on the world's food-producing capacity. Declining soil fertility associated with erosion is particularly acute in many tropical and subtropical areas, such as Africa. The high cost of fertilizer, and the resulting limited use of such fertilizer, is also contributing to a long-term decline in soil productivity in many areas.

Worldwide, large amounts of organic carbon are stored in soils, and agricultural activities have contributed to the loss of organic carbon from soils. Some of this carbon has been buried with sediments eroded from farm fields, and some has been broken down and may have contributed to increased atmospheric CO_2. Improved farm management practices that conserve soil also help restore soil organic matter, and are being promoted as a way to offset emissions of carbon to the atmosphere.

The study of landscape patterns at local and regional levels, using vegetation and soil maps, almost always reveals human disturbance. As human settlement increases, environmental managers worry about our impact on ecosystems. Although we now know that these activities can have global impacts, such as altering the global carbon budget, the processes that drive them are almost always largely regulated by local-scale conditions. For example, a well-drained soil on a hillside may have very different carbon-storing properties from a moist soil at the bottom of the hill.

Ecosystems

An **ecosystem** includes all living organisms in an area and the physical environment with which they interact (Figure 4-19). An ecosystem can cover an area as small as a field or a pond. On any scale of analysis, large or small, certain fundamental elements exist:

- *Producers*—green plants and other organisms that produce food for themselves and for consumers that eat them.
- *Consumers*—organisms that eat producers, other consumers, or both.
- *Decomposers*—small organisms, such as bacteria, fungi, insects, and worms, that digest and recycle dead organisms.
- *Abiotic materials and energy necessary for production and consumption to occur*—water, mineral nutrients, gases such as oxygen and carbon dioxide, and energy (light and heat).

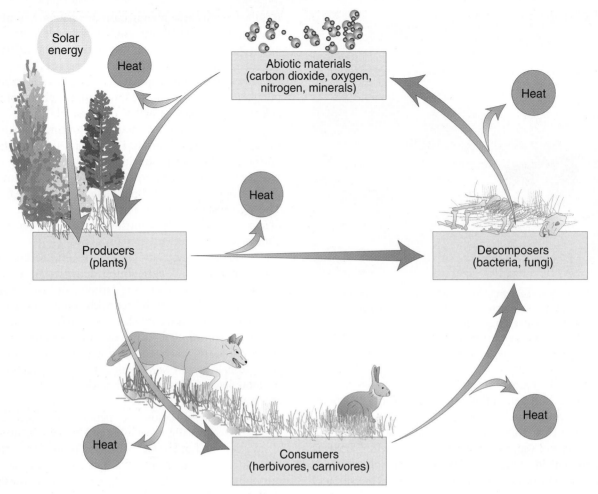

Figure 4-19 Ecosystems. Ecosystems consist of plants, animals, decomposers, and the physical environment with which they interact. Energy is supplied to an ecosystem from the Sun, passed through it via a food chain, and dissipated as heat.

Ecosystem Processes

Energy from the Sun is the starting point for understanding the operation of an ecosystem. Sunlight makes photosynthesis possible, and green plants produce food in the form of carbohydrates. The food is distributed through an ecosystem by way of a **food chain.**

Plant-eating animals, known as **herbivores,** begin the food chain by consuming plants (Figure 4-20). **Carnivores,** which are meat-eating animals, eat herbivores and may in turn be eaten by other carnivores or by **omnivores,** such as humans, who eat both plants and animals. Most of the food that animals consume is used to keep their bodies functioning. Animals excrete some of the food, and when they die, their bodies are rich in stored-up nutrients. Decomposers attack animal excretions and dead bodies. These small organisms return chemical nutrients to the soil and the atmosphere. Some decomposers provide nutrients to new plant growth, completing the cycle.

Each step in the food chain is called a **trophic level.** At each trophic level, food is passed from one level to

the next, but most of the energy is lost. As a rule of thumb, about 10 percent of the energy consumed as food at a given trophic level is converted to new biomass, and the remaining 90 percent is burned off and dissipated as heat, although this proportion varies substantially. Because of this loss of energy, the amount of biomass decreases as we go from the first trophic level—the green plants—to higher levels. This is why we find large numbers of herbivores such as mice and rabbits in natural systems, but relatively few large carnivores like wolves and lions. There simply isn't enough energy to support them.

Some chemicals in the environment, especially persistent pesticides such as DDT, create problems passing through a food chain. These chemicals do not break down easily in the environment, so they accumulate in animal tissues. If a pesticide accumulates in, rather than being excreted from, an animal, then at each trophic level the concentration of that pesticide increases in a process called **biomagnification.** Consider this example: If a falcon weighing 1 kilogram consumes 50 smaller birds, each weighing 0.2 kilogram, and if

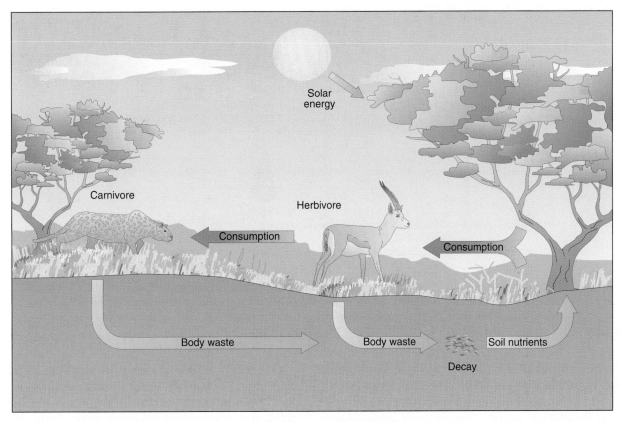

Figure 4-20 A food chain. Green plants are the primary producers of food, which is eaten by the primary consumers: herbivores and omnivores. Carnivores, or secondary consumers, derive their energy from other animals rather than directly from plants. At each step of the food chain, energy is lost as heat, and waste materials are broken down by decomposers.

each of these smaller birds consumes in its lifetime 2 kilograms of plant-eating insects, the falcon indirectly consumes 100 kilograms of insects. If each of these insects consumed during its lifetime 10 times its weight in plants, then the falcon has indirectly consumed 1,000 kilograms of plants! Therefore, even a very small concentration of pesticides at a low trophic level, such as a plant, can become a high concentration in a large carnivore. In some birds of prey such as eagles and falcons, this biomagnification causes eggs to fracture, killing chicks as they incubate. Because of these effects, DDT and similar pesticides have been banned or severely restricted in much of the world, and the pesticides used today break down relatively rapidly to reduce this problem.

Plant and animal success in ecosystems Within any particular ecosystem, living things compete for resources such as nutrients and water. The more successful plants and animals in this competition will dominate that environment.

Success of a species means that it survives, thrives, or even comes to dominate an ecosystem for a long period. Humans are an example: surviving for the first 150,000 years of our existence as a species, then thriving over the last few thousand years, and today dominating Earth (with dramatic consequences for other species).

The success of one species over another is the result of competition. In the case of plants, this competition is for light, water, nutrients, and space. Although plants require all these factors to grow, in any ecosystem one factor usually is restricted, which forces competition and adaptation. For example, in an arid environment, plants compete for scant water but do not need to compete for the abundant sunlight. In a humid environment, water is abundant, but plants must compete for sunlight. An area with adequate water and light may have poor soils, so plants must compete for nutrients. Through evolution, the plants that have adapted their life forms, physiological characteristics, and reproductive mechanisms that allow them to succeed in particular environments have survived. The plants that compete best in an environment dominate the ground cover there.

What adaptations are effective in competing for space and nutrients? In a humid area, height is effective: Plants that are able to grow tall capture the most light and shade out their competitors. An ecosystem with extreme cold, drought, fire, or a short growing season favors organisms that have adapted to survive or even benefit from the stress. For example, some species have adapted by developing the ability to take advantage of occasional fires. After a fire has felled other vegetation and heated the soil, long-dormant seeds respond to the flash heating and germinate quickly.

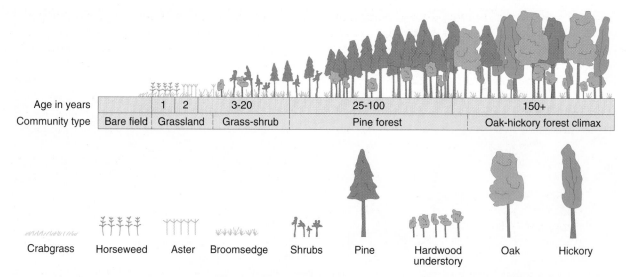

Age in years		1	2	3-20	25-100	150+
Community type	Bare field	Grassland		Grass-shrub	Pine forest	Oak-hickory forest climax

Crabgrass   Horseweed   Aster   Broomsedge   Shrubs   Pine   Hardwood understory   Oak   Hickory

Figure 4-21 Succession sequence. Clearing a field sets the stage for the succession sequence shown here, from left to right. This example is typical of conditions in the southeastern United States. In each successional stage, modification of the local microenvironment allows a new assemblage of plants and animals to become established. The sequence begins with pioneer species such as crabgrass and horseweed and culminates in the climax hickory forest.

They occupy the burned site, completing their life cycle and spreading new seed before the other vegetation recovers from the fire. Once the other vegetation regains dominance, the fire-resistant seeds lie dormant, awaiting the next fire opportunity.

Community succession As plants grow and multiply, they alter their environment, sometimes changing it to the point that it becomes more suitable for the growth of other kinds of plants. Thus, the plants of the community may change over time until a stable plant population emerges. This community remains until something, perhaps a fire, changes the physical environment and then different plants again begin to grow. These changes in the makeup of the plant community are called **succession.** A common example of community succession occurs when agricultural land is abandoned in eastern North America (Figure 4-21):

- In the first couple of growing seasons, the fields sprout fast-growing herbaceous weeds and other pioneer species. These spread seeds widely and tolerate bright sunshine.
- Over the next few growing seasons, the weeds and pioneer species are succeeded by slower-growing perennial shrubs and trees, such as pines, which prefer bright sunshine and gradually shade out the smaller plants beneath them.
- Hardwoods that tolerate shade grow beneath the pines, eventually overtaking them in height. As the pines die away, a forest of shade-tolerant species dominates the canopy. They are replaced by their own seedlings as they die, thus leading to a stable association of plants in equilibrium with local soil and climate, called a **climax community.**

The length of time necessary for succession to take place varies, depending on the growth rates and lifespans of plants, rates of seed introduction and establishment, and whether soil development proceeds along with plant growth. If the seeds of the late-successional species are present at the time succession begins and growth rates are fast, a well-developed forest may be established in a few decades. The illustration shows a more common situation, where a mature forest is achieved in 150–200 years. But if succession also involves soil development through weathering and the accumulation of organic matter, it may take thousands of years for a site to be converted from bare rock to a mature forest.

In environments subject to frequent disturbance, such as fire, windstorm, disease outbreaks, floods, and volcanic eruptions, it may take a long time to establish a climax community. Vegetation may actually change continually but never reach a climax.

Earth's recent history has been one of striking change, including dramatic climate changes and human transformation of vast areas. Even environments that have not been severely altered by humans are subject to natural disturbances such as storms, fire, or outbreaks of plant or animal diseases. These environmental changes have profoundly influenced vegetation distributions in many areas, favoring plants that are adapted to dynamic conditions. Very little of Earth's surface can be regarded as stable or undisturbed.

Biodiversity

Biodiversity is the diversity of species present in any environment. Biodiversity is important because it brings multiple food options to living things, thereby improving each one's chance for survival. Biodiversity enhances the ability of an ecosystem to function and thus process nutrients and provide food to the animals that depend on it. Experiments have shown, for example, that diverse grassland communities produce more biomass than communities that have only a few species. Biodiversity has many benefits to society including nutrient processing; providing food, medicines, and other materials that we need; and carbon sequestration in some environments.

Earth's biodiversity is vast: There are an estimated 10 million plant and animal species. Each of these has specific needs. For plants, critical needs include water, light, and nutrient availability; effective seed dispersal, good germination conditions; and the absence of soil disturbance. For animals, the most critical requirement is food supply, meaning a sufficient quantity of specific plants or animals to feed upon. In a world of rapidly changing environments, the critical needs of many species increasingly are not met.

Habitat loss is so widespread that a significant portion of the world's species may be threatened with extinction. The largest single cause of extinction is land-use change, resulting from increasing human settlement worldwide. About 37 percent of Earth's land area (excluding Antarctica) now is cropland or permanent pasture. Between 1960 and 2000, the land area of such agricultural uses increased about 10 percent, whereas the world's forest area decreased about 5 percent. Hunting, species introductions (such as invasive species that outcompete indigenous species), and pollution also contribute to extinction. In Europe and North America, radical changes in the landscape have taken place over centuries of human habitation. Although a few extinctions were noted, little attention was paid to the broader, unknown impacts of those landscape changes. Rapid change now is occurring in the Amazon basin at the is now occurring, while the science of tropical environments is still young. Biologists working in the Amazon discover new species virtually every time they look, raising concern that deforestation is causing the extinction of species even before we discover them.

Biosphere reserves To reduce the impact of land-use change on species diversity, protected areas are being established by national governments and by international agencies. One important international effort is the United Nations Biosphere Reserve Program. Biosphere reserves are typically designed to include a core area in which land use activities are highly restricted, surrounded by a zone in which resources can be used for human purposes but in a way that maintains the environmental integrity of the core. Of course, individual states have their own protection strategies as well. Worldwide, approximately 6.1 percent of all land area is protected in some way.

The size of individual protected areas is important. A single animal may range over only a few square kilometers, but that small area cannot ensure species survival. A reserve must be large enough to sustain enough individuals to allow genetic diversity in the breeding population. It is difficult to decide how large a protected area should be; this disagreement can become part of the political tug-of-war among conflicting potential land users.

Today conservationists are beginning to recognize that setting aside a few patches of relatively "natural" landscape will not be sufficient to preserve Earth's remaining biodiversity. Rather, we must also consider ways of promoting biodiversity in the managed landscape—on farms, in commercial forests, and even in urban areas. One reason is the sheer size of human-dominated land areas in comparison to the small patches that have been protected. The land available for protection is insufficient to retain biodiversity across the landscape. Another reason is the belief that biodiversity in itself provides important goods for humans, perhaps through enhancing biological productivity or promoting ecological stability. If biodiversity contributes to these services, then we need to maintain that biodiversity where we live and work—not just in remote preserves.

Biodiversity is a barometer of the consequences of human modification of the environment. A commitment to biodiversity means a commitment to a broad range of environmental policies—not just preserving wilderness areas, but maintaining diversity everywhere.

Biomes: Global Patterns in the Biosphere

Earth's ecosystems are grouped into **biomes** characterized by particular plant and animal types, usually named for a region's climate or dominant vegetation type. Biomes typically contain many ecosystems. A terrestrial biome's nature reflects two especially visible features: climate and vegetation. Underlying a biome's label is a diverse community of characteristic plants and animals.

The global distribution of natural terrestrial biomes imitates very closely the distribution of climate regions described in Chapter 2 (Figure 4-22; also refer to Figure 2-33). This correspondence reflects a close link between climate and vegetation. To grow, plants depend on the two most important elements of a climate: energy and water. The actual distribution of vegetation is somewhat different than shown in Figure 4-22 because of human disturbance, especially deforestation and agriculture.

We will now examine the vegetation and climates that distinguish major biomes: forest, savanna, scrubland,

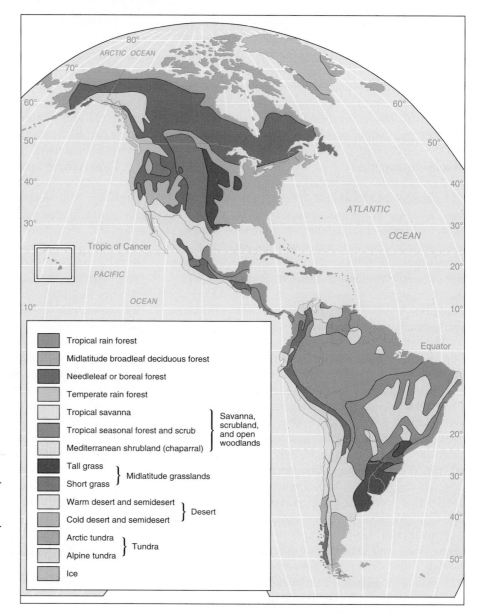

Figure 4-22 The major terrestrial biomes. This map shows vegetation that would exist if humans did not disturb the vegetation, as in agricultural regions. Map generalization at this scale obscures many local variations in vegetation types, especially in mountains and populated areas.

woodland, grassland, desert, and tundra. Figure 4-23 is a matrix showing some of the relationships among temperature, water availability, and vegetation characteristics. The distribution of animal species is strongly influenced by the distribution of vegetation; thus, our descriptions of biomes will focus on plants.

Forest Biomes

Forest biomes are dominated by trees: tall, woody-stemmed perennials (plants that live for at least several years) with spreading canopies. Forests occur over a wide range of humid and subhumid climates, where ample water is available most of the year. We group forests into four broad categories: *tropical rain forest*, *midlatitude broadleaf deciduous forest*, *needleleaf or boreal forest*, and *temperate rain forest*.

Tropical rain forest biome Tropical forests and woodlands dominate the humid equatorial environments

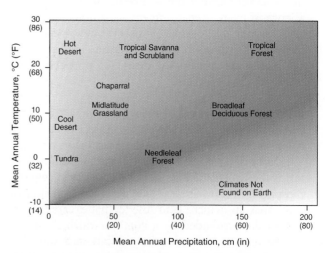

Figure 4-23 Generalized relation between vegetation type and climate. The amount of precipitation necessary to support a given amount of vegetation (such as forest relative to grassland) is greater in warm climates, where evapotranspiration is higher, than in cold climates. Cold climates with high rainfall amounts are not found on Earth.

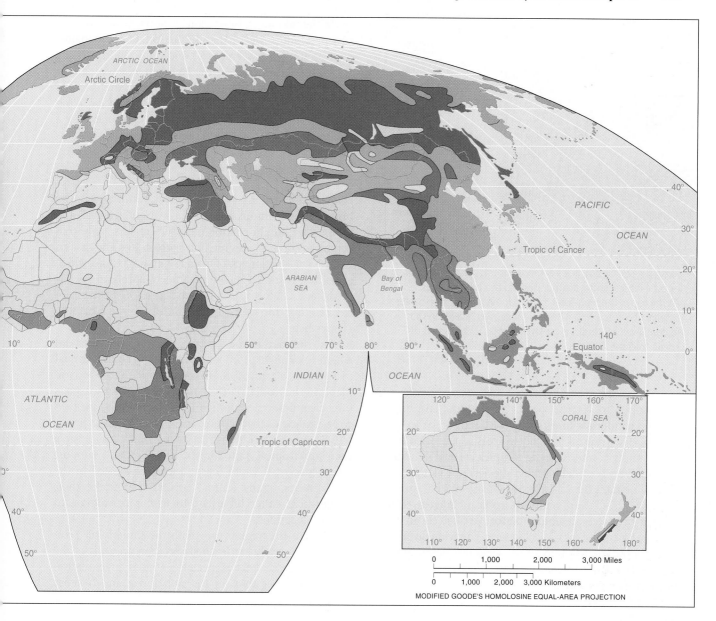

ARCTIC OCEAN

Arctic Circle

PACIFIC

OCEAN

Tropic of Cancer

ARABIAN
SEA

Bay of
Bengal

140°

Equator

ATLANTIC

OCEAN

INDIAN   OCEAN

Tropic of Capricorn

CORAL SEA

MODIFIED GOODE'S HOMOLOSINE EQUAL-AREA PROJECTION

of Central America, the Amazon River basin, equatorial Africa, and Southeast Asia (Figure 4-24). Tall, broadleaved trees retain their leaves all year. A **tropical rain forest** has a top layer, or canopy, and two more layers beneath (Figure 4-25). Each layer has different dominant species and associated animal communities.

The highest canopy consists of very tall trees. Underneath is the dense middle canopy. Most photosynthesis occurs in these two upper layers, where sunlight is most intense. The middle canopy blocks most light, keeping the forest floor in twilight, so plants at the lowest levels must tolerate deep shade to survive. Climbing vines are common in many tropical rain forests, as are plants that grow on the surfaces of other plants to gain height advantage without having their own long trunks or stems reaching to the ground.

Tropical rain forests are noted for biodiversity, having hundreds of tree species within a single hectare (about 2.5 acres). In contrast, midlatitude forests may

Figure 4-24 Tropical rain forest. Rain forest vegetation is generally dense and multistoried, as exemplified by this scene from Costa Rica.

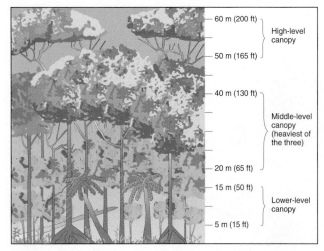

Figure 4-25 Canopy layers in a forest. Three levels of forest canopy can be distinguished in the tropical forest. At the top is the high-level canopy, in which the tallest trees emerge above the rest. The densest canopy layer is in the middle levels. A third, lower-level canopy is composed of shade-tolerant species.

have only 5 to 15 tree species per hectare, and high-latitude coniferous forests may have only three to five species. The tropical diversity of trees is paralleled by a great variety of animals. The complex vertical structure adds a diversity of habitats and species. A walk through a zoo illustrates the vast number of large animal species of the world that are native to tropical climates. Non-human primates are found almost exclusively in the tropics, especially in tropical forests. The wide diversity of life in the tropical rain forest places this biome at the center of controversy over deforestation, species extinctions, and biodiversity.

In the rain forest biome, much nutrient storage is in a living biomass rather than in the soil. Organic matter decays rapidly in warm, wet conditions, and nutrients are not easily retained in the soil. When leaves fall on the forest floor and begin to decompose, the released nutrients are quickly absorbed and recycled by living plants. This is yet another reason why deforestation is a major concern. If forests are removed, their store of nutrients may be lost quickly to leaching and runoff, leaving only the impoverished soil.

Midlatitude broadleaf deciduous forest biome   The **broadleaf deciduous forest** exists in subtropical and midlatitude humid environments where seasonally cold conditions limit plant growth (Figure 4-26). The eastern United States and much of China share this biome, which explains why some Chinese ornamental species are popular on lawns in the eastern United States. On the map you can see that this biome occurs mostly in the Northern Hemisphere.

During the summer growing season, long days and a high solar angle promote rapid growth, so that annual plants may attain 60 to 75 percent of the growth of plants in the tropics, even though the growing season is only five to seven months. These plants have evolved wide, flat leaves that capture as much sunlight as possible. Broadleaf deciduous forests are much less diverse than are tropical rain forests, but they still support many species. Because these forests are located in the midlatitudes, many animal species that make their home there either hibernate in winter or migrate to warmer locations.

Soils of this region vary with the underlying geology, but generally are leached less than are soils of the humid tropics. This is because annual precipitation usually is lower than in the rain forest and because

Figure 4-26 Mid-latitude forest in northern Michigan. This area is in the transition zone between midlatitude deciduous forest and northern coniferous forest in northern Michigan.

Carbon Emission Offsets

The complexity of the carbon cycle, with its many possible sources and sinks (places where carbon is taken out of the atmosphere, such as where it is absorbed by growing plants), means our emissions from fossil fuel combustion can be counterbalanced in a number of ways. It also means that from the standpoint of the CO_2 content of the atmosphere, it is just as helpful to reduce the amount of CO_2 we put in the atmosphere as it is to increase the amount that is taken out. This opens up a wide range of possibilities when it comes to managing the global carbon budget.

One approach to managing the environment involves creating markets on which some sort of commodity that affects environmental quality is bought and sold. In this way, environmental protection can be given a monetary value, and thus encouraged. With regard to carbon, emitters can buy "offsets," which are financial instruments created to help reduce carbon dioxide in the atmosphere. Offsets are sold in units of tons of carbon, with prices typically in the range of $5 to $30 per metric ton of CO_2. The value of carbon taken out of the atmosphere in various projects around the globe is sold on the Chicago Climate Exchange, as well as in markets elsewhere in the world.

When someone buys an offset, the money is used to encourage carbon storage or *sequestration*, or prevent emissions somewhere else. For example, planting forests and allowing them to grow takes carbon out of the atmosphere, and carbon offsets can be used to pay for these types of activities. Carbon offsets are used to plant forests, build renewable energy systems, and improve energy efficiency. At the present time, most offsets are purchased voluntarily by individuals or corporations that want to invest in a lower-carbon future, but it is possible that future CO_2 emission regulations might require some emitters to purchase offsets.

One important aspect of these systems is that it can be very difficult to accurately quantify the amounts of carbon that are moving in or out of a given part of the environment. And yet, because offsets are sold with prices of dollars per tons of CO_2, accurate measurement of carbon flows is essential.

To give a sense of the possible problems, consider the role of forests in carbon storage. It is often suggested that preservation of tropical forests will help slow global warming. However, the trade-offs are complex, and not every forest preservation action will reduce atmospheric carbon. Certainly, cutting down a forest releases carbon to the atmosphere. But will preserving a forest do the opposite? In a mature forest, young trees grow but old trees die and decay, so a mature forest does not necessarily take carbon out of the atmosphere because photosynthesis is balanced by respiration. In a young forest, on the other hand, most trees get larger year after year so there is a net increase in the carbon stored in biomass. If we were to cut down an old-growth forest and convert all the wood to furniture or other long-lived products and then allow a young forest to grow in the place of the former old growth, the net effect would be to remove carbon from the atmosphere as long as the lumber is preserved and not burned or allowed to decay. So it is not always the case that cutting down a forest releases carbon to the atmosphere. In recent years, a variety of proposals for actions that would actively manage the global carbon cycle—sometimes called "earth engineering"—have emerged. The possibility of income and profits from carbon storage increases interest in these activities. Ocean fertilization is one of the more controversial of earth engineering ideas. As we have already learned, the ocean takes up large amounts of atmospheric carbon. Some of this is converted to biomass through photosynthesis in the oceans, and some of that biomass settles to the bottom of the sea and is effectively stored there. In much of the world's oceans, the rate of photosynthesis can be increased by adding fairly modest amounts of key nutrients needed by marine organisms. Experiments have been carried out showing that addition of iron, for example, to surface waters can increase photosynthesis rates, although it is unclear whether the added biomass produced ultimately is deposited on the ocean floor. If the added biomass is respired instead, the effect on atmospheric CO_2 may be negligible. Of course actions like this would likely have unanticipated impacts on other aspects of marine ecosystems, some of which could be negative.

As more countries move toward regulating carbon emissions, the value of carbon offsets, and thus the likelihood of major investments in carbon sequestration, will increase. As interest in carbon offsets grows, an accurate understanding of carbon movements will become increasingly important.

lower temperatures slow chemical weathering. Lower temperatures also contribute to reduced decay rates of organic matter. Consequently, undisturbed soils typically have a substantial *O* horizon, ranging in thickness from a few centimeters (1 to 2 inches) to over 10 centimeters (4 or more inches). When these forests are cleared and the soil is brought under cultivation, the soils generally are quite fertile, but that fertility declines once the organic matter decays and is not replaced.

Needleleaf, or boreal, forest biome   In poleward portions of the midlatitudes, a needleleaf (coniferous) evergreen forest flourishes. The name "boreal forest" comes from the northern location of this vegetation (in Latin, *borealis* means "northern"). The **boreal forest** thrives in cold, continental midlatitude climates. Extensive boreal forests are restricted to the Northern Hemisphere, since there is little land in the Southern Hemisphere in latitudes 50° to 70°. The most extensive boreal forest regions are in central Canada and Siberia. Needleleaf forests also occur in some areas of seasonally dry soils.

During cold winters, low humidity and frozen ground cause moisture stress, but needleleaved trees survive because the leaves have low surface area in relation to their volume, and they are covered with a waxy coating to reduce water loss. This adaptation allows them to survive in colder climates that would kill broadleaved trees. Because they are evergreen, these trees photosynthesize whenever the temperature is warm enough, instead of having to wait until after the last frost to produce leaves. Most needleleaved trees are evergreen, but not all. The larch, for example, is a needleleaved deciduous tree.

Coniferous forests also occur in the southeastern United States, where sandy soils drain quickly and have limited water storage. During warm summers, these soils dry out, and the moisture-retaining properties of coniferous trees allow them to compete well against broadleaved trees. The ability to sprout from roots and grow rapidly following forest fires is an added advantage of many coniferous species in areas of summer aridity.

Needleleaves accumulate beneath coniferous forests, and as they decay, they produce strong acids in water that percolate down through the soil. This acidic water leaches the upper soil of all but the most insoluble minerals, leaving a light-colored, sandy upper soil layer. Some of the iron and aluminum leached from the upper soil accumulates lower in the soil profile. This produces a distinctive soil associated with coniferous forests called a *spodosol*, which you remember seeing in Figure 4-15.

Temperate rain forest biome   On the west coasts of North and South America, lush evergreen coniferous forests grow. These temperate rain forests are found in areas of marine west coast climate, where moderate temperatures and ample rainfall allow year-round plant growth. The evergreen trees grow faster in summer, but they can photosynthesize in winter, too, despite occasional freezing temperatures. The temperate rain forest is much less diverse than the tropical rain forest and is dominated by relatively few species.

Savanna, Scrubland, and Open Woodland Biomes

Savannas and open woodlands have trees spread widely enough for sunlight to support dense grasses and shrubs beneath them. This vegetation is common in climates that have a pronounced dry season. The term **savanna**

Human-Dominated Systems

In recent years, quantitative study of biogeochemical cycles at the global scale has determined the amounts of materials that are processed in various environments per unit of time. This research was done in part to understand the relative importance of different components of those cycles. One of the conclusions of this research is that a considerable amount of material processing—the conversion of substances from one form to another—is carried out by humans. In some cases humans do more processing than nature does.

Consider some examples. We have already seen that humans are playing such a major role in the global carbon cycle that the quantity of CO_2 in the atmosphere is steadily increasing, despite the vast amounts of carbon processed each year by the biosphere. The nitrogen cycle is another important cycle that is influenced by humans. In this cycle, nitrogen is taken from the atmosphere and converted to forms that can be used by plants and animals to build proteins and other molecules. This bioavailable nitrogen is then passed through food chains before being broken down and released back to the atmosphere. Humans *fix*

nitrogen (that is, convert it to biologically available forms; see Chapter 8) through fertilizer manufacture, and today more nitrogen is fixed by humans than by natural processes. Food production through photosynthesis is another critical biogeochemical process in which humans play a major role. About 8 percent of global photosynthesis takes place in agricultural lands. Humans play an even larger role in consuming plant matter, either directly as food for themselves and their domestic animals and as fiber for material such as lumber, or indirectly by controlling its characteristics or fate, such as grass grown on golf courses or crop residues left in the field. Estimates of the amount of biological production that is "appropriated" by humans range from about 3 to 40 percent, depending on what human uses are included in the estimate.

These conclusions are of great importance, not only for scientific and practical purposes but in describing the physical relationship between humans and the environment. This, in turn, has implications for the way we think about nature and our role in it. In short, it goes to the very heart of the human–environment relationship.

refers especially to vegetation characteristic of large seasonally dry areas in tropical Africa and South America. The rainy season brings green, lush grass, but during the dry season the grass dries out and only deeper-rooted trees continue to photosynthesize. If the dry season is pronounced, even the trees may lose their leaves, and in some areas dry-season deciduous trees are common.

Fire is common in the savanna. Although aided by the seasonal aridity, it is probably as often caused by humans as well as lightning. Fire does not destroy most grasses, which grow back quickly. Slower-growing trees might in time become dominant, shading out the grasses, but they are more susceptible to fire damage, so the savannas remain. Whether the savannas of Africa and South America are the result of human activity is often debated. Some expanses of African savanna occur in climates that support forests elsewhere, thus encouraging the hypothesis that human-caused fires have created the extensive African savannas.

Soils of tropical savannas usually are deeply weathered, as in the humid tropics, but they may be less leached of nutrients as a result of reduced rainfall. High temperatures and reduced plant growth combine to lower the organic content. Many soils of these regions are quite productive if sufficient irrigation water is available.

In more arid margins of tropical savannas, a seasonal forest or thorn-scrub vegetation occurs. Drought-resistant trees and shrubs dominate, as well as cacti, and grasses are less important. Savanna-like open woodlands (also called parklands) also are found in some semiarid midlatitude areas.

Mediterranean climates promote a distinctive shrubland vegetation type called *chaparral*. This is a shrub woodland dominated by hard-leaved trees and shrubs that withstand the severe summer aridity. Fire is common because of summer aridity and flammable waxy-leaf coatings. Many plant species in the chaparral are adapted to frequent fire, and some even require fire to maintain their continued presence in the landscape.

Midlatitude Grassland Biome

In Figure 4-22, extensive midlatitude grasslands are shown in deep red and brown. Major midlatitude grassland areas include interior Asia, the Great Plains of Canada and the United States, and central Argentina (Figure 4-27). The midlatitude semiarid climate of these areas features hot summers, cold winters, and moderate rainfall. In moister portions, the grass is usually taller (up to 2 meters, or 6 to 7 feet), and the biome is called **prairie**. In drier areas, such as Asia, the biome is called short-grass prairie, or **steppe**.

Grasses are well suited to this climate because they grow rapidly in the short season when temperature and moisture are favorable (generally spring and early summer). During dry or cold periods, above-ground parts

Figure 4-27 Tall-grass prairie in Kansas. A small remnant of formerly extensive tall-grass prairie in Kansas.

of these plants die back, but the roots become dormant and survive. This trait also allows grasses to survive fire and grow back rapidly, using available moisture at the expense of trees or shrubs that might invade. Many experts believe that occasional fires, either natural or deliberately set by humans, are at least partly responsible for establishing and maintaining extensive grasslands.

The soils of these midlatitude grasslands are very fertile and have a dark upper horizon with high nutrient content. Because of this fertility, many of the world's midlatitude grasslands are "breadbaskets": They are primary production areas for wheat and other cereal grains (Figure 4-28). In North America, virtually all original *tall-grass prairies* are farmed to produce wheat, corn, soybeans, and other small grains. Much of the *short-grass prairie* present at the time of European settlement remains, although it has been heavily grazed.

Desert Biome

In **deserts**, moisture is so scarce that large areas of bare ground exist, and the sparse vegetation is entirely adapted to moisture stress. Plants with such adaptations are called **xerophytes**. Some desert plants are drought-tolerant varieties of types common in more humid areas, such as grasses. Others, such as cacti, are almost exclusive to deserts. Most desert plants are adapted to limit evaporative losses, with thick, wax-coated leaves or with needles rather than leaves, meant to protect cacti from being consumed for the water they contain.

Another common adaptation is water-collecting root systems. These are extensive shallow roots to gather occasional rainwater or long tap roots to reach deep groundwater. Many desert plants survive long periods without moisture by becoming dormant, either

Figure 4-28 Former tall-grass prairie. Today most of the tall-grass prairie has been converted to row crops.

as a mature plant or as a seed. When occasional rains do occur, these dormant plants spring to life and flower in a few weeks before becoming dormant again. Animal life in the desert is dispersed because of the reduced rates of plant production, but it is still diverse. Many desert animals are nocturnal, foraging and hunting at night to avoid dehydration during the hot daytime hours.

Scant moisture and the consequently slow chemical activity leave many desert soils poorly developed. Leaching of soluble nutrients is limited, and usually the horizons in desert soils are developed through vertical movements of soluble minerals, controlled by rainfall and evaporation. When water enters the soil, it may dissolve some minerals, but as evaporation draws this water back to the surface, these minerals are redeposited. As a result, high concentrations of soluble minerals near the surface (especially salts) are common. Desert soils might be fertile if irrigated, but irrigation usually introduces more minerals to the soil, possibly making it salty.

Tundra Biome

In cold, high-latitude environments or in high mountains above the treeline, freezing and short growing seasons severely limit plant growth. During winter, water is locked up in the soil as ice. Dehydration and abrasion by blowing snow damage exposed upper parts

Fire and Forest Management in the Western United States

In the chaparral woodlands of California and many other forest ecosystems of the western United States, fire is an important natural part of ecosystems. Unfortunately, people have built homes and businesses in forested areas. Conflagrations like the 2006 fires near Sedona, Arizona, destroy entire communities, cause vast property damage, and occasionally take lives. Fire has also been an increasing problem in western Canada recently, and one major fire that started in the United States burned across the border into British Columbia. When a fire starts near a populated area, the natural reaction is to extinguish it at once. Paradoxically, this action to halt fire actually makes fire-prone conditions worse. In the chaparral, plant growth produces flammable biomass that does not decay as rapidly as it is produced, so it accumulates over time. The longer it accumulates, the more fuel is available for severe fires. If a fire starts in a recently burned area, little fuel is available, the fire

burns slowly at cooler temperatures, and it does not spread widely. But if a fire starts where one has not burned for a long time, a hot, dangerous fire is likely.

Consequently, many areas that routinely would burn naturally have not burned, and fuel accumulations—dry leaves and dead wood—are unusually large. When a fire starts during dry, windy conditions, it becomes very dangerous and uncontrollable. Its high temperatures make it much more destructive than fires during moist conditions.

Decades of firefighting in the western United States have not eliminated fires; they have only allowed fuel to accumulate. Normal, "natural" fires, low temperature and slow burning, have been eliminated and replaced by infrequent but catastrophic conflagrations. Fire managers now use controlled burns in some areas to reduce fuel accumulations.

of plants. Therefore a **tundra** is dominated by low, tender-stemmed plants and low, woody shrubs. These survive the cold by lying dormant below the wind, often buried in snow, growing only in the short, cool summer. Animal life in this biome is very limited in winter, but in summer, the tundra comes alive with insects, migratory birds, and grazers such as reindeer and caribou.

As in deserts, tundra soils often are poorly developed because of slow chemical activity. **Permafrost**, or soil/rock that has a temperature below 0°C (32°F) all the year, occurs where the mean annual temperature is below 0°C (32°F). Soil horizons are weakly developed and often disrupted by frost. In poorly drained areas, organic matter often accumulates rather than decays, and peat is formed. As was mentioned previously in this chapter, peat is an accumulation of plant debris and an early stage in the formation of coal.

Because of slow plant growth in this environment, recovery from disturbances can be very slow. This, combined with the disruption of permafrost caused by construction activities and vegetation disturbance, has led to concern about the long-term effects of oil exploration and related activities in Alaska, Siberia, and similar areas.

Natural and Human Effects on the Biosphere

Plants, soils, climate, and human activity each reflect the influence of the others. In the absence of humans, climate would have the strongest control on vegetation. Climate controls soil formation and plant growth so strongly that, with the exception of agricultural areas, world vegetation and soil patterns correspond closely to world climates. Humans, however, have had profound influences on ecosystems in most of the world's land areas, including the 37 percent of land area (excluding Antarctica) that is cropland or permanent pasture (Figure 4-29). Locally, topography and geology also exert strong influence on vegetation and soils.

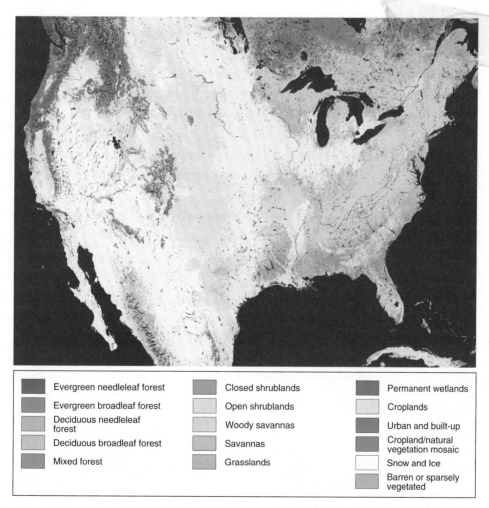

Evergreen needleleaf forest	Closed shrublands	Permanent wetlands
Evergreen broadleaf forest	Open shrublands	Croplands
Deciduous needleleaf forest	Woody savannas	Urban and built-up
Deciduous broadleaf forest	Savannas	Cropland/natural vegetation mosaic
Mixed forest	Grasslands	Snow and Ice
		Barren or sparsely vegetated

Figure 4-29 Land cover of part of North America, based on satellite imagery. Compare the extent of agricultural land with the distribution of natural vegetation types, as shown in Figure 4-22.

Regionally and globally, climate is in turn affected by ecological processes and soil-vegetation linkages:

- Plants regulate evapotranspiration from land surfaces and thus moderate the hydrologic cycle. If vegetation is cut back, thereby reducing the number of plants that draw water from the soil and transpire it to the atmosphere, moisture flow to the atmosphere may be reduced. The excess water runs off to the sea as river flow. In this manner, vegetation could modify the climate by affecting humidity.
- Soil eroded by wind from semiarid and arid areas may cloud the atmosphere with dust and modify energy exchanges there, thus modifying the climate.
- The biosphere stores and regulates carbon, helping to control the global carbon budget. If vegetation grows more quickly and absorbs atmospheric carbon dioxide (CO_2) faster than decaying organic matter can return carbon dioxide to the air, there is net storage of carbon in biomass. If the reverse is true, there is net release of carbon as carbon dioxide to the atmosphere. Thus, vegetation partly controls the level of this important greenhouse gas in the atmosphere.

Desertification is a good example of the impact of interactions among human and natural processes. Economic and social forces, especially population growth and demand for food, cause us to populate semiarid areas with domesticated animals or to replace native vegetation with plant crops. The reduction in vegetation cover causes increases in runoff and wind erosion that reduce the amount of water and nutrients present in the soil, further limiting vegetation growth. The loss of vegetation may also affect the atmosphere by increasing the amount of dust it contains or by reducing the amount of solar energy absorbed by the ground surface. Changes in dust or sunlight absorption are hypothesized to affect temperature and precipitation amounts, although such effects have not been confirmed by observation. These intimate relations among the soil, water, air, and living things lead some scientists to view Earth as a single interconnected dynamic system—almost a functioning organism—rather than just a set of cause-and-effect linkages. We can see all these subsystems interacting with each other in much the same way that different parts of a single living organism interact.

Chapter Review

Summary

Climate is the dominant control on local environmental processes, primarily through its influence on water availability and movement. Knowledge of the water budget is essential to understanding both physical and biological processes because the water budget is a critical regulator of ecosystem activity. This influence extends to human use of the landscape, through such issues as determining the natural vegetation cover or agricultural potential.

Carbon is the basis of life on Earth. It cycles through the atmosphere, oceans, biosphere, and soil. Photosynthesis transfers carbon from the atmosphere to the biosphere, and respiration returns it to the atmosphere. Plant growth is most prolific in warm, humid climates where water and energy are plentiful. Plant growth is less in dry or cool climates. Large amounts of carbon are exchanged between the atmosphere and the oceans, and the oceans are a major sink for carbon dioxide that is added to the atmosphere by fossil-fuel combustion.

The outermost layer of Earth's surface in land areas is soil, which is a mixture of mineral and organic matter formed by physical, chemical, and biological processes. Climate is a major regulator of soil development, through its control on water movement. Plant and animal activity in the soil produces organic matter and mixes the upper soil layers. In humid regions, water moves downward through the soil, carrying dissolved substances lower in the profile or removing them altogether. In arid regions, these substances are not so easily removed. Soil characteristics reflect these processes, and there is close correspondence between the world climate map and the world soil map.

Photosynthesis is the basis of food chains and ecological systems. Plants compete for water, sunlight, and nutrients, and plants that are best adapted to compete for limiting factors will dominate a given environment. Such adaptations help explain the world distribution of major vegetation types. Disturbances modify ecological communities, and succession following disturbance is a fundamental process of landscape change. Very large portions of the world's land surface have been modified by human activity, and humans are major players in most of the world's ecosystems.

The world vegetation map closely mirrors the world climate map. Ecologically diverse and complex forests occupy humid environments, storing most nutrients in their biomass. In arid and semiarid regions, sparse vegetation is adapted to moisture stress. Forests adapted to winter cold are in humid midlatitude climates, developing as broadleaf forests in warmer areas and coniferous forests in subarctic latitudes. In high-latitude climates, cold-tolerant short vegetation occupies areas that have a mild summer season. Vegetation is absent in ice-bound polar climates.

Overall, climate is the strongest control on *natural* vegetation, but humans have had profound influences on ecosystems in most of the world's land areas. Plants regulate evapotranspiration from land surfaces and thus moderate the hydrologic cycle. The biosphere stores and regulates carbon, helping to control the global carbon budget. Only a few thousand years ago, Earth was regulated by nonhuman processes, and people were merely negligible players. Today, however, humans are globally significant and locally dominant players in Earth's biogeochemical cycles. Environments in which human impacts are minor are increasingly scarce.

Key Terms

actual evapotranspiration
   (ACTET) p. 126
biodiversity p. 142
biogeochemical cycle p. 122
biomagnification p. 140
biomass p. 131
biome p. 143
boreal forest p. 148
broadleaf deciduous forest p. 146
carbon cycle p. 128
carnivore p. 140
climax community p. 142
desert p. 149

desertification p. 136
ecosystem p. 139
evapotranspiration (ET) p. 125
food chain p. 140
herbivore p. 140
hydrologic cycle p. 123
infiltration capacity p. 126
omnivore p. 140
parent material p. 132
permafrost p. 151
photosynthesis p. 122
potential evapotranspiration
   (POTET) p. 125
prairie p. 149

respiration p. 129
runoff p. 123
savanna p. 148
soil p. 131
soil fertility p. 136
soil horizon p. 132
soil order p. 133
steppe p. 149
succession p. 142
trophic level p. 140
tropical rain forest p. 145
tundra p. 151
water budget p. 123
xerophyte p. 149

Questions for Review and Discussion

1. Diagram the hydrologic cycle, using boxes and arrows. Proportion box sizes to indicate the relative amounts of water stored at each point in the cycle. Label the arrows linking the boxes with the appropriate hydrologic process (evapotranspiration, condensation, precipitation, runoff, and so on).

2. Examine the water-budget diagram in Figure 4-7 for a climate similar to the one in which you live. Explain the seasonal patterns of precipitation, potential evapotranspiration, actual evapotranspiration, soil-moisture use and recharge, runoff, and moisture deficit.

3. What are photosynthesis and respiration? How are they related? How do they vary seasonally in relation to temperature and water availability? How do they vary in relation to major climate types?

4. What is soil texture? How does it affect the movement and storage of water in the soil? How are soil-forming processes affected by water availability?

5. For each major climate type described in Chapter 2, describe the major vegetation type generally associated with that climate.

Thinking Geographically

1. Go to the U.S. Geological Survey's surface water data website (http://waterdata.usgs.gov/nwis/sw) and locate monthly flow data for a river near you. Download the data and make a bar graph showing mean monthly flow. Explain the variations in river flow in terms of seasonal variations in precipitation and evapotranspiration.

2. In your library, find the soil survey for a place near you where excavation is occurring, such as for a highway or building foundation. Determine the soil characteristics in the area. Then visit the excavation site and view the soils in the field, comparing what you see with the descriptions in the soil survey.

3. Consider a typical water budget for a midlatitude continental site in which soil-moisture use is important for part of the year. How is actual evapotranspiration affected by the amount of water stored in the soil and the rooting depth of the vegetation?

4. World maps of climate and vegetation have many similarities. If climate were to change significantly, do you think the vegetation map would also change? Why or why not?

5. How has the vegetation cover where you live changed in the past 200 years? How do you think this change may have affected the local water budget?

MapMaster™

Residents of some Pacific islands have recently experienced higher King tides that flood croplands and wash away homes. Shown here is upwelling water on the Amatuku Islet in Tuvalu. Some, including residents of the Carteret Islands, have had to move off their islands.

5

Population and Migration

The quiet arrival of five men on the island of Bougainville, Papua New Guinea, in May 2009 signaled the start of a new era of migration. They were evacuees from the nearby small Carteret Islands that are expected to sink beneath rising sea levels by 2015, a situation generally thought to be caused by climate change. Like many other Pacific islanders, the 3,000 people living on the Carteret Islands live close to sea level. In recent years, rising seas and stronger storms have washed away homes and crops. During the annual peak high, or "King," tide, Carteret and other small islands are flooded, with no safe highlands. "It gives you the scary feeling that you don't know what is going to happen to you, that any minute you will be floating," one resident told reporters. The floods also eroded their small fields, leaving soils more saline and less productive each time. As conditions worsened, the Papua New Guinea government and local charities finally began to move the islanders to a nearby larger island. Like new arrivals in many other countries, their new neighbors did not welcome the Carteret evacuees.

The Carteret Islands are merely the first of many planned evacuations as other islands succumb to a similar fate. Flooding of small low-lying islands and coastal areas is expected to produce many more "climate refugees." The small island nations of Tuvalu and Kiribati may well be uninhabitable by 2050. Villagers on Vanuatu's Tegua Island and other coastal areas have already begun moving inland to higher ground. And their numbers will add up. The Maldives is considering what it would do to evacuate 300,000 people, and populated coastlines around the world could add millions more to the growing number of people heading for drier land. Rising sea levels could displace as many as 162 million people by 2050. Moreover, climate change will produce other migration triggers as the pattern of droughts and floods shifts or becomes more severe. Climate induced events such as these may displace another 50 million people by mid-century. Whatever we call such migrations, they may permanently alter the distribution of the world's population.

A Look Ahead

The Distribution and Density of Human Settlement

About 90 percent of Earth's 6.8 billion people physically occupy only 20 percent of Earth's surface. The distribution and density of settlement is due largely to attributes of the physical environment, to factors of history, and to varying rates of population growth.

World Population Dynamics

Earth's population is rising, but the rate of increase is slowing. Most population increase is occurring in poor countries, and some people fear that Earth may be overpopulated or facing potential overpopulation.

Other Significant Demographic Patterns

Sex ratios vary greatly among national populations, and these variations may be explained as the result of both economic and cultural factors. The median age of the human population overall is rising, and this presents distinctly different challenges to rich countries and to poor countries.

Migration

All the world's modern peoples descended from a single stock that migrated away from Africa as long as 200,000 years ago. Since 1500, the great migrations that have made our world include migrations of Europeans to the Western Hemisphere, Asia, Oceania, and Africa; migrations of Africans, mostly to the Western Hemisphere; migrations of Indians; and migrations of Chinese.

Migration Today

Migration of people from poor countries to rich countries, of people from rural areas to cities (discussed in Chapter 10), and of people seeking greater political freedom continues today. Pressures of migration have become a political issue in many countries, rich and poor.

Effects of Emigration

Migration has important effects on the countries of origin. Brain drain describes the loss of skilled people, especially from developing countries.

This chapter examines the world's **population geography**—that is, the distribution of humankind across our planet. The relative distribution of humankind is constantly changing, and this is because of two factors. First is the fact that different countries have different internal population dynamics, such as the numbers of births and deaths and the consequent rates of increase or decrease. The second factor is human migration, which occurs whenever people move their homes from one place to another. This move might be across town or around the world. Migration affects not only the lives of the migrants but the places they leave and the places they move to. The **emigration,** or departure, of persons from one place might improve living conditions there. This is the case when emigrants send money earned from their new jobs back home. Different issues are raised by **immigration,** when people move to a place from somewhere else. People already living there might not welcome newcomers, perhaps because they think immigrants will worsen living conditions or disrupt social cohesion. Migration has affected the distribution of the world's peoples in the past, and continuing migration affects world affairs today.

This chapter will occasionally include ideas and information about **demography,** which is the study of individual populations in terms of specific group characteristics. These characteristics may include the distribution of ages within the group, the relative numbers of males and females, income levels, or any other characteristic. Demography means "describing people." What demographers and geographers know about human populations may come from several sources, including governments, non-governmental organizations, and academic studies. The two most basic kinds of information are censuses and estimates. Many governments periodically perform a census to count every person resident in their country. The U.S. Constitution requires a census every 10 years to reapportion legislative representatives among the states. Censuses also collect information about the population, such as their age, employment, and language spoken. They are enormous and expensive undertakings that often fail to count every person in a country. Where census data are missing or unreliable, demographers use estimates based on what they do know about the behavior of different populations. Well-trained social scientists can produce very accurate estimates, but all estimates must be interpreted carefully. This book uses census data where possible and scientific estimates as necessary.

The Distribution and Density of Human Settlement

In 2010 the population of Earth reached 6.8 billion, but people are not evenly distributed across the landscape (see the map on the rear endpaper). The map shows that large areas of Earth's surface are surprisingly uninhabited. About 75 percent of the total human population lives in the Northern Hemisphere between 20° and 60° north latitude, and even great expanses of that area are sparsely populated. About 90 percent of the population is concentrated on less than 20 percent of the land area.

The map reveals three major concentrations of population: (1) East Asia, where eastern China, the Koreas, and Japan total about 1.6 billion people; (2) South Asia, where India, Pakistan, and Bangladesh together account for about 1.5 billion; and (3) Europe from the Atlantic to the Ural Mountains, with almost 750 million. Southeast Asia forms a secondary concentration, with about 589 million people; the United States and Canada are home to 342 million.

These five population concentrations, plus parts of West Africa, Mexico, and areas along the eastern and western coasts of South America, are densely populated, with more than 25 people per square kilometer (64 per square mile). More than half of Earth's land area, by contrast, has fewer than 1 person per square kilometer (2.6 per square mile) (Figure 5-1). A surprising portion of Earth's land surface is virtually uninhabited: central and northern Asia; northern and western North America; and the vast interiors of South America, Africa, and Australia. The countries that occupy these spaces are enormous in area, but they contain relatively few people. The African country of Chad, for example, covers nearly 13 times the area of South Korea, but its population is less than one-quarter that of South Korea.

Figure 5-1 Low population density. The enormous island of Greenland (2.17 million square kilometers, or 836,000 square miles—more than three times the size of Texas) is inhabited by only 56,000 people. The fact that any people live here at all is a tribute to how human culture and technology allow us to adapt to all environments.

Figure 5-2 is a cartogram of world population. The size of each country reflects its population, not its land-surface area. The most populous countries are shown as the biggest, and the least populous as the smallest, no matter how big or small their land-surface areas. The single country of India, therefore, is drawn considerably larger than the entire continents of South America or Africa.

A number of factors combine to explain the distribution of the human population. In this section, we will consider the impact of the physical environment and history on population patterns. In the next section, we will take a close look at rates and distributions of world population increase. We will examine these factors as hypotheses of explanation for the population distribution.

Population Density

The uneven distribution of humans means that some places are more "crowded" than others, and other places are more "empty." Geographers use several measurements to compare population densities. **Arithmetic density** is the number of people per unit of area. Using this measure, the countries of the world range from a sparse 2 people per square kilometer (5 per square mile) in places like Mongolia and Western Sahara to dense city-states like Monaco, with 16,905 people per square kilometer, (43,767 per square mile) and Singapore, with its 7,013 people per square kilometer (18,157 per square mile) (Figure 5-3). The map on the rear endpaper reveals, however, that arithmetic density varies greatly even within individual countries. In Egypt, for example, the arithmetic density is nearly 2,000 people per square kilometer (5,178 per square mile) in the delta and valley of the Nile River, but it is only 3 persons per square kilometer (fewer than 8 per square mile) in the rest of the country, which is desert.

Some scholars believe that a country's **physiological density,** which is the density of population per unit of cropland, tells us more about its ability to support its population than its arithmetic density. This measure relates people to food production and assumes that food is distributed equally, which it is not. In this measure, the lowest physiological density is found in Australia, which has 2.34 hectares (5.78 acres) of cropland for every person living in the country. One of the highest physiological densities is found in tiny Singapore, which has approximately 10 square kilometers (3.9 square miles) of agricultural land for its 4.8 million people. Obviously, this measure does not take into consideration the international food trade.

Do these statistics tell us anything about a country's standard of living? No. In the past, physiological density may have influenced the well-being of primitive societies that were entirely dependent on local agriculture. Today, however, there are several reasons why measures of population density do not reveal national welfare. We will study world agriculture in greater detail in Chapter 8 and economic development

in Chapter 12, but we should note here that few countries actually farm all their potential cropland, and scholars differ about how much potential cropland each nation has. The amount of potential cropland can be increased by discovering crops that can thrive under new conditions, by treating soils in new ways, and by expanding irrigation. Chapter 4 noted, however, that the amount of potential cropland can be decreased by misusing and thus degrading soils or by expanding urban or industrial land uses.

Furthermore, the amount of land people have to sustain themselves is less important than the productivity of that land and how people use it. **Carrying capacity** is a theoretical concept to describe the maximum number of people that an area can sustain given its physical qualities, as well as the social, technological, and economic system that depends on it. A fertile region, for example, could presumably support a higher population density than an infertile region. Land used to grow crops for human consumption will feed more people than if the same land is used to raise cattle. Carrying capacity is affected by many other factors, too. Productivity depends on the level of agricultural technology and other inputs, including labor. Some countries enjoy technologically sophisticated and highly productive agriculture—even on infertile soils. Other countries suffer technological backwardness and low productivity—even on fertile soils. Political and economic considerations must be taken into account, as well. If, for example, most of the land in an area is owned by a small number of farmers who use it to grow tobacco rather than food, then its carrying capacity is reduced. On a global scale, there is intense concern about the carrying capacity of Earth itself. The number of people that Earth can support depends enormously on how we use our finite resources.

Finally, no countries are fully dependent upon local agriculture. Trade and circulation loosen societies from the constraints of their local environments and allow them to draw resources from around the world. Physiological density is irrelevant to the welfare of a rich city-state such as Singapore, which imports food in exchange for manufactured goods and services, or to any densely populated but industrial country, such as Japan or Belgium. Neither arithmetic density nor physiological density fully explains how the people at any place feed themselves today, or how they support themselves. Some of the most densely populated areas on Earth are rich, yet others are poor. Conversely, some sparsely populated regions are rich, while others are poor. We can draw no clear correlation between density and welfare.

Climate The clearest relationship between population distribution and environmental factors is that population densities are low in most of the world's cold areas and dry areas. Compare the map on the rear endpaper with Figures 2-32 and 2-33. The major exceptions to this rule are places where rivers flowing

Figure 5-2 A cartogram of world population. On this population cartogram, the size of each country reflects its population, not its actual land area. Notice that on a population cartogram the whole African continent is smaller than India, because it is less populous. South America is also relatively small.

through dry areas provide water for irrigation, such as India, Pakistan, and Egypt (Figure 5-4).

Earth's coldest areas support habitation only where mines or other special resources make it worthwhile for populations to work there, as in parts of central Asia or high in the mountains of South America. People will settle in harsh areas if it is profitable to do so.

Some warm and wet equatorial regions are also sparsely populated, such as the Amazon River Basin in

South America, the Congo River Basin in Africa, and the island of New Guinea. Despite the lush vegetation of these regions, conventional agriculture has not proved successful. Some soils are poor in nutrients. In other cases, agricultural crops cannot compete with other plants. A wide variety of plants and animals flourish in the warm, moist climate, but many of these are hostile to humans and their agriculture. For example, insects thrive in the warm climate of the Amazon,

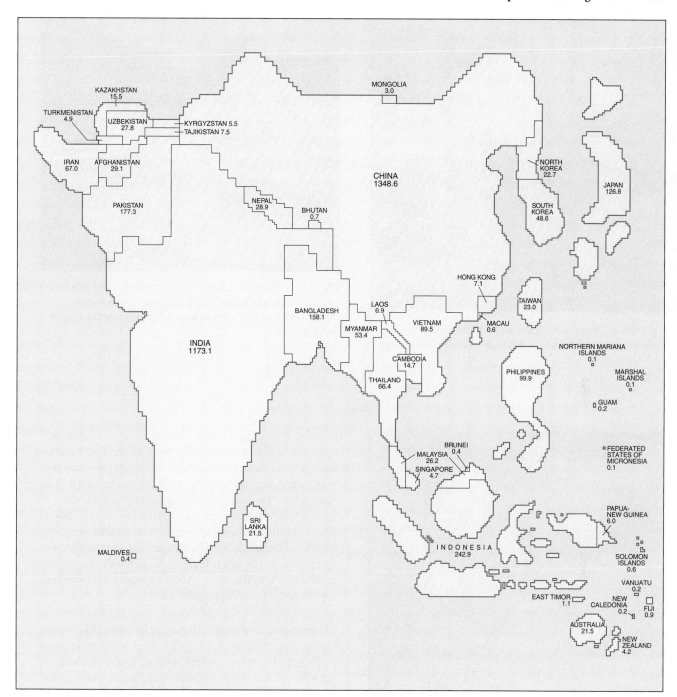

as do the diseases they carry. In addition, these environments support countless forms of parasites, microbes, and fungi that weaken and kill humans, wilt and blight their plants, eat crops alive in the fields, or quietly feast on them in granaries or storerooms.

Tropical Asia, a prime example of a warm, moist climate, is home to several great concentrations of people, but these areas are mostly in seasonal regions that are not wet year-round. Some of these regions, such as the island of Java or the Deccan Plateau of India, offer rich volcanic soils, and others are well adapted to the construction of flooded fields for growing rice (Figure 5-5). Rice yields the highest number of calories per acre of

any known crop. Historically, the high population densities in subtropical eastern China have been supported by a farming region that mixes wheat production, more dominant in northern China, and rice production, more prevalent in the south. Northern India's high population densities are fed by a similar mixed crop area that lies across the river valleys of northern India. In this case, wheat and other cereals are more common in the west, and rice is more common on the east.

In conclusion, most of the world's population is concentrated in areas of seasonal environments that are not too wet, too hot, too dry, or too cold. In fact, most of Earth's inhabitants live in warm midlatitude climatic

Figure 5-5 Rice terraces. The construction and intensive cultivation of rice terraces such as these on the Indian Ocean island of Bali support a very high human population density.

Figure 5-3 Singapore. Space is at a premium in the crowded streets of Singapore. Sidewalk cafes on Bugis Street spill out into the road.

Figure 5-4 The Nile Delta. This satellite photograph illustrates why the ancient Greek historian Herodotus called Egypt "the gift of the Nile." A thin strip of cultivable land winds northward through uninhabitable desert and then fans into a delta where it meets the Mediterranean Sea. Each year the Nile River floods and recedes, leaving a narrow plain of rich black mud, although construction of the High Dam at Aswan has significantly reduced this annual input of natural fertilizer.

regions, including western Europe, eastern China, and northern India (areas designated *C* in the Köppen climate classification system), or in tropical areas that experience distinct seasons, such as southeastern Asia (categories Am and Aw). The geographic incidence of these patterns is visible by comparing the maps of population density on the endpapers with the climate regions shown in Figure 2-33. The geographic explanation for this relates to the advantages these climates provide for productive agriculture, as shown in Figure 8-1, which in turn support growing populations.

Topography and soils   Topography often affects population distribution, although its effect is less prominent than that of climate. People tend to settle on flatlands because of the ease of cultivation, construction, and transportation. Thus, most of the densest population concentrations in the world are found on level terrain, often near navigable waters. (Compare the map on the rear endpaper with the map on the front endpaper.)

Flatness alone, however, is insufficient to attract a large population. Usually there is an association of other environmental attributes, such as fertile soil, available water supply, and moderate climate that helps to explain high population density. Flatlands in central Siberia, western Australia, and central Brazil are nearly unpopulated because other factors, such as the climate or soils, diminish the area's productivity.

Conversely, sloping land (hills and mountains) usually has a sparse population density, although this is not always the case. High density is found in many mountainous portions of South America, Japan,

New Guinea, Southeast Asia, central Europe, and elsewhere. Specialized attractions, such as mineral resources, often help explain such concentrations. Moreover, in most mountainous areas, the settlements actually are concentrated in valley bottoms, even though they may be small and constricted. Thus the connection between topography and population density is too inconsistent to validate a direct relationship.

Historically, people have settled areas of fertile and potentially productive soil unless there are powerful, negative factors against the choice. For example, floodplain soils are typically fine-textured and full of nutrients, so they can be used intensively for agriculture. Thus, most river floodplains are densely populated unless they are located in extremely cold or dry areas. People do not necessarily avoid areas of poor soils, however, because some poor soils can be enriched with fertilizers.

Overall, the complexity of human history and the variety of human adaptations teach that environmental conditions alone can never explain population densities.

History History helps explain the pattern of human settlement. The populations of China and the Indian subcontinent achieved productive agriculture and relative political stability thousands of years ago. These peoples domesticated many plants and animals early in their histories. **Domestication** is the process of taming and training animals and of sowing, caring for, and harvesting plants for human uses—largely for food. Intensive cropping and irrigation yielded generous food supplies, which supported rising populations. The western European population multiplied when world exploration and conquest brought Europe new food crops—notably the potato. Later, during the Industrial and Agricultural revolutions, European food production increased again, and so did its population. Migrants from Europe settled and brought European technology to more sparsely populated areas, so some of today's secondary population concentrations grew as extensions of Europe. These areas include eastern North America, coastal South America, South Africa, and Australia and New Zealand. It has been estimated that Europeans and their descendants increased from 22 percent of the world's population in 1800 to 35 percent by 1930.

Human population distribution can also be affected by the demarcation of cultural territories, particularly political territories. Two governments on opposite sides of an international border may have different environmental policies, and these can drastically alter the carrying capacity of the environment. Furthermore, a country may manipulate the distribution of its population. It might subsidize or command the settlement of harsh territories in order to occupy them effectively, or it might establish settlements along its borders for defense. The Soviet Union (1922–1991) provides an extreme example of brutal government population policies. It relocated or expelled some 12 million people and killed millions more.

World Population Dynamics

Earth's population numbered 1 billion around 1800, 2 billion by 1930, and 4 billion by 1975, and it is expected to reach 7 billion in 2013. World population grew by almost 80 million persons per year over the past decade (Figure 5-6). The rate of increase, however, is decelerating. In the period 1965–1970, the average annual population increase was 2.06 percent, but it fell to 1.73 percent for the period 1985–1990, to 1.35 percent in 1995–2000, and to 1.22 percent in 2000–2005. It has been registering 1.12 percent since 2005.

The rate of population increase represents a balance among several demographic statistics. First we will define each of these statistics, and then we will explain through the next several pages how changing relationships among these statistics cause Earth's total population, and the populations of individual countries, to rise or fall.

The **crude birth rate** is the annual number of live births per 1,000 people. The **crude death rate** is the annual number of deaths per 1,000 people. The difference between the number of births and the number of deaths is the **natural increase** if there are more births than deaths, or the **natural decrease** if there are more deaths than births. In the United States, for example, there were 4.33 million births over the course of one year, ending in 2008. The population at the start of that period was 301.2 million. This equates to a crude birth rate of 14.3 per 1,000 people. During the same period, there were 2.45 million deaths, yielding a crude death rate of 8.1 per 1,000 people. With more births than deaths, the United States experienced a natural increase of 6.2 per 1,000 people.

For individual countries or regions, we can also calculate the **net migration rate** by subtracting the number of people who emigrated out of the area and adding the number of people who immigrated into that area. The rate of natural increase (or decrease) plus the net migration rate (which may be negative) is known as the **demographic equation** (sometimes called the balancing

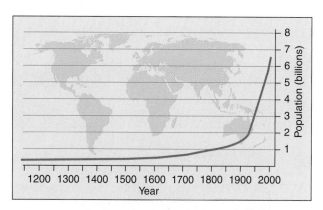

Figure 5-6 The increase of human population. The human population grew rapidly from about 1550 until about 1950, and then its growth accelerated to a still higher rate of increase after 1950. The total number of people is still increasing greatly today, but the rate of increase is slowing.

equation), and it measures a population's total growth or decrease. The result of this equation can then be divided by the total population of that area, resulting in a **growth rate,** which may be positive or negative. For example, Mexico currently has a birth rate of 20 births per 1,000 people and a death rate of 5 deaths per 1,000 people, which gives Mexico a rate of natural annual increase of 1.5 percent. When the net migration rate of -4 people per 1,000 of population is taken into consideration, however, Mexico's growth rate falls to 1.1 percent.

A great deal of attention is paid to each country's **total fertility rate (TFR),** which is the average number of children that would be born to each woman in a given society if, during her childbearing years (ages 15–49), she bore children at the current year's rate for women that age. This statistic carries great predictive value. A TFR of 2 would mean that parents will have replaced themselves when they die, eventually stabilizing the population size. Not all children survive to reach their own reproductive years, however, so a TFR of about 2.1 is the actual **replacement rate.** If a country's TFR falls below 2.1 and the country does not experience immigration, that country's population will ultimately decrease.

Population Projections

A **population projection** is a prediction of the future, assuming that the world's current population trends remain the same or else change in defined ways. Projecting population is an uncertain task because the smallest rise or fall in the percentage of increase today would increase or decrease the total population in the next century by hundreds of millions of people. Even if worldwide crude birth rates and TFRs continue to fall, the total number of people on Earth will continue to grow. This **population momentum** is due to the fact that the number of young women presently reaching childbearing age is larger than ever before.

Earth's population will go on rising, but the sooner the world's TFR falls to the replacement rate, the lower the total population will be if or when the population eventually stabilizes. Many observers insist that the sooner Earth reaches a constant population—**zero population growth** overall—the better we will be able to achieve a better standard of living for all. When will this happen? At what total population count?

The United Nations offers a range of projections in answer to this question. In 2010, the world TFR was about 2.55 but falling. If the world's average TFR remains high, falling only slightly to 2.51, then world population will grow to 10.7 billion by 2050. That's about 1 billion more people per decade! If TFRs fall further, to an average 2.02, slightly below replacement level, then world population will be 9.2 billion in 2050. Should TFRs fall well below replacement level, to 1.54, then Earth's population will be 7.8 billion by mid-century. Another way of interpreting the significance of this low-TFR scenario (1.54) is that it represents an average of about "one child" fewer per woman than the current TFR (2.55). As more couples choose to have, for example, two children instead of three, they are contributing to a significant decline in world growth rates.

Our need to consider different scenarios highlights the uncertainty of population projections (Figure 5-7). Based on current trends, it is likely that total world population will be close to the middle-range estimate of 9.2 billion by 2050. Predicting changes in world population, however, also requires that we consider changing death rates. Death rates have been falling globally for more than 50 years, adding to total population size. In contrast, deaths from HIV/AIDS have slowed population growth rates where the epidemic is worst. In southern Africa, life expectancy has dropped from 62 years in 1990 to 49 years today.

Even if population totals begin falling by 2100, many fear that it might be difficult to sustain acceptable

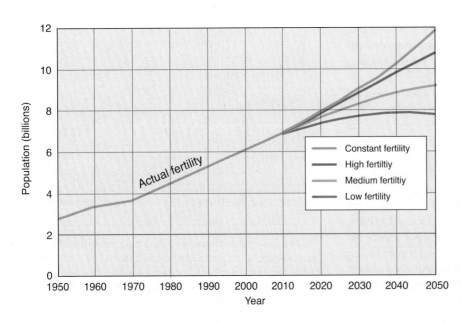

Figure 5-7 Population projections. Predictions of future population growth must assume certain population dynamics, especially the total fertility rate. This graph shows how different total fertility rates are likely to affect the size of Earth's population until 2050. The trend lines reflect scenarios for **constant fertility** (TFR remains at about 2.55), **high fertility** (TFR lowers to 2.5 by 2050), **medium fertility** (TFR drops to 2 by 2050), and **low fertility** (TFR of 1.5 by 2050).

economic and political conditions with billions more people on our planet, especially if it remains true that some of the densest concentrations of the population are in some of the poorest areas.

Regional Variation in Population Growth

The overall population projection conceals wide variations in the rate of population increase from region to region and country to country. For example, the annual growth rate in 2010 was 2.4 percent in sub-Saharan Africa, 1.19 percent in Latin America, and 1.33 percent in Southeast Asia, but only 0.52 percent in North America. Europe experienced slightly negative growth. Figure 5-8 shows how the rates of population increase vary from country to country. These different growth rates suggest that there will be significant shifts in the geographic distribution of the world population. By 2050, Africa's share of Earth's population is expected to increase from 14.5 in 2010 to 21.7 percent, but other regions will represent smaller shares even as their populations grow: Latin America's share will decrease from 8.6 to 8.4 percent, Asia from 60 to 57 percent, and North America's from 5.1 to 4.8 percent. Europe's population decline will also decrease its share of world population, falling from 11 to 7.2 percent by 2050.

Some scholars compute the **doubling time** for each country's population—that is, the number of years it would take the country's population to double at its present rate of increase. A number compounding at 2.8 percent annually (the rate of population increase in Yemen) doubles in only 27 years, but a population compounding at only 1 percent annually (as in the United States) takes 70 years to double.

Overall rates of population increase vary significantly between rich and poor countries. Almost all of the population increase projected to the year 2050 will be in countries that today are poor, those least able to support a larger population. The share of the world population living in the countries that are poor today will increase from 82 percent today to 86 percent by 2050. In these countries, economic growth is in a race with population growth. In order to achieve rising per capita (per person) incomes in these countries, the rate of economic growth must exceed the rate of population growth. If the rate of economic growth equals the rate of population growth, per capita incomes will remain static; but if economic growth rates fall below population growth rates, per capita incomes will fall. Complicating this situation further, the addition of several billion people to our planet over the next four decades will strain existing resources.

The Age Structure of the Population

Demographic analysis reveals that in countries with high fertility rates, the populations are young. This is often represented by a graphic device called a **population pyramid,** which represents two aspects of a population: age and gender (Figure 5-9). Young people are indicated at the base of the pyramid, so a wide base

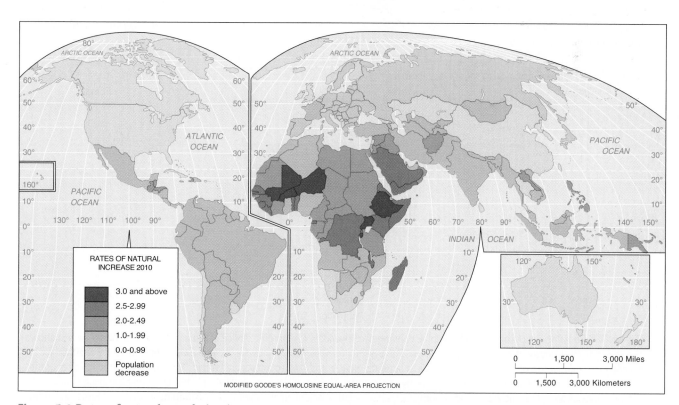

Figure 5-8 Rates of annual population increase. Most countries with the highest rates of population increase are in Africa and the Middle East.

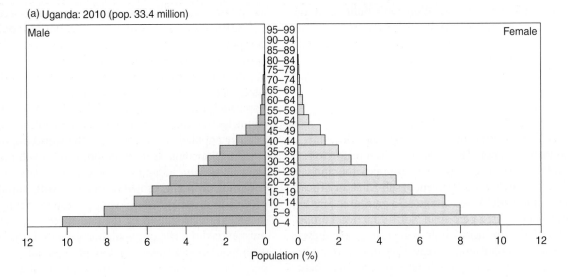

(a) Uganda: 2010 (pop. 33.4 million)

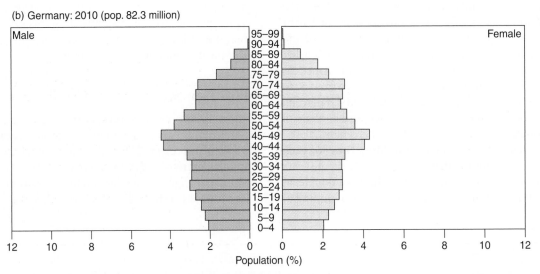

(b) Germany: 2010 (pop. 82.3 million)

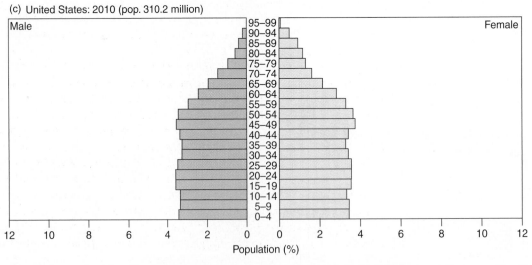

(c) United States: 2010 (pop. 310.2 million)

Figure 5-9 Population pyramids. These three drawings are examples of a graphic device called a population pyramid, which shows the age and sex structure of a country's population. The pyramid for Uganda (a) shows a broad base—indicating many children—tapering to a narrow top of fewer older people. Birth rates are high, but life expectancies are limited. This pyramid is typical of countries with high fertility rates. The pyramid for Germany, by contrast (b), shrinks at the base, indicating a birth rate that has been falling steadily since about 1970. Life expectancy is long. Germany's relative lack of elderly men is the continuing evidence of losses in World War II (1939–1945). The population pyramid of the United States (c) shows a bulge of people between 45 and 65 years of age—the "baby boom" of 1945–1965. Below this group, the figure shrinks to reveal the "baby bust," or Generation X, followed by a rise in births due to "boomers" reaching their childbearing years, a phenomenon called the "echo of the baby boom."

indicates large numbers of young people. As the individuals of the same age (called a *cohort*) grow older together through time, the horizontal bar that represents that cohort moves toward the top of the pyramid. Numbers for males and females are set on separate sides of the pyramid to show their relative numbers in each cohort. Demographers can project the future structure of a population based on this knowledge. For example, if we know how many people are 20 years old today, we can estimate how many people will be 50 years old 30 years in the future.

In 2010, the median age of the world's population was about 29, but the various countries and regions demonstrate great differences in the ages of their populations (Figure 5-10). Where birth rates are high, median ages are generally younger than where birth rates are low. The median ages of the populations in many countries in Latin America, Africa, and the Middle East are between 18 and 25 years. In contrast, the median ages of populations of wealthier countries are older. In Western Europe, for example, the median age is 40.2; in the United States it is 36.6.

The **dependency ratio** of a country suggests what proportion of its people are in their most productive years. It is defined as the ratio of the combined population of children younger than 15 years and elderly people over 64 years (both groups considered dependents) to the population of those between 15 and 64 years of age. It might be helpful to imagine a traditional family farm on which working-age adults are responsible for clothing, feeding, and housing the children and grandparents.

The larger the percentage of dependents, the greater the burden on those who are working to support those too young or too old to work. Even in societies with day care and schooling for children, and retirement funds and nursing homes for the old, the cost of these services is ultimately paid for by those who work. A dependency ratio will be high if a country has either a high percentage of young people (which is typical of the poor countries) or else a high percentage of elderly people (which is typical of the rich countries). If a country is poor, a highly dependent population compounds its financial difficulties. In Ethiopia, for example, which is a poor country, the dependency ratio is 0.86, and most of the dependents are youngsters. In the United States, by contrast, the ratio is only 0.50, and a growing share of the dependent population is made up of elderly people. Economists suggest that a country's age structure is a key to understanding its economic growth, which is usually highest when its dependency ratio is low.

In the countries with low median ages, a high percentage of the total population is drawing on national resources, but it has not yet reached its most productive working years. These countries are challenged to feed, clothe, and educate youngsters. As the youngsters reach maturity, the national economies must be able to provide jobs for them. If economic growth has not provided enough jobs, their frustration may break out in civil disorder. Economic development is urgently needed. We will see below how rising median ages challenge both rich and poor countries.

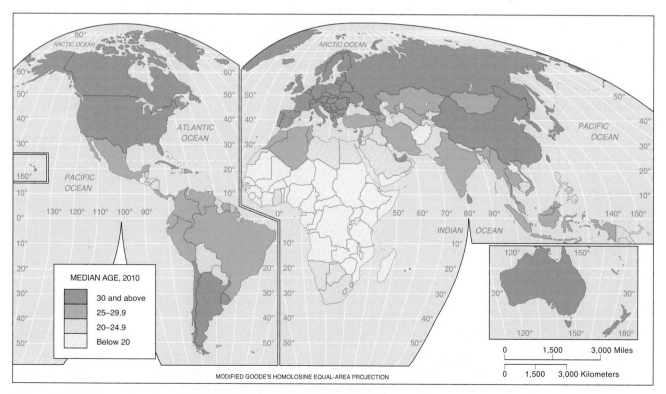

Figure 5-10 Median ages of national populations. Comparison of this figure with Figure 5-8 shows that populations are young where rates of population increase are high.

The Demographic Transition

Scholars who have studied population growth in the countries that today are rich have defined a model to describe these countries' historical experience. That model is called the **demographic transition.** The demographic transition model defines a pattern of growth that exhibits four distinct stages (Figure 5-11).

(1) In Stage One, both the crude birth rate and the crude death rate are high, so the population does not increase rapidly. All countries experienced this stage in the past, when the human population was a fragile number in constant danger of significant reduction, locally if not globally, by periodic epidemics.

(2) As incomes increase and medical science advances (which is usually, at first, simply an understanding of hygiene), crude death rates drop dramatically (Figure 5-12). The **infant mortality rate,** which is the number of infants per thousand who die before reaching 1 year of age, falls almost immediately. Geographically, high crude death rates and high infant mortality rates have very similar distributions.

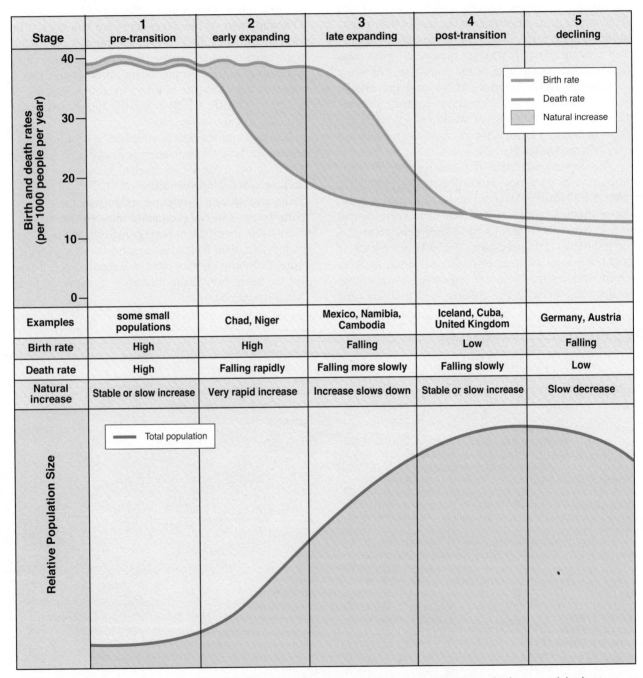

Figure 5-11 The demographic transition model. Shown here is the historical relationship between birth rates and death rates (above) and the resulting population growth (below) that together characterize the four stages of the model. The fifth stage, where birth rates fall below replacement level, currently describes some European countries experiencing natural population decrease. More countries, including China, are expected to enter the fifth stage in the coming decades. Note that this model does not account for migration.

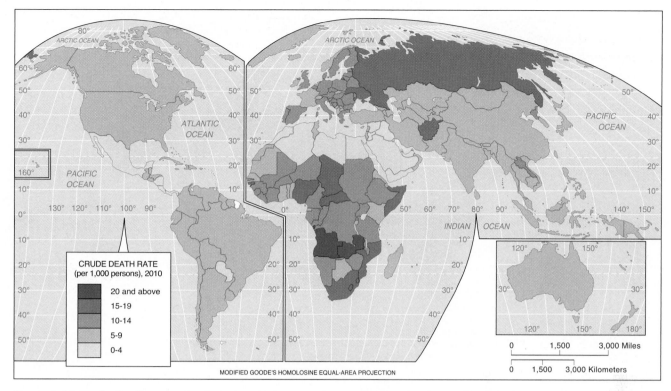

CRUDE DEATH RATE
(per 1,000 persons), 2010

20 and above
15-19
10-14
5-9
0-4

MODIFIED GOODE'S HOMOLOSINE EQUAL-AREA PROJECTION

Figure 5-12 Crude death rates. Death rates are high in very poor countries (sub-Saharan Africa), moderately high in wealthy countries with aging populations (Eastern Europe), and low in countries with very young populations and at least modest health-care availability (Latin America). Crude death rates are falling almost everywhere, but environmental disasters, local famine, epidemics, or war can still substantially increase a country's crude death rate.

Another factor historically lowering death rates is the improvement in the quantity and quality of food that resulted from the Agricultural Revolution (discussed in Chapter 8). Crude birth rates, however, remain high. Thus, in *Stage Two* of the demographic transition, crude death rates are falling, but crude birth rates remain high, so the rate of natural increase is high.

Several theories have been offered to explain the persistence of high birth rates during the second stage of the demographic transition. In traditional societies, children may be economic assets. They provide more hands to help in the fields, for example. Also, people traditionally expect their children to look after them in their own old age. If infant mortality rates are high, parents have many offspring to ensure that some survive into adulthood. Not until one or two generations have passed will adults accept the reality of lower infant mortality rates and start having fewer children.

(3) Eventually, crude birth rates begin to fall. In Europe, this decline in childbearing occurred along with economic growth, urbanization, and rising standards of living and education. Children came to be seen as expenses rather than as economic assets, and many parents began to have fewer children so that they could provide a higher material standard of

Figure 5-13 The cost of children. The cost of raising children now includes large expenses such as schooling, especially in developed countries where employers often seek college degree holders.

living for themselves and the children they did have (Figure 5-13). During *Stage Three*, birth rates remain higher than death rates, but both are dropping and slowly converging. Thus, the rate of natural increase is positive but slowing during this stage.

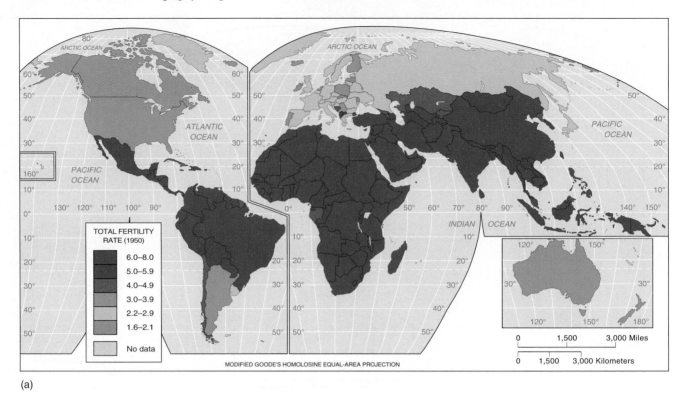

(a)

Figure 5-14 Falling total fertility rates. The "population explosion" that many observers feared in the 1950s and 1960s (a) is fizzling, although there was dramatic population growth in poorer countries. *(continued)*

(4) In *Stage Four*, death rates and birth rates converge at a low and relatively constant level, approximately 10 per 1,000. Population sizes stabilize as population momentum dissipates and TFRs hover at or below replacement level. Most of the wealthy, industrialized countries in the world today are nearing or in Stage Four of the demographic transition.

Demographers have noted that some countries are passing into a new phase in which birth rates drop well below replacement level while death rates remain low and relatively stable. The result is population decline, and many rich countries' populations will decrease without substantial immigration; some are doing so already. Italy, for example, has a TFR of only 1.3, driving the median age up to 43.3 (among the world's highest). Its native-born population is projected to fall 12 percent between 2000 and 2050, from 57 million to 50 million. Reasons cited by Italians for having fewer children typify the social changes in the rich countries: Women are pursuing careers, contraceptives are increasingly available, day-care services are limited, urban housing is short, abortion has been legalized, and people choose to enjoy material goods rather than to bear and raise children. Other social trends at work in lowering fertility rates are more years spent in education, postponement of marriage, cohabitation without marriage, and environmental and ethical considerations about population growth.

It is important to note that the demographic transition model was based on the population experience of wealthy Western European countries, where the transition took place over centuries. In the period from 1950 to 1980, scholars studying world population believed that the demographic transition model would eventually apply to all countries. They noted that birth rates and TFRs were still high in the poor countries, and they described these countries as being in the second stage of the demographic transition. Scholars concluded that when these countries achieved economic development, they too would progress to the fourth stage. Therefore, the 1974 World Population Conference in Bucharest, Romania, concluded, "Economic development is the best contraceptive."

Several characteristics of the demographic transition model still do apply in today's poor countries. Crude birth rates and TFRs are falling, but they remain relatively high (Figure 5-14). The average TFR for the countries of sub-Saharan Africa during the period 2000–2008, for example, was 5.4. In many developing countries, preferences for larger families persist, most likely due to a combination of economic factors and cultural preferences (Figure 5-15). It can even be argued that some African countries face labor shortages in agriculture, and because they lack capital for agricultural machinery, increases in rural labor forces could increase food output.

Furthermore, in even the world's poorest countries, infant mortality rates are dropping because antibiotics

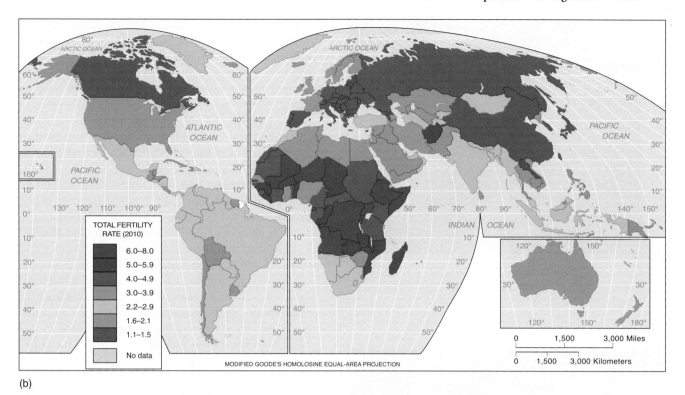

(b)

Figure 5-14 (*continued*) Total fertility rates remain high in many poor areas (b), but today women virtually everywhere are bearing fewer children.

and immunizations that today help many babies survive were not available in the past (Figure 5-16). Modern medicine also keeps people alive longer, and the combined effect of lowering infant mortalities and people living longer expands **life expectancy,** which is the average number of years that a newborn baby within a given population can expect to live (Figure 5-17). Today, life expectancies are lengthening fast in most of the world. The combined effect of lowering infant mortalities and lengthening life expectancies is to increase total population size.

Therefore, some circumstances in today's poor countries are comparable to the second stage or third stage of the demographic transition, as it was experienced in countries that are rich today.

Factors Affecting Fertility Rates

The demographic transition model describes a historical decrease in births known as the **fertility transition,** in which populations move from high to low birth rates. As described above, today's rich countries experienced a drop in fertility during the nineteenth century, and it was widely assumed that these changes were associated with a shift from rural family farming to urban wage-based employment. Today, total fertility rates are falling in many poorer countries but there are doubts about whether the demographic transition model applies to these countries as originally thought. Already by the time of the opening of the third U.N. Conference

on Population and Development, in Cairo, Egypt, in 1994, experts identified two trends occurring in the poor countries:

(1) TFRs have been dropping much faster in poor countries than they did in the countries that today are rich. From 1975 to 2010 the world TFR fell from 4.5 to 2.49. In East Asia, the decline was from 5.1 to 1.6; in Africa, the decline was from 6.6 to 5.4; and in Latin America, the decline was from 5 to 2.5. Over the course of just two generations, average family sizes have decreased.

(2) Falling TFRs are not necessarily associated with improving incomes and standards of living. Birth rates continue to fall in some countries where the economy is stagnating or even declining. Per capita incomes in many African economies have fallen, but TFRs have fallen, too. A few countries have even experienced rising TFRs as economies boom. These observations challenge the model's assumption that economic growth always leads to falling TFRs.

These statistical trends have convinced many scholars that today's decreases in TFRs in the poor countries are different from the fall of fertility rates that had occurred in today's rich countries. Therefore, identifying the new causes of falling TFRs may assist in policies meant to complete the fertility transition and slow population growth. Theories of explanation focus on many factors, especially government programs and policies, the availability of contraceptive technologies, the status

Figure 5-15 Children at work. Boys working at a brick kiln on the outskirts of Hyderabad, Pakistan, fill and turn brick moulds by hand.

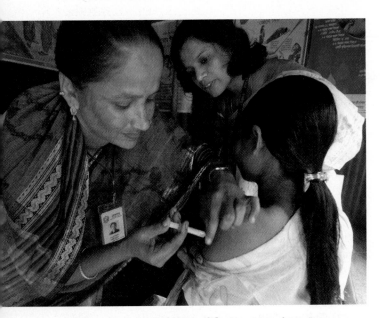

Figure 5-16 Lengthening life expectancies. A community health worker in Matlab, Bangladesh, gives vaccinations in her home-based clinic for children under five, pregnant women, and women of reproductive age and their children up to eighteen. Clinics like hers help to reach many who do not have access to physicians.

of women, and the diffusion of changing attitudes related to fertility.

(1) Government programs and policies During the demographic transition in today's rich countries, the concept of family planning was not quickly accepted. Early leaders in the population-control movement in Europe and the United States, such as Margaret Sanger (1883–1966), were even arrested for immorality or creating public disturbances. Today, by contrast, lowering the crude birth rate has become the focus of most

countries' efforts to reduce overall population increases. Over 90 percent of the people of developing countries live under governments committed to reducing the rate of population growth (Figure 5-18). Today, national public health services in many developing countries seek to change public attitudes in favor of smaller families. Family-planning programs seek out new clients and many governments have removed or lowered barriers to the availability of contraception.

The most dramatic and significant change in fertility has been in China, where the annual rate of natural population increase dropped from 2.9 percent in the early 1960s to just 0.5 percent in 2010. China is home to about 20 percent of the human population, so that decline alone accounts for much of the change in world trends. In 1979, China launched a so-called one-child policy, offering couples incentives to reduce childbearing, such as financial rewards and special privileges for small families, as well as penalties for exceeding the targets. There are numerous exceptions provided by the policy, however, so that only about a third of the population actually lives under a strict one-child rule and thus the policy, if followed, would create a TFR of about 1.47. The policies have helped to lower China's fertility rate to about 1.6, well below replacement level, although with enormous controversy about the government's tactics. Other factors, such as delayed marriage and changing ideas about family size, are likely contributing to this decline independently of government policy. If these trends continue, China's population will peak during the next two decades before decreasing.

India's population grew from 342 million at the country's independence in 1947 to 1.17 billion in 2010. If India's growth rate continues at 1.6 percent per year, it will overtake China as the world's most populous country before 2050. Today, the Indian government sponsors family-planning programs, and radio and television emphasize that small families are healthier and happier. A new social security program assures people that they will be taken care of by the government in their old age. Abortions are legal, but sterilization accounts for about 90 percent of India's family-planning program and is practiced by 37 percent of married Indian women. The TFR fell from 6 in 1951 to 2.8 in 2008. Programs in India's individual states also discourage large families. In Rajasthan, for example, village council members lose their seats if they have more than three children.

The Mexican population quintupled from 19.6 million in 1940 to 112 million in 2010 but today government family-planning clinics offer free contraception and sterilization, and a government slogan insists, "The small family lives better." The TFR fell from 7 in 1965 to 2.3 in 2010 but the total population will nevertheless continue to grow by 1 million per year for at least another 20 years.

South Korea began promoting family planning in the 1960s, fearing that overpopulation would impede its economic growth. A slogan at the time warned

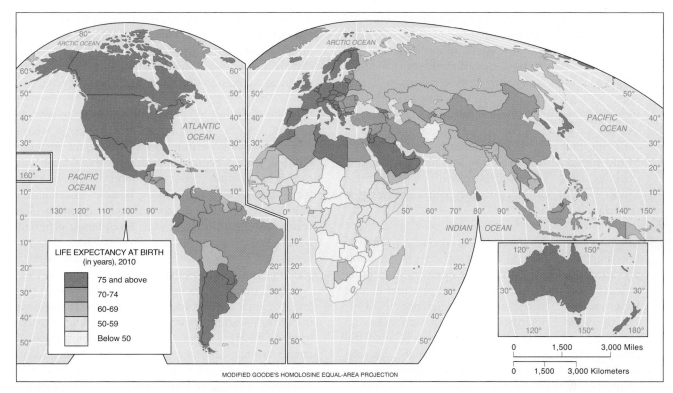

Figure 5-17 Life expectancies. People in North America, Europe, and the richest countries in Japan and Oceania can expect to live much longer lives than people in poor countries. Life expectancies are lengthening everywhere, but much of the human population still cannot be expected to reach what we would call "old age." Compare this map with Figure 5-12. Countries with high death rates have low life expectancies.

Figure 5-18 Family planning. This Indonesian government billboard suggests that smaller families can be healthier, happier, and better off. It is part of the government's family-planning initiative.

South Koreans, who averaged six children per family, that they would become "Beggars without family planning." Today, when the TFR is 1.3, the government still insists that "Even two are a lot."

(2) Availability of contraceptive technologies   The demographic transition occurred in Europe without modern contraceptive methods. Today, however, technology has provided new means of birth control. The United Nations has replaced the slogan that it proclaimed in Bucharest in 1974 ("Economic development is the best contraceptive") with a new slogan: "A contraceptive is the best contraceptive." Among married or cohabiting partners, the global rate of modern contraception use is 56.1 percent, but the rate of use and the preferred form of contraception varies widely across and within individual countries.

Female sterilization is the most widespread form of birth control in the world, used by almost one-fifth of married women. It is especially prevalent in parts of Latin America and Asia, including China. In Brazil, for example, more than 40 percent of all married women of reproductive age have been sterilized. Intrauterine devices are used by 15.5 percent of married women worldwide, most of them in China and other parts of Asia, as well as in developed countries. Almost 13 percent of women rely on birth control medications, whether pills, implants, or injections, and these methods are more common in Europe, North America, and Latin America than they are elsewhere. Contraceptive techniques also reduce abortion rates, but about 42 million abortions are still performed annually, and improperly performed abortions kill about 68,000 women each year. Efforts to develop male contraceptive medicines continue.

Cultural attitudes can affect contraceptive use. The Roman Catholic Church opposes all forms of birth control technology, and some other religions also preach against specific forms of birth control. Many religious groups adamantly oppose abortion. Other issues arise from the cost of manufacturing and distributing contraceptive devices and training users how to use the devices properly. Two-thirds of the users of condoms, diaphragms, and contraceptive sponges live in the industrialized world, and attitudes about family planning in the rich countries determine whether these countries offer assistance to the poor countries that may want it.

Geography Video
Staying Alive

(3) Status of women   Empowering women is key to lowering fertility rates. In societies where a woman's role is culturally or legally limited to being a wife and mother in the home, there are often enormous pressures to have large families. The 1994 Cairo Conference, the 1995 Fourth World Conference on Women, and follow-up conferences all emphasized that key factors in successful family planning include universal access to education and health care, and relative equality in status, employment opportunities, and full political rights for women. The 1995 conference declared, "women have the right to decide freely and responsibly on matters related to their sexuality" and must be able to do so "without coercion, discrimination and violence." Improving access to education and employment for young women has shown tremendous results in lowering fertility.

The status of women also affects access to birth control. For example, some countries ban sterilization on religious grounds. In numerous societies, women still lack political and economic rights, have limited access to education, and exercise little control over their own lives. A 2006 ruling of the Supreme Court of Colombia gave many women hope: The court ruled that Colombia's total ban on abortion violated international treaties ensuring rights to life and health. The court insisted that abortions must be allowed when the mother's life is in danger, when the fetus is expected to die, or in cases of rape or incest. To ban all abortions is to value the fetus ahead of the life of the mother. Other governments may follow this legal precedent.

In areas where male children are preferred to females, birth rates remain high for several reasons. If a couple's first child is female, they often continue having children until a male child is born. The fertility policies in some parts of China, for example, allow for a second child in the event that the first child is female. In these traditional societies, a girl is often regarded as just another mouth to feed and as a temporary family member who will eventually leave to serve her husband's relatives. Also, in many societies a woman's parents must provide her with a dowry, which is a substantial financial payment at the time of her marriage. A son, on the other hand, means more muscle for the farm work and someone to care for aged parents. In several traditional Asian and African religions, only sons can burn the necessary ritual offerings to ancestors, and only sons carry on the family name. As women win the right to work outside the home, as is happening in some societies, the financial preference for male children is reduced.

(4) The diffusion of changing attitudes related to fertility   During the demographic transition in today's rich countries, information about contraceptive methods diffused slowly because educational levels were low and mass communication was limited. Ideals about family size varied widely. Today, however, communication and the mass media in developing countries have accelerated the diffusion of ideas about family planning. Many governments, including those of Ghana, Nigeria, Gambia, Zimbabwe, and Kenya, broadcast family-planning messages on both radio and television.

Furthermore, public health officials have noted that even commercial television programs have a tremendous impact in lowering fertility rates by detaching sexuality from childbearing. Popular television programs transmit images, attitudes, and values of a modern, urban, middle-class existence in which families are small, affluent, and consumer oriented. Popular television has been credited with lowering fertility rates in India, Pakistan, the Philippines, Kenya, Brazil, and China.

Throughout the world, fertility rates are lower in urban areas than in rural areas. Moreover, the larger the city, the more likely women are to use contraception. Reasons for this may be that urban life changes the social and economic status of women. Their ideas about marriage and family size may change and family-planning services may be easier to reach; children cost more to raise in big cities, and space is at a premium. Today, half of the world's people live in cities, and that percentage is increasing (see Chapter 10). This development will probably continue to lower TFRs.

Changes in World Death Rates

A decrease in the crude birth rate is not the only factor affecting human population growth. Natural disasters and wars claim tens of thousands of lives around the world each year, but these have little effect on the global death rate. Decreases in the crude death rate, however, can increase life expectancy and population size. Across human history and around the world, there has been a gradual decrease of death rates first through improved nutrition and more recently with the prevention of infectious diseases. This long-term decrease in death rates is known as the **epidemiological transition,** and it figures importantly in the demographic transition model. (**Epidemiology** is the study of the incidence, distribution, and control of disease.) In developed countries, public health practices and medical advances since the nineteenth century have lowered the number of deaths caused by **pathogens,** disease-causing organisms that enter and multiply in the body. As a result, populations in Stage Four of the demographic transition model are more likely to die from degenerative diseases that occur as the human body ages. For example, in 1900, the leading causes of death in the United States were infectious or parasitic diseases: pneumonia, tuberculosis, and diarrhea. Today, the leadings causes of death are heart disease, cancer, and stroke.

Disease prevention techniques and medicines diffused quickly around the world during the twentieth century, reducing the deadliness of many common infections. As recently as the 1970s, scientists tended to see infectious diseases as a series of problems to conquer, ticking off victories like a to-do list. Smallpox was the greatest cause of death by infection in human history, but eradication campaigns begun after World War II had eliminated the disease by 1977. (Some countries, however, kept smallpox viruses and other pathogens for purposes of war.) Eradication efforts against polio and guinea worm disease continue. Despite these efforts, infectious and parasitic diseases take an enormous toll on populations in poorer countries.

The deadliest communicable diseases in the world today are respiratory infections such as pneumonia, which kill more than 4 million persons each year. It is the cause of about 7 percent of all deaths. It is possible that some of these deaths are attributable to an underlying infection caused by other communicable diseases.

The second leading communicable cause of death is diarrheal diseases, responsible for more than 3.7 percent

of all deaths. The most important of these is cholera. New strains of cholera appear regularly, and they are spread through contaminated water and poor sanitation. Cholera epidemics occur periodically in the developing world.

The third deadliest infectious disease, the human immunodeficiency virus (HIV), causes acquired immune-deficiency syndrome (AIDS). AIDS destroys the body's ability to fight infections, and it was responsible for about 2 million deaths in 2007, about 3.5 percent of all deaths. It is believed that the AIDS virus emerged in Africa in the 1950s, and possibly earlier; by 2007, an estimated 33.2 million people were living with HIV—many unaware that they were infected (Figure 5-19). Social and political leaders in some countries have denied the prevalence of the disease, thus giving it time to spread unchecked. South Africa is estimated to be the country with the greatest number of HIV-infected people (about 5.7 million), followed by Nigeria (2.6 million) and India (2.4 million). Globally, the rate of HIV infections has begun to slow, yet there were still about 2.7 million new infections in 2007. Two-thirds of all people infected are in sub-Saharan Africa, although infection rates have slowed in some African countries thanks to massive prevention programs. In several of these countries, AIDS-related illnesses are the leading cause of adult death, and AIDS has shortened life expectancy by 20 or more years. AIDS causes massive human suffering and also has a harmful economic impact. It weakens and kills adults in their prime years,

Geography
Video
Sowing Seeds
of Hunger

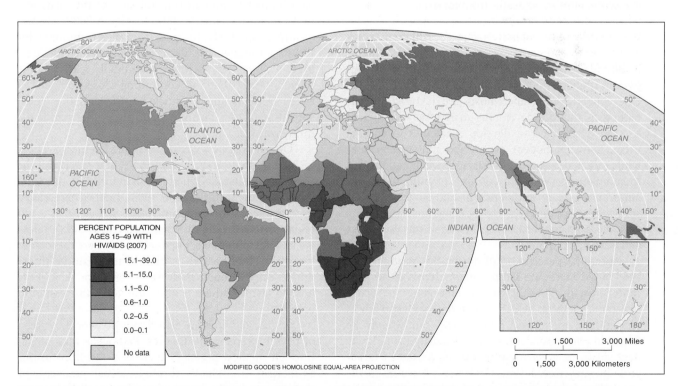

Figure 5-19 Population ages 15-49 estimated to be living with HIV/AIDS in 2007. Today, scientists think that human immunodeficiency virus (HIV) evolved in Africa and spread from other primates to humans there. It continues to spread around the world, but Africa is particularly afflicted.

leaving behind orphans and driving dependency ratios higher. New techniques of treating the disease offer promise, but their success is not yet proven. The high cost of prevention and treatment has been lowered through international aid and cooperation from pharmaceutical companies and national governments, but many countries lack the health systems necessary to reach all communities in need. Such programs have begun to make a small difference, as the total number of deaths from AIDS has slowly decreased since 2005. This is good news, but it means an even greater effort is needed to reduce new infections and to treat those already living with HIV.

Tuberculosis is the fourth leading communicable cause of death, causing about 2.5 percent of the total. Approximately one-third of the human population carries TB bacilli in their bodies, but in the vast majority of cases, the disease is latent and cannot be transmitted. New strains of drug-resistant tuberculosis, however, are multiplying; the fight against this disease continues.

Malaria is responsible for about 2 percent of human deaths. The mosquitoes that carry the disease live in a broad variety of climates, so malaria occurs widely, but about 80 percent of malaria-caused deaths occur in sub-Saharan Africa. Drugs are normally used to treat malaria, but drug-resistant strains are appearing in Southeast Asia and in Africa. In addition, mosquitoes are developing resistance to insecticides. Thus, health officials fear that the appearance of "supermosquitoes" armed with drug-resistant "supermalaria" could increase the number of deaths from malaria.

New and Reemergent Diseases Although communicable diseases are now less common and often less deadly in wealthier countries than in poor ones, this does not reduce the risks posed by new infectious disease. Malaria and tuberculosis are just two of the pathogens that are developing drug resistance. Scientists are challenged to develop medicines to combat them. Some pathogens evolve and mutate naturally, while others develop drug resistance when patients fail to complete their full prescriptions of antibiotic medicines. The use of antibiotics in animal feed and as sprays on fruits and vegetables during food processing also increases opportunities for organisms to evolve and develop resistance.

About 70 percent of all antibiotics produced in the United States are fed to healthy livestock to promote growth. This allows the emergence of drug-resistant bacteria, which are today widespread in commercial meats and poultry and can already be found in consumers' intestines. The drug Baytril may have given us a warning: It was once used in large-scale poultry production, but when its use in chickens reduced the effectiveness of a similar drug made for humans, Cipro, against infections, the United States banned the use of Baytril for poultry in 2005.

Only European governments, however, absolutely ban the nontherapeutic use of antibiotics in animals.

Other diseases are constantly evolving over time, sometimes reappearing in deadlier form. This sometimes happens when disease-causing pathogens are passed back and forth between humans and other animals. Many influenza strains, for example, originate in Asia because great numbers of ducks and pigs—common animal hosts for viruses—live there in close proximity to human beings. A Great Flu Pandemic swept the world in 1918–1919, killing as many as 40 million people, as did the pandemics of Asian flu in 1957 and the Hong Kong flu in 1968.

In 2002, a new viral infection, called SARS (severe acute respiratory syndrome), appeared in southern China, but the Chinese government at first withheld information about the disease to avoid embarrassment. This action abetted the disease's global spread. The epidemic subsided, but the discovery of the SARS virus in other animal species suggests how the virus may have been transmitted to humans. We may never eradicate the virus if it is harbored in wild animals.

In 1996, a new lethal avian virus, H5N1, appeared in poultry in China. It spread among poultry and then to humans who lived near or handled poultry. In 1997, the virus appeared in Hong Kong, and authorities ordered the destruction of all domestic poultry there. During 2003–2004, human cases multiplied throughout Southeast Asia, and through 2005 and 2006, the flu spread across Asia into Africa and Europe, killing people and devastating poultry stocks. The first identified fatal human-to-human transmission of the virus was in Thailand in January 2004. In March 2009, a strain of influenza, H1N1, jumped from pigs to humans for the first time. It was declared a pandemic by June, though it was less deadly than initially feared. Some scientists fear that the continuing close contact between humans and livestock in Asia and the global spread of factory farming could fuel other pandemic influenzas (see Chapter 8).

Threatening new viruses continue to emerge. These include not only HIV, SARS, H5N1, and H1N1 but the Ebola virus (first identified in Sudan in 1976), the Sabia virus (identified in Brazil in 1990), hantavirus (related to a virus first identified in Korea in the 1950s), and Nipah virus (first identified in Malaysia in 1998). These are all fatal to humans, but, luckily, most are difficult to transmit. The Ebola virus, for example, can be transmitted only through body secretions, and it kills its host so fast that there is little time for transmission to new victims.

Human activities are also responsible for the spread of new or reemergent diseases. Pathogens may spread through medical practices such as blood transfusions and organ transplants, through sexual practices, or through needle-sharing among drug users. As humans change the environments in which we live, we come into contact with new pathogens.

Environmental Disturbance and Disease

Human alteration of the environment can create breeding grounds for new viruses and increase the number of pathways viruses can take to new populations. As new human settlements put pressure on surrounding habitats, humans come into contact with unfamiliar species that may carry disease capable of jumping to human hosts. Settlers who clear forests often reduce the natural food sources used by forest mammals, which invade the new houses looking to eat. An outbreak of hantavirus occurred in the United States in 1993, when hungry rodents, driven into human settlements by rising waters, left droppings in Arizona kitchens. "Manmade malaria" occurs frequently around irrigation systems that contain large pools of standing water in open fields—ideal mosquito breeding grounds. Even simple deforestation at the edge of a city removes the canopy that normally reduces mosquito activity while leaving behind pockmarked land that fills with water. Dengue fever and Japanese encephalitis also spread through irrigation practices in mosquito habitats. Confined animal breeding, such as pig farms and poultry pens, is under intense scrutiny as the possible cauldron of recent viral outbreaks, including SARS and H1N1. Travel, of course, effectively introduces viruses and fresh hosts who may lack the locals' resistance, shuttling disease around the world. One of the worst scenarios for public health is a highly contagious infection entering the global air transportation network. Humans no longer benefit from relative isolation and the disease barrier of distance.

Large-scale environmental alteration is likely to change the opportunities for old and new diseases to appear. Climate change is increasing the portion of Earth that is hospitable to disease-carrying insects. Mosquitoes are already appearing at previously cooler higher latitudes and higher elevations. These changes may produce more infectious disease such as diarrhea. Climate change may also affect crop production leading to malnutrition, a health problem by itself, which also limits humans' ability to fight off infections. Heart and respiratory diseases may increase due to increased ground-level ozone. Human environmental alteration may, on the other hand, eliminate a pathogen's habitat, eradicate the pathogen, and prevent future epidemics. For example, the completion of Egypt's Aswan Dam in 1971 destroyed the floodwater habitat of the *Aedes aegypti* mosquitoes, carriers of Rift Valley fever virus. By 1980, Rift Valley fever had virtually disappeared from Egypt, although the dam provided an aquatic environment that spread Schistosomiasis.

Because environmental changes are highly localized in their effects, it is impossible to make accurate predictions about what will happen where. Some regions may experience relative relief from some disease burdens but many regions will experience a shift to new, unfamiliar, disease. Vulnerable populations, especially those with weak health systems, will have a difficult time coping with these unanticipated changes.

Degenerative diseases Most degenerative diseases are difficult to treat because they deteriorate bodily tissues or the operation of vital organs. There is little that can be done to reverse the effects of age on human health. Other degenerative diseases, such as heart disease and diabetes, may be attributable, in part, to the lifestyles, diet, and behavior of the sufferers, who are often adults. Degenerative diseases are leading causes of death in rich countries, but they are increasing in several developing countries, too. The World Health Organization (WHO—a United Nations Agency) estimated in 2006 that Earth's number of overweight people (1 billion) exceeded the number of undernourished people (about 800 million). This was a turning point in the history of humankind. Some 300 million people were obese—that is, in serious medical danger due to their weight—and that number has been rising fast. In the rich countries, causes of obesity include the switch to restaurant and take-out meals, which usually have higher calories than home-cooked meals, and increasingly sedentary lives lacking exercise. In developing countries, increased dietary fat—largely attributable to the spread of cheap, high-quality cooking oil—has helped to raise daily caloric intake by an average of

400 calories per capita since 1980. In developing countries, too, urbanization and sedentary lives take their toll. China alone counted 65 million obese people by 2006. Being overweight and obesity may trigger type 2 diabetes, from which some 30 million people in east and south Asia alone were already suffering by 2003, plus heart diseases. From 1986 to 2009, the estimated number of diabetics on Earth swelled from 30 million to 246 million. Type 1 diabetes, which arises almost entirely from genetic factors, accounts for only 5 to 10 percent of cases. In many countries, including the United States, life expectancies for the next generation may actually shorten.

Tobacco products are predicted to account for an increase of 50 percent in global cases of cancer by 2025. Tobacco now accounts for about 5 million annual deaths worldwide due to lung cancer and many other debilitating diseases, and more than one-half of the estimated 1.25 billion men and women who currently use tobacco are projected to die from the habit. WHO projects that tobacco will account for over 8 million deaths a year by 2030 if preventive steps are not taken now. The Convention on Tobacco Control, regulating crops, advertising, trade, and other aspects of marketing, was signed by 168 nations and came into force in 2005, but the United States has refused to ratify it, defending the "free speech" rights of the tobacco industry.

A reversible transition   The epidemiological transition is not a one-way street—it can move in reverse. Emergent and degenerative disease could increase death rates if left unchecked. Countries that attain lower death rates and reduce the prevalence of infectious disease may see those achievements slip away. There is the risk that new and reemergent disease will decimate even wealthy countries with advanced medicine. More likely, however, is that rapid societal change will limit infectious disease prevention programs and wreck health care systems. Intense political, social, and economic changes can bring about dire consequences for a population. War often worsens nutrition and disrupts health care systems, but the resulting increased death rates are overlooked because of combat-related deaths. Nonviolent reversals are also possible, as witnessed in the former Soviet Union.

Is Earth Overpopulated?

Many people fear that Earth may become overpopulated or that it is already. People disagree, however, on the meaning of the word *overpopulation*. The idea of overpopulation implies that there are too many people. But too many people for what? What criteria do we use to evaluate this concept? Do we mean too many people to feed and house? Earth has enough basic resources for a much larger world population, but this would require many more people to use resources sustainably, which is not the case in wealthier countries today. And how do we value environmental quality versus the needs of people? How much growth in both population and wealth can Earth sustain? On a global scale, these questions are too complex to yield a definite number beyond which Earth would be overpopulated. Such questions also relate to moral and political values that differ widely and often sharply.

Discussions of world population growth inevitably invoke the ideas of Thomas Robert Malthus (1766–1834), who asked whether humankind would always be able to feed itself. Malthus's statement of the relationship between population and food supply still demands our attention. Malthus was a professor of economy and also a clergyman, and in 1798 he published his *Essay on Population*. His essay, stated most simply, argued that food production increases arithmetically: 1–2–3–4–5 . . . units of wheat. Population, however, increases geometrically: 2–4–8–16–32 . . . people. Therefore, the amount of food available per person must decrease as the population increases.

The human population can be kept in balance with food supplies only through checks on population increase. Malthus defined two types: positive checks and preventive checks. *Positive checks* refer to premature deaths of all types, such as those caused by war, famine, and disease. *Preventive checks*, in contrast, are human actions designed to limit births, such as a decision by a young couple to delay marriage and childbearing. According to Malthus, couples should have a sense of responsibility for the economic welfare of themselves and their children. "There are perhaps few actions that

---

RAPID CHANGE

Demographic Collapse

Death rates have risen in Russia since about 1990, reversing its epidemiological transition. Increased death rates have been attributed to the physical and mental stresses of desperate economic conditions and unstable political circumstances since the fall of the Communist government and the dissolution of the U.S.S.R. (see Chapter 13). Advanced health care was readily available under the Soviet system, and its doctors pioneered many treatments for chronic and infectious disease. That system fell apart with the state that sponsored it. Russian income has fallen, use of tobacco and misuse of alcohol have spread, infectious diseases have multiplied, and life expectancy has fallen. At the same time, fertility rates are low (TFR is 1.4), and difficult economic conditions have driven up emigration rates among the educated. In the early twenty-first century, Russia's population has been decreasing by about 1 million per year.

These effects can be long lasting. Russia's population is projected to decline by 22 percent between 2008 and 2050. Other Eastern European countries face similar fates, including Bulgaria (−35 percent), Georgia (−28 percent), Ukraine (−28), Moldova (−23 percent), Serbia (−21 percent),

Passing time in Yekaterinburg. The end of the Communist era left many Russians out of work with little to do but drink.

Belarus (−20 percent), Romania (−20 percent), and Bosnia-Herzegovina (−20 percent). The difficulties facing these countries' readjustment after Communism will be exacerbated by a shrinking and aging population.

tend so directly to diminish the general happiness as to marry without the means of supporting children," he thundered. "He who commits this act, therefore, clearly offends against the will of God."

Malthus did not believe that preventive checks could control population growth. He pointed to high birth rates among the poor as evidence. Therefore, he predicted that the future of humankind would consist of endless cycles of war, pestilence, and famine. This is the pessimistic conclusion of the **Malthusian theory.** People who share Malthus's pessimism are often today called *neo-Malthusians* ("new Malthusians") or *Catastrophians.*

Malthus was replying to the French philosopher Jean Antoine Condorcet (1743–1794). Condorcet had argued that increases in productivity and improvements in technical ingenuity would keep pace with the needs of Earth's rising population, so per capita material welfare could actually rise. Condorcet also suggested that, given education and prosperity, most people would voluntarily limit their family size. People who share Condorcet's optimism are often referred to as *Cornucopians*; a cornucopia is a horn of plenty.

So far, Condorcet's predictions have proven more accurate than those of Malthus. The human population has sextupled since Malthus's time, yet the widespread starvation that he feared has not happened. Many people are hungry, but this deprivation does not result from insufficient food on Earth. People are hungry because the food is not priced within their reach, nor is it distributed to all. The problems are political and economic, not technological. Will these political problems ever be solved? We do not know. Chapter 8 details how humankind has managed to multiply the food supply. Could mounting numbers of people ever place demands on food and materials so great that per capita supplies will necessarily fall? That question will be answered by future developments in technology, and we cannot possibly foresee what those can be.

Other Significant Demographic Patterns

We have seen how variations in birth rates and death rates affect both overall world population growth and the growth rates in different parts of the world. Two other demographic statistics vary so significantly around the world that they demand geographic analysis. One is the variations in the sex ratios among national populations, and the other is the aging of the human population.

Sex Ratios in National Populations

It is considered a fact that under natural conditions, about 105 male children are born for every 100 females. The United Nations, however, estimated 107 males were born for every 100 females in 2010. There is ongoing debate to explain these differences, but cultural gender preferences for males appear to explain much of the difference. The higher birth rate for males is not new, and historical research suggests that in some cultures with strong preferences for male children, it was not uncommon for couples to abandon, neglect, or murder unwanted female infants. New technologies, such as ultrasound and amniocentesis, allow couples to know the sex of their child *in utero*, making it possible to selectively terminate pregnancies in favor of male births. While these practices may occur in many societies, there are large regional variations that appear in the sex ratio at birth. Sub-Saharan African birth ratios are slightly below normal (103 males for every 100 females); North American, Latin American, and European ratios are normal (105–106 males to 100 females); and Asian birth ratios average 109 males for every 100 females.

Asia's regional average is due almost entirely to birth ratios in China (120 to 100), India (108 to 100), and South Korea (110 to 100). Rates vary within these larger countries. India's rates vary across its states from normal ratios to one district that in 2001 had 133 male births for every 100 female. There is evidence that China's fertility policies may have exacerbated these practices, as couples grow desperate for male children. One study of abortions in southern China showed that 29 percent of abortions were to women who already had one daughter and, among these, more than two-thirds of the aborted fetuses were female. These imbalances in sex ratios cause problems later on in forming new families. Many governments, including those of China and India, have sought to ban or limit the practices associated with sex selection. China, however, still promotes abortion as a means of birth control.

An alternative theory that may partly explain the high ratios of male births in many poor countries is the prevalence of hepatitis B, which increases the chance of a male birth if the disease is carried by either the mother or father. Demographers disagree on the importance of this factor, with most accepting that higher male birth ratios are intentional.

After birth, sex ratios change. Women tend to live longer than men when both receive similar medical care. Therefore, a population should naturally have an approximately equal number of males and females, or perhaps slightly more females. In the world in the year 2010 there were 101.5 males for every 100 females, but sex ratios vary widely. Obviously, countries with a strong male bias at birth will see that trend continue as children age. China, for example, has a sex ratio of 107.7 males to 100 females in the general population. Other factors that can reduce one's life expectancy, thereby affecting sex ratios, are often gender biased in favor of men. Although men typically fight war, war also affects women as victims while also depriving them of basic needs taken for the war effort. In times of relative calm, women's access to sufficient food and medicine may be limited because of gender discrimination.

Limited access to education and employment means male-headed households may spend less on women's basic needs. In wealthier countries, where access to advanced medicine is higher and life expectancies are longer, there are more women than men (94 males to 100 females). This may be due to men's higher exposure to high-risk jobs and lifestyles, including excessive drinking and smoking. So, too, do lower birth rates in these countries mean women are less frequently pregnant and so are exposed to fewer of the associated risks.

Sex ratios favor women in both the richest and some of the poorest countries, reflecting different factors working across cultures and countries. In some countries of Southeast Asia, women outnumber men. In South Asia, however, particularly India and Pakistan, the ratio of men to women is high. Sharp contrasts exist even within individual countries. In the northern Indian states of Punjab and Haryana, for instance, which are among the country's richest, there are 116 males per 100 females, whereas in the state of Kerala in southwestern India, there are 97 males per 100 females, a ratio comparable to that in Europe, North America, and Japan. Sub-Saharan Africa has a similar ratio of 99 males to 100 females. Several of the countries with the highest ratios of males to females are Arab countries, where women's rights are not always guaranteed. Some demographers think this suggests that societies that limit women's rights do not count all women in the census. If females are actually undercounted in these countries, then the reported higher ratios of males may not be accurate. Certainly women are better off in countries that acknowledge female deprivation and seek to remedy it. Efforts to achieve this can be influenced by providing women with education and a greater role in public life, including employment, voting, and civil society (see Chapter 11).

The Aging Human Population

Through the twentieth century, improvements in health and hygiene triggered a rise in the median age of Earth's population for the first time in history. Humankind's median age increased from 23.5 years in 1950 to 29.1 in 2010, and it is projected to reach 38.4 by 2050. The percentage of people age 60 or older is expected to rise from 11 percent in 2010 to 21.9 percent by 2050.

In the rich countries, around 20 percent of the populations are already age 60 or older; that figure may rise to nearly 33 percent by 2050. The existence of aging populations challenges the possibility of sustaining economic growth and equitably distributing wealth among generations. Higher percentages of national budgets are devoted to senior citizen centers than to elementary schools. Funds for medical research are dedicated to degenerative diseases such as cancer, heart disease, and Alzheimer's, which means that enormous sums of money are buying relatively short advances in longevity. To support elderly populations, young workers may have to increase their contributions to pension funds, or else pension benefits may have to be reduced. Alternatively, workers may have to work more years before retiring. Most rich countries are already raising the age at which people may collect retirement benefits.

A few countries with TFRs below the replacement rate, including France, Italy, Poland, Australia, Russia, and Singapore, are encouraging higher birth rates. Japan, for example, announced in 2006 that the national TFR had fallen to 1.29 and that the national population had already begun to decrease. Furthermore, the population is aging, as the percentage of its people over age 60 will increase from 30.4 in 2010 to 44.2 by 2050. Only 13.4 percent were 15 or younger in 2010, and that figure will fall to 11.2 percent by 2050. Therefore, Japan is encouraging women to bear children. Government-funded day care is available, and employers offer extended childcare leave. Nevertheless, social rigidities limit childbearing. Social mores insist on a traditional view of marriage and parenthood (in Japan, fewer than 2 percent of births are to unmarried mothers, compared to 40 percent in Britain, 50 percent in Scandinavian countries, and 34 percent in the United States), and Japanese law even discriminates against children born out of wedlock.

The U.S. population aged over 60 is projected to rise from 56.9 million in 2010 to 111 million by 2050. Because of immigration (discussed later in this chapter) and America's higher TFR, which is almost unique among rich countries, America's median age is only projected to rise from 36.8 in 2010 to 41.7 by 2050.

Historically, every country that got old was rich first, but today this is no longer true. In many of today's poor countries birth rates are falling. Therefore, their median ages will soon rise fast. About 8 percent of the populations of today's poor countries are over 60 years of age, but that percentage will rise to 20 percent by 2050. Mexico's TFR, for example, has fallen from 6.5 in 1970 to 2.04 today, so its median age of 25 in 2006 is expected to rise to 44 by 2050. About two-thirds of Chinese population growth between 2005 and 2025 will occur in the over-65 category, a cohort likely to double in size to about 200 million people. By then, China's median age may be higher than America's. An aging China is already facing both strains on its pension system and shortages of young labor.

Aging populations will present tremendous challenges to the poor countries. National pension systems may be inadequate. In many poor countries, the elderly are today cared for by their families, but this situation will shift as the ratios of old to young rise. These countries will have to build up their national incomes and devise national pension systems. They have perhaps 50 years in which to do this.

One partial solution for both the rich and poor countries is the international migration of pensioners to poor countries. Tens of thousands of Americans

CONNECTIONS

The Economics of Aging

As the U.S. population ages and the balance between workers and retirees shifts, you might start planning for your own retirement, even if it is still many years away. The three traditional sources of income for U.S. retirees are pensions, personal savings, and social security.

Corporate pension plans multiplied through the twentieth century. Most were *defined-benefit plans* in which a worker received a set payment for life, based on a combination of salary and years of work. Clause 401(k) of The Tax Reform Act of 1978, however, encouraged *defined-contribution plans*, in which employers and workers themselves contribute fixed sums to savings accounts that receive favorable tax consideration. Most U.S. corporations have switched to defined-contribution plans.

Thus, private pensions—like health insurance—in the United States were linked to work. Today, however, as employers face global competitive pressures, and as the number of retirees grows relative to the number of workers, corporations have begun to reduce both their pension liabilities and their workers' health care coverage. In 1962, for example, General Motors had 464,000 U.S. employees and was paying benefits to 40,000 retirees and their spouses. By 2005, the corporation had 141,000 workers and paid benefits to 453,000 retirees. In 2009, it went bankrupt in part because it was struggling to cover its pension obligations. The telecommunications manufacturer Nortel announced in 2006 that it would shift its North American workers from a defined-benefit to a defined-contribution pension plan and eliminate all post-retirement health care benefits. At the same time, the corporation cut its North American workforce and hired thousands of new workers in Turkey and Mexico.

Other problems have arisen because not all workers have enjoyed satisfactory financial returns on the assets in their defined-contribution accounts, and the Pension Protection Act of 2006 only partially addressed the fact

that many corporate pension accounts have been substantially underfunded. About one-half of U.S. workers in the private sector still have no employer-sponsored retirement plan of any kind.

Local government pensions are underfunded as well, to an amount estimated to be hundreds of billions of dollars. Most state constitutions, however, guarantee these pensions, although experts wonder where the money to pay the pensioners will be found.

Many Americans find it hard to accumulate individual savings in a culture so rich in attractive consumer goods. The U.S. personal savings rate fell from about 8 percent of household income in the early 1990s to 1 percent in 2006. Other industrialized nations average about 7 to 15 percent savings rates. The U.S. trend began to reverse only with the deep recession that began in 2007, by which time far larger problems loomed. The financial meltdown of 2008 led to stock market losses that on September 29, 2008, made more than $1 trillion disappear from investor portfolios, much of which was retirement savings.

Social Security payments were never intended to allow retirees to live comfortably, just to prevent starvation, and Social Security will be strained as baby boomers reach retirement (refer to Figure 5-9). Suggestions to lower benefits meet strong resistance, as did the Bush administration's 2005 attempt to allow individuals to invest their savings in investments of their own choosing but then to eliminate the guaranteed aspects of the plan. Therefore, the United States is raising Social Security taxes by raising the base amount of a person's income that is taxed, although most Americans already pay more in Social Security taxes than in income taxes. The age at which workers may retire and receive benefits is also steadily increasing. Law current in 2006 has the retirement age for Social Security and Medicare slowly rising to 67 (from the current 65) by the year 2027.

---

have already retired to Mexico and Costa Rica, where they invest and spend income, while imposing few costs on the society.

Migration

Human beings do not stay put; they never have. Wanderings and migrations of people have distributed and redistributed populations throughout history and even prehistory. Significant redistributions continue.

Geographers who analyze human movements divide the causes for those movements into **push factors** and **pull factors.** Push factors drive people away from wherever they are. Push factors include starvation, disasters,

and political and religious persecution. Pull factors attract people to new destinations. Pull factors include economic opportunity and the promise of religious and political liberty. Physical geography can also serve as a push or pull factor. For example, we can say that migrants from northern cold regions have found the warmth of the U.S. Sunbelt attractive and that environmental disruption in the African Sahel, by contrast, has been found intolerable by emigrants from that region. The environment itself never actually pushes or pulls; people decide to migrate after considering environmental factors.

People choose a particular destination because of what they think they will find there, not because of the reality of conditions at that place. Therefore, migrants

Figure 5-20 Disappointed migrants. A Bangladeshi migrant worker looks out of the window of a building where he is housed, with other migrants, in overcrowded and unsanitary conditions in Male, Maldives. Many here are victims of scams where they are promised high paying jobs in the Maldives. On arrival as undocumented workers they are easily taken advantage of by unscrupulous business owners.

can be surprised by the reality of what they find at their new homes. Sometimes migrants find success beyond their expectations, but in other cases they are disappointed (Figure 5-20).

Migration has not always been voluntary. Some migrations have been forced. Millions of people suffered the tragedy of migrating in slavery, regardless of push or pull considerations of their own. Others have been forced to flee war, persecution, and catastrophes in their homelands. For example, millions of citizens of the Democratic Republic of the Congo have fled the civil war that recently raged in their country; many have fled abroad, but many more are displaced within their own country.

Many migrants fulfill whatever requirements are formally asked of them to settle in their host countries, but today great numbers of people are crossing borders without completing legal papers. These people are **undocumented immigrants,** and they might be called illegal or irregular immigrants, depending on their status in different countries. Even without papers, some governments and employers welcome these migrants as laborers, but many people vilify them. These immigrants raise many concerns for the countries that host them, whether South Africa, France, or the United States, including fear of increased competition over jobs, housing or public resources like schools. Other arguments are couched in cultural terms, as we explore below.

Movements of people can often be explained by studying flows of trade or by mapping the most convenient transport routes away from places people want to leave. People also follow information. Successful migrants write home, and then they may help new arrivals find employment and financial assistance. In this way, linkages are forged that produce **chain migration.** These linkages often depend upon information that travels

between formerly imperial powers and their former colonies, as between France and several African countries or between the United States and the Philippines. In these cases, potential migrants absorb some of the culture of their destination even before they leave home. Information flows can be astonishingly place specific. An estimated 80 percent of the Hispanics living in metropolitan Washington, D.C., are from the tiny Central American country of El Salvador—many of them undocumented. Nobody knows exactly how word first spread throughout El Salvador to trigger this migration.

Many migrants intend to stay in their new location only until they can save enough money to return to their homeland at a higher standard of living. These **sojourners** include men and women who often leave their families in their home country. Many sojourners eventually decide to stay, at which time they send for their families. The United States has always attracted great numbers of sojourners. In various years between 1890 and 1910 the number of Italians returning to Italy was as high as 75 percent of the number of Italians arriving in the United States. Even today, many Mexicans, Caribbean peoples, and Africans shuttle back and forth between their homelands and the United States.

The ease of international travel and the number of sojourners today has allowed migrants from specific villages in poor countries to form urban neighborhoods in rich countries. For example, when in 1982 a terrible fire killed 24 people in a tenement building in Los Angeles, authorities were astonished to learn that several hundred residents of the building were all neighbors from one small village, El Salitre, in the Mexican state of Zacatecas. Committees of the sojourners in the rich country not only send money home, but they decide on what expenditures will be made in their hometown—whether to build a new school or a new water-treatment plant, for example. A migrant's sustained interest in their place of origin reflects another facet of easier travel, which is that some migrants plan to go home some day. **Return migration** may occur after some period of work allowing them to save enough money to support them when they return. These plans may be seasonal or take decades. Their plans have important implications for understanding migration patterns and are often misinterpreted by host societies, who assume the migration is permanent.

Some migrations have numbered in the millions, but some small migrations have been disproportionately significant because the migrants have played key roles in transforming the government, language, economics, or social customs in their new homelands. As we shall see, the impact of the international migrations of Indians and Chinese, for example, or of Caribbean blacks into the United States, are not fully realized by their numbers alone.

In several cases, the impact of international migrants in their new homes has harmed the welfare of the natives, or **indigenous peoples,** of that region. The United Nations estimates that today there are 370 million indigenous people in 70 countries, defined as descendants of the earliest remaining inhabitants of a land who have been ruled by another people arriving after them. These include Native American groups, the Bushmen of South Africa, and the Lumad of the Philippines. The United Nations has voted that they should be free from discrimination and that they have the right to govern themselves.

Prehistoric Human Migrations

Recent advances in studying human genetics confirms archaeological evidence that all humans are descendents of a single population living in Africa about 150,000 to 200,000 years ago (Figure 5-21). From there, humans migrated near and far, peopling Earth. Some reached Asia (probably across the Arabian Peninsula) between 56,000 and 73,000 years ago and Europe between 39,000 and 51,000 years ago. Scientists are not certain about the timing of the first peopling of the Americas. Both archaeological and DNA evidence suggest a series of migrations, with some traveling on foot across a land bridge that once linked what is now Siberia and Alaska, and perhaps others coming by boat. The discovery in 2005 of human footprints dating from about 40,000 years ago, preserved in ancient volcanic ash near Puebla, Mexico, pushed back estimates of the earliest human presence in the Americas. Other discoveries raise the possibility of an even earlier occupation beginning around 50,000 years ago.

Races New studies of the information in DNA indicate that humans as a species are more than 99 percent genetically identical to each other. There are often fewer genetic differences between members of different race groups than between two members of the same race group. Nevertheless, scientists and the public have long used secondary biological characteristics to divide humankind into **races.** The criteria for these subdivi-

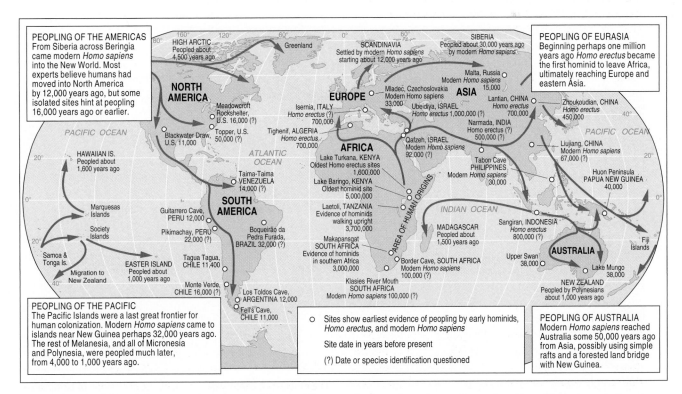

Figure 5-21 Prehistoric human migrations. This map illustrates the latest hypotheses about migrations of humans out of Africa to settle around Earth.

sions have been external features such as eyefolds or skin color, but they also include crude assertions about cultural and character types. On this basis, various classification systems used since the eighteenth century have lumped human populations into different races, such as European (Caucasian or white), Asian, and African (black). Later, scientists tried to use blood characteristics as a basis for a different racial classification. None of these schemes proved scientifically valid nor did they explain much in the way of human societies. Individuals and populations are more varied and diverse than simple racial categories.

The results of the Human Genome Project and other discoveries in the past few years have shown that there is less than one percent of genetic difference among humans. This confirmed what many scientists had already concluded: Most biological variation among humans occurs independently of superficial racial characteristics. Biologically, racial characteristics are like any other trait inherited by descendents. Racial traits are just a few of the many genetic traits that emerged, by chance, within specific human populations at different times and in different geographic regions. When human populations came into more frequent contact after 1500, these traits were incorrectly interpreted as a reason for cultural differences. Europeans presumed that their material advantage over other visibly different people was proof of their racial superiority. These supposed racial differences were used as a justification for European imperial domination of other regions. **Racism,** the belief in the inherent superiority of one race over another, has unfortunately played a major role in defining human relations during last five hundred years of increased migration and intercultural contact.

The Migrations of Peoples Since 1500

For purposes of understanding contemporary geography, we will begin our study of human migrations 500 years ago, before the European voyages of exploration and conquest. In 1500, Earth's many peoples and cultures were relatively isolated from one another. The world we know is the product of their migrations and mingling. The most significant population transfers during these years have been migrations of Europeans to carve enclaves of settlement around the world, the forced migration of enslaved blacks out of Africa, and migrations of other groups instigated by European expansion.

European migration to the Americas   The replacement of natives by newcomers was greatest in the Western Hemisphere (Figure 5-22). The Native American population fell victim to slavery, warfare, and most important, to European diseases, including smallpox, influenza, measles, and typhus.

We have only estimates, ranging from 8 million to 112 million, of the Native American populations at the time of the European conquests, but the numbers of deaths were appalling. There may have been 25 million Native Americans in Mexico before the arrival of the Spaniards; by 1600 that number had declined to about 1.2 million. The population of the Inka Empire in South

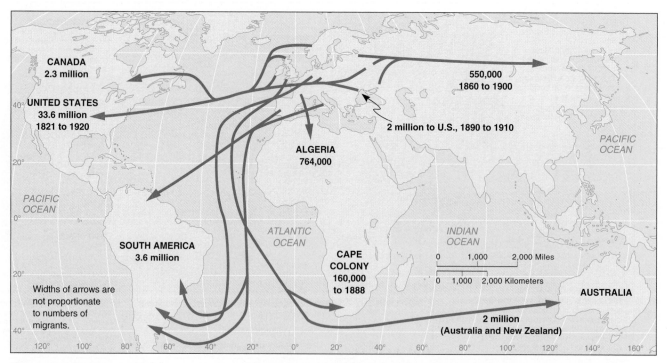

Figure 5-22 European migration in the nineteenth century. The nineteenth century saw the greatest migrations of Europeans. Europeans regarded the Western Hemisphere as a New World open to European settlement, and their massive migrations forever changed hemispheric demographics.

America plummeted from about 13 million in 1492 to 2 million by 1600. (You may be familiar with the spelling "Inca," which is Spanish, whereas the spelling "Inka" is Native American Quechua. For reasons discussed in Chapter 7, Inka is preferred today.) The estimated 5 million Native Americans in Brazil has shrunk to only 300,000 today. In what is today the United States and Canada, the Native American population fell from about 2 million in 1492 to a low of 530,000 in 1900.

North America received some 40 million Europeans before 1950. Today they and their descendants make up the majority of the population. The number of self-identified Native Americans in the United States more than tripled between 1960 and the 2008 estimate of 3.6 million, perhaps reflecting growing cultural pride, but they still constitute just 1.2 percent of the national population. Canada's 1.17 million "Aboriginal people"—First Nations, Métis, and Inuit—constitute 3.8 percent of Canada's population.

The United States and Canada have mixed policies of both assimilating Native Americans and segregating them (historically, sometimes against their will) on reservations, where native cultures may be preserved. In both countries, Native Americans retain preferred hunting and fishing rights over large geographic areas in addition to their reservations, over which they exercise more substantial, but never complete, control. Canada has granted the Inuit people (Eskimos) political domain over roughly one-fifth of the country, the eastern half of the former Northwest Territories, now the territory of Nunavut. Continuing litigation of other extensive Native claims embroils national politics. In the United States, Native American lands represent 2.2 percent of the total land area.

About 4.5 million Europeans, mostly from the Mediterranean area, migrated to what came to be called Latin America, a region that developed a complex racial structure. Ibero American (descended from inhabitants of Spain and Portugal) societies were composed as a pyramid of varying proportions, with a large base of Native Americans and blacks; a lesser number of people of mixed race; and a minority of whites ruling at the top. Numerous racial categories were invented with new terminology, such as mulatto, for a person of mixed white–black ancestry, for one of mixed black–Native American a *zambo*, and one of mixed white–Native American a *mestizo*. These terms are often considered derogatory today. Several Latin American colonies, and later the successor independent states, institutionalized racial status rankings, and discrimination has persisted, despite the fact noted above that all concepts of race are ambiguous. Native Americans lost their rights in their own homelands, even though they may have constituted the majority of the population. Only since the mid-1980s have some South American countries, including Peru, Venezuela, Colombia, and Brazil, set aside lands for Native Americans (Figure 5-23).

To varying degrees, racial animosities linger throughout Latin America. In Venezuela, for example, hostility

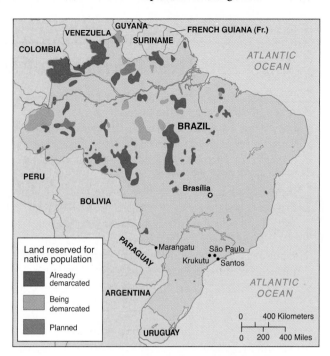

Figure 5-23 Lands reserved for the native population in Brazil. Brazil has only recently adopted the North American model of setting aside lands for indigenous peoples.

lingers between the *pardos*, brown-skinned people of Amerindian or African ancestry, and the *mantuanos*, people of European appearance and pretensions. Hugo Chávez rode this distinction to political power, winning election to the presidency in 1998 while openly referring to himself as "an Indian" against the "rotten white elites."

Native Americans still make up one-third to one-half or more of the populations of Mexico, Guatemala, Ecuador, Bolivia, and Peru (Figure 5-24). In Mexico,

Figure 5-24 Sacsayhuaman. Peruvians gather at the Sacsayhuaman Incan ruins to celebrate the Inti Raymi Festival, the Festival of the Sun.

The East–West Exchange of Disease

When the Europeans first came to the Western Hemisphere, they brought diseases that were new to the native populations. Smallpox, flu, and measles were among the viruses that significantly reduced the Native American populations soon after first contact. Native Americans had no immunity to these diseases, so they were decimated by them. Did the Europeans take new diseases back to Europe that decimated the European populations? No. Why not? Why was the direction of this terrible trade so one way?

One possible reason for the virulence of Eastern Hemisphere diseases in the Western Hemisphere is that the peoples of the Eastern Hemisphere had, through history, domesticated many more animals than had the peoples of the Americas. Many disease-causing pathogens are passed back and forth between humans and other animals with which we live in close proximity, and we and the other animal sufferers develop resistance to these pathogens. The peoples of the Americas had domesticated very few animals, so they had developed little resistance to any diseases. This helps explain why they fell victim to the new diseases arriving from the east. The only disease that, it seems, the Americans gave to the Europeans was syphilis, which a member of Columbus's crew took back to Europe.

---

the Native Americans have played the most conspicuous political role, and Native Americans are waking to power elsewhere. When Alejandro Toledo won election as Peru's first Native American president in 2001, he staged an inauguration ceremony at the Inka city of Machu Picchu presided over by a native holy man. In 2005, Bolivians elected an Aymara Indian president, Evo Morales, the first indigenous chief executive in Bolivia's 180-year history.

Among the southernmost midlatitude states, Chile claims to be 95 percent mestizo, but the Native Americans were practically exterminated in Uruguay and Argentina. In Central America most people are mestizos. Costa Rica is the only country in Central America in which whites form the majority.

The African diaspora   Africans have emigrated, either freely or in slavery, for centuries. These migrations are known collectively as the **African diaspora,** from the Greek word for "scattering," used first by the Jews to describe their worldwide migrations. The forced migration of enslaved black Africans was far and away the largest single factor in their diffusion (Figure 5-25). They were taken to the New World and also to the Middle East and India as part of the trade network in the Arab world and Indian Ocean. The trade diminished slowly, lingering longest in the Arab countries. (It was finally banned in Saudi Arabia in 1962 and in Mauritania in 1980.) The absence of a large black population across this enormous region today can be explained by the high mortality rate, by assimilation, and by the practice of castrating slaves. In central India, however, there are still communities of black descendants of African slaves.

International agencies insist that forms of slavery persist in Mauritania and that Arabs from northern Sudan continue to enslave the blacks of southern Sudan.

The transport of slaves to the Western Hemisphere European transportation of black slaves to the Western Hemisphere had a dramatic impact on population geography. In the fifteenth century, Europeans took a few hundred slaves from West Africa to Europe and the Atlantic Islands, and in the 1520s, Europeans began transporting slaves to the Western Hemisphere (Figure 5-26). Black Africans served as a labor supply to replace the dying Native Americans.

During the period of the transatlantic slave trade, which ran from 1526 until 1870, about 10 to 12 million slaves reached the Western Hemisphere from Africa. The sources of the slaves reflected patterns of European colonialism in Africa. For example, the Portuguese African colonies of Angola and Mozambique provided the slaves of Portuguese-ruled Brazil, whereas West Africa supplied most of the slaves for North America. DNA samples from Western Hemisphere blacks are today providing evidence of their ancestors' African homelands.

The pattern of resettlement in the Western Hemisphere reflected the uses of slave labor. Slaves were brought to tropical and semitropical plantations in the U.S. Southeast, to the islands of the West Indies, and throughout Latin America, to the *tierra caliente* (hot land) at elevations of less than 914 meters (3,000 feet). In Mexico, by contrast, the slaves were put to work in the cities and mines. Mexico never had a significant rural black population, and eventually Mexico's black population died out or mixed with the other groups almost entirely. Mexico abolished slavery at independence in 1821.

Blacks in the Americas today   Descendants of slaves now constitute the majority of the population on most West Indian islands and in Belize. In South America, blacks still occupy, for the most part, the lowland areas.

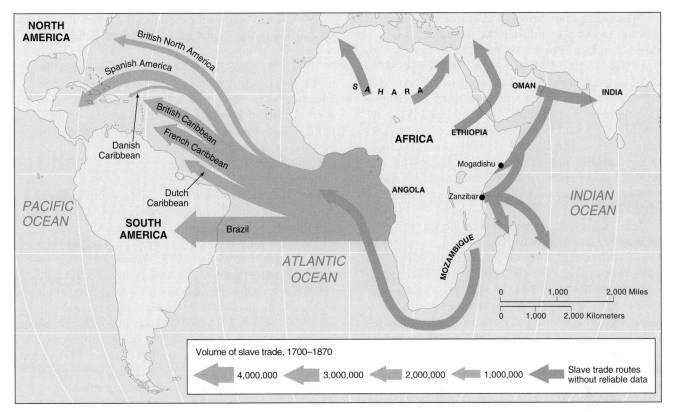

Figure 5-25 The movements of slaves out of Africa. Arab slave dealers first transported black slaves north and east, throughout the region from Spain to India. Transport of slaves to the Western Hemisphere started later and stopped earlier. Only about 5 percent of the slaves brought to the Western Hemisphere were brought to the area that became the United States.

Figure 5-26 A slave market. Tens of thousands of Africans passed through this old slave market on Goré Island off the coast of today's Senegal on their way to the Western Hemisphere. In 1992, Pope John Paul II stood in the doorway in the background, known as the Door of Tears, and blessed them and their descendants. Today, the Senegalese government is restoring the island's historic fortifications as a museum and is building an international conference center.

After centuries of intermarriage, tropical South American countries have significant minorities of black and mixed-race populations.

In the United States, about 13 percent of the population is black—38 million people—giving the country one of the world's largest black national populations. The black population was originally concentrated in the southeast, but even after slavery was abolished, blacks faced economic hardship and deep discrimination. Blacks began to migrate to the industrial cities of the north just before the time of World War I. Most importantly for the migrants, it was a dual migration: from south to north, but also from rural to urban conditions. Between 1910 and 1970, some 6.5 million African Americans headed northward, and the percentage living in the south fell from 70 percent to 50 percent. However, the civil rights struggles of the 1960s created new opportunities for blacks in the south, and partly for that reason but also for reasons not fully understood, African Americans have been migrating back to the south since the mid-1970s—almost exclusively to urban areas. Whereas 52 percent of African Americans lived in the south in 1980, 56 percent did in 2008. The size of black populations in many western, midwestern, and northern New England states remains very small.

West Indian and Caribbean blacks steadily migrated into the United States during the twentieth century—almost 3 million between 1960 and 2000 alone. A high percentage of these people brought capital or job skills, and they and their descendants played a role in twentieth-century U.S. politics and culture disproportionate to their numbers. For example, the first black U.S. Secretary of State, Colin Powell, was the son of immigrants from Jamaica. The flow of migrants and sojourners to the United States, combined with the flow of U.S. tourists to the Caribbean, has caused a degree of acculturation sometimes referred to as the Americanization of the Caribbean.

Since the 1990s, immigration to the United States from Africa has sharply increased, and the number of blacks with recent roots in sub-Saharan Africa has more than tripled. About 55,000 Africans have legally migrated to the United States each year—more than were brought each year during the transatlantic slave trade. The proportion of U.S. blacks who are foreign-born rose from 4.9 percent in 1990 to 8.0 percent by 2008. Black African immigrants seem to have brought skills or to have succeeded in the United States, so statistical descriptions of America's foreign-born blacks and of their children reveal high educational attainment, high income levels, substantial representation in professional and managerial occupations, and low levels of unemployment.

Some native-born American blacks wish to reserve the term *African American* for the descendants of slaves and to describe the newcomers by their countries of origin. U.S. President Barack Obama, however, born to a Kenyan father and an American mother, insists, "I'm African, I trace half of my heritage to Africa directly, and I'm American."

European migration to Asia and Oceania   The Spanish, French, Dutch, Germans, Portuguese, British, and Americans all created empires in Asia, but few people from these countries settled permanently in the Asian colonies. In no cases did any of these peoples come to make up significant proportions of mainland Asian populations. The most important European settlers in Asia and Oceania can conveniently be divided into those who migrated by land—the Russians—and those who migrated by sea—settlers from the British Isles who went to Australia and New Zealand.

Russian expansion   After the Russians rose up against their Mongol rulers in the fourteenth century, they sent a steady stream of settlers to the east, over the Ural Mountains into Asia. Between 1860 and 1914, almost 1 million Russians resettled in the east, and more continued to do so within the political framework of the U.S.S.R. In 1990, the population of Kazakhstan, which was ostensibly a republic of the Kazak people, was 38 percent ethnic Russian. Russians migrated particularly into the cities in the non-Russian republics.

In 1990, Tashkent, then the capital of the Uzbek Republic and today the capital of independent Uzbekistan, was primarily Russian.

When the U.S.S.R. dissolved in 1991, some of the newly independent republics imposed rigid language and citizenship requirements, triggering a migration of some 6 million Russians into Russia. Many of these families had not lived in Russia for generations. The position of Russians who remain in the newly independent Republics is still being defined.

Australia and New Zealand   More than 2 million Europeans, mostly from the British Isles, moved to Australia and New Zealand in the nineteenth century, and another wave of 3 million Europeans (half British) arrived between 1945 and 1973. They all but exterminated the native peoples. The latest censuses count only 150,000 Aborigines in Australia (of a total population of 20 million) and 400,000 Maoris among New Zealand's 3.8 million people. Extensive litigation in both countries has recently returned lands to the native peoples and awarded them substantial payments.

In both Australia and New Zealand, commercial farming and ranching thrived from the beginning of white settlement. Wool, wheat, and later frozen meat were sent to markets within the British Empire. In Australia, white settlement in the dry areas was hazardous and only intermittent until spectacular mineral discoveries attracted people inland. In general, however, the white populations of both Australia and New Zealand have always been highly urbanized along or near the coasts (Figure 5-27).

About 88 percent of Australia's 20 million people are of British and European ancestry, but the country abandoned an explicit "white Australia" policy in 1973. Today 40 percent of Australia's 100,000 annual immigrants come from Asia, making 7 percent of the national population of Asian ancestry. A full 20 percent of Australians in 2006 were foreign-born.

European migration to Africa   Europeans migrated to Africa and settled in substantial numbers only where environmental conditions made possible a Mediterranean or European style of life and commercial agriculture.

East Africa, Northwest Africa, and West Africa   The East African highlands attracted white settlement after the opening of the Suez Canal in 1867. Europeans claimed the fertile lands in today's Kenya, Tanzania, and Uganda and turned much of the native population into workers for the white-owned farms or enterprises. Tens of thousands of whites remain scattered throughout East Africa, but the black-majority governments have rescinded their legal privileges.

In northwest Africa, too, European colonial governments reserved the best lands for European settlers. During the French rule of Algeria from 1830 to 1962, some 1.2 million Europeans settled there—about as many as

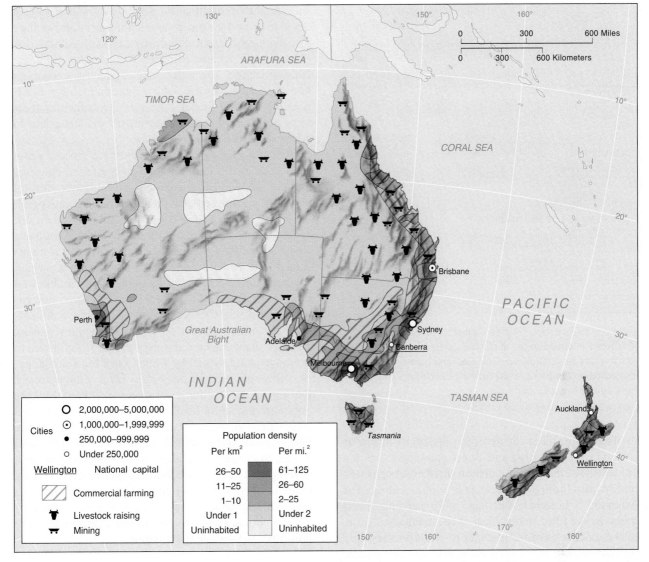

Figure 5-27 Population distribution in Australia and New Zealand. Vast stretches of inner Australia are virtually uninhabited.

in all 26 countries of tropical Africa combined. When Algeria, Morocco, Tunisia, and Libya gained their independence, most white-owned enterprises were nationalized or confiscated. Few Europeans remain in these countries.

West Africa never experienced substantial white settlement. European powers were at first interested in exporting slaves and natural resources. Later, a fairly small number of white administrators established residences to secure their countries' colonial claims. Today whites serve as business executives, technicians, or even government advisers, but they make up insignificant percentages of the populations.

Southern Africa Southern Africa includes large areas that Europeans found well suited to the kinds of agriculture they brought with them. Several colonies did achieve considerable prosperity—for the whites—but always based on repression and exploitation of the

native black majority. As these colonies won independence, white racist regimes yielded to black-dominated governments (Figure 5-28).

European colonial powers competed for what is now the Republic of South Africa because its climate was suitable for European-style agricultural and colonial settlement and because of its mineral resources. In 1671, the Dutch established the Cape Town colony as a staging point for voyages around Africa to Asia, and numerous Dutch farmers, called Boers, came to settle. When Britain seized that strategic colony in 1806, many Boers fled north, but the British eventually defeated them and conquered all of South Africa, establishing the Union of South Africa in 1910. That state was ruled by its white minority. International opposition to the government's policy of strict racial segregation, called **apartheid,** caused the government to declare itself a republic, totally free of any allegiance to Britain, in 1961.

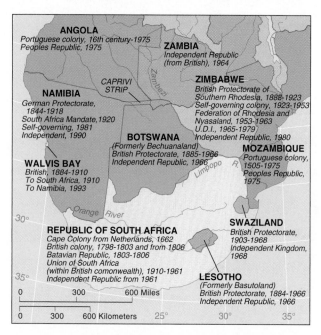

Figure 5-28 Southern Africa. Substantial numbers of Europeans settled in the midlatitude areas of southern Africa when these lands were colonies, and today's independent countries are still working to achieve rising standards of living for all.

The government divided the population into four groups: whites (British and Boers combined, totaling 13 percent of the total population in 1996); Asians primarily from India (3 percent); people of mixed race (or "colored," 9 percent); and blacks (75 percent). Each group held a fixed status. Most of the Asians had come to the eastern province of Natal to work on the sugar plantations, but many eventually rose into the middle class or even to substantial wealth.

Blacks were excluded from political participation. The apartheid government planned to move them into segregated *Bantustans* (homelands), on the basis of ethnic roots. The Bantustans covered only 13 percent of the territory and were areas of low agricultural potential and without other resources. The blacks obviously would never be able to support themselves in these lands, but would have to sell their labor within South Africa, where they would be immigrant workers without political rights. By the late 1980s, the system of apartheid had begun to collapse under the pressures of international sanctions and internal rebellion. A new governmental system was devised, and free elections open to people of all races were held in 1994. Nelson Mandela, a black leader who had been imprisoned for 27 years during the struggle for civil rights, won the presidency—and the Nobel Peace Prize.

The government still faces tremendous pressures. Blacks understandably want the life of conspicuous affluence that many South African whites long enjoyed, but South Africa does not have the wealth to provide such a lifestyle for all its people. Furthermore, South Africa is

considerably richer than its neighbors, so an estimated 2 to 8 million Africans from neighboring countries have illegally migrated into South Africa. South Africa's first Black Home Affairs Minister Mangosuthu Buthelezi complained that "the presence of illegal aliens impacts on housing, health services, education, crime, drugs, transmissible diseases—need I go on?" Nonetheless, these were problems that his government was struggling to address regardless of immigration.

The histories of some other southern African states echo that of the Republic of South Africa. The Portuguese colonies of Angola and Mozambique did not have many white settlers when they became independent in 1975, and most of those who did reside there left at that point. Botswana also had few white inhabitants when it gained independence in 1966; and when Namibia gained independence from South Africa in 1990, whites made up only 5 percent of its population. Zambia won independence in 1964 with a black-dominated government, which in 1975 nationalized all private landholdings and other enterprises. Today few whites remain. After Zimbabwe achieved independence from Britain in 1980, farmland owned by whites since the colonial period was supposed to be sold to black Zimbabweans. During the 1990s, the government of Robert Mugabe began seizing some land, giving it to poor blacks and some elites to secure his political power. His policies have devastated the country. Most recipients of the land lacked the tools, skills and money necessary to start farming and food production has plummeted, leaving many hungry. Mugabe's policies also led to hyperinflation, making Zimbabwe's currency worthless. International political pressure and domestic opposition have done little to improve the situation.

The migration of Indians When the British ruled the Indian subcontinent, they encouraged emigration to other parts of their empire. In many of the places where these Indian migrants went, they rose to positions that were socially and economically above the natives, as noted already in South Africa.

In British East Africa, Indians generally made up less than 5 percent of the populations, but when Uganda, Kenya, Malawi, and Tanzania achieved independence in the 1960s, Indians dominated the new countries' economies. Indians' high visibility made them targets of the majority blacks' frustrations, and Uganda even expelled more than 80,000 Indians. Several East African countries have recently tried to entice Indians to return, and Indians are regaining economic strength in Uganda and Kenya.

Indians settled in Myanmar (formerly known as Burma), and although several hundred thousand remain, they have faced discrimination since Myanmar achieved independence in 1948. Singapore is about 6 percent Indian. The Central American country of Belize is about 2 percent Indian. Indians still form key elements of the populations in several West Indian countries. Jamaica is only about 3 percent Indian, but Trinidad and Tobago is about 40 percent Indian, as is Suriname. Guyana on the South American mainland is almost 50 percent Indian. The population of the Pacific Ocean island of Fiji is almost equally split between Indians and indigenous Fijians.

Today the 20 million nonresident Indians make up one of the world's most widespread diasporas, and in many places they have prospered. Annual Indian migration to the United States is almost 50,000. Links are so strong between the U.S. computer industry and the industry in India that President Bill Clinton, while on a visit, referred to the Indian city of Bangalore (today's Bengalooru) as "Silicon Valley East," after California's concentration of high-tech industries. First-generation Indian Americans enjoy the highest median family income of any foreign-born group.

Several reasons have been cited for Indians' international success. One is that they have been able to take advantage of the network of commercial outposts established by the former British Empire. A second reason is their facility with English, which is increasingly the world language of business. Some Indians have themselves cited the cosmopolitan character of Indian culture as a third explanatory factor in their international success. The population of India is so diverse in terms of skin colors, languages, religions, and other cultural attributes that many Indians learn during childhood how to cooperate with people of all backgrounds. A fourth factor is that Indian migrants help one another in their new countries. If an Indian cannot borrow money from a local commercial bank to start a new business, local fellow Indians often raise the necessary capital. Hard work and frugality are additional factors often cited by Indian entrepreneurs themselves, but these characteristics also often describe groups that still do not enjoy the success that the Indians have.

Nonresident Indians constitute a valuable resource for India itself, because they have been able to transfer investment funds, technology, and marketing and financial expertise to India. The Indian government grants them privileges not available to other foreigners. For example, they may own land in India and move funds freely.

The Overseas Chinese From early in the nineteenth century until about 1930, waves of southern Chinese left their homeland to work in the British, French, and Dutch Asian empires. Their success mirrored that of the Indians. At the time the nations of Southeast Asia gained their independence, their economies were firmly in the hands of small Chinese minorities, who were often linked across international borders by bonds of family, clan, and common home province. The Chinese commercial supremacy and their international links still survive, so that today these 50 million **overseas Chinese** constitute a formidable economic power throughout Southeast Asia (Figure 5-29).

Figure 5-29 The migrations of Overseas Chinese. The number after each country is the percentage of its population that is Chinese. Even where the Overseas Chinese make up small minorities of the population, they exercise financial influence far beyond their numbers. Overseas Chinese control the economies of most of these countries.

Just as the migrant Indians faced native ethnic hostility, so have the Chinese. For example, ethnic resentments dominate the history of Malaysia. When the British ruled Malaysia as a colony, they founded the cities of Singapore and Kuala Lumpur. The populations of both cities came to be overwhelmingly Chinese, although the majority of the population of the colony as a whole was Malay. Malaysia was granted independence in 1957, but animosity was so great between Chinese and Malays that in 1965 Singapore broke off as an independent Chinese-dominated city-state. Even in Malaysia, however, the Chinese minority still controls the economy, although the Malay-dominated government blocks Chinese and Indians from advancement in government positions and encourages Malay entrepreneurship. The population of Malaysia today is about 68 percent Malay, 25 percent Chinese, and 7 percent Indian.

In Indonesia, the 7.3 million Chinese control an estimated 80 percent of the country's private industry, and the Indonesian government requires them to invest in businesses owned by other ethnic groups. A 1965 communist coup attempt, supposedly backed by China, ignited anti-Chinese sentiment, and many ethnic Chinese were killed. Anti-Chinese violence flared again during Indonesia's political and financial instability in the late 1990s and has lasted into the twenty-first century, making unclear the future of the Chinese in Indonesia. The loss of their skills and capital would devastate the Indonesian economy.

Overseas Chinese figure highly in their adopted countries around the world. They play an elite role in the Philippines and comprise almost the entire commercial infrastructure of Thailand. Ambitious Chinese also established themselves in New South Wales, Australia; along the west coast of the Western Hemisphere from British Columbia to Peru; and in small but significant numbers in the Caribbean. Tens of thousands of Chinese migrated to California in the mid-nineteenth century, and a new wave of Chinese migration to the United States began in 1965. Today, Chinese Americans make up 1 percent of the U.S. population, up from 0.4 percent in 1980. Tens of thousands of Hong Kong Chinese emigrated before that city reverted to Chinese rule in 1997; many headed for Australia or Canada.

The overseas Chinese have proved to be a valuable resource for their homeland, just as the nonresident Indians have. Overseas Chinese have raised investment capital, invested in homeland industries, introduced new technologies to China, and distributed Chinese products worldwide.

Migration Today

Human migration has by no means come to an end. The United Nations estimates that in 2010, 214 million people were living in countries other than the ones in which they were born. Large-scale migrations still make daily news. The United Nations Universal Declaration of Human Rights affirms anyone's right to leave his or her homeland to seek a better life elsewhere, but it cannot guarantee that there will be anyplace willing to take anyone. As in the past, the major push and pull factors behind contemporary migration are economic and political. People are trying to move from poor countries to rich countries and from politically repressive countries to more democratic countries. In addition, millions of people are fleeing civil and international warfare. Pressures for migration are growing, and they already constitute one of the world's greatest political and economic issues.

Many migrants may be poor, uneducated, or unskilled, but usually they are also enterprising, in good health, and of working age. They seek opportunity, and it takes courage, stamina, and determination to pull up roots and head to a foreign and perhaps hostile land. In 1993, Mexico's President Carlos Salinas de Gortari said, "I don't want any more migration of Mexicans to the U.S., especially those migrants who are very courageous and take the impressive risk of going to the U.S. looking for jobs. That is precisely the kind of person I want here."

Other migrants are highly educated and skilled, and even the richest countries compete to attract them. Many were born in poor countries, and they received excellent educations either at elite institutions in their homelands or in rich countries. The poor countries want these people to return.

The first half of this chapter noted that populations are declining in many of today's rich countries, and some observers may ask why migrants from today's poor countries cannot simply migrate to fill "vacancies" in these countries. New migrants would provide the young labor force that rich countries need in order to sustain the pension benefits of aging populations. The United Nations has estimated the numbers of immigrants each rich country would have to accept each year until 2050 in order for its total population to stabilize rather than fall. These numbers range from 29,000 for France to some 350,000 for Germany. Britain would need a million immigrants per year until 2025 to maintain its 2005 dependency ratio. The barriers to such mass migration, however, are cultural, economic, and political.

Forced Migration

There is no population more vulnerable than those who are forced to flee their homes, often with few resources and little time to prepare or to seek shelter in some unknown place. Although war, tyranny, and persecution have always produced refugees, refugees were not legally protected until the twentieth century. Since the 1951 Geneva Convention, a **refugee** has been defined as someone outside his or her home country with a well-founded fear of being persecuted in that country for reasons of race, religion, nationality, membership of a particular social group, or political opinion. States are obligated to accept refugees and grant them safety, called **asylum,** although some states routinely fail to do so. Through recent years, civil wars and international wars displaced millions of people and sent many streaming over international borders. The United Nations High Commissioner for Refugees (UNHCR) estimated almost 10 million refugees and an additional 22 million people as "of concern" in 2009 (Figure 5-30).

Some countries have broadened their definitions of the social groups to which they will grant asylum. For example, the United States, Canada, and several European countries have recognized persecution in the form of sexual abuse as a potential basis for refugee status. The European Parliament has accepted the legal principle that in some countries, women are customarily treated so badly that they constitute a persecuted social group. Both the United States and Canada have also offered asylum to women from countries where, in the Canadian government's words, customs place women "in a more vulnerable position than men."

Forced migrants are not always able to reach a safe country where they can claim asylum. Many are displaced within their home country and, therefore, not protected as refugees. These **internally displaced persons (IDPs)** are at the mercy of their home governments, which are often either the cause of the

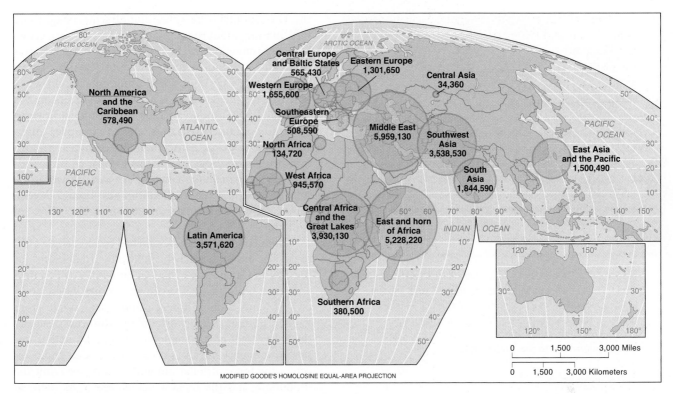

Figure 5-30 Persons of concern who fall under the mandate of UNHCR. This map shows the numbers of refugees, asylum-seekers, and internally displaced persons as of January 1, 2008.

displacement or are unable to assist the displaced. In 2009, the UNHCR was working to protect or assist 13.7 million IDPs, and almost that many more could not be helped often because their host governments barred outside agencies from getting involved in internal affairs.

While many asylum-seekers are genuine refugees, some are driven by the desperation of poverty in their homeland to seek a new life in another country. Of course, all migrants hope to improve their standard of living, but refugee status is meant for those suffering from direct persecution. It can be hard for would-be host countries to identify proper refugees from so-called "economic refugees." In just the five years from 2007 to 2009, European and other industrialized countries received 1.1 million applications from asylum-seekers. Although this is just a fraction of the refugee population in the world, these countries have millions of other migrants trying to enter, often illegally. Unfortunately, because refugees must request help from immigration officers, they become the target of frustrated governments trying to lower immigration rates. Even if economic refugees do not qualify as formal refugees, they are driven by very real and compelling conditions to leave their homes. Wealthier countries have hoped that small investments in economic development in these regions will stem the tide of economically motivated immigrants.

As described in the introduction to this chapter, some persons are compelled to flee rapidly deteriorating environmental conditions. Desertification and salinization destroys cropland. Flooding forces people to seek dry land. Volcanoes and mudslides wipe out whole communities. Some are displaced within their countries, including the Hurricane Katrina evacuees from Louisiana during 2005. Others may be displaced across borders, such as the residents of Montserrat who left after a volcanic eruption destroyed large parts of the island. Despite the term "environmental refugee," which aims to describe this type of forced migration, there is no international agreement that provides for their needs.

The Impact of International Migration

The impact of international migration is greater than the numbers might suggest. Immigrants are often greeted with apprehension, and immigration has become an explosive political issue in many receiving countries. This is for several reasons. (1) Official statistics substantially underestimate actual numbers, and there is evidence that undocumented migration is rising everywhere. (2) Migrants are usually in the peak years of fertility. Therefore, they are playing an increasing role in the total population growth in the rich target countries.

Figure 5-31 "Keep out." This high fence, seen here from the U.S. side between San Diego, California, and Tijuana, Mexico, winds its way across much of the U.S.–Mexican border in the U.S. Southwest. While the fence may be an extreme measure, some see it as the only way to control the flow of migrants into the United States.

(3) Migrant settlements are generally concentrated in a few places within a country. This usually adds to the immigrants' visibility and increases the perception of cultural differences. (4) Many migrants today seem to be assimilating to the cultures and habits of their new homelands slowly—if at all; this may be either because they are discriminated against or because they cling to their native traditions.

The peoples of the rich and democratic countries are beginning to exhibit "compassion fatigue," and doors are closing all around the world (Figure 5-31). In 2006, only 35 percent of Germans and 49 percent of Americans agreed that immigrants were "a good influence" on their countries. (However, 77 percent of Canadians thought so.)

Moreover, whenever unemployment rises, immigrants who were once grudgingly tolerated because they were willing to work in low-wage or dirty jobs suddenly come to be seen as competitors and become targets of wrath. The European countries are growing more selective; they welcome educated and skilled immigrants, but not poor or unskilled immigrants.

Every demographic impetus toward migration will be multiplied as populations rise in poor countries. Either substantial flows of investment capital and technology will be directed to the poor countries and create new job opportunities in them, or else migration from the poor countries will be stimulated by the combination of large increases in population, excess labor supply, rising social and political turbulence, and persistent or worsening inequalities between rich and poor countries. Furthermore, global communications media and social networks strengthen the pull of

the rich countries by spreading alluring images and information about them. Many poor people are growing desperate, and tragic stories of attempted migration can be found in the rich countries' newspapers almost every day: Would-be immigrants are drowned when flimsy boats capsize in the Mediterranean Sea or Straits of Florida, or they are found suffocated in cargo containers arriving on rich countries' highways or at airports. Altogether, an estimated 2.5 million people are today under the sway or in the hands of traffickers in human beings who abuse would-be immigrants in sexual exploitation, economic slavery, forced labor, debt enslavement, or other forms of exploitation.

Some countries may be relieved to see people emigrate. The emigrants may be politically troublesome, or the country may hope that its emigrants will return with new skills, make investments in their homelands, or just send money home. Money sent home from workers abroad, called **workers' remittances,** is several times the total of international foreign aid and is important support for several countries today. India receives the highest dollar amount ($22 billion), but the Philippines has the highest percentage of its national population working abroad (10 percent). Haiti receives the highest percentage of its national income (25 percent) from remittances.

Geography Vi
Cash Flow Feve

Many rich countries accept foreign workers, but they do not grant them citizenship or the rights that citizenship would bring. The host countries accept them only as sojourners, and it may be unclear whether the migrant workers themselves hope to stay. The conditions under which the host countries grant citizenship, or even guarantee full civil and legal rights to aliens, have become matters of international concern. Unscrupulous employers often profit by exploiting aliens. Illegal aliens especially have faced hazards in search of work, they are seldom unionized, and they often accept lower wages and worse working conditions than legal residents or nationals will.

Migration to Europe

Some 19.5 million non-European legal immigrants lived in the 27 countries of the European Union in 2008 (see Chapter 13), representing about 3.9 percent of the total EU population: 5.8 percent in Germany, 3.8 percent in France, 2.6 percent in the Netherlands, 3.9 percent in the United Kingdom, and 4.2 percent in Italy (Figure 5-32). Europe has not always easily absorbed immigrants. The European countries have not historically thought of themselves as multicultural, and even though many immigrants come from former colonies, many Europeans fear the arrival of increasing numbers of Asians and Africans. In a 1997 poll, 33 percent of EU citizens described themselves as "a little racist"; an additional 33 percent described themselves

Figure 5-32 Catering to new tastes. Immigrants bring with them food preferences that differ from their host societies' regular fare. Here an Indian sweet shop owner displays his products on Ealing Road in London.

Figure 5-33 Riots among immigrants to France. Burning cars lit the skies of Toulouse, Paris, and other cities throughout France as largely immigrant youths rioted in the suburbs in 2005. At least 10 percent of the children in France are Muslim, and Muslim youngsters suffer six times the national unemployment rate and five times the national incarceration rate for all young people in France. Similar riots occurred in 2006 and 2007.

as "very racist" or "quite racist." Compared to the Europeans around them, immigrant groups are often less educated, occupy substandard housing, and receive inferior services.

Furthermore, many immigrants from the Middle East and North Africa are Muslim; as we shall discuss in Chapter 7, their cultural differences with Europeans are often expressed as religious ones. Generalizations about Europe's immigrant Muslims, however, require many qualifications. Variations among their cultural and national backgrounds, and among the countries to which they migrate, may play a bigger part in their ability to integrate than their religion does. Britain's mainly South Asian Muslims have far more in common with other Britons than they do with France's North African migrants or Germany's Turks. Overall, however, Europe's Muslims are much more likely than non-Muslims to be jobless and ill-educated, harassed by police, and discontented (Figure 5-33). If they feel equally estranged from both the Muslim lands and their new homes (or, in the case of the second generation, from the European land of their birth), they may turn to Islam in search of an identity. European secularism sharpens that tendency. Thus, some immigrants do not want to acculturate to European cultures.

Europeans are sharply curtailing the granting of asylum. Most European countries have decreed, for example, that people who have passed through one or more countries on their way from their homelands are

no longer refugees but are merely "country shopping" for the most favorable new site to settle. Therefore, they will not accept anyone claiming refugee status who came overland through a third country. European states have also lengthened the lists of countries from which visitors' visas are necessary, have begun fining airlines for bringing people without visas, and often send people back to the country they were last in. The reduction of legal migration to Europe has increased illegal migration, which may have reached 500,000 per year. So many desperate would-be migrants suffer while sailing toward Europe in unseaworthy vessels that in 2006, Malta's Interior Minister Tonio Borg asked, "Can we look the other way when dead bodies are coming on our shores, as if they were dead fish in polluted water?" A European fleet patrols the Mediterranean to turn ships back to Africa.

European states have also redefined the criteria for citizenship. The member countries of the European Union have adopted a common European citizenship, rights that apply to all citizens of EU countries, regardless of where they live in the European Union. One purpose of EU citizenship is to encourage Europeans to migrate within the EU in search of jobs so employers will hire fewer immigrants from outside the European Union. Individual countries can still formulate their own rules for citizenship and many seek to limit non-European immigration. For example, France traditionally accepted the idea that an individual's citizenship is based on his or her place of birth, a concept called *jus soli* ("law of soil"). The United States also applies *jus soli*. France, however, has made it more difficult for children born in France of foreign parents to become French citizens. Some other countries, including Germany, have traditionally defined citizenship on the basis of the citizenship of one's ancestors. This concept is called *jus sanguinis* ("law of blood"). Germany has changed its conditions for claims to German citizenship to include *jus soli*, allowing some children born to foreign parents to become citizens. In Ireland, 79 percent of voters chose to end *jus soli* without at least one Irish parent following high-profile cases involving pregnant tourists arriving to give birth to children who would automatically become citizens.

Sealing borders will not solve the problem of existing immigrants and their European-born children and grandchildren. European governments are directing new attention to what the Dutch call *inburgering*—that is, the forming of citizens. Governments are requiring examinations in language, history, culture and traditions, and the rights and duties of citizenship. The Dutch require potential immigrants to watch a film about Dutch life that includes scenes of a topless woman and of men kissing. European governments are demanding that immigrant women and immigrants' daughters enjoy full rights not always guaranteed in the immigrants' homelands. Some Europeans ask whether such laws are discrimination or a last-ditch attempt at Europeans' own cultural survival.

Migrations of Asians

An estimated 5 million Asians work abroad, and in many cases, it is doubtful whether many will return to their homelands. Meanwhile, their remittances to their homelands are an important element in the economies of their home countries.

The absence of these workers lowers the local unemployment rates in their homelands, but it also constitutes a loss of important skills needed back home. Although the home country might not have the resources to reward these workers adequately, it frequently needs their contributions. A young Filipina woman, for example, may be needed in her hometown as a schoolteacher, but by staying there, she would earn $40 per month. If she is earning $150 per month as a maid in Singapore, she is better off, but the Philippines suffers. An estimated 60,000

Filipina women—many college graduates—work as maids in Singapore.

The oil wealth of some Arab states has drawn many Asian workers who seem little disposed ever to go home. In 2006, the populations of the United Arab Emirates and of Dubai were less than 20 percent native-born; that of Kuwait was only 35 percent native-born. The expatriate workers in these rich countries often come from poorer Muslim lands, and they are generally more susceptible to radical interpretations of Islam. As long as the Middle East remains an international hot spot, and as long as Arab oil supplies are important to world prosperity, any destabilizing migration into Arab oil-rich countries is of world concern.

Japan, of all the world's richest countries, limits immigration the most strictly. Between 1945 and 2000, fewer than 250,000 foreigners acquired Japanese citizenship, including those who married Japanese. The most significant minority in Japan is people of Korean ancestry—about 0.3 percent of the total population. These people are a reminder that Japan ruled Korea from 1910 until 1945, and they suffer discrimination. However, with Japan's low TFR and high median age, Japan would need to import 600,000 workers per year to keep its working population stable between 2000 and 2050. By that time, the population would be one-third foreign-born. The government has encouraged childbearing and broadened the category of jobs that foreigners in Japan (today about 1.1 million) could hold. The Minister of Justice has urged Japanese "aggressively to carry out the smooth acceptance of foreigners."

Migration to the United States and Canada

The single largest migration flow for the past 150 years has been migration to the United States. In recent years, the United States has accepted 706,000 to 1,064,000 legal immigrants per year. Some 12 million undocumented or illegal immigrants may also live in the United States.

In 2008, the Census Bureau found 12.5 percent of the U.S. population to be foreign-born, up from 7.9 percent in 1990. The nation's overall birth rate is low, so immigration is responsible for about one-third of U.S. population growth. If immigration continues at current levels, the foreign-born percentage of the population may account for nearly two-thirds of the population growth projected to the year 2050.

The source areas of U.S. immigrants have changed throughout history, thus changing the makeup of the national population. Immigration was totally unrestricted until the late nineteenth century, when the government enacted rules explicitly designed to keep out Chinese and other Asians. Later, in the 1920s, the government issued quotas for various nationality groups. The first quotas favored British, Irish, and

Germans, who had formed, along with African Americans, the bulk of the population in 1890. By the 1920s, however, southern and eastern Europeans had already established a strong presence. After the 1920s, a maximum quota was set on immigrants from the Eastern Hemisphere (in practice, Europe), but there was in theory no ceiling on immigrants from the Western Hemisphere. In 1965, the United States changed the rules again, in a way that tended to favor Latin Americans and Asians (Figure 5-34). In 1965, European

immigrants outnumbered non-Europeans 9 to 1, but within 20 years, that proportion was reversed.

The latest changes in the source areas of immigrants will again bring long-term consequences to national demography (Table 5-1). In 2008, nearly 66 percent of all Americans were non-Hispanic whites; that percentage may fall to less than 50 percent by 2050. The number and percentage of the total population that is Asian (alone or in combination with another race)—13 million, or 4.4 percent in 2008—is projected

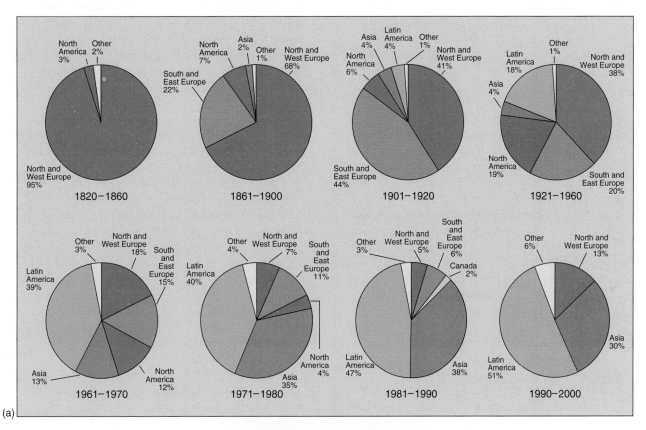

Figure 5-34 Sources of immigrants to the United States and foreign-born population, by region of birth. The graphs in part (a) show how the source areas of immigrants to the United States have changed through history. The percentages of immigrants from northern and western Europe have dropped steadily, as the percentages from Latin America and Asia have risen. These trends have affected the composition of the national population. Graph (b) shows what percentage of the total foreign-born population in 2002 had come from each region.

Race, Culture, and the U.S. Census

All categorizations of people in the U.S. Census are self-descriptions. Respondents are asked to report basic information about themselves and others in their household. Most census questions are straightforward, asking the name, age, and sex of each person. Census questions about race and ethnicity are more complicated, however, because respondents must choose from predefined categories that may not match their self-description. The meaning of these categories is often caught up in how respondents perceive their cultural identity.

In 1990 the U.S. Census Bureau asked respondents to identify their race by choosing from one of 15 options: "White"; "Black"; "American Indian"; "Eskimo"; "Aleut"; nine types of "Asian or Pacific Islander"; and "Other." These categories had been used in previous decades to compile information needed to enforce the U.S. Voting Rights Act and Civil Rights Act. In the United States of the late twentieth-century, however, the idea that a person belonged to one and only one race seemed antiquated and biased. The census effectively ignored the fact that people can have mixed racial ancestry. Forcing respondents of mixed racial descent to chose one race over another was reminiscent of discredited racial theories used by segregationists to "purify" the white race. Racists believed that anyone having just "one drop" of black blood should be segregated from whites.

In 1997, therefore, the federal government began to allow people to classify themselves as being of more than one race by selecting two or more racial categories on the census. To avoid the diminution of the minority voice in certain government programs, however, anyone who counted himself or herself as being both of a minority race and also being white, would be counted in the totals for minorities. In the 2000 census, 6.8 million people (2.4 percent of the total) chose to identify themselves as being of more than one race. People under 18 years of age were twice as likely to be identified as being multiracial than were people 18 or older. This reflects increasing population mixing in the United States, but also, perhaps, because younger peoples' heroes may include proudly multi-racial Mariah Carey, Derek Jeter, Barack Obama. The 2010 census is expected to show a further increase in the number of respondents who select more than one race.

The U.S. government has also tried to count persons of Hispanic origin living in the United States. The history of this effort illustrates again the difficulties of describing and categorizing the population. In the 1940 census, the Bureau asked how many people used Spanish as their mother tongue, and in 1950 and 1960 it tried asking how many people had Spanish surnames. The government adopted a new category, "Hispanic," for the 1970 census. The definition of the term was later codified in U.S. law as persons whose origin or descent is Mexican, Puerto Rican, Cuban, Central or South American, or of some other Spanish speaking culture. The law also recognizes that "persons of Hispanic origin may be of any race." Hispanics do not all share a common "culture" (see Chapter 6) beyond sharing a Spanish-language background. Some people who consider themselves Hispanic do not speak Spanish. Nevertheless, information on Hispanic origin has been collected in each census since 1970.

The census asks whether one is Hispanic separately from one's race, raising further issues. For some people, their identity as a Hispanic is more important to them than any racial categorization. Some Hispanics do not see a difference between race and Hispanic origin. Some critics have noted that asking separate questions on race and Hispanic origin forces respondents to be race-conscious in a way that Hispanic society traditionally is not, or that they do not want to be. Partly in response to this feeling, some analyses divide the U.S. population into the categories "White," "Black," "American Indian," "Asian," and "Hispanic." Whenever you see statistics reported in this way, however, you must remember that the category "Hispanic" has been created by pulling out from each of the other racial groups those people who reported themselves as "Hispanic." The 2000 and 2010 censuses also added the terms "Latino" or "Spanish" to the Hispanic category in an effort to include all respondents with origins in Spanish speaking countries. These terms, however, exclude Brazilians, who are also from Latin America but whose native language is Portuguese.

As we will see in Chapter 6, not everyone can agree on exactly what racial and cultural identities really mean. The task of selecting supposedly objective categories to describe oneself is complicated by the historical and social significance of these identities, as well as personal experience. We must be mindful of these complex issues when we interpret census statistics, too.

to rise to 9.2 percent by 2050, largely through continuing immigration.

The rising percentage of the national population that is Hispanic, however, is more dramatic. The 1970 census counted about 9.6 million Latinos, a little less than 5 percent of the population. By 2000, Hispanics numbered 35.3 million, 12.5 percent of the total U.S. population. In 2006, Hispanics exceeded non-Hispanic blacks as the largest minority group in the 50 states, reaching 14.7 percent of the population (when the total

TABLE 5-1 Population (in Thousands), by Race and Hispanic or Latino Origin, for the United States, 2000 and 2050

Race and Hispanic Origin	CENSUS 2000		PROJECTION 2050	
	Count	Percentage	Count	Percentage
TOTAL POPULATION	**281,422**	**100.0**	**439,010**	**100.0**
One race	277,524	98.6	422,828	96.3
White	228,104	81.1	324,800	74.0
Black	35,704	12.7	56,944	13.0
American Indian and Alaska Native	2,664	0.9	5,462	1.2
Asian	10,589	3.8	34,399	7.8
Native Hawaiian and Other Pacific Islander	463	0.2	1,222	0.3
Two or more races	3,898	1.4	16,183	3.7
Race alone or in combination:[a]				
White	231,434	82.2	339,441	77.3
Black	37,104	13.2	65,703	15.0
American Indian and Alaska Native	4,225	1.5	8,592	2.0
Asian	12,007	4.3	40,586	9.2
Native Hawaiian and Other Pacific Islander	907	0.3	2,577	0.6
NOT HISPANIC	**246,116**	**87.5**	**306,218**	**69.8**
One race	242,710	86.2	292,876	66.7
White	195,575	69.5	203,347	46.3
Black	34,313	12.2	51,949	11.8
American Indian and Alaska Native	2,097	0.7	3,358	0.8
Asian	10,357	3.7	33,418	7.6
Native Hawaiian and Other Pacific Islander	367	0.1	803	0.2
Two or more races	3,406	1.2	13,342	3.0
Race alone or in combination:[a]				
White	198,476	70.5	215,411	49.1
Black	35,498	12.6	59,318	13.5
American Indian and Alaska Native	3,456	1.2	5,633	1.3
Asian	11,632	4.1	38,644	8.8
Native Hawaiian and Other Pacific Islander	752	0.3	1,827	0.4
HISPANIC	**35,306**	**12.5**	**132,792**	**30.2**
One race	34,814	12.4	129,951	29.6
White	32,529	11.6	121,453	27.7
Black	1,391	0.5	4,995	1.1
American Indian and Alaska Native	566	0.2	2,104	0.5
Asian	232	0.1	981	0.2
Native Hawaiian and Other Pacific Islander	95	0.0	419	0.1
Two or more races	491	0.2	2,841	0.6
Race alone or in combination:[a]				
White	32,959	11.7	124,030	28.3
Black	1,606	0.6	6,385	1.5
American Indian and Alaska Native	770	0.3	2,959	0.7
Asian	375	0.1	1,942	0.4
Native Hawaiian and Other Pacific Islander	155	0.1	750	0.2

[a]*In combination* means in combination with one or more other races. The sum of the five race groups adds to more than the total population because individuals may report more than one race.
Source: U.S. Census Bureau, 2008.

national population reached 300 million). Hispanics are projected to reach 30 percent of a total U.S. population of 439 million by 2050.

The increase in Hispanics is due to two factors. (1) Hispanics form the largest share of all immigrants. Through recent years, about 20 percent of all legal immigrants have in fact come from just one country, Mexico, and Mexicans today account for about 30 percent of all foreign-born U.S. residents. (2) Hispanic TFRs are the highest among U.S. groups (about 3). Furthermore, Hispanics have the lowest median age of any U.S. group (27), so population momentum means

Hispanics will make up an increasing share of the national population even if immigration had ended in 2010.

In 2000, two-thirds of all foreign-born individuals lived in just six states (California, New York, Florida, Texas, New Jersey, and Illinois). About two-thirds of new immigrants settled in just 10 metropolitan regions (New York, Los Angeles, San Francisco, Miami, Chicago, Washington, D.C., Houston, Dallas, San Diego, or Boston). In these areas, they often replace native-born residents, so the populations of New York and Los Angeles, for example, are today 40 percent foreign born. In 2005, Los Angeles elected Antonio Villaraigosa its first Latino mayor in 133 years.

Other features of current U.S. immigration legislation give preference to immigrants who bring either skills or capital for investment and make it easier for foreign students already in the United States to stay.

Canada has in recent years welcomed some 260,000 new immigrants each year, and foreign-born Canadians numbered 6.2 million in 2006, nearly 20 percent of the total population. Immigration now accounts for over half of total Canadian population growth. Canada has also experienced a recent shift in the source area of its immigrants. In the 1950s, more than 80 percent of Canadian immigrants were from Europe, but today that figure is less than 25 percent. About 60 percent of Canada's new citizens now arrive from Asia, many from former British colonies. In 1999, Adrienne Clarkson, an ethnic Chinese from Hong Kong, became Canada's governor general, the country's head of state, and she was replaced in 2005 by Haiti-born Michaëlle Jean. Canada has set a goal of increasing immigration gradually until inflows reach 1 percent of its population, up from about 0.85 percent in 2002.

In Canada, the concentration of the foreign-born population is even greater than it is in the United States. In 2006, 90 percent of all foreign-born individuals in Canada lived in the country's 33 metropolitan areas, with metropolitan Toronto, Vancouver, and Montreal together accounting for about two-thirds of them. A full 43 percent of Toronto's secondary school students in 2006 were born outside Canada, and 49 percent had a first language other than English. Thus, North American metropolitan areas have become increasingly "exotic" enclaves of ethnic diversity markedly different from both countries' non-metropolitan populations.

Canada selects its immigrants according to a system in which applicants are assigned points on the basis of characteristics such as their ability to speak English or French, workforce skills, family ties, and ethnic diversity. How these factors are weighed is a political issue in Canada, but Canada is today giving more weight to skills and less to reuniting families.

New arguments over immigration to the United States   Debates over immigration are nothing new in U.S. history, and the issue is once again current. Some Americans are questioning whether the United States should continue to even allow immigration. These arguments have become heated as the immigrant share of the population rises, as many Americans have become concerned with securing U.S. borders since September 11, 2001, and as the number of immigrants in the country increases. Some of the arguments have become frankly racist.

Some people argue that immigrants—legal or illegal—cost Americans money for public services. The substantial costs of immigration must indeed be borne by local governments in the areas where immigrants concentrate, exasperating local taxpayers. In California, for example, 59 percent of voters approved a 1994 ballot initiative to deny public health and education to illegal immigrants. Courts, however, ruled that step unconstitutional, so today Los Angeles spends about $2 billion per year educating illegal immigrants and an additional $350 million per year for their health care.

In 2010, Arizona passed a controversial law that made violations of federal immigration law a state crime. The law also permits Arizona police to arrest persons suspected of being illegal immigrants. The law is meant to target persons coming from Mexico and Central America. It is unclear how the police can enforce the law without violating the rights of legally resident Hispanics, who make up nearly 30 percent of Arizona's population.

Others counter that immigrants contribute to U.S. economic growth. The National Academy of Sciences concluded in 1997 that immigration does produce substantial economic benefits for the United States as a whole. The poverty rate is significantly higher for immigrants than for the native-born (21.6 percent versus 12.1 percent in 2004), but intergenerational differences among immigrants and their children demonstrate upward mobility. First-generation male Mexican immigrants, for example, earn only half as much as non-Hispanic white men, but the second generation has overtaken black men and earns three-quarters as much as whites. Furthermore, the newcomers and their children will make up the labor force to support the growing number of pensioners.

A great many immigrants hold arduous low-paying jobs, and this may slightly reduce the wages and job opportunities of low-skilled Americans. Some scholars argue that this adverse impact on the U.S. poor is immoral.

Congress has taken up the issue of what to do about the estimated 12 million undocumented immigrants already in the country, many of whom are holding jobs and paying taxes (Figure 5-35). Many are parents of

Figure 5-35 Immigration rallies. Protesters in Los Angeles, California, demanded immigration rights following the passage of Arizona's controversial new law. In May 2010, Los Angeles City Council began a boycott of Arizona businesses.

children born in the United States, who are thus citizens. Is it realistic to think of identifying, arresting, and deporting millions of people? Can the United States be a true democracy with so many of its settled inhabitants unable to vote? On the other hand, many people are reluctant to grant amnesty to those who came illegally. To do so would reward law breaking and discourage those around the world who are waiting and hoping to enter legally.

Immigration rules were overhauled in 1986, legalizing more than 3 million unlawful immigrants and supposedly banning the employment of illegal workers. One 2006 study, however, estimated that undocumented immigrants accounted for 24 percent of the country's agricultural workers, 14 percent of construction workers, and 9 percent of manufacturing workers. Former U.S. Attorney General Edwin Meese III criticized America's "failure of political will in enforcing laws against employers."

Most arguments over immigration are often based in economics, but immigration policy has political and cultural aspects as well. For example, should the United States absorb immigrants from Mexico deliberately, to preclude potential political turmoil there? Almost 10 percent of the total Mexican population lives in the United States, and their remittances sustain many Mexican communities.

Is there an "American culture," and can that culture be sustained as the racial and ethnic makeup of the population changes? The United States was dubbed the **melting pot** in a 1914 novel of that title, but various groups have maintained their cultural identities, and the traditional melting pot image cannot be sustained as demands rise for appreciation of a U.S. "multiculturalism." Perhaps the United States will redefine itself as a **cultural mosaic,** which is what Canada proudly

calls itself. As yet, however, many in the United States seem blind to the changing composition of the national population.

The United States must decide how widely to open its borders and on what criteria of admission. Even the question of the definition of citizenship emerged in national politics in 1996, when the Republican Party called for a constitutional amendment "declaring that children born in the United States of parents who are not legally present in the United States or who are not long-term residents are not automatically citizens." Perhaps the United States will discard its historic *jus soli* definition of citizenship. In any event, immigration policy will continue to be a political issue.

Effects of Emigration

Although migration questions are often raised by host countries as problems of immigration, many poor countries are losing their best-educated people and most skilled workers through emigration. This migration from less-developed to more-developed countries has been called the **brain drain.** In some cases, a country's schools are qualifying students for professional careers faster than its economy can absorb them. In other cases, the skilled or educated seek political freedom, and in still other cases the country's best students go abroad for training with the intent of returning home to help their own people, but they stay abroad and never return home. A 2005 study in the 30 rich countries that make up the Organization for Economic Cooperation and Development (OECD) revealed that from a quarter to a half of the college-educated citizens of poor countries including Ghana, Mozambique, Kenya, Uganda, and El Salvador lived abroad in an OECD country—a fraction that rises to more than 80 percent for Haiti and Jamaica.

The United States profits most from the drain of skilled laborers from other countries, and in fact, the nation's scientific establishment would suffer without them. About one-third of the U.S. Nobel Prize winners in the sciences have been foreign born. Thousands of senior scientists, engineers, and other professionals migrate to the United States each year, and about half of the young foreigners who come for education never return home. Almost half of the scientists and engineers working in California's Silicon Valley are foreign born.

The drain of scientists, engineers, and physicians from poor countries undermines the ability of these areas to develop. Any advanced country is reluctant to expel these skilled immigrants. That country would suffer its own brain drain, and the skilled surely would migrate to another advanced country anyway. How to keep these skilled professionals in their homelands, or to attract them back, is a problem that must be solved to avoid an ever-widening gap between the world's rich and poor countries.

Chapter Review

Summary

The population of Earth is about 6.8 billion people, and 90 percent of the population is concentrated on less than 20 percent of the land area. Major population concentrations are in east Asia, in south Asia, and in Europe. Secondary concentrations are in Southeast Asia and eastern North America. Much of Earth's land surface is virtually uninhabited.

Factors that explain this distribution include the physical environment, technology, history, and local differences in rates of population increase. Most of the population is concentrated in areas of mild midlatitude or seasonal humid tropical climates, but people will settle in harsh areas if it is profitable. In some places, local population densities may be high, but trade and circulation free members of societies from the constraints of their local environments.

The populations of China and the Indian subcontinent long ago achieved productive agricultural methods and relative political stability. Europeans multiplied when they gained material wealth and improved their food supplies as the result of world exploration and conquest, and later again during the Industrial and Agricultural revolutions. European migrants took technology to more sparsely populated areas.

The world population is increasing, but the rate of growth is slowing. The rate of increase in each country represents a balance among the crude birth rate, the crude death rate, and migration. Concern over population size focuses on the total fertility rate (TFR), which has been dropping in most countries. Most of the population increase is occurring in poor, not rich, countries. The history of population growth in today's rich countries traces a demographic transition. In the first stage, both the crude birth rate and the crude death rate were high, so population sizes were relatively stable. In a second stage, public health improved, so crude death rates dropped. Therefore, rates of population increase rose. In the third stage, crude birth rates dropped, so rates of natural population increase slowed. In the fourth stage, birth and death rates are about the same, and population sizes are once again relatively stable. Many rich countries' populations are not even reproducing themselves, signaling a new phase of population decline.

In today's poor countries, crude birth rates are dropping fast, even without significant increases in material well-being. Theories of explanation focus on the growing role of family-planning programs, the status of women, new contraceptive technologies, and changing attitudes about fertility. Family-planning efforts face obstacles in traditional cultures, religions, and in some cases the cost of devices.

The total population will continue to grow because the number of young women reaching their childbearing years is larger than ever before, but the sooner the world's TFR falls to the replacement rate, the lower the total population will be, at which point the projected numbers level off. Currently, the most probable estimate is that Earth's population will reach 10.7 billion by 2050.

Natural population decreases can result either from an increase in the death rate, usually disease, or by a lowering of the birth rate. Today, almost all poor countries are committed to family planning. A preference for large families is traditional in some areas, and having children may be economically rational for families in poor countries, but many people in poor countries want small families. The low status of women inhibits family planning in some societies. Urbanization usually lowers TFR.

The Malthusian theory argues that the rate of growth in the human population will always outpace the rate of growth in food supplies, triggering cycles of famine, disease, and war. Human creativity and technology have repeatedly increased food production and reduced disease-related deaths.

The natural ratio between females and males in a human population should be one to one, but, worldwide, men outnumber women. However, the sex ratio varies greatly among different countries. This may be due to a combination of economic and cultural factors.

The median age of humankind as a whole is rising. In rich countries, questions arise about the possibility of sustaining economic growth and about the equitable distribution of wealth among different generations. Many poor countries must quickly build up their national incomes and devise national welfare or social security programs.

Human migrations have redistributed populations throughout history, and significant migrations continue. Push factors drive people away from wherever they are, and pull factors attract them to new destinations, but some migrations have been forced. Sojourners intend to stay in their new location only temporarily.

The most significant population transfers during the past 500 years have been migrations of Europeans around the world, migrations of blacks out of Africa, and migrations of other groups instigated by European expansion.

It is sometimes difficult to differentiate political refugees, who are fleeing politically repressive countries, from economic refugees, who are trying to move from the poor countries to the rich countries. The brain drain of educated people and skilled workers from poor countries undermines these countries' ability to develop.

Key Terms

African diaspora p. 184
apartheid p. 187
arithmetic density p. 157

asylum p. 190
brain drain p. 199
carrying capacity p. 157
chain migration p. 180
crude birth rate p. 161

crude death rate p. 161
cultural mosaic p. 199
demographic equation p. 161
demographic transition p. 166
demography p. 156

Questions for Review and Discussion

1. Briefly describe the distribution of the human population.

2. Explain the demographic transition model and explain the new developments in demographics that challenge the continuing validity of that model.

3. What current developments may increase the world's death rate?

4. What have been the major currents of the African diaspora?

5. Who and where are the Overseas Chinese? Why are these people of such importance in the world today?

6. Differentiate a refugee from an economic refugee and an environmental refugee.

7. What is the Malthusian theory?

8. How does brain drain negatively affect emigration areas?

Thinking Geographically

1. How has physical geography affected the distribution of people in any given continent or region?

2. Investigate any country's family-planning efforts and experience.

3. Do you know any elderly people who have retired and are now collecting pensions? Where did they spend their working years? Where are they living now?

4. Do you know how many generations of your family have moved either to this country or within this country? What push and pull factors motivated them?

5. Has your local community seen net in-migration or net out-migration in recent years? If in, where from?

If out, where to? What push and pull factors did they consider?

6. Compare the cartogram in Figure 5-2, countries drawn according to their population size, with Figure 12-5, countries drawn according to their economic output, with a standard area world map. Do you think that one or the other of these criteria—population, output, or area—should determine the amount of time you, as a student, should devote to each country in your studies? Which should determine how much time on the evening news is devoted to each country? Can you suggest any other measures of each country's "importance"?

7. Are any major refugee streams reported in this week's headlines? If so, what are the push factors?

Log in to www.mygeoscienceplace.com for videos, animations, Map**Master**™ interactive maps, RSS feeds, case studies, and self-study quizzes to enhance your study of Population and Migration.

MapMaster™

An uncontacted tribe living in Brazil's Amazon forest. Tribe members reacted to the appearance of the photographer's plane in the sky.

6

Cultural Geography

Deep in the Amazon forest, there are human communities almost totally isolated from the rest of the world. Their existence may be known, but little else about them has been documented. They are referred to as uncontacted tribes. The aerial photographs shown here were circulated around the world to prove that such groups exist but are fast disappearing. Activists want to protect these people from what has happened to other uncontacted tribes. In 1996, for example, loggers harvesting mahogany in another part of the Amazon forest came into contact with a different tribe, the Murunahua. This first contact with people outside their tribe quickly spread diseases among the Murunahua against which they had no immunity, killing at least half their people.

This story has not changed much in the past 500 years. Since first contact with European settlers, untold numbers of native cultures in Brazil have disappeared due to disease, persecution, and assimilation. In trying to address this unfortunate legacy, Brazil's government tries to protect native groups by setting aside sections of the Amazon forest for them. The problem is that powerful economic interests want access to the Amazon forest's timber and mineral resources and illegally operate on Indian land. These companies have long denied that such tribes exist or have rights to the land. The Brazilian government struggles to balance its policy of opening the Amazon forest to economic development and human settlement with its responsibilities to the native peoples who live there. Brazil is not alone; small numbers of uncontacted and isolated tribes across the globe face similar prospects.

It seems almost impossible to think that people live without the material goods we associate with "the modern world." Yet make no mistake, uncontacted tribes live in the world of today. What may seem primitive or rudimentary to us must be recognized as a viable alternative form of human existence, one that has no need for packaged food, electricity, or brand names. But this does not mean that these people are "backward" or "uncivilized," terms used to justify the oppression of many native peoples. It is quite possible, however, that elements of their culture have remained relatively unchanged for a long time. Some have suggested, from examining the fly-over photographs, that some of their possessions show signs of indirect outside influence, perhaps ideas or objects that arrived through neighboring tribes. If so, the tribe fit these new things into the world, as they understand it. So, while it is possible that their language, religion, livelihoods, and social customs changed little over the past centuries, their culture remains an active process. It is what allows them to make sense of their experiences, whether familiar or radically new. Even if they had never seen an aircraft before, for example, they had to make sense of the low-altitude fly-over that took this picture.

There is an undeniable impulse in seeing these pictures to want to know more. We could learn a lot about the diversity of human cultures by contacting and studying this tribe's culture. It might even be possible that what we learn could be used to help protect such tribes—if they survived first contact. Some indigenous rights activists, such as Survival International, argue that the best protection for uncontacted tribes is to encourage governments to set aside the area they live in and leave them alone. There is still a lot that we can learn from uncontacted tribes, not the least of which is to ask why our culture is so different from theirs.

A Look Ahead

Cultural Evolution Contrasts with Cultural Diffusion

Cultural evolutionism studies the sources of cultural change that occur within individual cultures themselves, whereas diffusionism emphasizes how cultural elements spread out from the places they originate and are adopted by other peoples. Folk cultures preserve traditions, but popular cultures embrace changing norms.

Identity and Behavioral Geography

The behavior of any individual is partly unique to that person, but much of any person's behavior is specific to his or her cultural group. People often group themselves and interact with other groups according to racial, cultural, ethnic, or gendered identities, even though it may be difficult for an observer to describe these groupings "objectively." Geographers study how individuals and groups perceive and respond to their environments.

Culture Regions

The entire area throughout which a culture prevails is called a culture region or area. Culture regions may be defined on the basis of religion, language, diet, customs, economic development, or still other criteria. Aspects of culture realms may be visible in the landscape.

The Global Diffusion of European Culture

Much of today's world cultural geography is the legacy of the diffusion of European culture that accompanied European political and economic global supremacy from about 1500 until about 1950. Cultural trends in the United States became globally influential as U.S. economic and political power grew. As other countries grow in size and wealth, they contribute to new cultural influences.

Cultural Preservation and Hybridity

Many societies, especially in the developing world, have found that the diffusion of European culture has displaced their own traditions. Efforts to preserve or mix cultures vary widely around the world today.

Cultural Evolution Contrasts with Cultural Diffusion

In the most general sense, culture is what separates us from our natural condition and makes our species human. The word *culture* was introduced in Chapter 1 to describe everything about the way a people live: their clothes, diet, articles of use, and customs. An object of material culture is called an *artifact*, which means literally "a thing made by skill." Culture, however, includes patterns of behavior as well as possessions. The way a group does things—its interpersonal arrangements, family structure, educational methods, and so forth— are part of its culture, and so are the facts of a group's mental or creative life—its poetry, music, language, and religion. Geographers study the origin, diffusion, and extent of all aspects of cultures, as well as the places, regions, and landscapes they create and occupy.

Human cultures are never static; they are constantly changing, as people learn new techniques and develop new cultural traits. The forces that cause these changes may be divided into two general categories. Some changes evolve within the society itself, but others are triggered by contact with other societies. The uncontacted tribes mentioned earlier, for example, are very isolated, so their culture would change almost entirely through internal evolution. They might discover new uses for local plants through the years, or they might devise new religious ceremonies. **Evolutionism** is the idea that cultural change occurs from within the culture itself, as when a group invents a new word or tool. **Diffusionism,** by contrast, emphasizes how various aspects of cultures, such as words or tools, spread out from their places of origin and are adopted by other peoples. The process of spreading is called **cultural diffusion,** and the process of adopting some aspect of another culture is called **acculturation.** Cultural diffusion can be actively imposed, as when an outside power conquers a region and imposes its way of life, or it can be freely chosen, as when one group discovers and adopts some aspect of a different culture that it considers superior to its own.

This chapter will begin with an examination of two classic evolutionist theories. Both emphasize the role that technological and economic evolution play in the development of cultures and in the interaction between societies and the environment. Both theories have been used to suggest that certain developments in the past were inevitable, or even that certain events are inevitable in the future. This is followed by analysis of diffusionistic approaches to the study of culture.

The second section of this chapter examines the identities and behaviors that stem from cultural differences. As we will see in the third section, culture has an effect on the landscape, leaving its mark on places and regions. Through history, the diffusion of goods, ideas, and techniques among peoples has increased, and the original individual characteristics of world cultures have come to be overlain or commingled with new shared characteristics. This mingling has created new and original cultural combinations.

The fourth section of this chapter analyzes the worldwide diffusion of European culture that accompanied European global conquest and economic dominance. European political dominance has retreated, but it left behind cultural, political, and economic legacies. The United States inherited Europe's leadership, yet America's global role is being tested by other countries eager to add their own innovations to the "global marketplace of ideas." The final section examines how other societies adapt to or resist influential global cultures.

Theories of Cultural Evolution

Humankind has, for the most part, adapted itself to varying environmental conditions through cultural and technological evolution, not through biological evolution. For example, humans survive Minnesota's bitter cold winters by inventing furnaces and wearing warm clothes, not by evolving furry bodies. Our cultural or technological tools allow us to survive anywhere. Scientific research stations can be maintained even in the frozen wilderness of Antarctica or in the driest desert, although supplies may have to be brought to the stations' residents. Thus, humans exist in every biome.

The theory of human stages   One theory of cultural evolution argues that all cultures evolve through certain stages of development, and it defines these stages by the way in which the culture exploits the environment. This theory was first articulated by a Roman general, Marcus Tarentius Varro (116–27 B.C.), and it remained almost unchallenged until the nineteenth century.

Varro argued that humankind originally derived its food from things that people hunted or harvested naturally. He referred to humans in this stage of evolution as **hunter-gatherers.** Varro stated that humans then domesticated animals and moved into the evolutionary stage of **pastoral nomadism** (Figure 6-1). Pastoral nomads have no fixed residences, but drive their flocks from one place to another to find grazing lands and water. If their movements are regular and seasonal—as, for example, between summers on the mountain and winters in lowland pastures—their movements are called *transhumance.* The evolutionary stage of pastoral nomadism was in turn followed by settled agriculture. Agriculture was at first *subsistence agriculture,* which means that people raised food only for themselves, but subsistence agriculture slowly evolved into *commercial agriculture,* which means that people raised crops to sell (see Chapter 8). Following agriculture, the final stages of social and economic evolution were urbanization and industry.

Figure 6-1 Nomadic sheep herding in Mongolia. These summer pastures near the Tsast Uul mountains are used by nomads to feed their flock. The traditional Mongolian *ger*, or yurt, is a portable home that nomads reassemble as they move their herds.

Varro's theory was generally accepted until the German naturalist and explorer Alexander von Humboldt (1769–1859), whom we met in Chapter 1, pointed out that some native peoples of South America had never experienced a stage of pastoral nomadism. Therefore, Varro's theory could not be true of all societies everywhere. The different ways of life described by Varro might be "categories" of human societies at different times and places, but they are not "stages" on a single path of social evolution.

Today there are still human groups living in each category described in this theory of cultural evolution, which we will look at in greater detail in Chapter 8. Varro's theory that all human societies advance through the same series of stages, however, carries a danger. According to this unilinear theory, societies cannot just be "different" from each other; some must be "ahead" and others "behind." Such presumptions have been used by "advanced" peoples to justify replacing "more primitive" peoples. For example, in the nineteenth century many Americans argued that it was inevitable that the Native North Americans' way of life (pastoral nomadism and subsistence agriculture) would be destroyed and replaced by American farmers (ranching and commercialized agriculture). Thus, any activity causing this replacement was morally neutral. Other thinkers, however, argued that human behavior is based on conscious choices. Therefore, the results of human actions can never be scientifically inevitable and morally neutral.

Historical materialism   **Historical materialism** is a school of thought that tries to write a plot for human history based on the idea that human technology has increased humankind's control over the environment. Karl Marx (1818–1895) was the founder of historical materialism.

Marx said that humankind has progressively conquered the physical environment in order to improve its own material welfare. A contemporary ruler may or may not be wiser than one in the past, but a modern tractor can do more work than a horse dragging a wooden plow. Therefore, historical materialists assert, technological change drives historical progress. Marx insisted that all social evolution is rooted in technological evolution. This is because technology determines any society's economic system, and the economic system, in turn, determines the society's political and social life (Figure 6-2). Thus, as technology advances, technological change triggers changes in all these other aspects of society. Marx's argument that all aspects of a society are rooted in its technology and economic system has always aroused great controversy.

Marx was an optimist, and he agreed with Condorcet (see Chapter 5) that technology would eventually produce material abundance for all people. Goods would no longer be scarce, so people would no longer dispute their allocation through economic and political structures. Goods would be created and distributed "from each according to his abilities; to each according to his needs."

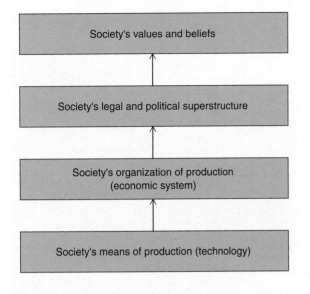

The Environment
(from which humankind must wrest material welfare)

Figure 6-2 The theory of historical materialism.
Historical materialists believe that any society's productive technology determines its economic system, and its economic system determines its political system. Ultimately, even a society's values and beliefs rest on its technological foundation. Technology is always improving, so all the other aspects of any society are constantly being dragged along and forced to change, too. Technology is the truly revolutionary force in society.

Some of what Marx wrote has turned out to be misguided predictions, and his ideas have been often reinterpreted, but much of what Marx wrote is generally accepted. Today no one doubts that science and technology progress almost constantly, and that these changes often reshape society's economic, political, and even philosophical condition. The cloning of a lamb and the birth of another with human genes in 1997, for example, challenged our laws and even our ideas of life. Many of the political movements that have used Marx's name, however, disgrace his memory. The communist parties that seized power in Russia in 1917 and imposed their ideology on other peoples in the twentieth century bear no relationship to the humanitarian society that Marx envisioned material abundance would bring.

Historical materialism contrasts with Malthusianism
Recently some economists and ecologists have attacked historical materialism from a new direction. Historical materialism maintains that in the past, technological progress offered rising standards of living to human populations: Historical materialists assume that this rate of technological advance will continue. This optimism characterizes some of our present-day leading scientists and industrialists, who would be classified among the Cornucopians discussed in Chapter 5.

Today, many concerned scientists and critics, however, belong to the opposing neo-Malthusian camp. They warn that there is no guarantee that technology

will continue to provide rising standards of living for the increasing human population. Confidence that technology will always solve problems of scarcity might lead humankind to disaster. Marx himself never imagined the explosive growth of the human population that has occurred since his death. During his time, the human population was only about 1.5 billion, and it was growing at an annual rate of less than 1 percent. The human population today is over 6.8 billion, growing at an annual rate of over 1 percent, though slowing. The neo-Malthusians ask whether the increase of the human population is, in fact, outpacing technological development or whether it might do so in the future.

These are some of the most controversial arguments of our time, and you will hear echoes of both the neo-Malthusian and the Cornucopian positions throughout this book. People of different political and economic philosophies offer different answers, and by the time you finish reading this textbook, you will have more information with which to form ideas of your own.

Cultures and Environments

Marx wrote that the key to understanding a society is the degree of control that society has over its environment. This degree of control is more important than the nature of the environment itself. Some writers, however, argue that variations in the physical environment itself explain even the variations among human cultures. The simplistic belief that human cultures or behaviors can be explained entirely as the result of the effects of the physical environment is called **environmental determinism.** The study of the ways societies adapt to environments, by contrast, is called **cultural ecology.**

Many writers have emphasized climate as a major control on human affairs. The Greek physician Hippocrates (460?–377? B.C.) argued in his book *Airs, Waters, Places* that civilization flourishes only under certain hospitable climatic conditions. Historian Arnold Toynbee (1889–1975) turned Hippocrates's idea upside down with his own *challenge-response theory.* Toynbee argued that people need the challenge of a difficult environment to put forth their best effort and to build a civilization. A rich environment encourages only sluggishness. Today, when we describe the invigorating effects of winter or a bracing wind, we are invoking the challenge-response idea.

Some political scientists have argued that societies in environments that are either particularly wet or particularly dry have necessarily developed totalitarian governments to organize and care for waterworks. Karl Wittfogel (1896–1988) described these ancient hydraulic civilizations, such as Mesopotamia, as requiring a powerful central authority capable of controlling scarce water for crop production. Such environmentally deterministic theories are intriguing, but exceptions can always be found. For example, the Dutch have managed

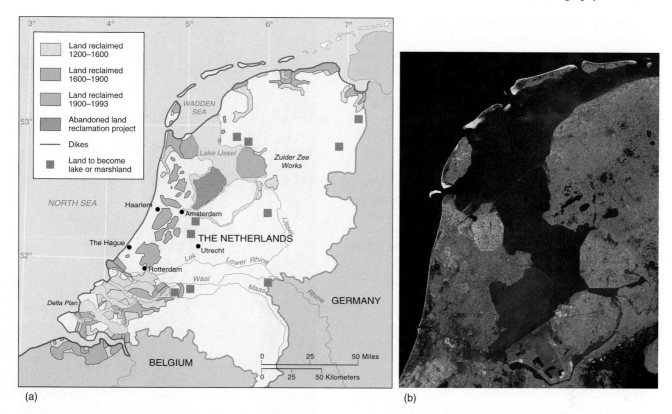

(a) (b)

Figure 6-3 The Netherlands versus the sea. More than half of the Netherlands lies below sea level, so most of the country would be underwater if it were not regularly drained and dammed off from the North Sea. Since the thirteenth century, the Dutch have reclaimed more than 4,500 square kilometers (1,800 square miles) of *polders*, which are pieces of land created by draining water. As a democracy, the management of these massive projects has responded to growing environmental concerns. The Dutch government has returned some of the wetland habitats and forests. In the infrared satellite photograph showing the northern area of the map (b), the vegetation shows as red, and more developed areas as blue-gray.

their waterworks democratically for hundreds of years (Figure 6-3).

The ability to find exceptions should warn us that no simple theory completely explains the relationship between the environment and human societies. Human affairs are not simple, and when we examine any world situation or historic event carefully, environmental determinism, or any other single-factor explanation, proves insufficient. Nineteenth-century French geographers proposed the theory of **possibilism** as an antidote to environmental determinism. Possibilism insists that the physical environment itself will neither suggest nor determine what people will attempt, but it may limit what people can profitably achieve. The choices and constraints involved in utilizing the natural environment are often as much cultural, economic, political, and social as they are technological. For example, Canadians have the technological capability to overcome environmental limitations and to grow bananas in greenhouses in Canada, but they choose not to do so. It is cheaper for them to buy bananas grown in Costa Rica.

Environmental determinism today Assumptions about the physical environment's influence in human affairs still affect economic and political debates. For

example, today many of Earth's tropical regions are poor, and there are three main hypotheses about why they are poor:

1. One theory emphasizes that the tropical environment handicaps human societies. Tropical heat multiplies the activity of organisms that are harmful to people and to agriculture, and it reduces human work efficiency. Tropical storms frequently are violent, and many tropical soils are poor.
2. Another school of thought, however, points out that many tropical areas were long held as colonies by countries in the temperate latitudes. This colonial experience left physical, cultural, and economic legacies that, combined with the continuing patterns and terms of world trade (discussed in Chapter 12), best explain tropical regions' current poverty.
3. A third point of view insists that the cultures of the peoples of the tropical regions hold back their economic growth. Perhaps their cultures do not encourage economic growth or these countries are poorly governed.

Each side in this debate may be partly correct. Which side you take will determine how you allocate responsibility for the current poverty. The first theory faults uncontrollable environmental factors; the second faults the rich, formerly colonial powers; and the third faults the peoples and governments of the tropics themselves. Furthermore, your assumptions will determine what solutions you suggest to the problem of poverty in the tropics. If the environment is to blame, nothing can be done. If the former colonial powers are to blame, perhaps they should help reverse the legacies of colonialism. If the tropical peoples themselves are to blame, they should change their habits.

All forms of environmentalist assumptions remain common in popular discussion and debate. Often they seem to be harmless figures of speech, but they are intellectually dangerous. They predispose conclusions and close the mind to alternative explanations. Watch carefully for these simplistic traps.

Cultural Diffusion

The isolation experienced by uncontacted tribes, like the one that introduces this chapter, is very different from the way most of us live today. Through history, global communication and transportation have increased, and trade and other cultural exchanges have multiplied. Most people no longer exploit their local environments and develop their cultures in isolation. They are interconnected by transportation and communication of goods, people, ideas, and capital. More people eat imported foods and combine them in imported recipes. They wear imported clothes in imported styles, and they fashion their built environments out of imported materials in styles that originated among distant peoples. Your daily life undoubtedly exemplifies this interconnectedness. Many articles that you use and other aspects of your activities draw on materials and ideas from around the world. To describe all this movement, we use the term *circulation*.

Almost everywhere today, cultural diffusion is more important than cultural evolution. Anthropologist Clifford Geertz (1926–2006) wrote in 1995, "The very notion of isolation lacks, these days, much application. There are very few places—there may not be any—where the noises of the all-over present are not heard, and most anthropologists work by now in places where such noises all but drown out local harmonies." What Geertz called "the noises of the all-over present" is cultural diffusion, and what he called "local harmonies" is cultural evolution. Indeed, what we describe as the "modern world" is one formed by ever-increasing and accelerating forms of cultural diffusion. As noted above, uncontacted tribes likely possess objects that diffused long distances. Even in closed-off and remote places, the chances are high that people there are wearing or carrying something that they did not make themselves (Figure 6-4). What happens *at* places depends more and more on what happens *among* places, and mapped patterns of economic or cultural factors can be understood only if we comprehend the patterns of movement that

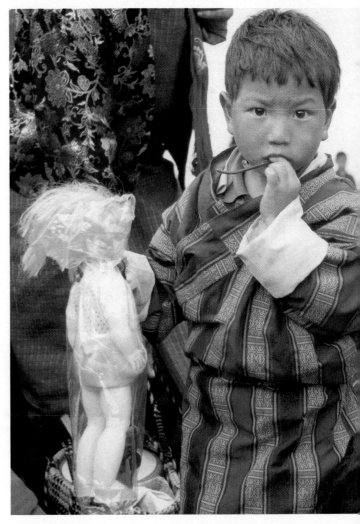

Figure 6-4 Manufactured clothes. This little boy wears a traditional robe, as is legally required in Bhutan, one of the world's most isolated countries. Bhutan struggles to preserve its traditional culture, and it allows only a limited number of foreigners to visit each year. The collar of a modern blue shirt, however, peeks out from underneath the robe.

create them and continuously rearrange them. Geography doesn't just *exist*; it *happens*.

The anthropologist Clark Wissler (1870–1947) called the places where cultures are developed *geographic culture centers*. He stated a principle, called the *age-area principle*, that if traits diffuse outward from a single geographic culture center, the traits found farthest away from that center must be the oldest traits. This description of cultural diffusion suggests concentric waves spreading out from a stone dropped into a pool. Carl O. Sauer (1889–1975) elaborated geographic studies of cultural diffusion. Chapter 1 explained that we call the place where a distinctive culture originates the hearth area of that culture, and it outlined the paths of cultural diffusion. Geographers can investigate the distributions of any cultural attributes; they are limited only by their imagination and curiosity.

Diffusion does not explain the distribution of all phenomena. Sometimes the same phenomenon occurs spontaneously and independently at two or more places. In the history of mathematics, for example, the idea of the zero and its use as the basis of a numerical place system was conceived in two places: among the Maya, a Native Central American people, and among the ancient Hindus of India or perhaps the Babylonians before them. Diffusionists once argued that Phoenicians must have sailed across the Atlantic from their homeland in the eastern Mediterranean thousands of years ago and taught the Maya how to use the zero. There is, however, no evidence for this assumption, and it underestimates the ingenuity of the Maya. To demonstrate diffusion, we must be able to illustrate its path from one culture to another.

Folk culture Today's rapid pace of cultural innovation and diffusion requires us to differentiate folk culture from popular culture. The term **folk culture** refers to a culture that preserves traditions. Folk groups are often bound by a distinctive religion, national background, or language, and they resist change. Most folk-culture groups are rural, and relative isolation helps these groups maintain their integrity. Folk-culture groups, however, also include urban neighborhoods of immigrants struggling to preserve their native cultures in their new homes. *Folk culture* suggests that any culture identified by the term is a lingering remnant of something that is embattled by the diffusion of other cultures.

Cultural geographers have identified culture hearths and routes of diffusion of a surprising number of folk cultures across the United States. Evidence of their presence and movement can be drawn from the material landscapes they produce. Geographers who have investigated the architecture of houses, for example, have differentiated among various house types and traced the diffusion paths of these styles westward with various immigrant groups (Figure 6-5). The groups brought these styles with them when they settled in the United States, such as the English "I" house, which subsequently evolved as people discovered new materials and techniques. Barns and other structures are also built in distinct architectural styles that reveal the origins of their builders. Folk geographic studies in the United States range from studies of folk songs, folk foods, folk medicine, and folklore to objects of folk material culture as diverse as locally produced pottery, clothing, tombstones, farm fencing, and even knives and guns.

In North America, the Amish provide an example of a folk culture. The Amish are a sect within the Mennonite Protestant religious denomination formed in Switzerland in the sixteenth century. Early Amish immigrants to North America settled in Pennsylvania and spread across the Midwest and into Canada. The Amish are notable because they wear plain clothing and shun modern education and technology. They prosper by specializing their

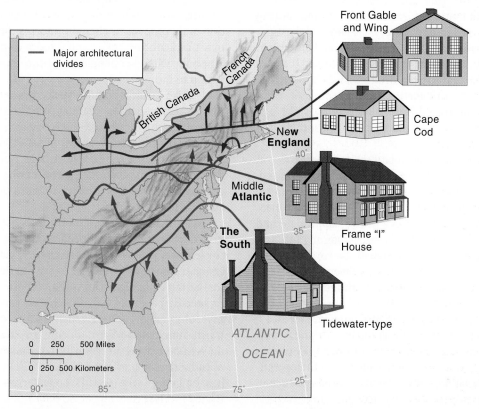

Figure 6-5 Paths of diffusion of house types. Geographers have identified four types of folk-house styles that originated on the U.S. Eastern Seaboard and diffused westward along with pioneer settlers. The illustrations show a typical house of each type.

Figure 6-6 Amish folk culture. Although Amish communities typically reject many technologies, such as automobiles, they may be receptive to others that are not motorized. Here an Amish woman on rollerblades pushes her child in a wheeled carrier through Lancaster County, Pennsylvania.

Figure 6-7 Heritage sites draw many tourists. The Dutch settlement of what is now New York's Hudson River Valley is remembered at historical sites such as Philipsburg Manor. Here a historical interpreter demonstrates the use of a spinning wheel from around 1750.

farm production and marketing their produce, but they severely curtail the choice of goods that they will accept in return (Figure 6-6). They have been able to preserve many of their cultural practices by remaining relatively isolated from surrounding communities.

Many African Americans have long felt that their traditional folk cultures were taken from them during the years they suffered in slavery, yet careful investigation has yielded a rich variety of cultural elements in some African American communities today that are traceable to specific African origins. These range from musical styles to the meticulous sweeping of front yards (a tradition traced from villages of coastal Nigeria to rural Georgia) to the patterns of speech discussed in Chapter 7.

The United States is relatively homogeneous in culture and is modern and technologically advanced. Cultural diffusion and change churn across the country rapidly and repeatedly. In such a country, the identification and study of folk cultures help us appreciate and cherish the richness of some of our remaining folk traditions. Chapter 5 noted that Canada, by contrast, more carefully nurtures and preserves its diversity as a "cultural mosaic." In many countries, **heritage sites** seek to preserve and commemorate dead or dying folk cultures and often become major tourist attractions (Figure 6-7).

Some newly independent countries are today emphasizing or even resurrecting traditional folk cultures as a way of enhancing national identity and community. When the nation of Eritrea won independence in 1993, for example, one of the new government's first acts was to open a national university mandated to re-search and define folk literature, folk music, and other distinct Eritrean traditions.

Popular culture   **Popular culture**, in contrast to folk culture, is the culture of people who embrace innovation and conform to changing norms. Popular culture may originate anywhere, and it tends to diffuse rapidly, especially wherever people have time, money, and inclination to indulge in it.

Popular material culture usually means mass culture—that is, items such as clothing, processed food, books, electronics, and household goods that are mass-produced for mass distribution. Whereas folk culture is often produced or done by the people at-large (folk singing and dancing, cooking, costumes, woodcarving, etc.), popular culture, by contrast, is usually produced by corporations and purchased.

Mass manufacturing lowers the cost of items, but in order for a mass-produced model to succeed, consumer taste must be homogeneous. This is not left to chance: consumer preferences are studied closely and marketing campaigns can reshape those preferences to fit specific products. Advertisements can even become part of popular culture, such as the expensive and much-anticipated ads run during the Super Bowl. Such mass taste setting necessarily requires some sacrifice of individuality and cultural identity. The United States has long been the world's largest consumer market. Even a product that appeals to only 1 percent of the U.S. market can potentially sell to over 3 million people!

GLOBAL AND LOCAL

Lahic

The village of Lahic is one of the world's most extraordinary examples of an isolated community that has created and maintained a distinctive local culture. Lahic is a village of approximately 2,000 people, perched high on a ragged cliff surrounded by the towering Caucasus Mountains in the Republic of Azerbaijan. Nobody knows when this village was first settled or who the first settlers were, but the unique local language provides a clue. It is a form of Persian (Iranian) that died out in Iran hundreds of years ago. Perhaps some ancient Iranian emperor sent settlers to Lahic to defend the empire's borders against the peoples to the North.

Still today, Lahic is connected to the outside world by only a rugged dirt path that winds its way around precipitous gorges and is closed by rain and snow much of the year. Wires from Baku, the capital of Azerbaijan some 128 kilometers (80 miles) away, brought electricity and telephone service to the village in the 1980s, but these services are provided only a few hours a week.

The desolate landscape of Lahic allows the people to raise livestock, but they cannot feed themselves, so long ago they developed a tradition of exchanging craftwork for food. In their isolation, the people of Lahic developed artisanal skills that have made their handcrafted products famous and prized around the world. Virtually every kind of handcrafted object has been made here and exported for hundreds of years: objects in copper and brass, carpets, hand carved and wrought iron, guns, cutlery, clothing, and leather goods. Every home is a workshop. Lahic crafts have long been treasured by collectors throughout the Middle East and Europe. Traders discovered them many centuries ago and sold them for high prices in bazaars in Baghdad, Shiraz, and other great Middle Eastern

Traditional artisan workmanship. This artisan in Lahic, Azerbaijan, demonstrates the skills that have earned renown for products from his isolated village and kept its economy thriving for hundreds of years.

cities. A display of copperware made in Lahic won a gold medal at the Paris World Exposition in 1878, and today Lahic crafts are on display in museums from London and Paris to Moscow and Istanbul.

A 1997 U.N. report on Azerbaijan noted that "Lahic . . . has kept almost perfectly intact a microcosm of cultural features which has preserved its communal harmony and social cohesion across the centuries. The spirit of the Middle Ages still lingers there."

These days, more young people are going off to school or to find new job opportunities in Baku and thereby abandoning their village's ancient crafts. How much longer can this time capsule survive?

This has lowered the cost of items and raised Americans' material standard of living. The substantial unity of the Canadian and U.S. markets has raised Canadian standards of living, too, but some Canadians believe that it has threatened Canadian cultural identity.

The popular culture of the United States exhibits a great degree of national homogeneity, but individual consumer goods win varying degrees of acceptance (called *market penetration*) in different regions of the country. The regions are often the vernacular regions, as mapped in Figure 1-8. The marketing managers of corporations are "applied cultural geographers" who specialize in this type of analysis. Carbonated soft drinks are a common trait of American popular culture, but there are regional and even local preferences. The soft drink Dr Pepper, for example, originated in the U.S. South, and it still enjoys its greatest acceptance there.

Even the word Americans use for soft drinks varies regionally. The generic term "coke" is most common in the U.S. South, where it was borrowed from Coca-Cola, an Atlanta-based company (Figure 6-8). Less common terms, like "fizzy" or "tonic," are used more locally, perhaps reflecting particular regional brands.

Geographers investigate the origins and diffusion paths of popular material culture and also of popular social culture. For example, people devise new ways of living, working, and playing, and they innovate in education and in employer–employee relations. Sport is an important part of popular culture, and popular culture regions can be defined by their sporting preferences (Figure 6-9). Regions differ in their popular entertainments—a movie that is a success in Seattle, for example, may flop in Houston. Cultural geographers want to know why. The radio stations in different regions

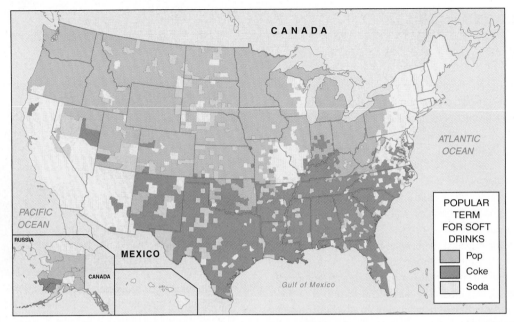

Figure 6-8 Coke or pop or soda? The word that Americans use for soft drinks varies regionally. This map was produced from the 300,000 votes cast by visitors to http://www.popvssoda.com, a website created by Alan McConchie, a graduate student at the University of British Columbia.

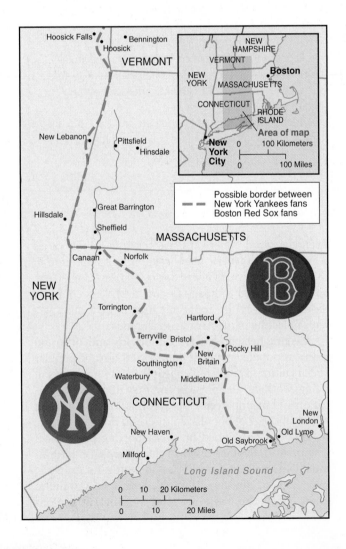

across the United States play varying mixes of country, gospel, rock, classical, and other popular kinds of music; DJs base their decisions on market research about local preferences. Geographers have investigated these and many more attributes of popular culture.

The marketing of popular culture can overwhelm folk culture. For example, how many of the characters and tales of traditional U.S. folk culture—or even of genuine historical heroes and heroines—have survived the onslaught of mass-produced and mass-marketed merchandise related to comic-strip characters? Who is most famous in the United States today: Paul Bunyan (a mythical folk hero), Johnny Appleseed (a real person, John Chapman, 1774–1845), or Iron Man a product of popular culture? The answer is probably Iron Man.

Figure 6-9 A "watershed" between territories of fans of two baseball teams. In 2006, researchers from Connecticut's Quinnipiac University set out to define the borderline between baseball fans loyal to the New York Yankees and fans loyal to the Boston Red Sox. Research techniques included interviews, checking sales of baseball caps and other merchandise with team logos, and even sales of locally made cookies with the different teams' logos. Rocky Hill, Connecticut, is the midpoint between the teams' homes of Yankee Stadium in the Bronx, New York, and Fenway Park in Boston, but the dividing line of team loyalties was found to lie slightly south of it. The choice of a boundary line simplifies the landscape, however, making two distinct areas instead of what is probably two overlapping areas of intensity. Do you think there are no Red Sox fans in New York and no Yankees fans in Boston?

Identity and Behavioral Geography

Identity is a frequently used term that has two common meanings. One meaning refers to a person's *individual identity*, a set of distinct characteristics that makes him or her unique. Some of these distinguishing features may be physical features or personality traits. For example, you have a friend who is good looking, sarcastic, and hard working. Another meaning of *identity* relates to your friend's *group identities*, which may be as a heterosexual, Japanese American lawyer. Group identities imply that group members share characteristics that differentiate them from other people. Unfortunately, other people often form incorrect assumptions about groups and apply these characteristics to all individuals in the group. This is known as *stereotyping* and may lead to biases such as sexism, racism, and elitism. Nonetheless, the group identities we use to describe ourselves can convey powerful ideas about our heritage, our peer group, and even our individual characteristics and values.

Group identities affect relations between people. Political leaders often use the group identity of nation as a basis for defining friend and enemy (see Chapter 11). The press often refers to the parties of disputes as having different racial, cultural, ethnic, or other identities. On closer inspection, however, explanations based solely on group identity prove too simplistic; people are driven by many needs and fears that often have little to do with identity. Thus, it is important to carefully consider group identities because they are how people understand themselves and others. When geographers study these behaviors, they are also interested in **spatial identities**—how certain identities attach to particular places or regions.

Grouping Humans by Culture, Ethnicity, Race, and Gender

It is easy to imagine that some groups form naturally and that humans are simply born into identity groups. This is incorrect even when identities appear to be based on biology, such as race and gender. Identities are human—in fact, cultural—inventions. Even our briefest encounter with other people is consumed with sorting out similarities and differences and assessing the meaning of these comparisons. Definitions of who belongs to which group can change over time and between cultures, as can the status of a group in society. Governments may even assign people to particular groups, having a powerful effect on their lives. Most importantly, people may not consider themselves part of a group that others would put them in on the basis of superficial appearances or behaviors.

Cultural or ethnic groups The definition of a cultural group may include a great number of shared characteristics or just a few. For example, all social scientists agree that language is an important attribute of human behavior. Two people who share a language share something very important. If, however, those two people hold different religious beliefs, feel patriotism for two different countries, and eat different diets, social scientists may insist that although those two people share one attribute of culture, they do not share one culture. The great number of English speakers who presently live around the world share few attributes other than language, so we would not say that they all share one common culture. If, however, two people do share a language, religious beliefs, political affiliation, and dietary preferences, then social scientists would agree that those two people share a culture.

The question of cultural identity seems relatively simple in dealing with isolated tribes. In large complex societies, however, there may be many groups. While many people identify with parts of the dominant or *mainstream culture* in their society, some also identify with smaller groups that are important to them. A **subculture** is a group that shares a smaller bundle of attributes with a larger, more diverse society. Its attributes are often related to a common folk culture or ancestry, and may include the experience of being marginalized by the larger society. For example, many Italian Americans, Chinese Americans, and African Americans share subsets of cultural attributes within the larger American culture. Durable subcultures are often based on spatial identities such as common origins, regional affiliation, or specific residential patterns. Another definition of subculture is one built around popular culture. Science fiction fans have become a worldwide subculture (Figure 6-10a). Football (soccer) subcultures in Europe, however, are intensely place based. Their loyal supporters often exemplify the word *fan*, which is short for *fanatic* (Figure 6-10b). Their team's home stadium is considered a shrine and a place of pilgrimage.

The term **ethnic group** or *ethnicity* is frequently used to describe a cultural or subcultural group. The word *ethnic* comes from the Greek for "people," and the definition of an ethnic group may depend upon almost any attribute of biology, culture, allegiance, or historic background. The word has historically been used in a pejorative sense: Its meanings have included "alien," "pagan," and often "primitive." Some social scientists nevertheless define ethnic groups and study the groups' characteristics or attributes. Ethnomusicology, for example, is the study of ethnic groups' music, and ethnobotany is the study of ethnic groups' knowledge of the uses of plants (Figure 6-11). It often implies a common ancestry or descent group that bears distinguishing cultural features. In this sense, the term ethnicity is often used to describe a type of political claim that we will discuss as nationalism in Chapter 11. **Ethnonationalism** is the claim that a particular culture group has special political rights over other groups.

(a)

Figure 6-11 Ethnomusicology. Music plays an important role in many cultures. The Kolge Duma Group dancers of the Western Highlands Province, Papua New Guinea, drum while performing ceremonial dances.

(b)

Figure 6-10 (a) Popular culture. More than 40 years since Star Trek first appeared, fans of the original television series attend the annual Comic convention in costume.
(b) Football fans. German football (soccer) fans cheering their team during the 2006 World Cup.

In the American context, the term ethnicity has come to mean subcultures, especially among immigrants and their children (as discussed in Chapter 5). As a spatial identity, geographers describe **ethnic enclaves** as urban, and increasingly suburban, neighborhoods with a high concentration of a particular ethnic

group. Although today such segregation may be self-imposed, in the past such segregation was enforced. In either case, these neighborhoods become part of a subculture's spatial identity. Even after those residential patterns dissolve, the neighborhood may still be known as "Little Italy" or "Chinatown," for example. Some of these enclaves endure, bolstered by new immigrants. Most importantly, the enclave provides a cultural, as well as economic, resource for new immigrants during their acculturation into the wider society.

Ethnocentrism is the term given to the tendency to judge other cultures by the standards and practices of one's own, and usually to judge them unfavorably. Practices in other cultures that may seem strange to us, however, may in fact be sensible and rational. Conversely, some aspects of our own culture may seem strange or even offensive to others. Most Americans, for example, assume that a person should be married to only one other person at a time, but any number of spouses are allowed in a series after divorce. This would shock many people. Americans, in turn, may be shocked to learn that Tibetan brothers are often married to the same woman. This is called *fraternal polyandry*. Through thousands of years, fraternal polyandry has made possible sustainable population increases and prevented the fragmentation of land-holdings in Tibet's poor mountain valleys. The social ramifications of fraternal polyandry confound most Americans. An American might ask how to identify the father of any given child. To a Tibetan, however, it makes no difference, and a Tibetan might consider the question prurient.

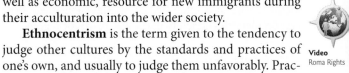

Video
Roma Rights

RAPID CHANGE

Who Killed the Record Store?

The cultural landscape can change quickly. Not long ago, the record store was a basic feature of any U.S. town. It was more than just a place to buy music. Going to the record store was an important cultural practice, especially for U.S. youth. Going to the record store alone, without parents, marked a coming of age for many young people. It was an exposure to new music, played in the store by a usually knowledgeable clerk who tried to play something good but relatively unknown. Staff and other shoppers might share suggestions for music, providing descriptions about not only the music but cultural cues about who listens to the music and why. A bulletin board listed local concerts, advertised used musical gear for sale, and sought new members for garage bands.

Going down to the record store meant wearing the right clothes, affecting the appropriate attitude, using the right language, and, above all, picking the right music to define oneself and one's peer group. As with youth today, there were different ways of being "cool," but each had its own set of "rules" and social expectations. In short, each musical style had is its own subculture. T-shirts, buttons,

The last of the independent record stores?

patches, jewelry, and other visible markings of one's identity were sold alongside the music.

Ironically, the anxiety over being authentic (not a "poser"!) was also a part of becoming a modern consumer. Sorting out one's identity through musical genres, clothing, and friends was a way to learn about one's own tastes, preferences, and lifestyle in relation to others. Even if your peer group and taste preferences changed, the experience was a valuable lesson in the social importance of melding behavioral traits to particular peer subcultures while also being in mainstream society.

But record stores, as described here, are fast disappearing. Small independent record stores struggle to stay open, but most cannot. Four in five national chain stores, such as Tower Records, have closed since 1991. Yet music remains a lynchpin of American popular and youth culture. So who killed the record store? Some blame mass marketers, like Best Buy, and discount marketers, like Target, who certainly took sales from record stores. Though record stores have been an important place in the cultural landscape, they are a business built around a rapidly changing technology. The same cultural preferences for selection, convenience, and portability that increased sales and diversified musical genres have more recently shifted demand toward digital downloads. Instead of paying for whole albums, consumers can buy just a few tracks and carry them everywhere. Virtual stores, such as iTunes, eliminate the need for physical stores. But the shift to digital didn't stop there. Digital technology makes it easy to "share" songs, and an estimated 40 billion tracks were illegally downloaded in 2008. That's 95 percent of all music downloads. Not all those tracks would translate to legal sales, but "sharing" does explain why fewer and fewer young people go to music stores anymore. The few independent record stores that remain open cater to a "graying" clientele. These trends have enormous economic implications for the music industry as well as perhaps profound repercussions on the role of music in popular culture.

Racial groups As we saw in Chapter 5, race is an ambiguous way to describe people because physical appearances tell us little about individual characteristics or the shared characteristics of cultures. Most people know that a person's race does not determine his or her personality. But many people still confuse race and culture. The physical features long used to classify races, especially skin color but also hair and facial features, can include people from very different cultures. Some cultures, meanwhile, are multiracial, especially some Muslim ones. Nonetheless, race is very important to many people. The experience of racial groups in different

societies shapes what race means in those contexts. In fact, much of what makes race meaningful in any society is the historical and contemporary experience with race-based preferences, laws, and attitudes. In the African American experience, slavery, discrimination, and segregation gave shape to a subculture among people forced to endure the legacies and bigotry of racism. More recently, this subculture has been defined by the role of African Americans in the civil rights movement, educational and professional success, and strides towards social equality, as well as many contributions to a wider American culture.

Gender and sexual identities   The role of men and women is not everywhere the same. The **gender roles**, those duties and behaviors associated with being a man or a woman, vary across cultures and even within complex societies. In Afghanistan, a Muslim country, women have traditionally worn a veil that covers their entire body, and they are not part of public life. In Turkey, another Muslim country, many women wear skirts, not veils, and a woman has led the government. Questions of gender often cause passionate debate because these identities are very personal for many people and cause some great anxiety. The same is true when **sexuality**, one's sexual orientation and behavior, differ from mainstream cultural expectations.

Part of the misunderstanding surrounding gender and sexuality is that they are culturally defined roles, whereas being male or female is a function of biology. Many cultures in the world today assume, even insist, that sex, gender, and sexuality are the same thing but of two types: men and women. In reality, these are different things. Biologically the sexes are typically distinguished by two sets of genetic and physical characteristics, but some people are born with characteristics of both. As described above, gender roles vary across history and cultures (see the Global and Local feature). Compare the role assigned to women in conservative Muslim cultures with the role of women in your community. But because sexual reproduction requires a typical female and male (technically, only male sperm), some have argued that heterosexuality is the basis for "natural" gender roles and sexuality. The requirements for sexual reproduction do not "naturally" preclude other sexual behavior, however. Homosexuality has been common through history and across cultures, although its cultural significance and acceptance has varied widely.

Gender and sexuality are also identities with spatial aspects. One of the most important distinctions is often that between public space and private spaces, especially the domestic sphere of home. What some cultures may tolerate behind closed doors, such as public displays of affection, homosexuality, or certain clothing styles, may be frowned upon in public streets or the workplace. The French government in 2004 banned the wearing of the Muslim headscarf by women and girls in public schools. The government argued that the scarf was a symbol of religious identity that had no place in a secular country. Many Muslim French women argued that the ban denied them their right to dress modestly. Gender roles and sexuality often imply specific places. Men and women, for example, are expected to use different changing rooms. Homosexuals sometimes go to what are considered or advertised as "gay" bars or neighborhoods to socialize with less chance of confrontation.

Any geography book will contain examples of ways of life that contrast with your own. None is necessarily "right" or the best for everybody. All people have to overcome the initial assumption that "different from" the way they do things themselves is "worse than" their way: People everywhere also can learn to appreciate and respect the integrity of other people's cultures. Indeed, life's most profound lessons usually appear only after we learn to see ourselves differently, from the position of another person or culture.

Behavioral Geography

Behavioral geography is a subfield of cultural geography that studies our perception of the world around us and how our perception influences our behavior. The political commentator Walter Lippmann (1889–1974) differentiated "the world outside" from "the pictures in our heads." The world outside is the way things really are, but the pictures in our heads may be based on preconceptions, misperceptions, or incomplete understanding. Geographers call these pictures in our heads **mental maps.** Geographers borrow from psychology the theory of **cognitive behavioralism**, which argues that people react to their environment as they perceive it. In other words, people make decisions on the basis of their mental maps. Thus, we cannot understand why people make the decisions they do or act as they do by studying only the real environment in which they acted. We must discover what was in their heads when they made their decisions.

For example, in 1898 Senator George Hoar of Massachusetts cast a key vote for the annexation of Hawaii because, according to his diary, he drew a line on a map from Alaska to southern California, saw that Hawaii fell inside that line, and concluded that its possession was necessary for the country's defense. In fact, Hawaii does not lie inside such a line. Check its position on a globe. No one has been able to discover what sort of map Senator Hoar was looking at (perhaps one with an inset map of Hawaii?), but his misperception affected history. We make choices based on our perceptions of the world, but we must act in the real world.

Where we choose to live, shop, or visit depends on our feelings about places—whether we think certain places are good or bad, beautiful or ugly, safe or dangerous (Figure 6-12). Geographers have investigated how we perceive environmental threats, either natural (earthquakes) or human-made (chemical or nuclear installations). How we perceive these threats affects our actions or reactions toward them. The geographic subfield of environmental hazards examines why, for example, people live on the slopes of volcanoes or next to rivers. The risk of eruptions or floods is given less consideration than the need to farm rich volcanic or alluvial soils.

Our individual notions of usefulness or of aesthetic value affect our evaluation of landscapes. Most urbanites, for example, agree that the Painted Desert in Arizona is beautiful, and it is a major tourist attraction.

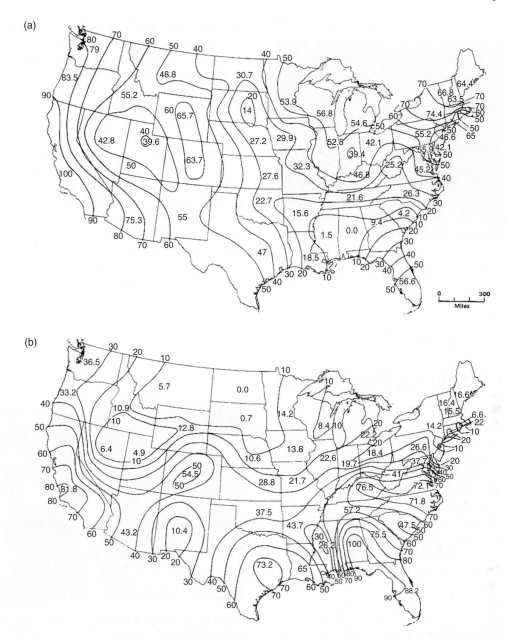

Figure 6-12 Residential desirability. These are the mental maps of places' residential desirability in the heads of students at the University of California at Berkeley (a) and the University of Alabama (b). High numbers indicate a positive image in the students' minds. Students at both places seemed to like where they were. Berkeley students gave coastal California a score of 100; Alabama students rated Alabama 100. The Alabamans, however, rated the region of Berkeley (60–70) much higher than the Californians rated Alabama (0.0). Both seemed to have positive images of Colorado and low images of the Dakotas. These preferences are not necessarily based on personal experience. We all have opinions of places we have never visited, but we might be surprised if we did visit them.

A farmer, however, might be appalled at such a landscape (Figure 6-13).

Studies that have grouped men, women, children, the elderly, the disabled, and individuals categorized in many other ways have taught us that each of these groups perceives the environment differently. Therefore, these groups may almost be said to live in different environments. This knowledge is useful in the design of new environments, such as housing. Housing for the hearing impaired, for example, should have open interior sight lines.

Differences in perception and behavior may be studied among different cultures. Human behavior is rooted in biology and physiology, but it is filtered through culture. **Proxemics** is the cross-cultural study of the use of space. People of different cultures balance the way their mind relies on input from their eyes, ears, and noses differently, even if they have the same physical sensory capacity. They do not agree, for instance, on crowding. In a room in which most Americans would feel uncomfortably crowded, most Middle Easterners would not. In designing a library, for example, it helps

Figure 6-13 Farmland in Iowa. This black soil photographed near Martinsburg, Iowa, is so rich that a farmer might find this a beautiful view. What is your aesthetic reaction?

to know that North Americans are more distracted by sounds than by sights, whereas Japanese are more distracted by sights than by sounds. Therefore, people talk in Japanese libraries, but the libraries have study cubicles. Proxemics affects human behavior at all levels from polite interpersonal behavior to the design of buildings or even whole cities.

Some scientists believe that humans exhibit **territoriality.** Many animals lay claim to territory and defend it against members of their own species. Most aspects of territorial behavior among animals have to do with spacing, protecting against overexploitation of that part of the local environment on which the species depends for successful reproduction. Applying these studies to human behavior is complicated by the interplay between human biology and culture. People defend their standing space in a crowd (they bump

GLOBAL AND LOCAL

Sworn Virgins of the Balkans

A century ago, travelers and anthropologists passing through the Balkan Peninsula took a great interest in persons born as women but who lived as men. The practice was uncommon but still aroused intrigue among outsiders. These women often dressed and acted as men, having publically vowed to remain virgins and to live as men for the rest of their lives. Among the strongly patriarchal (male dominated) cultures of southeast Europe, this "anomaly" appeared fantastical to foreign visitors whose own cultures had strict social conventions about gender roles and repressed attitudes about sexuality.

The explanation given at the time was a "cultural" one that responded to the often-harsh conditions of traditional life in the rugged mountains of northern Albania, southern Serbia, and eastern Montenegro. These *tobelija* ("one bound by vow" in Serbian) or *virgjinéshë* ("sworn virgin" in Albanian) could be found in all three religious communities in the area: Muslim, Catholic, and Orthodox Christian. They took on responsibilities assigned to men, especially caring for parents who had no sons. In these communities, only men could inherit property, which provided for basic needs in a remote village. Men were also responsible for heading the family and keeping good community relations. Vowing to live as men, respected by the community, allowed these women to fulfill otherwise male-only roles. A *tobelija* can dress as a man and carry a gun, but some wear only a man's hat or take a male nickname. These practices were interpreted as culture traits necessary for the family's survival and good for the community.

More recent interviews with these women have highlighted other reasons for their choice. Some cite the need to help raise and protect younger male siblings if parents died before the children reached adulthood. Others

Qamile Stema, one of the remaining sworn virgins in northern Albania.

mentioned their desire to work for a living and to avoid the constrictions of traditional marriage. Some of these women live as part of their extended family. A few live with other women for companionship but without the intimacy typically used to define homosexual relationships.

The practice is disappearing. The communist regimes that ruled these countries between World War II and the 1990s were focused on rapid modernization. Rural life was viewed as backwards and uncivilized. Schools and other programs were critical of many local cultural traits, including the destructive blood feuds that sometimes erupted between families. Even in remote mountain villages, the introduction of radio and television had profound effects on changing local gender roles. Only a few dozen sworn virgins are thought to remain, many quite old. In interviews, they say they understand the choice of young women to "come down from the mountains," meaning to abandon rural traditions.

back), urban gangs defend their turf, and nations defend their land; but sometimes they avoid conflict and instead cooperate. Assuming that humans are little different than animals ignores the importance of human culture, which allows us to solve problems creatively.

pMaster™
ed Thematic
ld: Cultural
r Cultural
ths

Culture Regions

A **culture region** or **area** is defined by a relatively continuous presence of one or a set of cultural traits. Such traits may be simple, such as bumper stickers for the Boston Red Sox, which imply a culture region of its fans. This region possesses a **culture core area,** in this case Boston, which is almost exclusively dominated by Red Sox fans—and those who aren't at least play along. The rest of New England may be described as a **culture domain,** where most, but hardly all, baseball fans support the Red Sox. Finally, there is a **culture realm,** in which there is some support for the Red Sox, but it is less than for other teams—that could include the whole country and parts of Canada! This means that culture areas overlap, as we saw in Figure 6-9.

Conventionally, geographers have identified culture regions according to more traditional identities. Religion is an important source of many cultural traits, so religion is often chosen as a criterion. The use of the phrase "the Christian World," for instance, implies that the prevalence of Christianity across a large region means there are important cultural traits shared by inhabitants there, even if not all of them identify as Christians. This region might significantly be contrasted with, for instance, "the Islamic World," which also includes non-Muslims. Each is a large culture realm, and Chapter 7 examines how the prevalence of each of those religions encourages other

similarities across those realms. Other criteria that may be chosen to define culture realms include language, diet, customs, or economic development. These topics will be discussed individually in the following chapters.

Culture regions must be considered carefully. First, cultures change over time and, thus, so do culture regions. It is important to think about the cultural diffusion processes that produced large culture regions. It is crucial to recognize the processes that are changing them today. Second, culture regions are not isolated. Global diffusion patterns and economic connections, not to mention colonial legacies, link together places that once had very different cultures. Third, the larger a culture region one considers, the fewer culture traits will actually apply. All large culture regions contain many specific cultures that may differ tremendously from one another. For example, Westerners may label Asia as a culture region. A Korean person, however, might insist that the many cultural traits that differentiate him from a Burmese person are more important than anything they share. Finally, culture regions are defined by traits that are similar in culturally meaningful ways. Other kinds of regions, such as climate regions and demographic regions, use criteria that are not culturally relevant.

Visual Clues to Culture Areas

Culture areas and regions can reveal themselves through visual clues, or visible traits, in the cultural landscape. These include the language of posted signs, the clothing the local people are wearing, and the goods available in local shops. In the built environment, building materials, architecture, and settlement patterns are all visible manifestations of cultures. People usually rely on local materials for building, so in one

CONNECTIONS

Is Latin America a Region? How Did It Get Its Name?

Some criteria of homogeneity justify labeling Central and South America "Latin America." "Latin" suggests that certain important aspects of the culture spring from the cultural tradition of ancient Rome (*Latium*), and, in fact, most of Latin America was long ruled by either Spain or Portugal, both of which are Latin countries. Most Latin Americans today speak Spanish or Portuguese—both languages descended from ancient Latin—and most of the people in Latin America belong to the Roman Catholic Church. However, neither Rome nor anywhere else in Italy played a historic role in Latin America. Why, then, don't we call the area Luso-Hispanic America (*Luso* meaning "Portuguese," because today's Portugal was the ancient Lusitania) or Ibero America? The more one thinks

about the term Latin America, the more mysterious its origin becomes.

The name Latin America originated as a geopolitical term (see Chapter 11). During the U.S. Civil War, France's Emperor Napoleon III thought he saw an opportunity to take over Mexico. France had never had any colonial interests there, but Napoleon III invaded on the pretext that France, a Latin nation, was avenging military humiliations suffered by the Spanish and was therefore "defending Latin honor." On July 3, 1862, Napoleon published a letter in the French newspaper *Le Moniteur,* introducing the term *Latin America* to justify what was really French imperialist aggression. Eventually the French were defeated and retreated, but Napoleon's sly propagandistic term has long survived him.

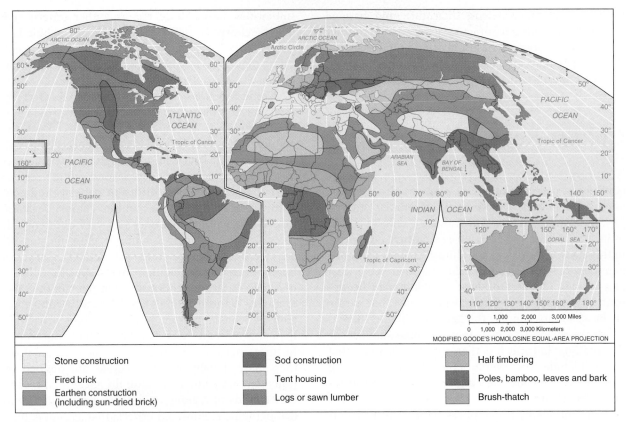

Figure 6-14 Traditional building materials. The materials that are locally available—stone, brick, bamboo, or wood—are used everywhere in traditional architecture, called vernacular architecture.

Legend:
- Stone construction
- Fired brick
- Earthen construction (including sun-dried brick)
- Sod construction
- Tent housing
- Logs or sawn lumber
- Half timbering
- Poles, bamboo, leaves and bark
- Brush-thatch

MODIFIED GOODE'S HOMOLOSINE EQUAL-AREA PROJECTION

Figure 6-15 The mosque at San, in the West African country of Mali. This building demonstrates that when people have mud and very little wood, they can still build extraordinary structures. It is in the distinctive Dyula architectural style, named for a trading people who diffused the style as they moved around West Africa. It is quite different from the Middle Eastern architectural styles that many Americans expect mosques to reflect. Compare it with Figure 7-24.

place stone may be the traditional building material, in another place brick, and in another wood (Figures 6-14 and 6-15). Innovations in the uses of these materials may diffuse across cultures.

Styles of architecture often represent adaptations to climatic conditions, so a particular style may be adopted in different regions with similar climates (Figure 6-16). Cultural preferences are sometimes so powerful, however, that an architectural style may diffuse beyond the limits of where its building materials can be found and even beyond the range where that architecture is comfortable. The style of the Italian Renaissance architect Andrea Palladio (1508–1580), for example, spread to England because of aesthetic preferences, despite the fact that Palladian buildings are uncomfortable in England's damp, cool climate. From England the preference diffused to America, even to areas that lacked both the appropriate climate and the necessary building materials. As Europe and the United States came to dominate other regions, Palladian architecture diffused throughout the world (Figure 6-17).

Public statuary and monuments may reveal local cultural values, and they may change as local values change. For example, many cities in the U.S. South erected monuments to commemorate the Confederate cause, but as African Americans gained equal rights and political power there, they insisted that some monuments be taken down. In some cases, new memorials were erected to remember slavery or those who died in the civil rights movement. Russia has marked the fall of communism by erecting public statues of its pre-Communist czars

Figure 6-16 Simla, India. The building in the foreground is a *bungalow*, which is a Hindi word for a low-sweeping single-story house with a roof extending out over a veranda. Such houses were first built in the mountain foothills of northern India, but the style was copied throughout the British Empire and eventually the United States. Today in English the word *bungalow* means almost any sort of small house. Simla was the summer capital of the British Indian Empire between 1864 and 1947. It was cooler up at Simla (2,200 meters/7,100 feet; notice the vegetation) than down at New Delhi in the plains, about 275 kilometers (170 miles) to the south. The medieval-style cathedral in the background seems an odd presence in northern India. It is not, however, a medieval building. It was built in the nineteenth century in the medieval style that was popular for Christian churches then (and still is today). Wherever the British went, they took their culture and traditions, including architectural styles.

(Figure 6-18). Statues also may reveal that a place is dominated by a cultural or political outsider. For example, several Eastern European countries fell under the domination of the U.S.S.R. after World War II and statues of Russian "liberators" were often set up in their public places. Since these countries have regained their freedom, however, the local people have replaced these statues with statues of genuine local heroes. Budapest, Hungary, for example, boasts new statues of ancient Hungarian royalty and Christian saints.

Settlement patterns The designs of settlements reflect cultural differences, and a trained observer can see in the look and layout of whole towns and cities the cultural backgrounds of their builders. Rural societies can be differentiated by the way that some cluster housing settlements, whereas others isolate settlements in individual farmsteads. City planning will be discussed in Chapter 10.

Figure 6-17 A Palladian building in the United States. This house, called the Morris-Jumel Mansion, was built in 1765 in Manhattan 10 miles north of Wall Street. The choice of the Palladian architectural style is highly symbolic. Standing virtually on the frontier of Western civilization, this style boldly—almost arrogantly—announced European conquest and the coming of European civilization to the New World. Symbolism and aesthetic taste can prevail even over comfort, for although this style is appropriate for the Italian climate, in New York it would have been impossible to keep warm. Furthermore, in Italy it would have been built of stone, yet here it was built entirely of wood masquerading as stone.

Figure 6-18 Russia's Czar Alexander II. This statue of Czar Alexander II (who ruled 1855–1881) was unveiled in Moscow in 2005, 14 years after the fall of Communism, which had replaced the monarchy in 1917. The government that created the statute honored the czar for having emancipated the serfs and achieved judicial and military reforms.

In societies that cluster housing, farmers may choose to live together in clusters ranging from a few dozen homes up to thousands. There are no dwellings in the surrounding farmland, so the farmers journey out to work in the fields each day. The farm buildings are usually concentrated together with the human settlement.

Such compact villages may be found in many forms: irregular; wandering along a principal street, river, or canal; clustered about a village common; or checkerboard (Figure 6-19). Clustering may reveal family or religious bonds, communal land ownership, or the need for common security against bandits or invaders. The government may deliberately cluster the population to supervise it or provide education or health care. Clustering may also have environmental reasons—people may cluster at water sources, for example, or on dry places when the surrounding land is swampy. Clustering is more common among farmers than among

(a) **Polsbroekerdam, Netherlands**

(b) **Holca, Mexico**

Building
Road
Crops
Wooded area

(c) **Finchingfield, England**

(d) **Ambohimandroso, Madagascar**

Building
Road
Crops
Wooded area

Figure 6-19 Village settlement patterns. The configuration of villages may reveal current or historical land uses. (a) Row villages line a common thoroughfare, in this case a canal, with back-lot farmland running perpendicular to the canal. (b) Checkerboard villages such as those in the Yucatan were laid out by the Spanish to ease the movement of horse-drawn wagons. (c) A village green, or commons, once provided a central pasture enclosed by houses. (d) Irregular villages have no clear center but include several small clusters of houses on hilltops overlooking farmland.

livestock ranchers, except in Africa, but it generally characterizes settlement across much of Europe, Latin America, Asia, Africa, and the Middle East.

Isolated farmsteads, by contrast, characterize areas of Anglo-America, Australia, New Zealand, and South Africa that were settled by Europeans, and also some parts of Japan and India. The conditions for isolated farmstead settlement usually include peace and security in the countryside; agricultural colonization of the region by individual pioneer families rather than by socially cohesive groups; agricultural private enterprise, as opposed to communalism; unit-block farms in which all of a farmer's land is in a block rather than in scattered parcels; and well-watered but well-drained land.

The history of the United States provides examples of both patterns of settlement. In the southern colonies in the seventeenth and eighteenth centuries, settlement was characterized by widely dispersed, relatively self-sufficient plantations. New England settlement, by contrast, reflects the social cohesion of the settlers' society. The people of New England were tightly bound in religious communities, and they advanced westward in tiers of adjacent, well-planned towns. The reason for the differences in the settlement patterns was not strictly economic; it reflected cultural preferences.

Figure 6-20 Barri Gòtic, Barcelona. European cities are notable for the presence of older cultural landscapes amid modern life. The Gothic Quarter in Barcelona includes buildings from the Roman and Medieval periods that line narrow streets busy with people moving through Spain's second largest city.

Forces That Stabilize the Pattern of Culture Regions

Despite the force of diffusion, a number of factors tend to fix the geography of culture regions. Culture leaves its mark on the landscape. The fixed pattern of activities, land uses, transport routes, and even individual buildings will guide, restrict, or predispose future patterns and activities. The construction of a factory, for example, represents a great investment of money, and once the factory is operating, it relies on a local workforce and develops ties to local suppliers. An industrial complex such as this cannot easily be picked up and moved. **Inertia** is the term for the force that keeps things stable. All of a people's fixed assets in place—railroads, pipelines, highways, airports, housing, and more—are called the **infrastructure**.

Historical geography is the subfield within geography that studies the geography of the past and how geographic distributions have changed. Historical geographers can sometimes read landscapes as if through time the landscape has been overlain with layer after layer of peoples using the land in different ways and organizing it for different purposes. The landscape is like an old manuscript on which a reader can discern earlier erased writings (Figure 6-20). The influence of past land uses was defined as *sequent occupance* studies by the geographer Derwent Whittlesey in 1929. An alternative approach to historical geography focuses on how the transformation, use, and organization of the landscape is continuously changing.

Culture includes a set of values and ways of doing things, and culture groups seldom get displaced or eliminated entirely. Culture is learned behavior, and cul-

tural norms are handed down through generations. Groups may be quick to take up new techniques or products, but tradition is a powerful force in human life, and imported ideas or lifestyles only slowly transform a people's entire cultural inheritance. This inertia is reinforced by the existence of territorially sovereign states and the enormous power that governments exert over their citizens to teach and to enforce norms of behavior. This power includes, as we have seen, the power to resurrect and promote folk traditions. Lee Kuan Yew, Senior Minister of Singapore, emphasized to an interviewer the role of culture in various Asian countries: "Culture is very deep-rooted; it's not tangible, but it's very real: the values and perceptions, attitudes, reference points, a map up here [he tapped his head], in the mind."

Today each local culture is a unique mix of what originated locally and what has been imported, and each unique culture is a local resource, just as surely as the minerals under the soil and the crops in the fields.

Most peoples value their culture, and they usually try to preserve key aspects of it. This influences the way they interact with other peoples. A people's self-consciousness as a culture may be codified in their religion or in their sense of their own history. The Jewish religion, for example, uses scriptures that refer to Jews' historical homeland. Jews also remember the persecution of their ancestors, most recently during the Holocaust. These are examples of a **historical consciousness.** Many Americans have difficulty understanding other people's historical consciousness, because one aspect of American culture is an optimistic denial that history can shackle future opportunity. Other peoples, however, nurture their traditional cultures and their historical consciousnesses. Most people, after all, closely pattern their parents' behavior, and neighbors may also impose expectations drawn from tradition.

One major theme in geography is the tension between forces of change and forces for stability. Cultures evolve, cultures diffuse, and peoples can transform themselves and their behavior, but cultures and culture realms also have elements of stability. Any cultural pattern or distribution maps the current balance between those forces.

Trade and Cultural Diffusion

Cultural isolation is usually accompanied by economic self-sufficiency, but trade diminishes people's cultural isolation at the same time as it expands their economic possibilities. Trade also leads to cultural diffusion. Every item in trade is a product of the culture that originates it, so economic exchange is one of the most important ways that ideas, beliefs, and practices circulate through space and time. Choosing goods is an act of self-definition, of social and cultural identity (Figure 6-21). Trade, economics, and culture are intertwined.

The study of how various peoples make their living, how economies develop, and what peoples trade is **economic geography**. Trade releases people from dependence on their local environment. It allows them to draw resources from around the world. Fewer and fewer people anywhere rely on their local environment for all of their needs, so the link between the resources of any local environment and the well-being of the

people who live there has been weakened. The Swiss, for example, live in an environment that is poor in natural resources, yet the Swiss have grown rich through trade and services.

As peoples come into contact with others and begin to trade, they usually first export only whatever they have in surplus and have no use for. They usually view imports as luxuries unnecessary to their way of life. Trade, however, triggers far-reaching cultural and economic changes.

For example, imagine three isolated communities in three different environments (Figure 6-22). Village A is located along a river plain. The people there catch fish, which they fry; they grow rice, which they eat as rice cakes; and they distill rice into sake (rice wine). Village B lies on the slope of nearby hills. The people there have domesticated grapes, which they have learned to distill into wine, and oats, which they make into oat bread. They have also domesticated goats, which clamber about the rugged hillsides. They milk the goats for dairy products, and they roast goat meat. The people in village C, up on a nearby plateau, grow corn for cornbread and to be distilled into whiskey and have domesticated cattle, which they eat as broiled steaks. Each village has developed other aspects of a unique culture, too, including its own language, religion, and customs, but let us focus on their diets as representative of their cultures.

The construction of a road and the commencement of trade among the three villages will probably trigger at least four significant changes:

1. Each village will have access to the products of each of the other villages. The people of village A, for example, will taste broiled steaks, and the people of village B will be introduced to fried fish. We might call this the simple *addition of cultural possibilities.*

Figure 6-21 New consumer goods change cultures. An Avon salesperson demonstrates a product to Tembe tribesmen in Brazil's Amazon region. The tribesmen obviously have their own cultural tradition about makeup, but the introduction of new cosmetic products will change these people's culture in ways that cannot be foreseen.

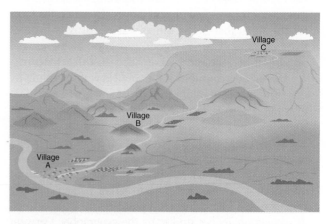

Figure 6-22 Isolated cultures yield to interaction. The inhabitants of three isolated villages in three different environments learn to use local resources, and they develop three distinct cultures. When a new road connects those villages, however, the lifestyles of all three villages may be transformed. Their cultural possibilities will multiply, and their economies will change.

2. New cultural combinations, or *cultural permutations*, will appear: fried steaks, broiled fish, roast beef, and more. A good example of cultural permutation is the appearance of pita fajitas on menus in Southern California. A pita fajita is a piece of pita bread, an item brought to southern California by immigrants from the Near East, stuffed with fajita ingredients, usually beef or chicken, which were introduced to Mexico by the Spaniards long ago and brought to Southern California by Mexican immigrants. Nobody knows who first devised this permutation. Today, many U.S. McDonald's restaurants offer sweet-and-sour sauce, which is of Chinese derivation, with Chicken McNuggets. People often find surprising new uses for objects created by other cultures and societies, even when they do not grasp the fundamental values or technologies that created the objects in the first place.

   Over time, all aspects of the cultures of all three villages will experience these two results—cultural addition plus cultural permutation. Residents of one village may convert to the religion of another, or perhaps the religions will blend. The languages, styles of architecture, music, games, clothing, and other customs of all three villages may add up and also permutate. Those who profit from a new trading system and exchange with other cultures often challenge and overturn local politics and traditions.

3. The residents of all three villages will see how imports can raise their standard of living. In order to pay for the imports they want, they will dedicate more effort to *producing items for trade—that is, for markets*. The change from self-sufficiency to production for markets is one of the greatest transitions in history. The existence of trade and markets gives each village, for the first time, an incentive to produce surpluses of its local goods. Agronomists (economists who specialize in agriculture) say that, in general, "The market produces the surplus." In other words, if farmers have a market where they can sell their surplus for a profit, they are likely to produce a surplus. Furthermore, when people specialize in producing an item, they become more efficient, and the quantities they produce increase. Therefore, the total amount of food—of all goods—produced in all three villages will probably increase.

   Production for the market affects cultures in another subtle way: The people may produce less of those local goods that cannot be exported, but more of those local goods that win broader markets. Traditional local items may be altered in order to increase exports. For example, many Native American tribes once made traditional craft items such as baskets, jewelry, and clothing for themselves only. Then they began to produce these items for tourists, and today they concentrate on producing those items and styles that tourists favor. They even alter traditional designs

if requested to do so by customers. Thus, people may slowly change their traditional culture.

As trade multiplies, more of what people produce in any one place is consumed elsewhere. Conversely, a growing percentage of the things people use are produced elsewhere. Eventually, if the terms of trade are favorable, people dedicate most of their efforts to producing export products, and they rely on imports even for their necessities. They have surrendered their self-sufficiency and become dependent on trade, and they surrender their cultural isolation and experience cultural change.

4. Village B will probably develop a market larger than those in villages A or C. This is because of its situation between the two other villages. Village B is the most convenient place in the pattern or network of exchange, called the *central place*, so it will probably grow to be the largest.

This scenario describes what happens when relatively isolated places open up to trade with their nearest neighbors. The growth of a global trade network linking small villages with the rest of the world took a long time to emerge. It developed as part of the modern world since 1500 in conjunction with colonialism and the migration streams described in Chapter 5.

World Trade and Cultural Diffusion Today

Chapter 12 examines world trade in detail, but we must note here that virtually all peoples have today experienced this evolution from self-sufficiency and cultural isolation to trade and cultural exchange. The share of any country's territory that is devoted to export production may be small, but the shares of the national population and the national income that are involved are rising everywhere. Even within individual countries, growing cities create markets for food from surrounding rural regions. This draws the rural population out of subsistence agriculture into commercial agriculture.

Regions or peoples have not always freely chosen to enter into the system of production and exchange. Some areas were forced into specialized production by colonialism. Peoples in other regions were forced to buy goods, and they had to develop exports in order to pay for these goods. Less coercive but no less effective are the intense marketing campaigns that associate products such as electronic equipment, foods, and fashion wear with a sense of material and social well being. These consumer goods assume importance beyond their strict utility value—what they do—and become culturally meaningful as status objects. As people demand the availability of more such goods, they become more deeply enmeshed in the web of trade and circulation.

Trade leads to specialization of production and to greater production, but it does not benefit all regions equally. Some regions prosper, and others fall behind.

Why international trade causes this to happen is a principal subject of research in economic geography. Many different reasons will be suggested throughout this book.

We noted earlier that items of popular culture achieve different market penetration throughout the United States, suggesting differing regional cultures. Similarly, the international marketing divisions of large corporations study why some products have worldwide appeal, whereas others are successful only in geographically restricted markets. Salespeople strive to break down cultural differences, but the integrity of each culture realm resists complete homogenization. Therefore, cross-cultural advertising and marketing of consumer goods present fascinating cultural–geographic questions. For example, Domino's Pizza has spread around the world, but the corporation has learned that Germans prefer smaller pies than North Americans do, and Japanese like pies topped with squid and sweet mayonnaise.

Wal-Mart, which is by some measures the largest and most successful corporation in world history, retreated from both Germany and South Korea in 2006, losing billions of dollars. The company opened in poor locations and was unable to meet consumer preferences in a market already dominated by domestic discount stores. Company executives admitted that they simply could not understand German and South Korean consumer cultures. Some German customers reported that they were confused by smiling cashiers who bagged their purchases—neither is common behavior for German clerks.

A few consumer products, such as Coca-Cola, have achieved almost global diffusion. Its appeal extends beyond taste. American soft drinks are popular in even the poorest countries (Figure 6-23). They are small and therefore more affordable artifacts from an exotic world. Heavily marketed American soft drinks are associated with hip and fun Western lifestyles, with movie stars and being "cool"; consuming this product conveys status and establishes a bond with those images.

Soft drinks are not popular everywhere, including in the state of Utah. Many of Utah's devout Mormons avoid caffeinated drinks and have not developed a preference for the decaffeinated brands. Similar motivations and geographic patterns are evident with cigarette smoking.

Consumer goods are only one aspect of culture. In some places, people wear American blue jeans, but they wear them to political demonstrations against the United States. They like blue jeans, but that does not mean that they like everything about U.S. culture. In fact, when popular culture from one country overwhelms the folk culture of another, this process may be viewed as offensive cultural aggression. We will examine below how this process feeds some people's animosity against the United States.

The Acceleration of Diffusion

In the past, travel and transportation were more difficult and expensive than they are today, whether we measure the cost in time, money, or any other unit. The friction of distance was so high that only a few things moved far, and those things moved slowly. Over the past 200 years, however, technology has reduced the friction of distance and accelerated the diffusion of cultural elements (Figure 6-24). We often hear that the world has "shrunk." As the cost and time of moving almost anything— people, food, energy, raw materials, finished goods, capital, information—have steadily fallen, things are not necessarily so fixed in place as they were in the past. Many activities have been significantly released from the constraints of any given location. As we will see in Chapter 12, these changes have allowed economic activities to

Figure 6-24 Container ships diffuse culture. The introduction of standard-sized shipping containers was a revolution in ocean freight delivery. Here, a container ship passes under the Golden Gate Bridge on its way to the Port of Oakland. In 2007, a similar ship hit the bridge, itself a cultural icon.

Figure 6-23 A global product. Some consumer goods practically blanket the Earth.

overcome the inertia that traditionally defined them. This key aspect of globalization has meant many industries have relocated to take advantage of other savings.

Originally, information could move only as fast as a person could carry it, but electronics disengaged communication from transportation. When Samuel Morse demonstrated the first intercity telegraph line in 1844, a Baltimore paper wrote, "This is indeed the annihilation of space." The telephone, invented in 1879, furthered the annihilation of space by allowing a person, figuratively, to be in two places at the same time. The annihilation of space has continued with computer modems, fax machines, electronic mail (e-mail), text-messaging, and social networks such as Twitter and Facebook. The Internet is at the center of the revolutionary changes in communication technology, but other **electronic networks** are also important. More people with personal electronic devices are plugged in everywhere with increasing regularity (Figure 6-25). Schoolchildren with personal computers can tap into networks of information that were unavailable to the world's leading scholars 25 years ago. Today the cost of global communication is a fraction of what it was even a decade ago. Between 2000 and 2006, the monthly cost of leasing a line for phone calls and data transmission from Los Angeles to Bengalooru, India (formerly Bangalore), fell from almost $60,000 to under $10,000.

pMaster™
ered Thematic
rld: Economic
ular Phones

The pace of technological change can be immediate. The advantages of mobile (cellular) phone technology have meant that some areas that never had good landline connections have become suddenly awash in personal phone service. Remote villages from northern Europe and villages in West Africa have been connected to the rest of globe overnight.

The compression of space compresses time. The latest music heard on radio stations in Los Angeles and New York will be heard in Nairobi, Kenya, and Montevideo, Uruguay, before the week is over. Hollywood blockbusters are frequently pirated and on sale in Eastern Europe before they've been released in U.S. theatres.

The activities of communicating with other people through an electronic network or even of playing a game alone with a computer create a new mental world—a new "place" where that activity is occurring, called *virtual reality*. That extension of reality through global electronic means of communication is **cyberspace,** a word coined by William Gibson in his 1985 science fiction thriller *Neuromancer*. The term may be applied widely. Many office workers, for example, collaborate through electronic networks without sitting down together in one office. One worker may be at home tending children, a second in a car on the road, and a third in an airport waiting room. Some people interact virtually on sites like Second Life, which presents users with a virtual world. Networks connect them in a virtual office in cyberspace, so office work is increasingly mobile.

Despite the technological wonder of global communication, not everyone in the world has equal access to new information technology. Africa's digital networks are still way behind demand. Only about one percent of Africans have high-speed internet access, compared to almost a quarter of Americans and 30 percent of Europeans. This **digital divide** has important ramifications for the diffusion patterns in and out of Africa.

Even among those with access to the rapid diffusion of world news and culture, most continue to be selective in what they pay attention to (Figure 6-26). Our own backgrounds; our education, perceptions, and prejudices; and the media to which we are exposed affect our understanding of other peoples and places and how we rank their relative importance. People remain most closely interested in local and national concerns, such as their state government or major domestic news. Language is one of the major impediments to true global communication. This is often cited as one reason that English is such a popular second-language among younger people around the world. This adds to a growing perception that global diffusion technologies are a one-way transmission of culture from the United States to other places. When was the last time you read a Chinese or Arabic website?

The Challenge of Change

Maps of culture areas, regions of economic specialization, or political jurisdictions reveal current distributions of specific human activities. They are a snapshot of human activities. Yet humans are dynamic, continuously

Figure 6-25 At home anywhere? The widespread adoption of mobile phones has allowed people to disengage from the world immediately around them. Family meals, highways, classrooms, even vacation spots are full of people distracted by what is going on somewhere else.

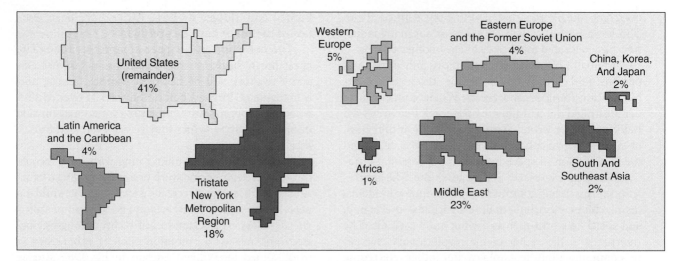

Figure 6-26 Relative world coverage in one newspaper. The relative sizes of world regions on this illustration correspond to the coverage devoted to each on the front page of the *New York Times* for two months in 2006. Clearly, editors select stories that reflect readers' interests, beginning with news about the United States and New York. The large amount of coverage of the Middle East reflects not only the war in that region, but also that New York's large and influential Jewish population seeks news about Israel. Some suggest that the decline of print newspapers is due to the medium's need to apportion limited content space among topics that will sell to specific audiences.

organizing and reorganizing, forming and reforming the world around them. These activities remake the places they inhabit and the larger spaces in which they operate. Distributions and patterns are disrupted and reshaped repeatedly, just as the patterns in a kaleidoscope are. A map of human activities at any time is comparable to a weather map. We cannot understand the activities if we do not know what forces are at work. For geographers to answer the question "Why there?" we must study activities and forces that overcome geographic inertia.

The geography of global manufacturing, which we will examine in detail in Chapter 12, exemplifies an activity that is redistributed continuously. The factors that redistribute manufacturing include new products; new technologies; new raw materials (the replacement of metals with plastics, for example); new sources of traditional raw materials; new technologies of manufacture; new governments with new policies regarding investment, taxation, or environmental preservation; growing and shrinking labor supplies; and the opening and development of new markets. When the executives of global corporations decide where in the world to build new factories, they balance these and still more factors: Each of these factors changes every day!

This textbook will detail many redistributions: Some religions are winning new converts, expanding geographically, and exerting new influence in world affairs, whereas other religions are withering away. Some languages are demonstrating the flexibility to adapt to new communication technology and are therefore gaining users at the expense of other languages. The relative rise and fall of nations' influence in world political or economic affairs affects how people view their

products, cultural artifacts, or lifestyles. Former French President François Mitterand said, "History has accelerated." So has geographic diffusion. Understanding how geographic diffusions have changed over time helps us to identify the processes that continue to reshape our world.

The Global Diffusion of European Culture

Despite the rich variety of indigenous local cultures around the globe, the world is increasingly coming to look like one place. In consumer goods, architecture, industrial technology, education, and housing, the Western model is pervasive. To ethnocentric Westerners who presume the superiority of their own culture, or to local people who have accepted Western culture, this may seem natural. To them this may be "modern" life, or "progress," or "development."

A neutral observer, however, might expect more diversity, more styles and models of development. Why are so many people around the world imitating Western examples and adopting aspects of Western culture? In much of the world, acculturation to the Western way of life is rapidly replacing both the positive and negative features of other cultures. It is the most pervasive example of cultural diffusion in world history. It illustrates all types, paths, and processes of diffusion that we have studied, and therefore an understanding of European cultural diffusion over the past 500 years is essential to understanding the world we live in today. In fact, much of what we today understand as globalization is a product of this history, though that may be changing in radically new ways.

Europe's Voyages of Contact

Europeans came to play a central role in world history and world geography because it was they who paved the way for the modern system of global interconnectedness. In the fifteenth century the great cultural centers of the world—the Inka and Aztec empires in the Western Hemisphere, the Mali and Songhai in Africa, the Mughal in India, Safavid Persia, the Ottoman Turks, the Chinese, and all of the lesser empires and culture realms—were still largely isolated from one another.

The European voyages of exploration and conquest connected the world (Figure 6-27). It was not inevitable that the Europeans would be the ones to do this. The Chinese were actually richer and more powerful than the Europeans, and earlier in the fifteenth century, they had launched fleets of exploration that had reached the east coast of Africa. We can hardly imagine how different world history and geography would be if the Chinese had continued their initiatives and gone on to explore and conquer Africa and the Western Hemisphere. But they did not. Instead, the Chinese government focused on internal affairs, and the Europeans continued with world exploration and conquest.

The first European initiative was Prince Henry of Portugal's conquest of the city of Ceuta on the north coast of Africa in 1415. He learned about trans-Saharan caravan routes and the riches of West Africa, and thus was inspired to finance studies of naval skills. Improvements in sailing technology enabled the Portuguese to sail down the west coast of Africa to reach beyond the Sahara Desert. Prince Henry, known as "The Navigator," launched the era of European seaborne colonial empires. Soon Spain, Holland, France, and England were racing to secure colonies. In contrast to these seaborne empires, Russia at the same time forged eastward overland from the Russian homeland in Europe west of the Ural Mountains. The Russian Empire pushed across Siberia to the Bering Strait, crossed over into Alaska, and eventually established colonies down the North American West Coast as far south as today's California.

This European outreach triggered the **Commercial Revolution**, between about 1650 and 1750. The development of the first oceangoing freighters that could carry heavy payloads over long distances allowed a tremendous expansion of trade. The evolution of superior ships was paralleled by the evolution of superior naval gunnery and of additional useful technologies such as clocks, which were perfected in order to determine longitude at sea. This first era of outreach ended with the death of the British explorer Captain James Cook in 1779. By then, Europeans were the first people in world history who could draw a fairly accurate outline map of all the world's continents and major islands.

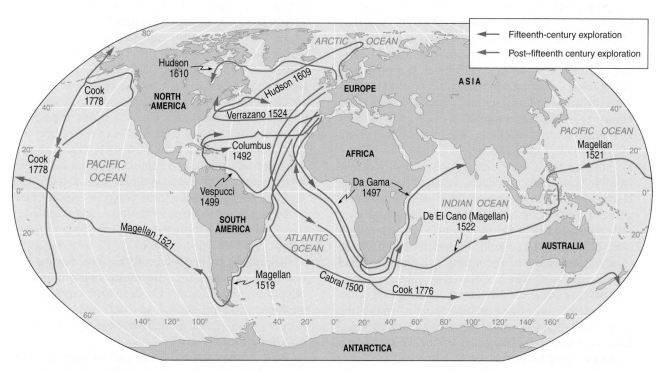

Figure 6-27 Major European voyages of exploration. Each of the European voyages of exploration was a daring enterprise. The Portuguese first established a string of bases around the coasts of Africa and Asia all across to Japan, where they established a trading post in 1543. Other European nations followed, and the race for discovery of new riches, of new converts to Christianity, and for help against Christianity's powerful Islamic foe was launched.

Expansion and cultural diffusion From the time of exploration, Europe did not actually originate every "modern" idea and then impose it on the rest of the world. Europeans learned from others, too, and then transplanted around the world the ideas that they had adopted from elsewhere. "Knowledge is power," said English philosopher Sir Francis Bacon (1561–1626), and because it was the Europeans and not the Chinese, the Inka, or any other group that had contacted all other civilizations, Europe became the clearinghouse of world information and products. Global diffusion was fixed hierarchically, with Europe as the apex. Europe became cosmopolitan—that is, familiar with many parts of the world—as it gained colonies. Other peoples, no matter how great their native civilizations, were pulled into the emerging Europe-centered trade network.

Chapter 5 described the migrations of peoples triggered by European expansion. This relocation diffusion brought people and cultures into contact, exchanging cultures and goods. Consider, too, what Europeans did with agricultural products. They took sugarcane from Asia and planted it in the Caribbean region. They replanted bananas from Southeast Asia to South America, cocoa from Mexico to Africa, rubber from South America to Southeast Asia, and coffee from Arabia to Central America. All these foods remain major products of international trade and important factors in the export economies of many countries.

Europeans not only relocated the production of many goods around the world, they also introduced many products into world trade, both for the European home market and for other overseas markets that they pioneered. Europeans introduced many Indian goods, for instance, into China, and South American products into Africa and Asia. An Englishman introduced tea plants from China into India, leading to the development of a tea industry in India and Ceylon (today known as Sri Lanka). Europeans created world markets and profited by controlling every stage: production, transportation, and marketing.

The story of Coca-Cola, one of the world's most familiar consumer products, exemplifies cultural blending. The drink was formulated in Atlanta, Georgia, and today symbolizes westernization so powerfully that the large-scale infusion of Western products into the non-Western world is often referred to as the "Coca-colonization of the world." The two original ingredients from which Coca-Cola takes its name, however, are coca, a Native South American Quechua word for the tree whose leaves supply a stimulating drug (used today to make cocaine and crack), and cola, a word in the language of the West African Mandingo people for the nut that supplied the other original stimulating ingredient. Westerners borrowed the knowledge of both of these ingredients from their far-flung hearth areas, combined them, and marketed the drink worldwide.

Thus, modern world culture is not exclusively a Western product. It is presumptuous for Westerners to think that it is. Even non-Westerners, however, often fail to see the many contributions of non-Western peoples because the route of global diffusion was through the Western powers.

Economic Growth Increased Europe's Power

Europe pulled ahead of the rest of the world economically as it underwent the tremendous transformation of the **Industrial Revolution.** Between about 1750 and 1850, Europe evolved from an agricultural and commercial society to an industrial society relying on mechanical power and complex machinery. We cannot fully explain why Europe experienced this transformation before any other part of the world, but we can list several factors that enabled Europe to industrialize. Europe's voyages of discovery and conquest resulted in an influx of precious metals and other sources of wealth that stimulated industry and a money economy. The expansion of trade encouraged the rise of new institutions of finance and credit. In the mid-sixteenth century the joint stock company was developed, allowing many investors to share both potential profit and risk in new enterprises. The creation of stock markets where stocks could readily be bought and sold granted capital new **liquidity,** which is easy conversion from one form of asset to another. This created, in the words of English writer Daniel Defoe (1660–1731), "strange unheard-of Engines of Interest, Discounts, Transfers, Tallies, Debentures, Shares, Projects."

In 1769, James Watt designed the steam engine, which multiplied the energy available to do work. Subsequent inventions and technical innovations in manufacturing, applied first to textiles and then across a broad spectrum of goods, dramatically increased productivity. Factories and industrial towns sprang up, canals and roads were built, and later the railway and the steamship expanded the capacity both to transport raw materials and to send manufactured goods to markets (Figure 6-28). New methods of manufacturing steel, chemicals, and machines played important parts in the vast changes. These innovations occurred first in Great Britain and subsequently spread to continental Europe and to North America.

Beginning in the eighteenth century, Europe also first experienced an **Agricultural Revolution.** This development, to be examined in detail in Chapter 8, both increased food production and released agricultural workers from the land, thereby creating a supply of labor for industry.

As a result of the Industrial and Agricultural revolutions, Europe and European settlements around the world drew far ahead of any other places and peoples in their productive capacity. As recently as 1800, the per capita incomes of the various regions of the world were

Figure 6-28 The first iron bridge. This is the world's first iron bridge, built over England's Severn River in 1779. It demonstrated iron's strength, and the bridge's bold design invited other uses for the new material. Its builder, John Wilkinson, launched the first iron boat in 1787, and when he died, he had himself buried in an iron coffin.

similar. If we index the Western European per capita income in the year 1800 as 100 units of wealth, then estimated per capita incomes in North America were 125, in China 107, and in the rest of the non-European world 94. By 1900, however, European and North American incomes were several times those of non-Western peoples.

Commercial contacts and economies At the beginning of the age of the European voyages, European demand for foreign products such as spices, sugar, fruits, and North American furs grew rapidly. Soon the Europeans were no longer content to trade with native peoples for these goods, and the Europeans themselves established overseas estates and plantations and applied large-scale techniques to specialized production.

European commercial plantations were at first concentrated along the coasts, but in the nineteenth century the railroad allowed penetration of the continental interiors and created access to superior agricultural lands or, later, as Europe industrialized, to mineral deposits. The world's railway network expanded from 200,000 kilometers (124,000 miles) in 1870 to over 1 million kilometers (621,000 miles) by 1900. European treaty ports (ports that by treaty had to be kept open for trade) and coastal footholds became inland empires (Figure 6-29). The development of the steamship also facilitated the transport of minerals, and increased quantities of minerals supplied Europe's multiplying factories. Between 1840 and 1870, the world's merchant

shipping rose from 10 million tons to 16 million tons, and then it doubled in the next 40 years.

New cities emerged in the non-European world as coordinating centers for these commercial activities and new ports sprang up along the seacoasts. The major seaport cities from the Straits of Gibraltar around Africa, across the Indian Ocean, and all throughout South Asia are the products of European contact (Figure 6-30). The same is true for most of the port cities of the Western Hemisphere. The railroads and associated commercial economies at first affected only a small percentage of the population, but over time an increasing share of the population and territory were drawn into the emerging global economy. In some countries, however, such as India, the modern commercial economy still overlies a traditional subsistence economy. The interaction between these economic activities is often unique and unexpected. For example, urban technology workers may travel to ancestral villages to help with the harvest, with all that implies for the exchange of cultural and material goods.

Political conquest In two waves of exploration and conquest—the first extending from 1415 to 1779 and the second occurring at the end of the nineteenth century—Europe (and, in the second period, the United States) dominated most of the rest of the world. European countries divided up the rest of the world in part to avoid expensive conflicts over land within Europe. Another reason for imperial conquest was the European nations' wish to protect their commercial investments in foreign lands and to control these lands as markets for themselves. Their ascendancy over the indigenous populations was guaranteed by their superior military power.

The United States and most of Latin America won independence between 1775 and 1825, but between 1875 and 1915 about one-quarter of Earth's land surface was distributed or redistributed as colonies among a half-dozen imperialist states. Of all the countries in the world today, the only ones never directly ruled by Europeans or by the United States are Turkey, Japan (although it was occupied by the United States from 1945 to 1952), Korea (which was ruled by Japan from 1910 to 1945 and remains split today), Thailand (left as a *buffer state* between the French and English empires), Afghanistan (a buffer between the English and Russian empires), China (which was nevertheless divided into foreign "spheres of influence"), and Mongolia (ruled by China).

Therefore, in most countries, European cultural attributes linger as a legacy of European rule and still predominate or mix with native pre-European traditions. European concepts of law, for example, drastically changed native societies, especially ideas of property rights and land ownership. Before the Europeans came, land was generally considered a good that was held in common for all members of the community. Local

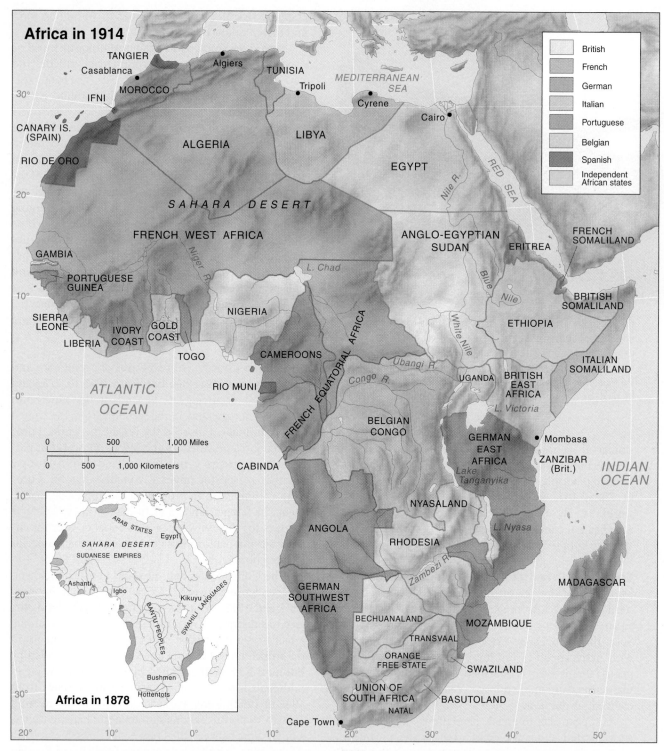

Figure 6-29 European partitioning of Africa. Europeans divided the African continent in order to prevent competitive war among themselves. British Prime Minister Lord Salisbury admitted, "We have been giving away mountains and rivers and lakes to each other, only hindered by the small impediment that we never knew exactly where they were." Treaty ports and trading stations along the African coast expanded into vast inland empires. Industrial Europe demanded African raw materials, and railroads allowed the Europeans to draw them out of the African interior. The native African peoples were not consulted in the political reapportioning, which was completed at a conference held in Berlin in 1884–1885.

political leaders apportioned land use and occupation by customs that brought the community together. Neither the leaders nor anyone else *owned* land. The idea that any individual could own land and single-handedly determine how to use it was largely unknown.

Europeans introduced their idea of "ownership" of an "estate" that could be bought, sold, or mortgaged by individual contract. Social cohesion was dissolved when land was no longer a common good and the regulation of its use was no longer a shared community affair.

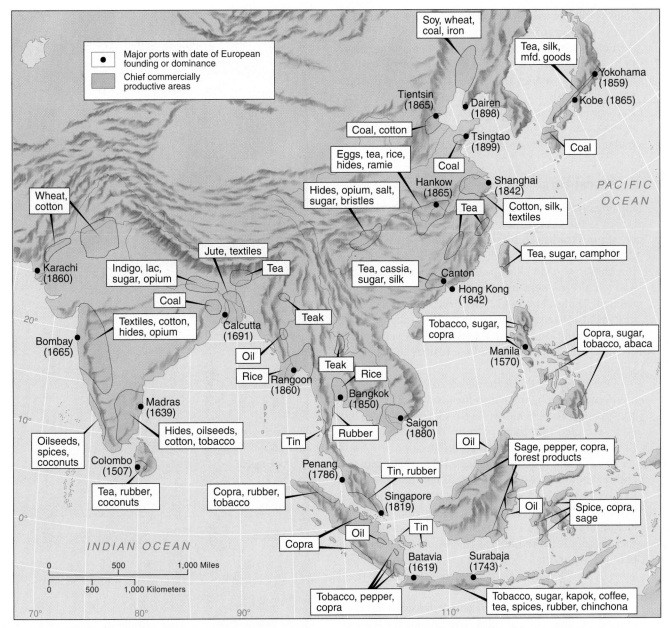

Figure 6-30 European intervention in Asia. Many of Asia's port cities were at first spigots that Europeans tapped into the continent to draw off Asia's wealth. As happened in Africa, however, the ports expanded into plantations for the production of goods valuable in international trade, and then the plantation regions grew into political colonies.

Native American chiefs, for instance, did not by their own customs have the right to transfer land out of tribal control, and they frequently did not actually understand what Europeans meant when they "bought" Native American lands. In the history of the United States, innumerable wars were sparked when Native Americans returned to hunt or harvest unoccupied land that the Europeans insisted the Natives had sold. Throughout Africa and Asia, the Europeans often simply assigned ownership to the local political leaders. This ended traditional egalitarian systems and created new classes of rich and poor. The descendants of many of these newly enriched leaders remain great landholders today throughout the Near East, for example.

Many traditional societies crumbled under the transformation of land from a public asset into a private commodity.

European law tended to transform labor into a commodity, too. Traditional communities were not idyllic; they restricted individual liberties, and slavery and serfdom were not unknown. These constraints, however, were often balanced by strong webs of responsibilities and rights that usually kept anyone from being entirely outcast and starving. The European idea of a self-regulating market for individual labor is more abstract and impersonal. It cut traditional ties of both rights and responsibilities. In addition, Europeans required the use of money as a universal measure of value.

Natives were forced to work for wages or to sell goods for money in order to pay taxes.

Europeans also brought their forms of administration, government, centralized state authority, written arrangements, uniformity, secularization, economic planning, public accounting and treasury control, central administration, and decision making. In many cases the civil services that the Europeans left behind remain the pride of new nations, as in India.

Cultural Imperialism

European rule was marked by **cultural imperialism**, which is the substitution or subordination of one set of cultural traditions by another, either by force or by degrading those who fail to acculturate and rewarding those who do. Europeans seldom doubted that native cultures were inferior and that native peoples needed "enlightenment." Therefore, the Europeans destroyed other ways of life, including religious and political traditions, physical artifacts such as art and architecture, and even records of history and science (Figure 6-31).

One reason for this is the nature of Christianity. It is a proselytizing religion, which means that its adherents try to convert others to their faith. The natives in many areas accepted that there is truth in all religions, and their toleration opened them to acculturation.

In addition, Europeans believed that their military and technological superiority presumed European superiority in all other aspects of life. Europeans did learn some things from those they conquered—farming techniques from Native Americans, for example—but they did not learn as much as they could have learned, because they often failed to appreciate the values, science, and technology of the civilizations that they conquered.

European cultural imperialism began with the systematic training of local elites. The missionary schools produced converts who proselytized among their own people, helping to eradicate the local culture. Later the government schools turned out bureaucrats and military officers who helped govern their own people. A second channel of transmission was *reference group behavior*. People who wish to belong to, or be identified with, a dominant group often abandon their traditions in order to adopt those of the dominant group. The returning slaves who carried the first wave of westernization to West Africa wore black woolen suits and starched collars in the tropical heat. In India, the native officer corps imitated English officers, complete with waxed mustaches.

Local elites also adopted Western ways because they were made to feel ashamed of their color and their culture. The rulers' racism and cultural imperialism meant that natives could succeed only by adopting the whites' ways. The later autobiographies of the new national leaders of every African and Asian country

Figure 6-31 A Mayan book. This page is from a book called the *Dresden Codex*, one of only four books known to survive from the considerable libraries of Central America's Mayan civilization. The rest were burned by the Spanish. This book is thought to be the oldest surviving book written in the Americas. It contains astronomical records for agricultural and religious purposes.

recount racial humiliation (Figure 6-32). The colonial school systems implanted in children's minds an image of the power and beneficence of the "Mother Country." Children in the Congo, for instance, knew more about Belgium than about their own land and

Figure 6-33 Self-westernization. In this historic photograph, Turkish president Mustafa Kemal Ataturk, whose surname means "father of the Turks," is teaching the Roman alphabet in an Istanbul public park. Ataturk recognized the technological superiority of the West, and he consciously turned his people toward westernization. One of his changes was the transformation of the writing of Turkish from the Arabic into the Roman alphabet.

Figure 6-32 Imperialism in the movies. This is a scene from the 1935 British movie *Saunders of the River*. In this Eurocentric and racist movie, Commissioner Saunders, a British imperialist officer, on the left, supervises and keeps peace among tribes in a British African territory. The tribes are depicted as dangerous children. In this key scene, Saunders prods a submissive native in the chest with his cane while lecturing on the superiority of British civilization. The native character was portrayed by Jomo Kenyatta, who was then a young actor but who later served as president of Kenya. What could Kenyatta have been thinking when this scene was filmed? What could he have thought years later while viewing the film in the presidential palace of independent Kenya?

peoples. Cristopher Monsod, the chairman of the Philippines Election Commission, remembers similarly having been taught from U.S. textbooks and says, joking, "Thanks to my American education, I know the capitals of North and South Dakota." History textbooks used in French African colonies opened with the words, "Our ancestors the Gauls. . . ." Schooling focused on each colony's ruler, so the native peoples scarcely knew that other countries existed. Residents of the Congo, for example, referred to all whites as Belgians.

Colonial intellectual institutions usurped the power to define local tradition. For example, after Europeans founded the Asiatic Society of Bengal in 1784, Europeans' study of Indian art and literature defined Indian culture. Europeans certified what were to be considered "classics" of Indian culture. European archaeological surveys determined what monuments were fit for description and preservation as part of "the Indian heritage." Educated Indians learned about their own culture through the mediation of European ideas and scholarship.

Self-westernization By the end of the nineteenth century, the elites of the entire non-Western world were taking the Europeans as their reference group. Even the three major nations that were never colonized— Turkey, China, and Japan—were all militarily humiliated, and traumatized by it. All three had had empires of their own, and their defeats forced them to reconsider all the assumptions on which their institutions and daily lives were based.

Turkey, China, and Japan all preempted or co-opted Western civilization, and to some degree this saved them from Western political rule. In Turkey, Mustafa Kemal (1881–1938) seized power and forced the country to undergo self-westernization (Figure 6-33). Many historians believe that the challenges he faced, the solutions he devised, and the degree of success he achieved defined a prototype for the leaders of all nations that emerged from colonialism after World War II. In China, the Republican Revolution of 1911 attempted to modernize the country, but China's subjection to the West continued. Later, another leader, Mao Zedong, applied an alternative brand of westernization— although in a uniquely Chinese form. The Japanese were forced by the United States to open their society to

Western trade in 1854, after centuries of near-total isolation. The Meiji Restoration in 1867 launched the self-westernization of Japan: The Japanese decided to become thoroughly Western but to retain control of the process. Japan sent students to the United States and Europe and adopted Western science, technology, and even many cultural traits so rapidly and so successfully that by 1904 the country was able to defeat a Western power—Russia—in war (Figure 6-34).

Japan was later defeated in its attempt to extend its empire in World War II, but its early military successes discredited the Western powers and encouraged non-Western colonial subjects to dream of freedom. Jawaharlal Nehru (1889–1964), the first prime minister of independent India, wrote in his memoirs of how, in 1904, when he was 15 years old, "Japanese victories stirred up my enthusiasm. . . . Nationalistic ideas filled my mind. I mused of Indian freedom. . . ." Chinese revolutionary Sun Yat-sen recorded, "We regarded the Russian defeat by Japan as the defeat of the West by the East. We regarded the Japanese victory as our own victory." Despite Japan's military defeat in 1945, it later developed into one of the world's dominant economic powers, and it continues to provide a non-Western model for economic development without complete surrender to Western culture.

The period of European relations with Africa and Asia that began 500 years ago is ending. Only two tiny Spanish holdings survive on the African continent—Ceuta and Melilla. The Western empires in Asia have been surrendered except for a few islands still held by France and by the United States. Great Britain yielded Hong Kong to China in 1997, and in 1999, Portugal returned the Chinese territory of Macau, which it had held since 1557.

The West and non-West since independence   The fixation with the West among the elite in the non-European world did not end when these countries won political independence. Those who assumed power in Africa and Asia were mostly educated in the West, and they demanded independence by quoting Western political writers whom they had read at Western universities. They often expressed anger that the West did not achieve the universal values it espoused and had taught them—racial equality, for example, and the equality of all people before the law. Their reference groups remained entirely European, not their own people. Nehru himself was eventually to say to U.S. Ambassador J. K. Galbraith, "You realize that I am the last Englishman to rule in India."

Few of the Western-educated elites developed indigenous models of development, since they continued the diffusion of Western models. They started building in the middle of their commercial cities and went on building outward. The technical term for the favored cities is *growth poles*. The leaders hoped to convert their whole national territories to modern Western societies, but they did not realize how difficult it would be to shift their countries from exporting cheap raw materials to profitable complex economies. We will revisit this problem in Chapter 12.

By the time the colonial rulers left, the traditional sources of social status in the societies, such as religion, family, and customs, had faded. The new politicians,

Figure 6-34 A Japanese copy of Western technology.   This is a Japanese woodblock cutaway view of a German battleship that visited Yokohama in 1873. This woodcut would later provide a virtual blueprint for the technologically less advanced Japanese Navy to copy.

bureaucrats, and business executives defined their status in the only way they knew, which was to exhibit Western material goods. Thus their homes, clothes, and cars mimicked Western status symbols. Even today, the parliament of Kenya forbids its members to wear African clothing but requires European-style suits and ties. This mimicry may inspire corruption in some of the world's poorest countries; their elite citizens pauperize their own countries and funnel money to banks or to buy prestigious properties in the rich countries. They flaunt their wealth in Paris and New York, and they vacation in villas along the French Riviera and in great mansions in London.

In many countries, the new rulers have practiced a sort of internal colonialism on their own people. For example, the rulers who assumed power in Latin America following the independence movements of the early nineteenth century were of European stock, and their descendants still dominate this region. Peoples of Native American or African backgrounds have been treated as subject groups, forced to adopt European culture, religion, and language, and subjected to discrimination. The situation is not very different in Africa and Asia. The "colonizers" are generally the westernized elite; the "colonized" are all those who do not belong to this group, which is often the majority of the population.

Westernization Today

The diffusion of Western culture continues today. Western culture diffuses from the top of societies, from the examples and activities of the local elites and international stars. Young people also diffuse it by adopting Western dress and lifestyles as status symbols. The rich and the young are everywhere the most cosmopolitan consumers, and most consumer items in international trade are artifacts of Western popular culture.

Even schools have become instruments of westernization. Their syllabuses emphasize modern, urban activities and values. The young sometimes emerge oblivious to their traditional culture or even despising it in favor of Western popular culture.

The media reinforce the message. Western television programs, movies, the Internet, advertisements, and videos penetrate millions of homes and implant Western values. Night after night, on television screens around the world, the images of "the good life" are images of the life among the wealthy in the United States. This imagery has dramatically changed behavior. Chapter 5 described how popular media can affect even national birth rates. People want the Western goods that they think will bring them long-term satisfaction, or at least immediate enjoyment.

Western media have been accused of entirely supplanting traditional culture, at least among urban middle-class adolescents. Sumner Redstone, the head of Viacom, the parent company of MTV, has said, "Kids on the street in Tokyo have more in common with kids on

Figure 6-35 Television makes a smaller world. Kenyans in the Kibera slum of Nairobi gather around a small television to watch the inauguration of U.S. President Barack Obama.

the street in London than they do with their parents." Historian William McNeill agrees that the diffusion of television "is a very deep transformation of human life. I would rank what is happening now with man's transition from a hunter and gatherer into a settled farmer. Television has replaced inherited culture" (Figure 6-35). Very similar arguments are made about the Internet's diffusion of Western culture.

Tourism provides still another channel of westernization. Westerners are attracted to "different" and "unspoiled" places but change the places they visit simply by their presence. The **commodification** of culture can turn complex local traditions into commercialized spectacles and shoddy souvenirs. Many local people abandon their own material culture and adopt that of the visitors (Figure 6-36).

Global flows of professionals and of professional education are another powerful force for the diffusion of westernization. The rich Western countries export professional services, and people from around the world attend Western schools for professional education. The elite and professionals of most countries have been educated in Europe or the United States. These graduates, acculturated to Western ways, return to hold influential roles in their societies.

Western architects, civil engineers, and urban planners—or non-Western individuals educated in Western schools—are transforming built environments. Non-Western countries' cultural landscapes are increasingly "modern," and there is no prototype other than the Western. For example, the largest homebuilders in both Thailand (Anant Asavabhokhin) and the Philippines (Manuel Villar), two of the richest men in those countries, worked for Los Angeles homebuilder Kaufman & Broad, Inc., before returning to their own countries with blueprints to build homes for the rising middle classes. Some countries save "traditional" landmarks

Figure 6-36 New cultural influences. Wireless laptops, and the content they carry, can be found in small villages like Bhaktapur, Nepal.

only as tourist attractions. For example, the government of Singapore bulldozed historic Bugis Street, which offered a mix of old shops and small bars, but then re-created it for tourists (see Figure 5-3).

Countries are being transformed by world flows of capital investment, and capital is invested according to the standards of the societies that export the capital. This activity will be examined in Chapter 12.

Under this barrage of westernization, many traditional cultures and social structures are being radically transformed, and these transformations are not always in accord with the people's conscious wishes. In some cases, whole cultures disappear, and their disappearance reduces cultural diversity and impoverishes all of us.

America's Role

Understanding the role of the United States in the diffusion of Western culture requires historical and geographic explanation. The United States was an unlikely world power when it began as a string of European colonies. In time, its economic and political self-interest caused it to throw off British rule, which was a success in part because of its geographic isolation from European powers. Westward expansion grew the U.S. economy and attracted immigrants, making the country one of the largest in the world. The United States pressed its economic and political

interests in Central and South America, and later the Pacific, but it remained a second-tier power on the world stage prior to World War I. After two disastrous wars wrecked the European powers, the United States was left as the preeminent economic and military powerhouse of the Western world. The resulting cultural and political influence of the United States was challenged only by the Soviet Union, which disintegrated during the 1980s.

Today, the U.S. economy represents approximately one-fifth of the world economy, and most Americans enjoy a high standard of material comfort. Although the United States was never an imperial power like those in Europe, it profited greatly from international trade. It exercises enormous influence over international affairs and its interests are embedded in international bodies, such as the United Nations and World Bank. Therefore, some poor and oppressed people throughout the world may inevitably resent the United States, whether or not American profit-making is the cause of their own suffering. U.S. investment and U.S. products are found virtually everywhere, and, conversely, the United States is the greatest market for many other countries' goods.

The diffusion of U.S. popular culture   Many of the world's most recognized brand names are American, and American corporations pursue sales abroad aggressively. Many people view the marketing of U.S. popular culture—that is, American salesmanship—as the deliberate destruction of their own traditional folk culture and its replacement by American popular culture. For example, from a distribution center in Singapore, the Disney Company floods the countries of Southeast Asia with more than 16,000 Disney products. Likewise, as more McDonald's restaurants open worldwide, they are regarded as a symbol of U.S. influence (Figure 6-37).

Figure 6-37 Ronald McDonald as a substitute for Uncle Sam. These Pakistani protestors of American foreign policy wrecked a local McDonald's restaurant. Ubiquitous and conspicuous symbols of U.S. private enterprises are often targeted during anti-American riots, even if the local business is a locally owned franchise, as it was in this case.

GLOBAL AND LOCAL

The Diffusion of "News"

Western media dominate the gathering and dissemination of news to such an extent that most people in non-Western lands learn about the affairs of other non-Western lands—and even about their own national affairs—through Western media.

Alternative sources of global news are appearing. The year 2005 saw the launch of Telesur, a television network aimed to provide an alternative to U.S.-based news and analysis for Latin America. Telesur is based in Caracas, Venezuela, and financed by Venezuela, Argentina, Cuba, Bolivia, Ecuador, and Nicaragua. The U.S. government has criticized the network's news coverage as biased.

Al-Jazeera, a Qatar-based television news service, is expanding its global coverage. When founded in 2001, al-Jazeera provided only Arab-language broadcasts, but in 2006, it launched an English-language news service. British journalist David Frost accepted a position with al-Jazeera, arguing, "We in the West have been broadcasting our views to the non-Western world for many years. It is only fair that these non-Western areas should have the chance to return the compliment." Then-U.S. Secretary of Defense Donald Rumsfeld criticized the network as offering an "anti-American worldview," but then-Secretary of State Condoleezza Rice chose to engage the network and appeared as a guest. Al-Jazeera's independence startles even many Muslim governments. It criticizes them, for example, for their lack of democracy and subjection of women. These topics are avoided in the state-owned media monopolies in most Muslim states.

The competitors to Al-Jazeera include the more moderate Al-Arabiya, based in the United Arab Emirates and backed by Saudi Arabia. When President Barack Obama granted Al-Arabiya the first interview after his inauguration, many viewed his choice as trying to promote healthy competition among news agencies. America's own government-subsidized Voice of America (VOA) has won tens of millions of listeners in 45 languages. Its charter states that the service should be "a reliable and authoritative source of news" and that it should be "accurate, objective and comprehensive," but some American critics demand that it presents exclusively a pro-American point of view.

Television remains an important medium for diffusing news in non-Western countries, where television viewership is often widespread. The growing use of the Internet in these countries

News presenters on Al-Jazeera dress in Western clothing and women appear without veils. These news anchors broadcast from the network's Kuala Lumpur broadcast studio.

has also increased demand for online news content. Newspapers and news agencies have quickly developed websites but so, too, have the news television agencies like Al-Jazeera. These sites offer content in the local language, but many online readers also read English-language news sources. Interestingly, the competition for online readers has caused Western media to tailor their news sites to a more international audience.

U.S. popular culture incorporates U.S. cultural and political values, so it can challenge traditions and initiate cultural change even unintentionally (Figure 6-38). For example, the U.S. women's magazine *Cosmopolitan* seeks profit, not revolution, yet its championing of women's rights and sexual freedom is revolutionary in many of the 60 countries in which it is distributed.

U.S. dominance of the world's television market has shrunk as the market has grown, but the very format

of television shows—dramatic or comedic shows, for example—are an American cultural product. U.S. movies still capture about 85 percent of total global box office receipts, and U.S. studios have profited more from international audiences than from domestic audiences every year since 1993. Foreign films shown in the United States, by contrast, capture only about 1 percent of the U.S. box office. Perhaps to appease non-U.S. audiences, the 2006 film *Superman Returns* had the hero

Figure 6-38 The diffusion of American values. Indian Muslim women look at Valentine's Day cards at a shop, on the eve of Valentine's Day in Ahmadabad, India. Social conservatives in India have threatened violence against young couples who exchange Valentine's messages, seeing the holiday as eroding moral values.

fighting for "Truth, Justice, and all that stuff," rather than the traditional "Truth, Justice, and the American Way." Many people, including many Americans, cited the winner of the 2006 Academy Award for best new song, "It's Hard Out Here for a Pimp," as evidence of America's immorality and decadence.

Many peoples see this onslaught of U.S. popular culture as cultural imperialism. Americans, however, generally accept the idea of *the democracy of the marketplace*. According to this theory, consumers' choice in purchasing goods is comparable to voters exercising their right to vote. People "vote" by buying products or tickets to performances. Thus, the market democratically reveals what people want. People watch Hollywood's latest movie because they want to, not because they are forced to. Americans interpret actions limiting consumer choice as comparable to restricting citizens' civil rights.

No one accepts this analogy completely—Americans debate it among themselves—but Americans do generally accept it more completely than do many other peoples. Many other peoples argue more strongly that the marketplace does not always protect values that must be preserved for the good of a society. The marketplace may not protect tradition; it may widen disparities in income or opportunity, and it may fail to protect the general welfare above economic results.

U.S. trade representatives, accepting the theory of the democracy of the marketplace, argue that when a government limits cultural imports, its purpose is not to protect culture but to protect markets for local producers. Other peoples say that Americans "just don't get it." Some governments insist that they have the right to censor or ban

films or products in the name of their people. The Chinese government, for example, reserves the right to determine whether U.S. cultural products are "catering to the tastes of the Chinese people," and it bans many American films. To Americans, this is simply preemptive censorship.

The global dominance of U.S. popular culture has triggered a backlash. In Iran, for example, a paramilitary unit known as the *Basij*, "those who are mobilized," patrol the streets battling prostitution, drugs, alcohol, and atheism, as well as objects and values they interpret as imported from the West, including stereos, pop music, videos, lipstick, and indecorous dress among women. One *Basij* commander, Ali Reza Afshar, has said, "This war goes to the root of our existence. While physically there is no loss of life, our young people are being felled by cultural bullets, and this cultural corruption makes our young impotent to rebuild the nation." Many of those "cultural bullets" say "Made in the U.S.A." on them. The case of Iran, however, illustrates that those who battle American cultural influence may have an agenda they themselves wish to impose upon their people, rather than to allow true freedom.

Even other rich countries resent U.S. cultural diffusion. France requires theaters to reserve 20 weeks of screen time per year for French feature films. Australia demands that 55 percent of a television broadcaster's schedule be filled with domestic programs. Canada insists that 60 percent of television programming be Canadian. In 2005, the United Nations Educational, Scientific, and Cultural Organization (UNESCO; see Chapter 13) approved a Convention on the Protection and Promotion of Cultural Diversity that allows each country to exclude its cultural policies, including media, from trade agreements. Governments also may use subsidies and quotas to promote their own cultures and to limit the access of other countries' cultural exports to their markets. Only the United States and Israel voted against this convention, arguing that it could allow governments to control culture, even through censorship, and to block the free flow of ideas and information.

U.S. cultural dominance is not just a matter of popular culture. U.S. educational institutions attract the world's future leaders. American universities educate the elite from around the world. In 2006, for example, 11 ministers in the cabinet of the government of Taiwan held U.S. degrees. U.S. intellectual, academic, and scientific journals set world standards.

It must be reemphasized that not every American cultural export is an original U.S. cultural product. The United States is the apex of global information diffusion, so U.S. global corporations find product ideas around the globe and then introduce them into other places, as demonstrated in the example of Coca-Cola. Endless additional examples could be cited: The U.S. corporation Häagen-Dazs developed in Argentina a dulce de leche ice cream that was soon the second most popular flavor in the United States and Europe. The U.S. corporation Nike has introduced around the world a shoe designed by Kenyans: It has a separate big toe, like a mitten, to simulate running barefoot. East German coaches developed inline skates for competitive ice skaters to practice without ice, but the popularity of these skates around the world today is a product of U.S. marketing. Even MTV is less distinctively American than when it began in 1981: It is increasing the local-content percentage of its programming everywhere. By 2010, MTV had channels in 162 countries, and it was watched in almost 632 million homes in 33 languages. MTV's chairman Tom Freston said that in opening new markets, the company "starts with expatriates [from America] to transfer company culture and operating principles," but then it surrenders control of programming to local executives. "We're always trying to fight the stereotype that MTV is importing American culture," says company president Sumner Redstone. "We aren't. To do so would be cultural imperialism." MTV is "cultivating and nurturing local" artists and shows.

The U.S. political ideals and influence U.S. political institutions and forms also sweep the Earth. As early as 1630, Puritan Governor John Winthrop had admonished New England settlers that "wee must Consider that wee shall be as a Citty upon a Hill, the eies of all people are uppon us." Puritan New England would set an example; it would demonstrate virtue, and surely all the world would eventually follow. Later, the United States was born in revolution, and the authors of the U.S. Constitution produced the modern world's first republic—that is, a government without a king. They trusted in the people to be able to rule themselves without one, and the Founders, too, saw themselves as setting an example that would sooner or later be followed around the world. In 1836, poet Ralph Waldo Emerson was asked to commemorate the first battle of America's War of Independence, the Battle of Concord Bridge, where patriot farmers had fired upon British soldiers. Emerson wrote, "By the rude bridge that arched the flood/Their flag to April's breeze unfurled/Here once the embattled farmers stood/And fired the shot heard round the world." Emerson was already voicing the popular belief that the American War of Independence had launched the idea of republican constitutional democracy—a "shot" that would eventually diffuse around the world, challenging and eventually bringing down all other forms of government (Figure 6-39). The idea that America would lead the world by example and, correspondingly, that its motivations in foreign affairs would always be benevolent, is often called *American exceptionalism.*

The freedom offered by U.S. political institutions, however imperfect, remains the envy of most people on

Figure 6-39 The Statue of Liberty in Tiananmen Square. This papier mâché copy of the Statue of Liberty (although with her arms incorrectly placed) was erected by Chinese students demanding democracy in Tiananmen Square in front of the imperial palace in Beijing in June 1989. Thus "the shot" fired at Concord Bridge in 1775 had diffused even to the site traditionally respected by Chinese as the very center of the world.

Earth. The American example of democracy, the rule of law, women's and minority rights, and other aspects of American life is still profoundly destabilizing in many places. By embodying, demonstrating, defending, and even *promulgating* these values, the United States attracts the hatred of people who feel threatened by, or who hate, these values.

In global political affairs, however, the role of the United States has been to a degree self-contradictory. On the one hand, the United States sets a revolutionary example, and it often justifies its actions with regard to other countries by citing the lofty principles of democracy and the right of nations to self-determination. At the same time, however, the United States has supported and continues to support many nondemocratic governments. During the Cold War, for example, the United States felt it necessary to accept as allies many governments that suppressed the rights of their own citizens.

The United States has also repeatedly intervened militarily in what would normally be regarded as the internal affairs of other countries (such as Iran, Afghanistan, and Iraq), citing its own interests as justification. U.S. foreign policy has often switched sides on international disputes, favoring, for example, Iran or Iraq in turn when there are disputes between those two nations. Some people call such behavior realism (pragmatic politics), but others call it hypocrisy. Whatever it might be labeled, this behavior fuels resentment of U.S. power.

In the late 1980s and early 1990s, the United States promoted democratization, most notably across Asia and Europe. In 1986, for example, the United States stopped supporting dictator Ferdinand Marcos in the Philippines when it became clear that tolerating his dictatorship was too high a price to pay for military bases there and for the "stability" of the Philippines that was in fact toppling into chaos. The United States supported democratic movements in South Korea, Taiwan, and China and, eventually, in 1989, the collapse of Communist government across Eastern Europe. In the Western Hemisphere, Africa, and the Near East, by contrast, the United States had less forcefully promoted democracy.

During the past decade, the U.S. government has begun to admit its role in Cold War plots that it long denied. In 2000, the U.S. Central Intelligence Agency admitted to its role in the overthrow of a democratically elected Iranian leader in 1953. It also confirmed its role in the coup against Chile's elected President Salvador Allende in 1973. Then-U.S. Secretary of State Colin L. Powell admitted that America's role in ushering in military dictator General Augusto Pinochet, who ruled for 17 years, was "not a part of American history we are proud of." The U.S. government hoped that by admitting its past, it would be able to rebuild relationships with countries harmed by U.S. interference.

The George W. Bush administration pursued a policy of encouraging democracy after September 11, 2001, believing that the diffusion of U.S. political ideals would improve national security. The few successes this policy achieved in a few small countries were overshadowed by the wars in Afghanistan and Iraq. Many criticized the idea that democracy could be imposed through military intervention. The revelations of abuses by American troops during the wars in Afghanistan and Iraq and at Guantánamo Bay have, for many global observers, jeopardized America's reputation as a model political culture.

Overall, the wealth and influence of the United States means that it has been the only country in history involved in the affairs of virtually every other country on Earth. People around the world consume U.S. products and debate U.S. government policies, and their homelands often host U.S. troops. U.S. ubiquity in the world places unique responsibility on Americans to understand other cultures and the impact of the United States on them.

Business for Diplomatic Action, a group of U.S. executives concerned about the world's growing disaffection toward America, publishes the *World Citizens Guide* for business travelers and students. The book reflects surveys of hundreds of non-American nationals working in U.S. offices worldwide, which asked how Americans could be better world citizens. The answers were overwhelming. Many people dislike America for four reasons: foreign policy, the negative effects of globalization, the vulgarity and violence of U.S. popular culture, and Americans' collective personality

(Americans are thought to "show no respect for others," and "Americans do not listen"). The Nations Brand Index, a ranking of global attitudes toward nations, ranks America 35th of 35 on its "cultural heritage scale." Clearly, the diffusion of America's best cultural innovators—Duke Ellington, Frank Lloyd Wright, Georgia O'Keeffe, and many others—is overshadowed by other exports.

One of the most reliable global opinion polls revealed that America's image in the world took a beating during the first decade of the new millennium. According to the Pew Research Center, "favorable" views of America in May 2008 had fallen in all 20 countries polled. Only 33 percent of Spaniards, 22 percent of Argentinians, 46 percent of Russians, 37 percent of Indonesians, and 12 percent of Turks viewed the United States favorably. These numbers improved dramatically after the election of President Barack Obama, especially with regard to foreign policy. Whereas in 2008 only 16 percent of the British were confident in Bush's foreign policy, some 86 percent expressed their confidence in Obama's leadership a year later. Such high expectations for the new U.S. president registered in all countries surveyed, except Israel.

Cultural Preservation and Hybridity

Today, many non-Western peoples are trying to defend or to revive their own cultures and values. **Cultural preservation** is the effort to document, repopularize, and rejuvenate traditional cultures. In societies that were colonized, however, European intervention was so profound that it is virtually impossible to reconstruct what existed before and to do so in a present world where change occurs so fast that it seems unstoppable. It is one thing to record songs and restore artifacts but something else to capture how people thought and felt before their culture changed beyond recognition. Much of what is today called "traditional" is in fact a result of European efforts to catalogue and describe what the peoples encountered during colonization, with the bias that necessarily entailed.

When broadcast television was first introduced in Malawi, many people feared an influx of Western culture, but Steven Mijiga, the postmaster general, said, "If you go into any bookstore, the books are all brought here from the West. Look at our religion, our legal system. They are all brought from the West." Today European museums hold African ritual objects for which Africans have forgotten the rituals. Many non-Western people study their history as a way of reclaiming it, along with their independent identity and their self-respect. Nigerian writer Chinua Achebe quotes the African proverb, "Until the lions produce historians, the stories of the hunt will glorify the hunters."

All non-Western national leaders today face a problem: They can neither re-create an idealized model of what their society was like in the past nor totally reject Western ideas and standards, so they must produce some synthesis of cultures (Figure 6-40). This requires difficult decisions and compromises. The Japanese, for example, adopted much of Western science and technology, but they have protected and maintained other aspects of traditional Japanese culture, such as an especially high regard for group effort and teamwork. The idea that cultures can be combined is called **hybridity,** but this tries to describe the preservation of two cultures at once, when the effort more likely results in cultural change to both.

If a nation is to have long-term success, cultural continuity must somehow be combined with attention to useful new imported ideas, practices, and technologies. Many of the problems that countries face today—such as urbanization, pollution, and cultural confusion—are what Europe faced first but has been unable to solve.

Some observers have spoken of the "exhaustion" of Western modernity, and they divide world history into three periods: (1) the *premodern world* of relative cultural isolation before 1500; (2) the *modern world* of Western expansion and cultural diffusion from 1500–1950; and (3) the *postmodern world*, in which more world societies are no longer passive receptacles of Western influence. Instead, they are active shoppers in a global cultural bazaar, picking and choosing what they want and then turning it into something of their own. As we noted previously in this chapter, new technology, such as the Internet, lowers the cost of producing and disseminating information, so virtually anyone can be an information producer. Movies made for just a few thousand dollars win international acclaim. The biannual Panafrican Film and Television Festival in Burkina Faso attracts entries from many countries, and many of the movies shown find global distributors. Some argue that the rise of new economies and world powers, especially India, China, and Brazil, will serve to diffuse their cultures just as trade and power did for European countries and America. Perhaps the film *Slumdog Millionaire* is what this future might look like: a British-produced film, with an Indian co-director and cast, about a boy from the Mumbai slums who goes on the Indian version of the U.S. game show "Who Wants to Be a Millionaire?" that wins big at the major film awards and the global box office . . . yet caused controversy at home (Figure 6-41).

Figure 6-40 Making national history. The 2006 Turkish film, *Valley of the Wolves: Iraq,* was hugely popular in theaters that chose to show it, such as this one in Istanbul. This Hollywood-style adventure film depicts a Turkish national hero who takes revenge on U.S. troops who humiliated his friend. Although the story is fictional, it unfolds amid recent events in Iraq.

Figure 6-41 Slumdog Millionaire. Residents of Mumbai's slums took offense at the portrayal of their lives in the film starring Anil Kapoor and protested outside his office.

Chapter Review

Summary

An isolated society depends entirely on its local environment for all of its needs, but the evolution of individual human cultures in isolation has, through history, yielded to the increasing interconnectedness of human societies.

Cultures diffuse, cultures change, and people can transform themselves and their behavior, but cultures and culture realms also have elements of stability. Any cultural pattern or distribution maps a current balance between the force of change and of stability. Cultures are constantly evolving through time, and these changes may result either from local initiatives and developments or else as the result of influences from other places. As global communication and transportation have increased, the balance of factors that explain the local activities and culture at any place has tipped steadily away from factors of evolution and toward factors of diffusion. What happens at places depends more and more on what happens *among* places.

Today's rapid pace of cultural innovation and diffusion requires us to make a distinction between folk culture and popular culture. Folk culture preserves traditions. Most folk culture groups are rural, and relative isolation helps these groups maintain their integrity, but folk culture groups also include urban neighborhoods of immigrants struggling to preserve their native cultures in their new homes. Popular culture, by contrast, is the culture of people who embrace innovation and conform to changing norms. Popular culture may originate anywhere, and it tends to diffuse rapidly, especially wherever people have time, money, and the inclination to indulge in it.

Humans identify as individuals and with each other. Identities may be self-ascribed or assigned by others. Human groups are often defined according to racial, cultural, ethnic, gendered, and sexual criteria. These criteria can vary widely, and their significance is interpreted culturally. Subcultures can form among people who participate in mainstream culture but also with others who share more specific identities. Individual behavior relates to their perceptions of their environment as well as the meanings they attach to what they perceive.

All human activities find territorial expression, and maps of culture areas, regions of economic specialization, or political jurisdictions reveal current distributions of human activities. Evidence of culture areas is often visible in the cultural landscape. Human activities, however, are dynamic, continuously organizing and reorganizing, forming and reforming. Therefore, their distributions and patterns are disrupted and reshaped repeatedly. Trade plays an important role in cultural diffusion.

Despite the rich variety of indigenous local cultures around the globe, the European cultural model is widespread. Europeans came to play a central role in world history and geography because they paved the way for the modern system of global interconnectedness. Global diffusion was fixed hierarchically, with Europe as the apex. Europe conquered most of the rest of the world, and European political domination imposed European concepts of government, law, property, and other aspects of culture. The spread of Western culture continues today, and many traditional cultures and social structures are being radically transformed. In some cases, whole cultures disappear.

Many non-European peoples are attempting to revive their own cultural history and values. Perhaps world cultural diffusion will be less hierarchical in the future.

Key Terms

acculturation p. 204
Agricultural Revolution p. 230
behavioral geography p. 216
cognitive behavioralism p. 216
Commercial Revolution p. 229
commodification p. 237
cultural diffusion p. 204
cultural ecology p. 206
cultural imperialism p. 234
cultural preservation p. 242
culture core area p. 219
culture domain p. 219
culture realm p. 219
culture region or area p. 219
cyberspace p. 227

diffusionism p. 204
digital divide p. 227
economic geography p. 224
electronic network p. 227
environmental
    determinism p. 206
ethnic enclave p. 214
ethnic group p. 213
ethnocentrism p. 214
ethnonationalism p. 213
evolutionism p. 204
folk culture p. 209
gender role p. 216
heritage site p. 210
historical
    consciousness p. 223
historical geography p. 223

historical materialism p. 205
hunter-gatherer p. 204
hybridity p. 243
Industrial Revolution p. 230
identity p. 213
inertia p. 223
infrastructure p. 223
liquidity p. 230
mental map p. 216
pastoral nomadism p. 204
popular culture p. 210
possibilism p. 207
proxemics p. 217
sexuality p. 216
spatial identity p. 213
subculture p. 213
territoriality p. 218

Questions for Review and Discussion

1. How did Alexander von Humboldt disprove Varro's theory of unilinear social evolution?

2. Many people believe that we have reached the end of a 500-year period of human history. What has characterized this period? What might happen next?

3. Define *historical consciousness* and give an example of how it can act as a force of inertia fixing cultural geography.

4. What four things usually happen when you build a road to connect three formerly isolated villages?

5. Define *cultural imperialism* and give specific examples of how much of the world's cultural geography is the legacy of European cultural imperialism.

6. According to the theory of historical materialism, why are political institutions almost always "behind the times"?

7. If everyone is an individual, then why are identities important? How do identities relate to different spaces and places?

8. Differentiate diffusionist influences on cultural development from evolutionist influences on cultural development and give specific examples of each. What forces or factors at work in the world today favor either diffusionist influences or evolutionist influences?

9. Differentiate folk culture from popular culture. Give a few examples of each that you know about personally.

Thinking Geographically

1. What were a few distinctive products of your region 50 years ago? Religious observances? Food products? Clothing or costumes? Architectural styles? Games? Have they been exported from the region? Are they still typically produced? Attend a local street fair or celebration. What aspects of the festival are different from a similar festival 100 miles away?

2. Investigate Survival International reports on uncontacted tribes. What similarities are there in their locations? What forces threaten their relative isolation?

3. Identify developments that are contributing to the formation of one global culture.

4. Clip a few photos from magazines and challenge your fellow students to identify the location on the basis of attributes of the cultural landscape.

5. How many cities are served by nonstop flights from your nearest airport?

6. Henry Louis Gates, Jr., has written, "... for those ...who value the survival of diverse human cultures in just the way naturalists value the diversity of flora and fauna—the march of international corporate capitalism is not without its mournful aspect. In Tibet, an intricate culture of Buddhist worship has been tragically disrupted and weakened by a brutal, deliberate state policy of cultural extirpation. In neighboring Nepal, a highly evolved practice of Hindu worship has been similarly disrupted and weakened, not by a hostile state regime but by the BBC World Service, Coca-Cola, Michael Jackson—in short, the encroachments of Western consumer culture." Do you agree that the processes he speaks of are similar or equivalent? Are they equally malevolent?

7. Explain the saying that what happens *at* places is increasingly dependent on what happens *among* places.

8. Find in today's newspaper an example of civil strife somewhere in the world. Can the strife be understood in terms of the identities or self-perceptions of the parties involved?

Log in to www.mygeoscienceplace.com for videos, animations, **MapMaster**™ interactive maps, RSS feeds, case studies, and self-study quizzes to enhance your study of Cultural Geography.

MapMaster™

The soaring arched dome of Hagia Sofia ("Holy Wisdom") in Istanbul remains a masterpiece of world architecture.

7

The Geography of Languages and Religions

*S*ince its construction in the sixth century, the building known as *Hagia Sophia has witnessed enormous cultural change. It was built during the reign of Justinian the Great, ruler of the Byzantine Empire, during the early centuries of Christianity. Marble and other materials, some weighing as much as 70 tons, were brought from all over the Empire to Istanbul to construct what would be for several centuries the largest church in the world. The freestanding arched dome measures 31 meters (102 feet) across. The building withstood numerous earthquakes and fires, each time ingeniously repaired to strengthen its delicate form. A perhaps more stunning change came in 1453 when the Ottoman Turkish Empire captured Istanbul. Their leader, Mehmed the Conqueror, ordered the renovation of Hagia Sophia to repair the damage incurred during the decline of Byzantium. He also converted the church into a mosque, removing the specifically Christian elements, such as the altar, and adding Islamic ones, including a minaret (a tower for the call to prayer). The Ottoman Empire made numerous changes to the building during their almost five centuries in Istanbul. Some of the mosaics showing Christian and Byzantine figures were covered over with geometric tiling and Koranic inscriptions. A religious school and other buildings were added to the grounds. The Ottomans later restored many of the Christian mosaics. When the Ottoman Empire came to an end in 1924, the first president of Turkey, Mustafa Kemal Ataturk, ordered that the building no longer be used as a mosque but as a museum. Today, fifteen centuries of art, architecture, and history from two major religions can be found under the magnificent dome of Hagia Sophia.*

Language and religion are two of the most important forces that define and bond human cultures. Their influences are so pervasive that many people take them for granted and cannot objectively observe their influences in the lives of others, or even in their own lives. Nevertheless, peoples who share either of these two cultural attributes often demonstrate consistencies in other aspects of their behavior, and often they can more easily cooperate with one another in other ways, such as in international affairs. Sharing these attributes may help them understand one another in ways that are unique and profound.

Each language and each religion originated in a distinct hearth, and although the carriers of each of these cultural attributes have diffused throughout the world, each language and each religion still predominates within a definable realm. Language and religion are two of the most important of all types of culture regions.

Defining Languages and Language Regions

Many social scientists believe that language is the single most important cultural index. A **language** is a set of words, plus their pronunciation and methods of combining them, that is used and understood to communicate within a group of people. Each language has a unique way of dealing with facts, ideas, and concepts; variations in languages result in variations in how people think about time and space and about things and processes. Exact translation from one language to another is virtually impossible. The language an individual speaks has an influence in structuring his or her perception and logic, and the comparative study of languages is one of the richest ways of understanding human psychology.

Any group of people who communicate exclusively with one another will soon develop "their own language," whether they are nuclear physicists or a social clique at a high school. In human history, languages developed among people who interacted regularly. The words communication and community share the same root: common. If human beings were to appear suddenly in today's highly interconnected world, one worldwide language might develop. The great variety of languages spoken today testifies to the relative isolation of groups in the past. The distribution of any language illustrates the pattern of dispersal of its original speakers, what Chapter 1 termed *relocation diffusion*, or their cultural impact on others, known as *contagious diffusion*.

The number of different languages recognized varies with the accepted definition of language. The term *language* is usually reserved for major patterns of difference in communication. Minor variations within languages are called **dialects,** but scholars do not agree on the amount of distinctiveness necessary for a pattern to be considered a language. Some scholars, for instance, accept Danish, Swedish, and Norwegian as distinct languages, but a speaker of any one of them can mostly understand the others. Therefore, other scholars insist that these are three dialects of one language. A **standard language** is the way any language is spoken and written according to formal rules of diction and grammar, although many regular speakers and writers of any language may not always follow all of the rules. A country's **official language** is the one in which official records are kept and government business is normally conducted.

A pidgin language is a system of communication that has grown up among people who do not share a common language, but who want to talk with each other. Pidgins are marginal or mixed languages, and they usually disappear after a few years or they evolve into a **creole.** A creole is a pidgin language that has survived long enough to become a mother tongue. That usually takes a generation or two. Examples of creoles include Gullah and Geechee, English-based languages that formed when enslaved West Africans were brought to work on plantations along the U.S. Southeast coast. There are numerous creoles based on French, including Haitian French Creole, Guyanese Creole, and Seychellois, a French-based creole spoken on the Seychelles Islands. There are also creoles based on Arabic, Dutch, Portuguese, and Spanish, among others. Many of these language groups are quite small.

A **lingua franca** is a second language held in common for international discourse. Today English is the world's leading lingua franca. Air controllers and pilots in international aviation, for example, all speak English. Other languages have served as lingua francas in the past. Latin long served Western civilization, and Swahili served throughout East Africa. Swahili developed among African peoples in communication with Arab traders, so it has many Arab words. Today it is an official language in Tanzania and Kenya.

Individual languages change through time, but religious classics or classics of literature can exert a powerful force for stabilization. In English, for example, the works of William Shakespeare and the 1612 King James translation of the Bible have molded the language, yet parts of even these works may be difficult for many English speakers to read today.

Linguistic Geography

The study of different dialects across space is called *dialect geography*, or *linguistic geography*. Dialects can contain differences in pronunciation, grammar and vocabulary. Dialects usually diverge more in the way they are spoken than in the way they are written. This is because writing is often widely dispersed, but sounds are localized only among a group of people who speak together, called a **speech community.** Probably no student of dialects has ever developed the ability of the fictional Henry Higgins in G. B. Shaw's play *Pygmalion*: "I can place any man within six miles. I can place him within two miles in London. Sometimes within two streets." To this day, however, there are dialect regions across the United States, although many are **accents,** a dialect difference in pronunciation only.

Sometimes researchers survey speech and draw lines around places where speakers use a linguistic feature in the same way. These boundary lines are called **isoglosses.** Figure 7-1 reproduces a map from the early twentieth-century *Atlas Linguistique de la France,* showing some of the isoglosses between the two major dialect areas in France.

Isoglosses frequently parallel physical landscape features, because physical features often act as barriers to human migration and contact, thereby limiting diffusion. For example, the Pyrenees mountains divide Spain from France, and the Pripet marshes separate Belarus from Ukraine. By contrast, languages often diffuse quickly across broad lowlands, and languages have historically diffused along river valleys or other routes of trade and transportation.

Researchers often map what is called a *geographic dialect continuum.* This is a chain of dialects or languages spoken across an area. The similarities between these dialects decrease with distance. Speakers at any point in the chain can understand people who live in adjacent areas, but they find it difficult to understand people who live farther along the chain. An example of

such a chain is the North Slavic continuum, which links the Czech, Slovak, Ukrainian, Polish, and Russian languages. Another chain includes the languages of northern Spain from Galician in the west to Catalonian in the east excluding Basque, which is an **isolate**—a language unrelated to its neighbors.

The World's Major Languages

Scholars disagree on exactly how many distinct languages there are, but most suggest a count between 6,000 and 7,000, at least 77 of which are spoken by 10 million or more people as a first language (Table 7-1). Over half of all the world's people, however, speak one of the 21 major languages listed in the table. The language with the most speakers is Chinese, with more than 1.2 billion native speakers. Among these are 845 million first-language speakers of the Mandarin Chinese dialect, spread across 20 countries. English is the primary language of about 328 million people worldwide, and it is either the only official language or one of several official languages in approximately 50 countries (Figure 7-2). Arabic derives special transnational importance as the language of the Koran, the sacred scriptures of Islam. The Koran has been translated, but Muslims are still encouraged to study the original. Arabic is the official language in roughly 20 countries today.

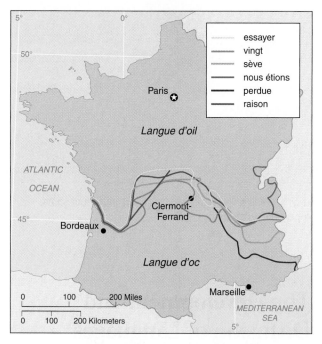

Figure 7-1 Dialect regions of France. These six isoglosses represent the dividing lines in France between regions to the north and to the south where people pronounce six individual words differently: *essayer,* "to try"; *vingt,* "twenty"; *sève,* "sap"; *nous étions,* "we were"; *perdue,* "lost"; *raison,* "reason". Clearly, this bundle of isoglosses represents a significant border between two spoken dialects.

TABLE 7-1 The World's Leading Languages and the Number of Native Speakers of Each (in Millions)

Language	Native Speakers
Mandarin Chinese	845
Spanish	329
English	328
Arabic	221
Hindi	182
Bengali	181
Portuguese	178
Russian	144
Japanese	122
German	90.3
Javanese	84.6
Lahnda	78.3
Wu Chinese	77.2
Telugu	69.8
Vietnamese	68.6
Marathi	68.1
French	67.8
Korean	66.3
Tamil	65.7
Italian	61.7
Urdu	60.6

Note: A native speaker is one for whom the language is his or her first language. Only the two largest Chinese dialects are shown.
Source: SIL, *Ethnologue,* 15th ed.

Figure 7-2 The official languages of the countries of the world. This map reveals the worldwide distribution of some European languages. This far-reaching distribution is one legacy of European imperialism.

In contrast to these widespread languages, some languages are extremely local. About 25 percent of all languages have fewer than 10,000 speakers each; 10 percent have fewer than 1,000 speakers. Linguists have discovered fully developed languages in New Guinea spoken by only a few hundred people living in certain valleys. These languages have developed in total isolation over long periods of time and are utterly incomprehensible to people just 20 miles away in the next valley.

The Development and Diffusion of Languages

Any isolated group of people develops a language of its own. This language describes everything that those people see or experience together. If groups of these people break away and disperse, then each group discovers new objects and ideas, and the people have to

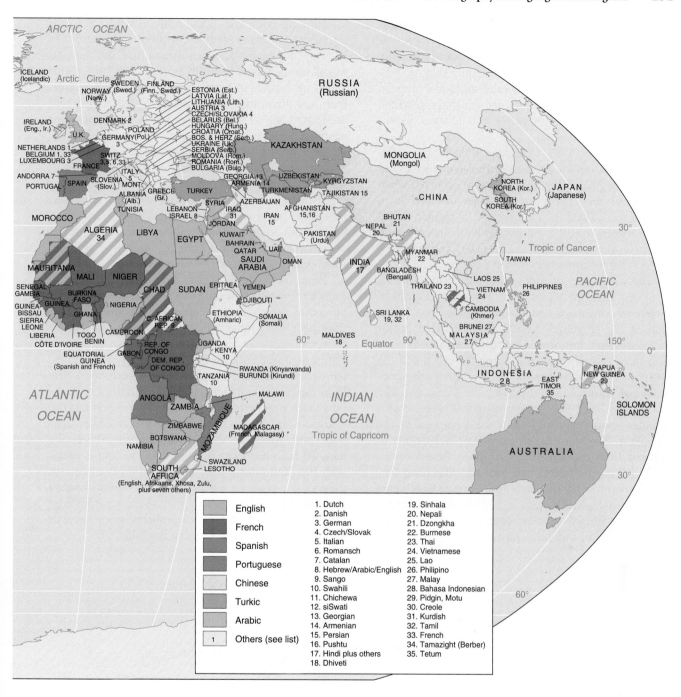

English

French

Spanish

Portuguese

Chinese

Turkic

Arabic

1 | **Others (see list)**

1. Dutch
2. Danish
3. German
4. Czech/Slovak
5. Italian
6. Romansch
7. Catalan
8. Hebrew/Arabic/English
9. Sango
10. Swahili
11. Chichewa
12. siSwati
13. Georgian
14. Armenian
15. Persian
16. Pushtu
17. Hindi plus others
18. Dhiveti

19. Sinhala
20. Nepali
21. Dzongkha
22. Burmese
23. Thai
24. Vietnamese
25. Lao
26. Philipino
27. Malay
28. Bahasa Indonesian
29. Pidgin, Motu
30. Creole
31. Kurdish
32. Tamil
33. French
34. Tamazight (Berber)
35. Tetum

make up new words for them. Other changes and errors, known as **drift,** appear within each of the now separate language groups. After hundreds of years, the descendants of each of these breakaway groups have their own language. Each descendant language has a vocabulary of its own, but each also retains a common core of words from that earliest shared language. Perhaps new words that you and your friends use will someday become standard English. When languages separate, they still have a *genetic relationship,* meaning they share a common ancestor language (and has nothing to do with DNA). The ancestor that is common to any group of several of today's languages is called a **root**

language, or **protolanguage.** The languages that are related by descent from a common protolanguage make up a **language family.**

The Indo-European Language Family

In 1786 the English philosopher Sir William Jones first pronounced his theory that a great variety of languages spoken across a tremendous expanse of Earth demonstrate similarities among themselves so numerous and precise that they cannot be attributed to chance and cannot be explained by borrowing. These languages, then, must descend from a common original language. The group of languages first identified by Sir William is called the *Indo-European family* of languages, and about half of the world's peoples today speak a language from this family (Figure 7-3). Sifting through the vocabularies of all Indo-European languages yields a common core vocabulary, which is the common ancestor of these languages, *proto-Indo-European*.

Jacob Grimm (1785–1863), one of the brothers who collected children's fairy tales, formulated rules to describe the regular shifts in sounds that occurred when the various Indo-European languages diverged from one another. There is, for example, a regular sound shift between words beginning with "p" in Latin and "f" in Germanic languages (as in "pater" and "father"). These rules are known as *Grimm's Law*.

The vocabulary of proto-Indo-European tells us a surprising amount about how proto-Indo-European society was organized and how the people lived. It also hints at the language's hearth. Reconstructed proto-Indo-European has words for distinct seasons (one with snow), woody trees (including the beech and the birch), bears, wolves, beavers, mice, salmon, eels, sparrows, and wasps. These things can be found together around the Black Sea, but proto-Indo-European also includes words borrowed from the languages of the Near East. The word for wine, for instance, seems to descend from the non-Indo-European Semitic word *wanju*. Thus, the hearth area for Indo-European languages was probably in today's Turkey, some 8,000 years ago. Archaeologists disagree as to whether proto-Indo-European diffused quickly, carried by warriors, or slowly, carried by individual farmers.

Hunting common Indo-European roots of words provides a fascinating study. The proto-Indo-European root *aiw*, for example, which means "life" or "the vital force," can be found in Hindi as *ayua*, "life," and it also shows up as *aetas* in Latin, *aion* in Greek, *ewig* in German, and the words *ever* and *age* in English. Words that are related appear in surprising places. *Maharajah*, for example, a Hindi word for a great ruler, may seem exotic, yet *maha* is a distant cousin of the English words "major" and "magnitude." *Rajah* is a distant cousin of the English "reign" and "royal." In this case, the cousins clearly look or sound alike, so they are called **cognates.** One must never assume any connection between two words from different languages, however, until it can be proven that they share a common root. The study of word origins and history is called **etymology.**

Proto-Indo-European provided the basic stock for all Indo-European languages, but that does not mean that one ethnic or racial group spread out to live where all these peoples are today. It is clear that the spread of Indo-European languages was in many places a contagious diffusion, a transmission of culture not people, accompanying the spread of agriculture. Sometimes a few Indo-Europeans conquered and imposed their language on a much larger group of people that previously had developed a language of their own—as occurred throughout Latin America and Africa. Sometimes peoples adopted Indo-European just to be able to communicate with Indo-Europeans. Language diffusion is a complex process. For example, not everyone who speaks English in the world today necessarily has an ancestor from England. Many readers of this book probably have parents or grandparents who could not read English.

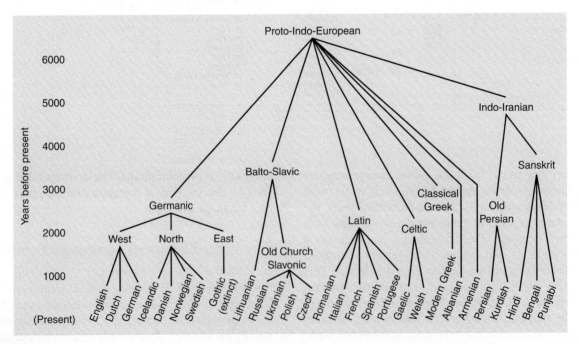

Figure 7-3 The Indo-European family of languages. The languages in this "family tree" are all descendants of proto-Indo-European. Common word roots can be found among them.

Also, cultures borrow both things and the words to name them from one another. Western culture gave Japan the word *erebata* along with the elevator itself. Japanese culture gave the West edible raw fish and a name for it: *sushi*.

Other Language Families

More people speak Indo-European languages than languages of any other family, and the Indo-European family has been studied more than any other. Researchers have benefited from written sources dating back 3,700 years—tablets in the extinct Indo-European language Hittite. These sources give researchers considerable confidence in including individual languages in the family and in reconstructing proto-Indo-European.

The classification of all Earth's languages into families is an enormous and difficult task. Comparative linguists do not agree whether it has yet been satisfactorily completed. Figure 7-4 reflects the most generally agreed-upon state of understanding of the geographic distribution of some of the language families (there are as many as 116).

The same principles apply to each of these language families as to the study of Indo-European. Scholars sift vocabularies of today's individual languages to compile vocabularies of a protolanguage. The protolanguage offers clues about the origins and culture of the people who originated that protolanguage, and the sound shifts help us trace how and when various individual languages and peoples broke off linguistically and geographically from the main branch of the group. We know, for example, that Finnish, Estonian, and Hungarian are related languages within the Uralic family, and their common ancestors long ago emigrated from the region of the Ural mountains in today's Russia. Most peoples of sub-Saharan Africa speak Bantu languages, whose origins can be traced to today's Nigeria about 500 B.C. The ancestors of the Turks of today's Turkey left their homeland near Asia's Altai Mountains 1,000 years ago, while their cousins—the Uzbeks, Kazaks, Kyrgyz, Azerbaijani, and Turkmen—stayed in Central Asia. Scholars believe that early migrants to Madagascar came from the South Pacific rather than from nearby Africa, because the language of Madagascar is related to those of the South Pacific.

Some comparative linguists believe that after all languages have been classified into families, protolanguages can be constructed for each language family and that the study of those protolanguages will eventually yield superprotolanguages. They argue that the Indo-European

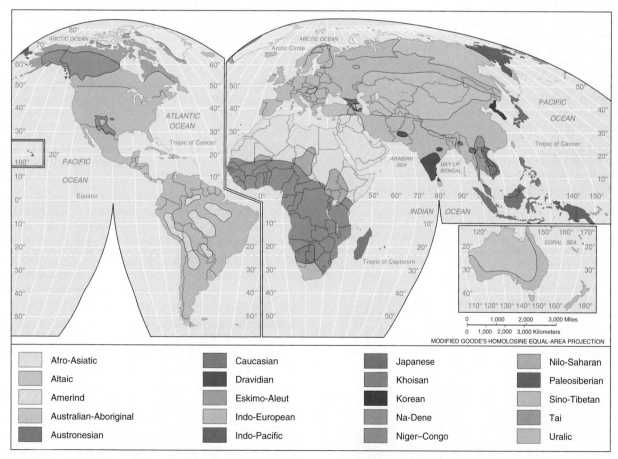

Afro-Asiatic | Caucasian | Japanese | Nilo-Saharan
Altaic | Dravidian | Khoisan | Paleosiberian
Amerind | Eskimo-Aleut | Korean | Sino-Tibetan
Australian-Aboriginal | Indo-European | Na-Dene | Tai
Austronesian | Indo-Pacific | Niger–Congo | Uralic

Figure 7-4 The world's major language families. Language families are groups of languages that have common linguistic ancestors. In some areas, languages other than those shown on this map have been adopted as official languages, even though they may be spoken only by minorities.

family, for instance, is only one of six branches of a larger group, which they call *Nostratic*. Theoretically, such efforts could reach back to a protohuman language, or Mother Tongue, that was spoken in Africa 100,000 years ago and then diffused around the globe.

A human gene that is specifically involved in language, the FOXP2 gene, was first discovered in 2000, when researchers found that a defect on a certain part of a person's chromosome is clearly linked to a form of extreme language difficulty (but not all language problems are related to such defects). The gene has since been found in 38,000-year-old Neanderthal bones. The gene had remained largely unaltered during the evolution of mammals, but it suddenly changed in early humans after the hominid line split off from the chimpanzee line of descent about 100,000 years ago. These discoveries suggest that we now have both linguistic and biological evidence that all humans can be traced back to one stock that evolved in Africa. The further back into prehistory that scholars attempt to reconstruct individual languages, however, the more controversial the work becomes. Some linguists believe that the time depth of such studies is too great to reconstruct more protolanguages, let alone one prehistoric mother tongue. Geography, linguistics, anthropology, and biology each contribute to this fascinating but controversial research into prehistory.

The Geography of Writing

The geography of languages is complicated by the geography of **orthography**, which is a system of writing. There have been two indisputably independent inventions of writing—by the Sumerians in Mesopotamia before 3000 B.C. and by the Central American Olmec peoples about 650 B.C. We think that Egyptian writing of 3000 B.C. and Chinese writing (by 1300 B.C.) may also have arisen independently, but probably all other peoples who have developed writing have borrowed, adapted, or at least been inspired by existing systems (Figure 7-5).

Most languages are written in alphabets, which are systems in which letters represent sounds. There are several alphabets in use today. The alphabet in which a language is written can reflect a historical diffusion. Modern Western European languages are written in the Roman alphabet because Western Europeans were converted to Christianity from Rome. In Eastern Europe, in contrast, Russian, Belarusian, Ukrainian, Serbian, Bulgarian, and a few other languages spoken in Russia are written in the Cyrillic alphabet. This is the Greek alphabet as it was augmented and taught by the missionary Saint Cyril (d. 869), who converted many of these peoples to Orthodox Christianity. When Bulgaria joined the European Union on January 1, 2007 (see Chapter 13), Bulgarian became the Union's first language in Cyrillic script. Scholars are now working to regularize a system of transliteration. Serbian and Croatian are one spoken language, but most Serbs write in Cyrillic, and Croats write in Roman letters. The Cyrillic script is closely associated with Slavic language groups that have traditionally practiced Orthodox Christianity, although there are Slavs who are not Orthodox and Orthodox Christians who use Roman script (Figure 7-6).

A similar alignment of orthography and religion now separates Pakistan from India. In everyday use, spoken Urdu, the language of Pakistan, is the same as spoken Hindi, the language of northern India. The Muslim Pakistanis write in Arabic script, but the Hindus of India write in Devanagari script. Several other Indian languages are also written in Devanagari.

Sometimes a government decides to change a country's orthography as a way of redefining its cultural and political identity. Romania, for example, switched from the Cyrillic to the Roman alphabet early in the twentieth century. This change reflected a choice to be a "Western European" nation. Montenegro is increasingly using the Roman alphabet instead of Cyrillic since its breakup with Serbia. The Turkish languages of Central Asia came to be written in Arabic when these people converted to Islam, as discussed later in this chapter. Early in the twentieth century, the Turks of the country of Turkey chose to replace Arabic with the Roman alphabet to signal a deliberate Western-style modernization (recall Figure 6-33). In 1939, Soviet dictator Joseph Stalin forced the Turkish peoples in Central Asia to replace Arabic script with Cyrillic script. His objective was to cut off these peoples from their cultural heritage and to intensify relatively minor linguistic differences among them. In 1992, however, with the collapse of the U.S.S.R., the newly independent republics of Kazakhstan, Kyrgyzstan, Uzbekistan, Turkmenistan, and Azerbaijan all chose to abandon Cyrillic script. They did not choose to revert to Arabic, but, instead, to adopt Roman. This reflected a repudiation of their previously colonial status and a commitment to join the cultural community of the West. A Russian law of 2002 prescribes the Cyrillic alphabet as mandatory for Russian and for the languages of all

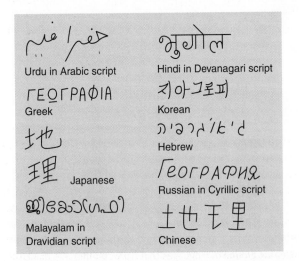

Figure 7-5 Examples of orthography. This figure shows the word *geography* written in several different orthographies.

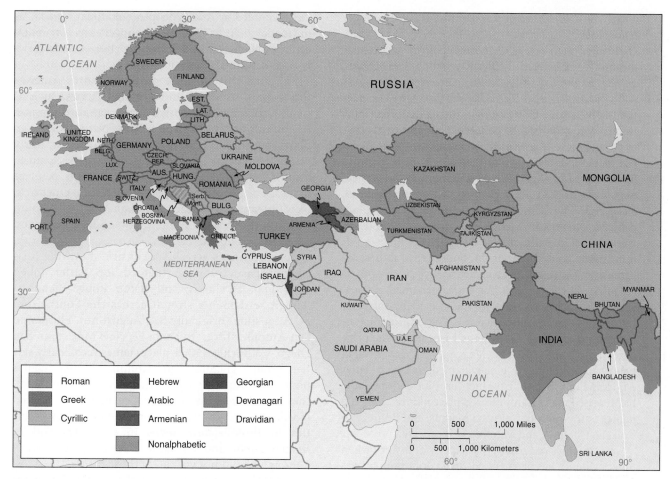

Figure 7-6 The distribution of alphabetic scripts in Eurasia. The distribution of scripts reveals the diffusion of religions and other cultural attributes.

ethnic republics within Russia; the rebellious region of Chechnya, however, defiantly uses the Roman alphabet as a symbol of cultural independence.

Not all languages are written in alphabets. The major nonalphabetic forms of writing are Chinese, in which each character typically represents a word or concept, and Japanese, in which characters can represent either whole words or syllables. Korean writing was once modeled on Chinese, but this was replaced by the *Hangul* alphabetic script. The Korean alphabet was made official following independence from Japan at the end of World War II. It is used in both South and North Korea, where it is known as *Choson'gul*. It is increasingly common for non-alphabetic languages to use Roman orthography for Western names or terms. Computer software has eased communication in non-alphabetic languages, but the demands of contemporary international communication are nevertheless pressuring these countries to adopt the Roman alphabet. The Vietnamese changed from Chinese writing to Roman orthography, introduced by Christian missionaries and made official under French colonial rule.

Many traditional Native American, African, and Asian languages never had written forms until Christian

or Islamic missionaries set out to translate the Bible or the Koran into these languages, using the missionaries' own orthographies. Somaliland, for example, first received a written language in 1972. These peoples may have benefited from having their languages written, but imposing any orthography on a people whose culture had been oral is an act of cultural imperialism. The transition from an oral to a literate culture is arguably the greatest transition in human cultural history.

Toponymy: Language on the Landscape

Toponymy is the study of place names. Place names record natural features as they exist in the present or as they were in the past, as well as something about the origins or values of a place's present or past inhabitants.

Some names describe environments as they are today (Oak Bay, British Columbia), whereas others describe environments as they were in the past. Many treeless towns in the North China plain, for example, are named Wang, which means oak tree. We assume that oaks grew there in the past but that either climate change or overharvesting has eliminated oaks from the landscape.

Names often reveal what the people in a particular location do or believe. If cities have names like St. Paul, we are in the Christian realm; conversely, Islamabad ("the place of Islam," the capital of Pakistan) indicates the Islamic realm. City names in Russia honored Communist heroes during the period of Communist government, but today many of those places have reassumed the names they had before the Communists came to power (Figure 7-7). St. Petersburg, for example, was renamed Leningrad through the Communist period, 1917–1991. Newly independent countries often replace colonial names. In Africa, for example, Rhodesia, named for Cecil Rhodes (1853–1902), an English administrator and financier in South Africa, became Zimbabwe (the name of an ancient native city there). Its colonial-era capital, Salisbury, was renamed Harare. South Africa has recently Africanized many formerly European-language place names. The city of Bloemfontein, for example, has become Mangaung, "the place of leopards," in the Sesotho language. The government may change Pretoria, originally named for Dutch settler Andreas Pretorius, to Tshwane, for a local river. India has reversed many Europeanized place names, changing names of the cities of Calcutta, Madras, and Bombay to Kolkata, Chennai, and Mumbai, respectively.

Toponymies can also reveal aspects of history when other evidence has been erased. Spain, for example, was long occupied by Arabic peoples, and today "guada," as in Guadalquivir or Guadeloupe, lingers as a corruption of the Arabic *wadi*, which means "river." *Guadalquivir* is *Wadi al Kabir*, "the great river."

The majority of place names in North America today are possessives and personal names, such as Jones Creek, but descriptive names are also common (the Red River, for example). Many Native American place names survive (Winnipeg is Cree for "dirty water"), but areas explored or settled by the Spanish, French, or English can be traced by trails of place names. The arc of French settlements from Quebec to New Orleans can be traced through Montreal (Mount Royal); Detroit (the "straits" between Lakes Erie and St. Clair); Fond du Lac, La Crosse, and Prairie du Chien (on routes across Wisconsin); down the Mississippi past Dubuque (named for settler Julien Dubuque); St. Louis; and Baton Rouge. Spanish-named settlements stretch across the South from St. Augustine (originally San Augustino), Florida; to San Antonio, Texas; to Santa Fe, New Mexico; to San Diego, California. Other place names recall early settler groups (Figure 7-8). Commemorative names honor historic personages. Alexander von Humboldt, the geographer introduced in Chapter 1, visited North America, and he is commemorated by counties in three states; towns in six states and in Saskatchewan; and several features in the U.S. West, although he did not visit all of those places.

Today all place names in the United States are recorded by the Board on Geographic Names, created in 1890. The board has a committee to certify foreign place names for American usage, such as all names used in this book.

Communist era name	Previous and now restored name
Andropov	Rybinsk
Brezhnev	Naberezhnye Chelny
Chernenko	Sharypovo
Frunze	Bishkek
Georgiu-Dezh	Lisky
Gorky	Nizhny Novgorod
Gotvald	Zmiev
Kalinin	Tver
Kuibyshev	Samara
Kirovbad	Gyanja
Leninabad	Khodjent
Leningrad	St. Petersburg
Mayakovsky	Bagdati
Ordzhonikidze	Vladikavkaz
Sverdlovsk	Yekaterinburg
Voroshilovgrad	Lugansk
Zhdanov	Mariupol

Figure 7-7 Politics changes place names. During the period of Communist government in the Soviet Union, all these historic cities' names were changed to honor Communist leaders. The cities have reassumed their earlier names.

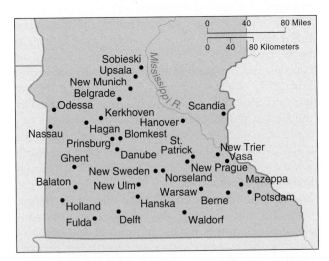

Figure 7-8 Minnesota place names. Some North American cities retain the names of explorers who just passed by, but these Minnesota place names supply a good clue to the origins of the immigrant settlers in the area. Each is a city in Europe or a hero to some national group. How many origins can you identify?

Linguistic Differentiation in the Modern World

Diffusion and differentiation of the major languages continues. Communication within individual speech communities (usually countries) tends to individualize each dialect from the common mother language, but at the same time, international communication among these groups tends to homogenize the dialects. For example, English has differentiated into many dialects. Today an English-speaking American, Nigerian, and Indian will not be able to understand one another completely without effort. Sometimes even the same words will convey different meanings when they are used by people from such diverse cultures as the United States, Nigeria, and India. In 1877, the linguist Henry Sweet (the model for Henry Higgins in *Pygmalion* and later the musical version, *My Fair Lady*) predicted that by 1977, even the English, Americans, and Australians would speak mutually incomprehensible languages, because of their isolation from one another. These dialects have not, however, differentiated to the degree Professor Sweet predicted. This is partly because English, American, and Australian cultures have not greatly differentiated from each other in their other aspects. Also, Sir Henry never could have foreseen the great increases in world communication, nor the fact that rising percentages of national populations would receive formal education, which tends to stabilize languages.

Other major languages continue to diffuse and differentiate. Both French and Spanish are widely used. The French and the Spaniards try to "purify" their languages of linguistic borrowing but also to add new words so that their languages stay useful in the modern world. The French Academy, a scholarly institution dating back to 1635, devises new French words for new concepts and items in international trade and discourse, and these terms become official for French. Despite the efforts of the Academy, however, the French spoken in the 35 Francophone countries and Quebec are different dialects. Spanish also experiences differentiation. In 1997 Mexico, the world's most populous Spanish-speaking nation, sponsored the first International Congress on Spanish. Scholars and writers from around the world discussed the national variations of Spanish. Other international conferences bring together Germans, Austrians, and Swiss to harmonize the spelling and grammar of German.

National Languages

There is no exact correspondence between languages and the countries of the world. A few languages, such as Icelandic and Japanese, are associated almost exclusively with one country, but several languages are shared by many countries. The relationship between languages and nationalism is very complicated.

In European history, language has been interpreted as the basis of nationalism. As early as 1601, Henri IV of France seized French-speaking territories from the Duke of Savoy and declared to his new subjects, "It stands to reason that since your native tongue is French, you should be subjects of the King of France." This logic, however, never stopped the French from seizing the territory of non-French speakers whenever they could.

In 1873, the Third International Statistical Congress recommended that all censuses henceforth include a question of language. The language that a person speaks at home was thought to be the only aspect of nationality that could objectively be counted and tabulated. Therefore, a person's language was "definitive" of that person's nationality. This idea gained in popularity and played a key role in creating feelings of competitive national identity that eventually helped cause World War I.

Language and nation building *Philological nationalism* is the idea that "mother tongues" have given birth to nations (Figure 7-9). This idea persists despite the fact that standard languages usually were, and still are,

Figure 7-9 Luther's German Bible. Martin Luther's 1534 translation of the Bible into German helped to standardize the modern German language, which had been divided by many regional dialects. This helped unify Germans around the idea of a German nation. German speakers can still read Luther's translation today.

The Rise of English

The map of the geography of languages is not static; it is always changing. The use of some languages has expanded as speakers of those languages diffuse throughout the world, rise in power and influence in world affairs, or win new adherents to their ideas. Latin served as a language of religion, scholarship, and some diplomacy in Europe after the Roman Empire. In the nineteenth century, French was the language of culture and diplomacy, a lingua franca for learned travelers. Today, English has quickly become a global lingua franca.

How did English acquire this status? English speakers spread around the world as part of the British Empire. British trade and colonies introduced English to every continent. Even after British rule, English remained the language of government, commerce, and education in some of the world's most populous countries, such as India, the United States, and Nigeria. Ironically, English became more common in these countries after independence, when their governments invested in public education. Since World War II, the political and economic power of the United States has helped to increase the use of English in global science and business activities. American cultural products, especially Hollywood films, inject American English into global popular culture.

Some writers have referred to the *Anglobalization* of the world. In 2006, French President Jacques Chirac stormed out of an international meeting at which a French business executive said, "I will speak English, the language of business." In the 1970s, Malaysia's then-Minister of Education Mahathir Mohammad expelled English from national schools and required the use of Malay, Chinese, or Tamil. In 2003, however, Mahathir Mohammad, who had become prime minister, reintroduced English for teaching all math and science classes, saying "We have to accept English whether we like it or not." Over 90 percent of secondary school students in the European Union's non-English-speaking countries

study English, whereas only 33 percent study French and 13 percent German. Chile wants all school children to be proficient in English, insisting that the use of English is an instrument of equality, and an estimated one-fifth of China's total population is studying English. Thus, by 2025, China's English speakers may outnumber all native English speakers in the world. Many multinational corporations have designated English as their corporate language, regardless of what the languages of their home countries might be. In computer terminology, there is virtually no language but English. Even web pages that display other languages use an underlying computer language based on English.

The genuine merits of English provide other reasons for its success: It has a huge vocabulary but a simple grammar. Furthermore, it is open to change and absorption of new words: foreign words, coinages, and grammatical shifts. Nevertheless, as English grows as a second language, it is losing ground as a first language. The percentages of Earth's population speaking Hindi, Urdu, Arabic, Spanish, and Chinese as first languages are all growing faster than the percentage speaking English as its first language.

Anyone anywhere can communicate via the Internet with people who speak his or her own language, but the Internet was born in the United States, so more than half of all web pages are in English. Depending on how one measures the Internet (by websites or web pages), the other most common languages are German, French, Japanese, and Spanish. English, however, accounts for about 97 percent of all pages linked to secure servers—that is, pages used in electronic business transactions. Technology is providing greater means of "localization," adjusting to language needs of users still uncomfortable in English or other major languages. Mobile phones can now provide menus for even small language communities. But technology is also easing access to English-language materials and increasing the popularity of learning English as a foreign language.

the product of deliberate efforts by centralizing governments to create national identities and loyal citizens. Standardized languages cannot emerge before mass schooling and mass literacy or, alternatively, universal service in a national army. Usually these centralizing pressures transform the language of a small percentage of the population, the political or cultural elite, into the national language. In France in 1789, for example, 50 percent of the population did not speak French at all, and only 12 to 13 percent spoke the standard language.

Several national languages have developed as a result of deliberate political pressure. In 1919, Hebrew was actually spoken only by about 20,000 people, but the Israeli state has nurtured it for nationalistic reasons. Today,

Israel's Academy of Hebrew monitors the language. Some national languages were virtually invented, such as Romanian in the nineteenth century. Other new states have cultivated new national languages to unite diverse populations. The Indonesian government, for example, has succeeded in codifying and establishing one common language: Bahasa Indonesian. Since the breakup of Yugoslavia into several countries, the new governments are encouraging language differentiation. Serbo-Croatian was one spoken language written in two alphabets, but today, because of political breakup, four languages are evolving: Croatian and Bosnian written in Roman, Serbian written in Cyrillic, and Montenegrin. As noted above, Montenegro's independence from Serbia in 2006

Language in New States

The emergence of new countries brings with it important questions about national identity. New states have a strong desire to distinguish themselves from the states they previously belonged to. East Timor faced a language problem when it won independence in 2002, after four centuries of Portuguese rule, 24 years of Indonesian occupation, and 2 years of U.N. administration. Only 57 percent of its population was literate in any language. The country's new constitution designates Tetum (an indigenous lingua franca) and Portuguese as official languages, plus English and Bahasa Indonesian as "working languages." The long-term survival of Tetum is questionable: Its written form is incomplete, and there are few written materials in it, but the Roman Catholic Church adopted Tetum for its services. English and Bahasa are used in business. About 25 percent of the population speaks the former colonial language, Portuguese, but the need to trade with and attract tourists from Australia puts a premium on learning English.

There is rarely a complete turnover in the population when a new state emerges. The Republic of Kosovo declared its independence from Serbia in 2008, and Albanian became the official language. Kosovo also possesses an important Serb minority, and Serbian was also made an official language. Most of Kosovo's population speaks Albanian, which is the basis of their cultural identity and was often suppressed by the Serbian government. For the last decade, the country was administered by the United Nations, which operated in English. Today, public signs are written in Albanian and often English in areas populated by Albanians. Although many Albanians have learned English, it is little spoken outside the cities. The smaller Serb areas speak Serbian almost exclusively; signs are written in Cyrillic. Kosovo once boasted a fair number of multilingual inhabitants, some of whom also spoke Turkish or another minority language. Young Albanians and Serbs are being raised in linguistic isolation from one another and increasingly rely on English if they interact at all.

Welcome to Kosova. Arrivals to the airport in Prishtina, Kosovo, are greeted in Albanian, the language of the majority of Kosovars, followed by English and Serbian. Even the spelling "Kosova" reflects the Albanian pronunciation, "Kosovë." This replaces the Serbian spelling "Kosovo," which is still used in the English language press.

has encouraged the rejuvenation of Montenegrin, which differs from Serbian. The governments are reviving the use of old words in official documents, and education ministries are printing checklists of forbidden words and fining people who use them. In 1996, President Tudjman of Croatia himself made a mistake. He welcomed U.S. President Clinton to Zagreb using the Serbian version of happy, *srecan*, rather than the Croat word *sretan*. This mistake was broadcast live, but the

government edited it out of later rebroadcasts of the event. The Czech and Slovak languages have been differentiating since the breakup of Czechoslovakia into the Czech Republic and Slovakia in 1992. Not all countries succeed in reviving or creating national languages. When Ireland achieved independence in 1922, the government tried to enforce the use of Irish rather than English, but the people themselves did not accept it. They continued to use English. Today, however, Irish nationalism has inspired the country to demand that the European Union recognize Irish as an official language.

A minority people in a country often clings to its language as a gesture of cultural independence. For example, from 1937 until the mid-1950s, the Spanish government tried to stamp out the Basque language by forbidding its use in all public places. Through the 1960s and 1970s, however, Basques won new rights, and in 1980, the first Basque Parliament was elected, with Basque as a second official language in Spain's Basque provinces. The 1992 European Charter for Regional and Minority Languages stresses the "value of interculturalism and multilingualism" and demands that treaty participants "promote regional or minority languages," encourage their use, create political links among their speakers, guarantee access to them in criminal and civil proceedings, and encourage their presence in television and radio. Some people object to the idea that governments should support minority languages and feel that everyone would be better off if they spoke the majority language in their country.

Language in postcolonial societies European colonial powers imposed their languages on their colonies, regardless of how many native languages were already spoken and whether many Europeans actually settled in any given colony. The conqueror's language became the language of government, administration and law, economic development, and usually education. When the colonies gained their political independence and had to choose official languages, many kept their former ruler's language. Figure 7-2 partly reflects the map of former European empires. Some former colonial peoples find the acceptance of a major international language useful. It facilitates ties of trade, cultural exchange, travel, and even diplomacy with its former ruler and with other countries that are former colonies of that ruler. The English-speaking countries of the Caribbean, for example, find English a useful tie among themselves.

In many countries, however, the official language is not actually spoken by most of the population. Figure 7-2 shows a few international languages covering most of the globe, and yet in some of these areas, the majority of the people do not actually speak that language. The difference between official languages and the languages actually spoken is particularly great in Africa. The principal official languages of Africa are those of the continent's former European rulers, but there is an extraordinary variety of native languages spoken, by some counts more than 2,000 (Figure 7-10). This large number reflects the minimal interaction of Africa's many native peoples before European conquest. Many of these languages do not have a written form, and only 40 or so have as many as 1 million speakers. Some African governments are reviving their use or study in order to rekindle appreciation of their people's pre-European heritage.

In some cases, the newly independent countries retained their former ruler's language to prevent arguments about which of the native languages should become the new national language. Many African countries have retained English or French for this reason. For instance, over 500 languages are spoken in Nigeria, and advocates of a national language could not decide among Hausa, Yoruba, or Igbo, three of the major native tongues. Therefore, English was chosen, although only about 20 percent of the population speaks it well and still fewer use it as their first language. Another reason for the continuing preferences for European languages is the high cost of schooling in the many indigenous languages. For example, Tanzania's choice of Swahili for primary education allowed the government to bridge linguistic divisions among the people, but it restricted the books available to Tanzanian schools.

Linguistic rivalries cause English to remain as an official language of India, although it is spoken by a small fraction of the total population. The leaders of India's independence movement intended Hindi one day to be India's official language, and Hindi is spoken by almost 400 million people. Hindi speakers, however, are concentrated in northern India, and other Indians are unwilling to accept the domination of Hindi. Whenever Hindi has been pressed on India's southern states, they have threatened to secede. India recognizes English as 1 of 16 official languages, and English remains the preeminent language of the upper classes and the upwardly mobile.

A former colony can rise to greater wealth and influence than its former ruler, in which case its dialect sets the international standard. This has happened with a few languages, including the U.S. version of English. Noah Webster published his *Spelling Book* in 1783, the same year that the United States won its political independence. Webster's books and dictionary popularized the spellings and pronunciations that now distinguish American from British English. Today, American English competes with British English for precedence worldwide. Philologists in Canada, Australia, South Africa, and other English-speaking countries have published dictionaries of their national variants. The French Academy rigorously defends its privilege to guard the French language, but in 2003, the French government accepted the Quebecois term *courriel* for e-mail, rather than the French *courrier électronique*.

Portuguese boasts about 200 million native speakers, most of whom live in Brazil, a country with 18 times Portugal's population. In 1990, the governments of Brazil, Angola, Mozambique, São Tomé and Príncipe, Guinea-Bissau, and Cape Verde—all former Portuguese colonies—accepted the Brazilian version of Portuguese as their standard. Brazilian is much simpler in spelling than is the Portuguese of Portugal, and it incorporates many Native American, African, and even American words.

The cultural legacy of imperialism explains why some languages are broadly distributed. But this is not

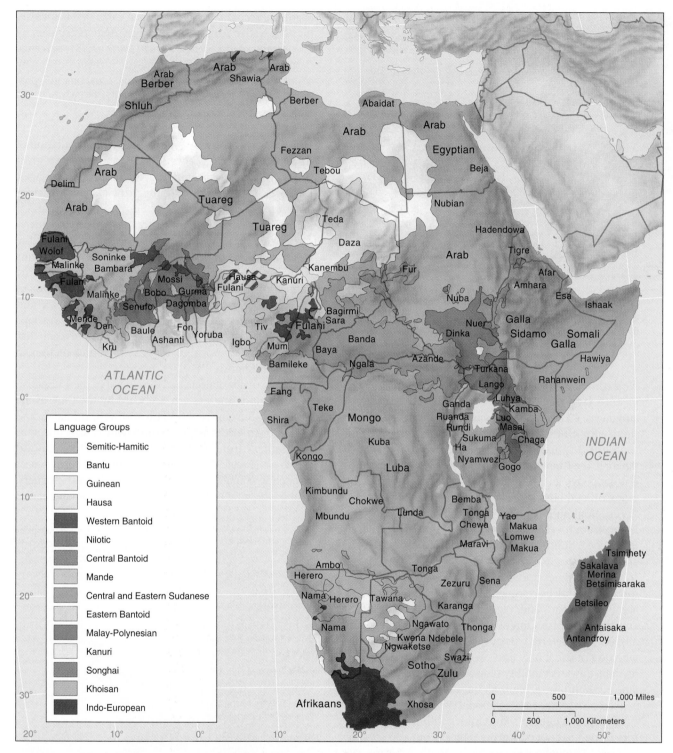

Figure 7-10 The languages of Africa. Figure 7-2 shows that most of the official languages in Africa are the languages of the former European rulers, but this map reveals major language groups and a sampling of the extraordinary number of native languages, many of which are actually spoken by a majority of the local population.

always directly related to the number of people who speak it. Table 7-1 lists, among the leading languages in terms of speakers, several Asian languages that are geographically restricted. Hindi and Bengali are examples. These languages are not widespread across the world map, because the speakers of these languages did not create vast empires.

Polyglot states   India is not the only country that grants legal equality to two or even more languages. The several states that do so are called **polyglot states.** If various languages predominate in distinct regions of a country, the country might accept each language as official in its region. Belgium, and even its capital city of Brussels, is legally divided into French and

Figure 7-11 Bilingual signs in Brussels. Seafood on display in Brussels beneath signs written in both Flemish and French. By law, all street signs in the capital of Belgium must be in both languages, reflecting the make-up of its population.

Flemish zones. In the center of Brussels, all signs are in both languages (Figure 7-11). The precedence of languages is hotly contested in Belgium because they signify the cultural differences between northern and southern Belgium. Several Belgian governments have fallen if they have been seen as giving preference to one language or the other. Canada is also officially polyglot, and language is a source of friction between the Anglophonic (English-speaking) majority and the Francophonic (French-speaking) minority centered in Quebec. South Africa's constitution names 11 official languages.

Polyglot states usually select one language as official for the central government and for communications among its regions and, regardless of what the country's constitution or laws might say, one language will generally be the preferred language of the country. Those who do not use it may find their opportunity restricted or their upward mobility blocked. In Tanzania, for example, Swahili is recognized, but knowledge of English is necessary to rise in the national power structure.

Different language communities within one country may be defined not only geographically but also socially, by class. Throughout Central and South America, the Native American populations and their cultures were smothered under European rule. Vestiges of this cultural imperialism remained even after independence. For example, the original constitution for Bolivia, written by Simón Bolívar, accepted as citizens only those who could speak Spanish. Today some countries are granting official recognition to indigenous languages as a way of enhancing national identity and community. Therefore, more and more countries today recognize two or more official languages—one international language plus one or more local indigenous languages. Today Bolivia, for

example, recognizes three official languages: Spanish, plus the Native American languages Aymará (spoken by 25 percent of the people) and Quechua (34 percent). Peru recognizes as official languages Spanish (spoken by about 80 percent of the population), Quechua (17 percent), Amaryá (2.3 percent), and smaller indigenous languages where they predominate.

Languages in the United States   The population of the United States has always been composed of a great variety of peoples speaking a great variety of languages. English was the language of the principal colonial ruler and of the greatest number of European settlers, and it has always served as a lingua franca. A distinct American English nevertheless evolved. From the days of the earliest settlers, the American language adopted terms from Native American languages. These included the names of native animals and plants unknown in Europe or Africa, products derived from them, and also place names. Each of the many immigrant groups has in turn learned American English, but each has also contributed vocabulary, grammar, and diction to American English.

Three major dialects had developed in the 13 English colonies along the Eastern Seaboard by the time of the American Revolutionary War: northern, midland, and southern American English. Toponymy reveals the paths that settlers from each seaboard region took toward the West (Figure 7-12). Users of the northern dialect frequently named places Brook, Notch, and Corners. People moving westward from the midland areas

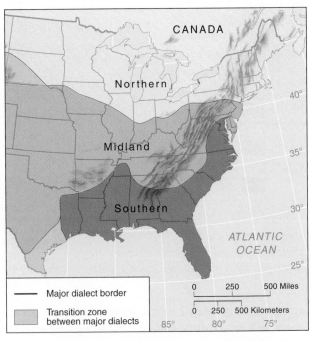

Figure 7-12 U.S. dialect regions. The three regional dialects of northern, midland, and southern American English were already developed in the eighteenth century, and as pioneers headed westward from each dialect region, they took place names with them.

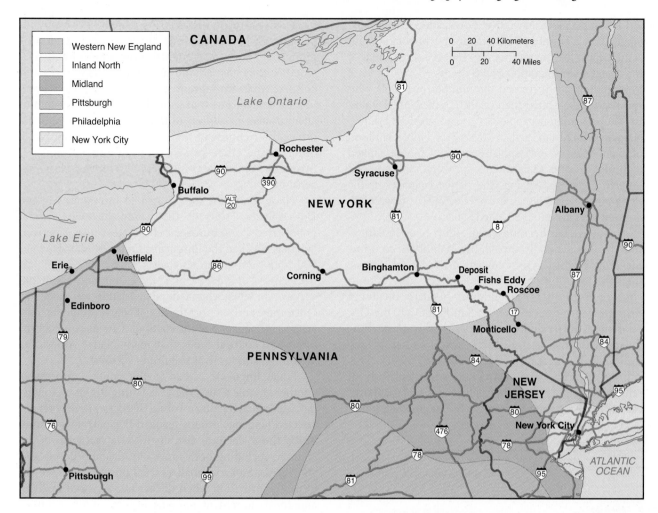

Figure 7-13 Pittsburgh and New York speech. Great Lakes commerce spread the Inland North dialect west to Chicago, so either in Rochester or in Chicago, you might hear a person ask, "What hee-appened?" "Cot" and "caught" sound rather like "cat" and "cawt." New Englanders often drop the "r." ("Pahk the cah in Hahvahd yahd.")

commonly named places Gap, Cove, Hollow, Knob, and Burgh. Southern settlers favored names using Bayou, Gully, and Store.

Regional variations in American English persist, both in grammar and in pronunciation, despite the nationwide reach of popular media. Television homogenizes national speech patterns much less than scholars originally thought it would (Figure 7-13).

Should American English be the official national language? The U.S. Constitution did not specify English as the official language, and many local governments through history and still today have found it useful to provide services and even keep official records in other languages (Figure 7-14).

In recent decades, however, a movement has grown to declare English the country's official language. In 1996, the U.S. House of Representatives approved such a law, and in 2006, the Senate did, but the House and Senate have never agreed on the same wording. Sixteen states have statutes or constitutional clauses declaring English their official language.

Figure 7-14 Bilingual signs. Some U.S. towns have added bilingual signs throughout their municipal buildings. The use of Spanish is widespread across the United States today.

This movement may reflect a dedication to a national language as a bonding force of a diverse population, but some people fear that it may also reflect a resentment of the changing immigration trends discussed in Chapter 5. Immigration has increased the number of people who speak a language other than English in the United States. In 2007, 55 million people claimed to speak at home a language other than English (Table 7-2). Fully 35 million of them (62 percent) spoke Spanish. Chinese ranked a distant second (2.5 million speakers), followed by French, German, Tagalog (the native language of the Philippines), and Vietnamese—reflecting the newest immigrant streams into the United States.

Language has become a civil rights issue in education. If, as is generally agreed, the inability to communicate in English is a handicap in the United States, then the teaching of English becomes a key route to equal opportunity for all children. In a case involving Chinese American children in San Francisco, the U.S. Supreme Court ruled that "students who do not understand English are effectively foreclosed from any meaningful education" (*Lau* v. *Nichols*, 1974). This ruling triggered a national concern to identify local school districts in which English was the students' second language and to sponsor bilingual educational programs in those districts. Arguments persist, however, over whether the only purpose of bilingual programs is to ease the students' transition to English or whether they also should preserve the languages and cultures that immigrant children bring to school. Some people argue that using the education system to acculturate all children to the English language denigrates the richness of their native inheritances, while others argue that preservation of other languages threatens to lock the youngsters into second-class citizenship.

These issues are not limited to immigrants to the United States. From the 1880s until the 1950s, federal policy tried to discourage or eliminate Native American languages. When the Europeans first arrived, more than 500 Native American languages were used throughout the territory of today's United States, but only about 200 survived in 1990, when federal legislation was passed to "encourage and support the use of Native American languages as languages of instruction." Language is such an important cultural index that many Native Americans view linguistic survival as cultural survival.

Equal linguistic access to the political process is another volatile issue. The U.S. Voting Rights Act mandates bilingual ballots in voting districts where voters of selected language groups reached five percent or more. Today those electoral districts include vast areas of the country, including both inner-city neighborhoods and rural areas. Some lawmakers tried unsuccessfully to remove the provision when the act was renewed in 2006, and the requirement remains U.S. law. Opponents of bilingual ballots argue that the right to vote comes with an obligation to use English. They also object to the cost of providing bilingual ballots. Proponents of the law point out that citizens have the right to vote regardless of their ability to use English. Some argue that English-only ballots would intentionally discourage minority voters, a form of election discrimination that the Voting Rights Act was designed to prevent.

TABLE 7-2 The Leading Non-English Languages in Use in the United States, 2007

Language	Number of Speakers
Spanish or Spanish Creole	34,547,077
Chinese	2,464,572
French (incl. Patois, Cajun)	1,984,824
Tagalog	1,480,429
Vietnamese	1,207,004
German	1,104,354
Korean	1,062,337
Russian	851,174
Italian	798,801
Arabic	767,319
African languages	699,518
Portuguese or Portuguese Creole	687,126
Polish	638,059
Hindi	532,911
Japanese	458,717
Persian	349,686
Urdu	344,942
Gujarathi	287,367
Serbo-Croatian	276,550
Armenian	221,865
Mon-Khmer, Cambodian	185,056
Miao, Hmong	181,069
Navajo	170,717
Yiddish	158,991
Laotian	149,045
Thai	144,405

Source: U.S. Census Bureau * Arabic speakers last counted in 2000

The Teachings, Origin, and Diffusion of the World's Major Religions

A religion is a system of beliefs regarding conduct in accordance with teachings found in sacred writings or declared by authoritative teachers. Most religions involve personal commitment to worship a god or gods, but the teachings of several Asian systems of belief, including Confucianism, Taoism, and Shintoism, are basically ethical and philosophical. That is, they focus on appropriate behavior (*orthopraxy*) rather than belief in a set of philosophical or theological arguments (*orthodoxy*). Several do not even address theological questions such as the nature of God or gods or goddesses, or life after

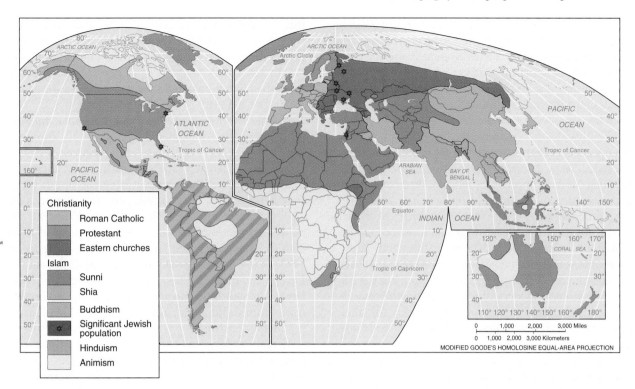

Figure 7-15 The world's major religions. This map shows the predominant faith or faiths in each region, but space restrictions limit it from showing minority faiths.

death. The strictest adherence to traditional beliefs is called **fundamentalism. Secularism,** by contrast, is a lifestyle or policy that purposely ignores or excludes religious considerations, usually because many religious claims cannot be scientifically tested or because religious values may be socially and politically divisive.

Religion is a major element of most cultures, so the geography of people holding and following a set of teachings affects politics, economics, agriculture, and diet, as well as many other aspects of human life. Most religions involve joining an organized body of fellow believers, and local administrative patterns, such as Roman Catholic dioceses, often organize and focus

socializing, community spirit, and, in many cases, education for their members.

Each religion originated in one place and spread out from there, but today communicants of various religions mingle around the globe, and religious affiliations cut across lines of politics, race, language, and economic status. Figure 7-15 shows the predominant faith or faiths in each region, but it cannot show minorities. At the most local scale, individual religious communities can be found within most of the world's large cities, sometimes clustering in identifiable neighborhoods.

No worldwide religious census is taken, but Table 7-3 provides an estimate of the adherents of each of the

TABLE 7-3 Adherents of the World's Major Religions (Mid-2008 Estimates, in Thousands)

Religion	Total	Africa	Asia	Latin America	Northern America	Europe	Oceania
Christians	2,254,535	465,880	364,106	536,162	277,089	583,802	27,496
Roman Catholics	1,130,418	159,776	128,901	474,595	83,210	275,209	8,727
Independents	411,142	92,928	179,166	42,381	74,085	21,104	1,478
Protestants	385,321	130,376	61,598	56,214	61,119	67,829	8,185
Orthodox	254,552	42,220	13,951	895	6,679	190,031	776
Anglicans	83,522	47,655	838	875	2,867	26,241	5,046
Buddhists	384,318	165	377,515	767	3,504	1,792	575
Confucians	6,418	300	6,346	500	—	18,300	53,300
Hindus	913,671	2,813	906,190	760	1,756	1,681	471
Jews	15,096	130	5,750	1,046	6,212	1,850	108
Muslims	1,434,081	392,636	992,850	1,830	5,556	40,749	460

Source: Adapted from the 2009 Encyclopaedia Britannica Book of the Year

world's principal religions. These figures cannot indicate the depth of anyone's belief nor the degree to which a person's religion actually affects his or her behavior. Also, several systems of belief listed are not exclusive; that is, a person may adhere to more than one. The following discussion will explain additional reasons why this table must be viewed with considerable caution.

This book cannot examine the fine points of each religion's message. For that, a text in comparative theology is needed. We will, however, briefly summarize the basic teachings of the world's major religions, because the distribution of people who hold those beliefs affects the distributions of other things: forms of government, dietary habits, women's rights, the organization of economies, and people's relationship with their environments.

Our coverage will identify the hearth region of each religion and review the story of its diffusion, but it is difficult for scholars to identify precise reasons why any religion diffuses. Believers in a religion will say that their religion diffused because that was God's or the gods' will. This kind of argument cannot be criticized or analyzed, for it is grounded in a faith that cannot be tested. A few religious groups **proselytize**—that is, try to convert others to their religious beliefs. Other religions do not proselytize. Some scholars differentiate *universalizing religions*, which seek new adherents through proselytization, from *ethnic religions*, whose adherents are typically born to a particular cultural group.

Judaism

Judaism has only about 15 million adherents, but it was the first of the widespread **monotheisms**—religions that preach the existence of only one God—to emerge in history. Many scholars suggest that **polytheism**, the worship of many Gods, is an older form of religion and that among some peoples polytheism developed into monotheism.

Judaism rests on a belief in a pact between God and the Jewish people that they would follow God's law as revealed in the *Pentateuch*—the first five books of the Old Testament section of the Bible. This covenant was granted to Abraham. Both Jews and Arabs claim descent from Abraham—Jews through his son Isaac, and Arabs through another son, Ishmael. Judaism is divided among a great variety of Orthodox (fundamentalist), Conservative, and Reform *sects*, or subdivisions.

Judaism developed historically in the Near East over many centuries. Under the Roman Empire, Jewish communities were established outside the Near East, but in A.D. 70 the Romans destroyed the temple in Jerusalem, and the Jews scattered in the Diaspora. During the Middle Ages, many European Christian rulers persecuted and expelled Jews, but Jews returned to Western Europe during the Enlightenment of the seventeenth and eighteenth centuries. On their return, they were required to live in segregated communities called *ghettos*. Legal emancipation came only in the nineteenth century. At that time, millions of Jews came to the United States from Eastern Europe, and migration from Russia and surrounding lands continues.

An estimated 1.7 percent of the U.S. population, more than 5 million people, were identified as Jewish in 2007. The majority of Jewish individuals live in New York State, other eastern states, and California, but Jewish communities are scattered throughout the country.

Israel as a Jewish state   During the Diaspora, many Jews visited Jerusalem if they could, but when the nation-state idea developed in Europe (see Chapter 11), many Jews came to accept **Zionism,** the belief that the Jews should have a homeland of their own in Palestine. When Israel came into existence in 1948, many Jews who had survived the Holocaust in Europe and many Mediterranean Jews migrated there. Israel has continued to welcome Jewish immigrants from around the world, although several Jewish sects repudiate Zionism.

Israel was intended to be a homeland for the Jewish people, but the degree to which Israel is "a Jewish state" is still being defined. In 1997, Israel declared that Arabs born in Jerusalem were aliens and thus subject to expulsion, but in 2000, the Israeli Supreme Court ruled that an Arab couple could not be barred from living in a community built solely for Jews. The Court declared, "We do not accept the conception that the values of the state of Israel as a Jewish state justify discrimination by the state between citizens on the basis of religion or nationality." In 2007, 20 percent of Israel's 7.2 million people were Arab. (These citizens are distinct from the Palestinians in the West Bank, Gaza, and East Jerusalem.) The balance of Jews and non-Jews in Israel has continuously shifted as a result of Israeli territorial expansion, immigration to Israel, and the birth rates of the various groups in Israel. In 1994, Israel withdrew from territory inhabited by Palestinians and that was put under the authority of the Palestine Liberation Organization (Figure 7-16). This was to lead to a future Palestinian state. That action increased the Jewish percentage in Israel's remaining territory, but Israel cannot remain distinctly Jewish unless it surrenders more territory occupied by non-Jews or expels non-Jews.

After a war in 1967, Israel held on to East Jerusalem, the Gaza Strip, and the West Bank of the Jordan River (a region previously part of Jordan). Israeli governments, spurred by Zionist extremists, continued to occupy and to build new settlements in these areas in violation of international law and despite admonitions against the occupation by the United States, the United Nations, and virtually all other countries and international agencies. In 1994, Israel granted self-rule to the Palestinians in Gaza. In 2005, it began to withdraw Jewish settlers, but continuing clashes between Palestinian groups and Israeli forces have left the situation troubled.

Figure 7-16 Israel and Palestinian territories. Israel was carved out of the Middle East in 1948, and it occupied larger areas during a war in 1967. Jewish settlers have moved into some of these occupied territories. In May 2000, Israeli troops pulled out of southern Lebanon, which they had occupied for 22 years. Israel has transferred about 20 percent of the West Bank to Palestinian control and another 22 percent to joint control, but the "Palestinian" areas are crisscrossed with Israeli-held roads and checkpoints.

In 2006, a quarter-million Jewish settlers still occupied the West Bank, where Israel began erecting an 805 km (500 mi) wall consisting of concrete, barbed

wire, electric fencing, motion detectors, trenches, and guard posts (Figure 7-17). The Israeli government said the wall was intended to lessen security risks, such as terrorist attacks, but it also separates land Israel seems to intend to retain from land that Israel may cede to Palestinians. This fence does not follow Israel's pre-1967 borders but encloses an additional 8 percent of the West Bank, so it incites Palestinian and Jordanian anger. While extending the wall, Israel has required Jewish settlers to abandon outpost settlements and return behind the wall. Thousands of settlers did return, although many struggled against their own (the Israeli) government. If the wall is a unilateral attempt to fix permanent borders, controversy will continue, because the demarcation deprives Palestinians of the best land in the area and leaves them with a jumble of unattached territories that will be difficult to coordinate into a successful Palestinian state. The Israeli Supreme Court has ruled that the wall's course should be determined by security needs alone, not the expansion of Jewish settlements.

The Palestine Liberation Organization (PLO), with both Muslims and Christians among its leaders (fewer than 8 percent of Palestinians are Christian), has sought to establish an independent Palestinian state since its founding in 1964. The governing functions of the PLO were transferred to the Palestinian National Authority after 1994. The Internet Corporation for Assigned Names and Numbers has assigned the Palestinian areas a domain code (.ps), and this action seems to suggest some unofficial international recognition of Palestine as a state. Thus, Palestine exists in cyberspace, but the creation of an independent Palestinian territorial entity remains the object of ongoing political confrontation and debate. The Palestinian people have fought Israeli occupation of lands they claim—often violently and even with terrorism—but the military force of Israel is so great that this contest of peoples has been an uneven one in which the "score" of dead and wounded, of destruction and persecution, enormously "favors" the Israelis.

Furthermore, anti-Israeli groups with their own militias not controlled by any national governments are based around Israel. Hamas, a group among the Palestinians, was once a part of the Palestinian National Authority but was barred from the government after it took over the Gaza Strip in 2007. Hezbollah, an Iranian-backed Shiite group in Lebanon, has representatives in the Parliament of Lebanon. Hamas and Hezbollah maintain their own independent militias and both deny the legality of the very existence of Israel. They have frequently engaged in armed conflict with Israel, with all sides causing civilian casualties. Israel invaded Lebanon again in 2006, laying waste to many Lebanese villages and destroying public infrastructure. Israel fought with Hamas in 2008 inside the Gaza Strip, one of the most densely populated areas on Earth. The process of achieving lasting peace within Israel and between Israel and its neighbors has been frustratingly slow and is marked by periodic retreats

Religious Fundamentalism and Political Terrorism

The people responsible for the September 11, 2001 attacks on the United States are often referred to as *Islamic fundamentalists*. Their actions were, in fact, repudiated by Muslim religious leaders throughout the world, as well as many Muslims themselves. But we must still ask what force of furious hatred this concept can carry for it to trigger the slaughter of thousands of innocent people, whatever one's religious beliefs. The 9/11 attacks were defined as acts of **terrorism**—that is, violent acts for political ends that intend to frighten and intimidate a civilian population beyond their immediate victims.

Religious fundamentalism of any kind sometimes provides meaning and direction to people who feel lost in a confusing world. Many people find comfort, or at least some kind of psychological "anchor," from strictly adhering to religious doctrines that are accepted as literal truth. These texts are followed absolutely, to the point that religious doctrine may override reason, judgment, and even conscience. Fundamentalism has led people to extraordinary acts of both good and evil. A "true believer" must encourage others to obey, too, or even coerce others, because other people's sins can corrupt the true believer, too. Sin must be punished and purged, even by force. When people feel that an issue is beyond matters of life and death, that the issue affects immortal souls, then they may commit acts other people would call "extremism."

The use of religion as a way to express repression and terror is not unique to Islam. The Christian Gospels teach peace, love, and forgiveness; but Christianity arguably has a worse history of persecution than Islam does. Christian history records centuries of persecution of Jews, mutual slaughter by Protestants and Roman Catholics, hunting for religious deviation by the Inquisition and other bodies, the murder of "witches," the Crusades, forced conversions, and other similar actions.

Today, Christianity is seldom associated with terrorism, but the actions taken by some Christians could be interpreted as such. The bombings of abortion clinics and murder of abortion providers by people who call themselves Christians, for example, demonstrate that some Christians continue to practice evil deeds to enforce what they see as the greater good. These Christian extremists believe that abortion is murder; therefore, the murder of abortion providers is logical to them. Such violence is seen as justified because it is viewed as the result of a choice between taking action and eternal damnation.

If religious texts written in primitive societies a thousand years or thousands of years ago are interpreted literally, the modern world can be terrifying. In recent decades, some Islamic fundamentalists have begun to argue that all of modern Western culture and society is sinful and evil. Believers of this fundamentalist perspective may argue that this world must be destroyed, or at least that its destruction is the consequence of its sin. Even some Christian fundamentalists hold this view. After the attacks on 9/11, the American Baptist minister Jerry Falwell argued, "God continues to lift the curtain [of His protection] and allow the enemies of America to give us probably what we deserve. I really believe that the pagans and the abortionists, and the feminists, and the gays and lesbians who are actively trying to make that an alternative lifestyle, the ACLU [American Civil Liberties Union—a civil rights organization], People for the American Way—all of them who have tried to secularize America—I point the finger in their face and say, 'You helped this happen.'" In other words, for Reverend Falwell, as for many other American Christian fundamentalists, the 9/11 massacre was divine retribution for American behavior.

Figure 7-17 Israel's security barrier. The project to physically divide Israel from the West Bank began in 1994, during negotiations that led to the transfer of territory to the Palestinian National Authority. The barrier is known as "the wall," but Israelis sometimes call it a fence. In many places, it divides Palestinians from their farmland and separates communities.

into violence, but a just settlement might encourage rising standards of living and security for everyone in the region.

Christianity

Christianity, the belief that God lived on Earth as Jesus Christ, emerged as a separate faith from Judaism, but part of the widespread appeal of Jesus' teachings rests on his emphasis on the nearness of the Kingdom of God and God's love for all. Jesus lived his entire short life in Judea, and after the death of Jesus, Saint Paul preached in many major cities in the eastern Mediterranean (Figure 7-18). Paul's letters (*epistles*) to early Christian groups make up much of the New Testament. Many Christians believe that proselytizing is a duty (Matthew 28:19).

The Bible records that the first non-Jew to convert to Christianity was an Ethiopian (Acts 8). Today, many scholars believe that man was probably from Makurra, in what is today the Sudan. The Christian church of Egypt, called the Coptic Church, was founded in Alexandria in A.D. 41, and Christianity diffused south from Egypt to today's Ethiopia by the fourth century. It has survived there despite having been cut off from the Mediterranean world by the later conversion of Egypt and the Sudan to Islam (to be discussed shortly). Jesus' disciple Thomas may have traveled to India, and although that is not certain, we do know that followers of Thomas had established Christian churches in India as early as the second century. Christianity also spread to the northeast, and very early in the fourth century Armenia became the first Christian nation. The Roman emperor Constantine (288?–337) converted to Christianity and favored the religion; and by the end of the fourth century, Christianity was the Roman Empire's official religion.

Under the Roman Empire, the Christian Church adopted a geographic organization that paralleled that of the empire's secular administration. Bishops ruled over dioceses. When the western half of the Empire fell to invaders in 476, the church organization survived as one of the few stabilizing and civilizing forces in Western Europe. Bishops, including the pope as bishop of Rome, assumed civil authority as well as ecclesiastical authority in areas where there was no other effective government. The first Western monastery was founded by Saint Benedict at Monte Cassino about 529, and as monasteries were established across Europe, the pope brought them under his protection. Throughout the West during the Dark Ages, the Church was the focus of life, and monasteries and cathedrals rose as the monuments of the civilization.

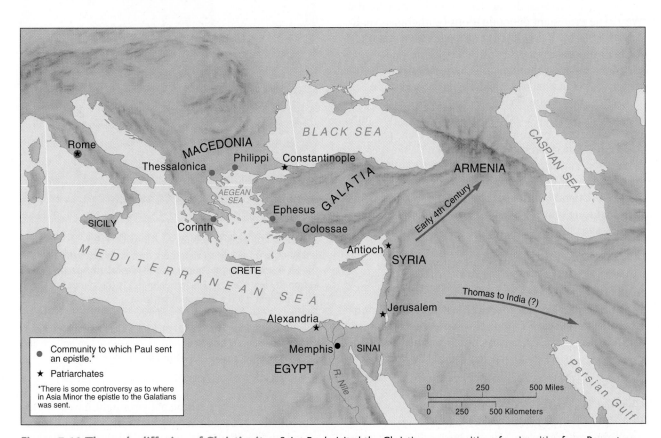

Figure 7-18 The early diffusion of Christianity. Saint Paul visited the Christian communities of major cities from Rome to Jerusalem. By A.D. 350, five patriarchs—co-equal leaders of the Church—had been recognized, with their seats at Alexandria, Jerusalem, Constantinople (today's Istanbul), Antioch (today's Antakya), and Rome. To the south, the Cathedral of Saint Mary was built at Axum (not shown) in A.D. 330, where the Christian emperors of Ethiopia were crowned until 1931.

In the eastern half of the Roman Empire, secular imperial control survived, with Constantinople (also called Byzantium, today's Istanbul) as its capital. The Patriarch of Constantinople headed the Eastern Christian Church, and missionaries from Constantinople (including Saint Cyril, mentioned earlier) converted the peoples of the Balkan Peninsula, Eastern Europe, Ukraine, Russia, and the Near East to Christianity. The Western and Eastern churches split finally when the Pope and patriarch excommunicated each other (officially declared each other to be outside the community of Christians) in 1054. Thus, Roman Catholic Western Europe came to be culturally divided from Orthodox Eastern Europe, although the mutual excommunications were revoked by the patriarch and the Pope in 1965. The Uniate Church of Ukraine bridges the gap by using the Cyrillic alphabet and following the Eastern rite, but recognizing some spiritual authority of the Pope. The fall of Constantinople to the Muslim Turks in 1453 opened Eastern Europe to Islam, but the office of the Patriarch of Constantinople survived, and all Orthodox Christians still recognize its spiritual leadership. Orthodox Christians are subdivided into national churches, such as the Serbian, Greek, and Russian, although U.S. and Canadian Orthodox Christians embrace historic national affiliations.

The Protestant Reformation of the fifteenth and sixteenth centuries split the Christian Church in the West. The Protestant denominations (Lutherans, Anglicans or Episcopalians, Baptists, and others) were named for their protest against the Church of Rome, and a variety of Protestant denominations and national denominations sprang up in Western and Northern Europe.

When Europeans explored and conquered other lands, they often cited religious conversion as their purpose. Christian missionaries were the chief agents of the partial Europeanization—of religion, language, social mores, and the acceptance of political authority—that made many natives more tractable to European rule. Christianity diffused rapidly and almost completely throughout the Western Hemisphere. Certainly, its message of all-embracing love carried strong appeal in lands in which many of the Gods had been portrayed as terrifying and cruel. Furthermore, the Christian concept of heaven after death is democratic and accessible to all classes of people.

Christians also have flexibly adapted some non-Christian practices. For example, in the fourth century, Western European Christian churches timed the celebration of Christmas to replace the celebration of the winter solstice among non-Christians. Christians still accept indigenous practices in order to bring peoples to Christianity. The 1992 Conference of Latin American Roman Catholic bishops pledged to incorporate the traditional religious symbols, rituals, and cosmologies of Latin America's Native Americans and blacks whenever "compatible with the clear sense of the faith" and "the general discipline of the church" (Figure 7-19).

Figure 7-19 Jesus as a Native American holy man. This image of Jesus as a Mescalero holy man dominates the altar at the St. Joseph Mission church on the Mescalero Apache Reservation in New Mexico. His left palm holds the symbol of the Sun; his right hand a deer-hoof rattle; at his feet an eagle feather, a grass brush, and bags of tobacco and cattail pollen used in ceremonies. He stands on Sierra Blanca, the Mescalero sacred mountain. The inscription, in Apache at the bottom and Greek at the top, reads, "Giver of life."

In Latin America, some natives might have adopted Christianity because they were astonished and demoralized by the Spaniards' invulnerability to the diseases that were wiping out the Native Americans. Neither the Spaniards nor the Native Americans understood that those diseases had been brought by the Spaniards themselves. The Spaniards convinced the natives to acquiesce to Spanish political authority and to seek both physical and spiritual salvation in the Spaniards' church. Similar beliefs prevailed among both the Native Americans and the whites in North America.

Christian missions to Asia were less successful. In 1542, exactly 50 years after Christians conquered Granada, the last Arab Muslim outpost in Europe, and after Columbus's first voyage to the West, Saint Francis Xavier arrived in Goa, India, to proselytize Asia. By 1549, Saint Francis was in Japan, which nearly converted to Christianity in the early seventeenth century but then expelled or persecuted Christians when it closed its doors to the world in the mid-seventeenth century (Figure 7-20). Christians had some initial success in China, but in 1723, Christianity was banned there as well.

Conversion to Christianity, and as will be seen later, to Islam, usually weakens the continuity of a people's cultural inheritance. This is because both Christians and Muslims believe that conversion is a spiritual rebirth. It is a deliberate repudiation of much of what came before, so religious conversion can destroy a people's historical culture more completely than can political conquest.

Christian sects and their distributions Roman Catholicism is the largest single denomination among Christians. It is headed by the Pope in Vatican City, a tiny independent city-state within Rome (Figure 7-21). Roman Catholicism has historically dominated in the Mediterranean basin, throughout Latin America, and wherever else Mediterranean peoples colonized or

Figure 7-21 Vatican City. Vatican City, made up of St. Peter's Church, Square, and the surrounding buildings and grounds, is one of the world's smallest independent states. It is entirely within the city of Rome, but it is ruled by the Pope, according to a treaty signed between the papacy and Italy in 1929. The official language here is ancient Latin, and in 2003, the Vatican produced a new Latin dictionary to deal with modern ideas and words.

converted, as in former French or Portuguese African colonies and in the Philippines and Vietnam.

Protestant denominations dominate in Northern Europe and wherever Northern Europeans have settled or converted: North America, Australia, and New Zealand, and the parts of Africa that were either formerly English colonies (today usually practicing Anglicans) or German colonies (Lutherans).

Several theological points differentiate Protestantism from Roman Catholicism, but among the most important is that Roman Catholics believe that Jesus gave Saint Peter unique responsibility for founding the Christian Church, that Saint Peter became the first bishop of Rome, and that the popes retain this special responsibility. The crossed keys on the Vatican flag represent Jesus' having given Saint Peter the keys to heaven. This belief that a church or priests intercede between God and humankind is called *sacerdotalism*.

Protestants deny sacerdotalism. Evangelical Protestants (from the Latin for "bringing good news") emphasize salvation by faith through personal conversion, the authority of Scripture and each individual's responsibility to read the Scriptures, and the importance of preaching, as contrasted with ritual. The evangelicals emphasize the ability of individuals to change their lives (with God's help), and this message offers to many people a new sense of personal empowerment. Thus, Evangelical Protestantism frequently brings a revolutionary force into traditionally rigid or stratified societies.

Today, Evangelical Protestantism is growing and spreading throughout the world. In some places, it is replacing non-Christian religions, but in many places,

Figure 7-20 The Basilica of Goa. The Basilica of Bom Jesus in Goa, India, is the final resting place of the missionary Saint Francis Xavier (1506–1552). Saint Francis is today the patron saint of Roman Catholic missionaries.

it is replacing Roman Catholicism. The population counted as Roman Catholic in Table 7-3 is the population baptized Roman Catholic. The Church insists that once a person has been baptized as a Roman Catholic, that person is always Roman Catholic, even if the person converts to another denomination or is excommunicated. Nevertheless, Evangelical Protestantism has replaced Roman Catholicism in several countries as the most widely practiced faith, and as a rule, the ardor of converts is strong. Evangelical Protestantism is also spreading in Pacific Asia (especially into South Korea and the Philippines), across Africa, and throughout Latin America. About one-quarter of the total Latin American population is Protestant today. This percentage is even higher in Brazil, which counts from baptism the world's largest Roman Catholic population, but where today full-time Protestant pastors outnumber Roman Catholic priests, and as many as 600,000 people convert to Evangelical Protestantism each year.

In 2000, more than 10,000 Protestant Evangelists from more than 200 countries and territories assembled in Amsterdam, the Netherlands. The "Amsterdam Declaration," which they adopted, urges evangelists to be sensitive to the societies in which they work and to avoid equating Christianity with any particular culture. It obliges evangelists to present the Christian gospel as authoritative, even while respecting people of other religions. A clause that encourages women to be gospel teachers could trigger change in many traditional societies that do not accept women in leadership roles and could hasten conversions from Roman Catholicism.

The spread of Evangelical Protestantism affects many other aspects of human geography. In most Latin American states, the Roman Catholic Church has traditionally enjoyed special privileges and a role in education, so its teachings have been enacted into law. Many Latin American countries, for example, have prohibited divorce. Today, however, a rising share of elected officials throughout Latin America is Protestant, and elections are explicitly referred to as "Holy Wars." The diffusion of Evangelical Protestantism is also at least partly responsible for slowing population growth. Evangelical Protestants are often as opposed to abortion as Roman Catholics are, but they are not always so adamantly opposed to other forms of birth control. Roman Catholicism has lost political power throughout Latin America, Quebec, Spain, Italy, the Philippines, and elsewhere, and presidents who are Protestant or *agnostic* (doubtful or noncommittal about the existence of God) have been elected, for example, in the Philippines, Guatemala, Colombia, and Chile. Governments have begun sponsoring family planning, distributing birth-control devices, legalizing divorce, and introducing sex education into school curricula. Chile, for example, legalized divorce in 2005 and in 2006 installed a new agnostic president, Michelle Bachelet, who "promised" rather than "swore" to uphold the Constitution.

Other possible results from the spread of Evangelical Protestantism have been hypothesized, and they bear watching through coming years. One is the spread of literacy. This follows from Protestants' individual responsibility to read and study Scripture. Literacy has risen in areas where Evangelical Protestantism has spread, but the exact cause-and-effect relationship is difficult to measure. If Evangelical Protestantism does raise literacy rates, this trend may carry further ramifications—greater political participation, for example. Still other possible consequences of the spread of Protestantism in the areas of women's rights and even economics will be discussed later in this chapter.

In addition to Roman Catholic, Protestant, and Orthodox Christians, many smaller sects include the Georgian and Armenian churches; the Maronite churches in the Near East; the Copts of Egypt; the Christians of India, Ethiopia, and China; and a few other groups. There are also many Christian movements that are nondenominational, meaning they are not members of one of the established church divisions. Another movement, the Latter Day Saints, includes the Mormon Church, formally known as The Church of Jesus Christ of Latter Day Saints. Mormonism, which follows the religious teachings of a nineteenth century religious figure, Joseph Smith, Jr., is one of the fastest-growing religions in the world. In 1950, there were 1 million Mormons, nearly half of whom lived in Utah. Today, the religion counts 12 million members around the world, fewer than 15 percent of whom live in Utah.

The future of Christianity   The geography of Christianity is undergoing significant changes. The observance of Christianity has increased throughout Eastern Europe with political liberalization since the fall of Communism. Nevertheless, within the lifetime of most living Christians, Christianity has shifted from being a religion of the Western industrialized nations to being a religion of Asia, Africa, and the Pacific. In Africa, for example, the number of Christians has grown, chiefly by conversion, from fewer than 10 million in 1900 to about 466 million in 2010 (Figure 7-22). In Latin America, the number of Christians has risen from 62 million in 1900 to 536 million today; in Asia from 19 million to 364 million. Today, there are more Lutherans in South Africa than in North America, and today more than half of the world's 83.5 million Anglicans live in Africa. South Korea is over 25 percent Christian—75 percent of them Protestant. The new state of East Timor became the second Asian nation to have a Christian majority, after the Philippines. In 2003, the World Council of Churches elected its first African Secretary General, Reverend Samuel Kobia, a Methodist from Kenya.

Christian missionary activity is increasing: There are today about 400,000 Protestant and Roman Catholic missionaries in the world, which is six times the number in 1900. About 100,000 of these missionaries are from Protestant churches in non-Western countries. Korea, for

Figure 7-22 The world's largest Christian church. The Basilica of Our Lady of Peace in Yamoussoukro, the Ivory Coast, is modeled after St. Peter's in Rome (see Figure 7-21), but it is larger, capable of holding more than 18,000 worshipers.

example, has sent almost 10,000 Protestant missionaries outside its boundaries. Many missionaries witness their faith only through good works, such as building educational and charitable institutions and providing humanitarian assistance. More aggressive efforts toward conversion, however, are meeting stiffening resistance from other religions; the divisions within Christianity aggravate the tensions. The spread of evangelical Christian missionaries throughout the Muslim world, for example, has provoked anger not only among Islamic clerics, but also among the native Christian groups. Trying to convert Muslims is often interpreted as a provocation that invites violence, and conversion from Islam is in many places punishable by death. Today the estimated 40 million Christians in countries ruled by Muslims find themselves an embattled minority facing economic decline, dwindling rights, and even physical jeopardy. This oppression and decline is in contrast to the rights (if not always full equality) that the surging Muslim minority enjoys in Western societies. Already in 2003, Vatican Foreign Minister Cardinal Jean-Louis Tauran noted, "There are too many majority Muslim countries where non-Muslims are second-class citizens. Just as Muslims can build their houses of prayer anywhere in the world, the faithful of other religions should be able to do so as well." In 2006, rage exploded among Muslims around the world when Pope Benedict XVI quoted a medieval belief that Muhammad had introduced "things only evil and inhuman."

Missionary activity by Americans has become a lightning rod for anti-American political sentiments. American missionaries have been killed in the Muslim Near East and Africa, and in India by Hindus. The U.S. National Association of Evangelicals has repeatedly reaffirmed its commitment to proselytizing, but it issued a "loving rebuke" to some evangelical leaders,

such as the Reverend Franklin Graham (who said Islam is "a very evil and wicked religion"), Reverend Jerry Falwell ("Muhammad was a terrorist"), and Reverend Jerry Vines (who described Muhammad as "demon-possessed") for remarks that "tarnish Christianity" and jeopardize the safety of missionaries and the indigenous Christians in some countries. In 2004, Reverend Richard Cizik, president of the association, criticized "unfortunate and particularly irresponsible" remarks that "complicate circumstances for foreign missionaries and Christian aid workers overseas who are already perceived, wrongly . . . as collaborators with U.S. intelligence agencies." Nevertheless, the close affiliations between Evangelical organizations and some U.S. political leaders is interpreted by many in the world as an explicit tie between missionary Christianity and American political power.

Pope John Paul II, who served 1978–2005, doubled the number of Roman Catholic saints. Virtually all of the saints he canonized were African and Asian. In 2002, he canonized Juan Diego, a Native American who supposedly sighted the Virgin (Mary) of Guadalupe in today's Mexico in 1531. Historians insist that Juan Diego never existed, but the pope hailed Diego as "the first indigenous saint of the American continent."

The Vatican maintains that the only purpose of any interfaith dialogue is to convert others to Roman Catholicism, and this conviction of Roman Catholicism's unique truth has triggered resentment and even political and social repercussions. For example, on a trip to India, John Paul II declared that "the peoples of Asia need Jesus Christ and his gospel." Prominent Hindu, Sikh, Muslim, and Buddhist priests walked out of the room. Similar statements made during visits to the Near East triggered strong objections and even street riots. Roman Catholic proselytizing in Russia—trying to convert Russians from their traditional Orthodox Church to that of Rome—has triggered political restrictions on Roman Catholicism, such as the denial of permission to build new churches. Roman Catholic Cardinal Walter Kasper's attendance as Pope Benedict XVI's representative at the World Summit of Religious Leaders in Moscow in July 2006, however, signaled a thaw in relations between the two churches.

Chapter 6 noted that any cultural system or artifact may adapt and permutate as it diffuses across local cultures, and the shift in Christianity's population center of gravity has inspired changes in Christian theology and practice. As Roman Catholic Bishop Bonifatius Hauxiku of Namibia has said, "Our African people have accepted Christ, but this Christ walks too much among them in a European garment." Local churches are adopting local foods for the sacred rite of communion, although, historically, Christians' need for wine and bread for this rite triggered the global diffusion of grape and wheat crops. Roman Catholic bishops around the world have recommended ordaining married men, a practice forbidden by the Vatican. The bishop of Nassau,

The Bahamas, emphasized that on such questions "the church should not be tied to cultural vestiges typical of the European experience."

Christian leaders struggle to express Christian belief in terms meaningful to a great diversity of cultures without abandoning essential Christian distinctiveness. In many areas, Christianity is challenged by **syncretic religions,** which combine practices and beliefs from two or more religions. The African and Caribbean religions of Voodoo and Santería, for example, identify African deities and spirits with Christian saints. When Mathieu Kerekou, a Christian, was sworn in as president of Benin in 1996, he left out the italicized words in the oath: "Before God, *the spirits of the ancestors*, the nation and before the people . . . I . . . swear to respect the constitution." The public of this largely Voodoo nation insisted he retake the oath with the italicized words. Chapter 5 noted the Native American ceremony at which Alejandro Toledo, a Christian, was sworn in as president of Peru in 2001. Many Brazilians practice a syncretic religion called Candomblé, and in 1999, Brazil's minister of culture declared a Candomblé temple a national monument. Other syncretic religions include Cao-Dai from Vietnam and Chondo-Kyo from Korea.

Islam

Through the centuries after Abraham and Ishmael, the Arabs fell away from monotheism to polytheism, but they were brought back to monotheism by Muhammad (c. 570–632). He founded the religion called Islam, which means "submission [to God's will]." "One who submits" is a Muslim. The five essential duties of a Muslim, called the *Five Pillars*, are belief in God, five daily prayers, generous giving of alms, fasting during one month (called *Ramadan*), and, if possible, a pilgrimage (*hajj*) to Mecca at least once in one's lifetime.

The Arabic word for the one God is *al-elah*, or Allah, a cognate of the Hebrew *eloh*, "God." The Muslim God is the God of Abraham, Moses, and Jesus, so Judaism, Christianity, and Islam are often called "the Abrahamic religions." The essential belief of a Muslim is stated in the expression "There is no God but God" ("la Ellaha Ela Allah"), with which Jews and Christians would agree. This Arabic statement, however, is often half-translated or mistranslated. For example, the U.S. Marine Corps Intelligence Service's "Iraq Culture Smart Cards," distributed to U.S. troops serving in Iraq, state that Muslims believe that "Allah is the one true god" (with a small "g," rather than, for example, "Allah is the Arabic word for God" or simply "Muslims believe in God"). Such mistranslation is not only incorrect but prejudicial. It alienates Christians and Jews from Muslims rather than reconciling them. Some Muslims believe this insult is deliberate.

Despite Muhammad's flight from his native city of Mecca in 622 (the *Hegira*), which was a temporary setback and the year from which Muslims date their calendar, he had converted and united most Arabs by his death. Just as Christians see their religion as building on Judaism, adding the New Testament to the Old, so Muhammad envisioned his teachings as a continued evolution of monotheism. Muslims believe that Muhammad was the last of God's prophets, who also include Adam, Noah, Abraham, Moses, and Jesus. Muhammad expected Christians and Jews to be among the first to embrace Islam. He first directed Muslims to face Jerusalem in prayer, but he later changed the direction toward Mecca (Figures 7-23 and 7-24).

As Islam diffused, the Arabic of the Koran became the language of ethnically different peoples throughout the Near East and across North Africa. They had not previously been Arabic speakers, but as they converted, all came to be known as Arabs. To the East, Persia (today's Iran) had its own ancient culture, and although the Persians converted to Islam, they retained their own language, Farsi, an Indo-European language. Therefore, today's Iranians are not Arabs, nor are any of the peoples to the east of Iran.

Within slightly more than a century after Muhammad's death, Islam stretched from the Atlantic coast of Spain across North Africa and through Southwest and Central Asia to the borders of China. An Arab army met a Chinese army in the Talas River Valley in today's Kyrgyzstan in 751, just 19 years after another Arab army

Figure 7-23 Mecca. This black-shrouded building is the Kaaba in Mecca, where, Muslims believe, the biblical patriarch Abraham and his son Ishmael built the first sanctuary dedicated to the one God. Through millennia, the temple was given over to the worship of idols, but Muhammad threw out the idols and rededicated the Kaaba. That building has since been replaced, but this is probably the longest continuously revered spot on Earth.

Figure 7-24 Dome of the Rock, Jerusalem.
Sacred to Jews, Christians, and Muslims, this site is believed to be where God prevented Abraham from sacrificing his son on this rock (Mount Moriah). King David later built the Jews' Temple here. Muslims believe that Muhammad ascended to heaven from this spot. The beautiful building with the golden dome is the Dome of the Rock (begun in A.D. 643), the oldest existing monument of Muslim architecture. Christians believe that Jesus ascended to heaven from the Mount of Olives in the background. In the foreground is the Western Wall, the only wall remaining of the Jewish Temple, which was destroyed by the Romans in A.D. 70.

had faced a Frankish army in Tours in today's France thousands of miles to the West. Within this vast realm, dominance eventually passed from the conquering Arabs to the non-Arab majority, including the Persians and the several Turkic peoples. Islam later expanded down into South and Southeast Asia (Figure 7-25). This culture realm greatly exceeded the Christian culture realm in extent, power, and riches for hundreds of years. Learned travelers crossed its length and breadth, and their descriptive writings constitute some of the greatest works of historical geography (Figure 7-26).

Islam denigrates the earlier cultures of its converts, just as we noted that Christianity can. Everything before Islam was, in Arabic, *jahiliya*, "from the age of ignorance." This leaves little room in these people's historical consciousness for their pre-Islamic past, so they often lack interest in it. For example, despite Persia's brilliant antique history, for contemporary Iranians the glory began with the coming of Islam. Pakistan is a new Muslim state, so even though the land contains ruins of civilizations thousands of years old, contemporary Pakistanis disdain them. Many people in Muslim countries view their own ancient cultural landscapes without interest and may even discourage tourists from viewing pre-Islamic ruins. Similarly, the study of the history and art of pharaonic Egypt is the result of European historical interest. The fact that Egyptians are interested in it today demonstrates that Egyptian nationalism, a newer cultural force, competes with Islam for the primary loyalty of the Egyptian people.

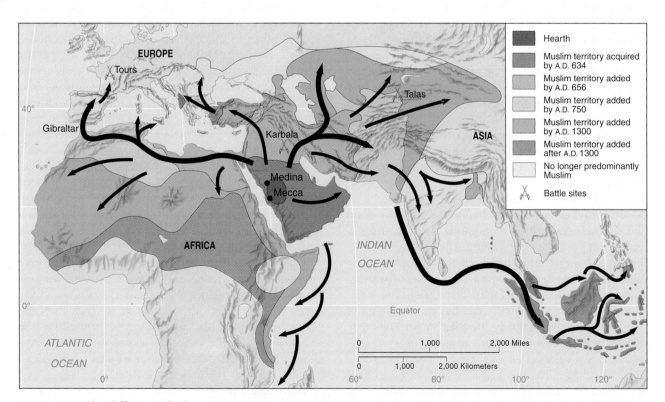

Figure 7-25 The diffusion of Islam. The first nine centuries of Islam were a period of almost continual expansion, and by 1800, Arab and Indian traders had spread the faith through many of the islands of Southeast Asia.

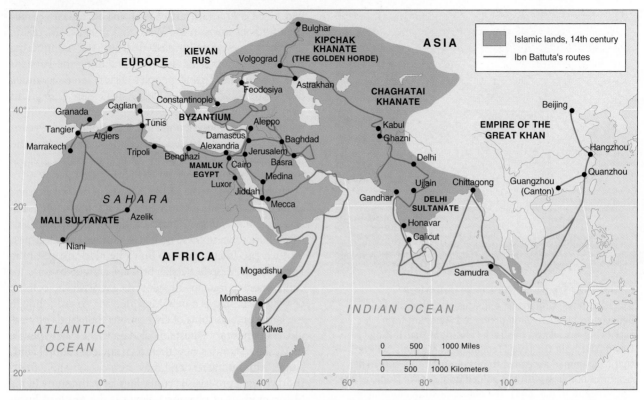

Figure 7-26 The travels of Ibn Battuta. In the fourteenth century, the Muslim scholar Ibn Battuta traveled from Marrakech to China, where he was happily received by rulers. Several even appointed him as a judge thousands of miles from his own home. His geographic writings are a superlative description of life across the vast Islamic realm of his day.

As Islam filtered south across the Sahara desert from North Africa, its message reached the black peoples of the Sahel and savanna. There it still competes with Christianity, which entered sub-Saharan Africa from the Europeans' coastal incursions. From Senegal to the Congo the coastal areas are Christian, and inland areas are Muslim. When several of these countries first won independence, the westernized political leaders were Christian (for example, President Léopold Senghor in Senegal and President Félix Houphouët-Boigny in the Ivory Coast), but the majority of the population was Muslim. Christian politicians continue to play prominent roles, but as democracy matures in these countries, Muslims often gain more power.

Within Europe, only Albania and Kosovo, long ruled by Turks, are predominantly Muslim. Bosnia and Herzegovina is nearly half Muslim. There are important Muslim minorities in most of the other countries in southeastern Europe. Islam in these formerly Communist countries is often a traditional or relatively weak religious identity that has a far less important role in public or daily life than in societies that interpret Islam more strictly.

Islamic sects   The two principal sects of Islam date back to a struggle over rule of the Islamic world that occurred shortly after Muhammad's death. Sunni Muslims accept the tradition (*Sunna*) of Muhammad

as authoritative and approve the historic order of Muhammad's first four successors, or *caliphs*. About 85 percent of Muslims worldwide today are Sunni. Most of the other 15 percent are Shia Muslims, or Shiites. They believe that Muhammad's son-in-law Ali was the rightful successor to the Prophet, and they commemorate the martyrdom of Muhammad's grandsons in a Muslim civil war battle at Karbala in today's southern Iraq in 680. Through the centuries, differences in ceremony and in law have further differentiated the Sunnis from the Shiites.

The ruling family of Saudi Arabia is of the Wahhabi sect, a puritanical movement that is the only modern separatist Sunni sect. Muhammad Ibn Abdal-Wahhab (1703–1792) sought to purify Islam from what he saw as the corruptions of mysticism, rationalism, and Shiite theology. Western collaboration with Wahabbites during World War I to defeat the Ottoman Empire gave birth to Saudi Arabia, a Wahhabite state whose immense oil wealth since World War II has allowed the aggressive diffusion of Wahhabism through the construction of schools and mosques throughout the Muslim realm. The diffusion of this fundamentalism has had political repercussions, as we shall see below.

Shiites form the majority in Iran, Azerbaijan, and Iraq and are important minorities in Kuwait, Lebanon, Bahrain, Syria, Saudi Arabia, and Pakistan. Animosity between Sunni and Shia can be fierce. Countries that contain significant shares of both sects are split by the enactment into national laws of either Sunni or Shiite interpretations of Muslim religious teachings, and several, including Lebanon and Pakistan, have suffered civil disturbances between the two groups. Ever since the overthrow of the government of Iraq in 2003, fighting between the historically dominant Sunni minority and the newly enfranchised Shiite majority has threatened U.S. peacekeeping and nation-building efforts (see Chapter 13). Some observers suggest that, in the long run, the United States has been only a pawn in a 650-year-long clash between Shiites and Sunnis. Today, a newly resurgent Shiite Iraq stands with Shiite Iran and Shiite Hezbollah in Lebanon against Sunni Pakistan (where Shiites are regularly referred to as "mosquitoes") and Saudi Arabia (where school textbooks refer to Shia Islam as "heresy"). Even the sacred *hajj* to Mecca has regularly been disrupted by violent clashes between Sunnis and Shiites.

The Muslim world today Today, Muslims are distributed from Morocco to Indonesia, north to the frontiers of Siberia and south to Zanzibar, with smaller communities scattered throughout the world. Even Rome boasts a splendid mosque. The leading Muslim states, in numbers, are Indonesia, Bangladesh, India, and Pakistan.

Muslims believe in proselytizing, as Christians do, and this necessarily injects an element of competition into the relationship between the two religious communities. Islam may evolve as Muslims move outside the Islamic realm, just as Christianity has evolved as it has diffused. For example, speaking at the dedication of a new mosque in Lyons, France, the mosque's imam (Islamic prayer leader), Algerian-born Chellali Benchellali, proclaimed that he stood "between two cultures. . . . I think this is the beginning of a French Islam." At a gathering of European imams, Muslim community leaders, held in Vienna in 2006, the Bosnian representatives insisted that "Muslims who live in Europe have the right—no, the duty—to develop their own European culture of Islam." Sayed Ghaemmagami, the Grand Mufti of the Shiites in Germany, argued that "the existence of an Islamic diaspora . . . requires new thinking about relations between Muslims and others, so Muslims should engage with their new countries and not set up parallel structures." Institutes for Islamic studies are proliferating across Europe. If European Muslims create an Islam that is open to democracy, sexual equality, and modernity, it will affect Islam around the world.

In other non-Arab regions, local Islamic practices differ. Many Indonesians, for example, retain ancient local Gods and goddesses and folk practices in a form of syncretism. These are declining, however, as a result of urbanization and Wahhabite missionary work.

Although many Americans tend to equate Islam with Arabs, Arabs actually constitute only a fraction of Muslims. There are more Muslims in Indonesia and Pakistan than in all the Arab countries combined.

Americans' association of Islam with Arabs is a political preconception, caused by the distinctive role of Islam in Arab countries and by U.S. news media focus on Arab lands as sources of oil and antagonists of Israel. There are historically important, if now small, communities of Christian Arabs scattered throughout the Middle East.

Hinduism and Sikhism

Hinduism is the most ancient religious tradition in Asia. The oldest Hindu sacred texts (the *Vedas*) date to about 1800 B.C., but the religion originated somewhere in Central Asia long before that. It entered the Indian subcontinent with the coming of Central Asian peoples about the time of the writing of the *Vedas*. Today, it is confined almost exclusively to India and Nepal, where it is the official state religion (Figure 7-27).

Hindus believe in one Supreme Consciousness, Brahman, whose aspects are realized in three deities: Brahma, the creator; Vishnu, the preserver; and Shiva, the destroyer. These three are coequal, and their functions are interchangeable. All other Hindu "Gods," saints, or spirits are emanations of Brahman.

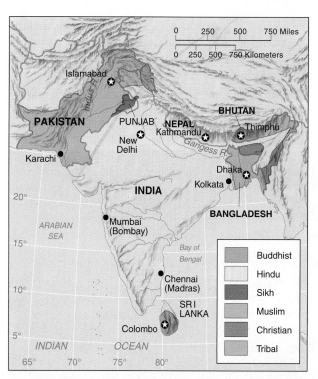

Figure 7-27 Religions in South Asia. Hinduism is today restricted almost exclusively to India, Nepal, and to places where Indians have migrated, including Jamaica, Trinidad, Guyana, and Fiji in the Pacific Ocean. Sikhs form a majority in the Indian state of Punjab, but smaller Sikh communities can be found throughout India and the former British empire. The regions labeled Christian converted from animism in the nineteenth century.

Hinduism classifies people in a hierarchy of classes called **castes.** The four main castes are (1) the Brahman, or priestly caste; (2) the Kshatriya, or warrior caste; (3) the Vaisya, or tradesman and farmer caste; and (4) the Sudra, or servant and laborer caste. Each of these castes is split into hundreds of subcastes, many of which are defined by occupation. People are expected to mix socially, marry, and stay in the caste into which they were born. A group of people called **untouchables** (*Dalit*) is considered so low that their status is below the formal structure of the four castes. To some degree, caste discrimination still structures Indian life, but it was legally abolished in the Indian Constitution of 1950, and an untouchable, K. R. Narayanan, was even sworn in as president of India in 1997. (India's president is elected by parliament, not by popular vote.)

Hindus believe in *reincarnation*—that is, individual rebirth after death. The caste into which you are born is not haphazard but depends upon your behavior in an earlier life. This teaching, called *karma*, discourages ambition because only by docilely keeping to your place in this life can you hope to enjoy a better position in your next life. The goal of Hindus is liberation from the cycle of death and rebirth.

Sikhism is an offshoot of Hinduism based on the teachings of Guru (teacher) Nanak (c. 1469–1539). Nanak tried to reconcile Hinduism and Islam, teaching monotheism and the realization of God through religious exercises and meditation. Nanak opposed the maintenance of a priesthood and the caste system. Under a series of gurus, the Sikhs had their own state in northern India, but they were eventually conquered by the British. Many Sikhs dream of the restoration of an independent Sikh state. The Sikh's holy temple is in the city of Amritsar in the Indian state of Punjab (Figure 7-28). This state is largely Sikh, and it might provide a territorial base for independence.

Buddhism

Siddhartha Gautama (c. 563–483 B.C.) was a Hindu prince born in what is today's Nepal who, through meditation, achieved the status and title of Buddha, or Enlightened One. He taught Four Noble Truths: (1) Life involves suffering, (2) the cause of suffering is desire, (3) elimination of desire ends suffering, and (4) desire can be eliminated by right thinking and behavior. This cessation of suffering is called *nirvana*, or total transcendence.

 Video Untouchable?

As Buddhism diffused out of India, sects and schools arose. The Theravada school ("doctrine of the elders") diffused to the south (Figure 7-29). This school centers on the idea of a monk striving for his own deliverance. Mahayana Buddhism (the "great vehicle" because it carries more people to nirvana) diffused northward. Mahayana idealizes the concept of the *bodhisattva*, someone who merits nirvana but postpones it until all others have achieved enlightenment. In Bhutan, Tibet, and Mongolia, Buddhism evolved a special form called *Lamaism*, which is known for its elaborate rituals and complex priestly hierarchy (Figure 7-30). Chinese

Figure 7-28 The Golden Temple in Amritsar. The Sikhs' 400-year-old Golden Temple in Amritsar, India, stands in the middle of a sacred lake (in Sanskrit, *amrita saras,* "pool of immortality"). The building was occupied by armed Sikh extremists in 1983; the following year, Indian troops stormed it and drove out the occupants in a bloody encounter. Sikh fanatics retaliated by assassinating India's Prime Minister Indira Gandhi.

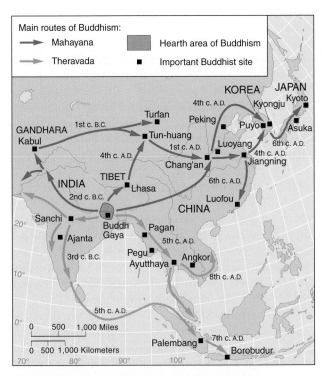

Figure 7-29 The early diffusion of Buddhism. Buddhism was localized near its hearth for hundreds of years, but it spread widely due to the support of the third-century Indian emperor Asoka. It has largely been abandoned in India.

Figure 7-30 The Dalai Lama. The Dalai Lama, the Buddhist spiritual leader of Tibet, was forced into exile by the government of China. He has won worldwide acclaim—including the Nobel Peace Prize in 1989—for his efforts to promote religious freedom.

Figure 7-31 Lao-tze and Confucius with the infant Buddha. This fourteenth-century Chinese painting shows Lao-tze (on the left) and Confucius protecting the infant Sakayumi, the future Buddha. Lao-tze's appearance reflects his legendary life spent wandering and communing with nature, in contrast to the courtly scholar Confucius. Images such as this, emphasizing the compatibility of Confucianism, Taoism, and Buddhism, became popular as Buddhism spread across East Asia.

Buddhists produced a new theory of spontaneous enlightenment, called *Ch'an*. This diffused into Japan as *zen*.

Buddhism has several hundred million followers, but its adherents are hard to count because it is not an exclusive system of belief. Its practice has declined in India, but today it is the state religion in Thailand and Sri Lanka, and it may achieve that status in Mongolia. Buddhist philosophy has also won considerable influence in the modern Western world. Much popular "New Age" philosophy derives from Buddhism.

Other Eastern Religions

Confucianism is a philosophical system based on the teachings of K'ung Fu-tzu (c. 551–479 B.C.). He taught a system of "right living" preserved in a collection of sayings, *The Analects*, which governed much of China's political and moral culture for 2,000 years.

Confucianists believe that people may attain heavenly harmony by cultivating knowledge, patience, sincerity, obedience, and the fulfillment of obligations between parents and children, subject and ruler. These moral precepts permeate life in many Eastern societies, and today the word *Confucian* is often popularly used as a synonym for these qualities. Confucianism influenced

Western philosophy and political theory at the time of the Enlightenment, when it appeared as the realization of Plato's utopian dream of a state ruled by philosophers.

Taoism is derived from the book *Tao-te Ching* (third century B.C.) attributed to Lao-tze. It advocates a contemplative life in accord with nature (Figure 7-31).

Shinto is the ancient religion native to Japan. Its rituals and customs involve reverence for ancestors, celebration of festivals, and pilgrimage to shrines, but there is no dogma or formulated code of morals. Shinto is nationalistic, and it traditionally recognized the emperor as divine. The Emperor Hirohito renounced this divinity in 1946, following Japan's defeat in World War II. Nevertheless, signs at the Yakasuni Shinto Shrine still proclaim, "Some 1,068 people, who were wrongly accused as war criminals by the Allied court were enshrined here. . . . War . . . was necessary in order for us to protect the independence of Japan and to prosper together with Asian neighbors." Periodic visits to Yakasuni by Japan's prime ministers roil antagonism between Japan and Japan's World War II enemies, especially China and the Koreas. In 2005, Chinese Foreign Minister Li Zhapxing asked, "The leader of a certain country is still worshipping war criminals.... What sort of behavior is this?" In 2006, Japan's new Prime Minister Shinzo Abe promised to end official visits to Yakasuni, although he has often gone privately in the past.

Animism and Shamanism

Animism is a belief in the ubiquity of spirits or spiritual forces (Figure 7-32). Animistic religions may be basically monotheistic, but they recognize hierarchies of divinities who assist God and personify natural forces. Millions of Africans believe in animism.

Animism is frequently accompanied by **shamanism.** A shaman is a medium who characteristically goes into auto-hypnotic trances, during which he or she is thought to be in mystical communion with the spirit world. Shamanism exists among the peoples of Siberia, the Inuit, some Native American tribes, in Southeast Asia, and in Oceania. Almost everywhere, both animism and shamanism are yielding to Muslim and Christian proselytizing. Shamanic knowledge often includes traditional medicines and cures drawn from local plants. The world's pharmaceutical companies have been seeking to learn shamans' ethnobotanical knowledge.

Figure 7-32 An animist altar. This animist altar stands in a Dogon village in the West African country of Mali. It is covered with a coating of cooked grains laid on as an offering to the spirits of the harvest.

The Political and Social Impact of the Geography of Religion

The distribution of religions affects the distributions of many other facets of life and culture. Religion is an important part of people's identities, both personal and cultural. It influences what they eat; their ideas of marriage and family; their interpretation of other cultures; when and how they work; their appreciation and use of nature; which types of behavior they encourage and discourage; and many other aspects of life. Religion's influence is so pervasive that it is difficult to isolate and identify precisely. Clearly, however, many people choose to obey their religion's teachings. The following pages catalog just a few of the many ways that the geography of religion influences the geography of other aspects of human lives and societies.

Religion and Politics

Many countries guarantee freedom of religion, and most governments observe a form of secularism. Governments nevertheless often favor one religion over others, either implicitly or explicitly. Therefore, the world political map partly fixes or stabilizes the map of world religions, just as it fixes the map of world languages.

Theocracy is a form of government where a church rules directly. The Vatican is a theocracy. Morocco may be a modified theocracy, because the king's legitimacy derives partly from his descent from Muhammad (Figure 7-33). Many theocracies have existed in history: Tibet, for example, and Massachusetts Bay Colony in British North America. Utah was once a Mormon theocracy, and although it is today increasingly cosmopolitan, no Utah politician can ignore Mormon Church leaders (Figure 7-34).

In Israel Israel has steadily enacted more religious strictures into law since its founding in 1948. Recent laws, for example, ban the production or sale of pork and prohibit activities on the Sabbath. Also, the government has become more Orthodox, recognizing, for example, conversions to Judaism performed only by Orthodox rabbis. This has distressed some U.S. Jews, 85 percent of whom practice Conservative or Reform Judaism. Many of the Russian Jews who have recently migrated to Israel are not Orthodox, and they press for secularization, including civil marriage, eliminating the "nationality" designation on identity cards, and the operation of public transit on the Sabbath—all policies opposed by the Orthodox.

In Christianity The Christian Bible records Jesus' words, "Render therefore unto Caesar the things which be Caesar's, and unto God the things which be God's"

Figure 7-33 King Muhammad VI of Morocco. King Mohammed VI alternates between wearing modern westernized clothes and the robes traditional to his role as hereditary religious leader.

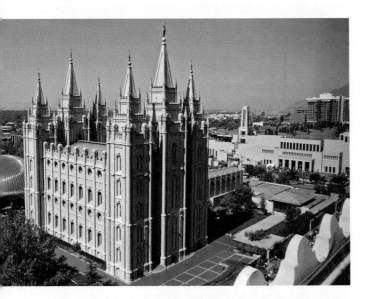

Figure 7-34 Temple Square of the Church of Jesus Christ of Latter-day Saints (often called the Mormon Church) in Salt Lake City. In 1847, this site was surrounded by hundreds of miles of wilderness, but Mormon leader Brigham Young recognized it as an oasis that Mormon industriousness could make bloom and where Mormons might escape the persecution they had suffered in the East. The building at left in the form of an overturned hull of a ship is the Tabernacle, and the neo-Gothic building in the center is the Temple.

(Luke 20:25). For 2,000 years, governments among Christian peoples have argued about exactly where to divide responsibility between religion and government, but they accept the idea that church and state, religion and politics, must be separated.

The separation between religion and the political order allowed Europe to develop secular government, secular knowledge, and a secular culture. Nevertheless, several countries are today explicitly Christian or at least support various Christian sects (called *established* churches) with public funds. These include Argentina, Peru, Ireland, Norway, Denmark, Iceland, the United Kingdom, Finland, and Germany (although in 1998, Gerhard Schroeder became the first-ever German chancellor to refuse to end his oath of office with the words "so help me God"). At the same time, other countries are disestablishing official churches. For example, Italy disestablished Roman Catholicism in 1984, as did Spain in 1988, and Colombia in 1991. Sweden disestablished Lutheranism in 2000, and the government of Greece stopped listing religious affiliation on state identity cards in 2000. Many countries in Eastern Europe have revived their national churches as a means of defining and emphasizing their national identities since the fall of Communist governments. The Uniate Church of Ukraine, for example, survived persecution by the Communists and revived in the late 1980s.

Liberal democracy, as it is known in the West, was shaped by 1,000 years of European history and, beyond that, by Europe's double heritage of Judeo–Christian religion and ethics and Greco–Roman statecraft and law. No such system ever originated in any other cultural tradition, and it remains to be seen whether such a system can survive when transplanted and adapted in another culture.

In Islam Theoretically, no distinction can be made between church and state in Islam, which teaches that the only purpose of government is to ensure that each person can lead a good Muslim life. Church and state should be the same, thus making Islam inherently political. (The centrality of religion in Islamic cultures is illustrated even in city planning; see Chapter 10.) Therefore, many countries with largely Muslim populations are today officially Islamic, including Mauritania, Afghanistan, Libya, Saudi Arabia, Yemen, Oman, Qatar, Bahrain, Iraq, Iran, Comoros, Maldives, Pakistan, Bangladesh, and Malaysia. These countries may be considered modified theocracies. Islamic states enact Islamic teachings into law, called **Sharia,** and establish Sharia courts to rule whether their secular laws conform to Islamic teaching. If the Sharia court rules against a secular law, that law is usually repealed. All legislation in Iran, for example, must be approved by a body of 12 religious judges, who enforce fundamentalist beliefs. Indonesia is officially a secular country, but it has recognized Sharia rulings on family law. Under Sharia law, and in many Muslim countries, *blasphemy* (an irreverent or impious act or utterance) is punishable, as it was in most Western Christian countries until the concept of "free speech" evolved. Furthermore, *apostasy* (abandonment of one's faith, or conversion) may even be punishable by death.

Fundamentalist Islamic political parties are contesting elections in several countries that are not now officially Islamic. Many observers fear that these parties see democracy as a means to an end—the creation of an Islamic state—rather than as a system to be valued for itself. Turkey is officially secular, but the government pays for mosques, supervises the (compulsory) religious education, specifies qualifications for imams, appoints all imams to mosques, and pays their salaries. The Turkish Constitution declares the army to be the official guardian of the country's secular status, and Turkey's army has repeatedly intervened in democratic processes in order to frustrate the rise of fundamentalists. Secular government probably survives in Egypt only through brutal repression of Muslim fundamentalists.

The role of education

Some observers insist that the rise of fundamentalist Islamic parties is due to the failure of the secular governments in many Muslim countries to provide basic services, such as health care and education. Many see education as secularists' strongest weapon against fundamentalism. Mohammed Charfi, who served as education minister in Tunisia from 1989 to 1994, presided over hundreds of Arabists, historians, philosophers, and other specialists to remove aspects of political Islam from Tunisian schoolbooks. "I left not a single schoolbook untouched," he wrote. "The reform is not against Islam, but rather designed as a modern version of Islam consonant with democracy. School is the best place to fight fundamentalism, and school is where it is born and dies. No one could be a fundamentalist if they read Spinoza, Voltaire, and Freud." The school curriculum in Jordan underwent a similar revision in 2005. In schools in Saudi Arabia, by contrast, religion occupies from one-quarter to one-third of class time, and textbooks still in use in 2006 teach that "every religion other than Islam is false" and even that "it is forbidden for a Muslim to be a loyal friend to someone who does not believe in God and His prophet." Fundamentalists of other faiths in other countries may hold such intolerant beliefs, but such beliefs are seldom taught in other nations' public schools.

In many Muslim lands, the only education available (and available only to males) is in Islamic schools, called *madrassas*, where modern mathematics and science, as well as humanistic concerns, are repudiated. Students spend hours poring over the Koran, and the top students are those who memorize the most passages. Fundamentalist governments, individuals, and Islamic foundations have financed madrassas in many countries in Africa, Southeast Asia, and other places where secular public education is weak, thus arousing fears of networks of anti-Western indoctrination.

Pakistan's President Pervez Musharraf has acknowledged that much of the backwardness of the Muslim world is rooted in the education systems. "Do we believe," he has asked, "that religious education alone is enough for governance, or do we want Pakistan to emerge as a dynamic Islamic state? The verdict of the masses is in favor of a progressive Islamic state." President Musharraf has promised educational reform.

The Muslim Umma

To a Muslim, the international community of all Muslims is called the *Umma*. To a fundamentalist, the division of this community into many political states is blasphemy. In 1998, the terrorist Osama bin Laden issued a *fatwa* (holy directive), saying, "The call to wage war against America was called because America has spearheaded the crusade against the Islamic community, sending tens of thousands of troops to the land of the two holy mosques . . . and . . . meddling in its affairs and its politics. . . ." Bin Laden referred to an "Islamic community" that spans many ethnicities and states and to the Al-Aqsa Mosque in Jerusalem and the Al-Haram Mosque in Mecca as being in the same "land." Bin Laden thus saw all the governments across the entire Muslim world (including Israel) as illegitimate. This belief made bin Laden, ultimately, a threat to all individual Muslim rulers and states (compare this to Figure 7-26). By his definition, the stationing of U.S. troops in the Near East is blasphemous, even if U.S. troops were first based on Saudi territory in 1991 to protect Saudi and Kuwaiti independence against Iraq. Bahrain and Oman have also provided bases for U.S. troops, but Americans' 2003 conquest and occupation of Iraq has fired many Muslims' indignation. Even though U.S. troops left Saudi Arabia in 2003, Saudi Arabia, Jordan, and other Muslim countries have continued to suffer disruption by al-Qaeda-linked bands.

Bangladesh is officially an Islamic country, but Islam there is generally characterized as moderate. Nevertheless, in August 2005, Bangladesh suffered an extraordinarily sophisticated, coordinated bombing attack. Approximately 450 bombs went off nearly simultaneously at military bases, airports, government and police offices, universities, markets, and 63 of 64 district courthouses. The bombers' intentions were clearly to warn people rather than to harm them (only 2 people were killed and 100 injured), but a fundamentalist group claimed responsibility in a pamphlet arguing that rigid Sharia law must be instituted. The situation in Bangladesh has become so dangerous for Americans that all Peace Corps volunteers were pulled from the country in 2006.

Osama bin Laden never represented the Islamic community in general. The Organization of the Islamic Conference, a coalition of 56 Islamic nations representing 1.2 billion Muslims, declared in October 2001 that the "terrorist acts [of 9/11] contradict the teaching of all religions and human and moral values." In 2003, Grand Sheik of Al-Azhar Mohammed Sayed Tantawi, considered by many to be the Sunni Muslim world's highest religious authority, said Muslims should "wholeheartedly open our arms to the people who want peace with us. . . . I do not subscribe to the idea of a clash of civilizations," he insisted, revealing his familiarity with popular Western thought.

The strife within the Islamic Umma between nationalism and internationalism, over questions of sacerdotalism, and over efforts to define the role of religion in modern life will occupy headlines for the foreseeable future.

The geography of religion in the United States

Several of the American colonies were founded as theocracies, and most retained established churches until the War for Independence. In forging the new United States, however, James Madison included in the Bill of Rights a prohibition against the federal government's establishing any religion—the world's first such clause: "Congress shall make no law respecting an establishment of religion, or prohibiting the free exercise thereof. . . ." The 1797 treaty between the United States and Tripoli—ratified by Congress—declares that "the government of the United States is not, in any sense, based on the Christian religion." Congress added the words "under God" to the pledge of allegiance in 1954, but ongoing litigation disputes the constitutionality of that action. American coins do insist "In God We Trust."

People of every faith can worship in the United States as they wish, and they can even proselytize. The consequence of the United States' separation of church and state has not been the death of religion, but, rather, the birth of what is arguably the most religious nation in the world. A 2007 study by the Pew Forum on Religion and Public Life found that 78 percent of Americans affiliated themselves with a Christian church; 4.7 percent with other religions; and 16 percent unaffiliated, meaning they were agnostic, atheist, or "nothing in particular." Overall, this is a high degree of religious affiliation compared to other developed countries. More unusual is the extent to which Americans change, drop, or add religious affiliation; 44 percent of adults have a different affiliation than the one in which they were raised.

Religious affiliation does not equate with being religiously active or having strong religious beliefs. Compared to the United States, for example, Norway has a higher rate of religious affiliation, but very few people go to church. About 40 percent of Americans attend religious services at least once a week; only 11 percent never do. Surveys in different countries do not always ask the same questions, but for comparison, about 2 percent of people in England regularly attend church, and fewer than 1 French person in 10 goes to church as often as once a year. In 2002, 49 percent of Danes, 52 percent of Norwegians, 55 percent of Swedes, and 64 percent of Czechs said God "does not matter" to them, whereas 82 percent of Americans said God was "very important." The U.S. Census Bureau does not study religions, but Figure 7-35 presents six maps of the variations of church membership across the country. Americans' religiosity, despite the constitutional separation of church and state, is, perhaps, particularly baffling to Islamic fundamentalists, for whom the unity of church and state is important.

Immigration is also changing the religious composition of the U.S. population. In one survey of immigrants in the late 1990s, 41 percent described themselves as Roman Catholic (compared to about 24 percent of the current national population), and 9 percent were Protestant (compared to about 51 percent of the current population). Nearly 30 percent describe themselves as something other than Protestant, Catholic, or Jewish; 7 percent claimed to be Muslim. In 1995, Paterson, New Jersey, with a total city population approximately 15 percent Muslim and a school-age population almost 25 percent Muslim, became the first U.S. city ever to close its public schools for a Muslim holiday. In November 2006, the people of Minnesota's Fifth Congressional District elected to send the first Muslim U.S. Representative to Congress: Keith Ellison. There is no evidence that his personal religion played a role in his election.

Churches in the United States, particularly in central-city, African American neighborhoods, often provide an array of social services, including counseling, meals, education, and health care. Thus, churches are potential building blocks of government community-assistance programs: Bills funding "faith-based" programs first passed Congress in 1996. Such programs expanded under Presidents George W. Bush and Barack Obama. Some critics of these programs contend that government is shaping the priorities of participating churches; others argue that church priorities are being funded with tax dollars. In a case over the funding of faith-based initiatives, the U.S. Supreme Court in 2007 barred taxpayers from challenging the government's use of public funds.

Indirect Religious Influences on Government

In all countries, the prevailing religion can dictate national ethics or morals. Religion so deeply affects what people assume to be natural or desirable human behavior that religious prejudices are taken for granted and unconsciously translated into laws. Religion should never be underestimated as an institutional, economic, and political force. Any organized religion has its own bureaucracy, its own sources of income, and, in the pulpit, its own channel of communication that often reaches more of the population than does government information (Figure 7-36).

In several countries, churches operate a school system parallel to that of the government. In the United States, religious schools are accredited, but not funded, by the government. The provinces of Canada vary on this subject, but in Ontario, for example, the provincial government finances both Roman Catholic schools and secular public schools. In virtually all countries except

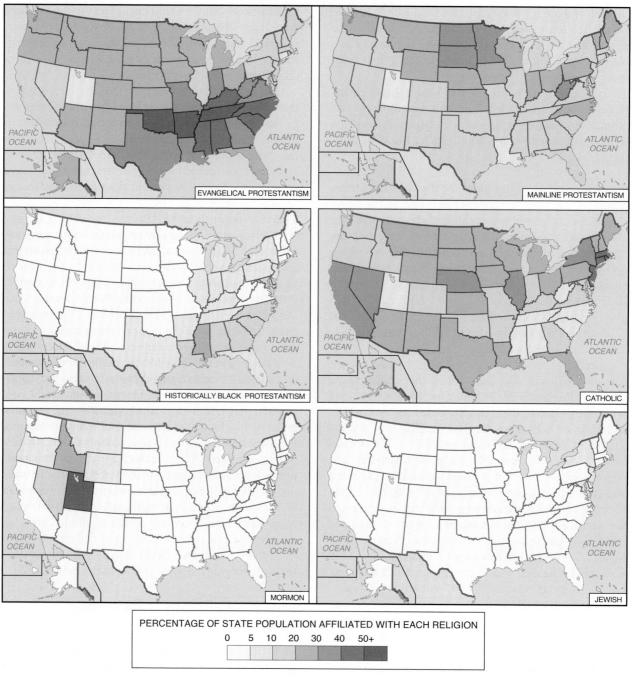

PERCENTAGE OF STATE POPULATION AFFILIATED WITH EACH RELIGION

0   5   10   20   30   40   50+

Figure 7-35 Church affiliation in the United States. These maps of church affiliation in the United States reveal regional geographies associated with settlement patterns and church activity. Evangelical Christianity is widespread in the United States but is most prevalent in the American South and Lower Midwest. Mainstream Protestant affiliations are also widespread but highest in the Upper Midwest, where there is a predominance of Lutherans, reflecting the Scandinavian and north German backgrounds of that region's pioneer settlers. The historically black Protestant tradition bears evidence of African American settlement in the southeastern United States, largely the result of slavery and a series of northward migrations. The distribution of Roman Catholics reflects, among other factors, the migration to the United States of Hispanics and southern Europeans. The distribution of Mormons reflects their westward migration to Utah and the surrounding states to escape persecution in the eastern United States (see Figure 7-34). The distribution of Jews in the United States has been historically concentrated in East Coast cities with major immigration ports. These maps do not pick up on smaller communities distributed across the country.

the most fundamentalist Muslim countries, the governments insist that religious schools cover such modern subjects as science and math along with religious studies.

Relations between the government and the leading religious organizations constitute a major political issue in most countries. A national church may sustain a suppressed nationalism, as in Ukraine or in Ireland. Religious leaders may lend legitimacy to rulers or, conversely, provide a rallying point for political opposition. The Polish Roman Catholic Church stood against the

Figure 7-36 The main square of Mexico City. Here in Mexico City, as in many other Latin American capitals, the cathedral (to the left) and the presidential palace (to the right) are given a central location on the main square.

Communist party and state as an alternative repository of Polish identity between 1945 and 1989. Attending mass was an act of political defiance. From 1959 until 1992, the constitution of Communist Cuba insisted that Cuba was officially atheist, yet the Roman Catholic Church has remained the only force outside the state-party machine with any significant structure or loyalty. After 1992, the government granted more freedom to churches, and Pope John Paul II visited Cuba in 1998. In Hong Kong,

Roman Catholic Bishop Joseph Zen fought the diminution of civil liberties; thereafter Hong Kong merged with China in 1997. The Vatican honored him by elevating him to the rank of cardinal in 2006. In 2003, then-U.S. Secretary of State Colin Powell singled out Archbishop Pius Ncube of Zimbabwe for his bravery in the face of government oppression there. Even in the United States, Roman Catholic Cardinal Roger Mahony of Los Angeles stood firm against a storm of criticism in 2006, when he announced that priests would defy legislation then under consideration in Congress that would have required churches to check the legal status of anyone to whom they gave assistance and that would have criminalized giving aid to illegal immigrants.

In many countries, a church or religion may form a political party, as the Christian Democrats have done in several European countries, or as various Islamic groups have done. India's Hindu nationalist party Bharatiya Janata has headed coalition governments.

In many countries, a religion can become a big landowner or financier, but religious organizations do not always maximize the productive capacity of their property. Real property that is restricted in ownership or in the purposes for which it can be used is called *mortmain*, literally "dead hand." Church property may also enjoy preferential tax rates or escape taxation entirely. Throughout Latin America, the Catholic Church is a principal landowner. In the United States, church-owned properties are usually concentrated in the cities. Because they are free from taxation, their concentration reduces cities' ability to raise property-tax income. In turn, however, the churches themselves may provide a wide range of services to a public beyond their own communicants.

CONNECTIONS

Liberation Theology

A philosophical dispute within the Roman Catholic Church has brought about revolutionary changes in politics throughout Latin America. **Liberation theology**—named after a 1968 book by Peruvian Father Gustavo Gutiérrez—puts the problems of overcoming poverty at the heart of Christian theology. It recommends political activism, using the church institutional framework to organize the population in the struggle for social justice and equality. Following this idea, many Latin American bishops have proposed national redistributions of land and income, provided legal services for the poor, and attacked elite power structures from their pulpits. Cuban leader Fidel Castro has praised liberation theology and supported liberation theologians throughout Latin America.

Liberation theology provoked political repercussions throughout Latin America. In some countries, it triggered militant opposition among conservatives threatened by changes on either religious or political grounds. For example, in El Salvador during the 1980s, nuns, priests, and even

Archbishop Oscar Romero were assassinated by right-wing elements of the military regime. The most conservative Latin American governments have favored Evangelicals as a counterforce to liberation theology. In Guatemala, for example, government death squads murdered politically active priests during the 1980s, and one Roman Catholic bishop was forced into exile. Meanwhile, Protestant sects, with government encouragement, have converted more than 30 percent of the population.

Popes John Paul II and Benedict XVI have tried to arrest liberation theology. John Paul II forbade Father Gutiérrez to write or teach for several years and appointed conservative bishops who closed church legal-services offices, land-rights offices, and liberal seminaries.

The arguments over liberation theology are ostensibly religious arguments restricted to church affairs, but in fact this theological dispute may activate politics in Latin America and other areas for years to come.

Religious Tensions on the Indian Subcontinent

The Indian subcontinent illustrates how religious divisiveness can cause problems both within and among countries. The British ruled the subcontinent (today's India, Pakistan, and Bangladesh) as one colony, "India," and they intended to grant it independence as one secular state. Many of the Muslims within this immense territory, however, feared Hindu rule and Muslim leaders demanded that the colony be divided so that they could create a Muslim majority state. When independence came in 1947, the Muslim population established the country of

Pakistan. It originally included a western territory (today's Pakistan) and a separate eastern territory (today's Bangladesh). In the months before the partition was realized, millions of Muslims fled toward areas that were designated to become Pakistan, and millions of Hindus fled to regions scheduled to become parts of India. Tens of thousands died in terrible acts of violence. Muslim–Hindu tensions were exacerbated by the fact that for hundreds of years in many areas, Muslim minorities had ruled Hindu majorities.

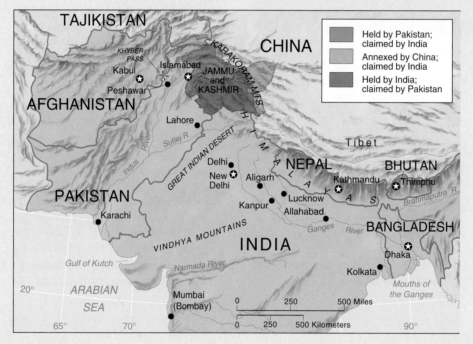

Border disputes in South Asia. Antagonism and border disputes persist between India and Pakistan and between India and China. These three nuclear powers contain about 40 percent of the human population.

In some countries, a community of citizens might be alienated from their national government if co-religionists in an adjacent country enjoy more complete religious freedom. For example, the Islamic vigor of Iran provided religious inspiration to the Muslim populations of the Central Asian Republics when they were under the rule of the Soviet Union.

France has long enforced a rigid separation of church and state, but the growth of Islam in France (today having at least 5 million adherents) has forced the government to rethink its role. In 2003, the government-sponsored elected Consultative Council of Islam was welcomed to the presidential palace, and, although the law forbids the government from building houses of worship, foreign subsidies to mosques and madrassas have been so great (mostly from Saudi Arabia) that local authorities are donating land for mosques.

States split by two or more religions When two religious communities compete to write a nation's laws, the country must devise compromises or else suffer internal conflict. Attempts to enforce Sharia in the Sudan, for example, have intensified conflict between the Arab Muslim–dominated north and the black Christian and animist south. Similarly, in Nigeria, several northern states have adopted Sharia, triggering rioting between Christians and Muslims. Christians have migrated out of Sharia states and Muslims into them. In the Philippines, tensions between Christians and Muslims have fostered revolts on several islands against the central government.

Sometimes states will enforce two separate systems of law, depending on the religion of the people involved in a case. Senegal and several other African states will allow Muslim men—but only Muslim men—more than one wife. In India, Muslims are allowed to follow

In 1971, eastern Pakistan broke off from Pakistan to form a separate Islamic country, today's Bangladesh, but religio-political turmoil continues in what is left of Pakistan. Sunnis (about 75 percent) scorn Shiites (about 20 percent), and tribal loyalties and identifications subdivide the population still further. Non-Muslims (about 10 percent of the population) are guaranteed 10 of the 307 seats in Parliament, but they do not represent geographic districts, so they have to campaign across the entire country.

India itself emerged as a secular state with a majority Hindu population. Between 10 and 13 percent of India's population, however, is Muslim, which is a greater number of Muslims than in Pakistan. India's Muslims were persecuted during India's wars with Pakistan in 1965

The Taj Mahal. A Muslim ruler built the Taj Mahal, a mausoleum in Agra, India, for his beloved wife in 1648. India's Muslim rulers were so rich and powerful that their name, Mughals, has come down to us in English as a word for a business executive: a mogul.

and again in 1971 and violence between Muslims and Hindus is occasionally intense. The rise of the Hindu nationalist party, Bharatiya Janata, has coincided with a growing climate of intolerance of minorities and the introduction of new Hindu-oriented syllabi in schools. Avul J. P. Abdul Kalam, a Muslim nuclear scientist, was inaugurated as India's president in 2002 (his family converted to Islam, but Hindus still consider him a Hindu with, in fact, Brahmin status), but growing hostility between Muslims and Hindus in India threatens domestic peace. Furthermore, the conversion of many untouchables to Islam or Buddhism further upsets India's internal religious balance.

Conflict continues on the Indian–Pakistani border. In 1947, India and Pakistan fought over the border state of Jammu and Kashmir (China controls a third part of the region). Its population was mostly Muslim, but the state was eventually split, with India taking about two-thirds. The population of Indian Kashmir today is about 60 percent Muslim, but India refuses to allow the people of Kashmir to vote on whether they want to be a part of India or Pakistan. It is important to India's image as a secular country to have a state with a Muslim majority. Fighting in Kashmir has claimed an estimated 70,000 lives from 1989 to 2006 and has been blamed for occasional terrorist attacks throughout India.

Water is another source of contention between India and Pakistan. A 1960 treaty regulates sharing of the water of the Indus River and its eastern tributaries that rise in China or in the Indian-controlled part of Kashmir. India, however, has threatened to revoke the treaty and cut off the water supply to Pakistan. Throughout much of the Indus watershed region, local aquifers are being depleted, water tables are falling, waterways are polluted, and soils are becoming saline from overuse of underground water supplies.

Muslim law in many matters of education, marriage, divorce, and property. The government of Malaysia enforces mandatory tithing (giving 10 percent of one's income to charity) among Muslims, about half of the country's total population.

In some countries, struggles that are ostensibly religious actually conceal other social divisions. Tension in Northern Ireland is, on the surface, a conflict between a Protestant majority and a Roman Catholic minority. That split, however, involves much more than theology. It is a competition for jobs and opportunity, and it is complicated by deep historic political and economic differences and the fact that both sides draw on outside support.

China recognizes five "approved" religions: Protestant Christianity, Catholic Christianity, Buddhism,

Islam, and Taoism. The government oversees religious activity carefully, and the Chinese Catholic Church does not recognize the authority of the Vatican.

Religion and women's rights Religions differ in the attitudes that their teachings advocate toward women, and national laws often reflect these religious teachings. In many cases, however, the actual practice of any religion varies from place to place. These alternate practices reflect variations in other aspects of the cultures among places. Therefore, it is difficult to determine exactly how much difference in women's rights can be attributed to the predominant local religion and, in turn, how much of that difference has been translated into local law.

For example, the Catholic Church is centralized, and it imposes a worldwide ban on the ordination of women.

The Vatican has excommunicated women ordained by bishops in Argentina and Austria. The worldwide Anglican Church, however, is less centralized. Its official head, the Archbishop of Canterbury, has approved the ordination of women, but local communities of Anglicans have different attitudes, and these reflect local variations in general social attitudes toward women's rights. In U.S. culture, for example, women's rights are protected in many areas, and the Episcopal Church (Anglican) ordains women. Reverend Katharine Jefferts Schori was even elected Presiding Bishop of the U.S. Episcopal Church in 2006. Anglican communicants from some other world regions, however—most notably Africa, where women's rights are not always assured—have refused to ordain women or to recognize ordained women.

As wealthier and more democratic countries have embraced women's rights and gay rights as well, some of the Protestant sects have splintered. In the United States, some Anglican congregations have broken away from the main Episcopal Church over these issues. In 2005, the Anglican Church of Nigeria, for example, deleted from its constitution all references to "the see of Canterbury." Protestant churches in Nigeria and other African countries have even spurned the financial aid that supported local schools and hospitals.

Variations in practice can also be found across the Islamic realm. Islam originated in an Arabic culture that granted women few rights, but the list of women's rights in the Koran was liberal for the time and place. Islam outlawed female infanticide, made the education of girls a sacred duty, and established a woman's right to own and inherit property. At the same time, Islam also fixed certain discriminatory practices. It teaches that a woman's testimony in court is half that of a man's and that men are entitled to four spouses, whereas women may have only one. In practice, women's rights vary throughout the countries where Islam predominates. Different countries interpret Sharia differently: In Dubai, for example, Sharia courts have ruled that a man may divorce a wife by sending her a text message over a mobile phone; Sharia courts in Singapore have ruled that he may not. Women are frequently secluded, or veiled in public, in Islamic countries, but this is compulsory only in Saudi Arabia and Iran. Also, only in the most fundamentalist Arab societies are women generally forbidden to work outside the home.

Future struggle is nevertheless almost inevitable between the expansion of women's rights in Islamic societies and the Islamic fundamentalist backlash that demands the repeal of laws that had banned polygyny (having more than one wife), permitted birth control, and given women the right to divorce in some countries. Almost everywhere, women struggle for equal access to education.

The Old Testament of the Bible contains restrictions on women's rights that are comparable to those found in the Koran, but few Jews or Christians today observe them. In Israel, however, the law denies women equality with men. For example, Israel has no civil marriage or divorce, and in Jewish law, only a husband can initiate a divorce.

It has been hypothesized that religious teachings about women's rights may affect women's role in politics, but this hypothesis is difficult to prove. Women have served as heads of government in Islamic countries, as well as in countries that are officially Christian (Catholic and Protestant), Jewish, and Buddhist. What sort of evidence for such a hypothesis could be sought? Reading different religions' scriptures alone does not prove whether those teachings are obeyed. Could we examine the percentage of national legislatures that are female? Geographers have noted that within the United States, the populations of Utah and of the states of the Southeast are, by some measures, the most committed to organized religion. The legislatures of these states have the smallest percentages of women legislators. Perhaps there is a cause-and-effect relationship between these facts, but clearly more research is needed on this topic.

Religion and Dietary Habits

Different religions teach different beliefs concerning the sacredness of plant and animal life, and this affects the farming and dietary practices of their adherents. Buddhists are generally vegetarians. Hindus believe that cattle are sacred; therefore, they will not eat beef, and they use cattle only as draft animals and as a source of milk. McDonald's restaurants in India feature lamb burgers and vegetarian sandwiches. Outside observers have tried to measure in economic terms the advantages and disadvantages of preserving India's cattle. Aging and unproductive cattle may overgraze the land or may compete with the human population for limited food supplies. Sacred cows roam the streets of Indian cities unmolested, befouling them and congesting them. On the other hand, the cattle provide dung as a fuel and manure, and those that die of natural causes are eaten by non-Hindus anyway. Whatever the balance of these economic arguments might be, to a religious Hindu this is a religious matter not open to economic debate.

Other religions prohibit certain foods. Jews and Muslims both refuse pork, so there are few hogs in Israel or anywhere else across the Islamic realm (Figure 7-37). Muslims are prohibited from drinking alcohol. Altogether, a significant share of humankind restricts its diet because of religious beliefs.

These restrictions affect even world trade in food. Several of the greatest importers of plants and seeds that are rich in protein (peas, beans, and lentils, for example) are countries that limit their consumption of meat for religious reasons. Annual per capita consumption of meat in India, for example, is only 2 kilograms (4.4 pounds), about one-tenth the U.S. figure, and India is a major importer of protein-rich vegetable products.

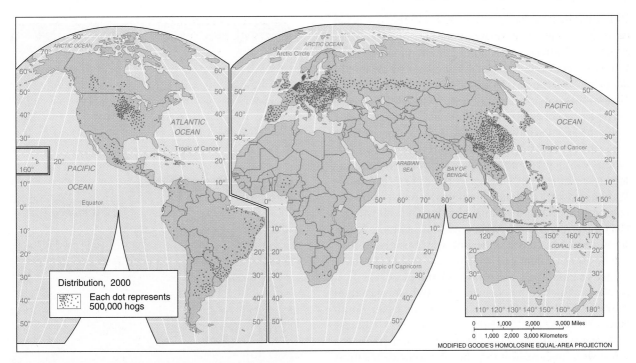

Figure 7-37 World distribution of hogs. The virtual absence of hogs across North Africa, the Near East, and all the way to the borders of China is not caused by any attribute of the physical environment but by the predominance of Islam and Judaism across this region. Both religions prohibit the consumption of pork.

Religion and Economics

The religion that predominates in any region may have a significant impact on that society's economy. Elaborate burial practices, for example, can drain a society of capital: Many peoples have buried their possessions along with their dead. Some societies lavish expenditure on houses of worship and other religious institutions. In some Buddhist societies, high percentages of the male labor force spend several years, or even their whole lives, as monks.

Furthermore, religious teachings may affect the way people view the accumulation of money. For example, the scriptures of most religions bar the charging of interest on a loan as taking unfair advantage of another person. Christian teachings, however, have evolved to differentiate between two reasons for borrowing money: for needs and for investment. It is considered sinful to charge interest on money borrowed to buy food, but not sinful to charge interest if the borrower invests the money for his or her own profit—as when the borrower buys a truck to increase the profit in his or her own business. Religious leaders in some societies do not accept such a distinction. Islamic financial systems forbid interest, but instead encourage risk sharing. Islamic banking systems have sprung up throughout North America and Europe that do not, for example, offer mortgages, but buy homes and devise rent-to-buy contracts for Muslim customers. In practice, financial structures and compromises have to be defined in each Islamic country.

Weber and the Protestant ethic German sociologist Max Weber linked Protestantism and capitalism in *The Protestant Ethic and the Spirit of Capitalism* (1904). According to Weber, Protestantism encourages individualism, and with the rise of Protestantism in Western Europe, acquisitiveness, as the result of exercising individual ability in the marketplace, became a recognized virtue. Today this ethic seems to characterize adherents of almost any religion, so it is usually referred to as the *work ethic*.

If, however, there is any true advantage in the Protestant Christian attitude toward moneymaking, then we could hypothesize that the worldwide expansion of Protestantism might encourage individualistic capitalism. This explains why the spread of Protestantism is often well-received in America's business press, which clearly accepts the hypothesis of a link.

The Catholic Church and capitalism In 1991, Pope John Paul II wrote, "On the level of individual nations and of international relations, the free market is the most efficient instrument for utilizing resources and effectively responding to needs." The Catholic Church's 1992 *Universal Catechism* notes, however, that markets do not always meet human needs, and it insists that governments regulate markets "according to a just hierarchy of values." It also demands that every worker receive a "just salary." The Church's attitudes toward markets, speculation, and worker–management relations may influence the behavior of individual Roman Catholics and of governments in Roman Catholic lands.

Religion and economics in Asia   The Confucian tradition in East Asian societies may exemplify another religious influence on economic development. Confucianism recommends societal leadership by an intelligent elite with the moral obligation to guide the people. Several East Asian leaders have praised their own societies' traditional communitarianism and decried Western Christian individualism. Confucianism enhances the status of jobs in government bureaucracies, so ambitious young graduates compete for positions with the Japanese Ministry of Finance or the Korean Economic Planning Board the way ambitious young Americans compete for jobs at investment banks. The ability to attract talent gives these Asian governments more legitimacy and competence in dealing with businesses than the U.S. government can bring to bear, which may have helped to build their successful economies.

Also, diligence, obedience, and high savings rates characterize the peoples of neo-Confucian Japan, South Korea, Taiwan, and Singapore, and of every Chinese settlement in the world. It is risky to hypothesize a cause-and-effect relationship, but the coincidence must be noted.

Religions, Science, and the Environment

Different religions teach different attitudes toward environmental issues, and cultural landscapes often reflect the religions of their creators. Most religions address the questions of how the world came into existence, what power created it, and what humankind's relationship should be to it and to other living things. Many major faiths not normally classified as animist nevertheless designate specific features of the environment as holy. Some religions revere rivers (Hindus revere the Ganges, Christians the Jordan); some revere high places (the Shinto revere Mount Fuji, Jews and Christians Mount Ararat, Hindus and Buddhists Mount Kailash) (Figure 7-38).

Early religious beliefs often formed as attempts to explain natural phenomena, such as the changes in the seasons or floods. Many religions celebrate natural events, time their annual ceremonies in accordance with astronomical events, or celebrate human affairs connected to the environment. Most religions celebrate harvest festivals of thanksgiving, for example.

Religions recognize different holidays and pace human activities through the year differently, which is often reflected in national laws across religious realms. Most Christian countries use the solar calendar. Jews use a calendar of 12 months, with an additional month added seven times in 19 years. Muslims use the lunar calendar. Muslims celebrate the Sabbath on Friday, Jews celebrate on Saturday, and most Christians on Sunday. Many governments prohibit work or other activities on that day. Muslims recognize an entire month as holy and accordingly curtail their activities.

Figure 7-38 Mount Fuji.   Volcanic Mount Fuji, Japan's highest peak at 3,776 meters (12,389 feet), is sacred to many Japanese. It has been celebrated for centuries in Japanese paintings and verse.

Most religious traditions award humankind a special role in the world, but, it has been argued, some religions suggest adaptive approaches to the physical world, while others suggest more exploitative approaches. Today, however, environmental degradation afflicts areas of all religious persuasions. Since 2006, more than 250 U.S. Evangelical Christian leaders have joined an initiative to fight global warming, saying, "millions of people could die in this century because of climate change, most of them our poorest global neighbors." Signers of the statement included presidents of Evangelical colleges, leaders of aid groups and churches, the Salvation Army, and pastors of megachurches. The leaders called for federal legislation that would require reductions in carbon dioxide emissions. They call on church members to purchase fuel-efficient vehicles, conserve energy use at home, find renewable energy sources, and support businesses that reduce pollution. The statement was only the first stage of an "Evangelical Climate Initiative" including television and radio announcements in states with influential legislators, informational campaigns in churches, and educational events at Christian colleges.

Some scholars have suggested that continuing advances in science, particularly in the life sciences, may be influenced by cultures' religious beliefs. Asian scientists, it has been argued, face less cultural resistance to their work than their colleagues in the West. In Roman Catholic areas and some fundamentalist Christian regions, for example, religious authorities argue against the destruction of microscopic embryos because in their teaching a fertilized egg is already a person. In 2006, Vatican spokesperson Cardinal Alfonso Lopez Trujillo stated that "excommunication will be applied to the women, doctors and researchers who eliminate embryos [and to the] politicians that approve the law." The independence of mind of Western scientists, however, was demonstrated by the Italian Dr. Cesare Galli, the first scientist to clone a horse, who said, "I can bear excommunication. I was raised as a Catholic . . . but I am able to make my own judgment on some issues and I do not need to be told by the Church what to do or to think."

In Confucian and Buddhist societies, by contrast, the defining moment of life is birth, not conception, and Buddhists and Hindus view life not as beginning with conception but as a cycle of reincarnations. One leading South Korean bioscientist said in 2004, "Cloning is a different way of thinking about the recycling of life. It's a Buddhist way of thinking."

Religious landscapes   Most landscapes reveal the predominant local religion. The architectural styles of houses of worship, monasteries, and similar institutions are often revealing—Muslim mosques, for example, usually have freestanding towers, called minarets.

Any religion's places of worship may become goals of pilgrimages, which attract visitors and play a major role in cultural diffusion (Figure 7-39). Fairs and markets grow up at these sites and along the pilgrimages' prescribed path. Muslims' hajj pilgrimages bring together people from around the world. At the local level, religious administrative patterns may organize and focus community spirit. More Roman Catholic New Yorkers can name the parish they live in than can name their city council district, and it is through the parish that they make friends and meet potential spouses. As noted earlier, the diffusion of religions can affect the diffusion of languages. Muslims everywhere are encouraged to read the Koran in its original Arabic, and the translation of the Bible into German by Martin Luther defined the German language and, arguably, the German nation.

Different religions' burial practices create different mortuary landscapes. Hindus and Buddhists cremate their dead, whereas Christians and Muslims usually bury their dead and erect monuments to them. Buddhist landscapes are marked by *stupas* (called *pagodas* east of India). These serve as temples today, but the form was originally devised as a funeral mound to contain the remains of a revered bodhisattva (Figure 7-40).

Religion means "rebinding" of people with their Gods and of people with one another. The geography of such binding will always be a key consideration of human geography.

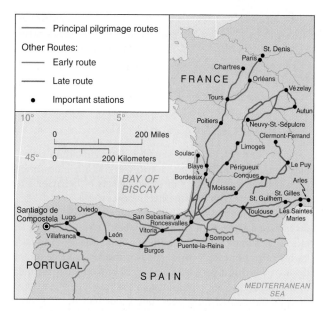

Figure 7-39 Traditional routes to Santiago.   The Cathedral of Santiago, in the northwest corner of Spain, is believed to hold the relics of Saint James. Pilgrims from throughout Europe and the world have sought it out. In medieval times these traditional routes and staging points played important roles in trade and in the diffusion of ideas. The routes are marked by a chain of historic churches, inns, universities, and trade fairs.

Figure 7-40 Great stupa in Nepal.   The eyes on this great stupa in Nepal emphasize that God sees everything.

Chapter Review

Summary

Language and religion define and bond human cultures. Peoples who share either of these two cultural attributes often demonstrate consistencies in other aspects of behavior and often cooperate with one another in other ways.

The distribution of any language illustrates the pattern of dispersal of its original speakers or their cultural impact on others. If human beings were to appear suddenly in today's highly interconnected world, one worldwide language might develop. The great variety of languages spoken today testifies to the relative isolation of groups in the past. There are more than 6,000 distinct languages, but about half of all people speak one of just 16 languages. The language with the most speakers is Mandarin Chinese, but English is the world's lingua franca. In contrast to these widespread languages, some languages are extremely local.

A few languages are associated almost exclusively with one country, but several languages are shared by many countries, and some polyglot countries officially recognize several languages. The idea that languages gave birth to nations persists despite the fact that standard languages usually were the product of self-conscious efforts at nation building by centralizing governments.

Today's world distribution of European languages as official languages partly reflects patterns of colonization. Europeans forced their languages on their colonies, and when the colonies gained their independence, many kept their former ruler's language. Across much of the world, a country's official language(s) may not be spoken by most of the population.

Orthography can also reflect a historic diffusion process. Toponymy records information about the environment or the beliefs of a region's inhabitants.

Languages can be categorized into families on the basis of common ancestry, and they continue to diffuse and differentiate today. The use of some languages is expanding because their speakers are diffusing throughout the world, gaining greater power and influence in world affairs, or winning new adherents to their ideas. Other languages are dying out.

The population of the United States has always been composed of a great variety of peoples speaking a great variety of languages. Regional differences persist, but some people think English should be the official national language.

Religion is a vital force in human affairs, and understanding the messages of different religions contributes substantially to understanding many other aspects of human life.

Judaism, Christianity, and Islam form one ancient religious tradition, originating in the Near East and diffusing from there. Christianity came to predominate in Europe, and it spread worldwide with European influence. Today most Christians are outside Europe and North America. Christianity continues to win new converts, but its practices and doctrines that derive from European culture are under pressure for change. New syncretic religions are thriving. Islam is newly influencing the politics of the countries in which it predominates.

East Asia is dominated by Hinduism and its offshoots Buddhism and Sikhism. Other Eastern religions include Confucianism, Taoism, and, in Japan, Shinto. The teachings of several of these systems of belief are exclusively ethical and psychological, and they do not address theological questions. Many people are adherents to more than one of these. Animism and shamanism have many followers in the less-developed parts of the world.

Religious teachings bear great influence on political organization. Theocracy is a form of government where the church rules directly, but religious teachings will often be translated into law in societies that claim to be secular. Religions' teachings differ regarding politics, women's rights, government, diet, economics, and environmental attitudes, so the geography of religious communities influences the geography of these other aspects of human culture. Christianity recognizes a distinction between church and state, but Islam teaches that the only purpose of government is to allow people to be good Muslims, and fundamentalist Islamic societies enforce Sharia law. The actual practice of any religion varies from place to place, often reflecting variations in other aspects of the cultures among places. Each society defines a balance among the teachings of its religion, other aspects of its cultural tradition, and the demands of modern life, but that balance may shift through time.

Key Terms

accent p. 249
caste p. 278
cognate p. 252
creole p. 248
dialect p. 248
drift p. 251
etymology p. 252
fundamentalism p. 265
isogloss p. 249
isolate p. 249

language p. 248
language family p. 251
liberation theology p. 285
lingua franca p. 248
monotheism p. 265
official language p. 248
orthography p. 254
polyglot state p. 261
polytheism p. 266
proselytize p. 265
root language, or protolanguage
   p. 251

secularism p. 265
shamanism p. 280
Sharia p. 281
speech community p. 248
standard language p. 248
syncretic religion p. 274
terrorism p. 268
theocracy p. 280
toponymy p. 255
untouchable p. 278
Zionism p. 266

Questions for Review and Discussion

1. Why is it that several former colonies retained their former rulers' language? What advantages and disadvantages did this offer?

2. Why are more countries today recognizing more official languages? What does this tell us about the internal politics of those countries?

3. Name a language that is expanding in use today and one that is shrinking. In each case, explain why.

4. What does it mean to say that Judaism, Christianity, and Islam are three religions but form one religious tradition?

5. What are the observed and hypothesized consequences of the conversion of much of Latin America from Roman Catholicism to Protestant Christianity?

6. How do religious teachings affect economic behavior? How do they affect the systems of national laws?

7. What is a lingua franca? What languages have served as lingua francas?

8. What is the caste system?

9. What is the Indo-European family of languages? Who first identified it? What is the geographic distribution of these languages? What is the history of the family?

10. Describe the diffusion of Islam from A.D. 622 to 751.

11. From what religious tradition did the Sikh religion spring? Where? When?

12. Where did Christianity diffuse before it diffused throughout Western Europe?

Thinking Geographically

1. How many languages are spoken in your community? Does the local school system have programs that are bilingual or to teach English as a second language? Where do your community's non-English speakers come from?

2. The Dutch have considered making English the official language for college-level education in the Netherlands. What would be the advantages of doing so?

3. If you speak two or more languages, can you think of specific ideas or images that may be difficult to translate from one to another? Try to explain this difficulty to your fellow students.

4. How many religions have temples, churches, or synagogues in your community? What, roughly, is the statistical breakdown of religious affiliations of the people of your community? Does this reflect the source areas of settlers to your region? Does each place of worship anchor a residential neighborhood of that religion's communicants? How many religiously oriented parochial schools exist in your community? With what religions are they associated?

5. Almost the whole world uses Arabic numerals, but the use of Arabic script is largely restricted to Islamic lands. Why?

6. Investigate the competition between Christianity and Islam for converts in an African country. Consider Nigeria, Senegal, or the Ivory Coast.

7. The fourteenth-century Islamic scholar Ibn Battuta could travel from Morocco to China unmolested—even being welcome across most of the region. What political, religious, and other criteria of territorial organization have fragmented that region today so that a contemporary traveler might have trouble retracing Ibn Battuta's travels?

8. What do the place names in your community reveal about its environment or history?

Log in to www.mygeoscienceplace.com for videos, animations, **MapMaster**™ interactive maps, RSS feeds, case studies, and self-study quizzes to enhance your study of The Geography of Languages and Religions.

MapMaster™

Food riots in Haiti, April, 2008. Household incomes for many in the developing world could not keep up with rapidly rising food prices since 2002. Riots and demonstrations, such as this one in Port-au-Prince, occurred in at least 60 countries as people demanded that their governments respond to growing hunger. On a global scale, the problem was not insufficient food production but a series of factors that drove up the price of food.

8

The Human Food Supply

In April 2008, dozens of hungry rioters tried to break into Haiti's presidential palace as thousands took to the streets to protest against the government. They were angry that their leaders had done little to address quickly rising food prices in their poor Caribbean country. The riots left five dead. Haiti's president tried to calm the public by replacing some of the government's other top leaders. In 60 other countries, such as Cameroon, Yemen, and Uzbekistan, food riots erupted for similar reasons. People could not afford to buy food. The effects were felt most sharply in poorer countries, but the prices affected wealthier countries, too. Some consumers in Italy could no longer afford pasta. In the United States, Sam's Club, a chain of warehouse-discount stores owned by Wal-Mart, restricted the amount of rice consumers could purchase as people began to hoard supplies.

The global trade in food was supposed to end crises like these. In some ways it has. The amount of food produced in the world is more than enough to feed everyone. But the global food trade has also made the poor more vulnerable to volatile prices. Consumers in poor countries cannot always afford food prices that are increasingly set by global markets. Sometimes, government policies can distort food prices or cause growers to grow less. For years, wealthier countries told developing countries to specialize their economies around goods for export and to use that income to buy food and other basic needs from world markets. The fields in many poor countries were converted to crops for export and no longer provide a complete meal. People in these countries now rely on food imports to eat. In 2007, the price of food traded on world markets skyrocketed as production levels began to fall while demand and shipping costs increased. Wheat and rice prices doubled in just a few months, and maize (corn) reached record highs. Yet poor countries received less and less income for their export products. Average people in poor countries could not afford the high cost of imported foods. Food aid and government interventions have provided short-term fixes, but the crisis added more than 100 million people to the world's already large numbers of hungry and undernourished.

Food is a basic human need. Its availability and quality have enormous effects on population growth and distribution (as we saw in Chapter 5). Acquiring food has always been a necessary human endeavor. It is the oldest economic activity in history. Even today, food production and distribution are key economic sectors in every country. Yet not everyone can acquire enough nourishment to lead full and healthy lives. Since farming began, humans have found creative ways to increase food supplies. The application of scientific knowledge to food production has yielded astonishing results, including the increase of productivity, plant and animal resistance to diseases and pests, and even food's inherent nutritional content. Some people, however, fear the scientific manipulation of something once thought so "natural" as food.

Furthermore, the political and economic context for the development of these possibilities will determine when they might be applied, where, and for whose benefit. Even today, producing enough food for everyone on Earth is not as much of a problem as is the distribution of food. There are great disparities in the map of food production and supplies. Some countries produce substantial surpluses of food, while others lack supplies sufficient to feed their entire populations a nutritious diet. Even in many of the countries with surpluses, some people go hungry. Problems of food production and distribution can be understood only as the result of interaction among environmental, technological, political, and economic factors.

Food Supplies Over the Past 200 Years

Chapter 5 examined a pessimistic prediction that Thomas Malthus made 200 years ago. Malthus argued that the population would always tend to increase faster than the food supply, so cycles of mass starvation would limit population increase.

Since Malthus published his theory, the human population has increased from 1 billion to 6.8 billion. The mass starvation he predicted, however, has not occurred. Chapter 5 noted that today there are more overweight people than undernourished people. Humankind has avoided starvation not by lowering the rate of population increase but by increasing the food supply faster than the population, so per capita supplies have actually increased. How has this been done?

New Crops and Cropland

Extensive areas that were scarcely utilized during Malthus's lifetime have been opened to productive agriculture (Figure 8-1). The vast prairies of North and South America, as well as food surplus regions in Australia and South Africa, have been developed

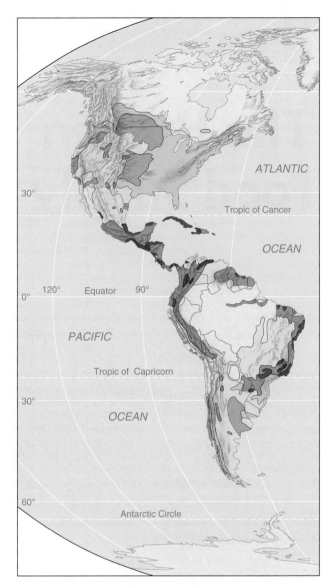

Figure 8-1 The world of agriculture. These are regions of different types of agricultural land use. The success of agriculture depends largely upon the possibilities of the physical environment, so we should not be.

since Malthus published his theory. Since the 1960s, there has been a net increase of 475 million hectares (1.2 billion acres) of agricultural land. Most of these lands were opened by irrigation.

A second factor is that many food crops have been transplanted to new areas where they have thrived, in some cases better than in their areas of origin (Figure 8-2). The world's first farmers raised barley and wheat in the Middle East about 10,000 years ago. Later, about 9,000 years ago, Mexicans began cultivating maize and beans, Chinese began cultivating rice roughly 8,000 years ago, and about 7,000 years ago, South Americans began growing potatoes. Natives in what is today the eastern United States raised sunflowers 5,000 years ago, and peoples in the highlands of New Guinea cultivated taro and banana between 7,000 and 10,000 years ago. It is unknown whether the first sub-Saharan African farmers

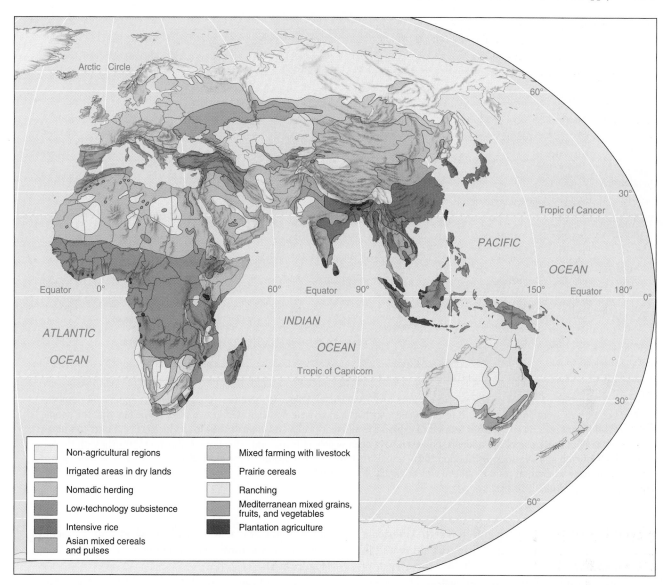

surprised that the categories mapped here largely coincide with the climates mapped in Figure 2-33 and the biomes mapped in Figure 4-22. "Nomadic herding" and "low-technology subsistence farming" cover impressive areas on this map, but their productivity is low. These areas are generally sparsely populated, as you can see by comparing this figure with the map on the rear endpaper.

began to cultivate millet, sorghum, and rice independently about 4,000 years ago, or whether this knowledge diffused from Middle Eastern cultures (Figure 8-3).

Even before Malthus wrote, the Western Hemisphere had contributed important food crops to the Eastern Hemisphere. For example, the potato is native to the Andes region of South America. It yields the second highest number of calories per acre of any crop, although its protein content is only 6 percent. By Malthus's day, it had already become a major food in northern Europe (Figure 8-4). Today, it is the world's fifth most important food crop, when measured by total tonnage harvested, after sugarcane, maize, wheat, and rice. China has recently recognized the potato's versatility and is today the world's largest producer. Russia is in second place, followed by India and the United States. The crop is becoming a mainstay throughout Africa and Asia.

Potato yields have increased due to **genetic engineering,** which is the manipulation of a species' genetic material. Two main techniques of genetic engineering are selective breeding, which has been carried on for millennia, and recombinant DNA, a new form of genetic modification discussed later in this chapter. New genetically modified potatoes are being introduced to the tropics, where they are ready to harvest only 40 to 90 days after planting, instead of 90 to 140 days. Potato planting continues to spread, and the value of the world's potato crop increases each year. The true treasure of the Andes was not the gold that the Spanish conquerors sought but the potatoes they trampled.

Maize is another Western Hemisphere native. "Corn" is the name English-speaking people have traditionally given to any major cereal crop in a given region. Therefore, in England corn means wheat;

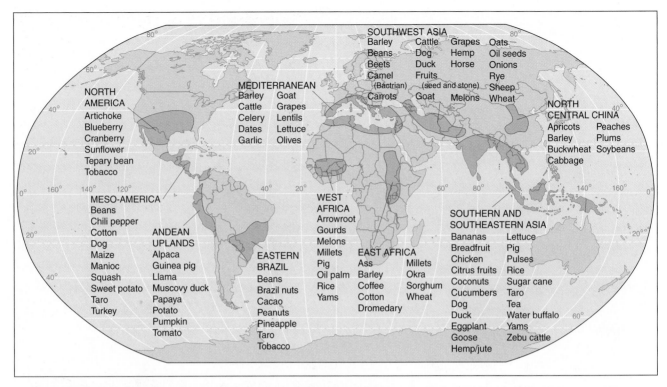

Figure 8-2 Regions of origin of the world's foods and livestock. This map identifies 10 regions where significant numbers of plants or animals were domesticated. A few examples, such as barley, were domesticated in two places. The fact that the people in so many different regions and biomes learned to exploit the different plants and animals they found locally is a tribute to human ingenuity. Many of these plants and animals, however, have been widely redistributed. Therefore, comparing this map with the following maps of crop and livestock distribution today reveals that an area that produces the most of any crop or animal today is not necessarily a place where that plant or animal was originally domesticated.

in Scotland and Ireland, oats; and in North America it means maize, or Indian corn. Maize was cultivated in Central and South America long before the arrival of Europeans, who introduced it into east-central North America, where almost one-half of the world's maize is grown today (Figure 8-5). Maize is susceptible to frost, so the length of the frost-free period limits its distribution. Maize is a staple for people in South America and Africa, but about 70 percent of the U.S. crop is fed to livestock. Maize's average protein content is 10 percent.

Still another transplant is manioc, or cassava, a plant grown for its edible roots. The plant originated in South America, and Portuguese traders introduced manioc into Africa in the sixteenth century. It has a very low protein content and should be supplemented with other high-protein foods, but it remains a staple across lowland Africa. You may be familiar with tapioca, which is made from manioc root.

Transportation and Storage

Improvements in worldwide transportation have allowed regional specialization in food production, and specialization can multiply productivity. Railroads, trucks, and cargo ships—many of them refrigerated—

move quantities of food with a speed and efficiency that could not have been imagined in Malthus's day (Figure 8-6).

Transportation also allows the shifting of food from surplus to deficit areas, better sustaining populations whose local crops may be insufficient. In the past, it was not uncommon for surplus food to rot in the sun just a few miles from a starving population because the food could not be transported. Improved transportation also allows us to maintain smaller inventories of food at any specific time and place—yet remain secure.

An increasing share of the world's food production is entering world trade—$1.04 trillion worth in 2008. Table 8-1 lists the largest importers and exporters of some of the major foods traded. Some of the major exporters of certain crops are not among that crop's major producers. This indicates that in those countries, food production for export is a highly commercialized business. For example, the United States produced only about 1.4 percent of the total world rice crop in 2005, but the United States accounted for 12.5 percent of all world rice exports because most rice harvested in Asia is consumed where it is produced. As much as one-sixth of world wheat production enters into world trade.

Storing food for later consumption has always been a part of farming, beginning with pots and baskets.

(a)

(b)

Figure 8-3 Millet (a) and sorghum (b). These common cereal crops are relatively unknown in European diets and agriculture but are important grains that provide food for large parts of the developing world.

MapMaster™
Tested Thematic
World: Economic
Production

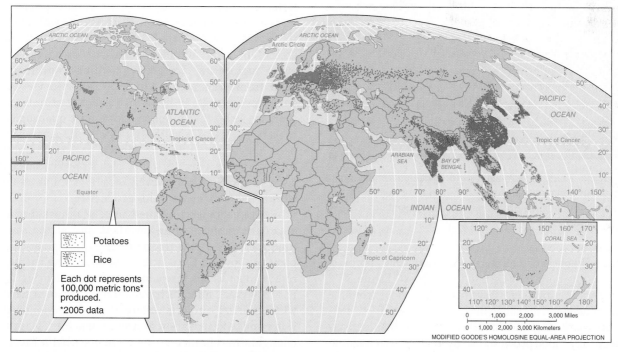

Figure 8-4 Current production of potatoes and rice. The potato is native to South America, but the country with the greatest production today is China. Rice remains the primary staple food produced in Asia, where it was first domesticated.

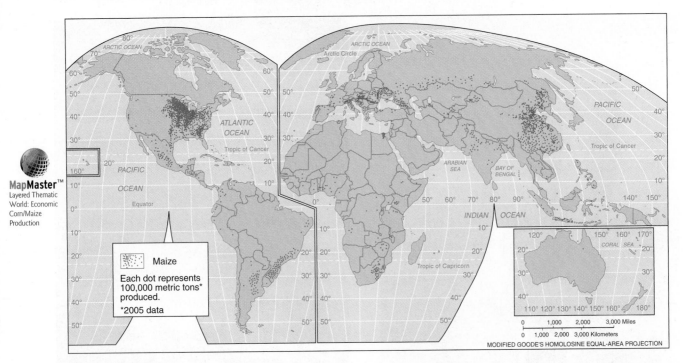

Figure 8-5 World maize production. Maize was first domesticated in Central America, but world production today is concentrated in North America.

Figure 8-6 Fresh food year-round. Modern techniques of food transportation and storage allow a grocery store in Alaska to offer fresh produce even when outside temperatures read –40°C (–40°F).

The technology of food storage has continually improved, decreasing both spoilage and loss to pests (Figure 8-7). Modern improvements have been continuous from the introduction of the silo in the nineteenth century to antiseptic packaging and freeze-drying. The Inka actually freeze dried food 500 years ago by leaving food outside at night to freeze, then scraping the ice off the next day. Modern freeze-drying was invented in the twentieth century.

Other Technological Advances

The addition of new crops and other technologies over the past 200 years is part of the continuing application of science to agriculture that began in the late eighteenth century, known as the **scientific revolution in agriculture.** New pesticides save crops from insects that once wiped out entire harvests, and new fertilizers multiply yields per acre. Parallel scientific research has been directed to livestock. As with crops, the developments include world redistribution, increases in total numbers, improved breeding, and even "engineering" for greater hardiness and higher yield from each animal. In 1950, the average U.S. dairy cow gave 2,339 liters (618 gallons) of milk per year; by 2007, that number had almost quadrupled, to 9,603 liters (2,537 gallons). In 1969, the average pig used for breeding in the United States produced 6.7 piglets per year. By 2000, piglet production had risen to 16. So far, however, only a fraction of the world's livestock herds is this productive.

Farm machinery invented since Malthus's day reduces the number of fieldworkers and at the same time increases yields by improving the regularity of plant spacing and the efficiency of harvesting. Heaters rescue many crops threatened by freezes. During the past 200 years, humans have literally re-formed Earth with drainage projects where there was too much water and with irrigation projects where there was not enough. Projects such as these are not new in theory, but they are achieved in a scale beyond anything that Malthus foresaw.

CONNECTIONS

Food Prices, Soybeans, Ethanol, and Deforestation

In 2008, food riots broke out among poor residents of Bangladesh, a new ethanol plant began operation in North Dakota, and 28,000 square kilometers (10,800 square miles) of forest were cleared for agriculture in Brazil. These events, in places far removed from each other, are intimately connected.

Consider the linkages: To replace petroleum fuels with biofuels like ethanol, new fuel production plants are needed around the world. Greater biofuel production puts greater demand on agricultural land to grow more fuel crops, causing deforestation in places like Brazil. Some farmers also shift their existing cropland to grow fuel crops instead of food crops, leading to food shortages in places like Bangladesh. On the supply side, food and petroleum are global commodities whose prices are affected by events around the world. Crops like sugar and corn can be used for food and for fuel, so the supply for one use is affected by the other use. On the demand side, as the price of oil goes up, so does demand for alternative biofuels made from these agricultural commodities.

Soybeans further complicate this supply and demand cycle. Biofuels are not the main use of soybeans, however they are still important. Soybeans have a high content of both oil and protein. Most of the soybean harvest goes into animal feed. Meat production depends increasingly on feeds made from corn and soybeans, and world demand for soybeans increased eightfold between 1961 and 2007. Growth has been especially rapid in Brazil. Per capita meat consumption in Brazil increased threefold, and when combined with population growth, this amounts to a sevenfold increase in total meat consumption.

Soybean production has added to demand for agricultural land in Brazil, creating competition between production for food and biofuel. Although they are used for different purposes, corn and soybeans have similar climatic requirements, and they are typically grown in the same regions. When the price of corn (maize) goes up, farmers tend to plant more land in corn and less in soybeans, causing the price of soybeans to rise.

In the United States, most biofuel production is in the form of ethanol made from corn, but soybean oil is also suited to biodiesel fuel production. Thus demand for soybeans is directly linked to the price of petroleum, as well as the indirect link through corn production. When oil prices rose dramatically in 2007 and 2008, the biofuel industry could offer high prices for corn and soybeans. As farmers shifted their cropland from food production to biofuel crops, it added to rising food prices. The higher cost of fuels also increased the cost of growing, processing, and transporting food. Combined, these factors had a global effect on food prices, which many of the world's poor could no longer afford, leading to food riots.

To complicate matters even more, the increasing demands on Brazil's agricultural land have contributed to deforestation. This occurs indirectly. Most of Brazil's soybean production is in the subtropical southern parts of the country, rather than the tropical Amazon basin. But when land in southern Brazil used for other purposes, such as pasture, is converted to soybean fields, then that increases the demand for pasture elsewhere, including the Amazon.

Brazil's soybean production has grown enormously over the last four decades.

Area of vast deforestation in Brazil

TABLE 8-1 Top Ten Importers and Exporters of Wheat, Rice, and Maize, 2007 (in thousands of metric tons; data from FAO)

Leading Exporters	Exports	Leading Importers	Imports
Wheat			
United States	33,442	Brazil	7,528
Canada	17,863	Italy	6,292
France	15,451	Egypt	5,911
Australia	14,980	Netherlands	5,290
Russian Federation	14,830	Japan	5,278
Argentina	10,878	Algeria	4,863
Kazakhstan	8,199	Spain	3,829
Germany	5,432	Belgium	3,744
China	3,369	Morocco	3,690
United Kingdom	2,114	Indonesia	3,423
Rice			
Thailand	9,196	Philippines	1,903
India	6,450	Indonesia	1,406
Vietnam	4,558	Senegal	1,073
Pakistan	3,129	Saudi Arabia	969
United States	2,987	South Africa	959
China	1,325	Benin	940
Egypt	1,223	Iran	935
Uruguay	799	Côte d'Ivoire	809
Italy	727	Malaysia	798
Argentina	451	North Korea	785
Maize			
United States	57,014	Japan	16,628
Argentina	14,990	South Korea	8,579
Brazil	10,933	Mexico	7,955
Hungary	4,976	Spain	6,675
China	4,917	China	4,530
France	4,749	Egypt	4,474
India	2,728	Netherlands	3,448
Paraguay	2,109	Iran	3,409
Ukraine	809	Colombia	3,323
Germany	712	Malaysia	2,658

Figure 8-7 New food-storage facilities. This officer of the U.S. Agency for International Development (AID) is overseeing the construction of a silo in Peru. Many countries produce enough food to nourish their people, yet much of the crop is wasted or lost to pests or spoilage because of inadequate storage and distribution facilities.

The Green Revolution

Technological advances in agriculture primarily suited European and North American farms. These crops and techniques were not appropriate for many climates and diets around the world. Starting in the 1940s, an intensive effort was launched to develop new grain varieties and farming methods and to establish them in developing countries. It focused on certain crops (rice, wheat) and certain techniques (breeding for response to fertilizer inputs). It was driven largely by private foundations such as the Ford and Rockefeller foundations. This focused effort to increase food production was known as the **green revolution** (and is unrelated to later innovations in green technology). Many projects continue. For example, specialists at Texas A & M University and at the International Rice Research Institute in the Philippines have produced higher-yielding strains of rice that are more resistant to disease and pests and that mature in only 110 days instead of 140. This allows for two or even three crops per year. New potato strains are developed at the International Potato Center in Lima, Peru. Wheat yields have multiplied from new hybrids. India, where 1.5 million people died in a 1943 famine, produced enough extra grain to become a net grain exporter in 1977, even though its population doubled between those years.

Productivity continues to rise: World yield of cereal crops per hectare more than doubled from an average of 1,653 kilograms per hectare (1,476 pounds per acre) in 1967 to 3,380 kilograms per hectare (3,018 pounds per acre) in 2007.

The increasing productivity of agriculture on a global level is the main argument against Malthus's prediction that population growth would outpace the food supply (see Chapter 5). There are many places, however, where new agricultural technologies have failed to increase yields or where increased yields have come at a high cost. The problems begin with seed and extend to the system surrounding its production. Higher-yield strains, in and of themselves, are often more reliant than traditional strains on additional **agricultural inputs,** which include all the factors that go into growing and harvesting a crop. High-yield strains, because they grow more edible plant material, often require much more water and fertilizer. Many strains are susceptible to devastating fungi and insects, and often can't outgrow weeds. These problems require the application of chemicals that solve one set of problems but might create more by poisoning the workers and the local environment. These inputs to maintain high yield strains are expensive if they are available at all. The application of these inputs require more hours of human labor, which may require farmers to hire more help. If the crop is successful, the total yield from the farm may again require additional workers or large machinery to harvest it. This, in turn, requires larger storage structures to keep it in good condition before spending money to transport it to market.

In some places, like the Philippines were rice is grown, small farmers at first thrived with the new crops despite the added costs. From the 1960 to 1970s, the Philippines went from being a net importer to a net exporter of rice. In other places, however, the high costs associated with irrigation equipment, chemicals, labor, storage, and transport meant only large farms could afford to invest in the new strains. Those large farms soon expanded, buying land from small farmers who went bankrupt. In some cases, the landless poor might be hired back to work the same crops for a wealthy farmer now grown richer. Meanwhile, the diversity of local crops that might be affordable and nutritious to the poor are largely replaced with a small number of crops headed for distant markets, replaced by cheap processed and unhealthy foods. The social and economic importance of farming in many countries means that these outcomes of new agricultural technology affect large segments of the population and have direct repercussions on national politics.

Agriculture Today

Humankind has thus managed to prevent the widespread starvation Malthus predicted. Not all of these developments, however, have taken hold equally everywhere. The following pages will first catalog regions of different types of present-day agriculture, and then we will examine the distribution of food supplies. What regions produce surpluses, and where are people hungry? Finally, we will address the question of whether and how we will be able to produce enough food for an increasing human population in the future. The pessimistic neo-Malthusian point of view introduced in Chapter 5 will again be set against the optimistic Cornucopian point of view.

Chapter 6 introduced the theory that human societies evolve through a series of stages. The theory was based, principally, on the evolution of agriculture, for humankind's first concern is to get enough to eat. Humans were at first hunter-gatherers before they domesticated animals and developed pastoral nomadism. Later, the domestication of plants encouraged settled agriculture, which is more productive. That is, settled agriculture yields a greater and more secure quantity of food. **Subsistence agriculture**—agriculture to feed oneself and family—was typically part of a communal agricultural system that traded foods locally or with nearby communities. These systems eventually yielded to **commercial agriculture**—growing food and raising animal products for sale.

A few hunter-gatherer bands still exist and provide evidence of how all humans lived in prehistoric times, before the invention of agriculture. Isolated groups inhabit the periphery of world settlement—the Arctic, the Amazon Basin in South America, central and southern Africa, and Papua New Guinea. Examples include the Bushmen of Namibia and some Aborigines of Australia. These groups are having increasing trouble gaining access to land and resources now used by expanding agricultural communities. Hunter-gatherers are under pressure to adopt settled agricultural practices and are sometimes forced to abandon their culture. In 2002, the government of Botswana forcibly rounded up and settled the last of the Basarwa, or San Bushmen—about 2,000 people. "Hunting and gathering," decreed President Festus Mogae, "is no longer a viable way of life." By 2006, few were left in the "protected area," and the government stood accused of genocide.

Subsistence Farming Contrasts with Commercial Farming

Chapter 6 discussed in general terms how societies change from self-sufficiency to participation in wider trade networks by examining three villages connected by a road. Subsistence agricultural villages rely on their own food production and are likely to raise many food crops that they will mostly consume themselves. Once they begin to trade food with other places, however, they dedicate more of their land and labor to the production of a smaller number of foods that can be exchanged more widely. Commercial farmers often specialize in one or two crops for sale at market, using the money they earn to import other foods from elsewhere. Thus, agricultural **polyculture,** the raising of a variety of crops, may yield to **monoculture,** specialized production of one crop.

There is no clear dichotomy between subsistence agriculture and commercial agriculture. They are the endpoints of a spectrum. The most self-sufficient hunter-gatherer may trade extra berries for fish, and a commercial wheat farmer in Kansas may also raise his or her own tomatoes. Nevertheless, many farmers today are called subsistence farmers because they do consume most of what they produce, and they produce most of what they consume. Subsistence farmers may sell some of their output, but surplus production is not their primary purpose, and in years of poor harvests, they may not even have a surplus.

The world is too interconnected, however, for there to be many completely self-sufficient subsistence farmers left anywhere. Farmers generally respond to economic incentives, so if a farmer has a market opportunity, he or she will try to produce something for it. In virtually all countries, urban populations offer markets for farmers to satisfy. The day after a dirt road or a railroad reaches the smallest village, many peasants are on their way to market something in the nearest city. The 800 million peasants of China are often referred to as subsistence farmers because most of what they raise is for their own consumption. They do, however, manage to raise and market enough of a surplus to feed China's 500 million urbanites and still have enough left over for China to achieve net exports of food.

Several characteristics differentiate subsistence agriculture from commercial agriculture. Subsistence farming usually relies on enormous investments of human labor but little animal or machine power. It is usually low in technology and, therefore, in productivity, and most of the food raised is consumed by the farmers themselves.

Modern commercial farming, by contrast, relies on substantial capital investment in machinery, chemicals, improved seeds, and livestock. The high price of farm machinery necessitates that it be used maximally, and this encourages an ever-increasing average farm size. Commercial farm products are seldom sold directly to consumers, but rather to large companies that process, package, store, distribute, and retail food. Some large companies are wholly organized around agriculture—agribusiness companies—and many of them, such as Ralston Purina and General Mills, are household names in North America. Another large agribusiness company, Cargill, has 159,000 employees in 68 countries; in 2009, the company earned $116.6 billion.

Most farms may be owned by individual families, but they are increasingly reliant on off-farm jobs and may rent out their land to other farmers. In the United States, 87 percent of farms are owned by families and individuals, but only 45 percent of farm operators do so as their primary occupation. Basic agriculture in the United States and other wealthy countries can no longer provide the small family farmer a middle class lifestyle. Instead, many small farm owners rent out their fields to other farmers who bundle enough acres together to make a living. For farmers in wealthy countries to survive, they must, like large agribusiness companies, achieve **economies of scale.** Economies of scale occur when one can achieve greater earnings per unit produced only by expanding the number of units produced. A farmer needs more or less the same tractor and other inputs to farm a small piece of land as to farm a large piece of land. The more land the farmer can work, the more money he or she can earn beyond the cost of his or her basic inputs. If the farmer can't afford more land he or she might have to find a job in town. If the farmer can rent land from others, he or she can, in effect, expand his or her farm until the input costs are covered.

Types of Agriculture

According to the United Nations Food and Agriculture Organization (FAO), in 2007, about 1.55 billion hectares (3.83 billion acres) of Earth's land area were devoted to crop production (12 percent of the total, excluding Antarctica); an additional 3.38 billion hectares (8.35 billion acres, 26 percent) were devoted to permanent pasture. Cereals (grains) and potatoes are the world's basic foods. Cereals have been grown since the dawn of history, and every major civilization has been founded on them as their principal source of food. Wheat, rice, maize, barley, sorghum, oats, rye, and millet are all members of the grass family of plants. They yield more food, both in bulk and in nutritive value, than most other crops. The preeminence of cereals as a food is also partly explained by the ease with which they can be produced, stored, and transported (Figure 8-8).

Most of Earth's "nonagricultural land" is too hot, too cold, or too dry for productive or profitable agriculture. Compare the distribution of major agricultural forms in Figure 8-1 with the climate regions shown in Figure 2-33. Agriculture has not been productive in these areas, nor have humans settled there in large numbers, as evident from comparing these regions with the population density map on the rear endpaper. In Figure 8-1, activities like nomadic herding and low-technology subsistence agriculture cover enormous areas, but these activities support very few people. In contrast, the world's largest population areas are supported by intensive rice farming or mixed crops. Also shown in Figure 8-1 is irrigated agriculture in dry lands, but this can cover a wide range of farming styles and

Figure 8-8 The shipment of grain. In the mid-nineteenth century, grain was still stored and transported in individual bags. This required enormous amounts of labor, and a lot of grain was lost to spillage. Late in the nineteenth century, however, methods were devised to store and ship grain as if it were a liquid. These techniques greatly reduced the cost and improved the efficiency of transportation.

population sizes, from the intensely worked and productive fields of California to subsistence oases in the Sahara (Figure 8-9).

Leaving aside nomadic herding and ranching, the farming regions in Figure 8-1 are defined by the interaction of five variables in each place: (1) the natural

Figure 8-9 Oasis in Algeria. The Grand Erg Occidental Taghit oasis in the Sahara Desert in Algeria supports a small population and feeds an occasional caravan passing through the surrounding sands.

environment, (2) the crops that are most productive in that environment, (3) the degree of technology used by local farmers, (4) the degree to which local farming is market-oriented, and (5) the degree to which crops are raised either for human consumption or to feed animals.

Another distinction to make in Figure 8-1 is that four agricultural regions are basically subsistence: nomadic herding, low-technology subsistence farming, intensive rice farming, and Asian mixed cereal and pulse farming. The other regions are predominantly commercial: mixed irrigated areas in dry lands, farming with livestock, prairie cereals, ranching, Mediterranean, and plantation agriculture.

Nomadic herding Pastoral nomads depend primarily on animals rather than crops, and pastoral nomadism may still be a successful low-technology adaptation to arid or semiarid environments. It exists across North Africa, the Middle East, and into parts of Central Asia. The Bedouins of North Africa and Saudi Arabia and the Masai of East Africa are examples of nomadic groups.

The animals upon which nomads depend may include horses, camels, goats, sheep, or cattle. The nomads will consume their animals or else sell them to sedentary farmers in exchange for grains. Sometimes, however, a nomadic group will plant crops at a fixed location and later return to harvest them. Pastoral nomads do not wander aimlessly; they have a strong sense of territoriality and often retrace the same routes.

There are probably no more than 12 to 15 million pastoral nomads today, and many of them are settling. Many governments are forcing nomads to settle in hopes of converting their land to crops. The government of Kenya, for example, is forcibly settling the Masai people. Some nomads are today allowed to move about only within limits fixed by governments. In the future, pastoral nomadism will probably be confined to areas that cannot be irrigated or that do not contain valuable mineral resources.

Low-technology subsistence farming The least intensive type of farming is shifting cultivation, or **swidden** cultivation, from a word meaning "to singe." It is usually called "cultivation" rather than "agriculture," because the word *agriculture* implies greater use of tools and animals and more extensive modifications of the landscape (Figure 8-10). Nevertheless, swidden cultivation is quite sufficient in tropical areas such as the Amazon region of South America, Central and West Africa, and Southeast Asia, including Indochina, Indonesia, and New Guinea.

Small patches of land are cleared with machetes and other bladed tools, trees are stripped and killed, and fires are set to clear the land for planting. The ashes contain a simple fertilizer, potash (potassium). Various subsistence crops are planted with the use of a simple digging stick or hoe. Polyculture is practiced even in a single swidden, which may contain a large variety of

Figure 8-10 Swidden burning. This fire was started to clear forest land for planting in Brazil. When the soil is exhausted, the site will be abandoned and reclaimed by forest.

intermingled crops. Soil nutrients are rapidly depleted, so the cleared land can be used to grow crops for only a short time, usually three years or less. Then it must be left for many years to regenerate, although it may be recultivated someday.

Shifting cultivation can support low levels of population without causing environmental damage, but it is being replaced by commercial logging, cattle ranching, and cultivation of cash crops. These activities, however, require cutting down much greater expanses of forest. Wholesale destruction of the rain forests reduces global biological diversity and contributes to global warming, as discussed in Chapter 4.

Intensive rice farming   The greatest number of farmers living in the large population concentrations of East, South, and Southeast Asia practice intensive subsistence agriculture. All usable land is planted, even tiny, irregular plots. Individual farmers' holdings are often fragmented as a result of subdividing holdings among children through generations. Most of the work is done by hand or with animals, largely because human labor is abundant, but capital for the purchase of equipment is scarce. Livestock are rarely permitted to graze on land that could be used to plant crops, and little grain is grown to feed the animals. The intensive agriculture region of Asia can be divided between areas where wet rice dominates and areas where other crops predominate. Wet rice is characteristic of Asia; dry rice is a grassy staple largely confined to West Africa. Rice is also grown in Egypt, Nigeria, southeastern Brazil, and parts of the American south.

About half the world's population subsists wholly or partially on wet rice. Wet rice occupies a relatively small percentage of the land, but it is the most important source of food in southeastern China, eastern India, and much of Southeast Asia. It has a protein content of 8 to 9 percent. Successful production of large yields of rice is an elaborate, intensive process. Wet rice should be grown on flat land, because the plants are submerged in water much of the time. Therefore, most wet rice cultivation is located in river valleys and deltas, although additional suitable land may be created by terracing the hillsides of river valleys (Figure 8-11). A farmer prepares the field using a plow drawn by water buffalo or oxen. The plowed land is flooded via dikes and canals that must be maintained to control the quantity of water in the field. A flooded field for rice cultivation is called a rice paddy in English, but has a variety of local names usually synonymous with field. Rice is either broadcast through the field, or else seedlings that have been nursed elsewhere are transplanted. Rice plants grow submerged in water for approximately three-fourths of the growing period, and they are harvested by hand.

In places with relatively warm winters, such as South China and Taiwan, two rice crops can be harvested per year from one field, a process known as **double cropping.** Most parts of India have dry winters, so double cropping there, where possible, usually involves alternating between rice in the wet summer and another cereal such as wheat in the drier winter.

Asian mixed cereals and pulses   Wet rice cannot be grown where summer precipitation levels are too low and winters too harsh. Therefore, other crops predominate in the interior of India and in northeast China. In milder parts of the mixed cereals and pulses region, crop rotation may allow double cropping. Wheat is the most important crop, followed by barley, which is the world's

Figure 8-11 Rice terraces. These rice terraces in China increase the amount of land available for rice cultivation by turning steep slopes into flat steps.

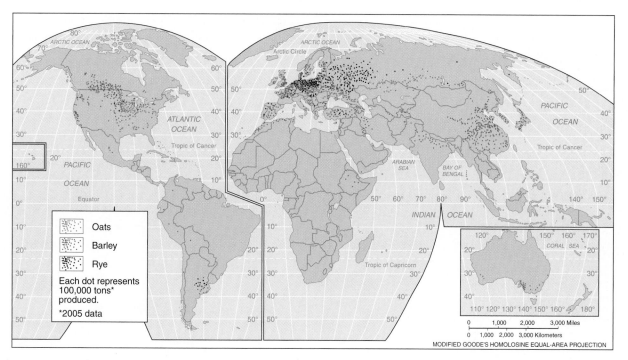

Figure 8-12 Production of barley, oats, and rye. Barley, oats, and rye can serve either as human food or as feed for livestock.

fourth most important cereal (Figure 8-12). Barley has the shortest growing season of all cereals and can be grown farther north, at higher altitudes, and in more arid regions than any other cereal. It is an important food crop in parts of Asia and in Ethiopia.

Pulses is the botanical name for a large family of herbs, shrubs, and trees, also called the pea or legume family. The seeds, which include beans, peas, lentils, peanuts, soybeans, and carob, are rich in protein and may form the principal source of dietary protein in poor regions or in regions where religious belief prohibits the consumption of meat. In China and Japan, tofu, made from soybeans, is a major food source. The root nodules of pulses have bacteria that absorb nitrogen out of the air and release it into the soil (a process called *fixing* nitrogen). This ability makes pulses valuable as temporary crops that reduce erosion and build up a soil's nitrogen content.

Mixed farming with livestock   Mixed farming with livestock involves growing crops and raising animals on the same farm. The intensity of land use is somewhere between intensive rice and extensive ranching. It is usually a commercial undertaking, for most of the crops are fed to animals rather than consumed directly by humans. Mixed farming allows the farmer to distribute the workload throughout the year (fields require less attention in the winter) and also to maintain his or her income by selling animals or animal products throughout the year. Crop rotation allows the farmers to maintain the soil fertility because different crops may deplete the soil of some nutrients but restore other nutrients.

Mixed farming is the dominant form of agriculture in much of the world today. Although it is still widespread in North America, farming there is evolving toward more specialized production. Farms that are truly mixed, including both feed production and animal production, remain common; but today maize and beans are increasingly grown on farms that no longer keep more than a few animals, and the feedstock is sold to feeding operations that produce the meat. Specialization has separated the two components of the system in both ownership and location. In the U.S. "maize belt," which stretches from Indiana across Iowa to the Dakotas, approximately half of the cropland is planted in maize. Some of this maize is consumed by people directly as oil and other processed foods, but most of it is fed to livestock (either on the farm or concentrated in feed-lots), including pigs, cattle, and poultry. The second most important crop in this region, soybeans, is grown almost exclusively as animal feed. The third most important crop in the United States by area harvested (although not by value) is hay, which is also an important component of mixed farming systems.

This farming is enormously productive. Maize production in the United States represented 42 percent of world total production in 2007; soybeans grown in the United States also accounted for 33 percent of the world total of that crop.

Prairie cereal farming   Crops on commercial grain farms are grown primarily for consumption by humans rather than by livestock. Large-scale commercial grain production is found in only a few countries, including

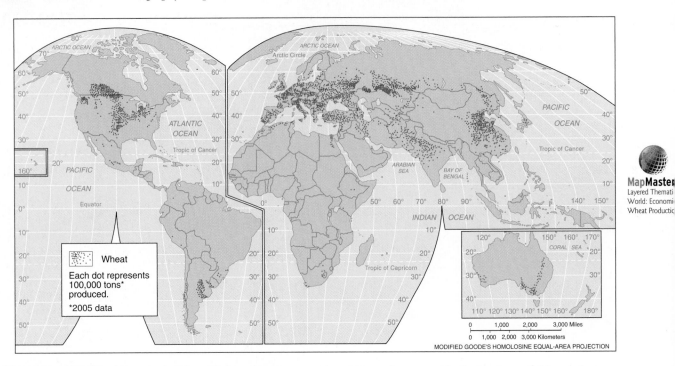

Figure 8-13 World wheat production. Today, wheat is grown almost everywhere except in the tropics, and it is a major commodity in international trade.

the United States, Canada, Argentina, Australia, Russia, Ukraine, and France. This production is mostly, but not exclusively, in regions that are too dry for mixed crops and livestock. Wheat is today the most important cereal in world food production. It supplies about 20 percent of the total calorie consumption of the human species and provides a staple for over one-third of the world's population (Figure 8-13). The protein content varies between 8 and 15 percent.

Wheat generally has more uses as human food than other grains, so it is more valuable. It can be stored relatively easily without spoiling and transported long distances both easily and economically. Hard-kerneled varieties yield flour with a high content of *gluten* (a protein substance) and are used to make breads; flour from soft-kerneled varieties is starchier and is used to make cakes and biscuits. *Durum* wheat is an especially hard-kerneled wheat used in pasta products. Wheat grain, *bran* (the fibrous outer coat of the wheat kernel that remains as a residue after milling), and the rest of the plant are all valuable livestock feed.

In North America, large-scale grain production is concentrated in three areas. In the winter wheat belt that extends through Kansas, Colorado, and Oklahoma, the crop is planted in the autumn and develops a strong root system before growth stops for the winter. The wheat survives the winter and is ripe by the beginning of summer. In the spring wheat belt of the Dakotas and Montana in the United States, and in southern Alberta, Saskatchewan, and Manitoba in Canada, the winter is usually too severe for winter wheat, so spring wheat is planted in the spring and

harvested in the late summer. A third important grain-growing region is the Palouse region of eastern Washington State. Large-scale grain production is highly mechanized and conducted on large farms, but the effort required is not uniform throughout the year. Therefore, some individuals or firms may own fields in each belt in order to keep their expensive machinery in use longer. Larger growers start working in Texas in late May, then move to Oklahoma in early summer and farther north through the season.

Ranching Ranching is the commercial grazing of livestock over extensive areas. It is best suited to arid or semiarid land, where the soil is too poor to support crops. Comparing again Figure 8-1 to the climate map in Figure 2-33, one can see ranching is most prevalent in dry and mid-latitude climates. Cattle raised on ranches are frequently sent for fattening to farms or to local feedlots before being moved to meat processors (Figure 8-14).

Spaniards introduced cattle to the Western Hemisphere in the sixteenth century, and the cattle thrived on the grazing lands in both North and South America. In the late nineteenth century, British capital developed the commercial cattle industry in Argentina in areas convenient for shipping to overseas markets. Today a large portion of the prairies of Argentina, southern Brazil, and Uruguay are devoted to grazing cattle and sheep.

The interior of Australia was opened for grazing at the same time, although sheep are grazed along with cattle. Reviewing Figure 5-28 reminds us that ranchlands are often sparsely populated, which helps to make

Figure 8-14 Cattle feedlot. Cattle that have been grazing over extensive areas are usually fattened at feedlots such as this before slaughter. Cattle are usually kept in their pens for several months.

Figure 8-15 Grapes in Chile. Chile long exported substantial quantities of grapes such as these, but Chileans now multiply the value of this rich agricultural harvest by converting the grapes to wine and exporting wine. A wine made entirely, or chiefly, from one type of grape is called a *varietal*; these Malbec grapes produce a fine red wine.

the large land areas needed for grazing relatively less expensive. Ranches in the Middle East, New Zealand, and South Africa are also more likely to raise sheep. Commercial ranching is found in other relatively developed regions of the world, but it is rare in Europe, except in Spain and Portugal.

Mediterranean agriculture Mediterranean agriculture is found in the distinctive Mediterranean climate regions—usually on the west coasts of continents at about 30° to 40° north and south latitude around the Mediterranean Sea, in Southern California, central Chile, the southwestern tip of South Africa, and southern Australia. The Mediterranean climate has hot, dry summers and cool, rainy winters. The lack of summer water and good grazing land hinders livestock raising, so Mediterranean-style farmers traditionally derive only a small percentage of income from animal products. Many of these areas are mountainous, or at least hilly, so some farmers do practice transhumance—keeping sheep and goats on the coastal plains in the winter and transferring them to pastures in the hills in the summer.

Most crops in Mediterranean lands are grown for human consumption, including most of the world's olives, grapes, and other fruits and vegetables. Hilly landscapes allow farmers to practice polyculture, with some crops facing the Sun, while others face away.

Two-thirds of the world's wine is produced in countries that border the Mediterranean Sea, especially Italy, France, and Spain, while other Mediterranean regions elsewhere (such as Southern California, Chile, and South Africa) produce most of the rest (Figure 8-15). About half of the land may be planted in cereals, especially wheat for pasta and bread. Seeds are sown in the

fall, and the crops harvested in early summer. Land is periodically left fallow to conserve moisture in the soil.

California has less land planted in cereals than is traditional in Mediterranean regions, because California growers concentrate on profitable citrus fruits, tree nuts, and fruits produced by deciduous trees (apples, peaches, plums, and similar fruit). Rapid urbanization along the coast has driven agriculture inland, where irrigation is required. Competition for water among farmers, urban and recreational areas, and environmental demands constitutes one of the most important political issues in the West.

Plantation farming A plantation is a large farm that specializes in the commercial production of one or two crops. Plantations can be found in tropical Latin America, Africa, and Asia where they originated as colonial agricultural estates. Wealthy and politically favored colonizers were given land on which they used enslaved or low-wage labor. In the American South, plantations grew rice, cotton, and tobacco until they were broken up following the Civil War. Independence for former colonies helped close plantations in some parts of the world. Latin American plantations persisted, often providing low wages and little food for local communities while plantation owners became incredibly wealthy. New plantations have been established, some owned by European or North American companies. Like the original plantations, crops are grown primarily for sale in rich countries. Latin American plantations grow coffee, sugarcane, and bananas, while Asian plantations may provide rubber and palm oil (Figure 8-16). **Land reform,**

Figure 8-16 Tapping a rubber tree in Indonesia. This woman is tapping a rubber tree on a plantation on the Indonesian island of Sumatra.

the redistribution of large land-holdings such as plantations to poor would-be farmers, has been a common objective in many of the revolutionary and nationalist political movements in poorer countries (see Chapter 11).

What Determines Agricultural Productivity?

The various types of agriculture we have described all concentrate on different sorts of crops, but we can learn something about the great range of productivity in agriculture around the world by looking at yields of cereal grain (Figure 8-17). This range of output is partly attributable to variations in the physical environment. For example, Western European countries enjoy mild climates, and their agriculture is highly productive, whereas the Sahara Desert region of northern Africa is generally unproductive. Surprising exceptions, however, can also be found. For example, several Middle Eastern Arab countries and the islands of Mauritius record some of the highest grain yields anywhere on Earth.

When we compare Figure 8-17 with Figure 8-18 and Figure 8-19, however, we see a considerable correlation between the variation in productivity and the application of fertilizers and the use of tractors. The countries that enjoy the highest crop yields rank high in at least one of these two important inputs. Investment in agricultural inputs seems to be a principal factor determining productivity. Many of the rich places that are able to invest in tractors and fertilizer can also be assumed to be the places where farmers

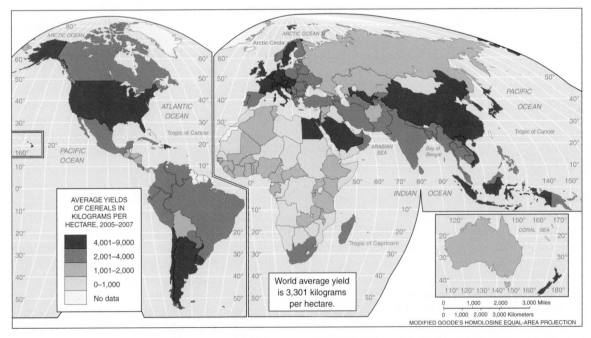

Figure 8-17 Agricultural productivity. Yields per hectare vary greatly around the world, and these variations are not always clearly related to natural environmental conditions. Other factors, especially capital investment (see Figure 8-18) and fertilizer use (see Figure 8-19), can greatly increase yields.

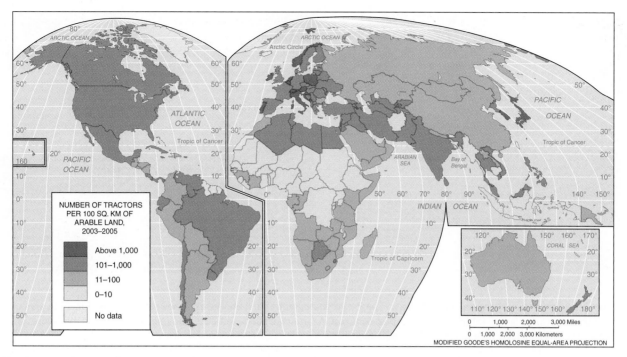

Figure 8-18 The use of tractors. Capital inputs in agriculture vary tremendously. The use of tractors is a good measure of modern capital-intensive farming. Does this map reproduce the pattern of Figure 8-17 and, therefore, help explain variations in yield?

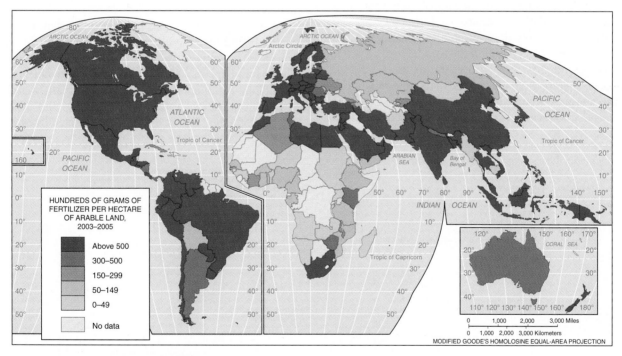

Figure 8-19 The application of fertilizer. Fertilizer is an important capital input in agriculture. Its application can override deficiencies in the local soil.

have high quality seeds. These countries also have advanced irrigation techniques, pesticides, and all the other technologies of modern agriculture. Therefore, we can conclude that capital and technological inputs are the principal determinants of agricultural productivity

today. Capital investment may be as important as the natural environment, or even more so. If investment in agriculture could be increased in the regions where there is little investment today, total world agricultural output would undoubtedly soar (Figure 8-20).

(a)

(b)

Figure 8-20 Extremes of capital investment. These two photographs illustrate the extremes of capital investment in agriculture. If the Senegalese farmer (a) had as much capital to invest in farm machinery, fertilizer, and improved seeds as the Nebraska farmers (b) do, who knows how much food he could raise?

Livestock Around the World

Humans began to domesticate animals about the same time as they did plants, some 8,000 to 10,000 years ago. Dogs were probably domesticated first, independently at a number of places around the world. Dogs helped in hunting, and they were also eaten for food. Looking back at Figure 8-2, one notes the domestication of other animals occurred at various locations in the centuries that followed. As humans learned to domesticate plants, they found it convenient to herd plant-eating animals and to pen them close to settlements. From then on, growing crops and raising livestock advanced side by side.

A wide range of animals have been domesticated as livestock, and their numbers totaled more than 24 billion in 2007—over three times the human population. These numbers included about 18 billion chickens, 3.5 billion ruminants (cud-chewing animals, mainly cattle, sheep, goats, buffalo, and camels), and about 918 million hogs (Figure 8-21). Domesticated livestock

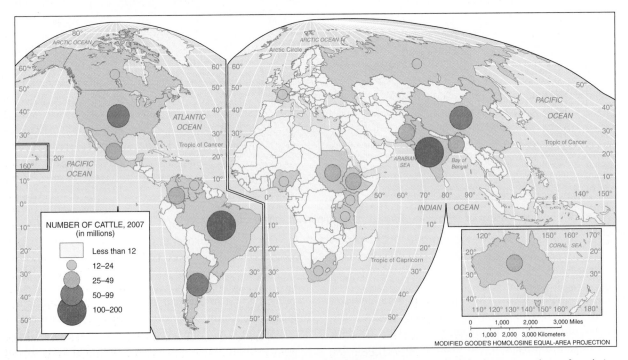

Figure 8-21 World distribution of cattle. Cattle are widespread, but these 20 countries had the largest numbers of cattle in 2007. Together, these countries account for over 70 percent of the world's cattle.

Figure 8-22 Masai cattle. Cattle represent wealth among the Masai people of Kenya.

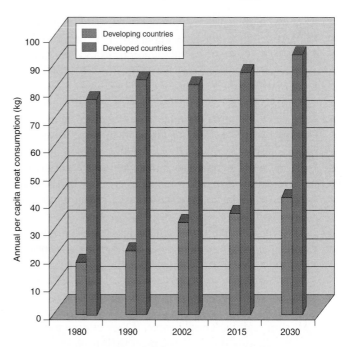

Figure 8-23 Meat consumption increases with rising incomes. The addition of millions more consumers to the global middle class in recent years led to a surge in demand for meats in the developing, poorer countries. Increased meat consumption is projected to increase and will require larger amounts of grain to be given as feed to animals.

provide high-quality protein, either as meat or milk, as well as supply hides, wool, and other raw materials. Livestock serve as draft animals that pull plows and wagons; large numbers of people in poor regions still depend on oxen and buffalo. In India alone, there are over 80 million draft animals. Among pastoral peoples, livestock represents wealth on the hoof (Figure 8-22).

The Direct and Indirect Consumption of Grain

Many domesticated animals eat grass, foliage, and other plant material that humans cannot digest. Some grains, however, called *feed grains*, are fed to livestock. Eventually, either the livestock or their products (milk, butter, cheese, and eggs, for example) are consumed by humans, so we say that humans consume feed grains indirectly. Over one-third of the world's total grain harvest is fed to livestock. This percentage is higher in richer countries. In the United States, for example, the figure is approaching 60 percent, whereas in India the figure is about 4 percent (Figure 8-23).

As a country develops economically, the population consumes a greater amount of grain, but an increasing percentage of that grain is consumed indirectly as meat and dairy products. The people of the United States and Canada, for example, consume as much as 907 kilograms (2,000 pounds) of grain per person per year, but only about 68 kilograms (150 pounds) of that amount is consumed directly as bread or cereal. The rest is consumed indirectly. In poor countries, on the other hand, only about 181 kilograms (400 pounds) of grain are available for each person per year, and it mostly is consumed directly. Meat is the center of a people's diet only in the richest countries, and some nutritionists argue that the consumption of so

much meat and dairy products leads to increases in debilitating diseases such as heart disease. Nevertheless, most people want and enjoy meat and dairy products. Therefore, rising incomes in some countries multiply the populations' appetites for grain consumed indirectly as meat.

Global per capita consumption of meat rose from 26 kilograms (57 pounds) in 1970 to 40 kilograms (88 pounds) in 2003. The total supply of meat tripled during the period from 1980 to 2002, driven primarily by consumer demands in the developing world. China's rapid economic growth accounts for more than half of the rising meat consumption among these countries. Demand has risen in developed countries, too, although not at the same pace. Many observers see this rise in meat consumption as good news, but neo-Malthusians believe that it threatens the food supplies of the poorest people. As long as rich people are willing to pay high prices for meat, feed grain for animals will be grown instead of the food grain that is more affordable to poor people. Cornucopians, however, argue that if grain prices rise, triggering a rise in the price of meat, grain production will multiply, increasing the supplies and thus lowering the price of grain again, making more available for human consumption. In any case, improving cereal yields might cause prices to continue to fall. We do not know which of these economic events will occur, but such future price variations will affect the diet of billions of people.

Some kinds of animals transform grain into meat more efficiently than others do, and humankind overall could greatly improve its food supply by concentrating on raising these types of animals. Chickens are the most efficient. They yield 0.45 kilogram (1 pound) of edible meat for every 0.91 to 1.8 kilograms (2 to 4 pounds) of grain they consume. Pigs produce 0.45 kilograms (1 pound) of meat for every 3.2 kilograms (7 pounds) of grain, and beef cattle 0.45 kilograms for every 6.8 kilograms (15 pounds) of grain. In addition, chickens reach maturity and can be consumed in just seven weeks—much faster than pigs or cattle (Figure 8-24). There are already more than twice as many chickens on Earth as people, and today many countries are concentrating on increasing their chicken populations. China, for example, increased its chicken population more than 500 percent between 1977 and 2007. In many societies, however, cattle are a status symbol, and this preference delays the switch to more productive livestock.

In some places, people avoid meat and dairy products because of religious prohibitions. Figure 7-37 noted the absence of pigs, for example, in Jewish and Muslim regions. These people, if they choose to follow their religions' dietary laws, meet their protein requirements from beef, lamb (mutton), fish, grains, or pulses (Figure 8-25).

Figure 8-24 Intensive livestock raising. Hogs are just one animal raised in close-quarters. Poultry, cattle and fish are similarly kept on factory farms, crowded into sheds, feed-lots, and tanks where they are fed until large enough for sale. The proximity of so many animals to each other and their own waste requires antibiotics and drugs to keep them well enough to reach maturity.

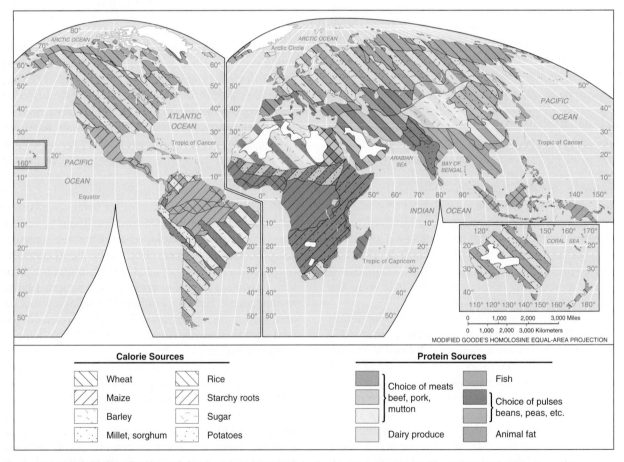

Figure 8-25 World distribution of diets. Everybody needs both calories and protein. The stripes and patterns on this map convey the predominant local source of calories, and the colors convey the predominant source of protein. The meat and dairy diet of Europe and of Europeans' settlements around the world stands out sharply. East and Southeast Asians rely more on fish, while pulses are important in South America, Africa, and in Southwest and South Asia.

Problems Associated with Animal Production

There are about 1.4 billion cattle on Earth today, and some scientists criticize these herds as inefficient and even harmful. The indictment against them is long. In the United States, runoff from feedlots causes serious water-quality problems. In sub-Saharan Africa, cattle contribute to desertification by stripping arid lands of fragile vegetation. In Central and South America, ranchers are cutting down rain forests for pasturage for cattle. Some agriculture scientists argue that the millions of acres of land dedicated to feedstocks should be rededicated to growing crops for direct human consumption.

In addition, cattle contribute to greenhouse gases. The bacteria that live in every cow's gut enable the animal to digest cellulose, a tough fiber found in grass and other plants that humans cannot digest. As a by-product, however, these bacteria produce methane, which is a greenhouse gas that worsens global warming. The amount of methane produced by livestock is estimated to be almost double the amount emitted from landfills (which give off gas from decomposition).

Other livestock also contribute to environmental devastation. Sheep and goats, for example, have overgrazed substantial areas in the Mediterranean basin, as well as parts of Africa and India. The global populations of hogs and chickens are rising faster than the cattle population, and their proportionate contributions to environmental problems, especially water pollution associated with feedlots, are increasing.

The pollution associated with "industrial" livestock farming is increasing in the United States. Industrial farms housing tens of thousands of hogs generate millions of gallons of waste each year. Hog waste is collected in cesspools because it must be treated before it can be used as fertilizer. It is then sprayed on fields or into the air in liquid form. Seepage from unlined waste ponds pollutes groundwater, and runoff pollutes local waters, too. Cesspools the size of football fields emit the toxic gases hydrogen sulfide and ammonia. Enormous dairy farms present the same problems. Furthermore, concentration of livestock increases their vulnerability to disease, as well as the chances of transferring diseases to humans. Chapter 5 noted the problems associated with the use of antibiotics in animal feed, as well as the potential for industrial animal farming to promote viral outbreaks. These huge facilities are truly factories, but they are not treated as such in the law or subject to the same environmental regulations.

Dairy Farming and the Principle of Value Added

For decades, most of the world's supply of cow's milk was produced in developed countries, but the FAO predicts that by 2015, an equal amount will be produced in developing countries. Developing countries have more than doubled their production of milk, partly through increased yields from each cow. India alone accounts for about one-quarter of this increase, followed by China and Brazil.

In the past, fresh milk was available only on farms and in nearby towns, but as the populations of more countries urbanize, they create a demand for large-scale commercial milk production. Milk is heavy and spoils quickly, so the cost of transporting it is high. Therefore, most cities receive their supplies from dairy farms in surrounding regions referred to as the cities' *milksheds*.

Some areas that are remote from urban centers but that have cool, damp climates unsuited to grain farming may specialize in dairy farming. Wisconsin in the United States and Switzerland in Europe are two such locations. Much of the milk from these regions, however, is first transformed into butter, cheese, or dried, evaporated, or condensed milk. These products are not only lighter and less perishable than milk, they are more valuable. The difference between the value of a raw material and the value of a product manufactured from that raw material is called the **value added by manufacturing.** Adding value to raw materials produces wealth. Wisconsin cheese and Swiss cheese and milk chocolate are exported worldwide, providing incomes to those two regions.

There are many steps and activities that transform the milk from a cow into a milk chocolate bar. In economic discussions, it is common to use the image that all products "flow" from their sources as raw materials to their ultimate consumers; the economic activities closer to the consumers are called **downstream activities.** Downstream activities add more value, and therefore they are profitable. Figure 8-15 notes how Chileans add value to their grapes by converting them to wine. We shall see in Chapter 9 and again in Chapter 12 that the possession of either agricultural or mineral raw materials may not necessarily enrich their possessor significantly. The key to wealth is adding value to raw materials and taking a share of that value added as profit. The geography of wealth is largely the geography of adding value to resources. Economists even speak of **psychological value-added**—that is, adding to the sales price of an object not by increasing its actual functionality or usefulness but by increasing its price through design, packaging, or "status" advertising.

Aquatic Food Supplies

Foods acquired from Earth's waters include fish, crustaceans, mollusks, aquatic mammals, amphibians, plants, and other aquatic life. Humans harvested about 158 million tons of aquatic foods in 2006, primarily fish. Aquatic food comprised less than 2 percent of the world's daily calories and about 8 percent of its protein.

The Economic Geography of Food and Land: von Thünen's "Isolated City" Model

Johann Heinrich von Thünen (1783–1850) was a Prussian aristocrat who wondered what product he could grow on his suburban estates and market most profitably. Thinking about this problem led him to question how various crops became distributed across any countryside. To answer his question, von Thünen devised one of geography's earliest models, which combines the economics of food markets and land use. He began with the idea of an imaginary city market in a perfectly flat plain with absolutely no variations on it. This is called an **isotropic plain.** Von Thünen drew a model pattern for the distribution of different land uses.

Von Thünen noted that parcels of land are put to different uses according to the economic value of the land, called its *rent value*. On an isotropic plain, transport costs are the only variable, so differences in rent reflect the transport costs from each farm by straight-line distance to the market—this assumes that one always has access to a straight road into the city. The greater the transport cost, the lower the rent that can be paid if the crop produced is to be competitive in the market. In addition, perishable products such as milk and fresh vegetables need to be produced near the market, whereas less perishable crops such as grain can be produced farther away.

Von Thünen deduced that a pattern of concentric zones of land use will form around a city market. The intensity of cultivation—that is, the amount of costly labor and inputs applied—will decline with distance from the market. Perishable crops that have the highest market price and the highest transport costs per unit of distance, such as

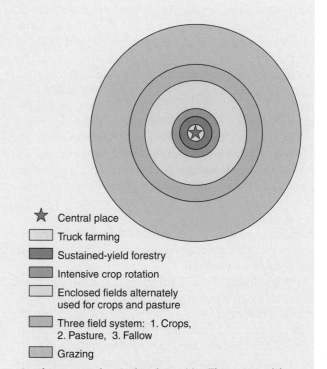

★ Central place
☐ Truck farming
☐ Sustained-yield forestry
☐ Intensive crop rotation
☐ Enclosed fields alternately used for crops and pasture
☐ Three field system: 1. Crops, 2. Pasture, 3. Fallow
☐ Grazing

Land use around an isolated city. Von Thünen's model shows rings of decreasingly intensive land use concentrically outward from the city. The ring of sustained yield forestry may be surprising, but forestry was once an intensive land use. The forest provided grazing for livestock, fuel, and raw materials for building and many other purposes.

Certain countries, however, draw more heavily than others on aquatic life. China gathers almost 36 percent of the world total. Aquatic foods are an important part of diets in many regions. In Japan, for example, aquatic foods account for about 10 percent of calories and 40 percent of protein. Worldwide, about two-thirds of aquatic foods are captured in the wild, and the remainder is produced by aquaculture, so-called fish farms on which humans raise a variety of freshwater or saltwater species (Figure 8-26).

Fish accounts for about two-thirds of all harvested aquatic life measured by weight, and most of this fish catch is marine. In addition to fish, humans harvest two other groups of marine species (Table 8-2). In 2006, world *marine capture production*—that is, basically, hunting and gathering at sea—totaled 82 million tons.

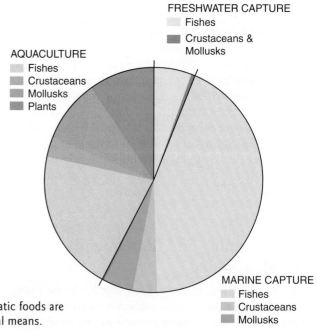

FRESHWATER CAPTURE
☐ Fishes
☐ Crustaceans & Mollusks

AQUACULTURE
☐ Fishes
☐ Crustaceans
☐ Mollusks
☐ Plants

MARINE CAPTURE
☐ Fishes
☐ Crustaceans
☐ Mollusks

Figure 8-26 Aquatic food sources. The main sources of aquatic foods are marine fishes caught in the wild and fishes raised by aquacultural means.

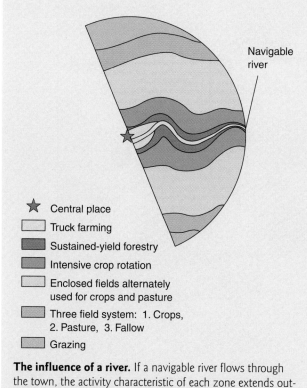

Navigable river

★ Central place

▢ Truck farming

▢ Sustained-yield forestry

▢ Intensive crop rotation

▢ Enclosed fields alternately used for crops and pasture

▢ Three field system: 1. Crops, 2. Pasture, 3. Fallow

▢ Grazing

The influence of a river. If a navigable river flows through the town, the activity characteristic of each zone extends outward along the river.

vegetables, will be grown closest to the market. Today, a hint of this zone around a city survives in New Jersey's nickname, The Garden State. New Jersey farms historically supplied the cities of Philadelphia and New York. Today, New Jersey is heavily industrialized in some parts, yet it also still specializes

in greenhouse products, dairy products, eggs, and tomatoes, which are typical market garden commodities. In von Thünen's model, fields farther away from the market will be dedicated to less perishable crops with lower transport costs per unit of distance. In the rings farthest outward from the cities, livestock grazing and similar extensive land uses still predominate.

Von Thünen elaborated his model by adding a navigable river flowing through the town. He noted that the areas along the river's banks would enjoy greater accessibility to the city market by boat. In other words, the cost distance to the city market was lower along the riverbank, so each of the zones would extend out along the river.

Some scholars have tried to apply von Thünen's model to entire countries. They have plotted the use of cropland in the United States, for instance, as if the Eastern Seaboard cities or Chicago were a central market. Trying to apply von Thünen's model to vast territories under contemporary conditions, however, strains the model's usefulness. The model focuses on distance from city markets, but in the real world, proximity to the market is only one consideration. Physical environmental conditions, governmental regulations, the economic system, the pattern and the regulation of the transport system, and still other factors must be considered. Also, transport costs have fallen dramatically relative to other costs of providing food. Food storage during shipping, such as refrigerated trucks, extends the distance that perishable foods can travel, providing access to less expensive production areas. Most of the fruits and vegetables now consumed in New York and Philadelphia, for example, come not from New Jersey but from California, Florida, Chile, and other distant places where production costs are lower and growing seasons longer.

The total *freshwater capture production* is only 10 million tons, of which 8.7 million tons are fish. Nearly all the freshwater catch is taken from the inland waters of Asia and Africa (Figure 8-27). Diadromous fish, such as salmon, move between saltwater and freshwater and may be fished in either location, but are less than 2 percent of all captured aquatic food.

Traditional Fisheries

Similar to the different agriculture regions described earlier, a **fishery** is an area where certain kinds of

fishing or fish farming are used to yield certain species. One of the basic distinctions among fisheries is between traditional, or artisanal fishing, and modern fishing. Traditional fishing is concentrated mainly in developing countries, particularly in Asia and Oceania (Figure 8-28a). Little capital is invested and the technology is basic, often nothing more than a person, a boat, and a fishing line. Traditional fishing provides low incomes, but these activities employ about 80 percent of the world's fishermen. Traditional fishing is an important food source in the lives of the majority of the populations of islands and

TABLE 8-2 Four Main Groups of Marine Catch				
	Demersal fish (bottom dwellers)	**Pelagic fish (surface dwellers)**	**Crustaceans**	**Mollusks**
Marine group members	Cod, haddock, sole, plaice	Herring, mackerel, anchovy, tuna, salmon	Lobsters, crabs, shrimp	Oyster, squid, clams, octopus
Tons of catch per year	70 million (combined with sharks, rays, and similar species)		5 million	7 million

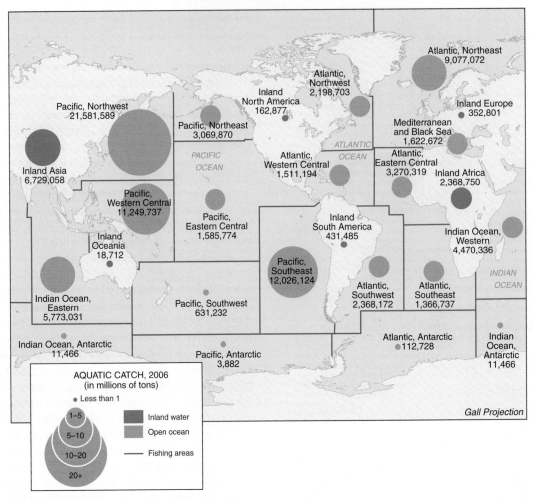

Figure 8-27 The world's major fisheries. This map shows each major fisheries' contribution to the total world catch. Many of these are currently fished to their maximum sustainable yield.

coastal areas, and it directly supports an estimated 30 million people.

Fishing is frequently a dangerous occupation, and it demands skills and local knowledge. Physical risks are high, and so are the risks in terms of income, which can fluctuate greatly. Many traditional fishing areas are characterized by close-knit communities with distinct customs and rituals. Traditional fishermen from Canada to Madagascar to Indonesia share certain hardships and experiences. Yields of fish from these traditional methods feed substantial numbers of people, but they are only a small fraction of the global catch. This way of life is rapidly being undermined and replaced by modern, highly capitalized fishing.

Modern Fishing

Large commercial fishing boats and fleets capture most of the aquatic life taken from the world's waters (Figure 8-28b). These boats often use fishing nets and large ones might pull in tons of fish each day. Between 1950 and 1989, the annual harvest from ocean fishing rose from about 20 million tons to 80 million tons

per year. Since then, yields have been driven higher by China while the total catch by the rest of the world has begun to taper off. In 2003, China, Peru, the United States, Japan, Indonesia, Chile, India, and Russia accounted for more than half of the total catch. These enormous yields come at a cost. Commercial fishing competes with traditional fishing both on the water by diminishing local stocks and at market by driving down prices. It also contributes to overfishing, described below.

The demand for fish and fish products continues to rise. Even as people eat more meat, they still create a demand for animal feed supplements, including fishmeal. Almost one-fourth of the global fish catch now goes into meal and oil, mostly to feed livestock and pets in the wealthy countries of Western Europe and North America. The Japanese, by contrast, want to consume more whole, fresh fish.

Overfishing and depletion of the seas It is difficult to regulate the exploitation of a resource that no individual or country owns because each country or individual fishing vessel harvests as much as it can.

Figure 8-28 Fishing activities compared. The relatively simple gear of traditional fishing activities (a) requires far less investment than commercial equipment (b). Traditional fishing, however, is affordable to poor coastal communities around the world. Traditional fishing increasingly must compete with commercial fishing in many fisheries.

This often leads to overexploitation and rapid depletion of the resource, a process called *depleting the common*. This has been the case with fish. The United Nations has sponsored periodic conferences on the world's fisheries, and most observers agree that several major fisheries are already fished to their sustainable limits or beyond, and that virtually every fishery is at risk. The major fishing nations have enlarged their fleets and developed new technologies to multiply their harvest: New electronic devices, for example, discover and track wholly new groups of fish. So-called factory ships are equipped with huge drag-nets that catch vast numbers of fish and other sea life indiscriminately. The FAO

notes that 20 million tons of catch include "bycatch," or species unintentionally caught in nets or on hook-lines. These include whales, sharks, porpoises, and other species important to aquatic ecosystems.

In 1999, the world's fishing nations agreed to reduce the size of their fishing fleets, and the World Summit on Sustainable Development held in South Africa in 2002 called for restoring stocks by 2015, but governments have been slow to act. Other efforts have targeted particular fisheries. The European Union in 2009 sought to ban Atlantic bluefin tuna following years of overfishing in the Mediterranean Sea. Annual catch totals of bluefin tuna peaked in 1996 at 50,000 tons and diminished to 30,000 tons in 2007. About 80 percent of the tuna from this fishery is exported to Japan, where bluefin is popular, especially in sushi.

We do not fully understand fish populations and their migrations. Systematic fisheries research began only in 1902, with the formation of the International Council for Exploration of the Sea, which studies North Atlantic species. Today, a great many international fishery commissions and advisory bodies exist. Some only consult, but others set quotas as well as conduct research. Some factors in changing fish populations are caused by human interference, but some are natural. Changes in water temperature and salinity can wipe out certain stocks, especially the small pelagic species such as sardines, anchovies, pilchard, and capelin.

Overfishing is not the only problem. Marine animals rely on coastal wetlands, mangrove swamps, or rivers for spawning grounds, but the world's wetlands and coasts are being destroyed by pollution and overdevelopment. The massive 2010 oil spill in the Gulf of Mexico will have long-term consequences for aquatic life in an area that previously supplied 20 percent of all U.S. catch. About one-half of the world's population lives within 200 kilometers (124 miles) of an ocean, contributing to the pollution that reaches the seas. Heavy metals such as mercury have contaminated fish and damaged the health of people who eat them. Sewage, fertilizers, and runoff from agriculture have overfed algae (tiny marine plants), causing them to grow so rapidly that they use up the oxygen that fish need to breathe. In turn, the collapse of oceanic food supplies puts greater pressure on inland, freshwater species.

The depletion of atmospheric ozone lets in ultraviolet radiation that harms sea life. Ultraviolet radiation may be responsible for the nearly 20 percent decrease of plankton (microscopic animal and plant life that form the lowest link of the food chain) near the surface of Antarctic waters, where most marine growth and reproduction take place. Global warming might also alter ocean currents that currently sustain many of the most productive fisheries.

International law and regulation of fishing
International talks focus on two categories of fish: species such as pollock in the Bering Sea and cod off

Canada's eastern coast, whose wanderings cause them to straddle territorial and international waters, and fish such as tuna, swordfish, and billfish, whose seasonal migrations cover thousands of miles. Coastal states dispute rights to these species with countries that operate long-range distant-waters fleets. The coastal states claim that factory ships in international waters are destroying the stocks in their territorial waters. Others argue that the crisis stems from coastal countries' mismanagement. Coastal countries want binding rules to regulate catches in international waters and countries that practice distant-water fishing want nonbinding guidelines drawn up regionally.

About 90 percent of the world's marine fish harvest is caught within 370 kilometers (200 nautical miles) of the coasts, and therefore many coastal countries have recently extended their claims out into the sea. The 1982 U.N. Convention on the Law of the Sea authorizes each coastal state to claim a 200-nautical-mile **exclusive economic zone (EEZ),** in which it controls both mining and fishing rights (see Chapter 11). Many countries have already restricted foreign fishing vessels in their waters.

Aquaculture   Global seafood consumption is on the rise, but the catch from the oceans can no longer keep up with demand. Therefore, humankind has begun to shift from fish capture to **aquaculture,** which involves herding or domesticating aquatic animals (Figure 8-29). By 2006, the world yield from aquaculture was over 66 million tons, a total that has been rapidly rising toward the total for wild capture production. More than half of all aquacultural yields are now from China. Asia as a region accounts for more than 90 percent of production. Fish farmers can monitor both the purity of the aquatic environment and what the fish are fed. Fish waste is a valuable fertilizer. Aquaculture technology is advancing (computer-controlled feeding is an example), and experiments with different species are revealing which are most productive in aquaculture. Mahimahi, for example, can be grown from fingerlings to 1 to 2 kilograms (2 to 5 pounds) in just five months.

About half of all aquaculture production is done in seawater, the rest in freshwater or brackish water. Although some fish are raised in closed ponds or tanks, many are raised in pens or cages immersed in open waters. Fish farms in Norway, for example, raise coldwater species in fjords, using net cages that keep the fish contained but allow water to circulate. Large-scale experiments have also begun with fish farming in huge cages in the seas. Aquaculture has already proven itself as a major frontier for the production of high-protein food.

Hunger and Food Security

Having enough to eat is a basic concern that everyone shares, although not everyone confronts food shortage on a regular basis or in the same way. More than 1 billion

Figure 8-29 Aquaculture of salmon. Hatcheries in New Zealand raise King Salmon, where salmon exports reached $30 million in 2008. Aquaculture is tightly regulated in New Zealand and salmon farms were only allowed in the 1970s. In about one year, these salmon grow to about 3.5 kilograms (7.7 pounds).

people in the world are hungry, meaning they do not have enough to eat, and about 90 percent of these people live in developing countries, over half in Asia and the Pacific Rim (Figure 8-30). The minimum amount of food that humans require varies with age but ranges from 1,600 to 2,100 kilocalories per day and must also include sufficient protein, vitamins, and micronutrients for normal bodily functions. **Hunger** is a deficiency in any or all of these basic requirements; it may be short in duration or chronic.

A person is **malnourished** when he or she has too few or too many nutrients. A person is **undernourished** when hunger physically affects the person, often making it difficult for the body to function properly, especially in fighting infections. Undernourishment endangers pregnancies, inhibits the production of breast milk, and slows growth during childhood. It impedes learning and weakens mental functions. Continuous undernourishment often leads to reduced height and weight, known as *stunting* and *wasting*. Extended undernourishment leads

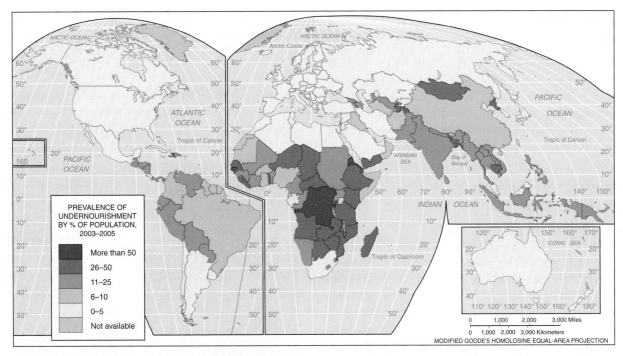

Figure 8-30 Prevalence of undernourishment. The amount of food needed per capita varies from country to country, depending on the age distribution of the population, activity levels, health, climate, and other factors. Nevertheless, this map reveals that in many countries, substantial percentages of the populations are undernourished.

to *starvation*, during which bodily organs cease to function, increasing mortality rates. In 2007, hunger-related diseases caused 60 percent of the 11 million deaths of children under age 5.

Access to food covers a broad spectrum of situations, all tied closely to social, economic, and political conditions. On one end of the spectrum, **famines** are food shortages that lead to extensive starvation in a population. Famines are often thought to have their cause in sudden environmental catastrophes, but most farmers can cope with a bad crop season. Famine occurs when all means of securing food fail, often because food costs are too high or access to food is disrupted by war or unrest. In fact, most famine results from political and economic disruptions rather than environmental crises alone. On the other end of the spectrum is **food security**, which the FAO defines as "a situation that exists when all people, at all times, have physical, social and economic access to sufficient, safe and nutritious food that meets their dietary needs and food preferences for an active and healthy life." Increasing food security means more than just increasing food production. Malnutrition, for example, can result from having too much food or eating unhealthy foods leading to obesity and disease. Food security means making sure nutritious foods are constantly available, affordable, and usable for all.

Food security must address trade, access, and consumption issues, which are closely entwined with global markets, government policies, and social attitudes. No country is completely self-sufficient in food. Most countries both import and export food, and a few countries are net exporters of food despite the fact that portions of their own populations are undernourished. Government policies have also hampered access to land for some groups. Unstable national economic policies may make food too expensive for many to afford even though there is enough food for sale. The number of hungry people in Zimbabwe increased in 2008, in part because the government cannot control currency inflation, sometimes doubling the cost of food overnight. In some cases, cultural preferences for some foods diminish food security for all, especially where expensive meat production for the few takes up cropland that could feed many. Food-supply problems are often problems of economics or politics, not problems of geography or technology. In poor countries, rich people have plenty to eat, while poor people are starving. Undernourishment is a problem in wealthy countries, too. For everyone on Earth to be adequately fed today as well as in the future as Earth's population continues to rise, we need both an increase in production and improvements in distribution; these must occur both within countries and among countries.

Peter Rosset, director of the Institute for Food and Development Policy, wrote in 1999, "There is no relationship between the prevalence of hunger in a given country and its population. . . . The world today produces more food per inhabitant than ever before.

Enough is available to provide 4.3 pounds to every person every day: two and a half pounds of grain, beans and nuts, about a pound of meat, milk and eggs, and another of fruits and vegetables—more than anyone could ever eat. The real problems are poverty and inequality. Too many people are too poor to buy the food that is available or lack land on which to grow it themselves." These are the problems we must examine in the following pages.

Problems in Increasing Food Production

Technology creates the potential for increasing food production, but it cannot guarantee that this will occur or that the food will be produced or distributed where it is needed. One of the biggest obstacles to food security is the use of farmland in poor countries for crops that do not feed the local population. Other potential problems in the effort to increase food supplies include lack of financial incentives for farmers and inequitable land ownership or unsuccessful land redistribution.

Commercial crops in poor countries Today, many farmers in poor countries do not concentrate on growing staple crops for local consumption. Instead, they raise cash crops for sale. Senegal, for example, exports peanuts, the Ivory Coast exports cocoa, Angola coffee, Zimbabwe tobacco, and Kenya tea and flowers, even though food availability in each of these countries is low (Figure 8-31).

Farmers often raise illicit drug crops for the same reason; it increases their income (Figure 8-32). Cash crops are often encouraged by governments that expect specialty

Figure 8-31 Flowers as a cash crop. Flowers raised for table or ceremonial decoration can earn farmers a lot of money, but the intensive labor requirements, specialized equipment, and high transport costs make it a capital intensive business. This greenhouse is located near Lake Naivasha, Kenya.

Video
The Trade Trap

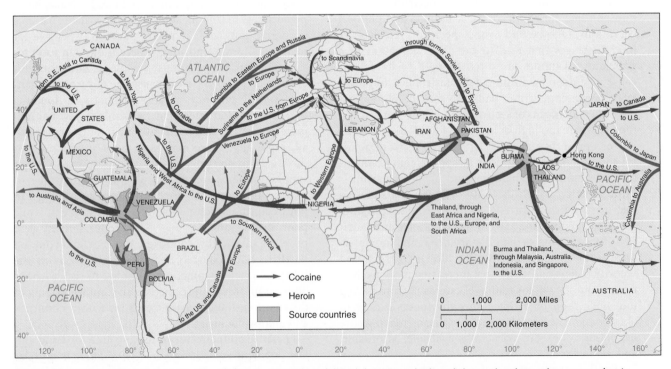

Figure 8-32 International trade in illegal drugs. The value of illegal drugs is so high and the market demand so strong that international trade flourishes despite international police efforts to stop it. The flow of opium from Afghanistan has increased since the U.S.-led war ousted the fundamentalist Taliban from rule there.

crops will earn their country enough money that they can buy more basic food on world markets than they could raise domestically. In that way, the country can achieve what is called *economic self-sufficiency* in food. Ideally, this would lead to a net increase in food security. Farmers who choose to depend on cash crops, however, face two risks: The price for their product could collapse because of overproduction elsewhere, or prices for the foods they will have to buy could rise faster than their own incomes.

Raising cash crops may increase farmers' disposable incomes rapidly, while maintaining food production. Guatemalan farmers raising snow peas for New Yorkers, for example, can make 14 times as much per acre as they would raising maize. Even after taking into account the increase in their costs, their return for each day of work more than doubles. A great share of the increased costs are for hired labor, which provides the local community with some income.

The rise in income may not, however, improve the lives and health of farm families and hired workers. The value of cash crops is often set by global markets, which tend to drive down prices and often fluctuate rapidly. Earnings are often too low to buy enough nutritious food to achieve food security. The benefits of cash crops are not equally shared. Wealthy landowners often have better access to global markets or can store crops until prices improve. Studies have revealed that in many poor countries, women control the production of food crops, but men control the production of cash crops. The men often spend the family's money on things such as alcohol and tobacco, which does not increase family welfare.

The cost of importing food has been rising faster than the value of most exported crops. This was the primary cause of the food riots described at the start of this chapter. World food prices began increasing dramatically after 2002, rising more than 60 percent by 2008. The reasons for this rise demonstrate the connections between food security and non-food factors. Major cereal crop output fell due to government policies in some countries and extreme weather in others. Rising oil prices increased the cost of shipping food, reflected in higher prices for consumers. Increased demand for agricultural products to make biofuels now consumes more than 4 percent of world cereal production. As consumers grow wealthier in some developing countries, especially China, India, and Brazil, they are increasing their food consumption, driving up prices. In response to these trends, some countries imposed trade restrictions that drove prices still higher. These price increases are estimated to have added more than 100 million more people to the number of hungry undernourished people in the world, reversing several decades of improving food security.

Many farmers lack financial incentives Chapter 6 referred to the agronomists' dictum: "The market produces the surplus." In other words, if a market exists for any surplus that farmers can produce, and if the political and economic system sufficiently rewards farmers for their effort, production of a surplus is all but assured (assuming that adequate technology is available).

In many countries, however, farmers have little incentive to increase food production, so they remain at the subsistence level. This is because many governments force farmers to sell their crops to the government at fixed low prices, charge high taxes on food exports, support high exchange rates for their national currencies, and impose high import duties on the tools and agricultural chemicals the farmers need. These policies lower the incentives to produce surplus crops; why would a farmer raise crops that would earn him less than the time, energy, and costly inputs spent raising them? The wealth that is squeezed from the farmers in these ways is often used to subsidize urban food supplies or to support urban civil service bureaucracies, military spending, or unprofitable state-run industries. In some countries, deteriorating conditions in the countryside trigger migration to the cities (a topic we shall examine in Chapter 10), and swelling urban populations increase the demand for subsidized food. When recent political changes have given farmers incentives to produce more food, the increases in production have been astonishing. India lifted some agricultural price controls in 1991, and since that year, grain production has risen about 4 percent annually. This rate is more than twice the rate of population increase, so per capita food supplies have improved markedly. When farmers enjoy higher profits, more rural jobs are created both in agriculture and in rural craft industries.

Systems of land ownership can discourage production In many countries, systems of land ownership reflect broad political and social inequalities. In Brazil, for example, about half of the country's arable territory is held by only three percent of all landowners. This concentration of land ownership spurs the landless to attempt to farm new claims in the forest regions, which degrades the environment while producing only disappointing yields. The government has announced land redistribution schemes, but it has been slow to act. Furthermore, to be successful, newly landed peasants require infrastructure (such as roads), technical support, and even assistance marketing and financing their production. Hundreds of landless peasants have died in clashes with armed landowners and police in Brazil in recent years, but, at the same time, the modern, highly capitalized sector of Brazilian agriculture achieved great success increasing exports of sugar, soybeans, and beef.

In many countries, the landholding system is a holdover from the country's colonial period. Colonial governments replaced communal systems with systems of private property, and the distribution of the property was highly inequitable. This created one class of large estate holders and another, much larger, class of landless poor. The large landowners often moved to the cities, becoming absentee landlords unwilling to make the investments necessary to maximize the return from their own agricultural holdings.

In the Philippines, for example, large blocks of land were granted to the Spanish colonial elite, and the natives on these land grants became tenant farmers. After a few generations, their descendants owned the best agricultural land. This land-grant system prevailed in all areas colonized by Spain, and it is at least partly responsible for political unrest in Central and South America today.

Indonesia presents a contrast to the former Spanish colonies. For more than 300 years under Dutch rule, only native Indonesians could own land, except for city lots. The Dutch, Arabs, Chinese, and English could rent agricultural land, but they could not acquire title to it. Thus, land ownership among Indonesian peasants remained much more widespread than it was in the nearby Spanish Philippines. Indonesian peasants who have inherited their land and expect to leave it to their children are much less likely to join peasant guerrilla movements than are landless Filipinos. In addition, individual farmers of small personal holdings are likely to maximize their productivity.

When land is leased to those who actually work it, the conditions of tenancy and of payment determine whether the farmers are encouraged to produce significant surpluses. If the landlord takes too large a share, surpluses will be small.

Most Communist regimes in Eastern Europe collectivized agriculture in large state-owned farms and in cooperative enterprises. Neither provided sufficient incentive to encourage farmers to maximize productivity. Since the fall of Communism, however, private farming is being encouraged. Giving or selling government assets to private individuals or investors is called **privatizing** an activity. Small private market gardens were always allowed in the former Soviet Union, and although these accounted for less than 2 percent of total Soviet farmland, they yielded 60 percent of the country's potato crop and approximately 30 percent of its vegetables, meat, eggs, and milk. Since the 1990s, the government of Russia has been privatizing agriculture. Poland ended food subsidies and restrictions on direct marketing in 1989; farmers' markets sprang up overnight, food shortages eased, and prices fell. The other formerly Communist Eastern European countries are privatizing the large communal landholdings. Vietnam ended collectivized agriculture in 1990, and the country swiftly became able to export rice.

China has experimented with several landholding systems in the past 50 years. When Communists came to power in China in 1949, they seized land from landowners (killing thousands of them in the process) and redistributed it to the peasants, but in the 1950s, the peasants' holdings were collectivized. In 1978, China relaxed its state-directed agricultural system.

The collectives were dismantled, and peasants were allowed decision-making power and profit incentives. Within 10 years, farm output rose 138 percent, and China turned from a net importer of food products into a net exporter. The peasants were granted 30-year leases, but not ownership, so they still could not sell it, rent it out, or borrow money to invest in improving productivity. These restrictions on peasant land rights slowed rural development and contributed to a growing gap between rural and urban incomes. To remedy this problem, China has recently abolished a wide range of fees and even the ancient land tax.

Video
Helping Ourse

Privatizing of industry and of other aspects of national economies is proceeding around the world; we will examine this important development in more detail in Chapter 12.

Agricultural productivity can also be hampered when title to land—legal ownership—is unclear, as is the case in many countries. Generations of peasants may have tilled small plots, but their families have never officially owned the land, and bureaucratic complications make it difficult to establish title now. It has been estimated that in poor countries only about 20 percent of all land has clear title. Lack of clear title discourages the occupants from making improvements on the land. Furthermore, as we have seen, successful farming today requires capital investment in irrigation pumps, chemicals, improved seeds, and machinery. Farmers can borrow money to make these investments only if the land can serve as *collateral* (a thing of value that can be seized by a creditor in case of nonpayment of a debt). Farmers without clear title to the land lack collateral. The satellite-based GPS and GIS have greatly reduced the cost of land registration, and new programs have been launched around the world. Establishing farmers' title to land triggers investment and increases productivity.

Unsuccessful land redistribution: Mexico   Mexico demonstrates how communal land ownership can restrict productivity because communal holdings cannot serve as collateral. More than one-half of the country's arable land is held in *ejidos,* a form of land tenure in which a peasant community collectively owns a piece of land and the natural resources and houses on it (Figure 8-33). Mexican law states that *ejidos* are "inalienable, nontransferable and nonattachable." They cannot be used as collateral, so banks will not extend loans. This system, combined with a government tradition of paying farmers low prices for their crops while at the same time subsidizing food for urban consumers, has kept farmers poor and productivity low. These conditions in turn intensify the migration of peasants to the cities. Mexico imports 35 percent of its total food supplies, including nearly half of its staples of maize, wheat, and beans, from the United States.

The government is struggling to overhaul the *ejido* system, but official recognition of *ejido* ownership

Figure 8-33 A Mexican *ejido*. All members of a Mexican *ejido*, or farming collective, gather to make decisions regarding planting, harvesting, and marketing their crops. *Ejidos* represent the traditional Mexican form of communal landholding.

was a principal issue of the Mexican Revolution (1910–1920), in which more than 1 million Mexicans died. Many Mexicans remain attached to the concept of the *ejido* system. In 1992, the government introduced a program of certifying individual land rights, but by 2000, scarcely 1 percent of *ejido* land had been sold. Since the NAFTA trade pact between the United States and Mexico was launched in 1994 (see Chapter 13), Mexican farmers have found it difficult to compete with U.S. farmers. The Mexican government has ended price guarantees without helping farmers market their produce, and it costs about twice as much to get a ton of wheat to Mexico City from Mexicali, near Mexico's northwest corner, as it does to get a ton of wheat to Mexico City from Kansas. Some Mexican farmers are switching to new commercial crops, or they are growing crops for large agribusiness companies. These companies provide the needed capital investment, as well as an assured market. The Mexican government has also experimented with guaranteeing loans made to the *ejidos*, thus encouraging capital investment and improving agricultural productivity.

Barriers to increasing production in Africa Many of the problems found in Mexico's agricultural economy can also be found in Africa. Landholding is often communal, so successful farmers cannot expand their holdings or borrow money to invest in greater productivity. Farmers cannot always afford fertilizers, leading to widespread soil depletion. Many African governments themselves hold ownership of agricultural land and lease it to farmers. In Zimbabwe, for example, the government nationalized numerous large white-owned private farms that were exporting food. The government relocated black settlers onto the properties but did not transfer ownership, so the farmers cannot borrow to invest in increasing productivity. Productivity has fallen, and Zimbabwe has lost export income.

In addition, farmers in many countries are required to sell their crops to state marketing boards. These

boards often pay the farmers very little. In Nigeria, for example, the government owns almost all land, leases it to farmers, and then pays the farmers fixed low prices for their crops. Many African state marketing boards have proved so corrupt that the farmers' returns for their work are virtually stolen from them. Some governments also have failed to invest in a rural infrastructure or in agricultural research to develop technologies or native crops suited to local conditions.

These conditions discourage farmers and reduce agricultural output. Many farmers migrate to the cities for work, but because land tenure is based on occupancy, they must leave their families in the villages. In some parts of rural Zimbabwe, 40 percent of the households have lost the father to the town. The overworked women, left to tend children as well as crops, can scarcely rise above subsistence. The United Nations Children's Fund (UNICEF) estimates that women grow 80 percent of Africa's food. AIDS has also taken a terrible toll on African farming, killing almost 10 million farmers by mid-2006, thus greatly reducing food output. Civil wars are still another factor that has reduced food production in some areas of the world, again particularly in Africa.

Growing international debts owed by African countries to banks in the developed world further undermine agricultural productivity. Debt weakens a country's currency and lowers the price of its exports on the world market, lowering earnings for farmers. A weak currency also makes buying farm inputs and food from the global market more expensive. Lenders often impose policies on governments that end useful assistance to farmers who grow food for local consumption.

Africa was feeding itself in 1960, but today the continent imports about 40 percent of its food supply. The reasons for inadequate food production in many countries are not necessarily environmental or even technological. Food production could be greatly increased if many governments could end their civil wars, change national economic policies that discriminate against farmers while subsidizing urban food supplies, stop discriminating against women, reduce import taxes or quotas on imported tools and chemicals, improve infrastructure (particularly roads and water supplies), reduce their debts, and reform land holding systems. All of these factors influence food production.

Rich Countries Subsidize Production and Export of Food

Many rich countries have erected figurative walls around their countries using high *tariffs* (taxes on imported goods) to protect their food producers against lower-priced imports from poor countries. This requires the urban consumers in rich countries to pay inflated prices for food. These inflated prices are indirect subsidies to rich countries' farmers. Many rich countries also give their farmers direct subsidy payments. In 2008, the European Union spent $150 billion in agricultural price supports, the United States about $23 billion, and Japan about $42 billion (Figure 8-34). As a result, farmers in many rich countries produce food surpluses, which the countries then export or sell on world markets at prices below the costs of production (a practice called *dumping*).

These subsidies distort the world geography of agriculture. They encourage production of agricultural surpluses in many rich countries and discourage increasing production in poor countries. As long as rich countries sell their surpluses on world markets at low prices, the governments of many poor countries neglect to invest in their own rural areas. Farmers in poor countries cannot export to rich countries or even compete in their own national markets, so they remain at the subsistence level, or they are driven off their farms entirely. International trade negotiations have often foundered on objections to agricultural subsidies and protectionism.

Why Do Some Rich Countries Subsidize Agriculture?

Many rich countries subsidize their farmers and erect import barriers for several reasons:

1. Some countries pursue self-sufficiency in food production as a national security measure. The governments want some defense against grain embargoes and crop failures in those countries that normally have surpluses to market. Subsidizing a country's own farmers offers some protection against these threats. Following U.S. threats to halt grain shipments to the Mideast in the 1970s, for example, Saudi Arabia spent heavily to improve its agriculture. Saudis give farmers subsidies to raise grain, import irrigation equipment, and even raise dairy cattle in air-conditioned sheds where outside temperatures exceed 48°C (120°F) (Figure 8-35).

2. Poorer countries want to keep their farmers from migrating to the cities in search of work. Rural–urban migration can overwhelm the ability of cities to absorb the new labor force.

3. Many rich countries, particularly in Europe, subsidize agriculture to preserve traditional agricultural communities. Some rich countries

Figure 8-35 Food for national security. The desert in the background is a typical landscape of Saudi Arabia, but Saudi Arabia has invested some of its oil income in irrigating the desert to achieve agricultural self-sufficiency. Large capital investment in agriculture allows Saudi Arabia to achieve such high productivity that it occasionally dumps surplus food on world markets. This bankrupts farmers in other countries with environments more naturally suited to agriculture. Thus, political and economic considerations can outweigh the role of physical conditions for agriculture.

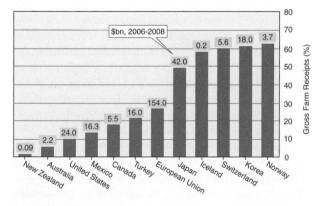

Figure 8-34 Farm subsidies. Farmers in rich countries are subsidized both by direct government payments and by taxpayers, who pay more than world prices for their food.

subsidize farming as part of national land-use plans. They enjoy agricultural landscapes and want to preserve green areas around cities, called *greenbelts*. Their urban citizens are willing to subsidize farms in order to preserve them.

4. Many rich countries' political systems favor farmers. The farmers in rich countries constitute a decreasing percentage of the national populations, yet, particularly in Europe and Japan, the electoral districts have never been redrawn to reflect the relative depopulation of the countryside and the urbanization of the populations. The U.S. Senate disproportionately favors farm-state populations, as we shall see in Chapter 11. Thus, farmers enjoy disproportionate political power over urbanites. The contrast is sharp between rich countries, in which the national governments generally drain wealth from the cities to support farming, and poor countries, in which governments often heavily tax farmers and subsidize the urban populations.

5. In rich countries, only a small percentage of what urban consumers pay for food actually goes to farmers; most of the cost represents value added by processing and packaging. In 1980, the cost of farm commodities in the United States—the raw material—represented 37 percent of the price of food; by 2000, that had fallen to 19 percent. Food processors and middlemen, including marketing and advertising, take an ever-larger share. For example, the cost of the maize in a box of cornflakes is only about 3 percent of the retail cost of the item; the box is worth more. The rest of the retail price represents costs for processing, packaging, distributing, and advertising the cereal. Advertising rates affect the retail price of cornflakes more than maize prices do. Therefore, most of the increase in urban food costs is attributable to these activities, not to farmers' profits. Most urban consumers remain unaware of how much the prices they pay for food are inflated to subsidize their nation's farmers. They may also be largely indifferent because the cost of food as a percentage of urban incomes in most rich countries is falling anyway, due to increases in agricultural productivity.

The American diet has actually become so full of rich and processed foods that the country has experienced periodic trade deficits in food products since 1998. That is, the United States has been importing more food—by value—than it has been exporting. This is astounding for a country so rich in agricultural resources. The reason for this is that America exports basic foodstuffs (wheat, for example), but Americans are importing food products of ever-higher value added (notably wines and out-of-season fruits and vegetables).

A myth exists that agricultural subsidies in rich countries go to struggling small family farms. In fact, the bulk of subsidies go to large multinational corporations (such as Nestlé, the world's largest food company) and to wealthy or well-connected individuals (such as members of the French Senate and even families of U.S. Congress members). The largest subsidy recipients in the United States are primarily large corporations. These subsidies are geographically concentrated, too. In the United States, just seven states received almost half of all subsidies in the period 1995–2006. Texas alone accounted for 9.1 percent of all government farm support. Overall, subsidies to U.S. producers in 2008 provided 7 percent of total farm incomes, down from 22 percent in the 1980s.

Rich countries periodically negotiate to end their subsidies, but they cannot even agree on what constitutes a subsidy. The United States heavily subsidizes water for California farmers but denies that this support is a "farm subsidy." Throughout California's Central Valley, sprinklers irrigate fields of alfalfa, cotton, and rice—crops more suited to wetlands than to a desert—and more than 50 percent of federally irrigated land in California is devoted to crops that are already in surplus. The issue of farm subsidies is complicated, and in each country, it is a volatile domestic political issue. The overall result, however, is to increase food surplus in rich countries and to discourage food production in poor countries. If rich countries stopped subsidizing their farmers and opened their markets to crops from developing countries, the farmers of poor countries might be able to sell more food to rich countries. This action would benefit poor countries more than direct foreign aid does.

In 2003, Amadou Toumani Touré and Blaise Compaoré, the presidents, respectively, of the African countries of Mali and Burkino Faso, published a joint plea to the American people. "Your farm subsidies," they wrote, "are strangling us." Rich countries' subsides to their own cotton producers totaled almost $6 billion in the production year 2002, "distorting cotton prices and depriving poor African countries" of their own ability to market this crop. During that year, they noted, "America's 25,000 cotton farmers received more in subsidies—some $3 billion—than the entire economic output of Burkino Faso. . . . Further, United States subsidies are concentrated on just 10 percent of its cotton farmers. Thus, the payments to about 2,500 relatively well-off farmers has the unintended but nevertheless real effect of impoverishing some 10 million rural poor people in West and Central Africa." Powerful U.S. congressional interests, however, continue to favor support for U.S. cotton growers, and subsidies totaled $2.7 billion in the United States alone in 2006. The year before, some U.S. cotton subsidies were ruled illegal under international treaties, but the U.S. Congress continued to approve the programs. Leaders of many of the world's poor countries have cited rich countries' agricultural protectionist policies as the reason why poor countries have balked at progress in international trade agreements (see Chapter 12).

Our study of world food production and availability illustrates many ways that the world's agricultural economy is manipulated and distorted. Thus, we cannot know how much more food the world's farmers would be capable of producing if these conditions were eliminated, or what the geography of food production would be.

Food Supplies in the Future

All the factors we have discussed have held off the specter of worldwide starvation. Is it possible that humankind is now, at last, at the end of its ability to increase food supplies? The answer to this question is a cautious "probably not." If demographers are correct in their projections of Earth's future population, people can be fed. Humankind has scarcely begun to maximize productivity with the best contemporary technology, and that leading technology has been applied to only a small portion of Earth.

Several of the other factors we have listed that have increased total production since Malthus's time still offer potential for advance. Spreading urbanization is replacing agriculture in many places, but more lands can still be farmed. In 1995, the FAO estimated that developing countries of Latin America and sub-Saharan Africa could more than quintuple their cropland; the countries of East Asia could double their cropland. Since then, sub-Saharan African countries have added 13 percent more cropland while Latin American countries added only five percent. New cropland is not the only way of increasing the food supply, however.

Irrigation can open new acreage or increase productivity on existing farms (Figure 8-36). In most dry areas where irrigation is most effective, however, water is already in short supply. Greater opportunities lie in increasing the efficiency of irrigation. Only about 37 percent of all irrigation water in the world, for instance, is actually taken up by crops; the rest is lost in runoff, evaporation, or leakage. Technological advances in irrigation could reduce the percentage lost. Drip irrigation, for example, is a method that pipes only as much water as crops need and delivers it directly to their roots. It reduces the average amount of water needed per irrigated acre, boosts crop yields, removes the threat of parasitic disease spread by irrigation canals, and reduces soil salination.

Research is continuing to explore the uses of *halophytes,* plants that thrive in saltwater. Samphire (Salicornia), for example, is a tasty vegetable that not only contains more and healthier oil than soybeans but also produces a valuable animal feed as a by-product. It thrives in saltwater, and it is now cultivated commercially in Mexico, Arizona, Egypt, Iran, Syria, and around the Arab Gulf. Interbreeding halophytes with conventional crops has made these crops more salt resistant, which means that they can grow in more diverse envi-

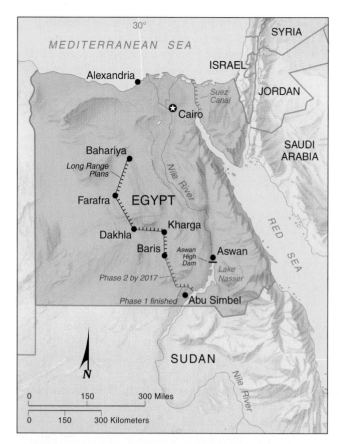

Figure 8-36 Egypt's "new Nile." Egypt has begun construction of an irrigation canal that will parallel the Nile River through its western desert.

ronments. Farmers are today harvesting lands once thought to be too salt soaked to support crops in Egypt, Israel, India, and Pakistan. Conventional crops may someday be grown in saltwater.

Techniques of storage and transport could still be improved in many areas. China loses at least one-quarter of its wheat harvest, for example, due to inadequate drying and storage, rats and mice, and poor transport. Improvements have been achieved, but in 1998, the U.N. World Food Program estimated that 40 percent of the world's crops were destroyed as they grew or before they left the field.

The Importance of Crop Diversity

The green revolution focused on the improvement of just a few of humankind's most important crops, and the success in improving these has raised concern that humankind is becoming too reliant on too few crops (Figure 8-37). If a new disease suddenly appeared and attacked any of these, it could destroy a significant percentage of humankind's total food supply. A model of such a disaster occurred in the U.S. maize crop in 1970. A new fungus suddenly appeared that was well matched to the T-cytoplasm that had been incorporated into 80 percent of the country's seed maize, and U.S. maize production fell 15 percent.

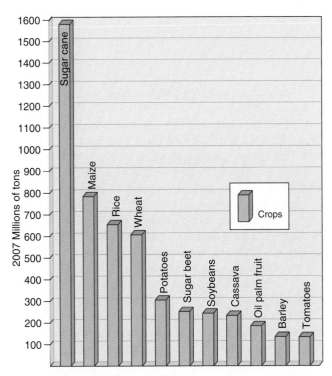

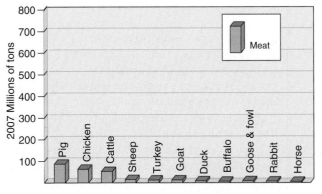

Figure 8-37 World food production. The great bulk of the human diet is today based on just a few crops and animal meats, but many others offer food potential.

To forestall a catastrophic loss of seed material, Norway, Iceland, Sweden, Denmark, and Finland have joined to build an International Seed Vault to hold seeds from all of the world's farm crops. Placed on the island of Svalbard, far to the North of mainland Norway, the vault holds samples inside concrete walls 70 meters (230 feet) under the permafrost, to be released only if all other seed sources are destroyed or exhausted (Figure 8-38). Some countries have begun their own seed vault programs.

A diversity of crops provides protection against catastrophe in case any one crop should fail, but modern commercial agriculture is increasingly specialized. Commercial agriculture has succeeded economically because each farm specializes in one or two crops grown over a large area to achieve economies of scale. Thousands of square miles in the U.S. Midwest are dominated by just a few varieties of maize or wheat. If diseases or pests attack these plants, an important agricultural region would be in crisis. In contrast, areas where multiple and different varieties of crops are grown may lose some of their harvest to a pest, but other crops will survive to provide sustenance. For example, ethnobotanists have reported more than 50 kinds of potato grown in a single village in the Andes.

New solutions to world hunger may arise from a diversity of crops. At least 75,000 plant species have edible parts, and many of them are nutritionally superior to plants cultivated commercially. Some of these crops provide superb nutrition yet thrive in climates hostile to commercial crops. The New Guinea winged bean, for instance, is entirely edible—roots, seeds, leaves, stems, and flowers. It grows rapidly, up to 4.5 meters

Figure 8-38 Svalbard Global Seed Vault. Opened in 2008, this underground facility on an island in the Arctic already holds more than 400,000 unique seed samples. The collection is meant to protect the genetic resources of the world's domesticated plants, which represent the careful work of farmers over thousands of years.

(15 feet) in a few weeks, and it offers a nutritional value as high as that of any known crop. Several Brazilian rain forest fruits—including acerola, camucamu, and açai—are both tasty and extraordinarily rich in vitamins and minerals. The National Academy of Sciences has recommended the cultivation of 36 other crops,

New Uses for Old Crops

The world's major crops, such as wheat, rice, potatoes, and maize, diffused rapidly under European colonialism. These foods dominate in many regions, but not everywhere. Some cultures have traditionally grown other crops that have advantages over today's major crops. With the globalization of the food trade, these lesser known crops are being rediscovered and produced in new places.

Chenopodium quinoa, for example, a member of the genus goosefoot, is a native of South America. It was domesticated in the Andes about 4,000 years ago. It is an annual, broadleaved pseudocereal (not actually a grass) usually standing about 0.91 to 1.83 meters (3 to 6 feet) high. Leafy flower clusters rise from the top of the plant. The dry, seedlike fruit is about 2 millimeters (0.08 inches) in diameter and enclosed in a hard, shiny, four-layered fruit wall. Quinoa seeds are flat and pale yellow, and they can be steamed, ground into flour, or fermented to make a mildly intoxicating beverage. The leaves are highly edible and often used for livestock feed. Quinoa is sometimes called Inka rice.

Quinoa has very high protein content, and it has an amino-acid profile that parallels the ideal standard set by the FAO. New varieties of this plant are being bred that are better suited to North American conditions, and quinoa has recently been appearing on North American menus as restaurants experiment with this food. Some stores also sell the grain-like material, which can be cooked and served like rice.

Amaranth is another traditional Inkan crop from the Andes that is becoming more popular in North America. A kind of herb, *Amaranthus caudatus* is high in protein and other nutrients and requires less water than major grain crops. Amaranth grains can be ground into flour, and the leaves of the plant are edible. These and numerous other traditionally cultivated plants and fruits hold great potential for providing healthy and sustainable food sources. The obstacles to developing their potential are primarily cultural—consumers' taste preferences, recipes, and cooking techniques continue to favor crops such as wheat, maize, and potato.

(a)

(b)

Quinoa (a) and amaranth (b). First domesticated in the Andes mountains, these crops require less water and provide more protein than cereal crops such as wheat.

Figure 8-39 A new food source? The fried spiders sold in Skuon, Cambodia are a local traditional food. Various cultures have divergent tastes and beliefs regarding what is considered edible or tasty.

Figure 8-40 A natural insecticide. The rows of potato plants on the left and right were devoured by the Colorado potato beetle. The plants in the center row, however, had a gene introduced into them from a common soil bacteria. Those plants produce a protein that acts as a natural insecticide. Biotechnology offers the promise of protecting many crops in this way.

including amaranthus, buffalo gourd, tamarugo, guar, mangosteen, and soursop. Most of us have undoubtedly never heard of these, yet each offers enormous potential for food. Agricultural exploration may offer many alternative foods, but what people eat, or refuse to eat, is to a great degree cultural (Figure 8-39).

The Scientific Revolution in Agriculture Continues

The scientific revolution in agriculture is not over. Some scientists argue that it has just begun and that new scientific advances will multiply future food yields. The economist Henry George (1839–1897) succinctly contrasted the rules of nature with the multiplication of resources through the application of human ingenuity. "Both the jayhawk and the man," he said, "eat chickens, but the more jayhawks, the fewer chickens, while the more men, the more chickens." This principle is key to understanding and counting all resources.

Biotechnology is the term given to a variety of new techniques for modifying organisms and their physiological processes for applied purposes. One aspect of biotechnology is **gene splicing,** or **recombinant DNA.** Scientists can now join the DNA of two organisms to produce a recombinant (or "recombined") DNA. Scientists can then introduce this recombinant DNA into another organism, thus permanently changing the genetic makeup of that organism and all its descendants. Such products are said to be **genetically modified (GM).** Since their commercial introduction in 1996, worldwide plantings of GM crops have ballooned from 1.7 million hectares (4.2 million acres) to over 125 million hectares (309 million acres) in 2008.

Another technique of biotechnology is **cloning,** the production of identical organisms by asexual reproduction from a single cell of a preexisting organism. Still other techniques of biotechnology develop new microorganisms for diverse purposes ranging from pharmaceuticals to cleaning up oil spills. Biotechnology can be applied to both plants and to animals.

Biotechnology offers genetically altered crops that can be custom-designed to fit the environment and produce bountiful harvests. It can even replace chemical pesticides (Figure 8-40). Breeding natural herbicides and pesticides into the plants themselves reduces the enormous quantities of chemicals and energy consumed in the manufacture, packaging, distribution, and application of chemical herbicides and pesticides. For example, a new GM maize significantly reduces the need for pesticides on that crop, and a new type of Russet Burbank potato contains its own protection against the Colorado potato beetle. In 1995, an estimated 5,000 barrels of oil went into the manufacture and distribution of insecticides to fight this one insect pest in the United States, and less than 5 percent of the insecticide reached the target insect.

The first GM food to reach U.S. groceries was the Flavr Savr tomato, in 1994, engineered to stay firm longer than regular tomatoes, but GM soybeans, cotton, potatoes, maize, and rapeseed soon followed. Most historic advances in the green revolution addressed improving yields of wheat, potatoes, and rice, but today many other traditional crops, including tropical crops such as sweet potatoes and cassava, are getting new attention (Figure 8-41). Today, lentils provide one of the world's leading sources of protein, and the GM Mason lentil is up to 18 percent more productive and is hardier than the traditional strain.

Figure 8-41 A leader in research on GM tropical crops. Dr. Florence Wambugu of Kenya has won world renown for developing a virus-resistant sweet potato that more than doubles yields for subsistence farmers. She spent her early life raising sweet potatoes on a subsistence farm. "I wasn't even supposed to be educated," she has said with a laugh. "My mother had to go to a tribal tribunal to get permission. I was a girl who was supposed to be married, and that was it." Today she heads a Nairobi-based scientific foundation and is an advisor to many scientific boards around the world.

Today's food has been improved not only in quantity but in quality. For example, GM pigs produce omega-3 fatty acids, which help reduce the risk of heart disease. Scientists have doubled the vitamin content of most vegetables and produced soybeans that are easier to digest and that also produce heart-healthy oils. Natural food advocates, however, argue that some GM foods are less flavorful and have unusual textures. They also question whether these foods will have unexpected negative effects on human health or the environment.

GM crops have even been bred to produce pharmaceuticals, in a process called biofarming or farmaceuticals. For example, DNA science has spawned a generation of drugs made from human antibodies, blood-borne proteins necessary for the immune system. The antibodies can be produced more cheaply, however, by splicing antibodies into the genetic fabric of plants, growing them in fields, and then extracting and purifying them. Hemoglobin is being bred into maize and soybeans. Maize-bred enzymes stimulate insulin production in diabetics, and safflower plants can even produce human insulin—an advance of increasing importance in a world of overweight diabetic humans. Bananas and potatoes can deliver a dose of vaccine; scientists have increased tomatoes' content of lycopene, which seems to prevent some types of cancer, and have introduced genes that produce beta-carotene, the precursor of vitamin A, into rice grains. A gene from *E. coli* bacteria allows rice to withstand drought, saltwater, and cold. Even a decaffeinated coffee bean has been invented.

The application of biotechnology to livestock itself is yielding hardier and more productive livestock. Biotechnologists are even improving animal feed. One product helps animals absorb phosphorous better, thus reducing the ecologically damaging phosphorous in animal waste. Cloning first achieved publicity with the birth of Dolly, a sheep, in 1997 (Figure 8-42). Dolly was the first mammal cloned from an adult cell, but since then mice, goats, monkeys, cats, horses, and rabbits have followed.

Resistance to Biotechnology

The U.S. National Academy of Sciences has found no evidence that GM foods are unsafe to eat, but the Academy does insist that rigorous monitoring of research must be maintained. History demonstrates that technological solutions to problems can trigger unexpected new problems, so some people fear that GM products may have unforeseen effects in nature (outcompeting and destroying natural species, for example) or on our own bodies.

RAPID CHANGE

Good-bye to the Banana?

Without a miracle of genetic engineering, bananas may be extinct within 10 years. The banana is one of our oldest crops, the first edible variety having been propagated about 10,000 years ago from a rare mutant of the wild banana, which, with a mass of hard seeds, is virtually inedible. All edible bananas are genetically decrepit sterile mutants—effectively clones of that first plant. Bananas are, therefore, unable to evolve to fight off new diseases.

Today we are seeing the global spread of black Sigatoka, a fungal disease that cuts yields up to 75 percent and reduces the reproductive lives of banana plants from 30 years to only two or three. The fungus reduced yields in Uganda 40 percent in one year, and it continues to spread across Brazil.

Genetic engineering may be the only way to save this popular fruit. Scientists first sequenced the banana genome and then modified it with rice genes that improved the banana's resistance to black Sigatoka and other diseases. Initial field tests have been planned for Uganda and Central America. Some producers fear that consumers will not accept a GM banana, but some 500 million people in Asia and Africa depend on bananas for up to one-half of their daily calories.

Figure 8-42 Dolly, the cloned sheep. On February 22, 1997, Scottish embryologist Ian Wilmut stunned the world by announcing that he had created an exact copy—a clone—of an adult Dorset sheep. The historic lamb, created from DNA extracted from a sheep's mammary gland, was named in honor of Dolly Parton. Dolly died February 14, 2003, of a lung infection.

In 2000, a GM maize that had been approved for cattle but not for human consumption found its way into commercially available taco shells, and unapproved GM rice was found in food supplies intended for human consumption in 2006. U.S. officials insisted that neither discovery threatened human health or the environment, but the discoveries triggered fear of unknown and unintended side effects.

The cloning of Dolly reminded some people of the novel *Frankenstein*, by Mary Shelley (1797–1851). In this classic, Dr. Frankenstein's confidence in science leads him to create a living creature that he hopes he can control but that turns on him and kills him. Some critics refer to GM products as *Frankenfood*. The idea survives in our imagination that science can create "monster problems" more serious than the problems that science solves. In 2001, scientists in Australia did, inadvertently, create a modified mousepox virus that destroys the animal's immune system. This scientific breakthrough may lead to a better understanding of AIDS, but it could also inspire new developments in biowarfare. Cloned animals have often exhibited severe and unpredictable health problems, including lung, heart, and immune system defects.

In 2005, Swiss voters supported a five-year ban on the farming of genetically modified crops, and some food manufacturers, supermarkets, and fast-food chains in several countries have stopped stocking food with GM ingredients. Opponents of GM foods have orchestrated demonstrations and even destroyed fields of experimental crops. Europeans insist that GM foods be labeled, but Americans oppose labeling as unnecessarily frightening. There are, however, no clear definitions of what GM means. For example, are hogs that have been fed GM maize necessarily GM hogs? Lack of agreement means that contradictions in distribution abound: Frito-Lay will not use GM maize in its corn-chips, yet its parent corporation, Pepsi-Cola, uses syrup from GM maize in its soda. McDonald's bans GM potatoes, but the company cooks potato fries in vegetable oil made from GM maize and soybeans. Some American opponents of GM foods have begun labeling things that are not GM as "natural" or "organic." To clarify labeling, in 2002, the United States defined standards for "organic" production and processing. Canada has lost potential international food sales by being slow to do so.

Delegates from 130 nations met in Montreal, Canada, in 2000 and adopted a treaty regulating trade in GM products. The Cartagena Protocol (after the city in Colombia where the talks started) is an outgrowth of the Convention on Biological Diversity forged in Rio de Janeiro, Brazil, in 1992. The new treaty went into effect in 2003, having been ratified by the requisite 50 signers. The treaty allows countries to bar imports of genetically altered seeds, microbes, animals, and crops if they deem the product is a threat to their environment. Labeling is required on products only in international shipments, not when the product is on store shelves. Exporters must obtain permission in advance from an importing country before the first shipment of a particular "living modified organism" is meant for release into the environment—like seeds, microbes, or fish that are to be put into a river. Advance notice and permission are not required for exports of commodities meant for eating or processing. The United States has never ratified the 1992 Biodiversity Convention, so it cannot be a party to the Cartagena Protocol. It will, nevertheless, have to abide by the terms of this treaty when exporting products.

To some degree, popular resistance to biotechnology is not entirely a technological question but also a religious one. Many people feel that nature is immutable or that tampering with it is sacrilegious. Prince Charles of the United Kingdom, for example, has said he will never eat any GM food. "That takes mankind into realms that belong to God, and to God alone," he insists. Monsignor Elio Sgreccia, however, president of the Vatican Bioethics Institute, has said, "We are open to the use of genetic technology in agriculture and with animals, as long as we don't do it with man. We believe that man has a primacy on this planet. . . . Nature is here for him."

Regardless of whatever happens in rich countries, many developing countries are betting on GM foods to increase their agricultural yields and food supplies. China, for example, is second only to the United States in biotechnology research, and it is forging ahead with productive and pest- and disease-resistant strains of rice, cotton, tomatoes, tobacco, sweet peppers, and green peppers, among other types of crops. China, India, and Indonesia are among the leaders in increasing plantings.

Other social obstacles remain in the path of technological advances in agriculture. For example, biotechnology is expensive. Will all farmers have access to this technology, or only those with sufficient capital? Some observers worry that the relatively few large corporations that own the rights to enhanced seeds and other advances will exercise increasing economic control over agriculture. Research and development in agricultural biotechnology are now occurring overwhelmingly in the private sector—for profit. The fruits of the earlier green revolution emerged almost entirely from public-sector laboratories and national breeding programs, allowing easy access for users. The new biotechnologies developed by private companies, however, are usually treated like inventions and shielded with patents. These new technologies are not diffusing easily, producing greater unevenness in the distribution of food production. This may enhance the productivity and profits of farmers in wealthy countries, but not for farmers in developing countries. What good is the new food technology if it only produces greater profits for some rather than more food for more people?

Another threat seems like an ironic paradox: If biotechnology truly unleashes the productivity the optimists foresee, then just a fraction of the farmers on Earth today will be able to feed everyone. Millions of farmers—hundreds of millions—will be unemployed. What will they do for a living? Multiplying our ability to produce food threatens to impoverish billions of people, to launch even greater migration to cities that are underequipped to absorb the influx, and thus to trigger civil unrest on scales never before contemplated in history—as well as, perhaps, to worsen the health problems associated with obesity outlined in Chapter 5. Will our political and economic systems be able to transform human societies fast enough to avoid chaos? Chapter 10 examines the pressures on cities already caused by the migration of farmers from the countryside, and Chapter 12 examines the constraints on economic development.

Climate Change and Food Security

Forecasts of climate change modify the optimistic scenarios of increased food production through new farmland, irrigation, or biotechnology. The FAO has estimated that by the year 2080, the world's poorest 40 countries (determined as of 2000) could lose 20 percent of their ability to raise food because of global warming. Food-growing capability would also lessen in the United States, France, Romania, Hungary, Belgium, the United Kingdom, the Netherlands, the Czech Republic, Ukraine, all of Central America and the Caribbean region, and India. In Canada, by contrast, agricultural production could double. Other "winners" would be Finland, Norway, New Zealand, Russia, and China.

The FAO anticipates that the effects of climate change on food security are both direct and indirect. Climate change may directly and negatively affect food supplies. Global warming will decrease crop and livestock yields due to heat stress while increasing transport and storage costs to keep food from spoiling. Changes in precipitation patterns and rising temperatures will produce new patterns of water shortages, reducing crop yields and herd sizes. Increased frequency and intensity of extreme weather may increase losses due to flooding and wind damage. Market and government responses to sudden food shortages may alter trade patterns or produce economic crises. Food producers, meanwhile, will lose their jobs unless they migrate to areas where new climate patterns are better suited to food production. Imagine the farming communities that have long defined the American heartland having to move northward or collapse altogether. The potential political conflicts stemming from these changes relate to almost every issue from immigration to national security. The degree to which these shifts decrease food security will depend, of course, on the extent of climate change. Food security will also depend on how human societies learn to adapt their agricultural practices to a changing climate.

Sustainable Agriculture

Sustainability describes humanity's use of Earth's limited resources in ways that can continue indefinitely. The question of whether we can sustain our use of resources without permanently depleting Earth's material supplies is at the heart of the debate between Malthusian and Cornucopian predictions. We will return to the sustainability concept several times in the coming chapters as we see, for example, that our unsustainable use of oil resources requires a search for new energy sources. Sustainability is not simply a question of how fast we use resources but whether their use degrades the larger environment.

Sustainable agriculture, therefore, is food production that can be continued indefinitely and that limits or even reverses environmental degradation. There are many definitions of what constitutes sustainable agriculture. One definition is provided by U.S. law, which describes sustainable agriculture as an "integrated system of plant and animal production" that will provide food and sustain the farm economically while also improving the environment and limiting the use of non-renewable and synthetic resources. Such practices should sustain local soil and water quality, for example, which are necessary for the continued success of the farm and the health of farming communities and consumers. Such a broad definition obviously touches on all aspects of food: production, distribution, commerce, and consumption.

The shift toward sustainable agriculture has come about as growers, governments, and consumers recognize the negative side-effects of modern commercial agriculture. Farmers were among the first to recognize what unsustainable agriculture can do to their farms. Overused soils can no longer provide profitable yields. Pesticides and herbicides are expensive, harm farm workers, and spoil water supplies. Fertilizers can improve crop yields significantly when used in appropriate amounts, but for many years the quantities of fertilizer applied in advanced countries have far exceeded the point of **diminishing returns.** Diminishing returns exist when, in adding equal amounts of one factor of production, each successive application yields a smaller increase in production than the application preceding it. Modern farming techniques apply more fertilizer to crops than is needed because farmers see greater economic risk in under-fertilizing than in over-fertilizing. However, excess fertilizer washes into groundwater, streams, and lakes, contaminating drinking water. Fertilizers also cause **eutrophication**, which is the rapid growth of aquatic plants, especially algae, that absorb oxygen and change the chemistry of a lake, inhibiting other life forms (Figure 8-43). New technologies allow farmers to more sustainably use fertilizer by adjusting applications to the minimum needed on different parts of their fields (Figure 8-44).

Governments, too, have long recognized that commercial farming can produce unsustainable outcomes. The expansion of farming into dry-land areas of the western U.S. led to an environmental catastrophe in the 1930s called the Dust Bowl, forcing thousands of farms to close. The government began to regulate the use of lands in the affected area and intervened to change farming practices. More controversially, the government bans

Figure 8-43 Dead Zones Created by Fertilizer Run-off. The eutrophic effects of fertilizer-fed algae are visible from the air over (a) the Gulf of Mexico and (b) Baltic Sea.

Figure 8-44 A GPS monitor in a tractor cab. GPS satellite navigation plus GIS field mapping and data analysis allow farmers to distribute seeds and apply fertilizers and pesticides mixed specifically for each square meter of a large field as the tractor drives across it.

on GM plants in Europe and elsewhere are considered a sustainable practice because they reduce risks to the natural environment.

Consumers are also driving the move toward sustainable agriculture by showing an interest in healthier and ethical foods. More and more consumers demand foods grown with few or no chemicals and meats raised on natural feed and without hormones. These foods are sometimes labeled "organic." The consumer market for organic foods has grown from $1 billion in 1990 to an estimated $23.6 billion in 2008. Most grocery stores now offer organic food options. Consumers and municipal authorities have also pushed for less packaging and more recyclable materials. Wal-Mart has repeatedly asked its suppliers to reduce packaging to save on material and shipping costs, and this has added environmental benefits. Some food retailers, responding to consumer demands, are addressing the sustainability of farming communities themselves. Starbucks Coffee, for example, now claims that 75 percent of its coffee beans are grown "responsibly": Coffee farmers are better paid for their crops, and farming practices better conserve the environment.

Sustainable agriculture is about food, which is the closest link between humans and the environment. The health of one depends upon the other. Recognizing this interdependence is crucial for improving the quality of our food supply. But it is equally about ensuring a sustained quantity of food production to feed a hungry planet. Today's farmers must be encouraged to use methods that improve environmental quality and assure the success of agricultural communities. Ultimately, sustainable agriculture is a means of improving our food supply today and into the future. The food security of future generations will depend upon how wisely we use our available resources.

Chapter Review

Summary

In the 200 years since Thomas Malthus wrote his theory, global starvation has not occurred. This has been because of the farming of more areas of the planet, the transplantation of many food crops, improvements in worldwide transportation and the technology of food storage, the introduction of higher-yielding and hardier strains of crops, new pesticides, and improvements in livestock and farm machinery. These developments are called the green revolution and the scientific revolution in agriculture.

Today, 10 principal types of agriculture can be identified. Four of them are basically subsistence (nomadic herding, low-technology subsistence farming, intensive rice farming, and Asian mixed cereal and pulse farming), and five are basically commercial (mixed farming with livestock, prairie cereals, ranching, Mediterranean, and plantation agriculture). Irrigated farming can be either subsistence or commercial. The richer a nation becomes, the more total grains it consumes, but the higher the percentage of grain consumption that is usually in the indirect form of meat and dairy products. Humankind overall could greatly increase food supplies by concentrating on raising the most efficient animals.

Some areas that are remote from urban centers may specialize in dairy farming and add value to milk by producing dairy products. Worldwide, there is growing demand for meat, but livestock consumes large amounts of grain. The harvest of aquatic life, primarily fish, is an important food source that is growing to meet increased demand. Some fisheries are overfished. Aquaculture or fish farming is an increasingly important source of aquatic life for many consumers.

About 1 billion people are hungry or undernourished, which can lead to increased mortality and famine. There is, however, more than enough food to feed everyone. Food security can be achieved only by addressing the economic, political, and social aspects of food production, distribution, and consumption. Problems in the effort to increase food supplies include cash cropping, lack of financial incentives for farmers, unsuccessful land redistribution, and inequitable land ownership.

Many rich countries have erected tariffs as protection against food imports, and they subsidize their own farmers. These subsidies contribute to production of agricultural surpluses in many rich countries and discourage production in poor countries. We do not know what quantities of food the world's farmers would be capable of producing or what the geography of food production would be if these conditions were eliminated.

Technology creates the potential for increasing food production, but technology is not readily available throughout the whole world, nor is it distributed in the most economic way. Biotechnology is providing new possibilities, but some observers fear possible injurious ramifications from tampering with plant and animal biology. Climate change may alter what foods can be grown where, while lowering agricultural productivity in general. Sustainable agriculture seeks to reform farming practices so that yields can continue indefinitely while minimizing farming's harmful effects on the larger environment.

Key Terms

agricultural input p. 303
aquaculture p. 320
biotechnology p. 331
cloning p. 331
commercial agriculture p. 303
diminishing returns p. 335
double cropping p. 306
downstream activity p. 315
economies of scale p. 304
ejido p. 324
eutrophication p. 335

exclusive economic zone (EEZ)
    p. 320
famine p. 321
fishery p. 317
food security p. 321
gene splicing (recombinant DNA)
    p. 321
genetic engineering p. 297
genetically modified (GM) p. 331
green revolution p. 302
hunger p. 320
isotropic plain p. 316
land reform p. 309

malnourished p. 320
monoculture p. 304
polyculture p. 304
privatize p. 324
psychological value-added p. 315
scientific revolution in agriculture
    p. 300
subsistence agriculture p. 303
sustainability p. 334
sustainable agriculture p. 334
swidden p. 305
undernourished p. 320
value added by manufacturing p. 315

Questions for Review and Discussion

1. What are the world's principal types of low-technology subsistence agriculture, and where can each be found?

2. What current conditions in many sub-Saharan African countries frustrate attempts to raise more food?

3. How have humans forestalled a "Malthusian" catastrophe for the past 200 years?

4. Why do rich countries subsidize agriculture? What effect does this have on the urban populations of rich countries, and what effect does it have on the farmers in poor countries?

5. Why are countries increasingly clashing over the rights to fishing at sea?

6. What is the difference between hunger and food security? What factors contribute to these outcomes?

Thinking Geographically

1. If fewer and fewer farmers can produce all the food the world needs, what will the rest of the people do for a living?

2. Why is humankind just now domesticating fish thousands of years after we managed to think of domesticating animals?

3. What percentages of the grain you consume do you consume directly? Indirectly?

4. How does your diet differ from that of your parents or grandparents? Compare the amount of meat or fish you eat. How much more processed is the food you buy? Compare the price of the ingredients of a cake with the cost of a store-bought cake.

5. Where are your foods grown or raised? What inputs go into making them, and where do they come from?

Log in to www.mygeoscienceplace.com for videos, animations, **MapMaster**™ interactive maps, RSS feeds, case studies, and self-study quizzes to enhance your study of The Human Food Supply.

MapMaster™

Salt flats in Salar de Uyuni, Bolivia. These deposits
are the world's richest known source of lithium.

9

Earth's Resources and Environmental Protection

S*alar de Uyuni, a vast salt flat in Bolivia, is home to the world's richest untapped deposits of lithium. Interest in, and demand for, this relatively lightweight metal has risen dramatically with the development of lithium ion batteries, which are used in a wide range of electronic devices, including cell phones and laptop computers. Interest is especially high because lithium ion batteries are the best technology at present for powering electric and hybrid vehicles. Bolivia, with about half the world's lithium resource, is looking to perhaps dominate or even control world lithium markets. The current president, Evo Morales, has asserted Bolivian control over natural resources by nationalizing natural gas and some mining companies, antagonizing foreign investors. Many imagine the lithium deposits forming a major export market that can boost the Bolivian economy, but there is always a risk that at some point, some other technology is likely to replace lithium as a power storage medium, and at that point, world demand for this metal will probably drop. In addition, high lithium prices have stimulated searches for resources elsewhere and substantial lithium resources were recently discovered in Afghanistan. Wealth generated from exploiting such deposits may be great, but is likely fleeting.*

Everything we consume is extracted from our planet and returned to it in one form or another. As we use Earth's resources of air, water, minerals, energy, plants, and animals, we simultaneously discharge our waste into the environment. As Earth's human population of 6.8 billion approaches 9 to10 billion, likely late in the twenty-first century, consumption of resources will increase. This expanded population will place tremendous stress on Earth's remaining resources and the ability of the planet's air, water, and land to accommodate human waste.

Consumption and waste vary among cultures and over time. Different cultures at different times have obtained energy from wood, coal, petroleum, natural gas, running water, nuclear energy, the wind, and the Sun. Issues of natural resource use and environmental quality must be understood in both their physical and human dimensions.

Resource management is exceedingly complex, because each resource varies geographically and physically. Resources also vary in value, depending on human factors: culture, technology, beliefs, politics, economics, and government style. Many resources are publicly controlled, so resource management is a political process.

The issue of lithium in Bolivia illustrates this complexity. Bolivia is a very poor country, and its economy is heavily dependent on natural resources. Struggles over resource issues have dominated Bolivian politics in recent years, especially issues of ownership and control of water and fossil fuels. Development of lithium offers an opportunity to generate income but also carries the risk of tying economic development to a very specialized technology, and world demand for lithium is both highly uncertain and beyond Bolivia's control. Similarly, development of mineral resources in Afghanistan would offer the opportunity of significant foreign exchange earnings, but would come at the risk of economic dependency on a limited number of resources and customers, and might not generate comparable income for farmers as they have historically earned from producing opium.

In this chapter, we explore the factors that affect the value of resources. These factors include the physical characteristics of resources and the natural systems in which they exist, the changing technology of resource use, and human value systems. We then consider how changing resource values affect what and how much of our resources we use. Finally, we look at environmental pollution and resource conflict and management.

What Is a Natural Resource?

A **natural resource** is anything created through natural processes that people use and value. Examples include plants, animals, coal, water, air, land, metals, sunlight, and wilderness. Natural resources are especially important to geographers, because they are the specific elements of the atmosphere, biosphere, hydrosphere, and lithosphere with which people interact. Natural resources can be distinguished from human-made resources, which are human creations or inventions such as money, factories, computers, information, and labor.

We often use natural resources without considering the broader consequences of doing so. For example, burning oil to generate heat or to power an automobile engine pollutes the atmosphere with exhaust gases. The nitrogen oxides emitted into the atmosphere contribute to acid rain, which then pollutes streams. Oil consumption also weighs heavily on international economic and political relations.

Characteristics of Resources

A substance is merely part of nature until a society has a use for it. Consequently, a natural resource is defined by the three elements of society:

- A society's *cultural values* influence people's decision that a commodity is desirable and acceptable to use.
- A society's level of *technology* must be high enough to use the resource.
- A society's *economic system* affects whether a resource is affordable and accessible.

Consider petroleum as an example of a natural resource in North America:

- *Cultural values.* North Americans want to drive private automobiles rather than use public transport such as trains.
- *Technology.* Petroleum is the preferred fuel in private automobiles because autos are easily powered by gasoline engines.
- *Economic system.* North Americans are willing to pay high enough prices for gasoline to justify removing petroleum from beneath the seafloor and importing it from distant places (Figure 9-1).

The same elements of society apply to the study of any example of a natural resource: rice to the Japanese, diamonds to South Africans, forests to Brazilians, and clean air to residents of Los Angeles. In every case, a combination of the three factors is necessary for a substance to be valued as a natural resource. Differences among societies in cultural values, the level of technology, and economic systems help geographers understand why a resource may be valuable in one place and ignored elsewhere.

Cultural values and natural resources To survive, humans need shelter, food, and clothing, and make use of a variety of resources to meet these needs. We can build homes of grass, wood, mud, stone, or brick. We can eat the flesh of fish, cattle, pigs, fish, or mice—or we can consume grains, fruit, and vegetables. We can make clothing from animal skins, cotton, silk, or polyester. Cultural values guide a process of identifying substances as resources to sustain life.

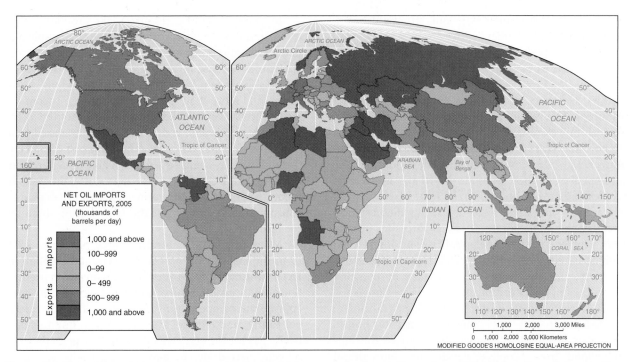

Figure 9-1 Net oil imports and exports. The largest oil importers are the wealthy nations of North America and Europe. Many developing nations are also net importers, and although the quantities they import are much smaller, the cost may be a higher percentage of the total national income for poor countries.

A swamp is a good example of how shifting cultural values can turn an unused feature into a resource. A century ago, swamps were seen in the United States as noxious, humid, buggy places where diseases thrived instead of a place that provides usable commodities. Swamps were valued only as places to dump waste or to convert into agricultural land. Eliminating swamps was good, because it removed the breeding ground for mosquitoes while simultaneously creating productive and valuable land.

During the twentieth century, cultural values changed in the United States. Scientists and environmentalists praised the value of swamps and documented their importance in controlling floods, providing habitat for wildlife, and reducing water pollution. Philosophers increasingly regarded nature as beautiful and praiseworthy. Changing public attitudes toward swamps is reflected in our vocabulary: instead of calling them *swamps*, now we use a more positive term, *wetlands*.

As a consequence of cultural change, wetlands now are a valued land resource, protected by law (Figure 9-2). We restore damaged wetlands, create new ones, and restrict activities that might harm them. Even in the Netherlands, where deep-rooted cultural and landscape traditions are based on draining wetlands to convert them into farmland, some of these same farmlands are now being converted back to wetlands to improve water quality and enhance species diversity.

⭐ **Technology and natural resources** The utility of a natural substance depends on the technological ability of a society to obtain it and to adapt it to that society's purposes. Metals are elements that can be formed into

Figure 9-2 Wetlands in southern Louisiana. Louisiana is losing wetlands at a rate of about 75 square kilometers (29 square miles) per year. In addition to providing important water quality and ecological functions, these wetlands help protect cities such as New Orleans from flooding.

materials that have high strength in relation to their weight, can generally withstand high temperatures, and are good conductors of heat and electricity. A metal ore is not a resource if the society lacks knowledge of how to recover its metal content and how to shape the metal into a useful object, such as a tool, structural beam, coin, or automobile fender.

Earth has many substances that we do not use today, because we lack the means to extract them or the knowledge of how to use them. Things that might become resources in the near future are **potential resources.** As a result of their high level of biodiversity, tropical rain forests are brimming with plants and animals that North Americans regard as potential resources. New medicines, pesticides, and foods might be developed from these substances. To the indigenous peoples of the Amazon rain forest, some of these plants and animals are already resources. But deforestation threatens the availability of these resources for indigenous peoples and their potential use by others even while it creates economic opportunities for Brazilians. By destroying the rain forest, we are diminishing Earth's pool of both current and potential resources.

Human needs can drive technological advances. People living in cold climates invented insulated homes and heating technology. The need to increase the supply of food drove people to develop new agricultural technology.

Because human need drives many technological advances, new technologies may emerge when a resource becomes scarce. New technology for reusing materials is being developed in part because space in landfills has become a scarce resource, especially in large urban areas of relatively developed wealthy countries, where consumption is highest. This space scarcity is stimulating development of new methods for reusing and recycling materials. Most waste currently is not reusable. But as we deplete the resource of landfill space, we will make more things recyclable and manufacture new products from waste, and waste materials themselves will become resources.

Economics and natural resources   Natural resources acquire a monetary value through exchange in a marketplace. The price of a substance in the marketplace, as well as the quantity that is bought and sold, is determined by **supply and demand.** Common sense tells us some principles of supply and demand:

- A commodity that requires less labor, machinery, and raw material to produce (for example, a bicycle) will sell for less than a commodity that is harder to produce (for example, an automobile).
- The greater the supply, the lower the price (for example, corn). The greater the demand, the higher the price (for example, Superbowl tickets).
- Consumers will pay more for a commodity if they strongly desire it (for example, a computer) than if they have only a moderate desire (for example, a textbook).
- If a product's price is low (for example, beer and hamburgers), consumers will demand more than if the cost is high (for example, champagne and prime rib).

In general, natural resources are produced, allocated, and consumed according to rules of supply and demand. Water is a good example. In areas where water is plentiful because of high rainfall and low demand—such as northern Minnesota—consumers pay a low price, little more than the cost of pumping it from the nearest well, river, or lake. But in the arid southwestern United States, water rights must be purchased, and scarce water must be carried hundreds of kilometers through pipelines and aqueducts. Thus, prices are generally higher. Because of these high costs, the government may subsidize water provision, and prices paid by consumers may not reflect the true cost of obtaining the resource.

Many important natural resource problems result from the inability of markets to account for pollution. **Externalities** are spill-over effects arising from production and/or consumption of goods and services for which appropriate compensation is not paid. They are external to the marketplace. The people who buy the electricity pay for the electricity directly. The price they pay reflects their desire to have the electricity and their willingness to pay for it. But people downwind from the power plant receive the pollution, whether they like it or not. The power plant does not directly pay for the privilege of discharging pollution to the atmosphere, although it does bear some of the costs of pollution control. This un-priced pollution thus constitutes an externality—a hidden cost. If the power plant were charged for its pollution and the people who receive it were compensated, perhaps the power plant would emit less pollution and thereby have less impact on the people exposed to it.

Example: Uranium   In every society, *level of technology, economic system,* and *cultural values* interact to determine which elements of the physical environment are resources and which are not. Uranium is a good example. Until the 1930s, uranium was a resource only because its salts made a pretty yellow glaze for pottery. After German physicists realized that the great energy stored in uranium atoms might be released by "splitting" their nuclei (nuclear fission), uranium gained value as weaponry during World War II:

- *Technology.*   The United States and its allies, fearing that Germany might develop a nuclear bomb, began what was known as the Manhattan Project, a crash program to develop the bomb technology first. Germany surrendered before the Manhattan Project achieved its goal, but the technology was eventually developed and used to win the war against Japan.
- *Economic system.*   After World War II, nuclear technology was applied to generating electrical power. Nuclear-generated electricity was slow to gain acceptance because it was more expensive than alternative power sources, but after Middle East

petroleum supplies were threatened during the 1970s, more nuclear power plants were built. The cost of nuclear power has had little to do with the cost of uranium, which is cheap and abundant. Recent increases in the price of non-nuclear-generated electricity associated with growing demand, concern about global warming, and rising prices of fossil fuels have renewed interest in nuclear power.

- *Cultural values.* With the construction of nuclear power plants, the public became increasingly concerned about the risks. Following power-plant accidents at Three Mile Island in Pennsylvania (1979) and Chernobyl in Ukraine (1986), government agencies regulated nuclear plant safety much more closely. Higher safety standards increased the cost of nuclear power at a time when conservation efforts had succeeded in reducing demand for electricity. Fear of the hazards of nuclear power has prevented growth of the industry, even though the hazards are probably less than those routinely accepted in other aspects of everyday life, such as the risk of death in an automobile or airplane accident. Orders for new power plants ceased in the United States during the 1980s. Renewed concern about security of radioactive materials may also reduce support for nuclear power.

 ## Substitutability

Many natural resources are valued for *specific properties*—coal for the heat it releases when burned, wood for its strength and beauty as a building material, fish as a source of protein, clean water for its healthiness. In most cases, several substances may serve the same purpose, so if one is scarce or expensive, another can be substituted (Figure 9-3). Copper is an excellent conductor of electricity, but it may be expensive relative to wire made of other metals that can be substituted. For information transmission, such as in computer networks, using light in a fiber-optic cable is more efficient than using electrons in a copper wire.

The **substitutability** of one substance for another is important in stabilizing resource prices and limiting problems caused by resource scarcity. If one commodity becomes scarce and expensive, cheaper alternatives usually are found. Such substitution is central to our ability to use resources over extended periods without exhausting them and without decline in our standard of living.

However, many resources have no substitutes. There is only one Old Faithful Geyser and only one species of sperm whale, so if we destroy Old Faithful or force extinction of the sperm whale, we have no substitutes waiting to be tapped. Other geysers and whales exist, but they are not the same as those we now know. The uniqueness of these resources is the essence of their value.

Figure 9-3 House construction in the United States. Wood is used in house construction in the United States mostly because it is relatively inexpensive and easy to work with. Here, oriented strand board, a manufactured wood product, is being used in exterior walls in new construction. This board is much cheaper than sawn lumber but has ample strength for covering exterior walls, floors, and roofs. The cost of this wood product is competitive with that of foam board made from petroleum that is an alternative to wood-based board.

Renewable and Nonrenewable Resources

In thinking about Earth's resources, we distinguish between those that are renewable and those that are not:

- **Nonrenewable resources** form so slowly that for practical purposes, they cannot be replaced when used. Examples include coal, oil, gas, and ores of uranium, aluminum, lead, copper, and iron.
- **Renewable resources** are replaced continually, at least within a human life span. Examples include solar energy, air, wind, water, trees, grain, livestock, and medicines made from plants.

Even a renewable resource can be *depleted*, or used to a point at which it can no longer be economically used. The only ones we cannot deplete are solar energy and its derivatives: wind and precipitation.

Geologic and Energy Resources

Geologic resources are substances that we derive from the lithosphere. They are basic materials that we use to construct roads and buildings, manufacture goods, and power transportation. They include rocks and minerals such as limestone and metal ores, as well as fossil fuels. Without these resources, modern industrial societies could not function. Our use of geologic resources in industry and commerce is governed primarily by technology and economics.

We value most minerals for their properties of strength, malleability (ability to be shaped), weight, and chemical reactivity rather than for their aesthetic characteristics. Few car owners care if the engine is made from aluminum or iron; what matters is that it is powerful, durable, and efficient. Few people care very much whether the roof of their house is made from slate or asphalt shingles, as long as the roof keeps out the rain and doesn't cost too much. Gold is the rare exception of a mineral valued mostly for its beauty (more than three-quarters of gold use is in jewelry), although even gold is increasingly demanded for industrial uses, especially electronics.

Because we value a mineral primarily for its mechanical or chemical properties, our use of mineral resources is continually changing as our technology and economy change. As new technological processes and products are invented, demand can suddenly increase for materials that had little use in the past. When these new processes and products replace older ones, demand for minerals that were previously important may be reduced. As a result of changes in consumer demand, remaining supplies, and prices, one mineral may become more favored while another is less desired.

Mineral Resources

The terms *Stone Age*, *Iron Age*, and *Bronze Age* indicate the importance of particular minerals at various times in the past. Minerals are as essential to civilization as plant, animal, water, and energy resources. They are present in virtually every product we manufacture, and though their value may be a small part of the total value of a finished good, those products could not be made without minerals (Figure 9-4).

Earth has 92 naturally occurring chemical elements, but most of Earth's crust is made up of eight elements: oxygen, silicon, aluminum, iron, calcium, sodium, magnesium, and potassium. These elements, as well as rarer ones, combine to form thousands of minerals, each with its own properties and distribution pattern throughout the world. Each mineral is potentially a resource if people find a use for it.

Metallic minerals, such as copper, lead, silicon, tin, aluminum, and iron, usually occur in ores, from which they must be extracted. Nonmetallic minerals, including building stone, graphite, rubies, sulfur, slate, and quartz, are generally easier to obtain because they are more plentiful and usually require less processing. Both metallic and nonmetallic minerals must be discovered, mined, transported, refined, and manufactured into useful goods.

Variations in Mineral Use

Historically, the use of particular metals and nonmetals has fluctuated between periods of high demand and price and periods of low demand and price. Discovery of a new resource could create a "rush" of people to the area of discovery. The "gold rushes" to California, Colorado, and Alaska are nineteenth-century examples. The period between 1970 and 1985 featured especially volatile mineral prices, as a result of rises and declines of industrial output and inflation. For instance, the price of copper doubled between 1973 and 1980, and then it fell nearly 40 percent between 1980 and 1985. Prices of most minerals were relatively stable through the 1990s, but since 2003, prices of some metals have risen dramatically in response to growing industrial production in China and India. For example, the price of copper tripled between 2003 and 2006, and between 2005 and 2008, the price of steel doubled. The prices of these commodities plunged in 2008, dropping to their 2004-05 levels in just a few months. Prices remained low in 2009 and 2010 because of the global economic recession. Historically, such spikes in mineral commodity prices have been short-lived, but sustained industrial growth in China, India, and elsewhere may prolong these price trends.

Mineral deposits are not uniformly distributed around the world. Most of the world's supply of particular minerals is concentrated in a handful of countries. For example, although aluminum is one of the most abundant elements on Earth, in 2009, just six countries—Australia, Brazil, China, Guinea, India, and Jamaica—produced about 87% of the world's total aluminum ore (bauxite) (Figure 9-5). Large countries such as the United States, Russia, Canada, China, and Australia are especially rich in metal and nonmetal mineral resources.

The concentration of mineral resources and production in a few countries favors the establishment of **cartels.** A cartel is a group of countries or firms that agree to control a particular market by limiting production in order to drive up prices. During the 1970s, when world demand for minerals was strong, a few cartels were able to control world markets for brief periods. But weak demand, falling prices, and political instability have limited the strength of cartels in recent years. In addition, the United States—the largest consumer of most minerals—has accumulated a stockpile of important minerals to protect against short-term reductions in supply caused by high prices, political instability, or hostile foreign governments.

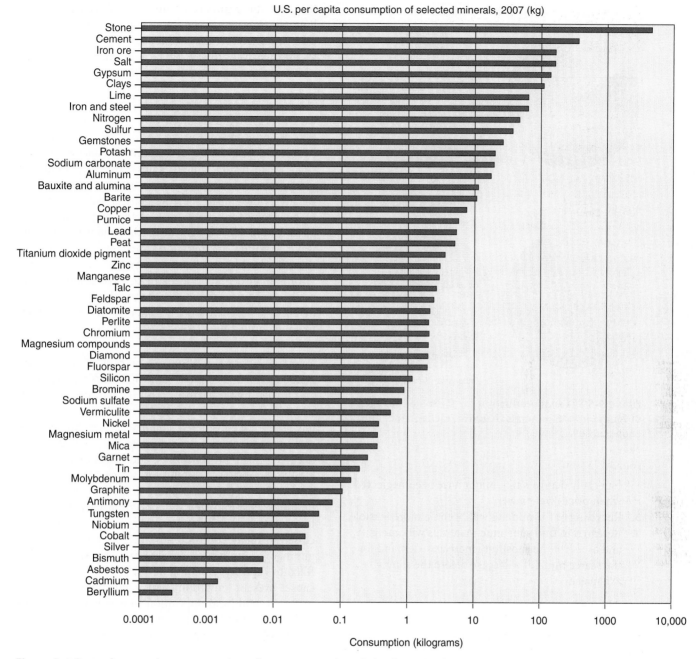

U.S. per capita consumption of selected minerals, 2007 (kg)

Figure 9-4 Annual per capita consumption of nonenergy minerals in the United States, 2007.

Depletion and Substitution

Fluctuations in the price of a mineral as a result of actions by a cartel, a political dispute, or a limited supply rarely continue for a long period of time. If high prices persist for several decades, technological innovations usually enable the substitution of cheaper minerals for more expensive ones. When the price of copper rose rapidly in the 1970s, for instance, plumbers began to substitute polyvinyl chloride (PVC). Today, PVC (a plastic made from petrochemicals) has largely replaced copper pipe for plumbing in new buildings. The likelihood that substitute products will be developed adds great risk to

enterprises of countries that become dependent on the production and sale of specific mineral commodities.

The substitution of one mineral for another has an important consequence: Even though world supply of a mineral resource may be limited, we will never run out of it. The reason is that if the supply of a resource dwindles relative to demand, its price will rise. The increase in price has four important consequences:

1. Demand for the mineral will decrease, slowing its rate of depletion.
2. Mining companies will have added incentive to locate and extract new deposits of the mineral,

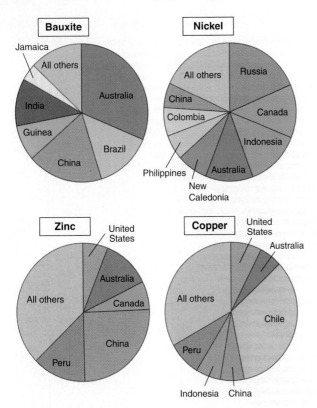

Figure 9-5 Share of production for the major producers of four common minerals: bauxite, nickel, zinc, and copper, 2009.

fear that these minerals will be depleted within your lifetime. However, if you know that a company has just invented a product that will replace one of these minerals inexpensively, this would be a sound investment!

Disposal and Recycling of Solid Waste

The average American throws away 2.1 kilograms (4.6 pounds) of solid waste per day, twice as much as people did in 1960. Paper accounts for one-third of all solid waste in the United States. Discarded food and yard waste accounts for another one-third (Figure 9-6). Relatively developed societies generate large quantities of packaging and containers made of paper, plastic, glass, and metal. We dispose of this solid waste in three ways: landfills, incineration, and recycling. Each of these methods poses significant problems, either in environmental degradation or in costs of disposal. Normally the choice of one disposal method over another means that costs are shifted from one group to another, making conflict inevitable.

Landfill disposal   About 55 percent of solid waste generated in the United States is trucked to landfills and buried under earth in **sanitary landfills** in which a layer of earth is bulldozed over the garbage each day. They are considered *sanitary* because burying the garbage reduces emissions of gases and odors from the decaying trash, prevents fires, and discourages vermin. Unlike air and water pollution, which are reduced by dispersal into the atmosphere and rivers, solid-waste pollution is minimized by concentrating the waste in thousands of landfills. However, landfills have been closed in many communities because they might contaminate groundwater, devalue property, or have been filled to capacity. Opening new landfills is difficult because present environmental regulations are more stringent, and local opposition to new landfills is usually overwhelming. The result in many areas has been a solid-waste crisis. Disposal sites are few and costly and

especially deposits that might have been neglected when prices were lower.
3. Recycling of the mineral will become more feasible.
4. Research to find substitute materials will intensify, and as use of the substitute increases, demand for the scarce mineral will cease before the supply is exhausted.

Although at current rates of use, the world would exhaust its remaining supply of lead in 23 years, zinc in 20 years, and copper in 33 years, there should not be any

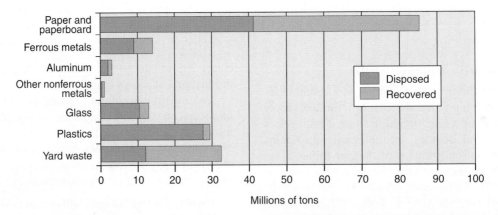

-6 Composition and disposition of U.S. solid waste, 2006. Paper and paperboard form the largest part of solid _ted in the United States. About one-half of all paper and paperboard waste is recovered and reused; much of this is _board boxes used for shipping. Recovery rates for aluminum are about one-third.

some communities must pay to use landfills elsewhere. San Francisco trucks solid waste to Altamont, California, 100 kilometers (60 miles) away. Passaic County, New Jersey, hauls waste 400 kilometers (250 miles) west, to Johnstown, Pennsylvania.

Incineration   One alternative to burying waste in landfills is incinerating it. Incineration reduces the bulk of trash by about three-fourths, and the remaining ash requires far less landfill space. Incineration also provides energy. The incinerator's heat boils water, producing steam that can be used to heat homes or to generate electricity by operating a turbine. More than 100 incinerators now burn and recover energy from about 12 percent of the trash generated in the United States. However, because solid waste is a mixture of many materials, it does not burn efficiently. Burning releases some toxic substances into the air, while some remain in the ash.

Recycling   Recycling solid waste reduces the need for landfills and incinerators and reuses natural resources that already have been extracted. Recycling simultaneously addresses both pollution and resource depletion. Most U.S. communities have instituted some form of mandatory or voluntary recycling, and about 32 percent of municipal solid waste in the United States is recycled.

Several barriers to recycling must be surmounted:

- *Waste separation.*   Solid waste comprises a variety of materials that must be separated to recycle (Figure 9-7). Metals containing iron can be pulled out of the pile magnetically; but paper, yard waste, food waste, plastics, and glass cannot be separated easily from each other. Consequently, many communities require consumers to separate solid waste themselves. Typically, consumers must place newspaper, glass, plastic, and aluminum in separate containers for pickup. Each type is collected on a different day or by a different truck, and is shipped to a specialized processor. The procedure is generally more expensive for the community than picking up all the trash together.
- *Consumer resistance.*   Separation for recycling is a nuisance for people in relatively rich developed countries who are used to throwing things away. To encourage recycling, some communities charge high fees to pick up nonrecyclables but take recyclables at little or no charge. Bottle and can laws requiring a deposit on beverage containers have been enacted in many states to encourage recycling and reduce roadside litter.
- *Lack of market.*   To succeed, recycled products must have a market. The lack of an assured market for many recycled products is perhaps the most difficult obstacle to increased recycling. Demand for recycled goods among industries and consumers is uncertain. For example, mixed plastics can be used as a substitute for wood in products such as picnic tables and playground equipment. The price of wood is not very different from that of recycled plastic; however, for aesthetic reasons, most consumers

Figure 9-7 Resource recovery. Resource recovery from solid waste requires sorting and separating different types of materials. One approach is to collect them together and separate them at a centralized facility like this one. Alternatively they can be separated at the source (home, school, work) and transported separately to reprocessing facilities.

prefer wood. This preference helps keep the market for recycled plastic small. To increase consumer acceptance, the quality of recycled material must also improve. Poor quality has resulted from the fact that recycling became mandatory before proper manufacturing methods had been developed.

- *Hidden costs.*   Recycling involves far more than just "melting down and reshaping." For example, to recycle paper, ink must be removed, which is an additional step compared to conventional papermaking and therefore an added cost. It is difficult to remove all of the ink, so recycled paper is unacceptable for some uses because it is too gray or speckled. Removing ink may create some pollution, though probably less than would be produced in making virgin paper. The relatively high cost of processing and the lower value of the products results in a limited market for the recycled materials.
- *Indirect losses.*   Trash burns only if it contains enough combustibles. If paper is recycled and yard waste is composted instead of being thrown in the trash, the trash may be difficult to burn. As recycling has increased, some communities have not had enough combustible waste to operate their incinerators.

Many companies are developing manufacturing methods and packaging that facilitate recycling. To reduce packaging volume, detergent is being sold in concentrated form, refillable containers are available for more products, and toner cartridges for some photocopy machines are built for reuse rather than for disposal. At the same time, however, the low cost of many materials means that recycling depends on cheap labor to separate materials prior to reuse.

Opportunities for recycling are growing, particularly as systems are developed to make it easier for consumers to recycle waste and for industry to recover recycled materials. The concentration of gold in computers, on a weight basis, is significantly greater than in gold ore. If sufficient quantities of old computers are brought together in one place, it becomes profitable to extract that gold from the computers.

One alternative is to require consumers to bear a greater share of the cost of waste generation. If those costs are clearly associated with the products that generate the wastes, then consumers have an incentive to reduce waste. For example, in 2002, the Irish government imposed a tax of about 17 cents on the use of disposable plastic shopping bags. Within a year, use of such bags dropped 95 percent, as consumers either reused bags or carried their purchases in other, reusable containers. The portion of plastic bags in litter dropped from 5 percent to 0.3 percent.

Landfills: An example of changing resource values

Garbage dumps are nothing new, as excavations of ancient cities reveal, but they are growing much more rapidly than in the past. Traditionally, food waste was fed to livestock, iron was used in durable goods, packaging was simple, and plastic products were unknown. Far fewer goods were manufactured. Our "throwaway culture" is a modern invention.

Garbage dumps have been sited in swamps and on other low-value land. For years, garbage simply was dumped and left uncovered. Rubber tires, deliberately burned in landfills to keep fires going, emitted offensive black smoke. Fires smoldered, rats thrived, flies buzzed, and homes situated downwind rarely enjoyed fresh air. Many coastal cities dumped their garbage at sea.

To reduce fire, vermin, and odor, cities have converted open dumps to sanitary landfills, which has reduced some environmental problems but aggravated others. Landfill toxic materials are leached by groundwater, polluting it. This consequence, plus problems such as health problems associated with chemical landfills in populated areas, heightened public concern. In response to these concerns, the U.S. government has passed several laws and regulations that control handling of toxic substances.

As landfills run out of space, new ones are difficult to create. People may support the creation of a facility, but they may not want it near their own homes. This attitude is sometimes called the NIMBY attitude—for

Figure 9-8 The Roosevelt Regional Landfill, Washington. This landfill is the site of an innovative program to convert waste to electricity, thereby reducing the need for combustion of fossil fuels for that purpose. In addition, the carbon dioxide produced from the landfill will be delivered to greenhouses that otherwise would burn propane to stimulate plant growth by increasing the CO_2 concentrations in the greenhouses.

"not in my back yard." Most people agree that landfills are needed, but few want them near their homes. Stricter regulations have forced some landfills to close and have prevented others from opening. The number of operating landfills in the United States has declined from about 30,000 in 1976 to fewer than 1,800 today. The few remaining landfills continue to grow as they accept the waste that formerly went to other sites.

At the Roosevelt Regional Landfill, an innovative project is operating in which garbage is burned to produce electricity (Figure 9-8). The CO_2 produced in the process is sold to nearby greenhouses that formerly burned propane to increase the CO_2 levels that accelerate plant growth in the greenhouses. This is encouraging: An industry that disposes of unwanted things also sells a commodity that people need. In this case, not only are we producing electricity but doing it in a way that reduces greenhouse gas emissions.

Modern facilities such as this are expensive, as is the process of collecting waste and transporting it to landfills. Unfortunately, in many parts of the world, the cost of collecting and disposing of solid waste is too high for society to bear. In our global economy, modern mass-produced materials are ubiquitous and inexpensive, so solid waste is becoming a major problem in many poor countries. The most visible example of this is the presence of plastic bags, which have replaced paper bags, baskets, and other biodegradable containers in rich and poor countries alike. Where solid waste collection is absent, plastic bags and other nondegradable waste accumulates wherever people congregate (Figure 9-9).

Today, landfills are a scarce resource that must be used carefully. To slow the rate of filling, many landfills

Figure 9-9 Plastic bags litter a beach in Mumbai, India. These bags are very inexpensive, which makes them easy to obtain, and they are difficult to recycle, making them a solid waste problem worldwide.

no longer accept grass clippings and leaves, which can be composted (decomposed harmlessly) at home, or any other materials that can be recycled.

Energy Resources

Earth has bountiful and varied sources of renewable energy that humans have been able to harness:

- Solar energy comes from the Sun.
- Hydroelectric power and wind power come from natural movements of water and air caused by solar energy and gravity.
- Geothermal energy comes from Earth's internal heat in volcanic areas, including California, Iceland, and New Zealand.

However, most of the world's energy comes from chemical energy stored in such substances as wood, coal, oil, natural gas, alcohol, and manure. Energy is released by burning these materials. People burn these substances to heat homes, run factories, generate electricity, and operate motor vehicles. Most of our energy comes from fossil fuels, but burning them reduces their supply—they are nonrenewable resources. We can continue to burn them for several more decades, but they are the major cause of global warming. Regardless of that consequence, eventually we must switch to new energy resources if we are to preserve our current standard of living.

Energy from Fossil Fuels

Oil, natural gas, and coal, known as **fossil fuels,** come from the remains of plants and animals buried millions of years ago. Through photosynthesis, plants convert solar energy to the chemical energy stored in their tissues. When plants die, this energy remains in their tissues. The energy may be released promptly if animals or decomposers consume the dead plants or if a fire burns the field or forest. Or the plants may become fossilized as coal, and the stored energy may wait millions of years to be released when the coal is mined and burned.

Oil and gas are also stored sunlight. When we burn these fossil fuels today, we are releasing the energy originally stored in microscopic plants (phytoplankton) millions of years ago. When animals eat plants, they store the energy that was incorporated into the plant tissues. As the sea's countless creatures died over thousands of years, their bodies sank to the bottom, creating organic sediment. Over time, this was converted to oil, accompanied by natural gas. Coal, oil, and natural gas still are being created, but the processes are so slow that from a human perspective, fossil fuels are nonrenewable resources: Once burned, they are gone forever as useful sources of energy.

From wood to coal, oil, and gas From the time that humans first lived in North America—probably at the end of the last ice age about 18,000 years ago—until the mid-1800s, wood was the most important source of energy. Prior to the arrival of European colonists, North American residents used wood almost exclusively for all of their needs, but because their total population was small, they did not significantly deplete the resource. When Europeans arrived in North America beginning in the seventeenth century, they harvested the forests for fuel and lumber and cleared the land for agriculture. By the end of the nineteenth century, most of the forests near populated areas of the eastern United States had been cut down, and fuel wood became very expensive. It was also inadequate for providing the large amounts of energy demanded by a growing industrial economy. Wood still provides the largest portion of energy in developing countries, though supplies are dwindling in areas of dense, rural populations, such as in East Africa and southern Asia.

Coal served as a substitute for fuel wood during the nineteenth century. Although large amounts of coal have been consumed, abundant supplies still remain in the United States and several other countries. Oil, another fossil fuel, was a minor resource until the diffusion of motor vehicles early in the 1900s. Today, it is the world's most important energy resource. A third fossil fuel, natural gas, once was burned off during oil drilling as a waste product because it was too difficult to handle, and markets for it were not established. In recent years, however, it has become an important energy source. Today, these three fossil fuels provide more than 85 percent of the world's energy and more than 90 percent in relatively developed countries (Figure 9-10).

For U.S. and Canadian industry today, the main energy resource is natural gas, followed by oil and coal. Some businesses directly burn coal in their operations,

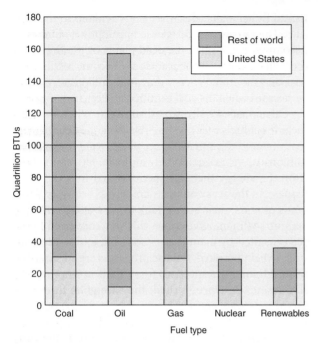

Figure 9-10 World energy sources, 2006. Fossil fuels make up the major proportion of global energy production; renewable and nuclear energy are relatively minor.

trucks, buses, airplanes, and most railroads. Only subways, streetcars, and some trains run on electricity, much of which is generated from burning fossil fuels.

Distribution of fossil fuels   Fossil fuels are not uniformly distributed beneath Earth's surface (see Table 9-1). Mineral *reserves* are known deposits that can be extracted profitably using current technology. Some regions have abundant reserves, whereas others have none. This distribution reflects how fossil fuels are formed. Coal forms in swampy areas, rich in plants. The lush tropical wetlands of 250 million years ago are today the coal beds of the world, relocated to midlatitudes by the ponderous movement of Earth's tectonic plates (see Chapter 3). Oil and natural gas form in seafloor sediment, but Earth's tectonic movements eventually elevate some seafloor above sea level to become land. Today, we drill for petroleum on both land and the seafloor.

Because minerals are available only in specific geologic environments, their distribution is very uneven. Table 9-1 reveals that, like the minerals shown in Figure 9-5, fossil fuel resources are highly concentrated in a few places. For each of the fossil fuels listed, the top five countries hold more than 60 percent of world reserves.

Relatively wealthy developed countries—which comprise about one-fourth of the world's population—possess more than 60 percent of the world's coal and more than 60 percent of natural gas. China is the leading producer of coal, but the United States is also a major coal producer and user. Russia is the leading producer of natural gas. Other than China and India, which have significant coal deposits, most countries in

while others rely on electricity generated primarily at coal-burning power plants. At home, electricity is used to operate diverse electrical devices, and in some homes to generate heat and hot water. Natural gas is the most common source for home heating, followed by petroleum and electricity. Nearly all transportation systems operate on petroleum products, including automobiles,

TABLE 9-1   Reserves of Oil, Natural Gas, and Coal

For each resource, only the top 15 countries are listed. The countries listed account for 92 percent of all oil reserves, 86 percent of all gas reserves, and 96 percent of all coal reserves. Data are for 2009, except coal data, which are for 2005. Oil reserves for Canada include tar sands.

Crude Oil (Billion Barrels)		Natural Gas (Trillion Cubic Feet)		Coal (Million Short Tons)	
Saudi Arabia	267	Russia	1,680	United States	263,781
Canada	178	Iran	992	Russia	173,074
Iran	136	Qatar	892	China	126,215
Iraq	115	Saudi Arabia	258	Australia	84,437
Kuwait	104	United States	238	India	62,278
Venezuela	99	United Arab Emirates	214	South Africa	52,911
United Arab Emirates	98	Nigeria	184	Ukraine	37,339
Russia	60	Venezuela	171	Kazakhstan	34,502
Libya	44	Algeria	159	Former Serbia and Montenegro	15,306
Nigeria	36	Iraq	112	Poland	8,270
Kazakhstan	30	Indonesia	106	Brazil	7,791
United States	21	Turkmenistan	94	Colombia	7,671
China	16	Kazakhstan	85	Germany	7,394
Qatar	15	Malaysia	83	Canada	7,251
Brazil	13	Norway	82	Czech Republic	4,962

Source: Energy Information Administration, *Oil and Gas Journal*

Africa, Asia, and Latin America have few reserves of coal or natural gas.

The distribution of oil is somewhat different. Two-thirds of the world's oil reserves are in the Middle East, including one-fourth in Saudi Arabia. Mexico and Venezuela also have extensive oil fields. In contrast, oil reserves in North America and Europe are relatively small, and production in those regions has passed its peak.

Compared to other continents, North America and Europe have much higher per capita oil consumption rates, so they account for nearly two-thirds of the world's energy consumption. The United States, with less than 5 percent of the world's population, consumes nearly 20 percent of the world's commercial energy. In North America and Europe, the high level of energy consumption supports a lifestyle rich in food, goods, services, comfort, education, and travel.

Because relatively rich developed countries consume more energy than they produce, they must import energy, especially oil, from developing countries. The United States imports roughly half of its needs, Western European countries more than half, and Japan more than 90 percent. U.S. dependency on foreign oil began in the 1950s, when oil companies determined that the cost of extracting domestic oil had become higher than for foreign sources. U.S. oil imports have increased from 14 percent of total consumption in 1954 to 62 percent in 2009. European countries and Japan increasingly depend on foreign oil because of limited domestic supplies.

Oil production and price history   Early in the twentieth century, the United States was an oil exporter, but its needs soon exceeded domestic supplies. Europe has always been a net importer of oil. In their search for cheaper oil, U.S. and Western European companies

drilled for oil in the Middle East and sold it inexpensively to consumers in relatively developed countries. Western companies set oil prices and paid the Middle Eastern governments a small percentage of their oil profits. To reduce dependency on Western companies, the countries possessing oil created the Organization of Petroleum Exporting Countries (OPEC) in 1960. Today, OPEC members include eight countries in the Middle East (Algeria, Iran, Iraq, Kuwait, Libya, Qatar, Saudi Arabia, and United Arab Emirates), plus Venezuela in South America, Gabon and Nigeria in Africa, and Indonesia in Asia.

In the 1950s and 1960s, Western oil consumption and dependence on oil exporting nations grew. By the early 1970s, 35 percent of U.S. oil consumed was imported—a low figure by modern standards but historically high at the time. The cost of those imports was low, typically around $3 per barrel (equivalent to about $15 in 2010). OPEC nations were eager to establish control of oil production and pricing, and the 1973 war between Israel and its Arab neighbors provided an opportunity for OPEC to assert itself. Angry at the United States and Western Europe for supporting Israel, OPEC's Arab members organized an oil embargo during the winter of 1973–1974. OPEC states either refused to sell oil to countries that had supported Israel or would sell only at significantly inflated prices. Soon gasoline supplies dwindled in relatively developed countries, and prices at U.S. gas pumps soared. Another oil crisis followed in 1979, triggered by a revolution in Iran. World oil prices went from $3 per barrel (160 liters, or 42 gallons) in 1974 to more than $35 in 1981 (Figure 9-11). To import oil, U.S. consumers spent $3 billion in 1970 and $80 billion in 1980, nearly 3 percent of the gross national income (GNI). This rapid price increase abruptly halted the rise in energy use in the United States and

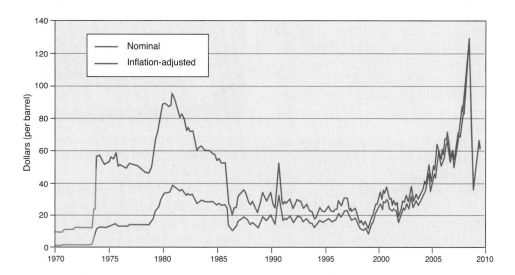

Figure 9-11 Crude oil prices, 1970–2009. Based on U.S. refiner acquisition cost for imported crude oil since 1973; price of Saudi Light Crude for 1970–1973. Crude oil prices jumped sharply in 1973–1974, 1979–1981, and 2006–2008 but dropped dramatically in the mid-1980s and 2009. Prices in mid-2008 were at record high levels, but by mid-2009, after adjustment for inflation, they were lower than they were in the early 1980s.

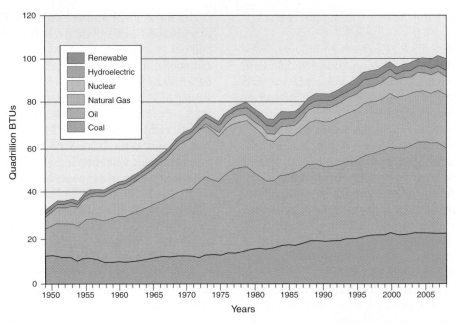

Figure 9-12 Trends in U.S. energy use, 1949–2008. Fossil fuels make up the major proportion of energy consumed in the United States. Use of oil and natural gas increased rapidly from the 1950s to the 1970s, but growth has been much slower since the energy crises of 1974 and 1979.

elsewhere and caused severe economic problems in relatively developed countries during the 1970s and 1980s (Figure 9-12).

In each of these oil crises, long lines formed at gas stations in the United States, and some motorists waited all night for fuel (Figure 9-13). In some cases, gasoline was rationed by license plate number; cars with licenses ending in an odd number could buy only on odd-numbered days. Some countries took more drastic action; the Netherlands banned all but emergency motor vehicle travel on Sundays.

Developing countries were especially hurt by the price rises. They had depended on low-cost oil imports to spur industrial growth and could not afford higher oil prices. Relatively wealthy developed countries somewhat lessened the impact of higher oil prices on their economies by encouraging OPEC countries to return some of their money by investing it in real estate, banks, and other assets in North America and Western Europe. Poorer nations could not offer this opportunity for reinvesting oil wealth.

No sooner had the world begun to adjust to high energy prices than the trend was reversed. Energy conservation measures in consuming countries reduced demand. Internal conflicts weakened OPEC's influence: Iraq and Iran fought a war that lasted eight years, and in the early 1980s, some OPEC members broke ranks and reduced oil prices. The price of oil plummeted from more than $30 to around $10 per barrel. Exporting countries flooded the world market with more oil in an unsuccessful attempt to maintain the same level of revenues as they had received during the 1970s. While supplies increased, demand in relatively developed countries remained lower than before the boycott. Oil prices briefly spiked during the

Figure 9-13 Gas line in 1970s. Cars lined up for gas in California during the oil crisis in 1979.

1991 Persian Gulf War but then quickly dropped back to about $20. In early 1998, following a significant drop in oil prices, OPEC members agreed to reduce production in an attempt to drive prices back up, and by late 2000 prices were again above $30 per barrel. They fell back to about $20 by late 2001, falling particularly sharply after September 11, 2001, but then rose again as tensions mounted in Iraq. The steep rise that began in 2003 was certainly influenced by supply disruptions in Iraq, but it also reflects growing worldwide demand for oil, especially in rapidly growing economies such as India and China. While there had been a significant reserve productive capacity for the last two decades of the twentieth century, that capacity has virtually disappeared so that any significant disruption of supply causes a rise in prices. In this environment, prices rose dramatically from 2003 to 2008, peaking at about $135 in July 2008. When the financial crisis struck in late 2008 and major economies around the world went into recession, oil prices dropped almost 75 percent in a six-month period.

The recent pattern of a rapid rise in oil prices followed by global economic recession is remarkably similar to that of the late 1970s and early 1980s, although the causes and circumstances are quite different. In the 1980s and 1990s, the world saw a prolonged period of relatively low oil prices. If rapid economic growth and expansion in numbers of automobiles continues in south and east Asia as it has in recent years, then oil prices are not likely to remain low for very long.

Future of fossil fuels   How much of the fossil fuels remain? Geologists can estimate fairly accurately the proven reserves of fossil fuels available in fields that have been explored, but there are significant differences of opinion on potential reserves of the fossil fuels—the amount in fields not yet discovered and explored, and the actual amount we will extract before shifting to other energy sources (see the Rapid Change box). The price spike of 2007–2008 shows that in the coming years, global oil markets are likely to be increasingly strained by growing demand and production limitations.

If we are to continue to base our economies on fossil fuels, what new sources of oil are available? Because much of the oil in accessible places on land has already

---

RAPID CHANGE

Peak Oil

If we divide Earth's current proven oil reserves (about 1,000 billion barrels) by current annual consumption (about 24 billion barrels), we get about 40 years of oil supply remaining. Rates of consumption will change, and new reserves will be discovered, but today's proven oil reserves will probably last only a few decades. Thus, unless large potential reserves are discovered, Earth's oil reserves will be significantly depleted during the twenty-first century, possibly in your lifetime.

Every discovery of new deposits extends the life of the resource. But extracting oil is becoming harder—and therefore more expensive. When geologists seek oil, they look first to accessible areas where geologic conditions favor accumulation. The largest, most accessible deposits already have been exploited. Newly discovered reserves generally are smaller and more remote, such as beneath the seafloor, where extraction is costly. Exploration cost also has increased because methods are more elaborate and the probability of finding new reserves is less.

At some point, the cost of finding and extracting oil will become so high that other sources of energy will be more attractive. When that happens, we will begin to shift our energy infrastructure toward the new energy source(s), and the demand for oil will begin to drop, reducing prices below what it costs to extract the oil. When that happens, oil production will reach a peak and begin a decline. These ideas were first developed by a geologist named M. King Hubbert, and their application to the current world oil situation has become known as **peak oil.**

If we accept that a peak must happen sometime, the key question is: when? It is a critical question because when it happens, there will be a fundamental transformation of our energy economy and all the machines and businesses that it. If the peak were to happen, say, 5 or 10 years from now, we want to know now so that we can prepare accordingly.

Some argue that the peak will occur very soon—in the next few years. These individuals, sometimes referred to as "peakists," point to declining production in major oil-producing regions such as the United States and the North Sea, and suggest that previous estimates of oil reserves may have been too high. Others, such as those associated with the oil industry, argue that undiscovered or undeveloped oil fields are many and that higher prices will spur development of these resources and provide oil for at least a few decades into the future.

Interestingly, each of these predictions can be self-fulfilling. If we believe that the peak will occur very soon, then we will avoid investing in things that use oil. For example, we wouldn't build a new gas station but instead we might build charging facilities in parking lots for plug-in electric vehicles. Once we begin to invest in alternative energy technologies, economies of scale will begin to make these technologies more affordable, and the transition away from oil will be under way. On the other hand, if we believe that oil will be with us for 20 or 30 years at least, then we will go ahead and build that gas station and the gasoline-powered cars that will use it, and people will pay the higher prices necessary to keep driving their cars. In this way, the debate over a technical question of "how much oil is left" actually becomes a policy debate with very real consequences. Not surprisingly, those concerned about greenhouse gas emissions and arguing for a change to new energy sources tend to be peakists, and those arguing for continuing a carbon-based economy tend to argue that there is plenty of oil still to be developed.

Figure 9-14 Deepwater Horizon. The oil rig exploded and sank in April 2010, causing the worst oil spill in US history.

been discovered and is being exploited today, future oil discoveries are likely to come from offshore areas and remote regions such as the Arctic. Offshore oil operations have moved into deeper water in recent years, and the risks of this expansion became clear when the Deepwater horizon oil rig exploded in the Gulf of Mexico in April of 2010 (Figure 9-14). The disaster produced the largest oil spill in U.S. history, and will likely result in more stringent restrictions and higher costs associated with future offshore oil operations.

In addition to oil resources in remote locations, unconventional sources of oil are being studied and developed, such as oil shale and tar sandstones. Oil shale is a "rock that burns" because of its tar-like content. Tar sandstones are saturated with a thick petroleum. Utah, Wyoming, and Colorado contain large amounts of oil in shale, more than 10 times the conventional oil reserves found in Saudi Arabia. The United States also has very large coal reserves, and technology that can convert coal to a liquid fuel is well developed. Canada has very large tar sandstone deposits that are currently being exploited. The largest operation is in northeast Alberta, where the sands are mined and oil is extracted. A new pipeline will carry this oil across the Rocky Mountains to a seaport on the Pacific near Vancouver, from which it can be shipped anywhere in the world.

The major concern about these unconventional resources is that converting them to liquid fuel that we can use in automobiles involves substantial carbon dioxide emissions. A unit of liquid fuel derived from tar sands, for example, would result in perhaps 40 percent more CO_2 emissions than a unit of fuel derived from conventional crude oil. In addition, mining large quantities of rock in order to extract the hydrocarbons

and then disposing of the waste is highly destructive to the environment. Because of these environmental concerns, it is unlikely that unconventional oil resources will make a major contribution to global energy needs in the near future.

Natural gas and coal: Short-term oil substitutes
In searching for alternatives to oil, we look first to the other two major fossil fuels—natural gas and coal. Natural gas is important in the United States as the clean-burning fuel of choice to heat more than half of the country's homes. A 1.6-million-kilometer (1-million-mile) pipeline network efficiently distributes gas from production areas in the Gulf Coast, Oklahoma, and Appalachia to the rest of the country. Demand for natural gas is increasing, in large part due to air pollution considerations.

Natural gas has its limitations, however. At current rates of consumption, natural gas reserves would last for 60 years, although potential reserves may be greater than for oil. Russia and the Middle East have about two-thirds of the world's natural gas reserves; the United States has only a few years of domestic gas reserves. Therefore, continued intensive use of gas will depend on increased imports.

Coal reserves are more abundant than oil or natural gas and, at current rates of use, can provide nearly 400 years of proven reserves to the world. The United States has a large percentage of the world's coal reserves. Coal is used mainly to produce electricity, and our use of electricity is expanding much faster than our use of other forms of energy. But several problems hinder expanded use of coal:

- *Air pollution.*   Uncontrolled burning of coal releases sulfur oxides, nitrogen oxides, hydrocarbons, carbon dioxide, and particulates into the atmosphere. The sulfur and nitrogen oxides are a major component of acid deposition. Acid deposition occurs when sulfur and nitrogen oxides combine with water to form acidic precipitation or when these acids are formed in the soil from pollutants that settle from the atmosphere. Many communities suffered from coal-polluted air earlier in this century and encouraged their industries to switch to natural gas and oil. The Clean Air Act now requires utilities to use low-sulfur coals or to install emission-control devices on smokestacks. Coal-fired power plants still pump large amounts of carbon dioxide into the atmosphere. In 2009, the U.S. Environmental Protection Agency recognized CO_2 as a pollutant that, as a greenhouse gas, "may reasonably be anticipated to endanger public health or welfare" and is thus subject to regulation.
- *Land and water impacts.*   Both surface mining and underground mining cause environmental damage, an externality caused by burning coal. The damage from surface mining is visible: Vegetation, soil, and

rock are stripped away to expose the coal. Today, surface-mined land must be restored after mining, but restoration may not leave the land as productive as it was before. Underground mining causes surface subsidence and can release acidic groundwater.

- *Limited uses.* Coal is a bulky solid, that cannot power cars, trucks, and buses without conversion to a liquid or gas. Coal-fired power plants can provide the power to run electric vehicles, however, which are expected to become more common in the next few years.

Nuclear and Renewable Energy Resources

Where can we turn for energy that is safe, economical, nonpolluting, and widely available; is not controlled by a handful of countries; and does not contribute to global warming? In the long run, we must look to energy sources that are renewable—or at least to resources such as nuclear energy that are so abundant they are not likely to be depleted. The two most promising energy sources are nuclear and solar. Other alternatives at present include biomass and hydroelectric power. We will now look at each of these sources.

Nuclear energy Nuclear power can be generated either by fission (splitting an atom into two or more parts) or by fusion (joining two atoms together). Most peaceful generation of nuclear power today relies on

fission of uranium, although plutonium also can be used. Someday we may use fusion to generate nuclear power, but as yet such technology is unavailable. Unlike solar or wind energy, nuclear energy is not renewable. It is derived from radioactive substances mined from the earth—but a tremendous amount of energy is available from a small amount of material. A kilogram of nuclear fuel contains more than 2 million times the energy of a kilogram of coal.

The peaceful use of nuclear reactors to generate electricity began in the 1950s, and today about 400 reactors are operating around the world. Nuclear power is used exclusively to generate electricity, a growing part of our energy needs. It supplies about one-third of all electricity in Europe. Japan (which has virtually no fossil fuels), South Korea, and Taiwan also rely on nuclear-generated electricity (Figure 9-15). Nuclear power generates approximately 20 percent of U.S. electricity (Figure 9-16). The United States derives a smaller portion of its energy from nuclear power than other wealthy developed nations, in part because of its more abundant coal reserves.

Like coal, nuclear power presents serious problems. These include potential accidents, the generation of radioactive waste, public opposition, and high cost:

- *Potential accidents.* A nuclear power plant cannot explode like a nuclear bomb, although it is possible to have a runaway chain reaction. The reactor can overheat, causing a meltdown, possible steam explosions, and a scattering of radioactive material

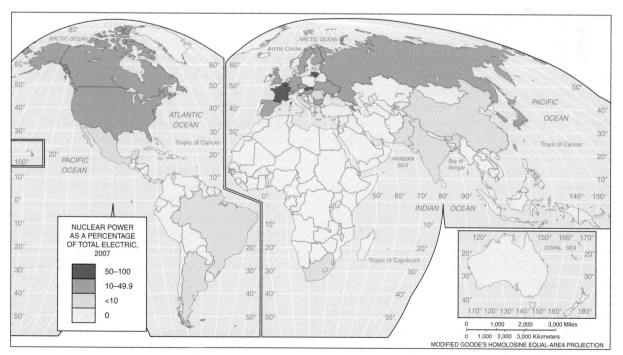

Figure 9-15 World nuclear power, 2007. Nuclear power plants are clustered in more developed wealthy regions, including Western Europe, North America, and Japan. Nuclear power has been especially attractive in the United States (which produces about 30 percent of the world's nuclear power) and in some European nations that lack abundant reserves of petroleum and coal.

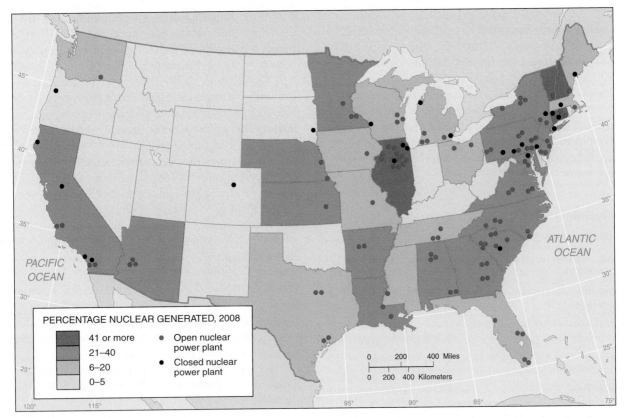

Figure 9-16 U.S. nuclear power. Nuclear power is an important source of electricity in several states. Some locations have more than one nuclear reactor.

into the atmosphere. An explosion did occur in 1986, in a nuclear power plant at Chernobyl, Ukraine, near the border with Belarus. (Both countries at the time were part of the Soviet Union.) The accident has caused over 3,500 deaths, including about 800 among emergency cleanup workers and potentially many more from subsequent cancers. Cancer rates remain elevated in southern Belarus. For example, childhood thyroid cancer rates in Belarus increased from fewer than five per year before the accident to more than 80 by 1994. The impact of this accident extended through Europe: Most European governments temporarily banned the sale of milk and fresh vegetables because of possible contamination by radioactive fallout. In addition to accidental radioactive releases, many people are concerned about potential terrorist attacks involving either nuclear weapons or "dirty" bombs composed of radioactive materials.

- *Radioactive waste.* The uranium-based fuel used in nuclear reactors eventually is broken down to the point that it is no longer usable. However, waste materials in this spent fuel remain highly radioactive for thousands of years. Spent fuel can also be reprocessed to extract plutonium, which can be used in nuclear bombs. Pipes, concrete, and water near the fuel also become "hot" with radioactivity.

This waste cannot be burned or chemically neutralized. It must be isolated for several thousand years until it loses its radioactivity. Currently, spent fuel generated in the United States is stored in cooling tanks at nuclear power plants, but these are nearly full. Work is under way to develop a long-term storage facility in Nevada, but many problems remain to be solved before waste can be moved from temporary storage sites to a permanent facility. Kazakhstan, which already has radioactive waste storage and disposal facilities dating from the Soviet era, is building a facility to receive nuclear waste from the European Union.

- *Public opposition.* Public concern about safety has been an obstacle to the diffusion of nuclear power since the technology first emerged. The accident at Chernobyl, as well as less damaging incidents in the United States and other countries, dramatically increased public concern about nuclear power (Figure 9-17). At Shoreham, Long Island, near New York City, a nuclear power plant was ready to operate and begin generating electricity when, under pressure from worried citizens, the government decided not to allow the plant to open and compensated the electric company for the loss of the plant. In 1992, Italian voters rejected future nuclear power development in that country, and public

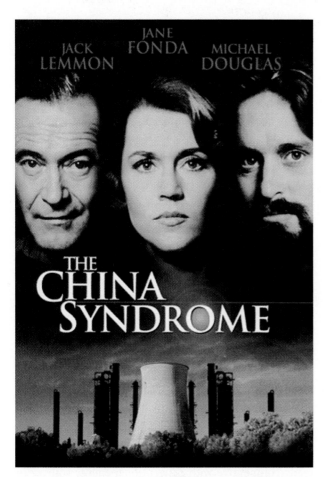

Figure 9-17 China Syndrome. This film, released on March 16, 1979, portrayed a fictional nuclear power plant near-meltdown that bore striking similarities to an incident just 12 days later at Three Mile Island, Pennsylvania. The film contributed to rising public fear of nuclear power.

opposition is similarly strong in Germany and Scandinavia. Even in France, where over 70 percent of electricity is generated from nuclear power, public opposition is a major barrier to new development. In addition, nuclear power plants cannot operate without a reliable source of cooling water, and summer heat waves in 2003 and 2006 caused critical cooling-water shortages in some areas of Western Europe.

- *High cost.* Nuclear power plants cost several billion dollars to build, primarily because of elaborate safety measures. Without double and triple backup systems, nuclear energy would be too dangerous to use. As a result, the cost of generating electricity is much higher from nuclear plants than from coal plants.

Biomass **Biomass fuel** derives mostly from burning wood but includes the processing of other plant material and animal waste. Energy is generated by either burning directly or converting substances to charcoal, alcohol, or methane gas. Biomass provides most of the

world's energy consumption for home heating and cooking, especially in the developing world. In China, some individual homes have fermentation tanks that convert waste to methane, which is used for cooking and heating.

Forms of biomass, such as sugarcane, corn, and soybeans, can be processed into motor vehicle fuels; Brazil in particular makes extensive use of biomass to fuel cars and trucks. When the price of oil rose rapidly in 2007 and 2008, the United States invested heavily in the production of biofuels using corn or other crops as the feedstock. Dozens of new ethanol-producing plants were built, spurred by substantial government subsidies. Between 2002 and 2009, U.S. ethanol production more than quadrupled. However, it takes a significant amount of land to produce the feedstock for biofuels, and a substantial increase in biofuel production would require much more agricultural land than is currently available. Also, the amount of fuel energy that is produced to manufacture ethanol from corn or biodiesel from soybeans is small, after one accounts for the amount of fuel consumed for tractor fuel and other needs in producing the corn or soybeans. Some analyses even indicate that more fuel is used to grow the crops than is produced when the crops are converted to fuel. When energy prices dropped dramatically in late 2008 and the price of corn remained relatively high, ethanol lost its competitive edge against gasoline made from fossil fuels. By 2009, the boom in biofuel production had stopped. If oil prices rise again then biofuel will probably come back to life, but concerns will remain about the wisdom of using crops to produce fuels when the net energy production is small.

The carbon and energy balances of biofuel production are potentially more favorable when based on crops that do not displace food products. There has been much discussion of switchgrass as a source of biofuels. Switchgrass is a perennial grass native to North America that can produce high yields without annual plowing and planting and the energy it requires. However, the biomass it produces is not as easily converted to biofuel as is that in crops such as corn or soybeans, and economically efficient commercial-scale processes that would use switchgrass as a feedstock have not yet been developed.

Biomass-to-energy conversion is much more appealing when it also serves to dispose of wastes. Opportunities for converting waste to biofuels may grow. In Bangladesh, for example, in 2003, a new pilot plant began generating electricity using the waste from 5,000 poultry farms. The concentration of U.S. meat production in large feeding operations (see Chapter 8) means that manure is already concentrated and managed instead of being scattered over the ground in barnyards. This increases the potential for using it to produce commercially valuable fuels. If the price of fossil fuels remains high, as seems likely, these new forms of renewable

eo
, Gas, Gas

energy production are likely to grow in importance. Used fats and oils are being converted to biodiesel in the United States, but the quantities available are not sufficient to meet our enormous demand for motor fuels.

Hydroelectric power   Flowing water has been a source of mechanical power since before recorded history. In the past, water was used to turn a wheel that could operate machines capable of grinding grain, sawing timber, and pumping water. Since the early 1900s, the energy of moving water has been used primarily to generate electricity, called **hydroelectric power.** It supplies about one-fourth of the world's electricity, which is more than any other source except for coal.

To generate hydroelectric power, water must abruptly change height, as at a dam. The falling water turns turbines that power electrical generators. A hydroelectric plant produces clean, inexpensive electricity, and a reservoir behind the dam can be used for flood control, drinking water, irrigation, and recreation.

Canada, China, Brazil, and the United States are the largest producers of hydroelectric power in the world. Together, North America and Europe generate about half the world's hydroelectric power. Hydropower supplies about 6 percent of total commercial energy production worldwide, but in the United States, it supplies less than 3 percent and in Canada 9 percent of commercial energy consumption. Most of the best sites for hydroelectric generation are already in use in the United States and Europe, but in many areas, there is considerable undeveloped potential. China, Brazil, Indonesia, Canada, and the Congo have especially great hydroelectric potential.

Opposition to construction of big dams and reservoirs is strong among environmentalists who fear the environmental damage they cause, such as loss of farmland or animal habitat. Hydroelectric dams may flood otherwise usable land and displace the people who lived on it. A dam converts a free-flowing stream to a lake, thus altering aquatic life. Many good sites for generating hydroelectric power remain in the world, but political considerations restrict their use, especially if the river flows through more than one country. For example, Turkey's recently built dam on the Euphrates River was strongly opposed by Syria and Iraq, through which the river also passes. They argue that the dam diverted too much water from the river and increased its salinity.

Despite these problems, hydroelectric power remains an attractive alternative to fossil- or nuclear-fueled power plants, largely because it generates no pollution. In rapidly industrializing countries such as China, increased electricity generation can mean an increase in the standard of living not only in industrial cities but in small towns where electricity provides clean energy for indoor cooking and lighting and operates pumps that help provide clean drinking water. The case of the Three Gorges Dam, which has drawn much international criticism for its negative impacts on the environment and human settlements, does have the advantage of providing a clean alternative to dirty coal-generated electricity.

Solar energy   Energy ultimately derived directly from the Sun offers much potential for providing the world's energy needs in future centuries. The Sun is a nonpolluting and perpetual source of energy that can be used directly or indirectly in forms such as wind and biomass.

At present, solar energy is used in two principal ways: thermal energy and photovoltaic electricity production. Solar thermal energy is heat collected from sunshine. Collection may be achieved by designing buildings to capture the maximum amount of solar energy. Alternatively, special collectors may be placed near a building or on the roof to gather sunlight. The heat absorbed by these collectors is then carried in water or other liquids to the places where it is needed.

Photovoltaic electric production is a direct conversion of solar energy to electricity in **photovoltaic cells.** Each cell generates a small electric current, and banks of them wired together can produce a large amount of electricity. Solar-generated electricity is now used widely, not only where conventional power is unavailable but also as a replacement for some conventional electric power. As more photovoltaic cells are produced, and as their technology and efficiency are improved, the cost of solar power will decline. Photovoltaic cells are already competitive with conventional energy sources in many new residential and commercial installations.

Video
Smokeless in China

Solar energy can be generated either at a central power station or in individual homes. Many countries are wired for central distribution, so central generation by utilities makes sense. But solar power technology now makes feasible individual home systems. An installation costing several thousand dollars provides a solar energy system that provides virtually all heat and electricity for a single home. Because the high installation price is offset by low monthly operating costs, home-based solar energy is economical for consumers who remain in the same house for many years. Individual solar energy users do not face rising electric bills from utilities that pass on their cost of purchasing fossil fuels and constructing facilities. The United States, Israel, and Japan lead in solar use at home, mostly for heating water. Solar energy is likely to become more attractive as other energy sources become more expensive.

Wind energy   Wind generation of electricity is one of the fastest-growing renewable energy technologies today. Significant improvements in turbine efficiency in recent years, combined with a growing demand for clean electricity, are making commercial-scale wind projects much more attractive. Large-scale wind-generating facilities are popping up across the landscape in many areas (Figure 9-18). The output of wind-generated electricity in the United States tripled between 2003 and 2007, although it still amounts to less than 2 percent of

but they likely will diminish in importance relative to other resources, especially renewable alternatives.

At present, the emerging energy sources are less versatile than oil and gas and are likely to find only specialized uses. Solar thermal energy might be used to heat buildings, whereas photovoltaic cells can power small electrical appliances. Centralized electric power plants—coal, nuclear, or hydroelectric—will probably remain the major source of energy for heavy users, such as large factories and shopping malls.

The United States has several options for meeting its energy needs. It might burn more coal, perhaps installing improved devices to protect against air pollution. Nuclear energy might win new proponents. Conservation and renewable energy sources also offer much potential. But none of these alternatives to coal and oil will grow significantly without the incentives provided by much higher energy prices. We can provide those incentives through government policy, or we can wait for increasing global demand for oil to drive prices up, forcing us to find alternatives. In any case, oil is certain to become increasingly expensive and less attractive as the basis of our economy.

If history is a reliable guide, the mechanism that will drive this transition to new energy supplies is the market. If the price of oil rises significantly, alternative technologies that are currently uncompetitive would become attractive. As these new technologies are more extensively used, their prices will decrease, further encouraging a shift away from oil.

Conservation is equivalent to discovering a new energy source, and in all likelihood, the greatest source of new energy resources will be conservation. More efficient use of energy means that we can produce more goods, operate more motor vehicles, and heat more buildings with the same amount of energy. Conservation is one factor in the slow growth of energy consumption since the 1970s. Sluggish demand for petroleum contributed to the decline in oil prices in the 1980s; similarly, slow growth in electricity demand contributed to the demise of the nuclear power industry.

In mid-2008, as oil prices reached record levels, interest in energy conservation and alternative technologies grew dramatically, especially in the United States. However, the economic downturn that struck in late 2008, combined with a major decline in oil prices, shifted the focus of concern away from energy. Nonetheless, many of the measures developed to promote economic development have also encouraged conservation, such as programs that subsidized replacement of older vehicles with newer more efficient ones, programs subsidizing home renovations to conserve energy, and subsidies for the development of renewable energy sources. In the long run, however, the conversion to renewable energy will have to be driven by the relative costs of fossil fuels, primarily oil, and renewable energy.

Figure 9-18 A wind farm in Illinois. Production of wind energy has grown rapidly in much of the world but still accounts for only a minor part of total electricity production.

eo
ack Time

the country's total electricity production. Wind energy, economical and clean as it is, is not without controversy, however. Environmentalists are divided: Some favor wind generation because it is renewable and pollution free, while others oppose it on the grounds of its adverse visual impacts and its hazards to birds. For example, North Carolina has prohibited construction of large wind turbines on its higher Appalachian ridges because of concerns about effects on the scenic beauty of the region. Nonetheless, the new growth in wind energy production signals a significant trend in the shift toward renewable energy.

Transition to new energy sources   The world contains a variety of energy sources. In addition to fossil fuels, people make use of hydroelectric, biomass, solar, and nuclear power. Other technologies are also in commercial use today, including geothermal, wind, and tidal power.

The emergence of new energy technologies suggests that we are beginning the transition from a fossil-based energy system to a new mix of energy supplies. Oil and gas are likely to remain important for several decades,

Air and Water Resources

In the preceding section, we considered renewable energy resources, which can be used year after year indefinitely. Renewable resources are limited in availability, however, and some renewable resources can be degraded in quality. Air and water resources are vital to humans; we need them for life and we also depend on them for waste removal. But in using air and water to carry away combustion products or animal wastes we degrade air and water quality. Burning fossil fuels pumps carbon dioxide into the atmosphere. The sulfur from burning oil and coal also enters the atmosphere and returns as acid deposition. Thus, the geography of resource consumption is also the geography of environmental pollution.

Air and water share two important properties: They are critical to human and other life on Earth, and they are useful places to deposit waste. An extreme environmentalist may argue that we should not discharge *any* waste into air and water, but in practice, we need to rely on air and water to remove and disperse *some* waste. Not all human actions harm the environment, and the air and water can accept some waste. For instance, when we wash chemicals into a river, the river may dilute them until their concentration is insignificant. **Pollution** (elevated levels of impurities in the environment caused by human activity) results when more waste is added than a resource can accommodate.

Pollution levels generally are greater where people are concentrated, especially in urban areas. When many people are clustered in a small area, the amount of waste they generate is more likely to exceed the air and water's capacity to accommodate it.

Air Pollution

The purity of air is paramount to life on Earth. Some air pollutants come from natural processes unrelated to human actions, such as dust, forest fire smoke, and volcanic discharges. Humans add to this by discharging into the atmosphere smoke and gas from burning fossil fuels, incinerators, evaporating solvents, and industrial processes.

The atmosphere is constantly stirred by temperature and pressure differences that mix vertically and horizontally. As it moves from one place to another, air carries with it various wastes. The more waste we discharge to the atmosphere, or the less the air circulates, the greater the concentration of pollution.

Average air at the surface contains about 78 percent nitrogen, 21 percent oxygen, and less than 1 percent argon. The remaining 0.04 percent of air's composition includes several *trace gases*. *Air pollution* is a human-caused concentration of trace substances at a greater level than occurs in average air. In addition to the carbon dioxide (CO_2) emitted by "clean" combustion of fossil fuels, the most common air pollutants include **carbon monoxide (CO), sulfur oxides** (SO_x, where the subscript x stands for the number of oxygen atoms), **nitrogen oxides** (NO_x), **hydrocarbons,** and **particulates** (very small particles of dust, ash, and other materials). Concentrations of these pollutants in the air can damage property and affect the health of people, other animals, and plants (Figure 9-19).

Each pollutant entering the atmosphere behaves differently. For example,

- SO_x and NO_x combine with water and fall to Earth as acid precipitation.
- *Particulates* in smoke are cleansed from the atmosphere quickly by gravity and precipitation.
- *Photochemical smog*—the product of *hydrocarbons,* as well as NO_x and sunlight—is created by chemical reactions that occur in the atmosphere itself.
- *Chlorofluorocarbons* (CFCs), chemicals used as refrigerants and in a variety of industrial applications, remain in the air long enough to be widely dispersed and carried into the upper atmosphere, where they damage Earth's protective ozone layer.

We are polluting our air in many ways today. We will focus on two important air pollution issues: acid deposition and urban air pollution.

Acid deposition   More commonly known as acid rain, **acid deposition** occurs when sulfur oxides and nitrogen oxides, produced mainly from burning fossil fuels, are discharged into the atmosphere. The sulfur and nitrogen oxides combine with oxygen and water in the atmosphere to produce sulfuric acid and nitric acid. When dissolved in water, the acids may fall as *acid precipitation*, or they may be deposited in dust. We use the term *acid deposition* to include both types of pollution.

Acid deposition seriously damages lakes and kills fish and plants. But the most severe damage of acid deposition is to the soil. The acid deposits harm the soil in a number of ways. Some of the acid deposited in soil is neutralized by calcium, magnesium, and other naturally present chemicals, but the amount of acid deposited can exceed the capacity of the soil's chemicals to neutralize the acid. If soil water grows too acidic, plant nutrients are leached away and are unavailable to plants. Acid deposition can dissolve aluminum in the soil, which can be toxic to plants and interfere with their nutrient uptake. Acids may also harm the soil-dwelling worms and insects that decompose organic matter.

Acid deposition is a regional problem, most severe in the densely populated industrial regions in Europe, eastern North America, and eastern Asia. It is a major source of damage to forests, especially in the industrialized and densely populated regions of the eastern United States, Central Europe, and eastern China, where high levels of SO_x discharges combine with

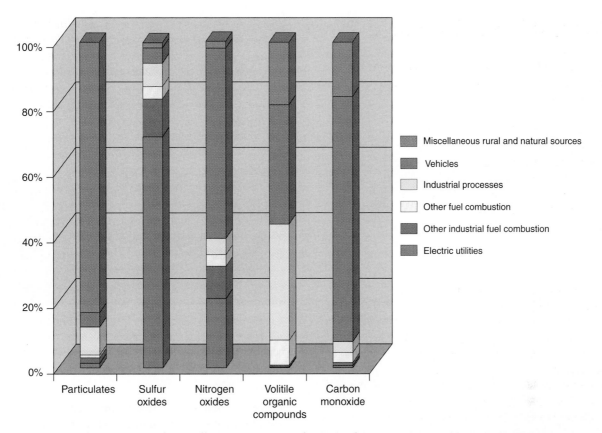

Figure 9-19 Major human-caused air pollution sources in the United States, 2003. Nationwide, particulates come mostly from soil erosion, but in urban areas they are derived from vehicles and industrial sources processes. Sulfur oxides are mostly derived from industrial processes. Vehicles, mostly cars and trucks, are the most important source of carbon monoxide as well as a major source of nitrogen oxides and hydrocarbons.

precipitation to form acids (Figure 9-20). Damage to individual forests varies widely, depending on its age, tree species, the capacity of the soil to neutralize acids, and interactions between trees and other organisms in the forest (Figure 9-21). Because the relationship between tree damage and high discharges of SO_x and NO_x has not been documented precisely, some governments are unwilling to impose the cost of controlling emissions on their industries and consumers.

Nonetheless, significant progress has been made. Since the early 1970s, the United States has reduced SO_x emissions about 47 percent. Over this same time period, emissions have been cut by larger percentages in other relatively wealthy developed countries, including 60 percent in Canada, 84 percent in France, 92 percent in Germany, and 92 percent in Sweden. NO_x emissions, which are more difficult than SO_x to control, have remained at about the same level in the United States during the past quarter century. Although precise figures are not available, SO_x emissions have probably increased in developing countries, especially China, which is responsible for nearly one-fourth of the world's coal combustion, the major source of SO_x emissions. China's sulfur emissions are a major source of acid deposition in Japan, again borne by prevailing westerly winds. In an attempt to reduce this problem, Japan is

helping pay for the installation of pollution-control equipment in China.

Urban air pollution Urban air pollution results when a large volume of emissions is discharged into a small area. The problem is aggravated in cities when the wind cannot disperse these pollutants. Urban air pollution has three basic components:

1. Proper burning in power plants and vehicles produces carbon dioxide (CO_2), but incomplete combustion produces carbon monoxide (CO). Breathing CO reduces the oxygen level in blood, impairing vision and alertness and threatening persons who have chronic respiratory problems.
2. Hydrocarbons also result from improper fuel combustion, as well as from evaporation of solvents as in paint. Hydrocarbons and NO_x in the presence of sunlight form **photochemical smog,** which causes respiratory problems and stinging in the eyes.
3. Particulates include dust and smoke particles. You can see particulates as a dark plume of smoke emitted from a smoke stack or a diesel truck—not a white plume, which is condensed water vapor. Many particles are too small to see, however.

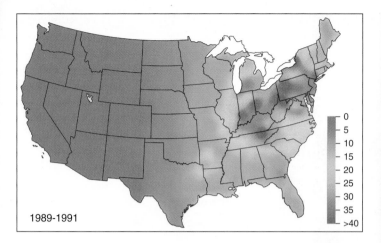

1989-1991

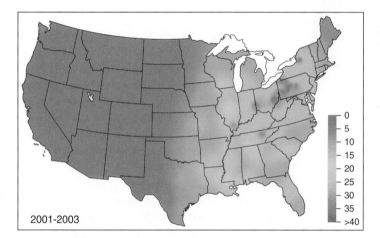

2001-2003

Figure 9-20 Acid deposition in North America. Acid deposition problems are caused by both nitrate and sulfate. These maps show wet sulfate deposition rates in the conterminous United States, in kg per hectare. The areas of North America most affected by acid deposition are in the Ohio Valley, eastern Great Lakes, and populated areas of the eastern United States. Acid deposition from sulfate, the larger component of acid deposition, has declined significantly as a result of emissions controls. Acid deposition from nitrate (not shown) has declined less, and has actually increased in some areas.

Three weather factors are critical to urban air pollution:

1. When the *wind* blows, it disperses pollutants. When the air is calm, pollutant concentrations build up.
2. Air *temperature* normally drops rapidly with increasing altitude. But over cities, conditions sometimes cause **temperature inversions,** in which warmer air lies above cooler air. This limits vertical circulation, trapping pollutants near the surface (Figure 9-22).
3. *Sunlight* is the catalyst for smog formation.

As a result of these three factors, the worst urban air pollution occurs under a stationary high-pressure cell,

Figure 9-21 Trees in the Black Forest, Germany, This forest, in the heart of Europe's industrial region, is suffering from decline probably caused by acid deposition.

where a combination of slight winds, temperature inversions formed by descending air, and clear skies allow pollutants to accumulate. A city that experiences frequent stationary highs, such as Denver, Colorado, has frequent pollution problems.

Mexico City is notorious for severe air pollution, especially in winter, when high pressure often dominates, and the surrounding mountains discourage dispersal of pollutants by wind (Figure 9-23). In the eastern United States, pollution problems are worst in summer and autumn, because stationary highs are most common then. In West Coast cities such as Los Angeles and San Francisco, the pollution "season" is also summer and autumn because inversions and bright sunshine are more persistent then.

Progress in controlling urban air pollution is mixed. In relatively wealthy, developed countries with strict regulations, air quality has improved. Controls on use of coal and improvements in automobile engines, manufacturing processes, and the generation of electricity have all contributed to higher-quality urban air.

To reduce auto emissions in the United States, for example, catalytic converters that oxidize unburned hydrocarbons have been attached to exhaust systems since the 1970s. As a result, carbon monoxide emissions have declined by more than three-fourths and nitrogen oxide and hydrocarbon emissions by more than 95 percent. Gains in relatively developed countries have been offset somewhat, though, by increased use of cars and trucks in recent years.

In many developing countries, urban air pollution is getting worse (Figure 9-24). Although ownership of motor vehicles is less common than in wealthy nations,

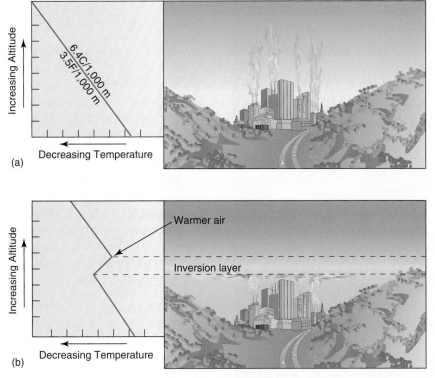

Figure 9-22 A normal temperature profile (a) and a temperature inversion (b) above a city. The temperature inversion traps pollutants.

the cars and trucks that do exist are older and lack pollution controls found on newer vehicles in relatively developed countries. Instead of relying on gas or electricity, many urban residents in developing countries burn wood, coal, and dung for cooking and heating. These smoky fires can create acute air pollution problems in poorly ventilated areas, even in the countryside. In developing countries, an estimated 4 million children die each year from acute respiratory problems, for the most part caused or aggravated by air pollution.

Water Resources

Water is our most immediate resource other than air—it is consumed daily and comprises about 70 percent of our bodies. Oceans occupy 71 percent of Earth's surface. From the sea we obtain fish, shellfish, oil, gas, sand, salt, and sulfur. The seafloor may someday yield manganese and cobalt and may become a burial site for nuclear waste. Countries with inadequate freshwater supplies, such as Saudi Arabia, desalinate seawater.

In the United States, about 1.5 billion cubic meters (400 billion gallons) of freshwater are pumped from the ground or from rivers and lakes each day. About 12 percent of the water is used in homes and businesses, 53 percent in industry (most of which is for electric power production), and 35 percent in agriculture. About two-thirds of the water we use is returned to rivers and lakes in liquid form, while the remaining one-third is evaporated, mostly from irrigated fields.

Water availability varies greatly around the world. South America has vast water resources (about 18 percent of all the freshwater runoff in the world flows to the sea via the Amazon River) and a relatively sparse population. Asia, on the other hand, is home to three-fifths of the world's people, and large parts of this vast continent are relatively dry, so per capita water availability is limited (Figure 9-25). Use of water for irrigation is especially significant in dry regions because the water is evaporated and not available for reuse (Figure 9-26).

Even in relatively well-watered parts of the world, use of water is reaching levels that strain available resources. As we saw in Chapter 4, many areas of humid climate have seasons in which precipitation does not meet the evapotranspiration needs of plants, let alone what humans remove from rivers or groundwater. Around the world, millions of reservoirs have been built to capture water in seasons with ample runoff and store it for use in other times of the year. However, not only are reservoirs expensive but they require displacement of people living where they are built, and they have negative effects on aquatic ecosystems. In some densely populated coastal regions of the northeastern United States and Western Europe, groundwater use is causing salty seawater to contaminate fresh groundwater and in some cases has made the groundwater unusable. And because we use some water to carry away our wastes, not all water in streams and lakes is available for irrigation, washing, or drinking. Thus providing water is an increasingly difficult problem nearly everywhere humans live. In Chapter 13 we will

Figure 9-23 Mexico City suffers from a combination of circumstances that all lead to significant air pollution problems. The city lies in a mountain basin that limits dispersion of pollutants. Automobiles are numerous and traffic is heavy, and there are few emission controls. In the early 1990s, the problem reached crisis proportions, and a pollution-control effort is now under way.

Figure 9-24 Satellite image of air pollution in north-eastern India and Bangladesh. The gray-colored haze seen in this image is common in winter, often remaining for weeks at a time.

discuss some of the approaches different societies are taking to meeting the water needs of human populations and natural ecosystems.

Water Pollution

Water is the "universal solvent." It can dissolve a wide range of substances, and it can transport bacteria, plants, fish, sediment, toxic chemicals, and trash of all kinds. As with air pollution, water pollution results when substances enter the water faster than they can be carried off, diluted, or decomposed. Water pollution is measured as the amount of waste being discharged in relation to a body of water's ability to handle the waste.

Pollutants have diverse sources. Some come from a *point source*—they enter a stream at a specific location, such as a wastewater discharge pipe. Others come from a *nonpoint source*—they come from a large diffuse area, as happens when organic matter or fertilizer washes from a field during a storm. Point-source pollutants are usually smaller in quantity and much easier to control. Nonpoint sources usually pollute in greater quantities and are much harder to control.

The *concentration* of pollution, such as sewage effluent, usually declines downstream from where the waste is discharged. This reduction occurs because the waste is diluted, and natural processes decompose pollutants and remove them from the water (Figure 9-27).

Reduction of oxygen in water Because aquatic plants and animals need oxygen, the **dissolved oxygen** concentration in water indicates the health of a stream and lake. The oxygen consumed in decomposing organic waste constitutes the **biochemical oxygen demand.** If a lake or stream contains an excessive amount of decomposing waste, the oxygen demand is too great and the water becomes oxygen starved, killing fish and other animals living in the water.

This condition often occurs when water becomes loaded with municipal sewage or industrial waste. The sewage and industrial pollutants consume so much oxygen that the water can become unlivable for normal plants, animals, and even fish. Similarly, when runoff carries fertilizer from a farm field into a stream or lake, the fertilizer nourishes excessive production of algae, which consume oxygen when they decompose.

Agriculture is the leading consumer of water, primarily for irrigation, and it is also the leading water polluter. Runoff from agricultural land discharges sediment, fertilizers, animal waste, and small quantities of chemicals into nearby streams. The fertilizer and animal waste, which are rich in nitrogen compounds, can nourish aquatic plants to the point of overproduction. Decomposition of these plants consumes so much oxygen from the water that fish may be unable to survive (Figure 9-28).

Wastewater and disease Wastewater comprises about 15 percent of the flow of U.S. rivers. Improved wastewater treatment is critical, especially in a world

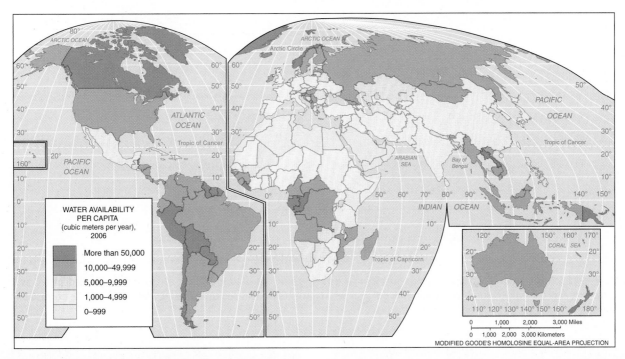

Figure 9-25 Per capita water availability. Water is most available in South America, Central Africa, Russia, North America, and Canada. These places have high amounts of runoff and relatively low populations. Major problems of inadequate water supplies occur in India, China, and other densely populated countries, as well as semiarid and arid regions.

Figure 9-26 The bed of the Yellow River (Huang He), exposed at low flow. Diversion of water from the river for irrigation has severely depleted its flow, and today the river barely reaches the sea for much of the year. Water is in critically short supply in much of northern China.

with a rapidly growing human population. Rich, industrial countries like the United States generate more wastewater than do poorer nations, but they also have greater capacity to treat this wastewater. In rich countries, strict legislation requires treatment facilities to be upgraded, making their rivers cleaner than a few decades ago. However, many of the ambitious water-quality goals of this country's Clean Water Act, which was first passed in 1972, have not yet been met.

In the developing world, untreated sewage often goes directly into rivers that also supply drinking water. A combination of poor general sanitation, nutrition, and medical care can make the drinking of water deadly. Waterborne diseases, such as cholera, typhoid, and dysentery, are major causes of death in developing countries. Because of improper sanitation, millions of people in Asia, Africa, and South America die each year from diarrhea. As people in these rapidly growing regions crowd into urban areas, drinking water becomes less safe, and waterborne pathogens flourish (Figure 9-29).

Chemical and toxic pollutants Any waste discharged onto or into the ground may pollute streams or groundwater. Pesticides applied to lawns and golf courses find their way into streams and groundwater. Landfills and underground tanks at gasoline stations can leak pollutants into groundwater, contaminating nearby wells, soil, and streams. Petroleum spilled from ocean tankers contaminates seawater. During times of flooding, normally secure tanks of chemicals may be dislodged and broken open and their contents mixed with the floodwaters.

Toxic substances are chemicals that are harmful even in very low concentrations. Contact with them may cause mutations, cancer, chronic ailments, and even immediate death. Major toxic substances include PCB oils from electrical equipment, cyanides, strong solvents, acids, caustics, and heavy metals such as mercury, cadmium, and zinc.

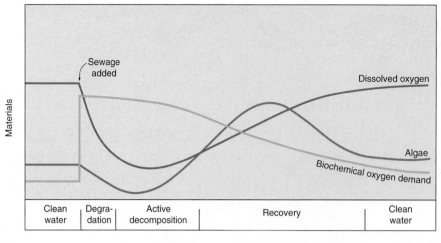

Distance Downstream ⟶

Figure 9-27 Water quality varies downstream from a point where sewage is added. Dissolved oxygen is depleted as organic matter is consumed, but it recovers further downstream. Nutrients are released to the water, and algae increase downstream as a result.

Figure 9-28 Water quality variations in lakes in the Minneapolis area revealed by remote sensing. This image was created using the relation between light reflectance from water and concentrations of nutrients and sediments. Lakes that are blue have generally good water quality, green and yellow lakes are intermediate in quality, and those that are red have poor water quality as indicated by nutrients and sediments. This map was created by a combination of GIS and remote sensing technologies.

During the 1950s and 1960s, toxic wastes were often buried, but by the 1970s, this method of isolating toxic wastes had proved inadequate because many waste sites were leaking. One of the most notorious was Love Canal, near Niagara Falls, New York, where several hundred families were exposed to chemicals released from a waste disposal site used by a chemical

Figure 9-29 Women washing and collecting water at a river in India. Less than 25 percent of the wastewater generated in India's 12 largest cities is collected by sewers. Most smaller cities have no wastewater treatment at all. It is estimated that between 5 and 10 million persons die each year from diarrhea caused by waterborne organisms in Asia, Africa, and Latin America.

company. The chemicals had been left by the Hooker Chemicals and Plastic Company, which had buried toxic wastes in metal drums in the 1930s. In 1953, a school and several hundred homes were built on the waste site. Eventually the metal drums were exposed, and beginning in 1976, residents noticed a strong stench and slime oozing from the drums. They began to suffer from high incidences of health problems, including liver ailments and nervous disorders. After four babies on the same block were born with birth defects, New York state officials relocated most of the families and began an expensive cleanup effort.

Love Canal is not unique; toxic wastes have been improperly dumped at thousands of sites. It was, however, very significant in raising public concern about groundwater pollution and stimulating action to combat it. In the United States, such polluted sites are slowly being cleaned up, but the cost is tremendous. As safe and legally approved toxic-waste disposal sites become increasingly hard to find, some European and North American firms have tried to transport their waste to West Africa and other areas where pollution regulations are less strict.

The U.S. government has spent well over $100 billion in construction of new sewage-treatment plants since the mid-1960s. Industry has also spent heavily to meet water-quality standards. In general water quality in the U.S. today is dramatically better than it was a half century ago; many streams and lakes that had become open sewers now are suitable for recreation. But the job is unfinished. Serious efforts are made to control chemical waste in relatively developed nations, but developing nations often give little attention to this problem, and wastes are freely released into streams.

Reducing Air and Water Pollution

The most often used pollution-control strategy is to remove pollutants from water or exhaust gases before they are discharged to the environment. Control of air pollution in the United States and other industrialized nations usually means installing equipment that either reduces the amount of waste gases generated or removes wastes from the exhaust. For example, particulate emissions from coal combustion are often controlled with devices called electrostatic precipitators, which use the principle of electric charge to attract particles or other systems that remove ashes from the smoke. SO_x emissions are controlled both by burning fuels that contain less sulfur and by using devices that remove sulfur from exhaust gases after combustion. Similarly, automobiles are fitted with catalytic converters that reduce harmful emissions in the exhaust and with computer-controlled fuel-injection and ignition systems that generate less pollution by increasing the efficiency of combustion.

CONNECTIONS

Meat Production and Water Pollution

Farm animals such as cattle, pigs, or fowl produce considerable amounts of waste that can constitute a significant source of water pollution. Many rural streams carry substantial loads of nitrogen, phosphorus, and partially decomposed wastes. In rural areas, agriculture is usually the dominant source of water pollution, and feedlots (where animals are kept and fed) are the most obvious and concentrated pollution source in the rural landscape. As just about anyone who lives in a farming region in eastern North America can confirm, pigs are especially noticeable in this regard.

In the late 1980s, a radical transformation began in the U.S. pork production industry. Prior to that time, the usual pig production system was a form of mixed agriculture combining grain and meat production. A typical farm may have included perhaps 100–200 hectares (250–500 acres) of land on which grain—primarily corn—was grown. Most of this was fed to a herd of perhaps 100–200 pigs kept in a feedlot, preferably downwind from the farmstead. But a new form of pig production is growing rapidly and replacing the traditional system. Huge industrial-scale feeding operations are springing up across the central and eastern United States. The new farms, if they can be called that, have pig populations in the thousands, fed from grain purchased on the grain market and not necessarily locally produced. These farms tend to be concentrated in specific areas, often supplying a single centralized slaughterhouse. For example, a slaughterhouse in Guymon, Oklahoma, processes 2 million pigs per year, mostly drawn from a few surrounding counties. Large pig-feeding operations enjoy economies of scale

that allow them to fatten animals much more cheaply than can be done on a traditional farm.

How does this concentration of pig-raising affect water quality? In many areas where new industrial-scale operations have sprung up, the impacts have been severe locally. An adult pig produces significantly more waste than an adult human. A facility feeding 5,000 pigs produces more sewage than a small city. And while facilities for collecting and treating the wastes are required, standards for design and operation are usually not as stringent as they are for municipal sewage-treatment plants. Wastes are often stored in lagoons, and in a few instances, the dams holding these wastes have failed, turning the rivers downstream into open sewers. Pollution from commercial-scale pig and chicken farms has been blamed for outbreaks of a mysterious organism called pfiesteria that is killing fish in many streams in Maryland, Virginia, and North Carolina.

Concentration of animal production, however, does not necessarily reduce water quality. If wastes are produced at a few centralized locations instead of many scattered small farms, it becomes easier to collect and treat. It is also easier—legally and politically as well as technically—to regulate and inspect waste-treatment systems. If industrial-scale feedlots are carefully regulated, they might provide an opportunity to manage a form of pollution that otherwise has been largely unmanageable. We do not yet have sufficient information to know whether the net effect of concentrated feeding operations is positive or negative. It is possible that across the landscape it is positive, while at the same time in the vicinity of some facilities local water quality is suffering.

The largest source of water pollution—sewage—is also controlled by cleaning water before it is discharged. In a typical sewage-treatment plant, large solid particles are screened from the water or allowed to settle and the remaining water is oxygenated to allow bacteria to break down organic matter. The water is then discharged to the environment. In advanced systems, some of the products of that breakdown, such as nitrogen and phosphorus, are removed. Most of the world's population living in developed countries is served by some kind of sewage collection and treatment system. In developing countries, however, sewage systems are less common. In Latin America, for example, only about 80 percent of the urban population is served by sewage collection systems, in Asia, about 60 percent have such service, and in Africa the figure is only 53 percent. In most poor countries the sewage collected in urban areas is not treated, but simply piped to rivers or the sea. In rural areas of developing countries, sanitation systems are uncommon.

Instead of using these so-called "end-of-pipe" approaches that remove pollutants from air or water *after* they are produced, some pollution can be controlled by simply not producing it in the first place. This is known as **pollution prevention,** as opposed to *pollution control*. Many opportunities exist for reducing or eliminating industrial pollution without sacrificing the quality of products or increasing manufacturers' costs. For example, toxic cleaning products are used in manufacturing to remove oil, grease, soldering residues, dust, coatings, and metal fragments. Trichloroethylene (TCE) is an excellent solvent, but it is also very toxic. Less toxic cleaning agents can be substituted, such as alcohol or detergents.

Another approach is to recycle polluting substances instead of discharging them into the environment. Some industrial processes, such as molding plastics and stamping sheet metal, generate recyclable scraps. Normally, larger scraps are recycled, but small fragments, such as dust from polishing or sanding, instead may be placed in landfills as solid waste or discharged in water. Capturing and recycling these fragments has dual benefits: It reduces pollution while increasing the supply of scrap for making new materials. Health benefits also may accrue if the dust presents a health hazard. In this way, environmental externalities are eliminated without increasing producers' costs.

In the early 1990s, a fundamental shift in pollution management began, from pollution control to pollution prevention. This change was spurred by several factors:

1. Despite pollution controls, significant water- and air-quality problems remained.
2. Polluters faced legal liability if they discharged substances that were later found to cause harm, so reducing discharges to the environment reduces potential later liability.
3. Pollution prevention often is cheaper than pollution control.
4. Publicizing its efforts to reduce pollution can improve a company's image and help it market its products, recruit employees, and negotiate with government regulators.

Pollution prevention has quickly been adopted by many industries, especially in the relatively rich developed countries of Europe and North America. Large corporations—which are especially visible to governments, consumers, and the general public—have led the way in reducing discharges into the environment. Companies increasingly recognize that the best way to prevent pollution is to design a product that can be manufactured with a minimum of toxic chemicals, used without generating pollution, and can be recycled easily when it is no longer wanted.

While industrialized countries have been successful in reducing pollution during the past few decades, pollution problems are increasing in many poor countries. Most of the differences in pollution between rich and poor countries result from differences in wealth and level of industrial development. Where electricity, clean-burning fuels, and pollution-control devices are widely available—as in Europe, North America, and Japan—urban air pollution mostly has been controlled. Problems are much more acute in places where people cook and heat with wood and drive old vehicles.

As manufacturing expands in developing countries, new facilities could be built with pollution control in mind. But pollution controls can be costly. Reducing the threat of large-scale pollution in developing nations requires development of pollution-prevention methods that reduce—not increase—costs.

Forests

As the world population grows and resources are used for more purposes, conflicts over resource use are inevitable. In some cases, the conflict concerns *who* has access to a resource in short supply, such as water in a desert region. But, increasingly, conflicts concern *how* a resource should be used. Should a valley be dammed to generate hydroelectricity, or should people continue to live there and practice their traditional way of life? Should a wilderness area be protected as a habitat for endangered species, or should it be opened for forestry, mining, or recreational development?

One resource may have competing uses, some of which may be incompatible with others. Balancing competing and incompatible uses requires careful management, and difficult choices must be made. In this section, we will review some of the important uses of Earth's forestlands and the reasons why these lands have long been, and will continue to be, controversial.

About 4.1 billion hectares (10.1 billion acres), or roughly one-third of Earth's land surface (excluding

Antarctica), is covered with forest and woodland. The amount of forestland has decreased substantially as a result of clearing land for agriculture, and experts estimate that forests may originally have covered as much as half of Earth's land surface.

A forest is part of an ecosystem in which vegetation plays a major role in biogeochemical cycles. By absorbing carbon dioxide and releasing oxygen, a forest is an essential part of a local climate; it also stores carbon that otherwise would be in the atmosphere as carbon dioxide. A forest reduces erosion, aids in flood prevention, and provides a place for recreation. A forest is drained by rivers that provide drinking water, irrigation, habitat, and waste removal. People may use a forest as a habitat and a place to obtain food, fuel, shelter, and medicine.

Because forests cover such a large portion of Earth's surface, often bordering populated areas, they have become centers of natural resource conflicts. Whereas agricultural land is typically under the control of individuals, corporations, or local political authorities, forestland is more often communally owned. And because of this communal or semicommunal ownership, many different people may have some level of legitimate claim to forestland. This further intensifies resource-use conflicts. So as you think about forests, remember that conflicts occur over other multiple-use resources, especially ones such as water and oceans that are commonly owned.

Forests as Fiber Resources

Forests, the wildlife they support, and the water that flows through them are all renewable resources. Conflicts occur between those who wish to harvest timber and those who wish to leave the trees standing (Figure 9-30).

Figure 9-30 Environmentalists and forests. These huge karri logs were cut in Western Australia to be chipped for paper. Environmentalists decry this destruction of old-growth forest, particularly when paper can easily be made from much smaller logs.

The United States, Russia, China, India, and Brazil are major timber-harvesting regions because their citizens demand wood products (Figure 9-31). Only a relatively small percentage of the world's remaining forestland is protected from cutting. Nearly all forests in the United States (except those in Alaska) have been cut at least once during the past 300 years, so very little original (virgin) forest remains. In Canada, much more original forest remains, especially in the north and west. The timber industry wants to cut remaining original forests because their large, straight trees yield high-quality lumber (Figure 9-32). Harvesting provides jobs and benefits lumber companies by keeping the price of timber low. Restrictions on cutting original-growth forests would possibly increase unemployment in regions where these forests are being harvested.

Environmentalists argue that when trees are cut, the forest no longer supports the same wildlife or maintains clean water as effectively as it did before. Cutting down a forest places its inhabitants at risk. During the late 1980s, the spotted owl was endangered in the original-growth forests of Oregon and Washington when timber interests wanted to cut the habitat. Environmentalists point out that U.S. laws prohibit the government from actions that could harm an endangered species. U.S. government attempts to find a compromise between the timber industry and environmentalists has proved elusive, because the two uses of the original-growth forest resource—commercial timbering and as habitat for the spotted owl—ultimately are incompatible.

Sustained yield management is a strategy that can maintain the productivity of a resource even as it is being used. In forestry, sustained yield management means that the number of trees harvested should not exceed the number replaced by new growth. The strategy also emphasizes harvesting in a manner that minimizes soil erosion, and therefore enhances the ability of the forest to regenerate. Sustained yield management is the official policy of the U.S. Forest Service, which manages the national forests. In practice, however, the Forest Service has allowed timber to be cut faster than mature trees are produced in national forests. As a result, we have a shortage of harvestable, mature timber in key areas such as the Pacific Northwest.

Similar conflicts regarding forest use are taking place in developing countries. For example, during the 1980s, tropical deforestation became a global issue, primarily because the rate of deforestation in the Amazon Basin was increasing rapidly. Environmentalists argued that once the forest was cut, it could never be restored to its original condition. Species would become extinct, the soil would be ruined, and indigenous cultures would be wiped out. But some involved in clearing the Amazon forests responded that environmentalists in the United States and Europe had no right to criticize Brazilian forest management when most of the forestland in the United States and Western Europe had already been cleared, some of it many centuries ago. Clearing the forests of eastern North

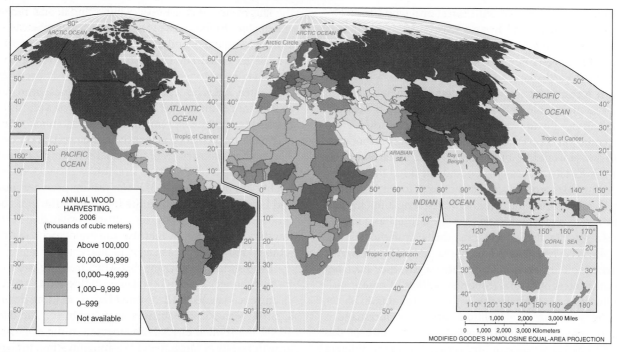

Figure 9-31 World timber harvesting. Most timber harvesting takes place in Asia, North America, and Brazil, and lesser amounts are harvested in Africa and Europe. Data from the former Soviet Union are lumped, with virtually all this harvesting occurring in Russia.

Figure 9-32 Deforestation in Java, Indonesia. Indonesia and Malaysia are the leading exporters of wood products in Asia.

America enabled agriculture that was key to the economic development of that region. Shouldn't other countries be able to develop their economies in the same way?

Ideally, the solution to conflicting uses of forests is sustainable management. Trees can be cut, but not to the extent that the future productivity of the forest is reduced. Some original forest should be protected to provide habitat for native species and prevent extinctions. Resource managers in northwestern North America and in the Amazon are working toward such compromise solutions.

Other Important Forest Uses

Recreation in forest areas   In many countries, especially wealthier ones, forests are important recreational resources. They are relatively undeveloped and thus have open space available for hiking, camping, and other outdoor activities. They offer a strong contrast to the noise, commotion, crowding, and pollution of cities. Their shade and water availability provide relief from summer heat. Many forest regions are located in mountainous areas that offer other amenities such as skiing, climbing, and the cooler weather of high elevations. And because they serve as habitats for game animals, they can be preferred areas for hunting and fishing.

In the United States and Canada, many important recreational centers are located in forested regions. Consider, for example, the coasts of Maine and Nova Scotia, the shores of the northern Great Lakes, or the mountains of the West. The most heavily used national parks, such as Acadia, Great Smoky, Yellowstone, Banff, and Yosemite, are all located in forest regions. Many of these parks are bordered by national forestlands, and tourists drawn to the parks also enjoy the forests outside of the parks. In fact, in the United States, recreational use of the national forests (measured in number of visitor hours) is more than double that of the national parks.

The tourists who use these forest areas appreciate them for their natural (or natural-like) landscapes: tall trees, clean rivers, wildlife. They are not interested in seeing swaths of land with only stumps, sharing the road with lumber trucks, or hearing the buzz of chain saws. Recreational use clearly conflicts with timber harvesting.

Forests and biodiversity   As we learned in Chapter 4, human occupation of Earth has resulted in significant threats to global biodiversity. Although damage to biodiversity is probably greater in agricultural lands than in forests, the remaining forests of the world contain some of the most important reserves of species diversity.

The tropical forests are especially important centers of diversity by supporting both larger numbers of tree species and complex vertical habitats. Foremost among these is the Amazon, a vast region of more than 4 million square kilometers (1.5 million square miles), containing literally millions of species. The Amazon has been a focus of attention in part because of the relatively high rates of deforestation taking place there. Since the 1980s, for example, the Brazilian Amazon has been cleared at rates of slightly less than 1 percent per year, an area roughly the size of the state of Connecticut. Worldwide, perhaps 40,000 square kilometers (15,000 square miles) of tropical rain forest have been cleared annually in recent years.

Concern over this loss is based partly on practical considerations, especially the potential medicinal uses of natural substances. Many important medicines are derived from forest species, and as mentioned in Chapter 6, indigenous peoples occupying these lands often have very good knowledge of such medicinal values. When the tropical forest is destroyed, these cultures may also be lost in the process, so both the species that provide medicines and the valuable source of information about them are at risk.

A moral argument in favor of biodiversity preservation is that humans, possessing unprecedented powers over nature and Earth, have a special responsibility to protect the natural environments and the species they contain from destruction by our own hands. This view holds that our great numbers and advanced technology do not bestow on us the right to eliminate other species from the planet.

Forests and carbon storage   Another long-term impact from cutting down forests may be the release of stored carbon into the atmosphere, through decay or burning of unusable parts of trees. Deforestation in the tropics may be responsible for as much as 10 percent of the increase in CO_2 in the atmosphere, as discussed in Chapter 2. Left alone, forest growth removes carbon from the atmosphere and stores it in the biomass of living trees and soil organic matter. We must therefore consider both deforestation and reforestation.

In much of the world, the rate of deforestation exceeds the rate of reforestation. Net deforestation is occurring, especially in the tropical rain forest, because people are cutting trees faster than they can be replanted or can regrow naturally. In some areas, cleared forests do revert to second-growth woodlands, while in others the land is converted to other uses more or less permanently. The timber industry does replant forests, especially in the midlatitudes. And in some areas, notably the eastern United States, forests now are growing in areas that were farmed or grazed until the early twentieth century, when these activities became no longer profitable. Forests can require more than a century to regrow, however, so these second-growth forests lack some important habitat properties of the old growth they replace. Nevertheless, as these forests grow, carbon is gradually removed from the atmosphere. The potential for removing carbon dioxide from the atmosphere in this way is considerable if forest growth is promoted worldwide. Some estimates indicate that as much as 15 percent of fossil-fuel emissions in the next half century could be trapped in biomass. Of course, this would preclude using these forests in some other ways.

Balancing Competing Interests

Political and economic relations are the key to any situation where competing interests battle for control of a scarce resource. Decisions on allocation of forests among various uses will be made both by governmental regulation and by the marketplace. In the case of government-owned lands in the United States and Canada, the political arena is where most decisions are made. The important interest groups include the lumber industry, environmentalists, recreational interests, and occasionally ranchers and the mining industry. Each of these has its own power base. Some may have strong allies in government, while others have based more mass support among voters. The fate of the remaining old-growth forest in the northwestern United States and southwestern Canada has been a particularly controversial issue because both of the principal competing interests—the lumber industry and environmentalists—are powerful groups with wide influence.

In market economies, the relative values that society places on different forest uses may be expressed through prices of resource commodities. If the price of lumber rises, there is more pressure to harvest timber and sell lumber. If cheaper substitutes are available for lumber, then the forest is more likely to be available for other uses. Similarly, if lumber from one region becomes expensive because of, perhaps, government restrictions on harvesting, then lumber from other regions will be demanded instead. Japanese demand for lumber leads to forest clearing in Southeast Asia and Alaska, while U.S. lumber interests argue that protecting old-growth forest there will only lead to a loss of business as foreign lumber is imported from Canada and elsewhere.

Government regulators sometimes make use of market-based principles, as well as political considerations, in deciding how to manage resources. Those who argue for restricting timber harvesting in favor of recreation and forest preservation, for example, note that the value of timber harvests in U.S. national forests in the early 1990s averaged about $1 billion per year. The national forests also provided about 300 million visitor days of recreation. If each of these visitor days were valued at $3, then the recreational value of the forests would be about the same as the value of timber harvested. Of course, both of these activities took place simultaneously in different parts of the forests. But as forest resources become increasingly scarce, such comparisons will likely lead managers to restrict logging further.

Similarly, other values may compete with lumber in economic terms. Since the Earth Summit in Rio de Janeiro in 1992, international negotiations aimed at reducing global carbon dioxide emissions have been under way. Industrialized nations consuming large amounts of fossil fuels are arguing that it would be cheaper to limit CO_2 releases by curtailing deforestation than by reducing fossil-fuel use. They are proposing to establish forest reserves in tropical countries as an alternative to reducing fossil-fuel use at home. In theory, this increases the value of those forests for carbon storage and reduces the supply of timber available for harvesting. We will discuss this further in Chapter 13.

Whether we consider forests, water and air resources, energy, minerals, farmland, or any other natural resource, it is clear that these resources are coming under increasing pressure worldwide, and the environmental impacts of resource use are increasing. As long as population grows and/or per capita resource use increases (and in general these trends have been ongoing for centuries), competition for increasingly scarce natural resources will increase.

In the early 1970s, an influential book called *The Limits to Growth*, by D. H. Meadows and others, argued that depletion of Earth's natural resources had placed us on a disastrous course. The report predicted that natural resource depletion, combined with population growth, would disrupt the world's ecosystems and economies and lead to mass starvation. If natural resource protection were not in place within 20 years, the report claimed, environmental systems would be permanently damaged, and everyone's standard of living would decline.

Few contemporary geographers accept the predictions made by *The Limits to Growth* a quarter century ago. Use of many resources has declined significantly since then. We may still deplete some resources, but substitutes and other strategies are available. Although pollution continues to degrade natural resources, industrial development has been made compatible with environmental protection in some locations. Nevertheless, many natural resource problems are likely to intensify considerably in the coming decades because of continued population growth and economic development. In addition, since *The Limits to Growth* was published, climate change has emerged as a central issue in global resource concerns, and some would argue that it is evidence we have already exceeded Earth's ability to support recent and future levels of resource use.

The developing countries face the greatest natural resource challenges. In places suffering from extreme poverty, sound management of resources for the future is difficult. Careful management of Earth's natural resources is even more difficult where population is increasing rapidly. The desire for economic growth, like that occurring in China, India, Brazil, and other rapidly developing nations, is a powerful incentive for increasing resource use and for valuing industrial uses of the environment above environmental protection. This growth has come at significant environmental costs, including air and water pollution. China, in particular, is beginning to recognize these costs and is taking significant steps to reduce pollution. China has also become a world leader in the production of renewable energy technology such as photovoltaic cells and wind turbines. Other developing regions, particularly sub-Saharan Africa, have yet to find ways to significantly raise standards of living and simultaneously protect the environment.

Chapter Review

Summary

A natural resource is an element of the physical environment that is useful to people. Cultural values determine how resources are used. Technological factors limit our use of some resources by determining the particular applications to which certain materials can be put. Economic factors such as resource prices and levels of affluence influence whether a resource is used, and how much. Renewable natural resources include air, water, soil, plants, and animals. Nonrenewable resources include fossil fuels and nonenergy minerals. Most resources are substitutable to some degree, so that if one resource is less available or more expensive another resource is available to take its place.

We depend on a great many different materials in our daily lives. Rich countries use large quantities of resources, causing depletion of some mineral resources. Mineral wastes and other materials accumulate in landfills. Recycling can help conserve landfill space as well as reduce resource use.

Over time, society has changed its use of energy resources from wood to coal, oil and gas. At present the world is dependent on fossil fuels for energy. The United States has abundant coal resources. Although the United States is a major oil producer, it must import most of its oil from other countries. Growing worldwide demand has caused large increases in energy prices. As fossil fuels are depleted, we will need to use energy more efficiently and develop other sources of energy. Nuclear power, renewable electricity, and energy conservation are promising new sources of energy.

Air pollution is a concentration of trace substances at a greater level than occurs in average air. Acid deposition and pollution of urban areas are particularly harmful forms of contemporary air pollution. Acid deposition is a regional problem that is most acute in areas that burn large amounts of coal, or have large numbers of automobiles and fossil-fuel-fired power plants. Air pollution is particularly severe in urban areas where there is a large concentration of pollution sources. Water pollution results from both point and nonpoint sources. Industrial facilities and municipal sewage plants are important point sources, while agricultural and urban runoff are significant nonpoint sources. Pollution prevention is a promising approach to reducing water pollution.

Forests are an example of a resource with many different uses, and conflicts over which uses are most important often arise. Among the important uses of forests are timber products such as lumber and paper, recreation, biodiversity preservation, and carbon storage. Sustained yield management is a strategy that attempts to balance the productive use of a resource while not depleting its supply.

Key Terms

acid deposition p. 360
biochemical oxygen demand p. 364
biomass fuel p. 357
carbon monoxide p. 360
cartel p. 344
dissolved oxygen p. 364
externality p. 342
fossil fuel p. 349

hydrocarbon p. 360
hydroelectric power p. 358
natural resource p. 340
nitrogen oxide (NO$_x$) p. 360
nonrenewable resource p. 343
particulate p. 360
peak oil p. 353
photochemical smog p. 361
photovoltaic cell p. 358
pollution p. 360

pollution prevention p. 368
potential resource p. 342
renewable resource p. 343
sanitary landfill p. 346
substitutability p. 343
sulfur oxide (SO$_x$) p. 360
supply and demand p. 342
sustained yield p. 369
temperature inversion p. 362
toxic substance p. 365

Questions for Review and Discussion

1. What is a resource? How do political–cultural, technologic, and economic factors determine whether substances in the environment become valuable resources?

2. What is a renewable resource? What is a nonrenewable resource? Give examples of each. Can any resources be considered either renewable or nonrenewable, depending on their use?

3. How has human use of coal, oil, and natural gas changed over the past 300 years? What changes are likely in the use of these fuels over the next few decades? What are some energy resources that are likely to be substituted for fossil fuels?

4. How do emission rates and rates of pollution dispersal interact to determine the severity of air and water pollution? What weather conditions aggravate air pollution?

5. How has water quality in U.S. rivers changed in the past few decades? What have been the major factors in this change? How does water pollution in wealthy nations differ from that in poor ones?

Thinking Geographically

1. Where does the drinking water come from where you live or attend school? What are the most significant potential pollution sources affecting your water supply? Your air supply?

2. If you drive, estimate the number of gallons of gasoline your automobile consumes in a year by dividing the number of miles you drive by your fuel efficiency in miles per gallon. If each gallon of gasoline contains about 2 kilograms of carbon, how many kilograms of carbon were emitted to the atmosphere by your automobile? Compare this with the number of hectares of growing forest it would take to store this carbon in biomass if a hectare of forest accumulates carbon at the rate of 750 kilograms per year.

3. Of the approximately 90 million new people added to the world's population each year, more than 80 million are in developing countries. But the per capita resource consumption of people in developing countries is a small fraction of that in rich nations. If population control is important for limiting global use of resources, should population control efforts be focused in developing nations or industrial nations? Why?

4. What are the most important issues of environmental quality where you live? Has environmental quality there improved or deteriorated in the past 30 years? What is your evidence?

PEARSON
mygeoscience place

Log in to www.mygeoscienceplace.com for videos, animations, **MapMaster**™ interactive maps, RSS feeds, case studies, and self-study quizzes to enhance your study of Earth's Resources and Environmental Protection.

MapMaster™

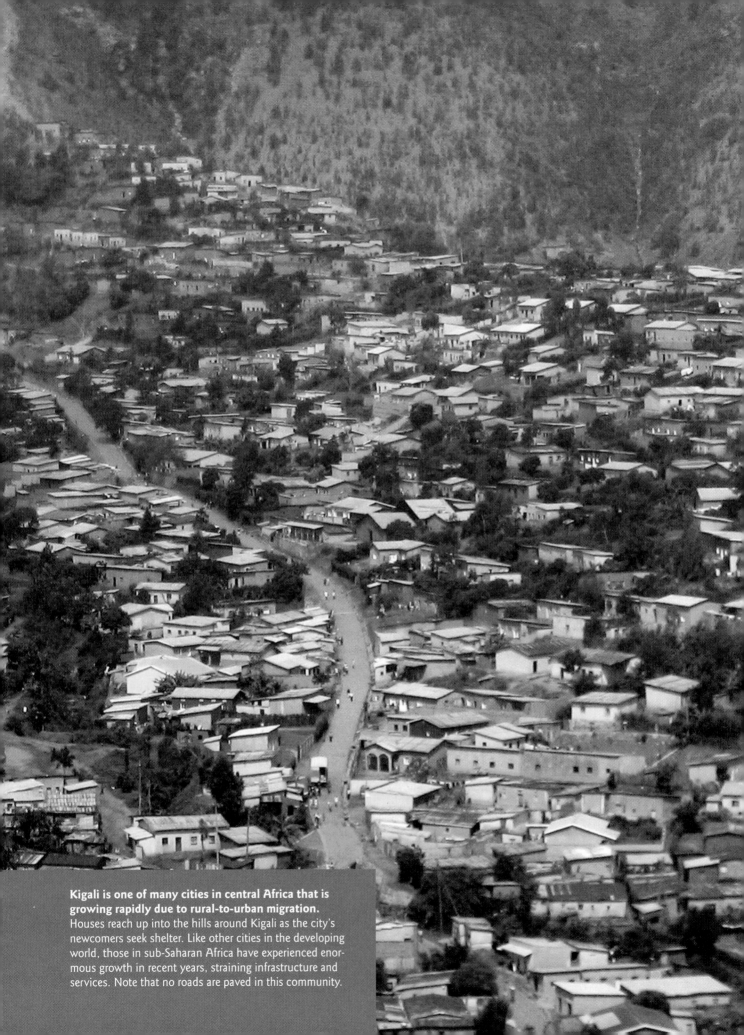

Kigali is one of many cities in central Africa that is growing rapidly due to rural-to-urban migration. Houses reach up into the hills around Kigali as the city's newcomers seek shelter. Like other cities in the developing world, those in sub-Saharan Africa have experienced enormous growth in recent years, straining infrastructure and services. Note that no roads are paved in this community.

10

Cities and Urbanization

A momentous day in human history occurred in 2007 when, for the first time, we became a predominantly urban species. Researchers estimated that day was May 23, 2007, when somewhere someone, like thousands of others that day, moved from countryside to city, tipping Earth's balance from a majority rural to majority urban population. We don't know the exact person, but we can make a reasonable guess based on current trends. From all that we know about urbanization today, the person who tipped the global balance toward cities did not have it easy.

The person was probably poor and moving with a family. Urban growth rates have been highest in developing countries, where urban populations have been growing at a rate of about 3 or 4 percent per year. The highest growth rates have been in sub-Saharan Africa, especially in central Africa. Rwanda's cities are growing at a rate of 7.3 percent. At this rate, its cities will double in size in less than 10 years, outpacing the ability for government to offer basic services to the largely poor newcomers.

Arriving poor in a city unprepared for new residents, the record-breaking person likely built a house out of scrap metal and wood that barely keeps out the rain and has no utilities. Their neighborhood has few or no public services, such as police, schools, or public transportation. This person's home may lie miles away from the city center, joining the 1 billion people who live in the world's slums on the outskirts of cities.

Finally, this person probably moved to the nearest city to escape violence and poverty in the countryside. Political unrest and violence are leading factors driving rapid urban growth, especially in sub-Saharan Africa. Most of the world's slums have been created during periods of civil war, when new arrivals far outpace affordable housing. Cities have become refuges for millions. Another factor driving rural-to-urban migration is conditions of poverty for farming communities. As we saw in Chapter 8, such problems are often caused by government policies and low prices set by global markets. Those who cannot make a living in the countryside seek employment in the city. The person who gave the world an urban majority in 2007 moved in search of a better life, if only a slight improvement over where he or she lived before. This person's family members will work hard to improve their condition, but much will depend on whether government policies improve economic opportunities or worsen them.

A Look Ahead

Urban Functions

Cities may have first developed for cultural reasons, but the economic reasons for their existence rose in importance over the centuries. Today, we can define basic and nonbasic sectors of urban economies and analyze their roles in the primary, secondary, and tertiary sectors of national economies.

The Locations of Cities

Some cities were founded to utilize advantageous sites, such as mines, while others exploit favorable situations, such as crossroads on transport routes.

World Urbanization

Urbanization is occurring everywhere, both because of natural increases of the urban populations and because of continuing migration of rural people to cities. Some governments are trying to regulate the growth of cities.

The Internal Geography of Cities

Any city's internal geography is defined by economic considerations, social considerations, and government actions interacting in the local culture.

Cities and Suburbs in the United States

U.S. metropolitan areas have witnessed explosive growth. Job opportunity and housing continue to grow at the periphery, but some central cities that were once hollowed out started growing again at the end of the twentieth century. The many local governments in metropolitan areas must devise ways of cooperating on common challenges.

Cities and the Environment

Cities are concentrations of resource use. The impacts of cities on the environment are felt well beyond their built-up areas. Reducing the urban ecological footprint is a goal for new development and a challenge for existing cities.

375

A **city** is a concentrated nonagricultural human settlement. Every settled society builds cities, because some essential functions of society are most conveniently performed at a location that is central to the surrounding countryside. Cities provide a variety of services, including government, education, trade, manufacturing, wholesaling and retailing, transportation and communication, entertainment, business, and defense and religious services. The surrounding region to which any city provides services, and upon which it draws for its needs, is called its **hinterland.**

In ancient Egyptian hieroglyphics—the earliest writing we can read—the ideogram meaning "city" consists of a cross enclosed in a circle. The cross represents the convergence of roads that bring in and redistribute people, goods, and ideas. The circle around the hieroglyph denotes a moat or a wall. Few modern cities have walls, but cities do have legal boundaries, and within those boundaries, a degree of self-government is usually exercised. The process of defining a city territory and establishing a government is called **incorporation.**

Sometimes several cities grow and merge together into vast urban areas called *conurbations.* In the northeastern United States, for instance, one great conurbation stretches all the way from Boston to Washington, D.C. This conurbation has been called *Megalopolis,* which is Greek for "great city." The world's largest urban areas are generally called *metropolises,* Greek for "mother cities" (Figure 10-1).

In several countries, one large city concentrates a high degree of the entire national population or of national political, intellectual, or economic life. These cities are called **primate cities.** Paris, for example, is the primate city of France, and Bangkok is the primate city of Thailand. Not all countries, however, have a primate city: The United States does not. Whether a country has a primate city depends on its national history and social and economic organization.

Today in all countries, urban populations are growing faster than rural populations. This process of concentrating populations in cities is called **urbanization.** The United Nations estimates that one-half of the world's population lived in urban areas by the year 2007, compared with 30 percent in 1950. Virtually all population growth expected in the coming decades will be concentrated in the urban areas of the world. The degree of urbanization, however, is not the same in all countries, primarily because the proportion of each country's population living in urban areas varies. But so, too, does the definition of "urban area" vary by country, ranging from settlements as small as 200 persons to as large as 30,000. Most countries use a minimum of 2,000 to 5,000 people (Figure 10-2). In the United States, the minimum threshold population is 2,500; in Canada, an urban population is defined as that population living in incorporated places of 1,000 or more and at densities of over 1,000 per square kilometer (2,590 per square mile). More than half of the

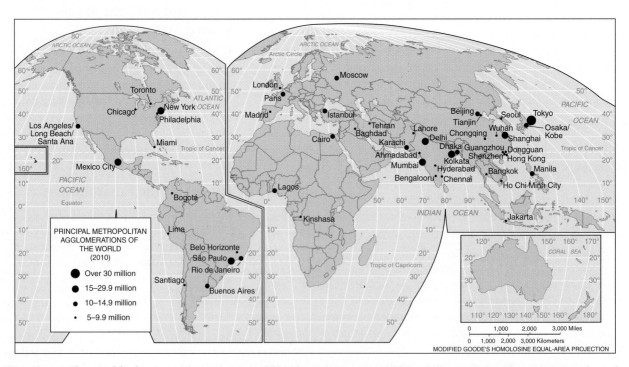

Figure 10-1 The world's largest cities. This map shows the 50 largest metropolitan agglomerations. The percentage of people living in urban areas is higher in already developed countries, but most of the world's large cities are in developing countries. The rapid growth of cities in these countries reflects both increases in overall national populations and migration into these cities from rural areas.

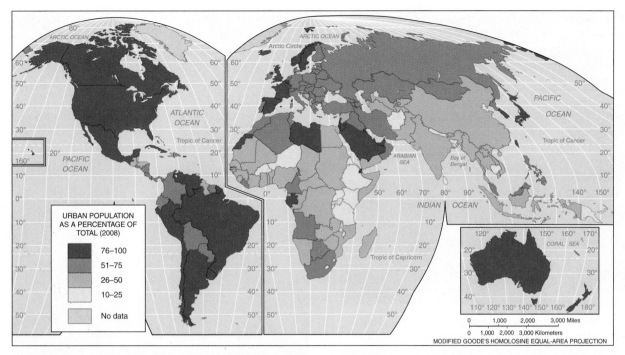

Figure 10-2 World urbanization. This map shows the percentage of each country's population living in urban areas, as defined by that country. Different countries use different definitions of "urban area," ranging from settlements as small as 200 persons to as large as 30,000, but most countries use a minimum of 2,000 to 5,000 people. In the United States, the minimum threshold population is 2,500; in Canada, an urban population is defined as that population living in incorporated places of 1,000 or more and at densities of over 1,000 per square kilometer (2,590 per square mile).

world's urban population live in cities smaller than 500,000. Only 8 percent live in **megacities** such as Istanbul, Kolkata, and Los Angeles that have populations of 10 million or more. One of the purposes of this chapter is to explain these patterns.

The geographic study of cities, **urban geography,** considers three topics:

1. The functions of cities and their economic role in organizing territory
2. Urbanization as it occurred in the past and as it is continuing in different countries today
3. The internal geography of cities—that is, the internal distribution of housing, industry, commerce, and other aspects of urban life across different cultures

This chapter will review each of these topics and then examine the growth and internal geography of U.S. metropolitan areas. This chapter concludes with a consideration of cities, natural resources, and the environment. This examination will illustrate many of the principles of urbanization in situations that will probably be familiar to you.

Urban Functions

The first cities appeared in today's Turkey and Iraq around 4000 B.C. (Table 10-1). Archaeological evidence suggests that settlements probably originated for cul-

TABLE 10-1 Historical Comparison of Select Cities		
City	**Year**	**Approximate Size**
Uruk, Mesopotamia (Iraq)	3700 B.C.E.	25,000
Memphis, Egypt	3100	30,000
Ur, Babylonia (Iraq)	2030	65,000
Babylon, Babylonia (Iraq)	612	200,000
Pataliputra (Patna), India	300	400,000
Changan (Xi'an), China	195	400,000
Rome	100 C.E.	450,000
Constantinople (Istanbul), Turkey	500	400,000
Changan (Xi'an), China	622	400,000
Baghdad, Iraq	775	700,000
Cordova, Spain	1000	450,000
Beijing, China	1500	672,000
New York	1925	7,774,000
Tokyo	1975	23,000,000

tural reasons rather than economic ones. The first permanent settlements may have started as places to bury the dead, or fixed sites for priests to perform ceremonies. Cities came to be embellished as centers of worship or even as the seats of the gods themselves (Figure 10-3). Many religions teach that the largest house of worship should be the tallest building in a city, and when commercial buildings first overtopped church spires in European and American cities in the

Figure 10-3 The Todaiji Temple in Nara, Japan. This temple is the world's largest wooden structure. It houses a colossal statue of Buddha 16 meters (53 feet) high. Both temple and statue were originally made in the eighth century, when Nara was Japan's capital city. Today, Nara is not a large city nor important in Japanese economic life, but much of it has been preserved as a treasure of Japanese history and culture.

Figure 10-4 Trinity Church at the head of Wall Street in New York City. Trinity Church's 86 meter (281 foot) high spire, erected in 1839, was for many years the highest point in New York City. Today, the spire is dwarfed by the skyscrapers of New York's financial district.

late nineteenth century, many observers found it symbolic of an unfortunate change in society's values (Figure 10-4).

The earliest settlements also may have served as places to house women and children while the men traveled in search of food. Household objects made by women, such as pots, tools, and clothing, provided a basis for the creation and transmission of a group's values and heritage. Today, settlements contain society's schools, libraries, museums, and archives—the repositories of knowledge and the vehicles for passing it to future generations.

Early settlements protected groups' land claims and food sources. Palaces arose to house the group's political leaders, and soldiers were permanently stationed there. Many settlements were surrounded by defensive walls. Long after the introduction of artillery, walls could still hold off an attacker until help arrived or the attacker ran out of food, so cities still built them. Paris, for example, surrounded the city with new fortifications as recently as the 1840s and did not completely remove them until 1932 (Figure 10-5). Today, cities are still the focus of military and political activities, but few retain walls except for historic interest (Figure 10-6). Most city walls were replaced in the nineteenth century by parks or grand boulevards.

The economic role of settlements may have begun simply as warehousing centers to store food, but as societies develop, production and trade join services as most cities' paramount activities. Cities bring people and activities together in one place for greater convenience.

This is called **agglomeration.** Agglomeration promotes the convenient *division of labor*, which is the separation of work into distinct processes and the apportionment of work among different individuals in order to increase productive efficiency. Craft workers flourished within city walls, and specialized occupations emerged, sustained by the peasantry of the hinterland. Settlements serve as convenient sites for trading, and local officials often regulated the terms of transactions, kept records, and created a currency system. Cities thus promote and administer the regional specialization of production throughout their hinterlands. With industrialization, cities become centers of production.

The Three Sectors of an Economy

Economic activities are generally divided into three sectors: primary, secondary, and tertiary. The names used for the three sectors indicate the degree to which each sector is removed from direct involvement with Earth's physical resources.

Workers in the **primary sector** extract resources directly from Earth. Most workers in this sector are in agriculture, but the sector also includes fishing, forestry, and

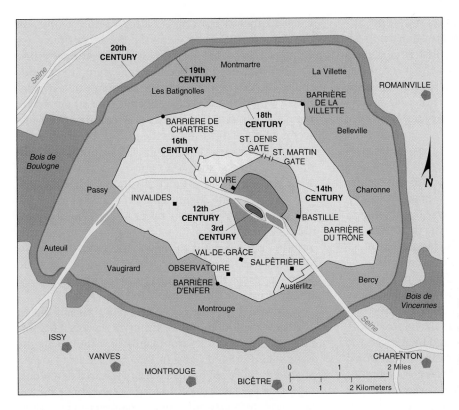

Figure 10-5 Concentric rings of growth. Paris periodically built concentric rings of protective walls to encompass new neighborhoods as the city expanded. The old gates of St. Denis and St. Martin have been preserved, but the walls have been torn down.

Figure 10-6 Quebec, Canada Two-hundred-year-old walls still surround Quebec, which stands above the St. Lawrence River and is the only walled city in North America.

mining. Workers in the **secondary sector** transform raw materials extracted by the primary sector into manufactured goods. Construction is included in this sector. All other jobs in an economy are within the **tertiary sector,** sometimes called the **service sector.** The tertiary sector includes a great range of occupations, from store clerk to

surgeon, from movie ticket seller to nuclear physicist, from dancer to political leader. This sector includes so many occupations that some scholars have tried to split it into a tertiary services sector and a *quaternary sector,* called an *information sector,* which includes advanced research and education. There is little consensus, however, as to how this sector is defined, and it is not regularly measured by statistical agencies.

As economies grow, the balance of employment and output shifts from the primary sector toward the secondary and tertiary sectors. Fewer people work on farms, and more people work in factories and offices. These activities normally require urban settings, and therefore, as an economy's secondary and tertiary sectors grow, its cities grow. This growth of the secondary and tertiary sectors is called *sectoral evolution,* and it will be examined in detail in Chapter 12.

The Economic Bases of Cities

Cities depend on their hinterlands for, at the very least, food. The cities must in turn provide services or "export" something to the outside. Many cities produce and export manufactured goods, but the exports of a city are not necessarily things that leave it. They may be things or services that people come to the city to buy. If people go to Houston for heart surgery, for example, then heart surgery is counted as an export of Houston. Vacations are an export of Miami Beach; gambling is an export of Las Vegas. In this economic sense, capital cities export government.

Some of the workers in a city produce the city's exports, but others serve the needs of the city's own residents. The part of a city's economy that is producing exports is called the **basic sector,** and that part of its economy serving the needs of the city itself is called the **nonbasic sector.** A city's basic and nonbasic sectors may be in the primary, secondary, or tertiary sectors of the economy as a whole.

Jobs in a city's basic sector create jobs in the city's nonbasic sector. For example, if a local factory makes a product that is sold around the world, the factory workers will spend their earnings by shopping locally, getting their hair cut locally, and purchasing other local goods and services. Each job in the basic sector actually supports several nonbasic sector jobs, because earnings from exports circulate and recirculate through the local economy. When a factory worker buys a shirt, the store clerk can get a haircut; the barber in turn might eat at a local restaurant, and so on. Thus, jobs in the basic sector have a **multiplier effect** on jobs in the nonbasic sector. This is why cities have long desired to attract basic sector companies, such as manufacturing firms, which spin off other basic and nonbasic jobs. A growing number of jobs also generates more tax revenues to pay for local government services, which in turn improves cities and attracts more employers.

Cities can be classified economically by examining each city's basic and nonbasic sectors and by comparing these sectors among different cities. New York, Seattle, and Los Angeles, for instance, each contains a number of dry cleaners and doctors. These people, for the most part, work within the nonbasic sectors. In terms of exports, however, workers in New York provide specialized financial services, workers in Seattle write computer software, and workers in Los Angeles make movies. A city's employment structure reveals its economic specialization. Any city that has an unusual concentration of workers in a specific job category must be exporting that product or service. A city that has a concentration of autoworkers, for example, presumably exports automobiles (Figure 10-7).

The Locations of Cities

Today, as in the past, the location of any city depends on a balance of site factors (characteristics of the place itself) and situation factors (its location relative to other places). Choice sites include defensive hilltops, oases, and the locations of mineral resources. Some mining towns virtually sit on top of valuable ore deposits but are otherwise isolated. At the time of the Industrial Revolution, cities often developed at waterfalls to exploit hydropower (Figure 10-8).

The locations of other cities more clearly result from advantageous geographic situations. Cities with the most convenient situations grow, which was illustrated with the example of three villages in Figure 6-22.

Cities frequently grow up at places where two different physical areas meet or along the border between two cultures. These are sometimes called **gateway cities.** Tombouctou has long been located along a key trade route, on the border between two physical environments (where the Niger River bends farthest north into the Sahara Desert) and also two cultures (nomadic Arabs to the North and settled black peoples to the South). Many cities spring up as transportation hubs or at bottlenecks, such as at a bridge across a river or where two political jurisdictions funnel trade through border checkpoints. Some cities grow at sites where the method of transportation necessarily changes. These are called *break-of-bulk* points. A seaport is an example. Another is the head of navigation of a river. If there is a waterfall on a navigable river, cargo has to be unloaded from ships and then reloaded beyond the waterfall up or downriver, or else be shifted to rail or truck.

Louisville, Kentucky, for example, was laid out in 1773 at the falls of the Ohio River. The river is navigable both up and downriver from Louisville. The opening of Louisville's Portland Canal in 1830 allowed ships to pass around the falls, but by then the city was well developed. It provided many services to its rich, developing hinterland and to westward-moving pioneers. The bridge across the Ohio River focused north-south traffic, and later the railroad lines also focused on the city.

If a situation is favorable, a great city may arise on an unfavorable site. The island city of Venice was originally located in a swampy lagoon because it provided a natural defense from invasion. This site also gave easy access to the Adriatic Sea that eventually allowed Venice to build a major commercial empire in the Mediterranean region. Today, it is a preeminent tourist attraction, yet Venice is sinking on the soft mud of the lagoon. Many of Asia's coastal cities were built by European merchants or conquerors at sites that provided access to the sea and that may have been defensible, but they sit on deltas or the swampy foreshores of tidal rivers. To this day, Karachi, Pakistan; Chennai and Kolkata, India (formerly called Madras and Calcutta); Colombo, Sri Lanka; Yangon, Myanmar; Bangkok, Thailand; Ho Chi Minh City, Vietnam; and Guangzhou, Shanghai, and Tianjin, China, all face formidable problems of drainage, water supply, construction, and health. New Orleans in the United States is a North American example.

One of human geography's great paradoxes is that one of the world's largest cities, Mexico City, is located on one of the world's most unfavorable places to build: a drained lakebed, in an earthquake zone, in a basin of interior drainage, at a high elevation in a dry climate. These conditions combine to cause physical instability, alternating flooding and lack of water, and air pollution. Both human lungs and internal combustion engines are inefficient when high altitude reduces oxygen levels by 23 percent, and air pollution is aggravated by local windstorms. Today's Mexico City, however, was the site of the Aztec capital Tenochtitlán at the time of

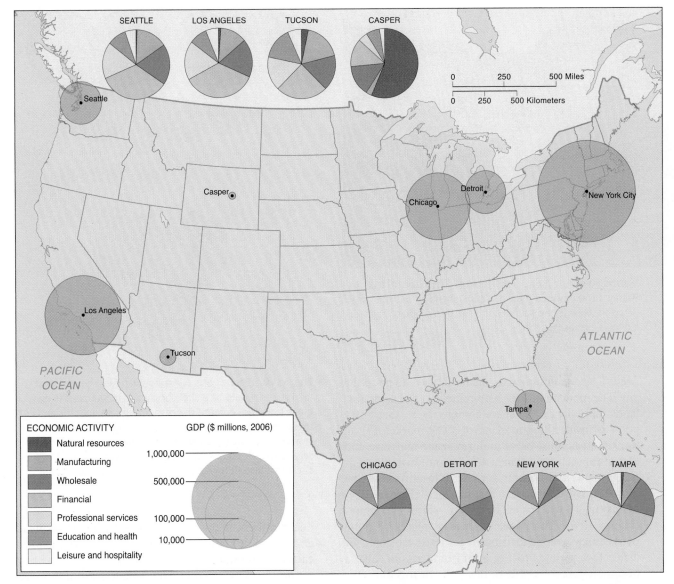

Figure 10-7 Economic activities in some U.S. cities. Urban economies vary in size and composition. The map shows a selection of large and small U.S. metropolitan areas and the value of their local economy (measured as gross domestic product; see Chapter 12). The pie charts provide a breakdown of each city's economy. Natural resources include agriculture, forestry, and mining, which are more important to cities that dominate agricultural and mineral-rich hinterlands. Manufacturing remains an important, though shrinking, component of many urban economies but tends to be larger in more populous cities. Wholesale and retail trade are key sectors in port or crossroad cities. Services are a growing part of all urban economies.

the arrival of the Spaniards. The Spaniards maintained the site as their capital, and so have modern Mexicans. Thus, this great city testifies to the power of history and geographic inertia.

Central Place Theory

The relationships between cities and their hinterlands have inspired a model of how cities are distributed across territory. Walter Christaller (1893–1969) began with the simplest imaginary landscape—an isotropic plain on which transportation cost is determined according to straight-line distance. Christaller then asked, "If cities are to serve as convenient centers for

exchange and other services across an isotropic plain, how will cities be distributed? What will be the pattern of towns and their hinterlands?"

To answer these questions, Christaller developed his **central place theory.** It is built on the idea that each city, town, village, or hamlet serves its hinterland as the central place to do business, defined as a market area. Central places of different sizes are also related to one another. The central places with the smallest market areas offer basic goods and services, but a city with its larger hinterland that includes smaller towns and villages offers more specialized goods and services. So a farmer might sell vegetables in the village, but would have to go to the town to buy animals or tools. If he

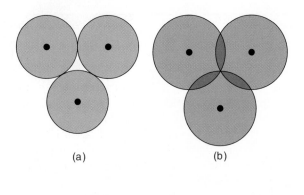

(a)   (b)

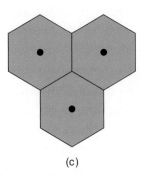

(c)

Figure 10-8 Paterson, New Jersey. These falls of the Passaic River powered the world's first planned industrial city, Paterson, New Jersey, founded in 1791 by a group of investors led by Alexander Hamilton.

Figure 10-9 Hexagons fill space efficiently. An ideal market area would be a circle because it minimizes the distance from the center to the edge. Circles do not efficiently fit together, however, because they leave gaps that are without service (a) or they overlap and some areas receive too much service (b). Hexagons are the geometric shape nearest to a circle that also fit together without gaps or overlap (c).

needs a lawyer or a new car, he might have to go to the city. The model has three requirements: (1) The hinterlands must divide the space completely, so that every point is inside the hinterland of some market; (2) all market areas must be of uniform shape and size; and (3) within each market area, the distance from the central place to the farthest peripheral location must be minimal. The only pattern that fills these three requirements is a pattern of hexagons—six-sided figures (Figure 10-9). Therefore, the central places of these market areas will be distributed across an isotropic, or featureless, plain as foci of a grid of hexagons. Like a bee's honeycomb, a grid of hexagons fit together efficiently without overlaps or gaps while minimizing the distance between center and edge.

Urban Hierarchies

Chapter 1 discussed diffusion up or down a hierarchy of cities. There are many small towns, fewer and more widely spaced medium-sized cities, and still fewer big cities. A hierarchy of cities exists because the more specialized a service or product is, the larger the number of potential customers that is needed for that product or service to be offered. No product or service can be offered without a minimum number of customers. This minimum demand is called the *threshold* for that product or service. Market area thresholds for inexpensive or basic goods are much smaller than market areas for expensive or rare goods. A coffee shop owner, for example, needs a minimum number of customers to earn a living. Most people periodically visit coffee shops, so the threshold of demand for a coffee shop is low. Each small town or city neighborhood can support at least

one. Similarly, many people frequently buy fresh bread and milk. Therefore, small groceries can be found in each neighborhood.

Few people, however, take tuba lessons or buy diamond bracelets. The threshold of demand sufficient to support tuba teachers or expensive jewelers is high, so only larger towns will be able to support these enterprises. On an isotropic plain on which people and buying power are equally spaced, coffee shops and grocers will be closely spaced; tuba teachers and jewelers will be widely spaced.

In real landscapes, other factors in addition to distance must be considered. Population densities vary and this affects the size of a market area. In places with high population density, grocery stores will be more numerous and located closer to each other than in low-density areas. Likewise, concentrations of wealth affect market area. If a place has a concentration of rich people, for example, then the density of jewelers will be greater there. This is true of any good or service: It takes many more poor people than wealthy people to meet the minimum threshold for a good or service.

A provider of a service can do one of two things to reach his or her necessary threshold of customers. One option is that the provider can be *itinerant* (travel from place to place). Alternatively, the provider can set up shop at one convenient place and wait for people to come. Convenience and accessibility are the principal purposes of cities. Agglomeration of services in cities saves travel costs, and it meets the thresholds necessary for more specialized goods and services.

The range of goods and services in a city offers small businesses the opportunity to rent pieces of equipment and to hire services or temporary employees only when they need them. This saves the businesses the cost of investing in equipment or hiring people full time. These available services or goods are called **external economies,** and their availability lowers initial costs for new business ventures. Thus, cities are incubators of new businesses. If a company grows large enough to justify buying its own equipment or hiring full-time employees, then the company has achieved **internal economies.**

The Patterns of Urban Hierarchies

Christaller's model of evenly spaced market towns with hexagonal hinterlands can be elaborated to represent an urban hierarchy of places. On an isotropic plain, the larger cities at the "top" of the urban hierarchy must be more widely spaced than the smaller market towns, which also must be evenly spaced, and their hinterlands must also be hexagonal. Thus, the hinterlands of villages are smaller than those of towns, and hamlet hinterlands are smaller than those of villages. Therefore, the distribution of the larger cities is represented by a grid of larger hexagons superimposed on several grids of smaller hexagons (Figure 10-10).

Many factors in the real world disrupt Christaller's model because the world is not an isotropic plain. The model has proved useful, however, in planning new cities in countries with unsettled lands. The land that the Dutch have claimed from the sea behind new dikes is similar to an isotropic plain. There, Dutch geographers have planned new market towns on hexagonal grids. Brazil has adapted the model to settle territories in Amazonia. The government built what it calls an "Agrovila" every 10 kilometers (6.2 miles), with a school, a health-care center, and a post office. These would be equivalent to the villages in the model. Every 40 kilometers (25 miles) the government placed an "Agropolis" that offers the services of an Agrovila, plus sawmills, stores, warehouses, banks, and other commercial services. Each 136 kilometers (85 miles) in Brazil a "Ruropolis," a city, was established for light industry. The ratio of 10, 40, and 136 kilometers cannot be reconciled with the proportions of a "pure" geometric grid, but they have been found useful by the Brazilian planners. The towns serve as foci for economic development. Venezuela has adopted a program similar to Brazil's.

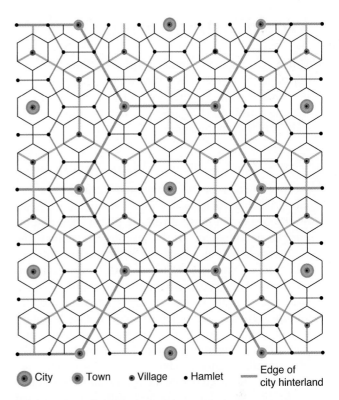

| ⊚ City | ⊚ Town | • Village | • Hamlet | —— Edge of city hinterland |

Figure 10-10 An urban hierarchy on an isotropic plain. The relationship between smaller and larger central places and their hinterlands can fit together in different ways. In the model shown here, four levels are shown. Notice that some smaller places are an equal distance from several larger central places. In this case, residents, have a choice of central places and may visit each for different needs.

When improvements in transportation allow people to travel farther to obtain services or goods, the smallest central places may lose their reason for existence and disappear. This is happening in the North American farming region from Kansas north into the Canadian prairies, where most small towns developed as commercial centers for farmers. They offered grocery stores, banks, hardware stores, farm implement dealers, automobile dealers, and feed stores. Technological advances, however, starting with the tractor, as well as advances in transportation and communication and other economic forces, have increased the size of farms but reduced the number of farmers. As a consequence, there is less need for the small-scale central places. Between 1960 and 1990 alone, over 70 percent of the 600 towns, villages, and hamlets in Canada's province of Saskatchewan lost the basic commercial functions necessary to sustain the communities. Farm areas are today punctuated by wholly abandoned towns, and the number of rural towns is expected to continue to decline. Small towns near metropolitan areas can survive or even prosper as urban workers seek to move there for more affordable housing or rural amenities. These towns become **exurbs,** the name given to settlements that make up the outermost ring of expanding metropolitan areas.

World Urbanization

Urban geographers compare urbanization as it occurred in the past and as it is continuing today. The rapidity of worldwide urbanization today presents many nations with both challenges and opportunities for the welfare of their populations.

The Rise of Modern Urban Societies

As noted earlier, cities have existed for millennia, but modern urban societies developed in cities at the heart of Europe's expanding networks of global trade and industrial development. In the seventeenth century, the Netherlands became the first modern urban country, with more than half its population living in towns and cities. The principal economic functions of Dutch cities were tertiary-sector functions. The cities arose as administrative and commercial foci of the Netherlands' global shipping, banking, and trading activity, and the urban populations could be supported by the Netherlands' highly productive agriculture. This urbanization occurred before the Industrial Revolution.

In Britain, a larger country with a more varied economy, urbanization occurred with industrialization. More than half of Britain's people lived in cities and towns by about 1900. Several developments over the previous 200 years had resulted in the concentration of Britain's population. These included the following:

1. Improvements in agricultural technology—part of the Agricultural Revolution—reduced the need for the number of agricultural workers. Landowners found it profitable to release employees and to evict tenants. The resulting rural depopulation is described in Oliver Goldsmith's poem "The Deserted Village" (1770), which laments that "rural mirth and manners are no more."

2. Displaced workers migrated to the cities. There, many were absorbed by the concurrent **labor-intensive activity** stage of the Industrial Revolution. An activity is labor intensive if it employs a high ratio of workers to the amount of capital invested in machinery. Other newcomers to the city found work in the tertiary sector. The largest class of urban workers was actually domestic servants—a tertiary-sector job. In 1910, they formed more than one-third of the total British labor force. A similar pattern could be found in the United States and other rich Western countries.

   Rural–urban dislocation caused appalling hardship. The history of urban life in nineteenth-century England records overcrowding in dreadful slums, malnutrition, starvation, crime, and early death. The descriptions of the miseries of England's slum populations in Charles Dickens's novels still haunt our imaginations. Cities have never easily absorbed all those who have flocked to them.

3. Population pressures were somewhat relieved by emigration, or else by the forcible exportation of criminals and debtors throughout the British Empire. The colonies of Georgia and Australia absorbed many of these deported people.

This British experience provided the world's first model for modern urbanization, and most of today's developed countries experienced similar histories. The growth of cities in Europe followed a similar path as industrialization and population growth fueled urban expansion. European cities grew in three ways. First, cities grew in absolute size. London grew from 676,000 in 1750 to 8.9 million in 1950. Rome increased from 157,000 to 1.7 million over the same period. They were growing because of the explosive European population growth during the middle stages of the demographic transition (described in Chapter 5). Second, there was also an increase in the number of cities with large populations. In 1750, there were two European cities with populations of 500,000 or more. In 1850, there were only three, but in 1950, there were more than 40 cities above a half million. Third, the European population has become more urbanized. Less than 3 percent of Europeans lived in urban areas in 1750. Today, about 71 percent of the European population lives in urban areas and 19 percent live in cities with populations of 750,000 or more.

The word *model* as used here does not mean that the British experience of urbanization was perfect or that it should be imitated. It imposed terrible hardship on millions of people. Britain provides a model only in the sense that Britain experienced these combined forces of industrialism, global commerce, and urbanization before anywhere else, so its experience can be compared with the current situation. In British history, the push of rural displacement and the pull of urban job opportunity were not coordinated. Governments have learned how to manipulate both forces in order to ease the process of urbanization, yet some aspects of contemporary urbanization still cause as much hardship as the experience did in Britain.

Urbanization Today

Today, urbanization is occurring in many places without economic development, especially in the world's poor countries. Burgeoning populations overwhelm the cities' ability to absorb the people and to put them to work. The incoming populations overload the cities' infrastructures of housing, education, internal transportation, water supply, and sewerage. By one estimate, 90 percent of sewage from urban areas in the developing world pours untreated into streams and oceans today. Reliable water, electricity, and telephone services are rare. From the tops of new skyscrapers in many modern cities, the view presents vast shantytowns of the desperately poor (Figure 10-11). Living conditions in these teeming cities are no better for the majority than those

eo
n Futures

Figure 10-11 La Paz, Bolivia. The tall buildings mark the city's downtown core. It occupies a steep canyon in which Spaniards founded this city in 1548 in order to avoid the chill winds of the plateau. The modern city has grown up and down the canyon to the east and west of the historic colonial core. Native American migrants to the city have settled up on the plateau, which is now a distinct city of over 1 million in population, called El Alto. The construction of housing and infrastructure has not been able to keep pace with El Alto's annual 10-percent population increase, producing vast slums.

that existed in Europe in the nineteenth century. The stresses of rapid urbanization in developing countries has been blamed for the breakdown of family life, recourse to drugs or religious extremism, and the spread of AIDS. The United Nations predicts that by 2030, about 60 percent of the world's population will live in cities, most of them in poor countries where infrastructure is inadequate for the population's demands.

The factors driving urbanization today are related to other geographic changes described in this book. Deteriorating conditions in the countryside often cause rapid urbanization. Global agricultural markets and misguided national policies (Chapter 8) have created large numbers of landless rural poor who move to cities. Civil wars and insurgent movements (Chapter 11) also make life difficult for rural communities in the developing world. Many internally displaced persons (Chapter 5) seek safety in urban areas. In many poor countries, rapid

urbanization is exacerbated by sharp cultural differences between rural and urban populations. Chapter 6 noted that many cities in today's poor countries evolved as outposts of international commerce grafted onto the local societies. Chapter 5 noted that urban populations may be cosmopolitan mixes of ethnic groups that are not native to the city and its immediate hinterland. Throughout Latin America, for example, the cities are predominantly white or mestizo, and measures intended to restrict urban migration are interpreted, often correctly, as racist discrimination against Native Americans. In East Africa, urban populations are often Asian, and in Southeast Asia urban populations are Chinese and Indian. These cultural and ethnic contrasts make it harder to deal with rapid urbanization.

In response to these obstacles, new arrivals to cities in the developing world often establish informal communities on open land at the edge of urban areas. These people build basic housing on land without clear title, meaning they do not legally own the land. So-called squatter settlements are often situated on marginal land such as steep hillsides and tidal basins. These settlements sometimes house migrants from the same rural area. Squatter settlements have different names around the world—for example, *gecekondu* in Turkey, *kampungs* in Kuala Lumpur, and *favelas* in Brazil. They lack public utilities and often pirate electricity from nearby powerlines. Over time, some squatter settlements are consolidated as homes are made more permanent and streets are filled in; a few are formally annexed by the city and receive public services. Some are rife with criminal activity while others provide affordable and functional housing to hard working families.

At least seven other circumstances differentiate urban growth today from that of the historic British experience:

1. The commercialization and mechanization of agriculture has accelerated, rapidly increasing the displacement of rural workers. For example, the introduction of mechanization to Brazilian sugar plantations reduced the labor force from 1.2 million in 1990 to fewer than 700,000 in 2000. Furthermore, modern agricultural technology requires capital investment, so the income gap between rich and poor widens.

2. National radio and television penetrate the countryside, spreading an image of the city as a place of opportunity. Many rural people feel pulled to the city.

3. In the past, disease and starvation among the poor kept urban death rates above urban birth rates. Without a steady influx of rural people, the town populations would have decreased. Today, however, the introduction of medical care and hygiene lowers urban death rates and triggers natural population increases. These factors compound the population growth that results from in-migration from the countryside.

4. Urban economies have changed in ways that make cities less able to employ the displaced rural population. Historically, urban economies offered entry-level job opportunities for the unskilled—jobs available for anyone with a strong back and a willingness to work. These jobs concentrated in domestic servitude, manufacturing, and construction. Today, however, domestic servitude has declined, and many of the jobs in manufacturing and construction demand skills. Manufacturing and construction are less labor intensive than they used to be; they have become increasingly **capital-intensive activities.** Machinery has replaced workers. Tertiary-sector jobs other than domestic servitude have multiplied, but most of these require job skills.

5. For most societies, the pressure-release valve of emigration has been stopped. Migration opportunities for the unskilled and for rural workers are decreasing.

6. The world economy is increasingly interdependent, so local governments have less control over local economies. Factory openings or closings in Bangkok are determined in Tokyo. Worker opportunity in Lima, Peru, is regulated from New York, and even job opportunity in New York is subject to international economic forces, as we will study in Chapter 12.

7. Many governments continue to favor urban projects over rural projects. One reason for this may be status. A shiny new hospital in the capital city, for example, may seem more impressive than a thousand new water pumps in poor villages. Another reason may be that the government fears urban rioting. Urban food riots have occurred in several African and Latin American countries. As noted in Chapter 8, many governments subsidize urban food supplies while discouraging their own farmers. These actions enhance the perceived opportunity in the cities without providing real opportunity.

Government Policies to Reduce the Pull of Urban Life

Governments could regulate the migration of people from the countrysides to the cities either by reducing the attractiveness of the cities or by improving the quality of rural life.

Forceful measures to limit urbanization are not new in human history. The Russians have been required to have internal passports since the days of Czar Peter the Great 300 years ago. Recently, the world has even witnessed brutal instances of compulsory "ruralization." When North Vietnam absorbed South Vietnam in 1976, the new rulers relocated millions of urbanites to rural areas to raise food using labor-intensive methods. The government argued that the cities had bloated on U.S. financial assistance. The government of Cambodia similarly relocated urban populations in the late 1970s.

Some governments use media to discourage migration to the city. Ghana, for instance, broadcasts to the villages films suggesting that life is better there than in the cities.

Many cities try to discourage newcomers by restricting housing and economic opportunity. They pass building codes, for example, that ban substandard housing. These codes are often ineffective in stopping the growth of slums, but they make squatters' settlements illegal. Therefore, the squatter-residents do not get city services and will not risk investing to improve the property. The city in turn cannot collect property taxes. In 2006, the United Nations Human Settlement Program estimated that these "illegal" communities, overwhelmingly slums, held 1 billion people—72 percent of the total populations in the cities of sub-Saharan Africa, 57 percent in Southern Asia, 35 percent in East Asia, 31 percent in Latin America, and some 25 percent elsewhere—70 to 95 percent of urban newcomers everywhere. Cities also try to restrict small businesses in residential areas, but backyard workshops may thrive anyway. Cities may discriminate against new urbanites by restricting education, housing permits, business licenses, or other job opportunities. Street vendors are chased from city centers, for example.

In some countries, frustrated city authorities have even bulldozed squatters' settlements after giving only one or two days of warning. This has happened in major cities throughout Latin America, Africa, and Asia, leaving hundreds of thousands of people homeless. In 1990, for example, bulldozers flanked by army troops, with air-force planes sweeping overhead, leveled the district of Maroko in Lagos, Nigeria, leaving about 300,000 people homeless. One survivor reported that the soldiers brought truckloads of coffins and said that if anyone was ready to die, the army was ready to bury him. In 2002, the government of India bulldozed shacks along the main airport road into central Kolkata so that World Bank delegates would not see slums on their way to a meeting in the city center. During 2005–2006, the Zimbabwean army virtually declared war on squatters and settlers in Harare, demolishing shanties and markets. Some 1.5 million people were scattered; over 10,000 poor people were rounded up and dispatched to rural areas. The urban poor were perceived as a threat to the 25-year tyrannical rule of the dictator Robert Mugabe, and the action was officially called "Operation Murambatsvina" ("Drive Out Rubbish").

Improving Rural Life

Instead of trying to drive people out of the cities, governments might invest in rural health care and education, housing, roads, and other infrastructure to raise

RAPID CHANGE

Urbanizing China

China had large urban populations long before modern Europe. These cities were small compared to the surrounding rural population and performed important administrative functions for a large empire. These early Chinese cities were spread across the interior and some grew to reach 1 million inhabitants by the eleventh century. In the nineteenth century, Britain forced China to allow foreign trade and with this came new coastal cities. "Treaty ports" were built along China's east coast and were dominated by European and American merchant interests. These coastal colonial cities grew in size and importance compared to the traditional inland cities, a reflection of the humiliating defeat of the Chinese empire.

It was not until after World War II, when the Communist Party created the People's Republic of China, that China's government encouraged industrialization in its cities, especially in the inland cities that were neglected during the colonial period. This led to rapid growth of urban populations, although China's population remained predominantly rural. Between 1949 and 1979, many Chinese cities doubled and tripled in size, and the urban share of China's population grew from about 10 to 19 percent. This changed in the late 1970s, when the communist government introduced a new economic policy focused on export industries. Emphasis then shifted again to coastal cities, including new towns such as Shenzen that were designated Special Economic Zones. These and other urban production centers became motors of economic development in China's recent growth. China's export economy soared during the last few decades and the percent of China's population living in cities more than doubled during this period.

China's cities would have grown even faster in recent years if the government had not limited migration from rural to urban areas. China has long enforced a household registration system under which people are not supposed to move without approval. China's authorities recognize that unrestrained urban growth will produce distinct social classes, zones of poverty, and potential unrest, all of which might weaken government rule. Despite the registration system, however, many people have moved to the cities. As many as 200 million Chinese peasants have drifted into the cities, where they are called *floaters*. Their residence in cities is technically illegal, so they have no right to education, medical care, and other government services. The floaters serve useful functions, such as bringing produce to the city or providing services not provided by central planning, but their continuing influx threatens to overburden China's urban infrastructure. The government has relaxed the registration system in order to create a unified national labor market, but the city governments themselves introduced new exclusionary techniques, such as granting residency permits only to those who can afford homes or to skilled workers. The wealth created in China's cities has produced a consumer economy for a growing middle-class that shops for western brand names and enjoys growing material comforts. And this attracts still more people from poorer rural districts.

Pearl River Tower in Guangzhou, China. This will be among the world's tallest buildings. Architects hope that it will produce more power from solar panels and wind turbines than the occupants will consume. Guangzhou is one of China's largest urban areas with a metropolitan population that will soon reach mega-city status, 10 million.

rural standards of living and keep people satisfied living in the countryside. This is usually addressed through programs to raise national food production. Rapid urban growth calls into question a country's agricultural policies that were discussed in Chapter 8. Mechanization and export-oriented cash crops tend to displace rural communities to cities. In 2006, the Chinese government significantly shifted its national budget toward rural roads, water and power supplies, schools, and hospitals.

A government can also organize labor-intensive investments in rural infrastructure (Figure 10-12). During the Depression in the 1930s, even the U.S. government Civilian Conservation Corps, Civil Works Administration, Public Works Administration, and Works Progress Administration all used labor-intensive methods to alleviate unemployment and at the same time build or repair national infrastructure.

The Economic Vitality of Cities

Most of the previous paragraphs about urbanization in developing countries may have seemed pessimistic. They presented urban growth as a problem for which solutions were needed. There is another side to the story, however: It is possible to view the backyard shops, the street hawkers, and the Chinese floaters as examples of opportunity and growth. These activities hint that urban migration, balanced between management and liberty, can provide a reservoir of vitality that can be harnessed for national growth.

The Peruvian economist Hernando De Soto, who co-chaired the U.N. Commission for the Legal

Empowerment of the Poor, has long noted that in many cities, the productive activities of a substantial share of the population do not appear in official accounts. The people may not have licenses to do what they are doing, they may be avoiding taxes, or for some other reason their activity escapes official notice. These activities make up the **informal, or underground, sector** of an economy. Every city in the world has such a sector, but the informal sector is particularly important in the cities of the poor countries. The International Labor Organization has estimated that informal employment is a full 72 percent of nonagricultural employment in sub-Saharan Africa, 65 percent in Asia, and 51 percent in Latin America.

Life in the extra-legal world is a constant risk. People lack title to property, so they build housing poorly; therefore, many die in earthquakes or other natural disasters. An estimated 4 billion people cannot create wealth or recuperate from disaster because of the lack of legal records. Neither capital nor credit will venture where there are no clear property rights, and seizure of property by the politically well-connected or powerful is a constant threat.

In an early study of Lima, Peru, De Soto found that the informal economy employed fully 60 percent of the population and produced 40 percent of all goods and services. The poor owned and controlled a public transportation network of private taxis and vans, plus land and housing worth billions of dollars. None of this, however, was legal, so it could not be taxed by the government or used as collateral by business owners. If the government simply legalized these activities, these assets would have liquidity and could provide collateral for investment and business enterprise. The government's refusal to recognize what was happening handicapped the country's economic growth and vitality.

In Peru from 1995 to 2001, more than 1.2 million households, including 6.3 million people, received title to the properties they were inhabiting. Title reform enabled more people to work outside the home and more family members to join the labor force, because now no one had to stay home to guard the property. The values of the newly registered properties have soared, and mortgage and consumer credit markets have developed. Studies in the Philippines found that 60 percent of Filipinos are holding real estate assets worth tens of billions of dollars outside the law, and thus illiquid. In Egypt 85 percent of the population lives in homes without property titles. In parts of West Africa, the figure is as low as 2 percent. De Soto estimated that at the start of the twenty-first century only 25 countries in the world had genuine contractual urban societies; the rest were informal.

A modern market economy cannot develop unless property rights are acknowledged and protected, and economic growth will probably occur most rapidly in the developing countries that ensure property rights. Therefore, De Soto has argued, governments must

Figure 10-12 Labor-intensive construction. Construction of even massive infrastructure projects such as this dam in India can begin with workers carrying baskets of dirt on their heads. In rich countries, however, labor-intensive construction methods such as this have yielded to capital-intensive methods.

formalize the spontaneous emergence of informal property. Chapter 8 already noted how productivity in agriculture can be raised by guaranteeing property rights. The same guarantees could develop in the cities. China has privatized landowning in the cities faster than in the countryside (see Chapter 8), and the privatization has contributed to faster economic growth in the cities. It created a middle class that is using its property as collateral to borrow money to launch enterprises. In both city and countryside, GIS and GPS greatly enhance the ability to record and register land holdings, so the spread of these techniques may spark economic growth and vitality around the world.

If governments view urban migration as a problem, they cannot see how urban immigrants' industriousness could be an asset for economic growth. Millions of people continue to choose to migrate to the cities, where they do survive or even thrive. There is something terrifically dynamic going on, and geographers, economists, and government officials at all levels are challenged to understand and measure it.

The Internal Geography of Cities

Urban geographers study not only the distribution of cities across the landscape but also their **urban form,** which is the distribution of land use and activities within cities. These distributions may be caused by economic forces, social factors, or deliberate actions of the government. In different countries and cultures, each of the three factors carries different relative weight.

Economic Forces

One of the first significant efforts to model, or explain, cities was carried out by urban geographers and sociologists in North America during the twentieth century. Many of these efforts were associated with a group of scholars known as the Chicago School, whose work began by looking at the complex interplay of economic and other forces in that city, especially among its diverse immigrant communities. Their models of land-use were primarily driven by economic forces. The first was E.W. Burgess's 1925 concentric zone model.

Figure 10-13 illustrates the *concentric zone model* of urban growth and land use. The core of the city, called the **central business district (CBD),** concentrates office buildings and retail shops. For businesses, accessibility is usually a principal determinant of a location's rent. The success of a department store, for instance, will depend partly on whether customers can reach the store easily. A city's most convenient and busiest intersections are most valuable for commerce. Landowners usually maximize the density of use on this valuable land by building up, so a traditional CBD is identifiable by tall buildings as well as crowded streets. Even within the CBD, clusters of functions appear. Lawyers' offices, for instance, cluster near courts or near the offices of their client firms. Retail stores of one type, such as jewelry stores, may cluster so that consumers can comparison shop. The CBD is surrounded by less intensive

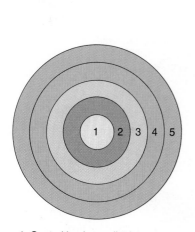

1 Central business district
2 Zone of transition
3 Zone of independent workers' homes
4 Zone of better residences
5 Commuters' zone

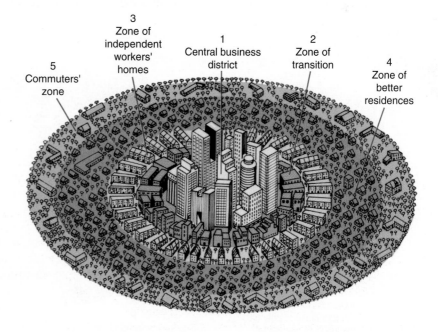

Figure 10-13 Concentric zone model of urban development. According to the concentric zone urban model, a city grows in a series of rings around the central business district.

business uses such as wholesaling, warehousing, and even light industry—that is, nonpolluting industries that require relatively small quantities of raw materials. Residential land use surrounds this urban core.

The concentric zone model can be modified by considering the effect of transportation routes, which affect accessibility. New means of transport—historically canals, then railroads and tramways—spread out radially from the heart of the city, although their paths are modified by topography. Industrial and residential growth take place in ribbons or fingers along these radial routes, and wedges of open land are usually left between these radial routes.

H. Hoyt proposed an alternative model in 1939, called the *sector model* (Figure 10-14). This model assumes that high-rent residential areas expand outward from the city center along new transportation routes such as streetcar and suburban commuter rail lines. Middle-income housing clusters around high-rent housing, and low-income housing lies adjacent to the areas of industry and associated transportation, such as freight railroad lines.

The third model, the *multiple-nuclei model*, shown in Figure 10-15, recognized the development of several nodes of growth within an expanding city area. It was proposed in 1945 by C. Harris and E. Ullman and drew from a combination of the concentric zone and sector models. The city's multiple nuclei may each concentrate on a different special function and each promotes further nearby development.

Harris updated this model to deal with sprawl that over the last half of the twentieth century had disseminated urban functions ever farther from downtown areas. His *peripheral model*, defined in 1997, shows how radial and circumferential highways continue to draw activities out of the central city and to disperse them around the region (Figure 10-16). The story of how this happens will be told shortly.

Social Factors in Residential Clustering

Social considerations play a role in urban residential clustering. The Chicago School models assumed that residential patterns were driven by economic opportunity. Economic growth in one neighborhood allowed residents to grow wealthier and move to better residential areas. As they left their old neighborhood, immigrants seeking affordable rent and entry-level work replaced them. This succession of residential groups was thought to explain the appearance of ethnic neighborhoods in large cities. Immigrants often live with people like themselves, which causes a clustering called **congregation.** Ethnic groups or immigrants of a common background, for example, may want certain services, such as grocery stores offering their traditional foods. Waves of immigration, as described in Chapter 5, have produced ethnic enclaves in many cities around the world, but these do not always last. As education and work opportunities open for the children of immigrants, they often move away, and these neighborhoods change in profile yet again.

In other cases, however, people live together because discrimination forces them to do so. These people suffer **segregation** from others. In any specific case, it may be difficult to determine the degree to which people are congregating or are victims of segregation. Chapter 7 noted that in the past, Jews were legally segregated in ghettoes, but today the word *ghetto* means any residential concentration of any one kind of people. The factors causing residential clustering of any group—of Chinese

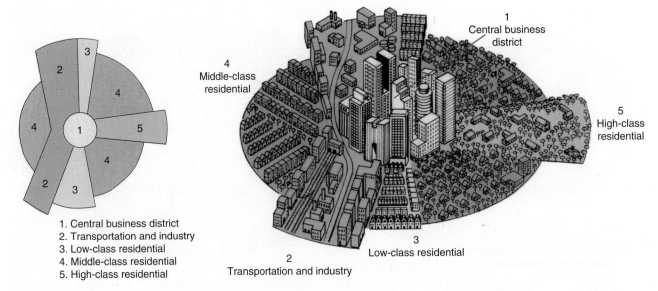

1. Central business district
2. Transportation and industry
3. Low-class residential
4. Middle-class residential
5. High-class residential

Figure 10-14 Sector model of urban development. In this model of urban form, a city grows out from the central business district in wedges, or corridors, of various land uses.

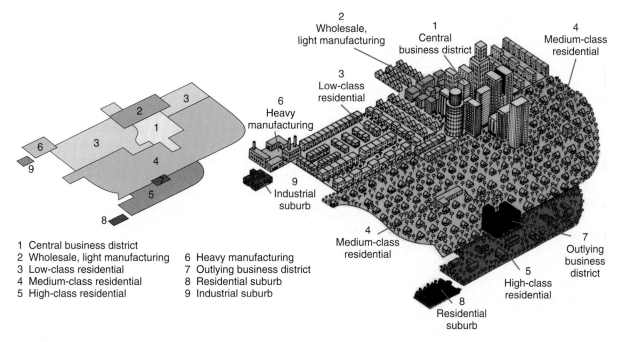

1 Central business district
2 Wholesale, light manufacturing
3 Low-class residential
4 Medium-class residential
5 High-class residential
6 Heavy manufacturing
7 Outlying business district
8 Residential suburb
9 Industrial suburb

Figure 10-15 Multiple-nuclei model of urban development. In this model of urban form, a city consists of a collection of individual nodes or centers around which different types of activities and people cluster.

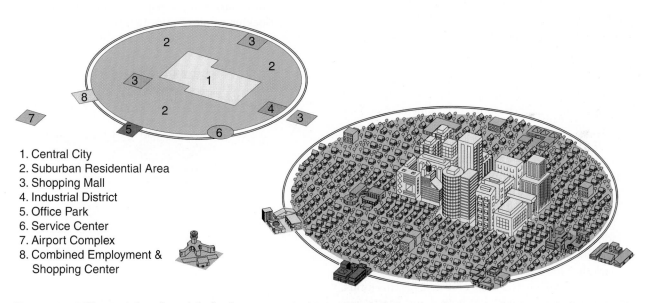

1. Central City
2. Suburban Residential Area
3. Shopping Mall
4. Industrial District
5. Office Park
6. Service Center
7. Airport Complex
8. Combined Employment & Shopping Center

Figure 10-16 The peripheral model of urban areas. In this model, which describes the growth of U.S. metropolitan areas in the last half of the twentieth century, activities disperse throughout a broad region.

Americans in "Chinatowns," of Italian Americans in "Little Italys," of Hispanic Americans in "barrios," or of African Americans in "ghettos"—must be evaluated carefully in each case. Sometimes the descriptive words placed in quotation marks in this paragraph are considered insulting, but they usually are not meant to be.

Religion is another social consideration that frequently causes clustering. People who share a religious faith may cluster around their house of worship, and in-migrants of that faith will seek that neighborhood.

Language communities frequently form, as do communities of the elderly and communities of gays and lesbians. Almost any factor of social bonding can encourage the creation of an identifiable residential neighborhood. Not surprisingly, these clusters are starting to reflect political preferences in the United States as homebuyers increasingly shop for neighbors, not just neighborhoods. The neighbors' yard signs, flags, cars, and bumper stickers are visible cues people use when house shopping.

Government's Role

Government may determine land use. Restricting or prescribing the use to which parcels of land may be put is called **zoning.** In U.S. history, local government has often been averse to planning, so zoning has more often been *restrictive* (dictating what cannot be done) than *prescriptive* (dictating what should be done). Industrial and commercial districts, for example, are usually kept away from residential neighborhoods. Each incorporated jurisdiction across vast conurbations—clusters of incorporated areas—exercises independent zoning power.

Today, most urban experts believe that the separation of land uses has been overemphasized. It may have been desirable to separate industry from housing when all industry was noisy, polluting, or smelly, but today separating homes from jobs may require excessive commuting. Most new planned communities emphasize the integration of residential and commercial activities, even including some light industry. They offer apartments above downtown shops and offices, for example, as in traditional small towns. The state of New Jersey subsidizes landowners who renovate downtown properties if the renovations create residential space on upper floors. This revitalizes downtowns and encourages the use of public transit—or even walking.

Government sometimes takes direct control over property in order to provide public functions, such as roads or schools. In the United States, this power is subject to a clause in the Fifth Amendment to the Constitution, called the *takings clause,* which states that "Private property shall not be taken for a public use, without just compensation." In other words, governments may take private land—this is called the right of **eminent domain**—but the governments must pay for it. In 1984, the Supreme Court extended the right of eminent domain to take land for any project "rationally related to a conceivable public purpose" (*Hawaii Housing Authority v. Midkiff*), and in 2005, the Court extended the meaning of "public purpose" still further to rule that fostering economic development is an appropriate use of eminent domain (*Kelo* v. *City of New London, Conn.*). Many people feel that local governments have abused eminent domain powers. The city council of Riviera Beach, Florida, for example, condemned 1,700 houses and apartments housing 5,100 people for a new development of shops, a hotel, a conference center, and yacht slips. Cypress, California, prevented a local church from building an annex in order to give the land to Costco. Such actions have roused complaints of political favoritism. The Supreme Court rulings have thrown the consideration over takings back to the state legislatures and courts, and most states have acted to restrict local governments' power of eminent domain.

Property owners are insisting that even short of complete seizure of land, some zoning restricts the use to which owners may put their land so severely that the zoning is in effect an unconstitutional taking. Some protest that both environmental and historic preservation

legislation are essentially takings. In 2010, a Minnesota court ruled that safety-related zoning restrictions on land around an airport constituted a takings that harmed the landowners. The takings clause is a fine point in the interpretation of law, but it is also important because it puts the public interest at odds with private property.

The Western tradition of urban and regional planning

The process of urban and regional planning applies many of the principles of urban geography to specific situations. The concept of designing an "ideal city" has challenged the best minds for centuries. The founder of Western city planning was probably the Greek thinker Hippodamus of Miletus. He laid out that city in today's Turkey according to a grid plan as early as 450 B.C.

Video
The Barcelor
Blueprint

The first modern attempt to formulate the needs of a city as a whole was the work of the British visionary Sir Ebenezer Howard. In *Garden Cities of Tomorrow* (1898), he outlined a plan to stop the unbounded growth of the industrial city and to restore it to a human scale. He wanted to relocate population into new medium-sized garden cities in the outlying countryside. These regional cities would be ringed by greenbelts of farmland and parks. All land would be municipally owned, and each town and its surrounding region would be planned as an interlocking whole (Figure 10-17).

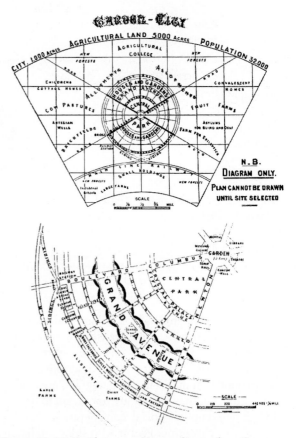

Figure 10-17 Sir Ebenezer Howard's garden city. Howard envisioned planned "garden cities" to disperse the concentrations of population in nineteenth-century cities.

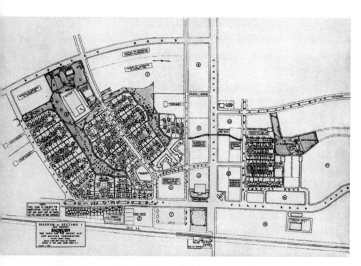

Figure 10-18 Plan for Radburn, New Jersey. This planned community, built in 1928, is still a desirable residential settlement. Key elements of the plan include plenty of park space and the separation of pedestrian walkways from car traffic and parking.

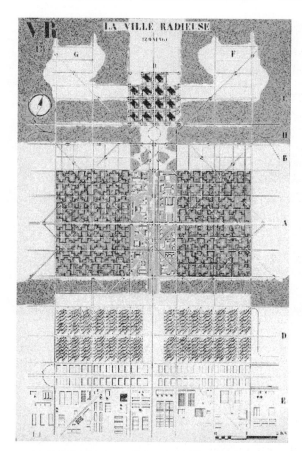

Figure 10-19 La ville radieuse (Radiant City). Le Corbusier's plan for his Radiant City featured high-density residential areas strictly segregated from other land uses.

Howard built two "garden cities" just north of London: Letchworth (1904) and Welwyn Garden City (1919). These inspired the Regional Planning Association of America, a private nonprofit organization, to construct two planned communities in the New York City area: Sunnyside Gardens, Queens (1924); and Radburn, New Jersey (1928) (Figure 10-18). Neither is a complete garden city, but both are harmoniously designed and have greatly influenced urban planning in the United States and Europe.

Probably the most important city planner of the twentieth century was the Swiss architect Charles Édouard Jeanneret-Gris (1887–1965), better known by his professional name, Le Corbusier. In a celebrated plan of 1922, he proposed to bulldoze the crowded, rundown historic core of Paris, preserving only the central monuments. In its place he wanted to build a *Radiant City* of tall glass offices and apartments, spaced so far apart that each tower would be surrounded by green space and have a fine, wide view (Figure 10-19). The concentration of facilities within high-rise slabs would liberate the city from its environment. It could be placed anywhere. Le Corbusier brought together two conceptions: the machine-made environment, standardized, technically perfect to the last degree and, to offset this, the natural environment, treated as open space, providing sunlight, air, greenery, and views. The degree of control and total design of the city was a hallmark of the "modernist city."

Paris was never torn down and rebuilt as a Radiant City, but Le Corbusier planned Chandigarh, a new capital for the Punjab State in India in 1950. The world's supreme modernist city, however, is Brasília, the capital of Brazil, designed by Lucio Costa in 1957 (Figure 10-20). Unfortunately, the city's gigantic scale demands a completely motorized population. That is the problem with excessive

Figure 10-20 Brasília, Brazil. Brasília, Brazil's capital since 1960, arose on a largely unpopulated open plateau 970 kilometers (603 miles) northwest of Rio de Janeiro. It was an effort to shift the nation's political, economic, and psychological focus toward the interior and away from the former colonial cities on the coast. As this picture reveals, it sometimes lacks water sufficient to keep the grass green or to operate the elaborate fountains. Slums surround the showplace buildings for 50 kilometers (31 miles) in every direction. Architect Lucio Costa admits, "Of course half the people in Brasília live in [slums]. Brasília was not designed to solve the problems of Brazil. It was bound to reflect them."

Figure 10-21 Canberra, Australia. Canberra became home to Australia's Parliament in 1927, but foreign missions and federal government departments did not move to the city until the 1950s.

openness—Corbusier's "city in a park" can become a city in a parking lot. The Australian capital at Canberra, planned by Walter Burley Griffin of Chicago, has less openness. Its layout allows walking and is generally considered superior (Figure 10-21).

In 1930, Le Corbusier planned the town of Nemours in Algeria. It has a geometric grouping of buildings that look like dominoes, and this plan set the international fashion for high-rise slabs for the next 50 years. These ideas were disseminated worldwide by the 1933 Athens Charter of the International Congress of Modern Architecture (CIAM). In the charter's codification, largely by Le Corbusier, the functions of the city—housing, work, recreation, and transport—provided the city's framework. The charter called for separation of high-rise development, industrial zones, parks and sports fields, and streets of different widths spaced for traffic at different speeds. These ideas diffused to dominate urban planning around the world.

Many people believe today that the widespread adoption of Le Corbusier's ideas produced a half century of monotony—not merely of detail and style but of insensitivity to essential differences between one place and another. The high-rise slabs ringing every big city in the world from Mexico City to Singapore look much alike.

Many urban planners worldwide have come to criticize the concept of high-rise living. Low-rise dwellings can achieve the same density of habitation as Le Corbusier's "towers in a park" can, and many people feel more content living in low-rise dwellings.

High-rise public-housing projects, it turned out, can breed a sense of alienation and helplessness, and many have been abandoned and razed across the United States.

The riots in Parisian suburbs in 2005 and 2007, visible in Figure 5-33, occurred in modernist, high-rise housing projects. Le Corbusier designed new towns on the outskirts of several French cities. The government called them "modern," but the housing was cheaply made, grim, and boring. When French people refused to live in these projects, the projects were dedicated to housing France's immigrant population. The immigrants' residential separation from traditional French life, as well as the problems of cultural assimilation discussed in Chapter 5, triggered repeated riots among residents. After a 1983 riot in such a project, President Francois Mitterand toured it and wrote that places like that "must disappear from our country." He insisted that it be torn down, and several such projects were dynamited in 1986, 2000, and 2004. Nevertheless, many still surround France's largest cities.

In 1961, a group of younger architects broke away from CIAM and proclaimed that architecture was more than the art of building. It was the art of transforming people's entire habitat. The School of Architecture at the University of California at Berkeley was reconstituted and renamed the School of Environmental Design. Today, we still work to design more humane urban environments (Figure 10-22).

Figure 10-22 Low-income housing in Newark, New Jersey. The abandoned high-rise public housing in the background (soon demolished) has been replaced by the low-rise development in the foreground. Low-rise housing can achieve equivalent density, and it has been demonstrated that it fosters more successful communities than high-rise projects do. Since 1992, the federal government has torn down more than 100,000 apartments nationwide in high-rise buildings and replaced them with low-rise housing.

Other Urban Models in Diverse Cultures

The concentric zone, sector, multiple-nuclei, and peripheral city models discussed earlier in this chapter were all devised to describe North American experience, but other models have been proposed to describe characteristics of other cultures. The three interacting processes identified earlier—economic factors (including transportation facilities), social factors (especially residential clustering and segregation), and governmental factors (most notably, planning)—affect all urban settlements, old and new, in all known cultures. These processes sort out the population and the land uses into distinct patterns that can be identified within any city.

The governments of most Western European countries, for example, have always been concerned with preserving the vibrancy and amenity value in their central cities. Therefore, they severely restrict suburbanization, as will be discussed shortly.

Latin American cities offer still another contrast to North American models. There, central business districts thrive (Figure 10-23). This is partly a result of continuing reliance on public transit and partly because high-income populations choose to live close to the central business district. A commercial spine such as a boulevard extends out from the central business district, and amenities such as opera houses, chic stores, and elegant parks follow this spine. Zones of more modest housing and value surround this elite zone, and the periphery is dominated by squatter settlements.

Western forms often overlie indigenous forms As we saw in Figure 6-19, distinct settlement patterns are landscape footprints of distinct cultures, but Chapter 6 also noted that many of the world's great cities came into being as a result of trading or political contact between the native peoples and Europeans. The cities that Europeans initiated and founded, such as those in Figure 6-30, were planned and built according to European notions of city planning. Their internal geography still shows port zones; enclaves of former European settlements; barracks; colonial government buildings; and racial, religious, and ethnic ghettoes based on the role each group played in the city's founding and during the colonial period (Figure 10-24).

In other cases, Western or modernizing interests built a new city alongside a preexisting native city. This was particularly common in those parts of Latin America (particularly today's Mexico and Peru), Asia, and North Africa where the local peoples had achieved significant urbanization before the coming of the Europeans. Today, the two cities often contrast sharply: Historic cores in forms traditional to the local culture

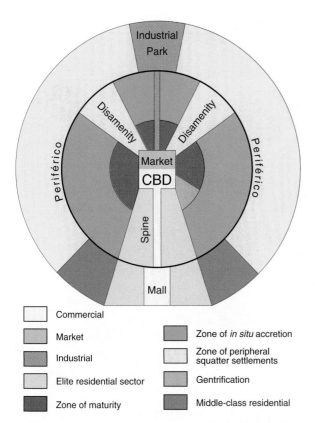

Commercial

Market

Industrial

Elite residential sector

Zone of maturity

Zone of *in situ* accretion

Zone of peripheral squatter settlements

Gentrification

Middle-class residential

Figure 10-23 Latin American model of urban growth. This model contrasts with the North American models. In Latin America an elite residential sector often follows a spine of high-value land use stretching out from the central business district.

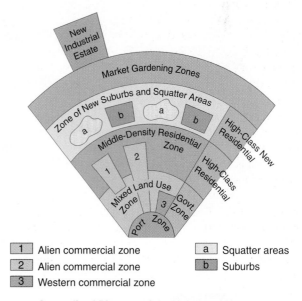

1 Alien commercial zone

2 Alien commercial zone

3 Western commercial zone

a Squatter areas

b Suburbs

Generalized Diagram of the Main Land Use Areas of the Large Southeast Asian City

Figure 10-24 Model of a Southeast Asian port city. Most Southeast Asian port cities are the product of Western influence, so they reflect Western design forms.

stand beside districts of modern commerce, retailing, industry, and associated residential areas. The modern or Western cities typically demonstrate grid layouts. In some cases a third element—vast slums of newcomers—surrounds both cities. The city of Fez, Morocco, for example, is clearly two cities (Figure 10-25): a modern Western-style city of broad, straight avenues lies on the plain below an older, traditional Islamic city. Western forms continue to be stamped on the world's cities today. Western architects, engineers, and urban planners, or non-Western individuals educated in Western schools, continue to transform even the older traditional built environments.

Individual cities, however, emerged at different times, for different reasons, and within different cultural contexts. Many great cities still boast historic cores that illustrate indigenous principles of urban planning, and they cannot all be squeezed into three or four simplified models.

Islamic urban form   The non-Western culture with the oldest and most articulate urban planning tradition is the Islamic culture. Traditional Islamic cities illustrate the role of culture in urban form. There are regional differences in cities across the Islamic realm, but most nevertheless show surprising similarity. These cities may seem chaotic to Westerners at first glance, especially to those accustomed to grid patterns, but they present an entirely rational structure. The structure develops from the basic needs of city dwellers but according to specific cultural influences. Among these are the central importance of religious obligations and the prominence of houses of worship in urban design. The resulting design characteristics can be identified from Seville, Granada, and Córdoba in Spain to Lahore in Pakistan, and elements of these principles can be found from Dar es Salaam in Tanzania to Davao in the Philippines.

The logic of traditional Islamic urban planning is announced in the Koran and has been codified by various schools of Islamic law. Certain basic regulations govern individual rights and the pursuit of the virtuous life in a densely crowded urban environment. For example, Islamic urban planning recognizes the need to maintain personal privacy; it specifies responsibilities in maintaining urban systems on which other people rely, such as keeping thoroughfares or wastewater channels clear; and it emphasizes the inner essence of things rather than their outward appearance. This last principle applies as much to the decoration of houses as to purely spiritual issues.

Take another look at Figure 10-25. At the heart of the traditional Islamic city stands the main mosque, the

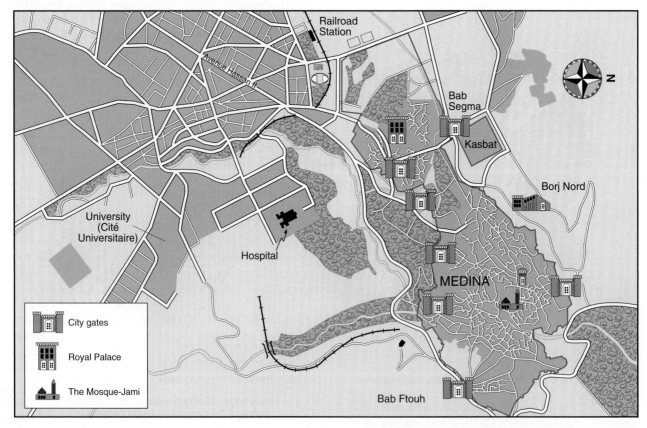

Figure 10-25 Fez, Morocco. The Moroccan city of Fez is really two contrasting cities. A modern Western city lies on the plain to the southwest (left in this figure) of an old Islamic city, the *Medina*. The new city has straight formal avenues, a railroad station, a university, and a modern hospital. The Medina contains old mosques and narrow constricted streets, and it is surrounded by city walls with great gates.

jami, which is typically the city's largest structure. Close to it are the main *suqs*, the street markets and enclosed shopping arcades. These arcades prefigured urban *galleries* in Europe and enclosed shopping malls in North America. Within the suqs, trades are diffused in relation to the mosque. The tradespeople who enjoy the highest prestige, such as booksellers and perfumers, are closest. Farthest away are those who perform the noxious and noisy trades, such as coppersmiths, blacksmiths, and cobblers. The neutral tradespeople, such as clothiers and jewelers, act as buffers.

An immense fortified *kasbat* is attached to the ramparts, on which are located several towers or gates. The *kasbat* was the place of refuge for the governor or sovereign. It had its own small mosques, baths, and shops, in addition to government buildings and barracks.

Everywhere else, the city is filled in with cellular courtyard houses tied together by winding lanes. Housing is grouped into quarters, or neighborhoods, that are defined according to occupation, religious sect, or ethnic group. The widest streets usually radiate outward from the core to the gates in the city wall. Slightly narrower streets serve the major quarters and define their boundaries, and still narrower third-order streets are used primarily by people who live in the neighborhood (Figure 10-26). Narrow streets provide vital shade, keep down dust and winds, and use little building land.

Interior courtyards of homes, often with trees and fountains, provide shade in hot climates, but, more importantly, they provide an interior and private focus for life sheltered from public gaze. This is true in Mediterranean architecture. The outside of a house may be plain, but the interior and courtyard may display lavish

Figure 10-27 The "French Quarter" of New Orleans. Here, as in Islamic countries and Mediterranean architecture in general, houses are not placed in the centers of gardens; courtyard gardens are placed in the centers of houses, not impressing the neighbors and passersby but providing cool private retreats.

wealth and decoration. The interior vividness parallels the Koranic emphasis on the richness of the inner self compared to a more modest outward appearance.

In the United States, New Orleans is the city closest to exemplifying these values (Figure 10-27). In the old Spanish sections of the city (misleadingly called today the French Quarter), houses generally present plain fronts to the street, but many enclose beautiful private courtyards. When Anglo-Saxons began to move to the city after its annexation by the United States in 1803, the Anglo-Saxon rich preferred to build grand homes in a new part of town called the Garden District. In the French Quarter, the courtyard gardens are inside the houses, but in the Garden District, the houses sit in the middle of their gardens. This is an example of how urban landscapes bear the impress of different cultures.

Cities and Suburbs in the United States

The dominant feature of the metropolitan form in the United States has been the explosive growth of cities across the countryside. Growing cities have spilled over their legal boundaries into areas called *suburbs*. Some suburbs are entirely residential, but others offer services for the surrounding residential population. In some cases, the suburbs are older cities that have been engulfed by the growth of a larger neighbor, but others are newly incorporated settlements. What defines an area as a suburb is its economic and social integration with a larger population nucleus nearby. *Town* and *village* are inexact terms that generally designate settlements smaller than cities, but the settlements may be incorporated.

Many of the developments described in the following discussion are now occurring elsewhere around the

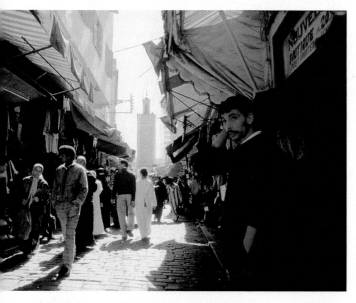

Figure 10-26 Casablanca, Morocco. In many traditional Islamic cities—as here, in Casablanca, Morocco—occasional straight streets provide views of minarets, towers for the call to prayer. People are constantly reminded of the importance of religion.

world, but they occurred first in the United States—largely because the nation's prosperity coincided with population growth and urban expansion.

The Growth of Suburbs

Early suburbs   Most large U.S. cities included manufacturing districts by the late nineteenth century. These were noisy and dirty, and they often attracted a working class, largely made up of immigrants, whom many long-established residents found to be unpleasantly "different." These biases pushed those who could afford city life to move to the lower-density neighborhoods further from the city center.

At the same time, a cultural preference for rural or small-town life pulled many people out of the city. Many Americans fell in love with the idea of "the country," and they favored a return to nature, to the land, or to open spaces—even if only a suburban yard. Therefore, when the railroads put older rural communities within commuting distance of the city, many people who had the time and the money necessary to commute to work from a home outside the city began to do so. In other cases, the wealthy built new towns (Figure 10-28).

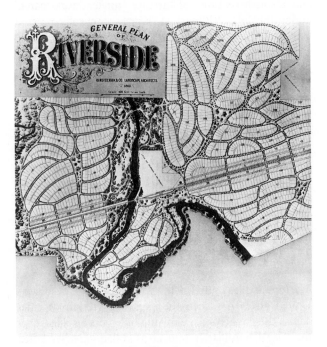

Figure 10-28 Riverside, Illinois. The designers of New York City's Central Park, Frederick Law Olmsted and Calvert Vaux, planned this real estate subdivision 14 kilometers (9 miles) from the center of Chicago in 1869. The plan included two straight business streets paralleling the railway into the city, but all residential streets were curved to slow traffic (before cars!). Open spaces contribute to the sense of breadth and calm enjoyed by "the more fortunate classes" for whom Riverside was designed. In 1992, Riverside residents refused federal financial assistance for traffic control because federal regulations would have required traffic intersections to be reengineered to 90° angles.

"Streetcar suburbs" sprang up when streetcar transportation was devised. Some of these planned suburbs eventually became completely built up, merged into other settlements, and lost their identities as they were absorbed into the expanding city.

The automobile ultimately opened the nation's landscape to suburban growth. For those who disliked urban life, the suburb was the solution. "We shall solve the city problem," wrote Henry Ford in 1922, "by leaving the city."

Government policies and suburban growth   The dispersion of housing to suburbs was slowed by the Great Depression in the 1930s and by World War II in the early 1940s. Following the war, however, government policies established a new balance of push-and-pull forces that encouraged the movement of investment, residents, and jobs out of the central cities into the suburbs.

The Federal Housing Administration (FHA) guaranteed loans, so down payments shrank to less than 10 percent of the house price. Suddenly, thousands of families could afford new houses. FHA benefits did not, however, apply equally everywhere or to everyone. The FHA favored the construction of new single-family houses in the suburbs over the rehabilitation of older houses or apartment buildings nearer the city's center. Also, the FHA opposed what it termed *inharmonious racial or nationality groups*. In some places, the presence of one non-white family on a block was enough to cut off the entire block from FHA loans. Thus, early government policy helped segregate the suburbs.

The government also granted tax and financial incentives to homeowners, including the deductibility of both mortgage interest payments and local property taxes from gross taxable income. These two benefits alone often made buying a new house cheaper than renting. In addition, the government protected homeowners from capital gains taxation—that is, taxes on any increase in the value of the house. By 2007, the tax loss to the U.S. Treasury of these three benefits totaled well over $100 billion per year. These subsidies to homeowners far exceed government spending on public housing. Furthermore, until 1980, the government allowed savings-and-loan institutions to pay higher interest on savers' money than commercial banks, because money in savings and loans was directed into housing.

The Veterans Administration Housing Program, begun in 1944 and called Homes for Heroes, pumped billions of additional federal dollars into housing programs. By 1947, the Levitt Company was completing 30 new single-family homes each day in Levittown, formerly a Long Island potato field, and similar developments were springing up on the outskirts of every other major U.S. city (Figure 10-29). Nationwide housing starts jumped from 114,000 in 1944 to 1,696,000 by 1950.

The suburbs brought homeownership to an increasing share of U.S. families. The percentage of U.S. housing

Figure 10-29 Levittown, Long Island. In building Levittown, 40 kilometers (25 miles) east of Manhattan, the Levitt family changed U.S. homebuilding techniques. The land was bulldozed and the trees removed, and then trucks dropped building materials at precise 18-meter (60-foot) intervals. Construction was divided into 27 distinct steps. At the peak of production, more than 30 houses were completed each day. Through the years, owners have personalized their homes so much that few visitors today can see that the houses were originally identical.

that was owner occupied rose from 44 percent in 1940 to 62 percent in 1960 and 69 percent in 2006, signaling middle-class status for a rising share of the population. Expanding homeownership has increased the number of citizens who have profited from the many homeowner subsidies, and it also has reduced the political possibility of rescinding them. Some scholars argue that the tax concessions were never necessary. Canada, Australia, and other countries achieved comparable levels of homeownership without offering such concessions.

At the end of the twentieth century and the beginning of the twenty-first century, housing prices in the United States rose rapidly during a period through which interest rates were low. At the same time, income and sales taxes rose, and property taxes fell from 40 percent of local tax receipts in 1970 to under 25 percent by 2007. These factors combined to encourage many people to consider their homes as investments—that is, to buy the most expensive home on which they could possibly meet mortgage payments, with the assumption that the value of the property would rise. Mortgage debt rose from 15 percent of gross domestic product in 1945 to 103 percent in 2007. This produced a dangerous situation that led to the financial crisis of 2008.

Over the decade leading up to 2008, relaxed mortgage and banking laws made more home loans available to more people. Loans were extended to new homebuyers, low-income families, and wealthier homeowners who wanted to buy larger homes. The surge of new buyers increased sales, and the prices of homes soared. Home-

owners were also able to borrow money based on the increased value of their homes. Meanwhile, some of the loans started to go bad. In some cases, irresponsible lenders had made complex loans to people who thought they could afford them because the loans had low interest rates for the first few years. Those loans were like time bombs, and after three or five years, homeowners were faced with much larger mortgage payments. Their paychecks that were going toward their home payments, however, were the same size because the wider economy was not growing. Many people suddenly found themselves unable to pay their mortgages.

Some borrowers defaulted on their mortgages, meaning they stopped repaying their loans. Lenders began foreclosing homes, which usually involves evicting the borrower and leaving the bank in possession of the home. The foreclosure rate more than doubled between 2006 and 2009, and it continued to rise in 2010 (Figure 10-30). Evicted families had trouble finding a place to live. It soon became clear that mortgage lenders and banks had invested heavily in mortgage loans, which were now considered potentially "toxic" investments. This started a financial crisis that mushroomed into recession during 2008 and 2009, as we will see in Chapter 12. The recession led to more job losses that led to more defaults and foreclosures. Large mortgage lenders disappeared overnight, and would-be homebuyers could not get loans. Home prices fell dramatically, and some people could not sell their homes without losing a lot of money. Other people were "upside down," meaning they owed more on their mortgage than their house was suddenly worth. About 1 in 11 mortgage borrowers were more than 90 days late on their payments in 2008. The U.S. government encouraged lenders to refinance some mortgages and backed these new loans with government guarantees but lenders were slow to respond. The long-term consequences of the housing crisis will be felt for years to come in many neighborhoods as old and new homeowners adjust to lower home values, new neighbors, or vacant houses. Suburban growth, however, is sure to continue after the crisis subsides.

Defining the expanding city As the population spread out during the twentieth century, new suburbs incorporated, and the Census Bureau devised a term for these sprawling conurbations: **metropolitan statistical area (MSA).** The Census Bureau defined a metropolitan area as "an integrated economic and social unit with a recognized large population nucleus." Thus, MSAs are the principal central cities and their suburban counties (except in New England, where the definitions are in terms of cities and towns). When two or more MSAs are next to each other, each is called a primary metropolitan statistical area (PMSA), and a group of adjoining MSAs are called a **consolidated metropolitan statistical area (CMSA).** By 2009, the nation's 366 metropolitan areas contained 84 percent of the total population. These MSAs covered about 20 percent of the country's land

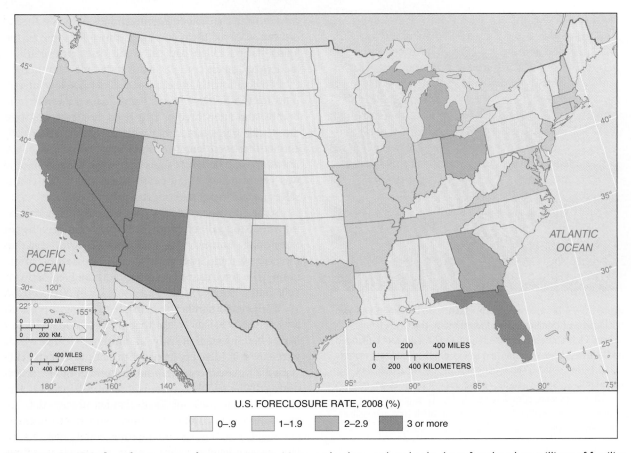

U.S. FORECLOSURE RATE, 2008 (%)

| 0–.9 | 1–1.9 | 2–2.9 | 3 or more |

Figure 10-30 U.S. foreclosure rates by state, 2008. Mortgage lenders, such as banks, have foreclosed on millions of families, usually forcing them to leave the property. Foreclosure rates are a measure of the percentage of households in mortgaged homes that were served with a foreclosure notice. The rates displayed here are for 2008. They range from a low of .04 percent in Vermont to over 7 percent in Nevada.

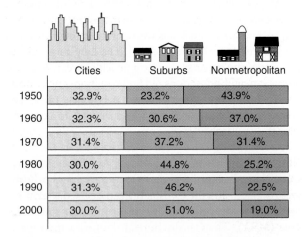

	Cities	Suburbs	Nonmetropolitan
1950	32.9%	23.2%	43.9%
1960	32.3%	30.6%	37.0%
1970	31.4%	37.2%	31.4%
1980	30.0%	44.8%	25.2%
1990	31.3%	46.2%	22.5%
2000	30.0%	51.0%	19.0%

Figure 10-31 Distribution of U.S. population. By 1970, a plurality of the U.S. population (37 percent) lived in the suburban portions of the MSAs, and by the century's end, a majority of Americans were suburban.

surface (Figure 10-31). Metropolitan areas accounted for a staggering 88 percent of the country's total economic activity in 2008.

In 2005, the Census Bureau first defined and collected data for 573 **micropolitan areas,** defined as developed regions with a core city of fewer than 50,000 people. These suburban areas cover the middle ground between rural and metropolitan living.

The suburban infrastructure The sprawl of single-family homes is expensive. It first requires roads for individualized transportation, which, in turn, demands energy. Heating and cooling individual homes is also energy intensive. Dispersed housing requires enormous investment in sewerage, water pipelines, telephone lines, and electrical wiring. The cost of providing infrastructure for 100 people in an apartment building is much less than the cost of providing it for the same 100 people spread out in 40 single-family homes over many hectares.

Infrastructure costs are further inflated by the fact that U.S. suburbs have not expanded contiguously outward from the city, like the waves from a stone tossed into a pond. Each developer wants to buy land as cheaply as possible and thus buys land beyond the edge of growth. This is called *leapfrogging*. The infrastructure network cannot be advanced in a regular pattern. Thus leapfrogging increases initial costs. Later, the leapfrogged areas are filled in, but then some initial infrastructure has to be rebuilt or upgraded to accommodate additional users. Many suburbs, for example, originally relied on individual-home wells for water, and their sewage was treated in individual-home septic tanks. As the suburbs matured and density increased, water supplies became polluted. Homeowners had to pay for wholly new public water mains and sewers. Suburban development also increased stream runoff, so financial losses from flooding have risen virtually every year.

The high infrastructure costs necessitated high property taxes, and these eventually provoked voter backlashes. The most notable was California's Proposition 13 of 1978, which limited property-tax increases. This has hobbled the government's ability to provide services no matter how much they are needed.

Suburbs take up a lot of space, and since 1945, U.S. urban areas have spread at a rate of about 405,000 hectares (1 million acres) per year. This growth has required that a good share of the country's most productive farmland be paved over, such as the Long Island potato fields covered by Levittown. The best agricultural hinterlands of many cities disappeared, and a rising percentage of the nation's food is now grown in conditions requiring expensive fertilizer or irrigation. In addition, the food has to be transported farther, consuming still more fuel, perhaps requiring refrigeration or special handling, and further boosting food prices.

Farmers have been virtually forced to sell their land to developers because property taxes are calculated on land's potential value, not its current-use value. A farmer who was making a small profit by farming could not afford to pay property taxes calculated on the land's potential value as housing. Even if a farmer could somehow meet the annual property-tax bill, the farmer's heirs would eventually have to sell the farm to pay inheritance taxes, which are also based on land's potential value. Only recently have these tax laws been changed to preserve greenbelts around some cities. In some cases, farmers now pay use-value taxes. Other farmers have sold development rights to conservation groups or to local governments, thus guaranteeing that the land will remain as farmland.

In the 1950s and 1960s, U.S. citizens did not worry about the costs of creating new suburbs. Between 1950 and 1973, median family income doubled in real terms. Demands for housing continued to mushroom with the baby boom and the splintering of families into separate households, partly as a result of the rising divorce rate. The average number of occupants of a U.S. household shrank from 3.67 in 1940 to 2.6 in 2008.

Today, a greater share of U.S. wealth is invested in housing and the necessary infrastructure than in any other nation. A high proportion of Americans enjoy private ownership of spacious, free-standing, well-equipped homes, and this investment has succeeded in bringing a sense of well-being to many Americans. It is becoming clear, however, that this development

has brought with it both high economic costs and steep social costs.

The Social Costs of Suburbs

The suburban lifestyle has imposed social costs not only on those who enjoy it but on many others, also. Americans generally sort themselves out residentially in such a distinct manner that sociologists and mass marketers can confidently construct an astonishingly accurate profile of people on the basis of their address alone. This residential segregation of the American people—racial, economic, and in many ways cultural—allows advertisers to target potential customers geographically. A great deal of demographic data on people's income, education, and other characteristics has been sorted by zip code, in order to fine-tune the selection of junk mail that arrives in mailboxes. GIS mapping and analysis of neighborhoods and metropolitan areas are skills in high demand by marketing and sales firms. Retailers want to know where the most probable potential customers for their product or service reside, what paths they take for work or relaxation and, thus, where the retailer might best locate a new store or service. One marketing consulting firm has categorized the country's zip codes and census tracts into 62 "lifestyle types," giving them catchy names such as Gray Power to describe communities of affluent retirees in Sunbelt cities; Towns and Gowns to describe college towns; and Latino America to describe neighborhoods of Latino middle-class families.

In low-density suburbs, local racial and social homogeneity have long been associated with conservative politics and social conformity. Property owners in many suburbs established *restrictive covenants*, which were legal agreements that the land would never be sold to people of a designated race or religious group. Such covenants are no longer legal, but they were common as late as the 1970s. Homeowners' associations, however, which regulate some aspects of property ownership, covered 20 million homes in 2003 (of the nation's 106 million) that house 50 million people, and they continue to grow. Most of these associations not only require that homeowners maintain their yards, but some dictate even when owners can put up holiday decorations or park their cars in their own driveways. Major changes in a home's structure or exterior appearance must be approved by the association, which can fine homeowners or even foreclose the "offending" homes—without the residents' knowing until they are evicted. Courts have accepted the argument that the regulation of "visual pollution" is analogous to the regulation of noise pollution, even though there may be greater disagreement on what constitutes visual pollution. In Portland, Oregon, for instance, the front of a house must be at least 15 percent windows and doors, and a garage door may take up no more than half the façade. In Arizona, a homeowner was sued for installing solar panels on his roof; a suburban couple in Washington State was sent to jail for painting their house mauve; and in December 2006, a couple in suburban Denver was fined for hanging a holiday wreath in the shape of a peace symbol on their front door. The stated purpose of these rules is to maintain neighborhood property values, but some Americans find them stifling.

The reputation of suburbs is perhaps unfair as they are, overall, quite diverse—there have long been poor suburbs as well as rich ones, and there are suburbs with varied ethnic and racial characteristics. In fact, many suburbs today are sites of intense ethnic and racial mixing for working-class and middle-class families seeking a trade-off between urban amenities and neighborhoods better suited for raising children. Denver and other cities that grew during the 2000s expect that the 2010 census will confirm what is already apparent: The suburbs are now a mixture of white, black, and Latino, as well as native-born and immigrant families.

The movement of jobs to the suburbs

The suburbs first expanded as bedroom communities for the middle-class workers who left their suburban families each morning to go to work in the city. Soon, however, the interstate highway system (authorized in 1956) and similar limited-access highways not only joined cities but also provided peripheral bypasses around them. These peripheral arteries provided access between suburbs and reduced the geographic advantage of a city's central business district. Developers put up suburban office buildings, and corporations built spacious office parks. Retailers soon began to build giant shopping malls at highway crossroads.

Manufacturing establishments abandoned the central city, too, for several reasons. Older heavy industrial buildings built for large equipment were too large or run down for new light-industrial needs. In addition, light industries relocated to escape central-city congestion, as well as the higher costs for energy, taxes, wages, and rent. Warehousing also relocated out from the inner-city railroad yards to the suburban highway interchanges. Metropolitan airports in the suburbs grew to provide jobs in both freight and passenger services.

In the 1950s, suburban growth was fed by young married couples who wanted to raise children away from the cities. By the early 1970s, however, the proliferation of jobs in the suburbs became the driving force for new housing. As the suburbs surpassed the central cities in employment, job opportunity became a pull factor for continuing suburbanization (Figure 10-32). Employment growth in U.S. suburbs has been greater than in central cities for every year since 1965. As jobs and housing have continued to expand outward, the extent of any metropolitan area has had to be continually redefined.

Some of the new exurbs, called *satellite cities*, or **edge cities**, boast greater retail sales and contain more office space than the old central cities. They even offer amenities formerly found exclusively in the central cities: art galleries, theaters, sports teams, and fine restaurants. These are the outlying nuclei of the periphery model that appear in the suburbs of metropolitan areas. Some edge cities outcompete the old city center for new jobs, investment, and even population density.

Changing commuting patterns and problems

When suburban workers commuted into and out of the city, radial mass-transit systems that focused on the central business district could serve transport needs tolerably well. Today, however, suburb-to-suburb commutes account for more than 50 percent of all U.S. metropolitan traffic. Only individualized transportation can serve populations that are spread out at low densities, and, as a matter of government policy, gasoline has never been taxed in the United States at rates comparable to those in other rich countries. Already in 1960, 64 percent of workers drove to work, and in 2008, over 86 percent did. More family members have been going to work, so the number of cars has increased faster than the local overall populations.

The Texas Transportation Institute has been studying traffic in 439 U.S. urban areas. The Institute reports that from 1982 until 2007, the annual delay per peak period (rush hour) grew from 14 hours to 36 hours. The number of urban areas with more than 40 hours of annual delay per peak traveler rose from 1 to 23. The total amount of delay reached 4.2 billion hours, and the amount of wasted fuel lost to engines idling in traffic jams rose to 2.8 billion gallons (at increasing prices per gallon). This accounting does not include the costs of air pollution or of highway accidents. A similar study in Canada in 2006 estimated costs of traffic jams in the nine largest cities there alone as some $3.7 billion. Commuting costs now consume more than 20 percent of the average U.S. household budget—as high as 23 percent in metropolitan Atlanta and Houston—and the percentage is rising. Older people often have difficulty driving, so as the American population ages, older people may become isolated or traffic accidents may increase.

The U.S. Department of Transportation has reported that the average American household used its cars and trucks for 496 shopping trips in 2001, with each trip averaging 7.02 miles (for a total of almost 3,500 miles). This is up considerably from 1990, when the average household took just 341 shopping trips, with an average length of 5.1 miles (totaling 1,700 miles). People are now taking more shopping trips than trips to and from work. No one knows how many of these trips are in fact discretionary, but suburban Americans seem to have little alternative to driving.

As driving has increased, concerns about energy consumption are rising. The United States consumes about 25 percent of the world's production of oil, most of it for transportation. Of the 520 million cars in the world in mid-2006, 200 million were driven in America. In the early 1970s, the United States imported about one-third of the energy it consumed, but by 2007, it imported 58 percent. Consumption levels are

(a)

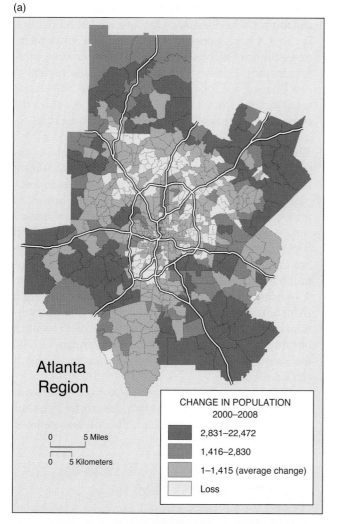

Atlanta
Region

0   5 Miles

0   5 Kilometers

CHANGE IN POPULATION
2000–2008

2,831–22,472

1,416–2,830

1–1,415 (average change)

Loss

(b)

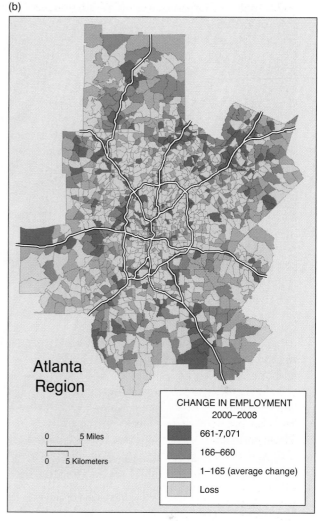

Atlanta
Region

0   5 Miles

0   5 Kilometers

CHANGE IN EMPLOYMENT
2000–2008

661–7,071

166–660

1–165 (average change)

Loss

Figure 10-32 Population and employment growth in metropolitan Atlanta. These maps show that in metropolitan Atlanta, Georgia, as in most other U.S. metropolitan areas, the greatest population growth (a) and job growth (b) are still occurring in the periphery. The city of Atlanta is at the center of the ring highway visible on the map. Central-city neighborhoods began losing population 40 years ago, and inner suburbs had already begun to lose population 30 years ago.

expected to begin falling due to the rising price of fuel, the availability of biofuels, and higher mileage requirements for new cars. The United States has only 3 percent of the world's petroleum reserves, and domestic production has been falling for decades. Meanwhile, worldwide demand for energy will rise rapidly, as China and India increase their consumption, thus keeping up the prices American businesses and consumers pay for gasoline and home heating oil, as well as jet fuel and other petroleum products such as plastics and pharmaceuticals.

Potential solutions to increased problems from traffic include mass transit, ramp metering, reserved high-occupancy lanes, improved methods of broadcasting information on traffic jams, better signal timing, and improved clearance of accidents (Figure 10-33). Stockholm, London, Singapore, and other world cities have reduced congestion by implementing fee systems that monitor traffic and automatically charge drivers according to the

time of day they are using busy roads and the fuel-efficiency of their vehicles. The sophistication of some in-vehicle GPS guidance systems would almost seem to obviate the need for drivers, as well as offer the possibility of collision-free highways, but the realization of such a possibility is years away. More immediately, GPS systems are helping drivers to avoid congestion and save fuel.

In an attempt to stimulate experimentation, the federal government has allowed states to use federal transportation funds for any means of transportation they choose. Salt Lake City, Utah, continues to build highways and to develop 400 hectares (988 acres) of land per month. The built-up area is expected to double in size, reaching 800 square miles by 2020, while the population is expected to increase by only 50 percent. Milwaukee, Wisconsin, by contrast, is using federal funds to tear down highways in an effort to revitalize the central city. The economic stimulus packages introduced

although many observers think that the system is ill suited to such a dispersed city with no central district. In 1999, Georgia created a new Regional Transportation Authority with power over all transportation projects in the 13-county Metropolitan Atlanta region. It can finance and operate public transport, take state money away from localities that refuse to enact its plans, and veto major projects such as new malls.

The "new urbanism" As suburbs continue to expand, several new developments have tried to re-create small-town community life, with walkable streets, compactness, downtown shopping and services, and even front porches. This effort has been called new urbanism. Most such developments try to free residents from dependence on the automobile (Figure 10-34). Celebration, Florida, was developed by the Disney Corporation in the 1990s. It tries to re-create small-town American life: Oak trees and white picket fences line the town's quiet streets, and all buildings are in traditional styles of architecture. Living in the town requires obeying the rules of the Celebration Company, a Disney subsidiary created to run the community instead of elected officials.

Some of these suburban centers form self-contained villages but offer mass transit into the central city, thus reproducing the advantages of the original planned

Figure 10-33 Road rage. As the frustration of sitting in traffic jams has increased on U.S. highways, the National Highway Traffic Safety Administration estimates that one-third of all fatal car crashes, in which almost 50,000 people die per year, could be attributed to "road rage," which is defined as "aggressive and even violent behavior by drivers caused by frustration or by the actions of other drivers." Road rage may win entry into the American Psychiatric Association's official *Diagnostic and Statistical Manual of Mental Disorders.*

in 2009 by the federal government to soften the recession dedicated over $19 billion the program's first year. Spending on roads in America can effectively inject money into every state and city because all U.S. urban areas depend on vehicular transportation to function.

Metropolitan areas are also experimenting with traffic monitoring and new methods of informing drivers of traffic jams. The technology for electric cars has not proved practicable. Hybrid cars, however, having both electric and internal combustion engines, are becoming increasingly popular on America's highways. Although the sales of hybrid and other high-efficiency vehicles are rising, they will remain a small fraction of all vehicles on the road for the near future. Several cities are rediscovering mass transit. Los Angeles built a 60-mile subway and light-rail system in the 1990s,

Figure 10-34 A new planned town. Valencia, California, is a new town 48 kilometers (30 miles) north of downtown Los Angeles that occupies a 6,070 hectare (15,000 acre) portion of a historic ranch owned by Newhall Land since 1875. The company has been planning and developing Valencia since 1965, and Valencia currently is home to approximately 48,000 residents in 17,000 homes. Its master plan balances residential uses with employment, education, recreation, open space, shopping, and services, so many Valencia residents live, work, play, and shop entirely within the community. There are 48,000 jobs within Valencia. Its residential villages, business parks, neighborhood retail and recreation centers, schools, parks, and shopping center are all interconnected by a 45-kilometer (28-mile) system of landscaped pedestrian walkways, bike trails, and bridges called *paseos*.

Public Space and Private Property

Cities have long been important political sites. In fact, the words *political, policy, police,* and *metropolis* are built on the Greek root-word *polis,* what the Greeks called the first cities. The polis was the center of authority, and citizens debated public laws and issues in the main square. These public spaces have always been a basic feature of democracy. One significant difference between traditional downtowns and most new villages and suburban malls is that the latter are private property. People cannot be banned from a traditional downtown, and the right to petition on a public sidewalk is constitutionally protected. People may, however, be banned from private property such as malls, and constitutional rights, such as political pamphleteering, may be restricted. In many communities today, malls, although private, are the only public gathering places, so if the mall owners are allowed to decide who may speak in them, mall owners can determine the public's access to ideas. Candidates for political office have been banned from busy malls owned by their opponents, and so have labor union organizers. In 2010, a man was ejected from a Denver mall for wearing a shirt that read "Yes We Cannabis." In a separate event, a California pastor was arrested for trying to talk to mall patrons about God.

Some communities are even locating schools in malls. The Landmark Shopping Center in northern Virginia, for example, includes the Fairfax County Mall School. The government insists that the students "feel comfortable" coming to the mall; they can shop during class breaks, and the mall offers opportunity not only for after-school employment but for "career-training programs" in retail sales. Mall restaurants replace costly school cafeterias, and heating and air-conditioning expenses are also reduced.

Private environments are patrolled by private security officers, one of the country's fastest-growing occupations. (Security guards constitute almost 2 percent of the nation's total labor force, which is triple the number of public police officers.) The privatization of space is a new form of economic and social segregation that carries profound influence throughout U.S. political and social life.

U.S. law recognizes that private properties can perform functions traditionally associated with government. This is called the *public function doctrine.* Yet the U.S. Supreme Court has ruled that state laws may protect free speech on private property, such as shopping malls (*Pruneyard* v. *Robins,* 1980). Some states, however, have struck down the right of public access under their state constitutions. Similar questions about public space versus private property arise with social networking sites such as Facebook, which can remove users according to its own policies. Some argue that these new virtual spaces are replacing our public spaces and so should uphold the right to free speech.

A security guard in a shopping mall. Does he enforce the law? Or the mall owner's rules? Can he arrest people? Evict them? What for?

suburbs, such as Riverside, Illinois, introduced in Figure 10-28. Metropolitan areas in Oregon, Washington, Minnesota, Colorado, and elsewhere are laying new light-rail tracks. Across the United States, the number of suburbs that enjoy rail service into the central city is growing, and in many cases, rail service is reviving (Figure 10-35). For example, in the 1990s, the Massachusetts Bay Transit Authority restored the public rail service system within 50 miles of Boston that

was terminated in the 1970s. The result was a boom in each community affected—both in new developments and in historic towns such as Newburyport, Massachusetts. Newark, New Jersey, is restoring a network of trolleys that honeycombs the city and reaches out into the suburbs. The system was abandoned in favor of private automobiles 60 years ago. Many of these commuter rail lines were bought by automobile companies that put them out of service to boost car dependency.

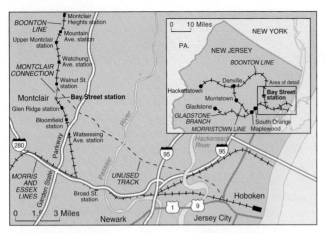

Figure 10-35 Improving metropolitan rail links.
A crucial link between two suburban rail stations in the New York metropolitan region that was opened in 2002 shortened commuting time, increased housing prices, and stimulated commercial development, with new stores and businesses opening all along the line. The convenience of suburban rail service—introduced in the nineteenth century—is enjoying new appreciation.

Suburbs as Sites of Change

Advances in information technology, rising incomes, population growth, and spending to develop infrastructure continue to drive residential and business development to the fringes of metropolitan areas. America's suburban population continues to grow faster than the central city population, so that by 2010, central cities housed only about one-quarter of the national population. The number of jobs continues to grow faster in suburban areas than in cities, too, so about 60 percent of metropolitan jobs are now in the suburbs. Furthermore, the rate of development of suburban land actually increased in the 2000s. The national population is growing at 1 percent per year, but land occupied by single-family houses was rising at 2 percent per year prior to the mortgage crisis. This slowdown will likely be followed by continued expansion at about the same rate.

Suburban airports are multiplying and expanding to serve personal and corporate aircraft, but this activity is launching new battles over land use and air and noise pollution. When Washington Dulles International Airport opened in 1962 in rural Virginia, it spawned a high-tech corridor that became the fastest-growing county in the country at the time. Dallas-Fort Worth Airport opened between the two metropolitan areas in 1974. The immediate airport area has all the facilities travelers expect—car rental, hotels, and cargo storage, as well as the company headquarters of American Airlines—with warehouses on all sides. Amazon built a huge distribution facility in Irving, 15 minutes away, and a new shopping mall is opening. Denver International Airport, covering 13,759 hectares (34,000 acres), opened 40 miles out of town in 1995, and by 2025, it is expected to be the center of a community of 500,000 people—almost as many as live in Denver itself. Working and living in the shadow of an airport has problems—height restrictions, noise, and traffic—but many businesses and people today feel the need to be near a runway.

The geography of retailing Highway congestion, more family members going to work, fear of crime, toll-free telephone numbers, credit cards, television and Internet shopping, and other factors are changing the geography of U.S. retailing. Many Americans are shopping from home by computer, telephone, or catalog. Elaborate catalogs fill mailboxes, and today almost two-thirds of the adult population buys at least occasionally from home. These developments introduce *virtual shopping*, which reduces the length of time before orders are filled, the quantity of goods sitting in inventories, and the number of middlemen standing between a product and the consumer. As we saw in Chapter 6, this shift to on-line sales has affected a wide variety of stores, from music and books to clothing.

Electronic commerce between businesses reached $3.1 trillion by 2007, and consumer e-commerce reached $251 billion. Stand-alone music retailers and travel agents are among the types of retailers being replaced by e-commerce; Amazon is an example of a company that has succeeded by lowering marketing, inventory, and warehousing costs (Figure 10-36). Its annual sales were growing at 20 percent per year before the recession and topped $22 billion in 2009. New technology significantly lowers warehousing and delivery costs. Company founder Jeff Bezos insists, "In the physical world it's the old saw: location, location, location. [But] the three most important things for us are technology, technology, technology."

U.S. retailers doubled shopping space per capita between 1970 and 1996, but as shopping at home increases, stores have begun to suffer. Almost 20 percent of the shopping malls in the United States that existed in 1990 were out of business by 2000. Malls are today being rebuilt as destinations for a wider range of activities than just shopping. Today, malls offer movies, skating rinks, games, and a greater variety of restaurants and other attractions (Figure 10-37). Even with such diversions, many of the 16 million sales jobs in the United States may be vulnerable to elimination.

Telecommuting It is easier to move information than to move people, so more people are working at home via the use of computer terminals. This is called **telecommuting.** At IBM, for example, at the end of 2006, a full 40 percent of the workforce had no official office. About 15 million employees in the United States work from home on computers, and several state governments have telecommuting programs. Some government bureaus have distributed telecommuting centers throughout the suburbs, complete with office equipment.

Figure 10-36 Amazon's computerized warehouse. Amazon's warehouse in Fernley, Nevada, a quarter-mile long and 200 yards wide, holds more than 3 million different books, CDs, toys, and other items. Computers send orders to workers' wireless receivers, weigh and sort packages, and apply shipping labels, and the facility can box 11,000 orders per hour. Consumers can order any two items out of the 3 million items and have them delivered quickly in one box—usually at a discount price.

Figure 10-37 New attractions at malls. The Mall of America located outside the Twin Cities in Bloomington, Minnesota, includes the Nickelodeon Universe Amusement Park. The Mall has more than 2,000,000 square feet (186,000 square meters) of shopping space plus millions more for other attractions.

Telecommuting eases the pressure on transport facilities, saves fuel, reduces air pollution, reduces the demand for office space, reduces absenteeism, lowers the fixed costs of a business enterprise, and has been shown to increase workers' productivity. It also allows employers to accommodate employees who want flexible work arrangements, thus opening employment opportunity to more people. Telecommuting tends to

reduce central-city employment, and we do not yet know what other ramifications increased telecommuting will mean for metropolitan geography.

Developments in the Central City

America's central cities suffered economic decline through most of the second half of the twentieth century. The drain of jobs from the central city and the concomitant development of the suburbs hollowed out many U.S. metropolitan areas. Many central business districts lost their purpose. Commercial, professional, and financial offices relocated to the suburbs, followed by upscale retailing. Many U.S. downtowns came to consist only of a government center, a convention center, and a few hotels; the streets were deserted after 6 P.M. The total populations of many central cities fell, particularly of older cities in the Northeast and Midwest. Between 1970 and 2008, population fell by the following amounts in major U.S. cities: St. Louis, 43 percent; Detroit, 47 percent; Cleveland, 47 percent; Pittsburgh, 43 percent; Baltimore, 30 percent; Cincinnati, 47 percent; Washington, 22 percent; Philadelphia, 26 percent; Chicago, 19 percent; and Boston, 4 percent. Visitors to the United States are astounded at the urban infrastructure that Americans seem simply to have abandoned: housing, sewerage, water pipes, roads and streets, industrial buildings, and more.

Economic decline Central cities can thrive as long as (1) their economies offer a complete range of job opportunities, ranging from entry-level jobs for the unskilled up to specialized jobs for skilled workers, and (2) family stability, education, and other social systems help urbanites ascend the socioeconomic ladder. In other words, cities do not have to retain their middle classes, but they have to offer the lower classes opportunity to become middle class. Unfortunately, the departure of the middle classes for the suburbs occurred at the same time as two other developments.

First, urban economies were transformed by the out-migration of entry-level jobs, particularly in manufacturing, construction, and warehousing. This out-migration broke the rungs of the ladder of upward mobility. In 1968, the economist John Kain first suggested that the removal of manufacturing jobs to the suburbs and the concentration of the poor in the central cities created a spatial mismatch between the suburban job opportunity and central-city low-income housing. This spatial mismatch, he argued, could explain the high unemployment found in the central cities. The **spatial mismatch hypothesis** has until today dominated analyses of inner-city unemployment—and therefore the formulation of potential solutions. We will discuss a new alternative hypothesis later in the chapter.

Second, at the very time that many unskilled jobs were leaving the central cities, new waves of unskilled workers—those lacking specialized training or advanced education—were pouring into them. The stream of new migrants to the cities included African Americans from the rural South, Latinos, and other immigrants. This influx continued long after the numbers of entry-level job opportunities began to shrink.

Much of the inner-city housing stock began to deteriorate. This was because government incentives still have made it profitable to give up a house in the city and move to the suburbs. Policymakers had thought that if new suburban houses were available to middle-class people from the cities, the urban poor could move into the older city dwellings. This succession, called *filtering*, would solve the housing problem for lower-income families. Much of the central-city population remaining behind, however, was financially incapable of maintaining the inherited housing stock. Therefore, many central-city neighborhoods deteriorated. The term *inner-city neighborhood* became a euphemism for slum.

The service economies At the end of the twentieth century, some central cities began to enjoy new growth in financial, information, and specialized technical services. In these areas, skyscrapers replaced rusty factories (Figure 10-38). New York's leading export for decades was garments, for instance, but today it is legal and financial services. In 1975, Baltimore's leading employer was the Bethlehem Steel Company; today it is Johns Hopkins University Medical Center. In 2009, almost 10 percent of all jobs in major metropolitan areas were in health care. This shift in job opportunity exemplifies a switch from *blue-collar jobs*, usually involving manual labor, to *white-collar jobs*, which are salaried or professional jobs that do not involve manual labor.

Analysts of urban economies suggest that cities increasingly compete for jobs and growth on the basis of lifestyle. Surveys have found, for example, that growth in the biotechnology and computing industries is clustered in a few metropolitan areas, including Austin, San Francisco, and Boston. These cities do not enjoy traditional economic advantages such as proximity to raw materials, cheap energy, or low costs of living. They do, however, share two features: a thriving arts scene (reflected statistically by a high number of artists, writers, and other arts workers) and a dense, highly diverse, and tolerant social character portrayed by, among other things, a high number of immigrants and gays and lesbians. Business journals periodically rank cities' relative attractiveness, and although the mix of criteria chosen is not scientific, a high ranking can boost a city's fortunes (see Table 10-2). Analyst Richard Florida identified "the three T's" of an economically successful city: tolerance, talent, and technology. These features attract the people who are crucial to economic

(a)

(b)

Figure 10-38 Pittsburgh transformed. No other U.S. city illustrates a transformed economy better than downtown Pittsburgh, which was transformed from a dirty and smoky industrial area (a) into a new park and gleaming service center, a "Golden Triangle" (b). This is where the Allegheny River (at left) meets the Monongahela (at right) to form the Ohio.

TABLE 10-2 Different Ways of Ranking Cities' Desirability	
Professor Richard Florida of the University of Toronto has ranked the following "city-regions" as the most "creative" in the United States:	
San Francisco, California	Houston, Texas
Austin, Texas	Washington, D.C./Baltimore, Maryland
San Diego, California	New York, New York
Boston, Massachusetts	Dallas, Texas
Seattle, Washington	Minneapolis-St. Paul, Minnesota
Raleigh-Durham, North Carolina	
By contrast, the AARP (formerly the American Association of Retired Persons) often compiles lists of America's best places to settle in retirement. In 2009, AARP offered a list of the 10 Best Places to Live the Simple Life. The empty nesters referred to in the text choose to move into central cities and enjoy their cultural attractions, but most of these smaller cities offer a relatively low cost of living and a relaxed pace of life. The increasing numbers of retired persons may initiate a revival of many smaller towns. Young upwardly mobile people might find some of these cities relatively boring!	
Tucson, Arizona	Northampton, Massachusetts
Greenville, South Carolina	Lexington, Kentucky
Montpelier, Vermont	Texas Hill Country, Texas
Logan, Utah	Oxford, Mississippi
Ames, Iowa	Walla Walla, Washington

success: creative workers, engineers and scientists who develop new products and industrial processes, and creative businesspeople, financiers, and other workers who start new businesses and improve old ones. Such people have the skills and the means to live wherever they choose, and they are attracted to cities that offer the amenities and broad quality of life they desire. *BusinessWeek* magazine has reported that two-thirds of college-educated adults aged 25 to 34 today decide first where to live and then where to work.

In some cities, the holders of the new white-collar jobs—sometimes called *yuppies*, young urban professionals—triggered a rediscovery and revival of urban life. They first occupied and restored select older residential neighborhoods in a process known as **gentrification**, but they soon began to convert even former industrial and warehouse buildings into residences. These conversions were usually in owner-occupied forms (condominiums or cooperatives) so that owners could profit from the tax advantages afforded to suburban homeowners. Some observers argued that the conversion of lofts to residences damaged the cities' chances for industrial resurgence, while others argued that the industries were never coming back, so the conversions were beneficial. Gentrified neighborhoods mix old and new architecture and a vibrant street life with upscale commercial activities such as bookshops, gourmet food and wine shops, and art galleries. Yuppies have been joined by *empty nesters*—that is, older people who had raised families in large suburban homes but found urban attractions and activities more appealing in later life. Historic preservation movements have assisted central cities by winning tax advantages for the reuse of older buildings. New zoning for mixed use revives the cities' days and nights. Even some businesses that left the central cities in the period 1950–1980 are moving back downtown, discouraged by suburban traffic tie-ups and attracted downtown by

relatively low rents and the availability of high-quality workers. Thus, rising educational levels, shrinking family size, and aging all contributed to some central cities' revival.

Most job and population growth is still in the suburbs, but signs of central-city revival are widespread. Census results reveal that Chicago, New York, Atlanta, and many other central cities are enjoying population growth for the first time in decades. Violent crime dropped 34 percent in America's 10 largest cities in the 1990s, and young people were increasingly choosing the urban lifestyle. In the newest editions of guidebooks to the best colleges, most of the top-rated schools are urban institutions. Higher education is a mainstay of many central cities' economies, and frequently the cities retain the graduates. Among the largest 100 U.S. cities, the 25 that began the 1990s with the highest percentage of college graduates ended the decade with even greater concentrations. These 25 cities saw the college-educated share of their population jump 6 percent, twice the average growth of the other 75 cities. Cities compete to attract and retain these educated workers.

In addition, many cities have cultivated the tourist sectors of their economies and now attract tourists with historic quarters, fine cuisine, shopping opportunities, performing arts, and art exhibitions and festivals. Memphis, Tennessee, for example, attracts 8–10 million tourists each year, about 14 percent of whom are foreigners (Figure 10-39). The city has borrowed exhibitions of artworks from around the world, and Memphis's role in American musical history draws many tourists to exhibitions and performances. Many cities are building new museums and performing-arts centers just to pull in suburbanites.

The role of immigrants As noted in Chapter 5, immigrants to both the United States and Canada concentrate in major cities, giving those cities a

Figure 10-39 Graceland. The home and grave of musician Elvis Presley remains one of the most popular tourist attractions in Memphis, Tennessee. Tourists coming principally to visit Graceland spend an estimated $40 million per year in Memphis. That sum equals $65 per city resident, and tourists' dollars have a high multiplier effect.

cosmopolitan sophistication. Immigrants have also played an important role in the cities' economic rejuvenation. Many bring capital or job skills, so for them a city's traditional advantages of agglomeration and external economies confirms the city's function as incubator of new businesses. In Los Angeles County, for example, corporations that employ fewer than 100 people offer more than half of all the county's jobs. Virtually all of the firms making clothing and textiles, toys, processed foods, furniture, and biomedical supplies are owned by foreign-born individuals. Furthermore, many immigrants have moved into declining neighborhoods and repaired deteriorating homes themselves—a process called investing *sweat equity* rather than money.

Latinos today outnumber blacks in some of the biggest cities in the United States—New York, Los Angeles, Houston, San Diego, Dallas, Phoenix, Chicago, Miami, and San Antonio. In Los Angeles, Miami, and San Antonio, Latinos outnumber non-Latino whites as

TABLE 10-3 The Diversity of Immigrants in U.S. Cities

Estimates of the U.S. population in 2008 reveal the importance of urban areas for immigrants. A striking 85 percent of all immigrants to the United States, regardless of citizenship, lived in major U.S. metropolitan areas. The table below lists the 10 U.S. metropolitan areas with the largest immigrant populations. The Miami area leads the country as having the highest percentage of immigrants in its population but the New York area has a larger total number of immigrants. Immigrants no longer live only in downtown areas. A rising share of immigrants live in the suburbs. Examining the birthplaces of contemporary immigrants in different U.S. cities bears witness to their diversity. The origin of Hispanic immigrants in the United States contrasts greatly between the predominantly Mexican-born immigrants found in Texas and California cities, the heavily Cuban immigrants of south Florida, and the Central American immigrants living in the Washington, D.C. area. The majority of immigrants in the San Francisco area are from Asia. Nearly one in four immigrants to Chicago is from Europe. Smaller cities show even more diversity in the number, origins, and residential patterns of immigrants.

U.S. Metropolitan Areas with the Largest Number of Immigrants

Rank	Metro Area	Immigrants	Immigrants as Percentage of Population	Mexico	Rest of Latin America	Asia	Africa	Europe
					Immigrants' Place of Birth (percentage)			
1	New York-Newark, NY-NJ-PA	5,328,033	28.0	5.7	44.5	26.8	3.8	18.2
2	Los Angeles-Long Beach-Santa Ana, CA	4,374,583	34.0	41.6	16.3	33.8	1.4	5.4
3	Miami-Fort Lauderdale-Pompano Beach, FL	1,995,037	36.8	2.9	82.7	5.1	1.1	6.3
4	Chicago-Naperville-Joliet, IL-IN-WI	1,689,617	17.7	40.9	7.5	24.4	2.5	23.5
5	San Francisco-Oakland-Fremont, CA	1,258,324	29.4	19.9	11.6	52.3	2.0	11.0
6	Houston, TX	1,237,719	21.6	47.8	21.7	21.3	3.2	4.5
7	Dallas-Fort Worth-Arlington, TX	1,121,321	17.8	57.1	10.7	21.5	4.7	4.3
8	Washington-Arlington-Alexandria, DC-VA-MD-WV	1,089,950	20.3	3.7	34.8	36.0	13.6	10.4
9	Riverside-San Bernardino-Ontario, CA	894,527	21.7	62.0	11.0	18.8	1.4	4.5
10	Boston-Cambridge, MA-NH	731,960	16.2	1.5	35.1	29.9	8.2	21.4
	All Large U.S. Metro Areas	**32,425,888**	**16.3**	**26.5**	**25.4**	**28.7**	**4.0**	**13.0**

Source: Brookings Institution Metropolitan Policy Program, based on U.S. Census data

well. The changing demographic mix in major U.S. cities carries political ramifications, as Latinos, Asians, and other minorities form new political alliances. The growing diversity of cultures and interest groups tends to submerge America's historic black–white dialogue in a new chorus of voices gaining political expression.

The shortcomings of service economies

Upscale urbanites account for only a fraction of the total inner-city population. The new white-collar jobs being created do not always equal the number of blue-collar jobs being lost. Although most high-paying jobs are in the service sector (doctors, lawyers, executives, sports stars, etc.), most service sector jobs are not high paying. Creative website designers, for example, earn higher incomes than janitors do, but usually janitors outnumber website designers. In New York City between 1989 and 1999, a time of national prosperity, the only net growth in the numbers of jobs was in jobs paying less than $25,000 per year, which was just about the national median for a full-time worker. The average weekly wages of workers in New York City's finance and insurance industries in 2006 was $8,323. This average figure conceals an extreme concentration of this income among the highest-paid bankers and stock brokers, but it contrasts vividly with the average weekly wage of $594 of workers in the fast-growing accommodation and food services industry and $803 in retail trade. Furthermore, workers who lose their blue-collar jobs often require retraining or education before they can capture one of the new opportunities.

The median household income in U.S. central cities hovers between 70 and 75 percent of the figure for the suburbs, and central-city unemployment rates hover about one-third above those in the suburbs. Furthermore, the percentage of jobs in most central cities held by commuters is rising—especially the percentage of the best jobs. Some 60 percent of the salaries earned in Washington, D.C., for example, go to suburbanites, and federal law protects them from D.C. taxes. In 1999, the New York state legislature, responding to suburban voting power, ended New York City's income tax on commuters. This gesture crippled the city's ability to provide services to them. The labor force participation rates of central-city populations—that is, the percentage of the population that is currently employed or even looking for a job—fall behind those of the nation and of the entire local metropolitan areas. For example, the labor force participation rate for the city of Detroit in 2008 was 55 percent, whereas for the Detroit metropolitan area it was 64 percent. This suggests that a smaller percentage of the city's population was working and paying taxes, and also that a higher percentage was receiving public assistance.

Security concerns in the twenty-first century also distress central cities. Central cities are the most likely targets of terrorism, their populations suffer increased stress levels, and, for the most part, the individual cities must bear the burden of high security costs.

U.S. cities are losing the middle class. Defining "middle" as between 80 and 150 percent of the median, the Brookings Institution found that the percentage of middle-income households in the 100 largest metropolitan areas had decreased from 30 percent in 1999 to 28 percent in 2009. Added to the disappearance of the middle class, poverty rose during the 2000s in U.S. cities. Suburban poverty rates also increased to 9.5 percent due to rising unemployment and wage cuts.

The urbanization of African Americans and segregation by race and income

Chapter 5 recounted how some 1.5 million African Americans migrated from the South to the cities of the North between 1910 and 1945, and an additional 6.5 million migrated north and west between 1945 and 1970. That second wave arrived just as the central cities were losing their ability to provide entry-level job opportunities. The migrants also often faced discrimination in employment and segregation in housing. Deteriorating conditions in new African American ghettoes eventually triggered civil unrest. A 1965 riot in the black Watts section of Los Angeles left 34 people dead and more than 1,000 injured and required military occupation of 119 square kilometers (46 square miles) to halt the violence. The Watts riot was followed by 150 major riots and hundreds of minor ones in cities across America that summer and the next three summers.

Since the 1960s, civil rights have made great advances, and a great many of the black urban inmigrants have achieved success. Others, however, have been left behind. If a family's first urban generation failed to find employment, skills, and upward mobility, the second and third generations may have failed also. Los Angeles erupted again in April 1992, for example, and in 2001, black neighborhoods in Cincinnati erupted in three days of rioting after the fatal shooting of the fifteenth unarmed black man by police since 1995. In 2003, 20 houses burned in Benton Harbor, Michigan, during riots that broke out after the death of a black motorist killed during a police chase.

Residential racial segregation continues. Statistics on America's 8.2 million individual residential blocks disclose that about one-third of blacks and more than half of whites live in blocks that are at least 90 percent of their own race.

During the 1990s, blacks' overall proximity to jobs improved slightly, but Census data from the 2000s reveal that no demographic group remained more physically isolated from jobs than blacks. Some decline in the spatial mismatch occurred during the 1990s and 2000s, and that decline was principally due to the residential movement of black households within metropolitan areas. Nevertheless, in nearly all metropolitan areas with significant black populations, the separation between residences and jobs was much higher for blacks than for whites.

The 2000 census revealed one more changing pattern: Whereas one of America's most dramatic trends from 1960 to 1990 was the movement of the American poor into urban neighborhoods of concentrated poverty, that trend reversed itself between 1990 and 2000, and poverty became less concentrated. Concentrations of poverty magnify the problems associated with poverty in general: crime, delinquency, joblessness, drug trafficking, breakdown of the family, and low-performing local schools. Furthermore, when a neighborhood has a concentration of poverty, middle- and working-class people often see it as "dangerous." They avoid the neighborhood, which becomes increasingly isolated, socially and economically.

In 1990, 10.4 million people, which was 15 percent of all poor people at the time, lived in census tracts in which at least 40 percent of the households were in poverty. By 2000, however, the number had declined to 7.9 million people (down 24 percent), about 10 percent of all poor people. Among blacks, the concentrated poor population fell from 4.9 million people (30 percent of poor blacks) to 3.1 million people (19 percent of poor blacks). The total number of poverty tracts fell from 3,417 in 1990 to 2,510 in 2000. Observers believe that the concentrations of people in poverty dispersed because public housing projects were torn down, central cities enjoyed some economic resurgence, immigrants revived some urban neighborhoods, and millions of poor people left urban slums for other neighborhoods.

People living in poverty had not disappeared between 1990 and 2000; their numbers had in fact increased, but some of them had moved to the inner ring of suburbs. Today, fewer central-city neighborhoods resemble the nightmarish scenes portrayed in the popular films *Fort Apache, the Bronx* (1981) or *Boyz N the Hood* (1991). Today's metropolitan reality has been captured better by rapper Eminem's fictionalized biography *8 Mile* (2002), in which poverty is located in the dreary inner suburbs of Detroit—the American city that in fact saw the most dramatic reduction of concentrated poverty 1990–2000, down by almost 75 percent. Suburban poverty is not pleasant, but, as a general rule, a mix of income and ethnic groups in suburbs reduces the isolation of the poor.

Only in pockets of the central cities are conditions at their worst, and even the worst slums of New York City, Chicago, Detroit, and Los Angeles do not compare with the conditions of life for many in Mexico City; Lagos, Nigeria; or Kolkata, India. The inner-city second- or third-generation living in deprivation, sometimes called the *underclass*, numbers less than 3 million people, which is less than 1 percent of the national population. Still, the conditions of deprivation and the lack of opportunity contrast starkly with the national self-image.

The network hypothesis   Scholars are formulating a new alternative to the spatial mismatch hypothesis to explain some urban unemployment. They have learned that the primary qualification employers seek in new unskilled workers is reliability. The best way to find reliable new employees is to ask current employees for recommendations. Therefore, networks of working family or friends provide unskilled urbanites seeking entry-level positions with information about jobs, sponsorship for jobs, and role models for work. Hiring is referential, not residential. Many urban black communities, after a generation or two outside the mainstream of labor opportunity, may lack these crucial links because the social networks conducive to upward mobility have deteriorated. Philip Kasinitz, a sociologist, has written, "The primary reason for ghetto unemployment is not the lack of nearby jobs but the absence of social networks that provide entry into the job market." The theory that it is the lack of these networks that causes unemployment may be called the **network hypothesis.** Scholars investigating this hypothesis note that significant numbers of inner-city jobs, even those in manufacturing, have often been captured by groups whose residences are distant, even though local residents cannot get work. For example, Kasinitz found that jobs in a manufacturing district of Brooklyn, New York, were captured by Hispanics who commuted from New Jersey, even though unemployment stayed high in contiguous black neighborhoods. Geographer Thomas Cooke concluded that "it is not possible to argue that census tract African American male unemployment rates are related to the number of local job opportunities."

These studies do not conclusively replace the spatial mismatch hypothesis, but they introduce a new partial factor of explanation. Assumptions about what causes a situation will determine approaches to altering that situation. In this case, our understanding of the causes of inner-city unemployment will determine what solutions we propose.

Efforts to Redistribute Jobs and Housing

The spatial mismatch hypothesis has dominated thinking about urban unemployment since the 1960s. Therefore, several government programs designed to deal with the problem of inner-city unemployment have addressed the spatial mismatch. Three approaches have been suggested: (1) Bring new blue-collar jobs to the cities, (2) move inner-city residents to the suburbs, or (3) transport inner-city workers to suburban jobs.

1. Bringing new blue-collar jobs to the cities involves efforts to reindustrialize the central cities by establishing **urban enterprise zones,** where manufacturers receive government subsidies. The Federal Budget Act of 1993 called for the designation of both urban and rural enterprise zones. Several states and cities had already designated zones and granted manufacturers assistance in their poorest communities. Some of these zones have enjoyed

success, exploiting the cities' advantage of existing infrastructure (water, sewers, and so forth), but the factors that historically have driven industries from the central cities are difficult to overcome.

In 2000, Congress granted tax credits for commercial projects that create jobs in low-income areas. Tax credits reduce development costs, thus encouraging riskier projects. Private investors pay lower taxes, and the developer passes the savings on to the community by, for example, lowering rent.

Many central cities are scarred by abandoned industrial facilities called **brownfields.** There are an estimated 450,000 brownfield sites across the United States, including industrial properties, old gas stations, vacant warehouses, former dry cleaning establishments, and abandoned residential buildings that may contain lead paint or asbestos. Developers have shunned these deserted industrial sites because of the liability for buried wastes and other pollution-related risks. It has been easier to buy undeveloped sites at the edges of cities. The most difficult brownfields to clean up, however, are often within 2 miles of the downtown core, so they offer the greatest potential for reclamation and rededication to new land uses. The U.S. Environmental Protection Agency provides grants for cleanups, and a 2003 survey of 244 cities found that federal help to renovate brownfields had cleaned 19,000 acres, boosting local tax revenue by as much as $1.5 billion and adding up to 570,000 jobs.

As many central-city districts were virtually abandoned, some scholars recommended that urban density shrink in a planned manner. In New York City's South Bronx and in parts of Detroit, Cleveland, Houston, and other central cities, decrepit apartment buildings have been replaced by new single-family dwellings that resemble typical suburban homes (Figure 10-40).

2. The second approach is to move poor people from the central city out into new subsidized suburban housing. The federal government actually built suburbs for people with low incomes in the 1930s: Greendale outside Milwaukee, Greenhills outside Cincinnati, and Greenbelt outside Washington, D.C. Today, however, most suburbs are zoned to exclude subsidized housing or even private apartments with young families. This means that local land-use laws effectively prohibit low-income residents. Municipal zoning instead tends to favor more expensive neighborhoods because these properties generate more property taxes to cover the costs of the services that residents require. Zoning also tends to prohibit partitioning one's home into accessory apartments. People with extra bedrooms or small guesthouses sometimes seek renters, providing a more affordable housing option. Most suburbs, however, ban this for fear of lowering overall property values.

For years, researchers have debated whether the poorest Americans would improve their personal situations if they were simply relocated from the worst neighborhoods and housing projects and dispersed among middle-class neighborhoods somehow to absorb a different set of cultural norms. The discussion has been largely theoretical, because an extensive resettling of the poor would be expensive, intrusive, and racially charged—in short, politically impossible.

Recent federal programs tried to break up the concentrations of subsidized housing in the central cities and build more in the suburbs. The government encouraged some suburban areas to accept a greater share of public housing by offering grants for other community facilities. The first experiment, known as Gautreaux, occurred as part of a court-ordered settlement of a racial discrimination

Figure 10-40 Central-city transformations. The new detached single-family homes in a neighborhood of New York City's South Bronx, shown here being visited by former president Bill Clinton, resemble typical suburban homes even more than do the homes in Figure 10-22. Here in the Bronx, low-income high-rise apartment buildings had been devastated by fires and virtually abandoned for 20 years.

lawsuit against the Chicago Housing Authority. From 1976 to 1998, the housing authority moved nearly 25,000 poor African Americans from crumbling city public housing to subsidized housing in suburbs where no more than 30 percent of the population was African American. Studies showed that heads of households were more likely to be employed than their inner-city counterparts, although they still earned poverty-level wages and their families received little in the way of extra counseling or social services. Their children did even better. They were much more likely than their inner-city peers to graduate from high school, go to college, and get good jobs.

In the 1990s, the federal government inaugurated "Moving to Opportunity," which gave vouchers to thousands of residents in poor city neighborhoods. The participants were required to move to neighborhoods with lower poverty levels. Nevertheless, they tended to move together into new clusters of households. These areas were characterized by the same problems of crime, single motherhood, and low educational expectations. So far, studies have found no statistically significant changes in employment rates for adults or educational attainment for their children. These results suggest that relocating households does not help people break the cycle of poverty created by discouraging peer groups, unhealthy environments, and weak economic opportunities. This would appear to confirm that social networks are at least as important as spatial mismatch in explaining poverty.

The devastation of New Orleans by Hurricane Katrina in 2005 has, unhappily, provided another chance for social scientists to collect evidence on whether relocating the poor is effective policy. The very poor make up a sizable portion of the nearly half-million evacuees spread across hundreds of towns and cities who are recipients of a large, focused package of federal and state aid. The evacuees have received rental vouchers and relocation allowances that they can spend on everything from job retraining to childcare. Many evacuees relocated to host cities such as Houston, but their outcomes have not yet been assessed.

3. The third approach to solving the spatial mismatch is to provide inner-city poor people with transportation to jobs in the suburbs. The American Automobile Association puts the average cost of car ownership at over $7,000 per year, which is more than many working Americans can afford. Therefore, several city governments, private employment agencies, and even suburban employers have instituted dedicated bus services for this purpose. Wisconsin and California are experimenting with interest-free loans to allow working poor to buy or maintain cars, and a federal-level program is called Bridges to Work.

These three efforts to solve the problem of inner-city unemployment build from an acceptance of spatial mismatch as the cause of central-city unemployment. The network hypothesis has only recently been offered as an alternative or complementary explanation. Therefore, few government programs have yet been devised to address the lack of social networks. Perhaps we will see proposals in the future.

Still another potential solution to inner-city unemployment among the unskilled is to educate and train the central-city population so that they can capture the skilled tertiary-sector jobs that do open in the central city. The quality of education in the central cities, however, is generally discouraging. Schools in upper-income suburbs usually have more money to spend on modern facilities than inner-city schools do. The latest experiments in public education suggest that the integration of students of varying economic statuses raises student achievement. In Wake County, North Carolina, for example, minority students have made rapid strides in standardized test scores since school officials began to use income as a prime factor in assigning students to schools, with the goal of limiting the proportion of low-income students in any school to no more than 40 percent. Other education districts across the country are experimenting with this model.

Governing Metropolitan Regions

Political geography is the subfield of geography that studies the interaction between political processes and the distributions of all other activities and transformations of the landscape. Political geography can be studied at any scale from local community politics to international boundary disputes and international law. Urban areas present special problems of interest to political geographers.

The legal boundaries of most U.S. cities were originally drawn to include some surrounding land for future growth, and when the city outgrew those boundaries, it annexed suburban areas. By the 1920s, however, the suburban populations had begun to incorporate themselves to avoid annexation (Figure 10-41). Suburbanites argued that their action upheld the U.S. tradition of local self-government, and many may have felt that by incorporating their own communities, they were escaping the problems (and people) of the old city, including high social costs and politics that were often corrupt. Only a few central cities—including Austin, Texas; Charlotte, North Carolina; and Oklahoma City—can still expand by annexing new suburban areas.

Today, autonomous municipal units form a legal retaining wall around almost every large city in the United States. Metropolitan areas cover a great number of municipalities and also myriad special district

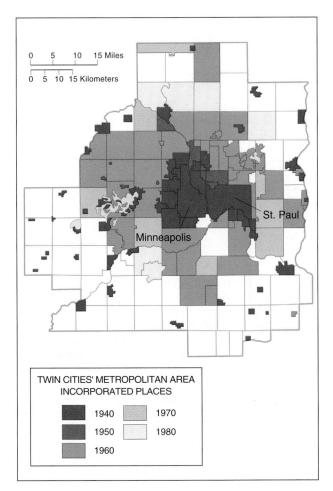

Figure 10-41 Metropolitan Minneapolis and St. Paul.
Minneapolis and St. Paul, Minnesota, were surrounded by tiers of independently incorporated municipalities, like the rings of growth of a tree. This pattern of suburban incorporation typifies U.S. metropolitan expansion.

governments, which are incorporated to deal with specific problems. For example, Nassau and Suffolk counties in suburban New York include two cities, 13 towns, 95 villages, 127 school districts, and more than 500 special districts, most of which exercise taxing powers for services from garbage collection to hydrant rental. The five-county Los Angeles metropolitan region contains 160 separate governments. Los Angeles County alone has 82, and even within the boundaries of the city of Los Angeles, there are seven independent city governments.

The boundaries among the many jurisdictions are not obvious to everyone, but each government has its own agenda and marks the landscape. These governments jostle for authority and tax dollars. Property taxes soar, but many metropolitan area residents have no idea which government is responsible for which service.

The governing bodies of many of the special districts are not chosen in elections in which each citizen exercises an equal vote. Instead, they are either appointed or

else chosen in elections in which votes are weighted in terms of payments for a service or use of a service, or by some other measure. As a result, the percentage of public funds spent by officials who are directly responsible to the voters shrinks. Furthermore, the boundaries of special districts may not conform to those of general-purpose governments but instead may overlap them. Overlapping boundaries multiply the difficulties in coordinating the provision of services. All these factors discourage voter turnout.

Decisions regarding metropolitan land use and the location of industries, recreation facilities, transport facilities, or new housing cannot be made in the best interests of the entire metropolitan population. Instead, they are made on the basis of competition among the local governments, each of which wants to enhance its own property-tax base by attracting commercial developments that pay high property taxes but demand little in the way of local services. For example, fast-growing Santa Clara County, California, has zoned for 250,000 new jobs but only 70,000 new homes. Each community hopes to let surrounding towns cope with the additional costs of schooling, pollution, and congestion (Figure 10-42).

Many metropolitan areas have created councils of governments (COGs). These are committees of officials representing each of the local governments in the region. COGs, however, exercise limited powers, but each local government can veto any proposed area-wide action. Some U.S. metropolitan regions are creating regional governments to address area-wide problems. State governments in Washington, Minnesota, Connecticut, Virginia, and elsewhere are devising new schemes to redistribute or equalize both the revenues and the costs of housing and schooling throughout metropolitan areas. Dade County, Florida, has assumed responsibility for many public services for Miami and the 29 other cities in the county, and county voters chose to rename the county Miami-Dade. In 2005, metropolitan Chicago created the Chicago Region Environmental and Transportation Efficiency Project (CREATE). The body plans transport facilities for the entire region, but it must deal with 272 independent municipalities. Other state governments are assuming increased responsibility for governing metropolitan areas. Some states are even drawing and applying state-level land-use plans (Figure 10-43).

The governing of metropolitan areas in Canada presents many of the same problems as in the United States. The province of Ontario created a government for metropolitan Toronto in 1953 that has since served as a model throughout North America.

Today, many city and suburban governments—backed by downtown business executives, environmentalists, farmers, and church leaders—are banding together to fight continued sprawl by pressuring state and federal gov-

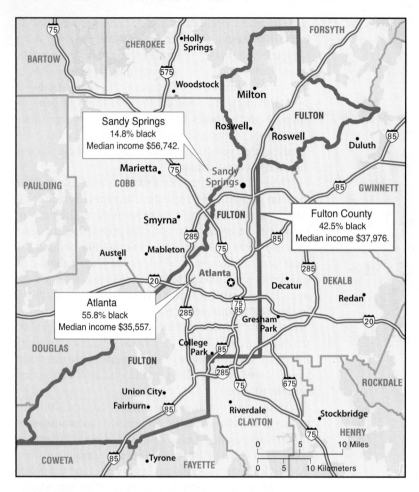

Figure 10-42 Remapping the suburbs. Questions of suburban incorporation continue to concern metropolitan politics. In 2005, Sandy Springs, Georgia, in suburban Atlanta, decided to incorporate itself to escape governance by Fulton County. That action encouraged residents of other parts of the county to incorporate their territories. If each succeeds and the county is left with no unincorporated territory, county government responsibilities will virtually disappear. Some observers have suggested that Sandy Springs residents were motivated by demographics of race and income. (New York Times Agency)

RAPID CHANGE

Controlling Sprawl in Metropolitan Portland, Oregon

Oregon has long championed laws that restrict urban sprawl. These concerns are shared by metropolitan Portland, Oregon, which is among the faster-growing cities in the United States. It has a "Metro" government that controls land use, conservation, and transportation for 24 cities and 3 counties holding 1.3 million people. Metro requires cities and towns of a certain size to bound their urban areas, thus keeping forests, farmlands, and open space free of development. The law has permitted new housing while protecting a rural character at the edges of cities. As a result, Portland is perhaps America's most "European" city, with a fine mass transit system, compact neighborhoods, and protected forests and farms outside the city limits. Critics point out that the law has pushed up housing prices by reducing the land available for building, but voters have turned down repeated initiatives that would have weakened the government's powers since the law was first passed in 1973.

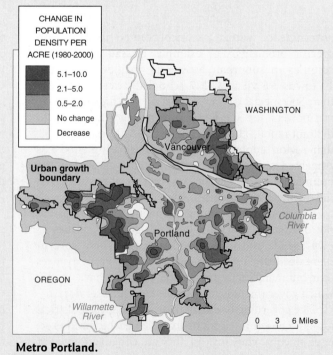

Metro Portland.

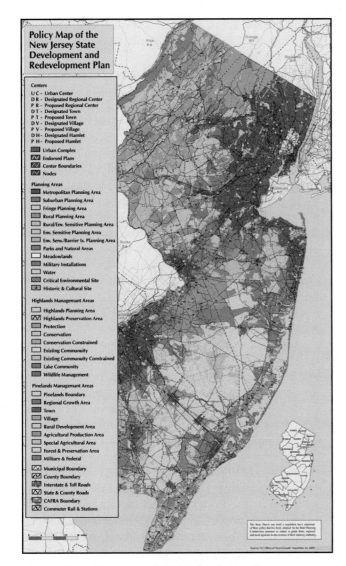

Figure 10-43 New Jersey's state land-use plan. New Jersey is among the states that are preparing land-use plans and realizing them through zoning and purchase.

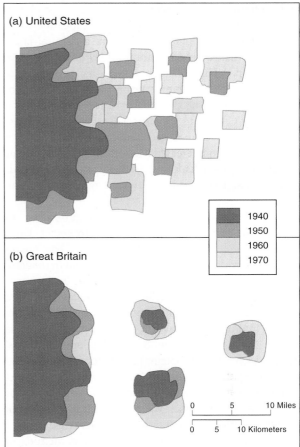

Figure 10-44 Suburban growth patterns in the United States and Great Britain. This schematic drawing contrasts (a) the leapfrogging typical of U.S. suburban growth with (b) the tighter, more controlled suburban growth typical of Great Britain. The British system saves open space and allows for better and cheaper planning and installation of infrastructure.

ernments to end the many subsidies that encourage suburbanization. The government of Ohio, for example, has cut back on building new roads, begun repairing old ones, and prevented new communities from offering tax breaks that might draw businesses from older core metropolises. In 1997, Maryland legislated to confine state spending on infrastructure to existing municipalities. Similar political initiatives have been launched in Minnesota, Pennsylvania, Indiana, and Oregon.

The weakness of metropolitan regional governments in North America contrasts sharply with models elsewhere (Figure 10-44). In the United States, states and municipalities are given primary responsibility for land use. Municipalities often want to encourage population growth because it adds to the local economy. Zoning laws, therefore, often favor developers who leapfrog each other to find cheaper land. In most European and Latin American countries, either strong metropolitan governments exist or else the national governments themselves oversee land use and the growth of metropolitan regions. Britain has for a long time used the authority of the central government to limit urban growth and preserve agricultural land near cities. Such policies generally concentrate activities in city centers and regulate land use very strictly to prevent urban sprawl. Also in contrast to the North American model is the fact that homeownership is relatively low, and single-family suburban lots are less common. Europeans are more content to rent smaller dwellings, which take up less land and slow the outward growth of cities.

Cities and the Environment

As humans become a majority urban species, we are more closely bound to environments we have made for ourselves. Rapid urban growth, driven by demographics and economics, has transformed the face of the planet. Chapter 1 differentiated a natural landscape, one without evidence of human activity, from a cultural landscape, one that reveals human modification of the local environment. That chapter noted that evidence of human activity is so ubiquitous on Earth that we might be justified in saying that all Earth is today a cultural landscape. Cities are an especially concentrated assembly of the technologies that modify the natural environment. Concrete, glass, electrical devices, and air conditioning overshadow the small patches of manicured grass, trees, and ponds that remain in urban parks.

Rapid world urbanization has required us to rethink our common definition of *environment*. Until now, in many people's minds, the word has meant Earth's natural landscapes. The phrase *environmental preservation* triggered images of struggles to preserve forests and wetlands, beaches, and grasslands. Today, however, the transformation of Earth's surface and the actual circumstances in which most people live force us to direct more of our attention to urban environments. The environmental problems of settlements are appearing on the political agendas of most countries and even of international agencies. Urban environmental problems increasingly impact the health and livelihoods of people in both the rich and the poor countries.

For example, unexpected recurrent epidemics of cholera have focused attention on health threats in human settlements. It had been assumed that the threat of cholera had been eliminated through improvements in water supplies, sanitation, sewage treatment, and food safety, but urban crowding and overloading of the urban infrastructure have brought the problem back. Other urban environmental problems of growing concern include poor or inadequate housing, unsafe water supplies and inadequate sewage treatment, air quality, exposure to chemical hazards in industrialized areas, physical hazards including road accidents, the collection and management of solid waste, and noise.

Cities transform natural landscapes not only within the built-up areas but for considerable distances around them. Ecologist William Rees has called the total area of land required to sustain a city its *ecological footprint*, and the United Nations has singled out four regional impacts of particular concern: uncontrolled and unplanned city expansion, liquid-waste disposal, solid-waste disposal, and acid precipitation. Cities concentrate demands for the products of fertile land, watersheds, and forests. Mining and the extraction of bulky, low-value substances that go into building materials, roads, foundations, and other parts of the built environment also disrupt local ecology, as does the waste matter dumped as a result of excavations. The United Nations notes that there is considerable potential for creating new jobs in reducing resource use and wastes and in recycling or reusing the wastes that are generated.

The challenge is great. We must transform our increasingly "cultural" landscape into a healthy and sustainable physical environment. Several countries have begun "low-carbon" settlements that employ "green" building technologies and renewable energy sources. The hope is to find sustainable means of building cities while lowering their ecological footprint. The Danish island town of Samso invested in various renewable energy sources, especially wind turbines, and succeeded in becoming a net exporter of energy. However, the level of energy consumption by the town's residents did not much change and what little power residents did conserve was offset by the needs of new electronics. Some countries are considering building eco-cities from scratch. The city of Dongtan was planned as a zero-carbon addition to Shanghai, but it was never built. Abu Dhabi has plans to build Masdar City, a 6-square-kilometer (2.3 square mile) carbon-neutral project (Figure 10-45). Incheon, South

Figure 10-45 Conceptual rendering of Masdar City. Plans for this "city of the future" place it near the city of Abu Dhabi in the United Arab Emirates. It is intended to be a model for sustainable zero-carbon urban development.

Figure 10-46 Rooftop ecology. Part of the solution for "greening" existing cities is to take advantage of the spaces that exist. The organic garden shown here is atop an office building in downtown Bangkok, Thailand. The garden cools the building, lessening the need for air conditioning, saving electricity, and reducing carbon emissions.

Korea, hopes to build an even bigger project. Whatever these mega-projects achieve, real change has to begin in the world's existing cities, where people already live. One place where people are already making changes is their rooftop. Lighter-colored roofing materials reduce heat

absorption and require less cooling. Solar panels convert sunlight into electricity or hot water. Rooftop gardens produce food while cooling lower floors (Figure 10-46). These and other simple innovations are part of how humans are changing how they live in urban environments.

Chapter Review

Summary

Every settled society builds cities because some essential functions of society are most conveniently performed at a central location for a surrounding countryside. The region to which any city provides services and upon which the city draws for its needs is its hinterland. Some cities' locations are the result of site characteristics; others are more the result of convenient situations. Urban environments often suffer from heat, noise, air pollution, and water and waste-management problems.

In many traditional societies, cities' ritualistic and political importance has outweighed their economic importance. As societies develop, cities' economic role usually becomes paramount. Cities offer agglomeration for the division of labor, and they promote and administer the regional and global specialization of production. Cities can become centers of industrial production and tertiary-sector functions. Some workers in cities produce basic sector exports, whereas workers in nonbasic sectors serve the needs of the cities' residents. Jobs in the basic sectors multiply jobs in the

nonbasic sectors. Cities can be classified economically by examining each city's basic and nonbasic sectors and by comparing these sectors among different cities. Central place theory models how cities and hinterlands are distributed across an isotropic plain.

Britain provided the first model for urbanization, but urbanization imposed hardship on millions of people. Today, urbanization is occurring in many places without economic development. People migrate from the countryside because of the balance of push and pull factors. Governments can regulate this migration both by reducing the attractiveness of cities and by improving rural life. Urban migrants' industriousness could be an asset for economic growth.

Urban geographers also study the distribution of activities within cities, which may be caused by economic forces, social factors, or government actions.

The concept of designing an "ideal city" has challenged planners for centuries. Howard's garden cities plan tried to stop the growth of the industrial city and to restore it to a human scale. Le Corbusier proposed cities of tall office buildings and apartments spaced far apart and surrounded by green space.

Many of the world's cities were planned and built according to European notions, and this is still reflected in their internal geography. Western forms continue to be stamped on the world's cities. Many great cities' historic cores illustrate indigenous principles of urban planning. Islamic cities present one alternative to Western forms.

In the United States, growing cities have spilled into suburbs. This dispersion has resulted from government policies, cultural choices, and other economic and social forces. Both housing and jobs have dispersed, and commuter traffic has been redirected. The central cities suffered economic decline, but some are reviving with new tertiary-sector activities. Attempts have been made to coordinate governmental activities across conurbations.

The fact that humans are now a majority urban species requires us to consider cities as our primary environment. More can be done to lessen the impact of cities on the natural environment. Conserving energy and water will be major challenges in the coming centuries.

Key Terms

agglomeration p. 378
basic sector p. 380
brownfield p. 413
capital-intensive activity p. 386
central business district (CBD) p. 389
central place theory p. 381
city p. 376
congregation p. 390
consolidated metropolitan statistical area (CMSA) p. 399
edge city p. 402
eminent domain p. 392
external economy p. 383

exurb p. 383
gateway city p. 380
gentrification p. 409
hinterland p. 376
incorporation p. 376
informal, or underground, sector p. 388
internal economy p. 383
labor-intensive activity p. 384
megacity p. 377
metropolitan statistical area (MSA) p. 399
micropolitan areas p. 400
multiplier effect p. 380
network hypothesis p. 412

nonbasic sector p. 380
political geography p. 414
primary sector p. 378
primate city p. 376
secondary sector p. 379
segregation p. 390
service sector p. 379
spatial mismatch hypothesis p. 408
telecommuting p. 406
tertiary sector p. 379
urban enterprise zone p. 412
urban form p. 389
urban geography p. 377
urbanization p. 376
zoning p. 392

Questions for Review and Discussion

1. Define and explain the three sectors of an economy.

2. What are the basic and nonbasic sectors of an urban economy?

3. How are cities distributed across a landscape?

4. In what ways does world urbanization today differ from the model provided by the historical experience of England?

5. How can governments slow urbanization?

6. In what ways does U.S. tax policy subsidize people who buy houses?

7. Contrast modern Western urban-planning ideas with the ideas of traditional Islam.

Thinking Geographically

1. What are the site characteristics of your city? Why was the site selected?

2. What are the situational relationships of your city? What principal transport routes converge on your city?

3. What range of external economies is available in your town's Yellow Pages?

4. Who are the biggest employers in your town? Which enterprises do the biggest dollar volume of business? What are your town's basic economic activities?

5. Are any planned suburbs or exurbs located around your town? What forms of transportation do their residents depend on?

6. How many local general-purpose governments and special-purpose governments are in your metropolitan

region, or the metropolitan region nearest to you? Study a map of all the forms of independent government in the area. How are the special-district government ruling boards chosen? What kind of regional planning body or government does the region have?

7. Some people enjoy living in cities; some people flee cities. What are the benefits and the drawbacks of living in either cities or suburbs?

8. What would you do to stop rural–urban migration in any given poor country?

9. Quality-of-life measurements can affect a city's prosperity. What are the basic measurements of urban or regional quality of life? How would your community score?

Log in to www.mygeoscienceplace.com for videos, animations, **MapMaster**™ interactive maps, RSS feeds, case studies, and self-study quizzes to enhance your study of Cities and Urbanization.

MapMaster™

One of two Russian miniature submarines that dove beneath the ice at the North Pole. Using a robotic arm, the subs' crew planted a titanium Russian flag on the sea floor.

11

A World of States

When submarines carrying two Russian politicians reached the ocean floor at the North Pole in August 2007, it was an unprecedented feat that should have been a milestone in exploration. Instead, it touched off a diplomatic crisis because one of the subs set down a Russian flag, forever marking the area with rustproof titanium. The act looked a lot like the days of European imperialism, when explorers would plant a flag on the shore of some "undiscovered territory" to claim it for their country. Today, planting a flag does not establish a country's possession of territory. Russia does claim that the North Pole is Russian territory, but Canada and Denmark also claim it. In fact, all of the coastal countries claim large portions of the Arctic Ocean as their own territory. These countries want to stake their claims to the Arctic as global warming melts back the ice that used to make navigation and exploration impossible. This is opening up the Arctic to intense competition for what is thought to be more than 20 percent of the world's undiscovered oil and gas. The countries also desire to control commercial shipping lanes where it was once thought impossible.

Canadian, Danish, and Russian scientists are mapping a topographic feature under the Arctic seas called the Lomonosov Ridge. Each of these countries wants to prove that the ridge is an extension of its continental shelf. As we will see in this chapter, a country can claim a large part of a surrounding ocean on the basis that its territory extends underwater on a continental shelf. At stake are the rights to unknown but potentially rich deposits of minerals, oil, and gas, as well as control over Arctic navigation. A Canadian and Danish team is trying to demonstrate that the Lomonosov Ridge is an extension of the North American continent, whereas the Russians are trying to prove that the ridge is an extension of the Siberian continental shelf. A settlement may be many years away. Meanwhile, oil companies have already paid billions of dollars for exploration rights. And in 2009, two commercial ships successfully used a new route from South Korea to the Netherlands through the Arctic region without needing icebreaking ships to open a path—about one-third shorter than the usual route. Competition for economic resources in the Arctic will lead to changes in the territories and borders of the surrounding countries, requiring cartographers to update maps once again.

A Look Ahead

The Development of the Nation-State Idea

A nation-state is a territory that includes a unified group of people with their own government. The nation-state idea originated in Europe and diffused from there—or was applied—to virtually all the rest of Earth. Collapsing empires have created many new states.

A Changing World Political Map

Relatively few states on today's world political map are nation-states. A map of nation-states could be achieved only by redrawing the map, by expelling people from some countries into others, or by forging national identities in the populations of the existing states.

The Internal Organization of States

States maintain their borders and subdivide their territory for governmental purposes. Democratic states have a variety of ways to design representative government, but these techniques can be manipulated for political ends. Individual rights and freedoms vary between and within states.

Relations Among States

States may cooperate peacefully with other states or they may engage each other in conflict. The world contains areas outside the jurisdiction of any one state. Globalization challenges the control that states have over their territory.

Figure 11-1 The ages of the world's states. The current political partitioning of the world is very recent. In many states, citizens remember the day their country achieved independence.

The world political map is probably the most familiar of all maps, because Earth's division into countries, or states, is the most important territorial organizing principle of human activities. **States** or countries are independent political units that claim exclusive jurisdiction over defined territories and over all of the people and activities within them. The governments are not always able to exercise this jurisdiction completely, but states can encourage or even force patterns of human activities to conform to the political map. Chapters 6 and 7 noted that many aspects of culture are affected by the political partitioning of the world. Patriotism, which is a strong emotional attachment to one's country, is itself a powerful cultural attribute. A country can be, in many ways, a fixed culture region.

Several of the countries shown on today's world political map are not, however, effectively organized (Figure 11-1). A great many people would like to see other countries on the map and boundaries moved, often because of cultural differences. In some countries, much of the territory is not ruled by the central governments. The existing governments try to consolidate

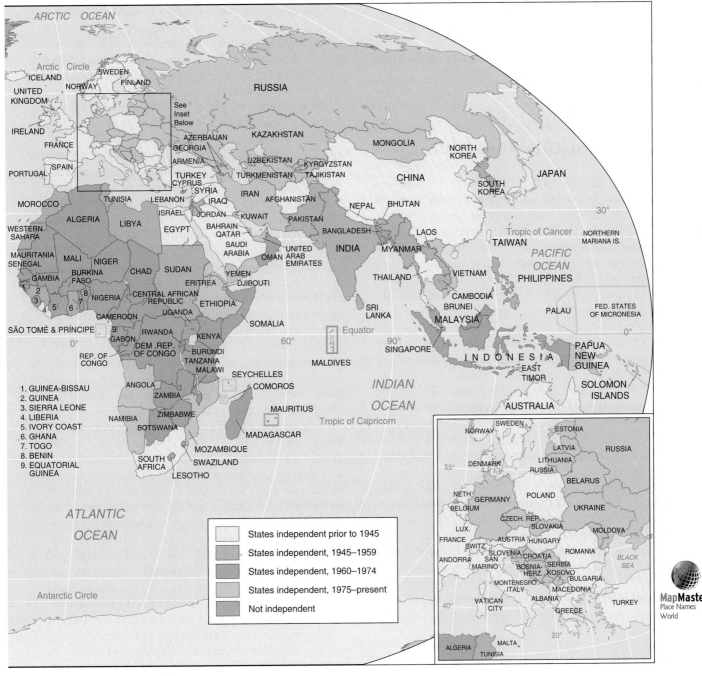

ARCTIC OCEAN

Arctic Circle
ICELAND
UNITED
KINGDOM
IRELAND
FRANCE
PORTUGAL SPAIN
MOROCCO
WESTERN
SAHARA
MAURITANIA
SENEGAL MALI NIGER
GAMBIA
BURKINA
FASO
NIGERIA
SÃO TOMÉ & PRÍNCIPE
REP. OF
CONGO
ATLANTIC

OCEAN

1. GUINEA-BISSAU
2. GUINEA
3. SIERRA LEONE
4. LIBERIA
5. IVORY COAST
6. GHANA
7. TOGO
8. BENIN
9. EQUATORIAL
   GUINEA

Antarctic Circle

SWEDEN
NORWAY FINLAND
AZERBAIJAN
GEORGIA
ARMENIA
TURKEY
CYPRUS
TUNISIA
ALGERIA LIBYA
LEBANON
ISRAEL
EGYPT
JORDAN
BAHRAIN
QATAR
SAUDI
ARABIA
CHAD SUDAN
ERITREA
DJIBOUTI
YEMEN
CENTRAL AFRICAN
REPUBLIC
CAMEROON
UGANDA
ETHIOPIA
GABON
DEM .REP.
OF CONGO
RWANDA
BURUNDI
TANZANIA
MALAWI
ANGOLA
ZAMBIA
ZIMBABWE
NAMIBIA
BOTSWANA
SOUTH
AFRICA
LESOTHO
MOZAMBIQUE
SWAZILAND
MADAGASCAR
KENYA
SOMALIA
SEYCHELLES
COMOROS
MAURITIUS

RUSSIA
KAZAKHSTAN
MONGOLIA
UZBEKISTAN KYRGYZSTAN
TURKMENISTAN TAJIKISTAN
SYRIA IRAN AFGHANISTAN
IRAQ
KUWAIT
UNITED
ARAB
EMIRATES
OMAN
PAKISTAN
INDIA
NEPAL BHUTAN
BANGLADESH
MYANMAR
THAILAND
SRI
LANKA
MALDIVES

See
Inset
Below

NORTH
KOREA
CHINA
SOUTH
KOREA
JAPAN
TAIWAN
LAOS
VIETNAM
CAMBODIA
BRUNEI
MALAYSIA
SINGAPORE
INDONESIA
EAST
TIMOR

Tropic of Cancer
PACIFIC
OCEAN
PHILIPPINES
NORTHERN
MARIANA IS.
PALAU
FED. STATES
OF MICRONESIA
Equator
INDIAN
OCEAN
Tropic of Capricorn
PAPUA
NEW
GUINEA
SOLOMON
ISLANDS
AUSTRALIA

States independent prior to 1945
States independent, 1945–1959
States independent, 1960–1974
States independent, 1975–present
Not independent

NORWAY SWEDEN
ESTONIA
DENMARK
LATVIA
LITHUANIA
RUSSIA
RUSSIA
BELARUS
NETH.
GERMANY POLAND
BELGIUM
UKRAINE
LUX.
CZECH. REP.
SLOVAKIA
FRANCE
AUSTRIA HUNGARY MOLDOVA
SWITZ.
ANDORRA SLOVENIA CROATIA ROMANIA
SAN
MARINO
BOSNIA- SERBIA
HERZ. KOSOVO
MONTENEGRO BULGARIA
ITALY MACEDONIA
VATICAN ALBANIA
CITY
GREECE TURKEY
ALGERIA
MALTA
TUNISIA
BLACK
SEA

MapMaster™
Place Names
World

their control over all their territory, but not all are able to do so.

The idea that the whole world should be divided up into countries seems natural to us, but it is a relatively new concept in human history, and Figure 11-1 shows how young many of today's countries really are. This chapter will explain how the idea originated in Europe and how it was diffused worldwide as part of European conquest. The neatness of the units on today's world political map nevertheless obscures the fact that states are not fixed or settled but change over time. Maps suggest that borders are clearly demarcated and that they correctly divide different activities and peoples on their two sides, but in fact some activities and peoples are divided

by borders. The map also suggests that the areas within those borders are politically homogeneous, which is also false. Governments are only more or less successful in organizing their territory, and no territory can be sealed off. Border wars among countries continue to afflict the world, and civil wars within countries are even more common.

After we have explored the state as a geographic entity, this chapter will examine the political geography of states' internal organization, including the kinds of political regimes that control the state and how they maintain their borders and subdivide their territories. We will then consider relations between states, including the regions that are not controlled by any one state.

The ways in which countries try to organize their territories economically—that is, to build national economies and to trade with other countries—will be the subject of Chapter 12. We will return to the forms of cooperation among states in Chapter 13.

The Development of the Nation-State Idea

The idea that a state has the exclusive right to rule over a demarcated space and all the people and resources within it is known as **sovereignty.** In medieval Europe, sovereignty was fragmented among various politically powerful people rather than definite territories. Kings and queens were often weak because their sovereignty required good relations with lesser nobility who actually owned most of the land. The Church, which had survived the collapse of the Roman Empire, was very important in medieval Europe. The Pope exercised the right to rule on religious matters over all Christians, including rulers, and interfered in most political matters. Cities were also becoming sites of great wealth, independent from feudal obligations but pursuing their own interests at the expense of monarchs. All this changed when, in France, kings claimed sovereignty over everything within a *territory*. The earlier Merovingian kings (fifth through eighth centuries), for example, had called themselves Kings of the Franks (the people of France), but the later Capetians (tenth through fourteenth centuries) could now call themselves Kings of France, a territory with everything and everyone in it. They had established the modern form of sovereignty, and the idea was adopted by other states in Europe. Historians often mark the 1648 Treaty of Westphalia, which ended Europe's Thirty Years' War, as establishing in Europe the general principles of territorially sovereign states and of noninterference by rulers in states other than their own.

The Idea of the Nation

A state is a territory on a map, but a **nation** is a cultural entity, defined by the people themselves. It is a group of people who want to have their own government and rule themselves. The feeling of nationality may be based on a common religion or language, but it does not have to be. For example, Switzerland has four official languages, and its people practice different religions. A group sharing a sense of national identity may or may not share any other attributes. Some nations define themselves in exclusive terms, based on ethnicity or race. The Croatian nation is defined as Catholic Slavs who speak Croatian; the Japanese are defined as members of the Yamato people who speak Japanese and are nominally Shinto-Buddhist. Both countries contain minority groups who are not members of the dominant national group. Nationalism is a cultural concept in its own right.

Nationalism is one expression of **political community,** which is a willingness to join together and form a government to solve common problems. Today, in most places, the state is the most powerful level of political community, but Chapter 13 will ask whether worldwide concern over environmental pollution or terrorism may nurture a sense of global political community. The evolution of nationalism is part of the story of the evolution of *legitimacy*, the question of who has the right to rule any group. Upon what sort of consent of the governed, if any, is rule based? Even the most totalitarian states today claim to represent the people.

European emperors and kings ruled over earthly matters with the blessing of the Church—by divine right—even after the establishment of the territorial state. Religion was a powerful legitimation of sovereign power that most people obeyed. Individuals were not expected to personally identify with the state. Kings could often assign and reassign thrones among themselves and redraw the political map without significant protest by the people. France's King Louis XIV (who reigned 1643–1715) insisted flatly, "I am the state." The Church, however, as protector of people's souls, demanded a greater degree of personal commitment than the sovereign did. Most people at the time identified as Christians more than as nationals.

The Protestant Reformation challenged this traditional church–state accommodation. Martin Luther preached that every person was his or her own priest and therefore carried individual responsibility for his or her own soul. This belief undermined the religious authority of the Church and, by extension, the divine right of kings. In 1581, the Dutch, who were then subjects of the king of Spain, adopted an Act of Abjuration that renounced (abjured) the theory of divine right and argued that a king had an obligation to rule for the welfare of the people. If he did not, then the people could abjure his rule over them. This turned the notion of divine right upside down. It suggested that the king, or any government, served the people or nation.

The English themselves already had risen up against their king, Charles I; defeated his forces in battle; tried him for crimes against "his" people; and executed him in 1649. The subsequent English Bill of Rights (1689) recognized that sovereignty did not lie in the king but in the people. The U.S. Declaration of Independence of 1776 claimed the right of the people against the abusive rule of King George III. Unlike the English or Dutch, the new American "nation" was not a unified cultural identity but a political community in the making. What makes a nation can vary widely.

The Nation-State Idea

The Swiss philosopher Jean Jacques Rousseau (1712–1778) laid the foundation for people's allegiance to the state. Rousseau believed that in nature, people were merely physical beings, but when they united in a *social contract*, they were capable of perfectibility. For Rousseau, politics was a means to moral redemption. Rousseau's ideas swayed France, and the French Revolution gave birth to the French nation. The 1789 Declaration of the Rights of Man stated: "The principle of all sovereignty resides essentially in the nation; no body nor individual may exercise any authority that does not proceed directly from the nation."

The nation demanded personal dedication and allegiance from its citizens. States rule over everyone in their territory. Therefore, the perfect state was idealized as a **nation-state,** a state ruling over a territory that contains all the people who are culturally joined as a nation. The idea of the nation-state assumed that nations develop first and that each nation then achieves a territorial state of its own.

In reality, true nation-states are rare. Perhaps Iceland's millennium of relative isolation provides the closest example of a match between one people and a state. Instead, some scholars have argued that several of the nation-states had historic **core areas,** or historic homelands, but then grew to encompass areas that included other cultures (Figure 11-2). France, for example, has long been focused on the region of Paris and Ethiopia on the highlands of

the horn of East Africa. In many other cases, however, core concentrations of settlement or activity developed only after the state had come into existence. In many cases, states lack a singular or dominant culturally defined nation. Instead, the nation is defined as those people subject to the constitution and laws of a state.

The particular circumstances of each state differ, and so each develops as a political community with a unique political culture. A **political culture** is the set of unwritten rules or the unwritten ways in which written rules are interpreted and actually enforced. Political communities differ widely, for example, in whether they honor elders, rich people, or religious leaders; in their tolerance of bribery; and in the rigidity or laxity with which they enforce laws. Political culture reflects other aspects of a people's culture, such as their religion. The key to the nation-state idea is that people who share a national identity will share the same political culture and thus form a government that better represents their interests.

Not all nations share one political culture. One scholar in the 1990s examined political life in Italy and found distinct regional differences in political culture, which he defined as "civic community"—patterns of social cooperation based on tolerance, trust, and

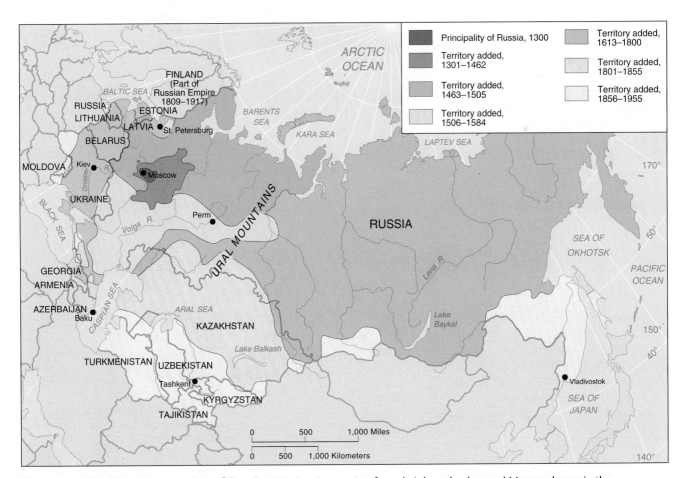

Figure 11-2 The historic expansion of Russia. Russians' expansion from their homeland around Moscow began in the fourteenth century, and by 1647, Russian cavalrymen had reached the Pacific. During the seventeenth and eighteenth centuries, Russians pushed south to the Black Sea. In the nineteenth and early twentieth centuries, czarist power advanced still further into Eastern Europe and Central Asia.

widespread norms of active citizen participation (Figure 11-3). In Italy, the distribution of civic community among the regions was already evident as long ago as the thirteenth century. Comparing the civic communities in Figure 11-3 with their economic productivity in Figure 12-22 suggests that patterns of political culture explain both a region's capacity for democratic self-government and its capacity for economic growth. Political leaders in regions lacking civic community may lack the fundamental building blocks upon which stable democracy can be built, and civic community may be difficult to create where it does not exist. Weak civic community is a common problem in many states. As we describe below, the "perfect" nation-state is more an idea than a reality.

The European Nation-States

The Napoleonic Wars that followed the French Revolution carried the idea of nationalism across Europe. Armies had formerly been composed of hired professionals, but now entire male populations had to serve. Military service demanded loyalty to the state and the nation as a whole and instilled a patriotic political culture among the citizens. Indeed, armies were called "the school of the nation." Although Napoleon was defeated in 1815, the code of laws that he had imposed left a widespread legacy in Europe. It swept away aristocratic privileges and strengthened the middle class.

The idea of nationalism matured in Europe during the nineteenth century. There were many distinct cultural groups within Europe's larger countries, such as the Austro-Hungarian Empire. These groups became the basis for nationalist movements that wanted greater rights or even independence. Other nationalist movements aspired to greatness and laid claim to the maximum extent of territory in which their people had ever lived, sometimes more! These assertions led to overlapping territorial claims and a series of wars.

After World War I, U.S. President Woodrow Wilson advanced the ideal of the nation-state, which he called **national self-determination,** or the right of a nation to govern itself. The victors redrew the map of Europe to break up the defeated German, Austro-Hungarian, and Ottoman empires (but not the victors' own empires) and to grant self-determination to several new European nation-states, including Poland (Figure 11-4). Russia's new Communist ruler, Vladimir Lenin, had criticized the czarist Russian empire as "a prison-house of nations." He and his successor, Stalin, reorganized the empire under a new totalitarian government disguised

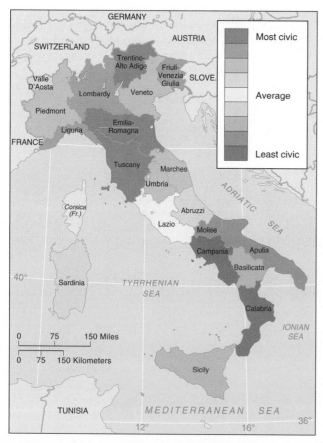

Figure 11-3 Variations in civic community within Italy. This map reveals the sharp differences in *civic community* that the scholars Robert Putnam and colleagues found in each of the regions of Italy.

Figure 11-4 Europe in 1920. After World War I, the defeated German, Austro-Hungarian, Russian, and Ottoman empires were dismembered, and several new states appeared. Some of them, such as Yugoslavia and Czechoslovakia, did not represent nations but were composed of diverse populations. The new Poland was unsatisfied with the eastern border it was originally awarded, and it seized more territory in a war against Russia.

as a union of nations—the Union of Soviet Socialist Republics, or Soviet Union—but that Union split apart in 1991.

It can be argued that several European nations existed before they achieved their own independent territory and governments—that is, their own states. Few states, however, have ever achieved a complete match between the spatial extent of the cultural nation and the territory of the physical state. The European map was redrawn again after World War II (Figure 11-5). However, rebellious groups in several states still claim parts of their neighbors' territories. Territorial claims on a neighbor are called **irredenta,** from the Italian word for "unredeemed." For example, many Austrians still claim South Tyrol from Italy, and many Hungarians claim the province of Transylvania from Romania, as shown on Figure 11-4. In both these cases, however, the state governments officially accept existing borders. By contrast, the Bolivian government still formally claims the coastline that Chile took from it after an attack in 1879. States often prioritize irredenta in national politics to stimulate patriotic loyalty or justify war.

The Collapse of Empires

At the same time as the idea of nationalism was maturing in Europe, the Europeans were actually enlarging their empires over peoples outside Europe (Figure 11-6). European imperialists were not willing to recognize that their colonial subjects had national rights. They argued

Figure 11-5 Europe in 1946. After World War II, the Soviet Union expanded considerably by absorbing territory from its western neighbors. It retook the territory that Poland had won in the earlier war and gave Poland some German territories, thereby effectively shifting Poland about 240 kilometers (150 miles) to the west. Yugoslavia took territory at the head of the Adriatic Sea from Italy, but the port of Trieste, which Yugoslavia wanted, remained Italian. Compare this map with the current political map (Figure 11-1) to see the changes that have occurred since the end of World War II.

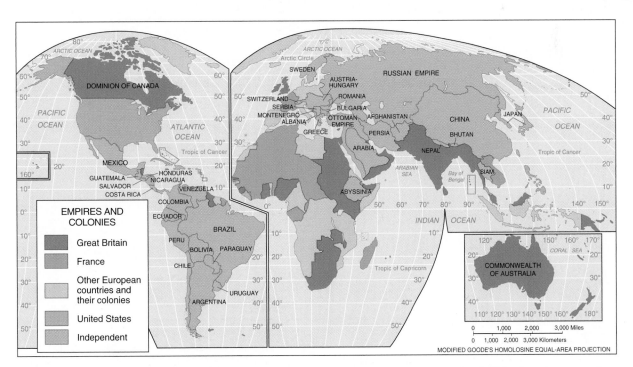

Figure 11-6 The world in 1900. In 1900, there were far fewer independent actors in international affairs than there are today. Compare this map with today's political map, Figure 11-1.

that non-Europeans were inferior, or "not yet ready" for political independence. The argument that native forms of government everywhere were "backward" compared to European forms was a rationalization for European conquest. Some intellectuals helped to justify brutal imperial policies. British sociologist Herbert Spencer (1820–1903) extended Darwin's theory of evolution to insist that "Nature's law" called for "the survival of the fittest," even among individuals, cultures, and whole peoples. Spencer's theory is called **social Darwinism.**

Individual colonies seldom matched pre-colonial political or cultural communities. When the European imperialists carved up the world among themselves, they drew borders that ignored any existing political organization among the native peoples. Some of these new **superimposed boundaries** split native political communities, whereas others combined two or more in one colony (Figure 11-7).

The colonies were administered by bureaucracies made up in some cases of Europeans, in other cases of

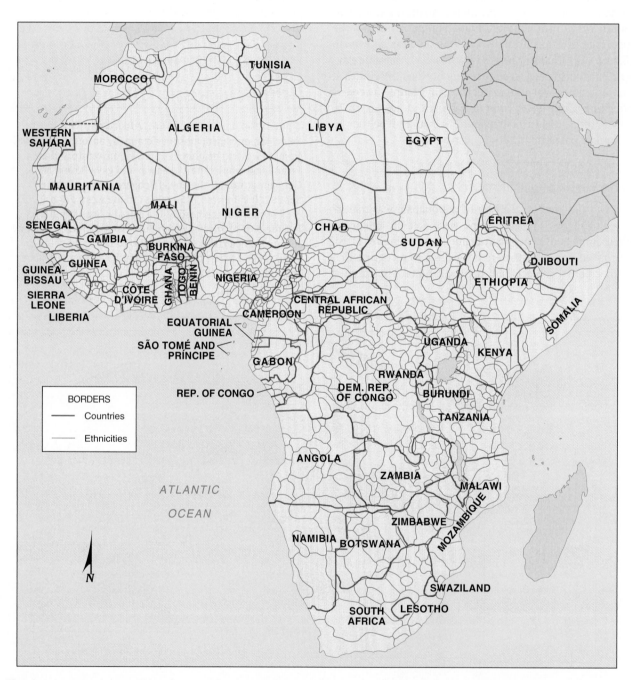

Figure 11-7 Superimposed borders in Africa. This map superimposes the borders of today's African states on the territories of just a few of Africa's many indigenous peoples, just as the European colonial powers superimposed these borders at the Berlin Conference of 1884–1885. When the colonies attained independence, the new states seldom contained unified nations.

acculturated natives, and in still others of foreign peoples imported by the Europeans. The Europeans often used native rulers as intermediaries between themselves and the people, especially if a colony included several groups. This form of government was called **indirect rule.** In many cases, the imperialist powers even combined groups into new tribes and appointed new kings or, in India, maharajahs. These new groupings soon gained the force of "tradition," making it difficult to later define a genuine pre-colonial community. Indirect rule hindered the native peoples from uniting against the imperialists. Later, the only feeling that brought native groups together in a sense of unified nationalism was often a shared indignation against the colonial power. It has been argued that still today the principal basis of some countries' nationalism is hatred of foreigners, of "traditional enemies," or even of internal minorities. Some countries' rulers skillfully create and manipulate these hatreds.

The twentieth century saw the fragmentation of empires that Europeans had built up over hundreds of years. After World War I, the Europeans did not offer national self-determination to their subject peoples outside Europe. The winners just took the losers' colonies while retaining their own colonies. After World War II, European powers could no longer afford to hold on to these territories, nor could they any longer justify colonialism. When the colonies received their political independence, their first new leaders were often those that had benefited from colonialism and they had a vested interest in maintaining existing borders, as well as in economic relations with the former colonial power. Therefore, today's world political map is not a map of nations that have achieved statehood. It is a remnant of colonialism. The retreat of empires also left behind many tiny, independent nation-states, including most of those in Table 11-1. Some, such as Bahrain, thrive economically, but others rely on foreign subsidies. Few are militarily defensible. Small, weak, or poor states often negotiate defense or economic treaties among themselves, with more powerful neighbors or their former colonial masters. The last few remnants of empires can be seen in Figure 11-8. We continue to live with the effects of a few former empires, such as the British, French, and Ottoman.

British Empire to Commonwealth

At its peak in 1900, the British Empire covered over one-quarter of Earth's land surface. This empire has gradually been transformed into a loose association of independent states called the Commonwealth of Nations. Most of the United Kingdom's former colonies have chosen to join, and today the Commonwealth has 53 members. The British monarch remains head of state in 15 countries in addition to the United Kingdom, including Canada, New Zealand, and Jamaica, but most former British colonies have become republics. The British legal system, its language and education system, the Anglican Church, and other cultural traditions linger in each of the countries. The Commonwealth offers a framework for consultation and cooperation for the achievement of common ends, where they exist; and its achievements have enticed Mozambique, a former Portuguese colony, to join.

Britain retains possession of the Falkland Islands ("Malvinas" in Spanish) against claims by Argentina, which even invaded the islands in 1982; and Britain retains Gibraltar against claims to it by Spain. Bermuda is a Crown Colony with its own internal representative government. All three colonies have repeatedly voted to maintain their status.

Northern Ireland's status is open to question. The United Kingdom gave independence to 26 of Ireland's 32 counties in 1921, but the 6 counties of Northern Ireland (Ulster) remain within the United Kingdom. The struggle between Catholic Irish nationalists who wanted to leave the United Kingdom and protestant unionists who wanted to stay with the United Kingdom led to almost nonstop violence beginning in the 1960s. A peace agreement in 1998 provided for a local government, but the province will remain part of the United Kingdom unless a majority of its citizens votes otherwise.

TABLE 11-1 A Few of the World's Smallest Independent States					
	Area (km²)	Area (mi²)	Population (in Thousands)	GDP (in Billions of U.S. Dollars)	Year of Independence
Bahrain	741	286	728	21.2	1971
Barbados	430	166	285	3.7	1966
Grenada	344	133	91	0.6	1974
Maldives	298	115	396	1.7	1965
Monaco	2	0.77	33	0.98	1297
Nauru	21	8.1	13	0.06	1968
San Marino	61	23.6	30	1.1	301
Tuvalu	26	10	12	0.015	1978
Vatican City	0.438	0.169	0.826	(not available)	1929 (as state)

Source: U.S. Central Intelligence Agency.

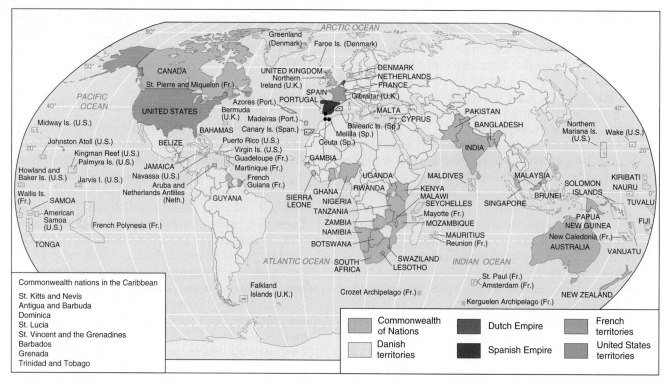

Figure 11-8 Remnants of empires. A few empires linger, whereas the breakup of others caused the creation of cooperative organizations. Some empires' former colonies maintain less formal ties not shown here.

The French Empire

France built up an empire in the seventeenth and eighteenth centuries but lost most of its colonies during an eighteenth-century rivalry with Britain. During the nineteenth century France built a second empire, beginning with the occupation of Algeria in 1830. France eventually ruled large parts of western and northern Africa and Southeast Asia, as well as Madagascar, Syria, and French Guiana. The French believed that their subject peoples would mature not to independence but to full participation in France itself. The four colonies of Martinique, Guadeloupe, Réunion, and French Guiana were organized as Overseas Departments of France, and each was granted representation in the French National Assembly in 1946. Algeria won independence in 1962, after a long and bloody war. France granted independence to most of its other colonies in the 1950s, but it continued alliances. France's former African colonies receive aid and still even host French military forces.

The present French Republic encompasses, in addition to mainland France and the four Overseas Departments, two Territorial Collectivities and four Overseas Territories. France also possesses a number of islands, including St. Pierre and Miquelon off the south coast of Newfoundland, a remnant of France's once-vast North American holdings. New Caledonia in the southwestern Pacific has major nickel reserves, and in 1998 its people agreed to defer a vote on independence for at least 15 years. France's other territories are generally poor and sparsely populated.

The Successor States of the Ottoman Empire

The Ottoman Empire was a vast entity extending from southeast Europe to the Indian Ocean, including much of what is today called the Balkans and the Middle East. Although the Ottoman family was Muslim and Turkish, their Empire contained and tolerated a wide range of different linguistic and religious communities. In the nineteenth century, some of these groups began to form nationalist movements and sought greater freedom. Serbian, Bulgarian, Greek, Kurdish, Arab, and Albanian nationalists often rebelled against the Empire. When the Ottoman Empire collapsed and a new Turkish nation was born in the 1920s, most of the Empire in the Middle East was divided between France and Britain. New states, including Iraq and Israel, eventually gained independence, but the map of today's states was drawn according to the interests of the colonial powers rather than the sentiments of the local people. The region is plagued by disputed borders; rulers whose power is based on foreign interests rather than popular support; religious animosities; and political and economic intervention by international oil companies and banks. Several wars have troubled the region since World War II, frustrating many hopes for democracy and economic development. As in many other former colonies, these areas host political movements that would like to change the map, often violently.

A Changing World Political Map

Today's world political map is made up of territorial states that each possesses political sovereignty. Most political theories assume that peace and stability are not possible unless all territory is part of one and only one state. The United Nations, and before it the League of Nations, was created to ensure the stability of a system of independent sovereign states by helping to resolve conflicts over territory and other disagreements between states. It also helps coordinate state cooperation over large parts of the world that are not part of sovereign states, including Antarctica and the High Seas. (This topic is discussed later in this chapter.) Yet states do not always cooperate and the world map of states can be altered, especially by acts of war. Changes may also come about within states as internal political forces struggle to define and control each state. This is because states are subject not only to challenges from other states but also from people within their borders. This section will examine the **centripetal forces** that bind states together and the **centrifugal forces** that tend to pull states apart.

States have to contend with both centripetal and centrifugal forces, which have specific geographic characteristics affecting how governments deal with disputes. Disputes arise, for example, over demands by minority groups in one region for greater participation in central government or more cultural rights. If the government accommodates their political demands, then this is a centripetal force that changes how the country is put together. If they are excluded, they may rebel and pull the country apart; a centrifugal force. Disputes also arise over the distribution of resources and wealth between different parts of a country. Governments may share resource wealth more equally, thus pulling the country together. If the resources go to only one group, it may cause other groups to rebel. Sometimes multiple claims against a state occur. Today in Iraq, for example, the Kurds in the north of the country want to keep the wealth that comes from oil fields in their region. The central government is concerned that this will fund Kurdish demands for greater independence from Iraq, which the Kurds have expressed before. In order to encourage the Iraqi Kurdish region to cooperate and remain part of the state of Iraq, the central government has to share enough of the oil wealth with the Kurds for them to see benefits in being part of Iraq and not rebelling.

Sometimes governments choose to fight centrifugal forces, such as rebellious minority groups who seek a change in borders and territory, even if it solves a difficult problem. Governments fear establishing precedent that would invite other groups to further carve up state territory. Allowing one region to leave the state might encourage other centrifugal forces to tear apart the rest of the state. Despite inevitable problems, political maps continue to be redrawn. Europe, in recent years, saw the unification of East and West Germany in 1990, after the country was split apart following World War II. Czechoslovakia split into two countries—the Czech Republic and Slovakia—in 1993. Yugoslavia broke up in 1991 (Figure 11-9). Montenegro split off from Serbia in 2006, and Kosovo declared independence in 2008. The island of Cyprus, meanwhile, has taken small steps toward easing the division between its northern and southern parts.

Few borders have been redrawn in Africa or Asia since de-colonization. African countries could have chosen to redraw their borders at the time they received their independence, but because many of these countries are quite young, they are still trying to promote centripetal forces that will bind them together. At a meeting of the Organization of African Unity (OAU) held in Cairo, Egypt, in 1964, the African states pledged themselves to respect the existing international borders. They made this pledge even though they resented the borders as a colonial legacy and found it difficult to govern within them. In 1994, however, Rwanda's president, Pasteur Bizimungu, demanded that neighboring Congo (then called Zaire) surrender territories occupied by the Banyamulenge, a people related to the Tutsi of Rwanda. President Bizimungu called for "a Berlin II [conference], so that we can reflect on the disorders created by Berlin I [where colonial powers divided Africa]," which are visible in Figure 6-29. In 2000, Edem Kodjo, former prime minister of Togo and Secretary General of the OAU, said, "A certain number of principles that date from the independence era are already being shattered. . . . The intangibility of borders is likely to be next, because many of our states are inviable, either for economic or political reasons, or because of their population or ethnic composition." Thus, 40 years after the Cairo Conference, African leaders began considering redrawing the map. Some African states may either break up (for example, Somalia) or else recognize subnational communities by adopting federal forms of government (discussed later in this chapter).

In Asia, the U.N.-assisted independence of East Timor brought an end to Indonesian occupation in 2002. Further remapping in Asia might include: independence for Palestine, re-unification of the Koreas, or the break-up of Iraq, Afghanistan, Pakistan, or even China (as will be noted later in this chapter). Other potential new countries that may be created in the coming years include Bermuda (a British territory), Greenland (a division of Denmark), New Caledonia (a territory of France), and Western Sahara (under the control of Morocco but in dispute).

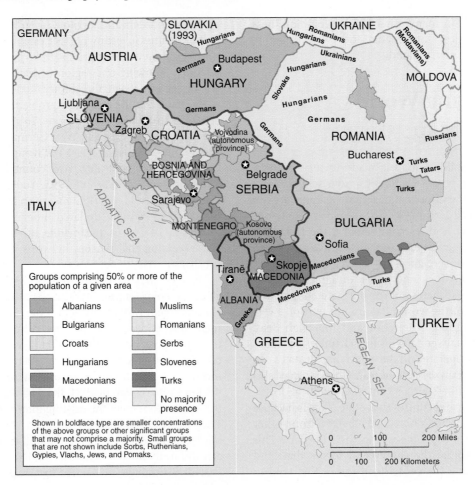

Figure 11-9 (a) Nations and ethnic groups in the Balkans in 1990. During the nineteenth century, nationalist movements in the Balkans rebelled against the Ottoman and Austro-Hungarian empires that ruled the region. Both empires were multiethnic and the settlement patterns were quite diverse. After World War I, Yugoslavia was formed out of the "South Slav" peoples. Some nationalists in these countries still wanted separate states and Yugoslavia dissolved in 1991 when the constituent republics of Slovenia, Croatia, Bosnia-Herzegovina, and Macedonia declared their independence. Serbia and Montenegro stayed together for several years, but Montenegro broke off in 2006. (For the borders current in 2010, see Figure 11-1.)

Beginning in 1992, Serb militias in Bosnia-Herzegovina, backed by troops from Serbia, began ethnically cleansing Bosnian Muslims (green in this map) from territories claimed to be part of a new Bosnian Serb state. Croatia eventually ethnically cleansed the Serbs from its territory and resettled the area with Croats. In 1998, Serbia opened an offensive to drive ethnic Albanians out of Kosovo, the province of Serbia that has an Albanian population. The Albanians returned under NATO and United Nations protections. Kosovo declared independence in 2008. In 2001, Macedonia, by contrast, peacefully granted Albanians cultural rights.

Centripetal Forces

Although the sovereign state is a basic part of the world political map, many states today are still trying to weld their populations into nations by forging national identities. This reverses the theoretical order in which the formation of a nation precedes the achievement of statehood. The nation-state ideal imagined that traditional cultures give rise to nations, which achieve modern states, not the other way around. Instead, there are many examples of how governments try to change human activities in their territories such as cultural identities and economic activities, to better "fit" the state. We explored some of these centripetal forces in Chapter 7, which noted that governments can affect the geography of religion and language. In addition, governments can affect the geography of political

community, of law, of land use, and as will be seen in Chapter 12, of economic activity. Countries can encourage the circulation of people, goods, and ideas within their territories, and they can restrict or discourage circulation across their borders. Countries can never control these activities entirely, but modern governments exercise incomparably greater power over human activities than governments did in the past.

We have already emphasized that many units on today's world political map are relatively new. No one can predict how long any world political map will remain fixed, but existing governments try to stabilize it. The history of the United States recounts many struggles to weld one nation out of many diverse groups. In Europe, a few nation builders faced their task self-consciously. The Polish hero and later president Joseph Pilsudski

Figure 11-9 (b) Bosnia and Herzegovina today. An accord concluded at Dayton, Ohio, in 1995 provided that Bosnia and Herzegovina would be a single country divided in two parts: One part was a Bosniak (Muslim)-Croat Federation and the other a Serb Republic. The entities have their own governments while the central government remains weak. The Dayton Accords guaranteed that refugees and displaced persons could return to their homes, and that war criminals would be put on trial. Neither has been fully implemented.

Figure 11-10 Masai people.   The westernized Kenyan on the left appears to regard these Masai men visiting downtown Nairobi with bewilderment (and, it seems, hostility). Kenya is forcibly acculturating the Masai to modern norms, and Masai are prohibited from dressing like this on the streets of Nairobi today.

(1867–1935) said flatly, "It is the state that makes the nation, not the nation the state." Italy was politically unified in the 1860s, and after the unification the statesman Massimo D'Azeglio observed: "We have made Italy; now we must make Italians." As illustrated in Figure 11-3, Italian politicians have not completely succeeded.

European experiences were echoed by Julius Nyerere, the first president of Tanzania (served 1961–1985). He commented on the African states: "These new countries are artificial units, geographical expressions carved on a map by European imperialists. These are the units we have tried to turn into nations." Many Asian states face the same difficulty, and even many Latin American countries that have been independent for over 150 years still have not earned the loyalty of their entire population.

The destruction of traditional loyalties   Before some states can forge their populations into nations, they often try to undermine the traditional loyalties within their borders. These loyalties may be based on kinship and tribe, religious authorities, or former rulers. Changing these loyalties can be profoundly destructive psychologically to at least the first generation of the new citizens. Typical is Kenya's campaign to abolish the ancient traditions and distinctiveness of the Masai people, once one of Africa's most powerful nations, and to integrate them into the modern state. The Masai must go to

school and conform to a new style of life dictated by the government. They must surrender their age-old lifestyle of nomadic cattle herding, settle down, and take up farming (Figure 11-10).

Some countries have absorbed the leaders of their traditional groups into the new political structure (Figure 11-11). Cameroon, for example, still allows the 17 traditional kings within its borders to exercise judgment over minor crimes, and it defers to their ceremonial importance. In Zimbabwe local chiefs appoint 10 members to the 150-seat parliament. In the 1990s South Africa's president Nelson Mandela tried to educate and train the country's 300 Zulu chiefs in law and administration. His efforts were criticized as a centralization of power, but the chiefs' powers had in fact been created by the British colonial authority. This struggle provides a fine example of the confusion of pre-colonial tradition with imperialist legacies. Members of many African royal families retain positions of power and influence in the modern states. Mandela himself was a traditional prince, and Ghanaian diplomat Kofi Annan, U.N. Secretary General from 1997–2006, is from a family of traditional chiefs. In Malaysia, a paramount ruler is elected every five years from among the nine traditional Malay sultans. The reduction of these sultans' power in 1993 symbolized Malaysia's emergence from a traditional monarchical country to a modern democratic country.

India abolished the privileges of its many maharajahs in the 1970s, although many of them subsequently won elective offices. India remains the world's largest

Figure 11-11 The Alake (king) of Abeokuta (seated) and his son. These two men represent the transition from tradition to modern statehood in Africa. The son of the Alake of Abeokuta was educated in Britain and returned home to serve as the chief justice of the new state of Nigeria.

experiment in bringing diverse groups of people together under democratic political structures. India is afflicted by centrifugal forces based on religion, ethnicity, caste, and language. Several of these distinctions are regional and might encourage separatism, but the Indian civil service, a legacy of British rule, remains a strong centripetal force.

Urbanization, which is occurring rapidly everywhere, often dissolves traditional rural-based identities. This development may reduce traditional loyalties, but it does not assure loyalty to the state.

Instruments of nation building Removing traditional loyalties is not sufficient to create a stable political community. Different countries rely on different instruments of nation building, and the choice reflects differences in other aspects of their cultures. Among the instruments most often used are religion, the armed forces, education, symbols, media, political parties, and labor unions.

Religion Religion historically played an important role in legitimating political rulers and unifying people around them. Today, national identities are often based on a strong sense of religious heritage. Chapter 7 noted countries in which a national church was a building block of nationalism: Ireland, Ukraine, and Poland. Orthodox churches are traditionally national churches, and some Protestant denominations are also rooted in specific nations: Lutheranism in Germany, Presbyterianism in Scotland, and Anglicanism in England. Chapter 7 noted that the emperor of Japan renounced his claim to divinity after World War II, but in 2000, Japanese Prime Minister Yoshiro Mori described Japan as "a divine nation with the emperor at its core." When the remark provoked a storm of criticism, the prime minister explained that he had meant only that the nation should be more spiritual. Many Russians view Orthodox Christianity as an essential element of Russian identity, and Russia severely restricts missionary activity of other faiths. Religion was also a unifying force in overthrowing the Iranian monarch in 1979 and establishing the Islamic Republic of Iran.

Armed forces In many countries, the armed forces are sometimes called the school of the nation and claim a high proportion of the people for service. The armed forces may consume a large percentage of the national budget, but defense against external enemies is not always their principal function. The army holds the state together, either by force or just by training and socializing young people as loyal citizens. In many cases, the armed forces provide a stable source of income and a disciplined labor force for building infrastructure. The military forces may even exercise an independent role in the government. In Chile, for example, the armed forces are guaranteed their own seats in the national legislature. The Turkish constitution grants the army a special role in preserving the government. Even the U.S. armed forces have been assigned nation-building tasks. In 1947, President Harry Truman ordered the racial integration of the U.S. military. This was brought about successfully decades before the society at large was ready to integrate.

Education National school systems are another tool with which states create nations. Rousseau wrote, "Education must give souls a national formation, and direct their opinions and tastes in such a way that they will be patriotic by inclination, by passion, by necessity." Today in France, it is said, the minister of education in Paris can look at his watch at any time of day and say precisely what all French schoolchildren are studying. In 2003, the United Nations administrator of Bosnia and Herzegovina mandated the integration in local high schools of Serbs, Croats, and Muslims, children of the groups who had fought so hard to live apart. It was hoped that integration in school could dispel age-old antagonisms.

Most countries have national curricula and textbooks. The interpretation of national history in those texts is a political issue, and a change in the government can alter that interpretation. After East Timor broke away from Indonesia in 1999, its education system moved towards instruction in Portuguese, one of the official languages of the new state. Indonesian history and geography were removed from existing textbooks until new books were ready. Education in Kosovo similarly reflected the political changes of recent years. In the 1990s, Albanian-language schools were closed by Serbia. Since the war in 1999, the Kosovo school system instructs students in Albanian, which is a problem for its minority Serbian-speaking population. History, culture, and geography lessons are now taught from the point of view of a newly independent Kosovo.

Schools instill in youngsters the society's values and traditions, its political and social culture. This is called the process of **enculturation,** or **socialization.** National curricula propagate national culture, literature, music, and artistic traditions. National history and geography lessons leave the rest of the world outside the focus of concern.

Manipulating symbols Each country has its own set of national symbols, called an **iconography,** and these items are circulated widely, beginning with schoolchildren. The national flag is one of the most important of these icons, and several countries have laws against its desecration (Figure 11-12). A nation's national anthem is another icon. In a unique instance of international solidarity, however, the nonpolitical prayer for God's blessing, "God Bless Africa," serves Namibia, Tanzania, Zambia, Zimbabwe, and South Africa. It was written by a South African schoolteacher and taken up by the African National Congress as its anthem.

The map is another icon of great power. Children can be taught to respect—even to cherish—the size, shape, topography, resources, and variety of their land. The U.S. weather map has demonstrated appeal on television, and news editors know that the weather map is one of any paper's most popular features. Few people actually read the map or even care about the weather in other areas of the country, but the map itself is an icon that wins allegiance.

Iconography also includes the pomp and circumstance of national ceremony—the celebration of holidays, national costumes, the designs on national postage stamps, and sports competitions. In 2003, the U.S. Treasury introduced a new $20 bill redesigned to foil counterfeiters and also specifically to add "symbols of freedom," meaning more eagles and flags. Many countries ban public displays of the iconography of past regimes, including fascist symbols in Germany, Hungary, and Romania and Communist symbols in several formerly Communist countries.

Media National media usually can reach and sway more of a national population than any other shared experience, even the army or the school system. One-third of India's people, for example, are illiterate, but 80 to 90 percent of them can be reached by state-controlled radio and television. Government media monopolies may inform the population, educate it, win its allegiance, or command it. Forms of media that spill over international borders can exert a powerful centripetal force. For example, one reason that East Germany never created an East German nation, despite 40 years of trying, was that the government could not prevent East Germans from watching West German television.

Political parties Political parties can politicize and mobilize diverse populations, recruit people, and give them a sense of participation. They can broaden the government's base of support, but they can also be divisive. If people identify themselves more closely with their tribe or faith than with their country, political parties can become the means of promoting narrow communal interests, or, worse, of fomenting ethnic grievances. In Uganda, President Yoweri Museveni, who seized power in 1986, long banned political parties because he considered them divisive, but he allowed them to form in 2003. In some countries, political parties reach out to minority groups to give them a voice in government, lowering the potential that they might rebel. Minority groups sometimes form their own parties. In Serbia, there are separate political parties representing ethnic Croats, Bosniaks, Hungarians, and Albanians, while most major parties represent Serbian nationalist parties.

Labor unions Labor unions may serve as still another building block of a nation. Political parties representing laborers or workers even govern in some countries. Traditionally, they have represented the interests of people who are not well represented by parties dominated by economic or political elites. In totalitarian states, the government might sponsor official labor unions and curtail the formation of independent unions. In Poland, under Communist Party rule, the independent Solidarity labor movement nevertheless overwhelmed the Communist Party candidates for parliament and deprived the

Figure 11-12 The flag of Montenegro. Montenegro adopted a new flag in 2004, even before it achieved full independence in 2006. The gold border contains a red field, centered in which is the nation's coat of arms. The coat of arms was taken from that of the nineteenth-century Montenegrin royal family.

Communists of any pretense of legitimacy. Independent unions also played a key role in bringing down the Communist government in the former Soviet Union.

Each of these seven institutions—religious institutions, armed forces, schools, symbolism, media, political parties, and labor unions—may be either a centrifugal force or a centripetal force in any country. Geographers study their presence or absence, the mix or balance among them that is unique to each country, and their interaction with other aspects of each country's culture. Furthermore, each institution has a geography within each country: One or more institutions may be equally influential across the entire territory, be concentrated in only one region, or spill over international borders. In many countries, for example, including the United States and Canada, the political parties show regional strengths. In several African countries, the army draws disproportionately upon one ethnic group or region for troops.

Centrifugal Forces

Not all states succeed in unifying their people into one nation. National unity can be threatened either by globalization or by regionalism. In a sense, globalization erodes the borders around a country and regionalism makes new borders within a country. Globalization includes forms of cross-border cooperation and interaction that decrease the sovereignty of the nation-state. This may take the form of partnerships such as military alliances or free-trade treaties that may expose a nation to problems in other countries. Today many global social and economic developments exert centrifugal force against the centripetal forces of nationalism. (Later in this chapter, and in Chapters 12 and 13, we will examine some of these global forces that weaken states but do not tear them apart.) **Regionalism,** on the other hand, tends to internally divide a country into many parts according to cultural and economic differences with a state. Some regions may even try to forcibly separate territory in hopes of making a new state.

When the entire population of a state is not bound by a shared sense of nationalism but rather is split by regional or other allegiances, then that state is said to suffer **subnationalism.** In some states, people grant their primary allegiance to a national or regional ethnic group that may be opposed to the existing state or its borders. These subnational identifications may be strong enough to foster a **separatist movement** that seeks to exercise its self-determination and form a new country, often leading to civil war. Some movements hope to ultimately achieve **partition,** making a new state by drawing a border around territory taken from the existing state. The Bosnian Serb politicians who started the war in that country wanted to partition Bosnia along supposed ethnic lines to form their own country. Separatist questions can be answered without violence, too. Canada's French-speaking population in Quebec have twice voted in referenda to decide whether to stay part of Canada, which they chose to do in 1980 and again in 1995.

In some cases, a national or ethnic group's membership may extend beyond the state's borders, producing the irredenta discussed above. Some groups may be inspired to alter territories and erase borders in pursuit of **unification** in an enlarged or even new state. A group's original division may itself have been the result of a partition to serve other interests. For example, the Kurds were promised a national homeland at the end of World War I, but instead they were split among several new countries to help Britain control Iraq (Figure 11-13). During the twentieth century, Kurdish nationalists in each country have rebelled against the partitioning states in hopes of unifying Kurds. After the United States defeated Iraq in the Iraq War in 1991, it allowed the Kurds in northern Iraq to establish a de facto independent state there. This action has frustrated the Turks, who are trying to squelch demands of the Kurds in Turkey for independence. The Turkish army has chased its Kurdish rebels over the border and into Iraq on several occasions. Since the U.S. invasion of Iraq in 2003, this has created problems for Turkish relations with the United States, despite a military alliance between them.

Civil wars demonstrate that many countries on the world map are not nation-states (Figure 11-14). Rebels in most civil wars claim to be fighting to correct religious, ethnic, or political wrongs, but those claims may also relate to economic motives. In some cases, separatists might feel aggrieved against governments that made them poor or failed to help raise the standard of living in their region. Basic issues, such as access to farmland, can drive peasants to take up arms. Another possible factor for civil war is economic dependence on commodities that can be plundered by rebel groups, giving them incentive and funding to continue their fight. Prime examples of wars of plunder include the wars in Nigeria (over oil), Colombia (drugs), and Sierra Leone (diamonds, as dramatized in the 2006 film *Blood Diamond*). The way to end such civil wars is for countries to develop away from dependence on commodity production or for international markets to embargo, or ban, sales of commodities by rebels. The United Nations did embargo the sale of diamonds from Sierra Leone on world markets in 2000. This does little to quell regional grievances, however.

Many observers fear a world in which the governments of some states cannot control their territories. Colombia, for example, suffered a civil war that took over 35,000 lives and displaced more than 1.5 million people in the 1990s. By the end of the decade, the government admitted that it did not exercise sovereignty over its total area, and it recognized guerrilla insurgencies' control over two areas. The United States, however, fearing increases in drug supplies shipped from those areas, supplied forces and arms to encourage Colombia to reconquer them.

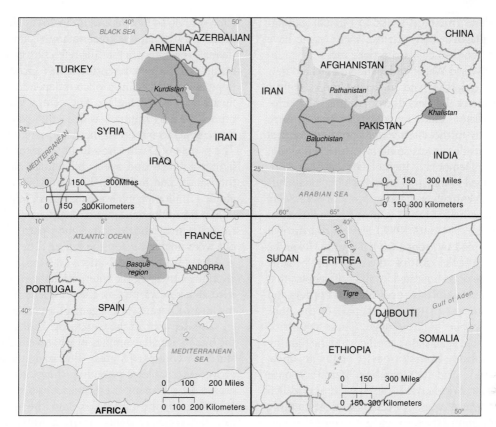

Figure 11-13 Nations without states. These are a few of the nations that do not have states of their own but may achieve statehood in the future. The present states, however, may succeed in dissolving these nations. Note that most of these nations overlap two or more existing states, so pressures by these submerged nations to achieve statehood are international issues.

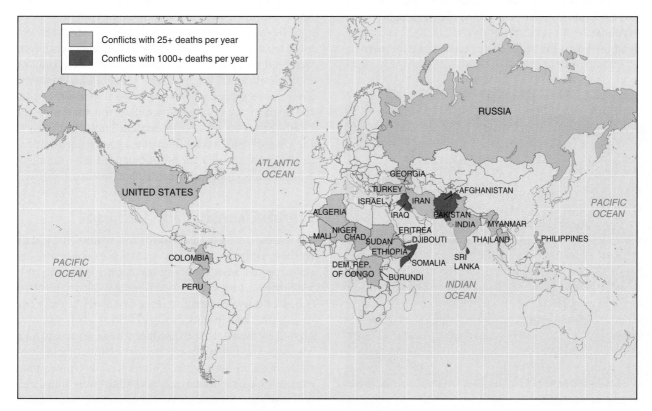

Figure 11-14 Conflicts around the world. In 2008, there were 36 armed conflicts across 27 countries with each conflict resulting in 25 or more battle-related deaths during the year. Five of these countries had 1,000 or more battle-related deaths. Most of these conflicts are between the government and a separatist group.

Subnationalism and civil wars in Africa Subnationalism has plagued many African states; some have suffered almost endless civil strife since receiving independence after World War II. Independence created new conflicts over leadership and jockeying for economic control of resources. Borders drawn arbitrarily by colonial powers created states in which multiple groups claimed authority in the absence of a common political community. Outside powers sometimes intervened, but not always. The Democratic Republic of the Congo, for example, dissolved in war soon after receiving independence in 1960. Eventually a dictator, Mobutu Sese Seko, seized power and held it for 32 years. (The country was called Zaire from 1971 until 1997.) He was overthrown by a rebellion aided by troops from Burundi, Rwanda, Uganda, and Angola. All of these troops became involved because of complicated loyalties of peoples that overlap all of these countries' superimposed borders and a quest for resources in the eastern part of the huge Democratic Republic of the Congo. These countries' armies soon turned against one another, and the fighting eventually involved troops from Congo, Rwanda, Uganda, Burundi, Sudan, Zimbabwe, Zambia, South Africa, Namibia, Angola, and Tanzania. It was called "Africa's First World War." A fresh round of fighting began in 2008, around the town of Kivu. These conflicts have left hundreds of thousands dead, mostly civilians, and displaced millions more (Figure 11-15).

In the Sudan, the central government, controlled by Muslim Arabs in the north, persecuted a black animist and Christian population in the south that fought for independence. A ceasefire was reached in 2005, according to whose terms the south could secede if the southerners choose to do so by referendum in 2011. The south, however, holds the nation's great untapped oil wealth, so it remains to be seen whether the referendum will peacefully occur. Meanwhile, resentment has grown among the population of Darfur, a once-independent Islamic sultanate in western Sudan: The central government recruited Arab militias called the *janjaweed* ("evil horsemen") to put down these non-Arab African tribes. The people of Darfur are Muslim, but the *janjaweed* are lighter skinned and speak of killing "blacks." Tens of thousands died in Darfur between 2003 and 2007, and about 3 million people were displaced. *Janjaweed* attacks even crossed the border into neighboring Chad. In 2006, troops sent by the African Union patrolled the region but were unable to stop the killing. They were joined a year later by a United Nations mission, yet the conflict continues. In 2008, the International Criminal Court issued an arrest warrant for the president of Sudan, charging him with war crimes against the civilians of Darfur.

The Republic of the Congo, Sri Lanka, Somalia, and the Sudan are all examples of what we today call **failed states**, or **collapsed states**—that is, countries that have proved incapable of providing their citizens with economic development or even peace and security.

Subnationalism helps explain the number of authoritarian governments in Africa. Authoritarian rulers argue that iron rule is the only alternative to the tribalism that could tear their countries apart, but this argument is at least partly a rationalization for them to hold on to power. Several have been accused of deliberately fomenting and manipulating subnational strife.

The existence of many failed states in Africa, as well as in Asia and in Latin America, created conditions that lead to civil wars, guerilla insurgencies, and the multiplication of independent rogue military forces. These have proved capable of projecting violence into the heart of the developed world, as the attacks on the United States on 9/11 demonstrated. Since those attacks, the United States has demonstrated greater involvement in the problems of nation-building and state-building, which means confronting centrifugal forces.

Ethnic Cleansing and Genocide

An extreme strategy used to make a nation "fit" a state territory is to expel from the state those people who are not accepted as members of the nation, or to exterminate them. Forcible and violent expulsion is often called **ethnic cleansing,** a term that entered daily use during the war in Bosnia (1992–1995). This war was started by Bosnian Serb leaders whose army expelled Muslims and Croats from territory claimed for a separatist Bosnian Serb state. This policy was more than expulsion; tens of thousands of civilians were murdered and any sign of Muslim or Croat culture was demolished, including mosques, churches, libraries, museums and cemeteries (Figure 11-16). **Genocide** is the practice of intentionally trying to eliminate a national,

Figure 11-15 Civilian victims of war. Internally displaced persons fleeing the town of Goma in eastern Democratic Republic of the Congo, where rebel and government forces were fighting during October 2008.

CONNECTIONS

Geopolitics

Chapter 10 defined *political geography* as the study of the interaction between political processes and the distributions of all other activities and transformations of the landscape. The term was coined by German geographer Friedrich Ratzel as the title of a book he wrote in 1897. Ratzel's studies focused on the political geography of states, and he suggested that each nation needed *Lebensraum*, "room to live." As national populations increased in size and productivity, he wrote, nations might need more Lebensraum.

Ratzel's idea of Lebensraum was adopted by some of his contemporaries who were at that time drawing analogies between states and living things. They argued that states are a form of biological organism. To explain why some states thrive while others perish, these writers combined Ratzel's idea of Lebensraum with Spencer's theory of social Darwinism, and they concluded that competition among states for territory resulted in "the survival of the fittest." In 1899, the Swedish political scientist Rudolf Kjellen coined the term **geopolitics** to describe the effects of natural geographic facts on states and their relations. He imagined that there were scientific principles at work

in "the natural and necessary trend towards expansion as a means of self-preservation." Geography, he argued, forces states to conquer other states. In 1924, a German professor named Karl Haushofer founded the Institute for Geopolitics in Munich, where he taught that it is natural for strong states to expand at the expense of the weak. These ideas impressed Adolf Hitler, and when Hitler became ruler of Germany, he quoted geopolitical theories to justify Nazi aggression in World War II.

The term **geopolitics** still loosely applies to the role of geography in international relations and military strategies, but political geographers do not regard it as a science. Chapter 12 will demonstrate that any state's possession of extensive territory—or even of natural resources—does not strictly determine the welfare of its people. Already in 1899, the British scientist Sir William Crookes had refuted geopolitical theories. His book *The Wheat Problem* suggested that technological progress can replace territorial aggression in raising a nation's standard of living. Instead of conquering national frontiers, Crookes coined the metaphor "scientific frontiers." Even he would be amazed to learn of the technological advances in agriculture described in Chapter 8.

Figure 11-16 The ruined national library of the Sarajevo, Bosnia, University. "Ethnic cleansing" of a landscape includes the deliberate destruction of libraries and other cultural sites.

ethnic, racial, or religious group. It is prohibited by international law. Genocide and ethnic cleansing are not mutually exclusive. In fact, the ethnic cleansing campaign in Bosnia led to genocide. These policies are tragic and abominable, but both have actually been implemented when groups have attempted to carry the logic of territorial nationalism to its inhuman extreme.

Ethnic cleansing and genocide occurred in southern Europe during and after World War I. The Turks massacred Armenians in 1915 and later expelled Greeks from Asia Minor. Greece responded by expelling Turks. During the Holocaust of the 1930s and 1940s, Nazi Germany attempted to exterminate Jews and other minority groups in Europe. The Nazi regime and its allies killed approximately 6 million Jews and millions more non-Jews whom the Nazis regarded as racially inferior. After Germany's defeat in World War II, German civilians were expelled from Poland and Czechoslovakia. Millions of refugees fled from Pakistan to India or vice versa at the partition in 1947.

Mass expulsions continue. In 1989, Bulgaria expelled about 100,000 ethnic Turks. In 1991, 1 million Kurds fled or were driven out of Iraq. In 1993, tens of thousands of people of Nepalese origin were driven out of Bhutan. In Tutsi-dominated Rwanda in 1994, Hutu nationalists rose up and slaughtered more than 500,000 Tutsis. The next year, the Tutsis regained the upper hand, driving several hundred thousand Hutus into Congo. Next door, Burundi has suffered ethnic violence ever since it won independence from Belgium in 1962. The Tutsi minority (15 percent) traditionally ruled the Hutu majority (85 percent), and in 1993, the first democratically elected Hutu president was assassinated by a Tutsi military uprising. Civil war raged, and more than 200,000 people were killed and an additional million were

Video
Srebrenica-
Looking for
Justice

CONNECTIONS

Nigeria's Oil Curse

Nigeria may be the world's most spectacular—and dangerous—example of a failed state.

In the late nineteenth century, the British established the Royal Niger Company to exploit resources in the Niger River Delta. Expanding inland, the British found themselves ruling some 250 ethnic groups that had never before coexisted in a single state. The British divided Nigeria into administrative zones along ethnic and religious lines. After Nigeria achieved independence in 1960, the Muslim north, which is poor but has half of the country's total population, gained supremacy over the army. Through a succession of military dictatorships, it dominated and plundered the fertile and oil-rich but disunited south, whose largest ethnic groups—the Yoruba in the west and Igbo in the east—together represented just 39 percent of the population. The Igbo declared their region to be the independent state of Biafra in 1967, and during the three years that the central government fought to restore its authority, fatalities were estimated to number more than 1 million. The workings of the civilian government that came to power in 1999 have been characterized by wholesale bribery and murder of government officials and candidates.

From 1980 to 2005, Nigeria earned more than $300 billion in oil revenues, but annual per capita income plummeted from $1,000 to $390. Billions of dollars have been squandered or stolen outright by the country's leaders. The country appears to be *de*-developing, and terrorist Osama bin Laden has called it "ripe for liberation."

Nigeria has the largest petroleum reserves of any African country, but its oil industry is characterized by crime, neglect, pollution, and corruption. Tapping into pipelines and siphoning oil into tankers hidden in the swamps of the Niger River Delta is widespread, causing losses, pollution, and catastrophic fires. Owing to the abysmal state of Nigeria's few refineries, the country must import gasoline. Fuel shortages are endemic. Nevertheless, Nigeria is the fifth-largest supplier of oil to the United States, and in 2010,

the White House declared the oil of Africa to be an "essential subregional linchpin" for U.S. interests suggesting that the United States would use military force, if necessary, to ensure Nigeria's stability.

A hastily erected façade of modernity is disintegrating and leaving Nigerian city-dwellers, in particular, struggling to survive in desolation. Lagos, Africa's largest city, with 13 million people, is a chaotic jumble of crumbling roads, falling power lines, electricity blackouts, and crime.

Since 1999, at least 15,000 Nigerians died in ethnic, religious, or communal fighting, and approximately 3 million people were internally displaced. Nigerian Nobel Prize–winning author Wole Soyinka has joined those Nigerians calling for a national conference that would bring together representatives of the country's groups and debate a new constitution. Only such a debate, Mr. Soyinka argued, would enable Nigerians "to make sense of our remaining together." "What do we want as a nation?" Mr. Soyinka wondered aloud. "We have never been able to decide for ourselves."

International investment. Royal Dutch Shell's manicured oil plant in Bonny, Nigeria, presents a sharp contrast to the squalid poverty of the neighboring traditional fishing village.

internally displaced. The Arab majority of Mauritania continues to expel the country's minority blacks into Senegal. In the Caucasus region since the 1990s, Abkhazia, Ossetia, and Naggorno-Karabakh all suffered ethnic killings and expulsions involving more than a million people combined. Ethnic conflict in South Ossetia led to war between Georgia and Russia in 2008.

The Soviet Union had forcibly relocated large numbers of its citizens during the twentieth century. Censuses taken in the Soviet Union in the late 1980s counted some 60 million "displaced Soviets" living outside the regions of their ancestors. When the Soviet Union broke apart, not all of the newly independent states guaranteed full

civil rights to their minorities. Ethnic Russians, especially, were subject to discrimination in some countries, forcing them to move. Millions of people have migrated into the homelands of their own ethnic origins.

The Internal Organization of States

A state has a government that is responsible for the country and its citizens. How governments operate varies widely around the world, but some main types can be identified. States are also responsible for managing their territory and maintaining borders. States subdivide their territory for more efficient administration or political representation. This section will examine some of the patterns of states' internal features and domestic politics.

Types of Regimes

The world today is a far cry from the age of monarchies and empires. It is no longer common for governments to justify their sovereignty on divine right or tradition. Instead, democracy, the rule of the people, has become the international standard, even if it is difficult to achieve. Democracy as we appreciate it today is more advanced and encompasses a wide set of basic rights. Democratic regimes are supposed to protect the rights of individuals, and citizens are to choose their leaders through competitive elections. Laws written by parliamentary assemblies, made up of elected representatives, govern democracies. People living in genuinely democratic countries assume that most governments work for the good of the citizens.

Democracy is an old idea, and as government by the people it appeared in ancient Greece and again in medieval Iceland. Modern democracy is much younger. Scholars recognize three waves of democratization, when today's countries became democracies. The first occurred in the early nineteenth century following the American and French revolutions and included parts of Western Europe. The second wave followed the decades after World War II when more of

Europe and some former colonies also established democratic constitutions. For example, India, now the world's largest democracy, held its first elections in 1952, just five years after gaining independence. Scholars think that a third wave followed the end of the Cold War, when countries in Eastern Europe and parts of Asia and South America became democratic. During these waves, however, some countries moved in and out of the democratic camp as militaries and dictators struggled to keep the government out of the hands of popularly elected representatives.

Virtually all governments today govern in the name of the people, but some are more genuinely democratic than others. The world has many imperfect or just plain fake democracies. An **autocratic** government, or autocracy, is run according to the interests of the ruler or ruling elite, rather than the people (Figure 11-17). Dictatorships are an extreme form of autocracy. In some cases, an apparently democratic country may have an *authoritarian* regime, meaning that the rulers use coercion or force to limit opposition and dissent. For instance, some countries have presidential elections with only one viable candidate, the others are silenced by government-owned media; candidates running for parliament and their supporters are beaten, arrested on dubious charges, or even killed; and constitutions are routinely rewritten to prolong the terms and powers of presidents. In many countries, authoritarianism is tempered by populism. Popularly elected presidents of both Peru and Venezuela, for example, actually seemed

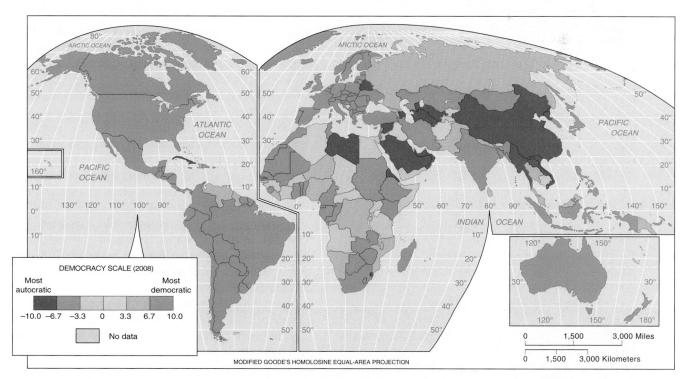

Figure 11-17 Regime Types Around the World. Governments may be characterized by their relative level of democratic features (for example, competitive elections and restrictions on leadership) and autocratic features (for example, unfair elections and unchecked government power). (Data from Center for Systematic Peace, POLITY IV dataset)

to rise in popularity when they unconstitutionally closed their countries' parliaments and fired their Supreme Courts.

Even in some democratic countries, the operation of government is often hidden from the public and the state exists largely to benefit the elite. These conditions are made worse when citizens are poor and ignorant, too busy trying to survive or too illiterate to see what their government is doing. These countries are often mired in corruption, meaning the government helps only those who can afford to buy off public officials for favors or special treatment (Table 11-2). The governments of some countries have even been called **kleptocracies**—government by theft. The national wealth may be funneled off and even taken out of the country by a privileged few, while most of the people remain in poverty. In these cases foreign aid, or any system of providing assistance to the poor through their governments, cannot guarantee that the needy will receive the assistance. For example, General Sani Abacha ruled Nigeria from 1993 until 1998, and in 2000, the Swiss government found $670 million in one Swiss bank account in his name, $600 million in another account in Luxembourg, and similar sums in other accounts in France and Germany. In 2002, U.S. oil companies revealed that Kazakhstan's President Nursultan Nazarbayev regularly accepted large bribes in connection with dispensing his country's oil concessions. The presidents or rulers of many of the poorest countries in Latin America, Africa, and Asia are among the world's richest individuals.

Authoritarian rulers often insist that their people "are not ready for democracy." Eritrean President Isaias Afewerki said in 2003, "We have not yet institutionalized social discipline, so the possibility of chaos is still here. . . . No one in Africa has succeeded in copying a Western political system, which took the West hundreds of years to develop. Throughout Africa, you have either political or criminal violence. Therefore, we will have to manage the creation of political parties, so that they don't become the means of religious and ethnic divisions. . . ."

Are there, in fact, preconditions for democracy, and are there states where the people are not ready for it? Without those conditions, do authoritarian rulers have the right to rule, presumably for the good of the people, to guide the people toward democracy? Farouk Adam Khan, former prosecutor-general of Pakistan, led a military coup that failed in 1973. During the following five years he spent in prison, he read the great works of Western political philosophy, including *The Federalist Papers* and John Stuart Mill's *On Liberty*. Later he said, "Every single ingredient that the authors of those books say is required for a civil society—education, a moral code, a sense of nationhood; you name it, we haven't got it! Just look at our history. . . . We need someone who will not compromise in order to build a state."

Western democracies have their problems, too. In some countries, citizen participation rates have fallen so low that observers wonder whether elections are valid. Italian courts invalidated seven national referenda in 2000, because only 32 percent of potential voters had turned out. Turn-out for the European Parliament elections has been declining and slipped to 43 percent in 2009. In the United States, an average of just 43.6 percent of the voting-age population has turned out for congressional elections since 1990, putting the United States in 139th place among the 163 countries in which turnouts were tabulated. The U.S. presidential election had a turnout of 69 percent in 1964 but fell steadily to 58.2 percent in 2008. British Prime Minister Tony Blair has argued that "failure to vote is the mark of a satisfied citizen," but many observers insist that this remark interprets the death of democracy as a vote for the status quo.

The early twenty-first century has demonstrated that there are areas where citizens may not want democracy. Many Arab rulers in the Middle East, for example, are autocratic, corrupt, and heavy-handed, but they may be more liberal, tolerant, and pluralistic than any form of democracy that would likely replace them. Elections

TABLE 11-2 Perceptions of Corruption Around the World

Each year, Transparency International, an organization based in Germany, publishes information about the levels of corruption in different countries around the world. Corruption is difficult to measure, and governments do not publish statistics about their own corruption. Transparency International uses opinion polls of experts and businesses around the world to track perceptions of government corruption. These data from 2009 show the 10 governments perceived as being the least corrupt and the 10 perceived as most corrupt (some country ranks are tied).

Rank	Perceived as Least Corrupt
1	New Zealand
2	Denmark
3	Singapore
3	Sweden
5	Switzerland
6	Finland
6	Netherlands
8	Australia
8	Canada
8	Iceland
⋮	
19	United States
⋮	
168	Haiti
168	Iran
168	Turkmenistan
174	Uzbekistan
175	Chad
176	Iraq
176	Sudan
178	Myanmar
179	Afghanistan
180	Somalia
	Perceived as most corrupt

in many Mideast countries could put dictators or religious fundamentalists in power. Thus, paradoxically, a more democratic country can be a more repressive one. A majority of Iraqis, for example, probably favor restrictions of the rights of women and minorities. Achievement of a democratic state is more than just holding elections. Unless majority rule is accompanied by legal protections, tolerance, and respect for minorities, the result can be populist oppression.

Furthermore, in any democracy, the demographic description of those who actually vote may differ from a description of the people at-large. In the United States, for example, the higher one's age and income, the more likely one votes. Popular public opinion polling, by contrast, usually includes all adults and income levels. Therefore, governmental activities that respond to the voters' preferences do not always reflect the population's wishes as reckoned in opinion polls.

Political geographers are interested in the ways that cultural and economic factors influence regime types, but they are also interested in how states organize and rule over their territories.

The Shapes of States

The physical shape of a state may affect its ability to consolidate its territory and control circulation of people and goods across its borders (Figure 11-18). A circle would be the most efficient shape on an isotropic plain because a circular state would have the shortest possible border in relation to its territory, and that shape would allow all places to be reached from the center with the least travel. States with shapes the closest to this model are sometimes called *compact states*. Bulgaria, Poland, and Zimbabwe are examples. *Prorupted states* are nearly compact, but they have at least one narrow extension of territory. Namibia and Thailand are examples. If these extensions reach out to navigable waterways, the extensions are called *corridors* (corridors' special importance

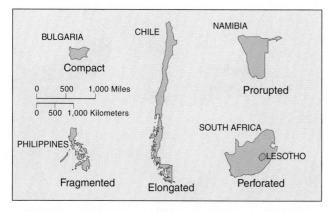

Figure 11-18 Shapes of states. Examples are shown of states that are compact, prorupted, elongated, fragmented, and perforated. The five states are drawn to the same scale. In general, compactness is an asset, because it fosters good communication and integration among all regions of the country.

to landlocked states is discussed later in this chapter). *Elongated states* are long and thin, such as Chile or Norway, and *fragmented states* consist of several isolated bits of territory. *Archipelago states*, made up of strings of islands, such as Japan or the Philippines, are fragmented states, but Azerbaijan is also fragmented, having a territory, Nakhichevan, separated from the rest of the country by Armenia. Nakhichevan is called an *exclave*. Still other states, called *perforated states*, are interrupted by *enclaves*, which are the territory of another state enclosed entirely within them. South Africa, for example, is perforated by Lesotho, and Italy is perforated by the Vatican and by San Marino.

The shape of a state's territory may influence the government's ability to organize that territory, but this is not always true. A topographic barrier such as a mountain chain may effectively divide even a compact state. Bolivia and Switzerland, for example, are compact in shape, but mountain chains disrupt their interiors. For some of their regions, trade across international borders is easier than trade with other regions of their own country. The people throughout an archipelago state, by contrast, may be linked successfully by shipping. Before drawing any conclusions about political control from the shape of a state alone, one must consider the distribution of topographic features of the state's population and resources, and whether any centrifugal forces, such as economic or cultural ties, straddle the state's borders.

International Borders

The line that marks the territorial limit of a state are a border or boundary. Seas and oceans form natural boundaries around some states, but most states share at least one land border with another state. Borders are often classified as one of three types: physical, geometric, or cultural.

Physical boundaries are defined by features such as rivers or mountains. They are often preferred by states because they are obvious and permanent parts of the landscape. The Oder and Neisse rivers were designated as the new boundary between Poland and German after World War II. Most of the boundary between Chile and Argentina is defined as the highest peaks of the Andes Mountains. These features are not necessarily "natural" dividing lines between people, however. Rivers bind the peoples on their two banks as much as they divide them, as well as connecting people upstream and downstream. Mountain borderlines can be drawn either from one peak to the next or else up and down the mountain valleys that reach between the peaks. Mountain valleys can actually provide foci for political organization. The early Swiss united the valleys on either side of mountain passes. The passes did not divide peoples but instead gave them a common interest.

Geometric boundaries are usually defined using straight lines of longitude or latitude or running between geographic coordinates. These borders are often used in otherwise featureless areas. Geometric lines mostly divide the Sahara Desert. In contrast, boundaries drawn with respect to existing patterns of human activity are known as **cultural boundaries.** These borders are typically based on the ethnic or national patterns that existed prior to drawing the boundary. The distribution pattern of different linguistic or religious community is rarely a clear line; trying to designate cultural boundaries often leads to violence. This was the case in Bosnia, where Bosnian Serb politicians tried to draw forcibly a cultural boundary around their "historic homeland," which had always contained other ethnic groups.

The function of international borders depends on the relationship between the states it divides. Some borders are themselves a matter of dispute. Guatemala recognized the independence of Belize in 1991 but claimed most of its territory until a treaty was signed in 2002. In 1992, El Salvador and Honduras accepted a World Court decision about their border, over which they had argued since independence in 1861. India, Pakistan, and China have fought over their borders several times, and they still have not agreed (see the map in the box on page 288). The location of an international border is sometimes visible from the air, even if there are no border structures. Borders may be evident by contrasting land uses on the two sides, by different types of landholdings, or by discontinuity in the transportation networks (Figure 11-19).

An international border across a sparsely populated area may not be marked or supervised at all, and its exact location may become a matter of dispute only if valuable resources are discovered in the region. This has happened along the inland borders of South American countries. In 1941, Peru attacked Ecuador and annexed, or took over, 55 percent of Ecuador's territory. The discovery of oil in the former Ecuadorian territory

of what is today northern Peru triggered renewed fighting in 1995, but the states agreed on a border demarcation in 1998. Venezuela claims about two-thirds of Guyana, and Suriname claims another 10 percent. Both Suriname and Guyana claim the mouth of the Corentyne River, where oil explorations look promising. The jungle border between Brazil and Venezuela is also disputed. Similar disputes have occurred in Middle Eastern deserts. Russia and Japan have not signed a treaty to end World War II, because they dispute possession of four islands. The Russians claim that the four islands in question are part of the Kurile chain occupied by Russian troops during the war and surrendered by Japan; the Japanese claim that they are not.

Some international borders are not disputed, but they are defensive, lined with minefields and military watchtowers. The Berlin Wall, which divided the city from 1961 to 1989, was a deadly border for those caught trying to cross it (see Chapter 13). The United States has built sections of high wall along the border with Mexico to keep undocumented migrants from coming across, and Chapter 7 discussed Israel's new wall barrier. Some states also want to control the passage of information over their borders. This was easier to do in the age of newspapers. Satellite television and shortwave radio can broadcast messages to just about anywhere. Landlines and cellular telephones are more easily blocked. North Korea, China, and Iran are among the countries that have recently limited their citizens' access to the Internet. New technologies can adapt faster than governments. During the Iranian government attacks on election protestors in 2009, citizens used social media networks to broadcast news and images of the violence to the outside world (Figure 11-20).

Figure 11-19 The U.S.–Mexican border at Tijuana. The United States (on left) has built extensive walls and fences along its southern border where Mexican cities abut U.S. territory.

Figure 11-20 The whole world is watching . . . on Twitter. In 2009, Iran discovered the power of social media networks, which allowed election protesters to send instant messages and images around the world.

Figure 11-21 The busiest border crossing in the world. Cars line up to cross into the United States at the San Ysidro port of entry.

Other international borders, by contrast, are relatively open and free, promoting complementary economic activities on the two sides. Even though part of the U.S.–Mexican border is defensive, trade between the countries was valued at more than $350 million in 2008, much of it crossing the boundary on trucks. The San Ysidro crossing between San Diego and Tijuana is the busiest in the world (Figure 11-21). In 2008, there were more than 25 million cars and 7 million pedestrians crossing into the United States. The cities along both sides of the U.S.–Mexican border have developed complementary manufacturing and transportation economies. Among most countries in Western Europe, promoting economic cooperation has meant eliminating the old border checkpoints to allow goods and people to move freely. As European countries have become more closely integrated, the need for border enforcement against illicit goods and illegal migration has been relocated to Europe's "outside" borders, the Mediterranean Sea and Eastern Europe.

Territorial Subdivision and Systems of Representation

All countries subdivide their territory, and each country has a unique balance of powers between its national government and its local governments. Countries in which political power is concentrated at the national level are referred to as **unitary governments,** or centralized governments. Those countries in which power is shared with the subunits are called **federal governments,** or decentralized governments. Today, most national governments are unitary. Federal governments have sometimes been designed to acknowledge regional identities, such as in Switzerland. Federal structures have also been used to localize governing authority in large countries, such as Canada, the United States, and Brazil. Strong local identities are a reason

GLOBAL AND LOCAL

U.S. Border Security

Congress responded to concern about American security after 9/11 by creating the new cabinet-level Department of Homeland Security. It exercises control over 46 federal agencies, including the Customs Bureau and the Coast Guard. The new department's responsibilities include both border security and internal security.

It is difficult to monitor borders as open and free as those of the United States. On an average day, some 1.3 million people, 348,000 private vehicles, 38,000 trucks and railcars, 2,600 aircraft, and nearly 16,000 freight containers arrive at U.S. borders. Many Americans are demanding improved inspection and tracking of this massive influx of people and goods. The United States has imposed new restrictions on the granting of visas to foreign visitors, tourists, and students, and on international shipping: Every one of the world shipping industry's 1.2 million merchant mariners will be required to carry a new biometric identity card. American ports may install scanners that can scan every container for radiation and also gamma-ray screeners,

which check for odd-sized objects. The systems can help identify suspicious cargo while keeping detailed records of what passes through the port.

Not all border security functions are conducted at the border itself. The United States has stationed customs inspectors in other countries to inspect shipments heading to the United States before they leave from the countries of origin. Immigration and customs officers check all international arrivals at airports located in the middle of the country. U.S. agents similarly review airplane traffic coming into the United States before passengers even board their plane in a foreign country. Inside the United States, immigration officers conduct routine traffic stops at points far away from borders, along roads heading away from border regions. Some local police and sheriff offices in non-border states have been granted authority to enforce U.S. immigration laws, which siphons off local law enforcement resources to enforce federal law. In these and other ways, U.S. border security is no longer confined to the boundary-line itself.

that Spain and the United Kingdom have decentralized national powers to give regional governments more responsibilities. At the same time, few newly independent or poor countries have adopted federalism, perhaps because they fear that if centrifugal regional identities were officially recognized, they could pull the country apart. Ethiopia, however, adopted a new federal constitution based on regional self-determination for 11 ethnic groups in 1994.

Unitary government   In unitary states, the central government theoretically has the power to redraw the boundaries of the subunits. This offers flexibility to accommodate geographic shifts in national population distribution or economic growth. It may be useful to redraw internal boundaries from time to time in response to changing needs, for example, by extending the boundaries of cities as they expand into the surrounding countryside. Interests become vested in any given pattern, however, so unitary governments seldom redraw internal boundaries. For example, the 50 U.S. states are each internally unitary, but they seldom redraw the boundaries of their counties because doing so would disrupt the provision of many local services, including schools and water.

China is constitutionally "a unitary multinational state." Five principal minorities and about 50 smaller national groups exercise some degree of local autonomy, and Hong Kong and Macau have been guaranteed special political and economic privileges for the next 40 years (Figure 11-22). The Turkish-speaking Muslim Uighur people of Xinjiang have rebelled periodically; they may be envious of the political autonomy enjoyed by similar groups in Uzbekistan, Kyrgyzstan, Turkmenistan, and Kazakhstan. China has encouraged neighboring countries to repress Uighur independence movements in their own countries. Meanwhile, settlement of ethnic Chinese in Xinjiang has made the Uighurs a minority there.

Similarly, the Mongols of Inner Mongolia (a part of China) may envy the Mongols in Mongolia (an independent country) and wish to secede from China, but Chinese settlement has reduced the Mongols to a minority in "their own" territory. Nevertheless, political instability may cause the Chinese multinational state to break apart.

Federal government   A federal government assumes that diverse regions should retain some local autonomy and speak with separate voices in the central government. Federalism also allows each area to serve as a laboratory for legislation that can be adopted elsewhere if it works. Wisconsin, for instance, might devise a successful school program that other states might copy, New Mexico might create a highway program, or Maine might develop environmental legislation. In other ways, subunits behave like independent units. Their policies, for example, may perpetuate economic or social inequality. Each U.S. state boasts an extensive iconography: state flags, birds, flowers, and so forth. Texas (where schoolchildren pledge allegiance to the state flag) and Hawaii have histories as independent countries.

The U.S. federal government has increasingly preempted state action in recent decades. Examples include federal laws governing consumers, food labeling, class-action lawsuits, and the "No Child Left Behind" education law. The Supreme Court has held that federal anti-marijuana legislation overrides the referenda in 12 states that approved marijuana's medical use. Such federal preemption has stimulated individualized responses from the states. Massachusetts offers near-universal health care coverage and allows same-sex marriage. California imposes limits on the emissions of greenhouse gases. California, New Jersey, Maryland, and Connecticut have allocated money for stem cell research. On the other end of the political spectrum, Idaho, Georgia, and 17 other states have constitutional amendments banning same-sex marriage, and Florida, Mississippi, and Utah even ban child adoption by same-sex couples. Historically, issues that enthrall the states eventually get sorted out nationally in federal legislation.

It is harder for national governments to change subunit boundaries in a federal system. The U.S. Constitution protects states from boundary changes against their will (Article IV, Section 3), although Virginia split into two states during the Civil War in 1863. Strong political currents within California have proposed breaking it into two or three states, but this would be legally almost impossible. It is possible, however, for federal governments to become more dominant over their subunits. Some people think this is happening in the United States, as federal law covers more and more issues that were once decided by states. Public interest in national issues is reflected in the fact that voter turnouts are highest in years with presidential elections than other years. Members of Congress also raise increasing percentages of their campaign funds outside their own constituencies. More than half of the U.S. senators, for example, raise more than 50 percent of their funds outside their home states. So whose interests do they represent—those of their constituency or those of a national list of donors?

Some countries' federal patterns reflect polyethnicity (for example, Russia, Myanmar, and India). Some countries' internal units are older than the federal framework, and they participate in the federal system only on the condition that they may retain certain powers (for example, the original 13 United States and Germany). Canada was formed in 1867 as a federation of its provinces, which had been settled as English and French colonies. The country is today officially bilingual, although not all Canadians speak both languages. There are also native languages spoken by the indigenous populations of Canada and

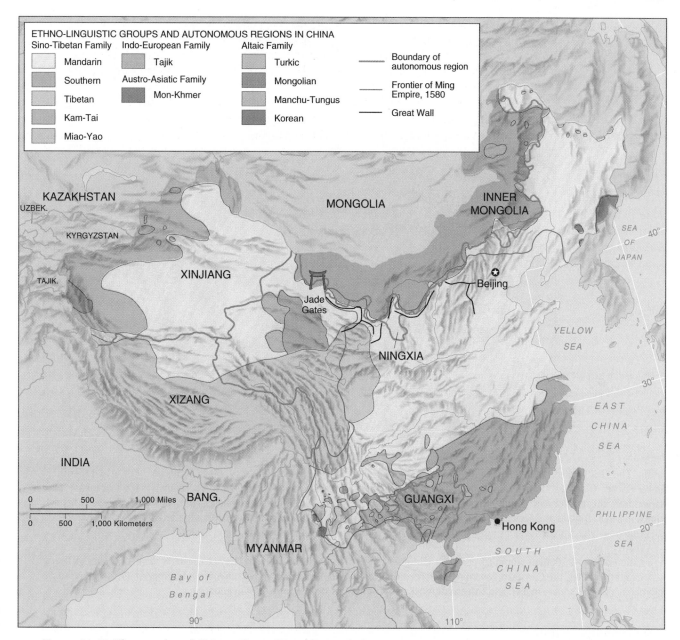

Figure 11-22 The peoples of China. The traditional limits of Chinese territory were the borders of the Ming Empire—the Jade Gate on the route to Central Asia and the Great Wall against the Mongols. Through the centuries, however, Chinese rule has intermittently expanded beyond these borders to include non-Chinese peoples. China's contemporary constitution guarantees some political and cultural autonomy to minority peoples in the five peripheral regions that are labeled plus Hong Kong. These regions constitute 42 percent of China's territory but hold only about 8 percent of the Republic's population.

that are most common in the northern territories (Figure 11-23).

Federalism may also serve countries that are relaxing central control. A new Italian Constitution adopted in 2001 devolves considerable power from the central government to the provinces. Belgium is decentralizing into three parts: Flanders, Wallonia, and the city of Brussels. The United Kingdom devolved some powers to new parliaments in Scotland and in Wales in 1999. In these European examples, some powers of the national governments are devolving down to the regions at the same time as others are being surrendered up to the multinational European Union (discussed in Chapter 13), so the

national governments retain fewer powers. In states torn between two religions, such as Nigeria, the adoption of Sharia law in some of the subunits can aggravate antagonisms, and if loosening the federal ties proves insufficient, a federal state may dissolve in civil war.

Special-purpose territorial subdivisions Several countries divide their territory into economic-planning regions, which can gather almost as much power as the legal constituent subunits. This has happened in France. In the United States, several federal agencies, including the Federal Power Commission, the National Labor Relations Board, the Bureau of Reclamation, the

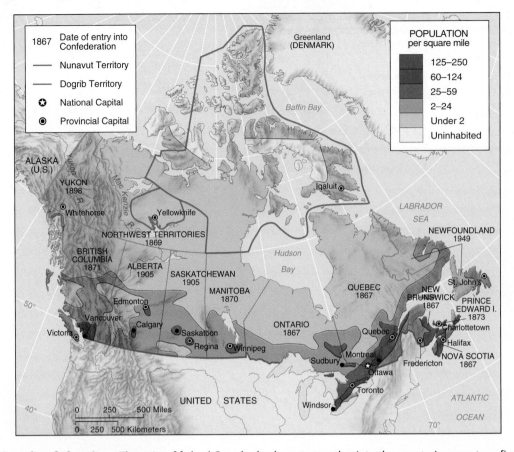

Figure 11-23 The Canadian federation. The units of federal Canada slowly came together into the country's present configuration. The 1867 British North America Act united the provinces of Upper Canada (Ontario), Nova Scotia, New Brunswick, and Lower Canada (Quebec). In 1869, the Northwest Territories were purchased from the Hudson's Bay Company; Manitoba was carved from this territory and admitted into the confederation in 1870. British Columbia joined in 1871, and Prince Edward Island joined in 1873. Alberta and Saskatchewan were formed out of previously provisional districts and admitted in 1905. Newfoundland, previously an independent country, joined in 1949 and took the name Newfoundland and Labrador in 2001.

The territory of Nunavut formally came into existence April 1, 1999. It covers one-fifth of Canada's land area but contains a population of only 27,000, of whom 85 percent are Inuit. The city that had been Frobisher Bay changed its name to Iqaluit as the new capital. In 2003, the Canadian government granted the Dogrib people local self-government in a territory of 39,000 square kilometers (15,058 square miles), and three additional self-government treaties are under negotiation.

Federal Trade Commission, the Federal Communications Commission, and the Federal Reserve System, have each subdivided the country into a different pattern of districts for its own uses. Few of these patterns conform to the pattern of the states, so when two districts of a federal agency enforce different policies in two parts of one state, the state government's own powers can be confused or undermined. The United States is also subdivided by 91 federal district courts and, above them, 11 federal courts of appeal (Figure 11-24). Different interpretations of the law among these different courts are sometimes resolved through appeal to the U.S. Supreme Court.

When Vladimir Putin became president of Russia in 2000, one of his first official acts was to divide Russia into seven new federal zones, each with an administrator appointed by himself, and to transfer power to these zonal administrators. Thus, Russia's

89 elected regional governors lost much of their power overnight.

Federal territories Most federal states contain territories that are not included in any of the subunits but that are administered directly from the federal center with, perhaps, some form of local government. These territories may constitute a significant share of a state's total area, including capital districts, colonies, strategic frontier areas, and federal territories.

Territories usually have only limited local government until some presumed future time when they will be ready for full statehood. If the territories lack resources, they may remain territories indefinitely. Most of the land area of the United States was once federal territory, but the territories were organized, populated, and finally admitted into the Union as

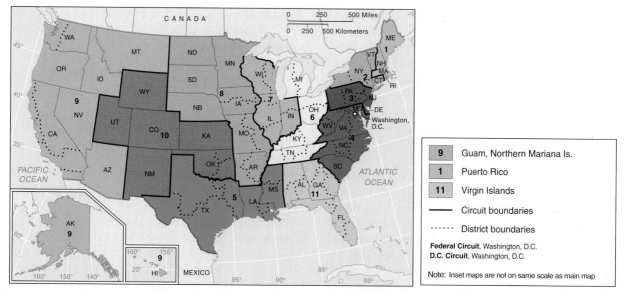

Figure 11-24 The pattern of U.S. federal courts. Any two of these courts may interpret federal law differently from one another. Therefore, federal law is not uniform throughout the country until the disagreement is resolved by the Supreme Court.

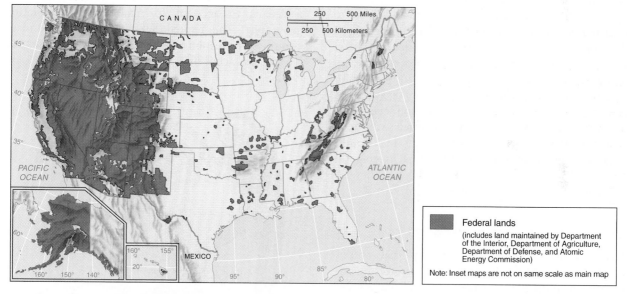

Figure 11-25 U.S. federal properties. The U.S. government still owns a substantial share of the total national territory. The largest land trade in U.S. history occurred in 1998. The state of Utah surrendered to the federal government 152,580 hectares (376,739 acres—an area almost half the size of Rhode Island) that had originally been given to the state to finance education; the land was scattered through national forests, parks, and monuments. In return, the federal government gave Utah $50 million in cash plus 58,725 hectares (145,000 acres) outside national parks and areas proposed for wilderness.

states. By contrast, today some 40 percent of Canada remains in territorial status. Mexico includes two territories that may become states upon reaching populations of 120,000 and demonstrating "the resources necessary to provide for their political existence."

The U.S. government still owns about one-third of the total national land area (Figure 11-25). The 1976 Federal Land Policy and Management Act requires the federal government to receive full value for any lands traded away, and Congress still wrestles with the question of how the federal government should exploit or preserve its lands.

Capital cities In some countries, the federal capital is also the capital of a component state (for example, Bern is capital of both its own canton and of Switzerland). Austria's federal capital, Vienna, is a federal state. Some federal capitals are governed directly by the federal government (for example, Australia's capital territory, Mexico's Distrito Federal). Residents of Washington, D.C., complain that they have no voice in the federal government despite the fact that it is more populous than Wyoming.

Several countries have moved their capitals because they believed that their old capitals were international cities grafted onto national life rather than truly representative of national life. A new capital, it is believed, especially if built inland, will symbolize a rebirth and rededication of a national spirit. In 1918, the Russians moved their capital from St. Petersburg inland to Moscow; in 1923, the Turks moved theirs from Istanbul to Ankara; Brazilians moved theirs inland to Brasília in 1960; Tanzanians moved theirs inland to Dodoma in 1975; Nigerians moved theirs to Abuja in 1991. Newly independent Kazakhstan moved its capital from Almaty (which was thought to be too Russian) to Akmola (today called Astana) in 1997. The Japanese, by contrast, moved their capital from inland Kyoto to coastal Tokyo in 1868 to symbolize Japan's opening to the world. Capitals are frequently designed as showplaces of national pride to impress both citizens and foreigners. In 1999, the offices of Malaysia's prime minister were moved to the newly built city of Putrajaya, next to the country's new scientific and computer-research center, Cyberjaya. Kuala Lumpur will, however, remain the official capital until 2012. In 2005, the Burmese government announced it was moving the capital to Pyinmana, renamed Nay Pyi Taw, about 250 miles north of the current capital, Yangon.

Districting and Redistricting

The subfield of political geography that studies elections and voting patterns is called **electoral geography.** Of particular interest is how democracies design geographic areas, called districts, as the basis for selecting representatives. Parliamentary bodies such as the British House of Commons and the U.S. House of Representatives are composed of representatives, each one elected by residents of a different district. Determining how many representatives each part of a country should have is known as apportionment. In the United States, the Constitution requires a decennial census for the purpose of balancing the number of representatives from each state in approximate proportion to its population. States are then responsible for districting: drawing districts that will elect a representative.

Districts are not drawn arbitrarily and can be the subject of intense political struggles because moving district boundaries can determine the outcomes of elections. This can occur in either one of two ways. First, if the electoral districts are unequal in population, then the ballots cast by some voters outweigh those cast by others. Second, district lines can be drawn in ways that include or exclude specific groups of voters so that one group gains an unfair advantage. This is called **gerrymandering** (Figure 11-26).

The electoral systems in many countries tolerate inequalities in numerical representation. Some countries grant rural populations a disproportionate amount of representation in the government. In the United States, this was long true at the level of the state and local general-purpose governments, but in 1962, the Supreme Court ruled that for these governments, the number of inhabitants per legislator in each district must be "substantially equal." In 1962, Tennessee was still electing state legislators on the basis of a 1901 apportionment. The population had urbanized, so that by 1962, one rural vote was equal to 19 urban votes. Many other states' legislatures contained similarly disproportionate representation, but nationwide redistricting now follows each decennial census.

Typically, redistricting occurs in most states after a census to account for changes in population geography. The United Kingdom has a boundary commission responsible for periodically adjusting boundaries according to criteria set by national law.

Figure 11-26 The original gerrymander. The original gerrymander was a district created in Massachusetts in 1812 to concentrate the Federalist Party vote and thus to restrict the number of Federalists elected to the state senate. The district configuration was at first likened to a salamander, but later it picked up the name of Governor Eldridge Gerry, the Anti-Federalist who signed the districting law. ("It looks like a salamander." "No, by golly; it's a Gerrymander!")

In the United States, most state legislatures are responsible for drawing new boundaries, giving the party in control at the state level considerable control over determining the outcome of the next Congressional election. More recently, some states have begun redistricting after each election. In 2003, Republican legislators in Texas redistricted that state shortly after winning the statehouse in the 2002 elections. The U.S. Supreme Court upheld the legality of the Texas legislature's action in 2006. The result is that voters in some areas no longer know what district they live in but parties have found ways of ensuring their victory.

Representation in the U.S. federal government The U.S. Supreme Court has upheld the principle of one-person/one-vote in all elections. Yet the U.S. federal government itself does not achieve that ideal. Neither the House of Representatives, nor the Senate, nor the mechanism for electing the president is based strictly on a population count. The Constitution specifies that "each State shall have at least one Representative" (Article I, Section 3). Several states have so few residents that without this constitutional protection, they would have to share a representative. There are 435 seats in the House of Representatives. Therefore, assigning one representative to each state leaves 385 seats to be allocated among the states by population. The Constitution's mandates cannot be reconciled to a one-person/one-vote ideal. After the 2000 reapportionments, the 496,304 people of Wyoming had one

representative, but each representative from California represented a district of about 640,000 people, 1.3 times as many (Figure 11-27).

Each state is also represented by two senators, no matter how few residents it has. Therefore, California's senators represented 68 times the number of people represented by senators from Wyoming. In both the House and Senate, people from Wyoming get "more representation" than people in California.

The president, in turn, is elected indirectly in the electoral college, in which each state has as many electors as it has total seats in Congress. The candidate who wins a majority of a state's popular vote wins all that state's electoral college votes. In 1888, Grover Cleveland won the popular vote over Benjamin Harrison, but Harrison won the electoral vote, and Harrison became president. The presidential election of 2000 repeated the experience of 1888. The Democratic candidate, Al Gore, received 500,000 votes more than George W. Bush, but Bush won most of the "overrepresented" rural states, giving him more votes in the electoral college and thus the presidency.

The national population of the United States is becoming increasingly geographically concentrated, so the system of representation is becoming increasingly arithmetically undemocratic. By 2000, California's population was greater than that of the 22 least populous states combined, yet those states had twice California's representation in the electoral college. It is projected that by 2030, a full one-quarter of all

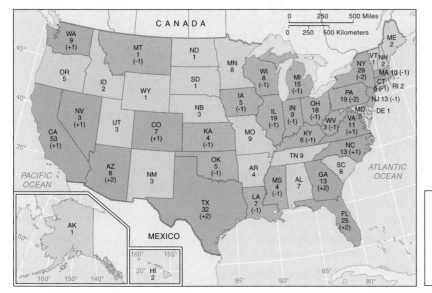

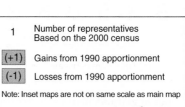

Figure 11-27 Reapportionment among the states after the 2000 Census. After the 2000 Census, the seats in the U.S. House of Representatives were reapportioned as shown on this map. Fast-growing states, mainly in the South and West, gained seats, but those with proportionately declining populations, mainly in the North and East, lost representation. Some observers feel that the United States has grown so populous that people are alienated from the Congress. The House of Representatives capped its own number at 435 in 1910, when the U.S. population was 90 million, but if the ratio of members of the House today were the same as it was in 1790, the House would have more than 9,000 members.

Constant Redistricting and Gerrymandering in the United States

Districts in the United States are supposed to reflect "communities of interest," meaning that people in the same geographic area share common concerns. Regardless of the original districts drawn in the United States, redistricting for the U.S. House of Representatives has been taken over by partisan interests except in those states that have only one representative or that have established independent or bipartisan commissions (that is, Iowa, Maine, Arizona, Hawaii, Idaho, New Jersey, and Washington). The stakes are high for political parties to capture seats in the U.S. House because doing so determines a good deal of where federal spending is directed. Majorities in the House can also control legislative agendas.

The basic means of gerrymandering is to move district boundaries so that people who are likely to vote for one party are either concentrated, to help them win the district, or dispersed, to ensure that they cannot win a district. In the United States, past elections are used as a guide by political parties to see where their support and opposition live. In the past, redistricting was easy to do because populations moved less often. As people move more frequently and communities change more rapidly, a person's political support is often inferred from his or her race, income, and other demographic characteristics. Social and economic data from the U.S. Census are used along with other survey data. Keeping track of all this

information on a block-by-block level was once nearly impossible. Today, the use of geographic information systems (see Chapter 1) has made analysis of all these data much easier. A GIS makes it easier to gerrymander, because anyone who has the right software can call up any district onscreen, shift a boundary, and get an instantaneous readout of what the total population, voting behavior, racial composition, and other characteristics would be of the newly drawn district (see the figure below). Redistricting that once took months or years to calculate with low precision can be optimized in a short period after each election.

Observers argue that easier gerrymandering threatens democracy, as incumbents and parties gerrymander to secure their own reelection. Voters are polarized; widening percentages of victory are interpreted as evidence of gerrymandered districts. Elections are less and less competitive, eroding one of the foundational ideas of democracy. In the 2004 congressional elections, only 5 percent of the 435 contests were decided by fewer than 10 percentage points, and only 5 incumbents were defeated. Nevertheless, more voters seem willing to shift their support between parties to punish poor performance. In 2006, the Democratic Party gained 31 seats in the House, in part because voters had grown frustrated with Republican policies.

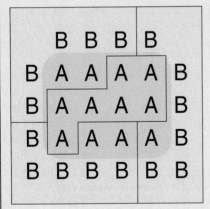

(a) Opponent concentration: A's control one district

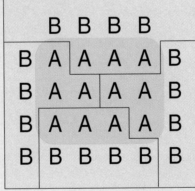

(b) Opponent dispersion: A's control no district

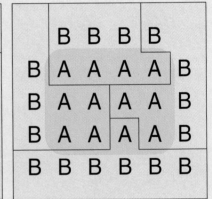

(c) Even division: A's and B's each control two districts

Types of gerrymandering. In this figure, the B's have an absolute majority in each case, so they have the power to draw the district lines. Drawing (a) illustrates concentration of the opponents (the A's), so the A candidate wins that one district with an unnecessarily large majority, but A's cannot win anywhere else. Drawing (b) splits up the A's so that they do not form a majority anywhere and cannot elect even one representative. In (c), each group controls two districts.

Americans will reside in just California, Texas, or Florida. Eliminating the electoral college would require the approval of the low-population states that profit from it, so the possibility is slim. In 2005, when U.S. President George W. Bush told Russian President Vladimir Putin that Russia should have a more democratic system, Putin asked, "Do you mean like the electoral college?" President Bush abruptly changed the subject.

The fact that the government is not truly numerically democratic causes inequitable distribution of federal spending. As a rule, the more-populous states subsidize the less-populous.

Redistricting and civil rights

A 1982 amendment to the Federal Voting Rights Act of 1965 banned any redistricting that would have a negative impact on the political representation enjoyed by minority communities. The law insists that when any court is examining a case of redistricting, the court must consider "the extent to which members of a protected class have been elected to office." Several states interpreted this to mean that they were mandated to create districts in which the majority of the voters belonged to racial minorities. These districts are called *majority–minority districts*, and they are drawn by practicing the opponent concentration technique illustrated in the feature above. Such majority–minority districts have been drawn at every level of government across the nation, from city council seats to congressional districts. In 1986, the Supreme Court defined a three-part test for assessing voter dilution that seemed to encourage the formation of majority–minority districts. The ruling stated: "First, the minority group must be able to demonstrate that it is sufficiently large and geographically compact to constitute a majority of a single-member district. . . . Second, the minority group must be able to show that it is politically cohesive. . . . Third, the minority must be able to demonstrate that the white majority votes sufficiently as a bloc to enable it . . . usually [to] defeat the minority's preferred candidate" (*Thornburg* v. *Gingles*).

Throughout the country, however, these majority-minority districts were challenged as representing racism—favoring a designated group rather than oppressing it, but racism nonetheless. Some people, however, argue that the United States must be color-conscious for a while as a way of making up for past discrimination and segregation. In 1993, the U.S. Supreme Court ruled that creating legislative districts with people of the same race who are otherwise separated "bears an uncomfortable resemblance to political apartheid . . . and reinforces the perception that members of the same racial group—regardless of their age, education, economic status or the community in which they live—think alike, share the same political interests and will prefer the same candidates at the polls" (*Shaw* v. *Reno*). Grouping voters by their ascribed identities and trying to win their votes is

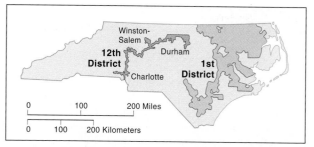

Figure 11-28 Efforts to gerrymander districts in North Carolina. In the early 1990s, North Carolina legislators designed these two congressional districts in order to elect black representatives to Congress. The U.S. Supreme Court upheld the constitutionality of the First District, but, in 1996, it ruled that the Twelfth District was drawn in a manner that was too "race conscious." In 1997, the North Carolina legislature redrew the Twelfth District in a way that would scarcely be distinguishable on a map at this scale. That new pattern was finally approved by the U.S. Supreme Court in 2001, just as North Carolina set about redrawing all of its congressional districts on the basis of the new data from the 2000 census!

commonly called *identity politics*—in contrast to *issue politics*, which argues public matters—but racial stereotyping in districting is legally impermissible. In examining one gerrymandered district, North Carolina's Twelfth Congressional District, the Supreme Court ruled that a conscious concentration of black voters did not automatically make a district unconstitutional, as long as the state's primary motivation was political (to create a Democratic district) rather than racial (to create a black district) (Figure 11-28). In other words, party-conscious gerrymandering is constitutional, but race-conscious gerrymandering is not (*Easley* v. *Cromartie*, 2001).

In 2006, the Supreme Court upheld the right of Texas to have redistricted in 2003; however, it criticized the redistricting scheme for having violated the Voting Rights Act, thus showing that the Voting Rights Act (extended for 25 years in 2006) remains a tool for minorities who can show that their right to equal participation in the political process has been impaired. Nevertheless, Chief Justice Roberts wrote a strongly worded dissenting opinion, which Justice Alito signed, insisting "It is a sordid business, this divvying us up by race." Arguments over gerrymandering continue in America's legislatures; in 2003, the U.S. Supreme Court upheld the federal courts' right to assume the activist responsibility of stepping in to draw legislative boundaries when state processes fail (*Branch* v. *Smith* and *Smith* v. *Branch*) (Figure 11-29).

Individual Rights and Freedoms

Countries differ greatly in the restrictions they place on their citizens' freedom and in the degree to which they encourage and achieve full development of their citizens' potential. Predominant attitudes toward some

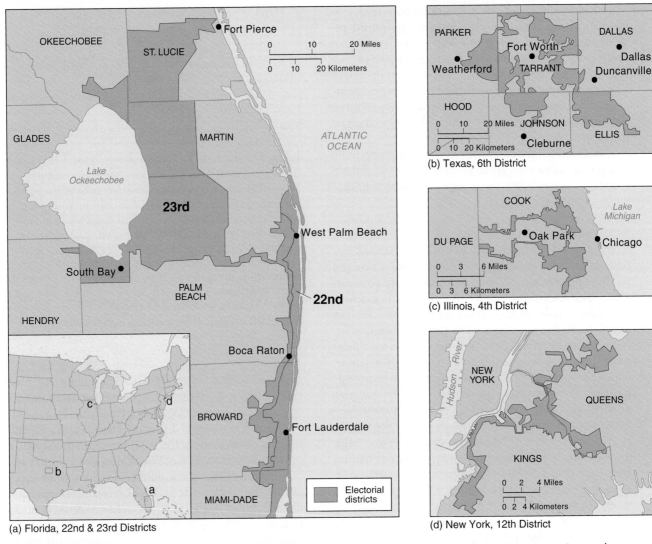

Figure 11-29 Some U.S. congressional districts in 2003. These are a few "gerrymanders" that sent representatives to the U.S. House of Representatives in 2006. In each case, the districts were designed to manipulate the outcomes of elections held in the districts.

issues, such as women's rights, may be determined by the underlying culture that may cross international boundaries. Arab culture, for example, as noted in Chapter 7, seems relatively insensitive to women's rights. Nevertheless, sovereign states are the units within which laws promote or restrict individual rights, so we can meaningfully analyze and compare the policies of different countries.

A society in which the most capable people can rise to the top on merit alone is called a **meritocracy,** which most countries claim to be. Rigid social stratification, in which people are unable to advance educationally or economically, can be unhealthy for a democratic society because it discourages participation. Hurdles to individual advancement may include stratification of the population by caste, race, or employment restrictions. The modern labor movement has won important victories for working people, but in some countries, labor unions may protect their members' jobs at the expense of opportunity for other individuals. In the United

States, for example, some labor unions have been bastions of racism.

Many countries of diverse populations have *affirmative action* programs designed to lift to national standards of achievement those segments of the population that have suffered a historic lack of opportunity. Laws in both Malaysia and Indonesia, for example, favor ethnic Malays over the more entrepreneurially successful Chinese. Nigerian law recognizes efforts of school and job-development programs to "reflect the federal character." This is a code for federally mandated ethnic quotas. Some European countries try to achieve diversity in groups in terms of members' age, as well as ethnic composition and sexual orientation. India has an elaborate system of preferences for lower-class "scheduled" castes and tribes, and Brazil enforces new race-based preferences in universities, the civil service, and the private sector. Affirmative action programs have been crafted in the United States in both the public and private sectors. In upholding race-consciousness among other admissions criteria for U.S. universities, the Supreme Court replaced the traditional appeal to historical justice as a rationale for affirmative action with diversity in itself as a key value in American life (*Grutter* v. *Bollinger* and *Gratz* v. *Bollinger*, 2003).

Other forms of discrimination reflect not only group preferences and biases but negatively affect an individual's opportunities to live a full and productive life. Sexism is by far the most pervasive form of discrimination in the world. Almost everywhere, women are worse off than men. They have less power, less autonomy, less money, harder work, and more responsibility. Educational opportunities for girls and women are in many places restricted by law or by custom, which limit their economic, social, and political advancement. Access to quality education is also a basic problem for both girls and boys in many countries. Because these problems deeply affect an individual's quality of life, we will examine them more closely in Chapter 13, which discusses human development. In consideration of one's rights, however, we must note that these forms of discrimination are incompatible with democratic values, which promote freedom over repression.

Relations Among States

Political geography includes not only an examination of how states are defined and internally organized but also how states relate to each other internationally. States cooperate on numerous issues through a variety of institutions. Yet states sometimes have competing interests they cannot resolve and that may lead to armed conflict. There are also parts of the Earth's surface that do not belong to states but raise important questions about how states cooperate. All states are finding it difficult to contend with globalizing forces.

Patterns of Cooperation

States are independent political units, but they typically require cooperation with their neighbors to succeed. At the very least, states must recognize each other's right to exist or they will find themselves consumed by conflict and instability. This fundamental form of cooperation is called **recognition,** which is when one state formally acknowledges the existence of another. When Kosovo declared its independence from Serbia in 2008, it was immediately recognized by several powerful governments, including the United States, France, and the United Kingdom, as well as other countries. Most countries have not yet granted official recognition to Kosovo, however, and still consider the province a part of Serbia's sovereign territory. The state of Israel is not recognized by several dozen countries, including many Arab and Muslim countries. Most countries enjoy recognition by most or all other countries, making it possible for them to

establish a formal relationship and cooperate on matters of common interest.

Formal relationships between two countries are carried out by heads of state or their representatives in acts of *diplomacy*. In many areas of cooperation, issues are worked out by diplomats who are either specially appointed by the head of state or who are part of a professional diplomatic staff. It is through these relationships that most cooperative work is accomplished, whether by discussing areas of common concern or making clear their differences and grievances. These personal meetings are important to establishing good relations because sensitive matters can be discussed in confidence, giving governments a chance to cooperate rather than confront each other. The possible number of diplomatic relations is enormous when one considers that there are more than 190 countries in the world.

The geography of diplomacy is unusual. The tradition of meeting in person means government officials and diplomats must be allowed to move freely and to carry messages and packages that are not subject to the usual inspection by border guards. Diplomats usually receive diplomatic immunity from the host country, meaning they cannot be prosecuted under local laws. Good diplomatic relations usually lead to two countries establishing a **permanent mission** in each other's countries, each with diplomats and staff that live there. Embassies, which house permanent missions, are not technically sovereign territory of the home country but they are considered "inviolable" and immune from entrance or searches by the host state. Individuals facing persecution have often sought safe haven in an embassy. In 2009, political opponents of Honduran president Manuel Zelaya forcibly removed him from office by putting him on an airplane leaving the country. He returned secretly to the capital later in 2009 to try to reclaim his position but feared for his life and immediately sought haven in the Brazilian embassy, which was willing to give him shelter. The embassy was then surrounded by Honduran soldiers, but they did not enter. Doing so would have jeopardized Honduran relations with Brazil and all other countries that have embassies in Honduras.

Countries can establish formal agreements with each other, which are usually called *treaties*. A treaty is a specific agreement that binds states to cooperate in set ways on various issues. Treaties cover everything from war and peace, borders, and trade to environmental protections, agreements that we will examine more closely in Chapter 13. Some *bilateral* treaties pertain to only two countries, while *multilateral* treaties are agreements among many states. The earliest known peace treaty is from thirteenth century B.C. Anatolia, a copy of which hangs in the United Nations. Contemporary treaties usually restrain states from exercising their full sovereign rights. Since 1997, for example, 156 countries have signed a treaty banning the production and use of landmines, which were once part of most militaries' weapons. The countries recognized that these devices do far more harm to civilians than good for national security and so they were willing to get rid of their landmines if other countries agreed, too. Treaties that create fairness, or a "level playing field," can entice countries to give up sovereign rights in exchange for a preferable outcome they could not achieve without cooperation. Some countries purposely avoid treaties, stating that their national interests outweigh the benefits of cooperation. In the case of the landmine ban treaty, 37 countries have not agreed, including China, India, Iran, Russia, and the United States. The absence of some of the world's largest countries certainly weakens the landmine ban treaty.

Geographies of Conflict

States do not always cooperate, and often they simply ignore each other. On some issues, however, states object to the actions or policies of another state. A situation of disagreement between states is a *conflict*, although this may not involve warfare or violence. Some conflicts are resolved through treaties and other conflicts simmer without solution but without escalating, either. Just as international cooperation may extend to cover a wide range of issues, conflicts may arise from any number of problems. Conflicts over borders, territory, and resources are conventionally the most likely conflicts to lead to *war*, which is a condition of formal hostility and armed violence between the militaries of two or more states. Overall, war is rare in the relationship between any two states in the world, but it is more common among neighboring states and especially in some troubled regions.

In 2008, researchers identified 36 ongoing conflicts, of which 5 had 1,000 or more persons killed during that year, as shown in Figure 11-14. Most of these conflicts could be defined as *civil war*, an armed conflict among militarized groups within one country including the government. Some countries have multiple conflicts occurring at the same time. Civil wars are often long running and cost the lives of many civilians as armed groups fight for control of the state. The war in Sri Lanka, for example, between the government and the Tamil rebels lasted from 1983 until 2009, during which time tens of thousands of civilians and many more soldiers were killed. Some wars are both international and civil wars. The 2003 invasion of Iraq included major military operations by the United States and the United Kingdom but grew into a war that also included the new government of Iraq and insurgent groups.

Jurisdiction over Earth's Open Spaces

One of the most important issues facing states is how to balance competing interests over Earth's open spaces. Open spaces include the Arctic, Antarctica, the world's oceans and seas, airspace, and outer space.

The Arctic and Antarctica   Eight countries have territory north of the Arctic Circle: the United States, Canada, Denmark, Finland, Iceland, Norway, Sweden, and Russia. In 1996, these countries formed the Arctic Council, which pledged to cooperate in monitoring Arctic pollution and in protecting the region's plant and animal life. Canada claims dominion over the uninhabited Arctic islands north of Canada all the way to the North Pole, but the United States has challenged Canada's claims. As noted at the start of this chapter, the melting Arctic icecap is producing a scramble among these countries to lay claim to larger areas of the north.

Antarctica has never had any permanent inhabitants, but seven states claim overlapping sovereignty. Australia, New Zealand, Chile, and Argentina, the world's most southerly states, claim sovereignty on the basis of proximity. The claims of Britain, Norway, and France are based on explorations (Figure 11-30). Neither the United States nor Russia, which support most of the scientific research there, make any territorial claims, nor do they recognize other nations' claims. Forty-six states are parties to a 1959 Antarctic Treaty that resolves that "Antarctica shall continue forever to be used exclusively for peaceful purposes and shall not become the scene or object of international discord." The treaty prohibits military installations, but it opens the continent to all nations for scientific research and study. The treaty also prohibits signatories from harming the Antarctic environment.

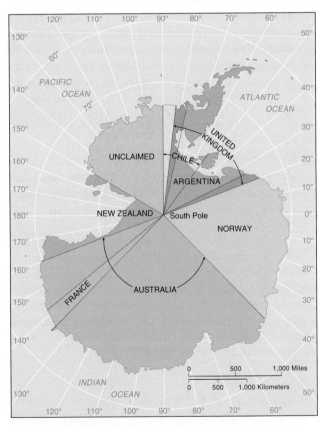

Figure 11-30 Claims to Antarctica. By treaty, these claims to Antarctica are not enforced.

Despite the terms of this treaty, however, some countries, led by Malaysia, demand that exploration should be undertaken for Antarctic resources and that any resources found should be exploited for the good of all nations. Other nations, led by France and Australia, insist that Antarctica should be set aside as a wilderness preserve.

Territorial waters The world's states have divided up all the land on Earth (except Antarctica), but international agreement negotiates the question of how far out to sea the territorial claim of a country can reach.

The Dutch jurist Hugo Grotius (1583–1645) first argued that the world's seas are open space, *mare liberum*, which no political unit can claim. Each coastal state, however, has claimed coastal waters for defensive purposes. A later Dutch jurist, Cornelius van Bynkershoek, accepted this in his book *De Domina Maris* (1702). The limit of sovereignty was set at three nautical miles from shore, which was generally accepted for more than 200 years. All distances at sea are measured in international nautical miles of 1 minute of latitude—1,852 meters (6,076 feet). Territorial sea is measured from the low-tide mark, and any disputes between states arising from irregular coastlines or islands are resolved by bilateral negotiations. The 1958 International Conference on the Law of the Sea set rules for ascertaining the seaward extension of land borders (Figure 11-31).

The United States accepted a 3-mile limit in 1793, but in 1945, President Truman broke that international covenant and claimed sole right to the riches of the continental shelf out to 200 nautical miles (about 230 miles or 370 kilometers). A **continental shelf** is an area of relatively shallow water that surrounds most continents before the continental slope drops more sharply to the deep-sea floor. The 1958 conference set a water depth of 200 meters (656 feet) as the definition of the outer edge of a continental shelf (Figure 11-32).

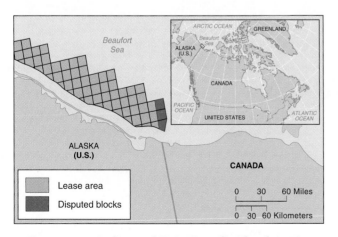

Figure 11-31 A disputed U.S.–Canadian border. The United States and Canada dispute the seaward extension of their border here in the Arctic, and the United States has auctioned off the rights to explore for oil in the disputed tracts of the Beaufort Sea.

The Truman Proclamation claimed shelf mineral rights, but it did not claim control over fishing or shipping in the seas over the shelf beyond the 3 nautical mile territorial limit. It nevertheless triggered extended claims to fishing rights by other nations. In 1976, the United States extended its own claims to exclusive fishing rights up to 200 nautical miles out from its shores. As noted in Chapter 8, most of the world's marine fish harvest is caught within 200 nautical miles of the coasts.

The 1982 United Nations Convention on the Law of the Sea In 1982, the United Nations proposed a new Law of the Sea Treaty. Sixty countries ratified it by 1993, and it came into force among the ratifying parties in 1994. The Law of the Sea Treaty authorizes each coastal state to claim a 200 nautical mile exclusive economic zone (EEZ), in which it controls both mining and fishing rights. Therefore, possession of even a small island in the ocean grants a zone of 326,000 square kilometers (126,000 square miles) of sea around that island. This might help explain why France retains many small island colonies. France claims a full 10.4 percent of the world total EEZ. The United States claims 8.4 percent, New Zealand 5.8 percent, and Indonesia 4.7 percent. No other state claims more than 4 percent. The potential for discoveries of oil in the world's seas has triggered an international scramble to claim even tiny uninhabited outcrops of rock. For example, China, Japan, and South Korea lay conflicting claims to areas of the East China Sea. Vietnam, China, Taiwan, the Philippines, Thailand, and Malaysia extend overlapping claims in the South China Sea.

Another clause of the 1982 treaty guarantees ships from one state **innocent passage** through the territorial waters of other states, provided that they do so peacefully. The extension of countries' territorial waters, however, means that countries now claim many of the world's narrow waterways. They remain open in times of peace, but they can be closed in time of war (Figure 11-33).

The United States at first refused to sign the Law of the Sea Treaty, in protest against provisions calling for internationalization of seabed mineral resources and the creation of an International Seabed Authority to control mining. The United States insisted on rights for free enterprise. Several of the treaty's other provisions, however, did serve U.S. interests, so U.S. President Reagan announced that the United States would regard all but the seabed provisions as law, even though the country had not signed the treaty. For example, one provision allows nations to declare a 12 nautical mile territorial limit. Therefore, in accordance with the treaty, on December 28, 1988, the United States extended its territorial limit to 12 nautical miles. Given the length of the U.S. coastline, this seaward extension enlarged the territory of the United States by some 478,000 square kilometers (185,000 square miles).

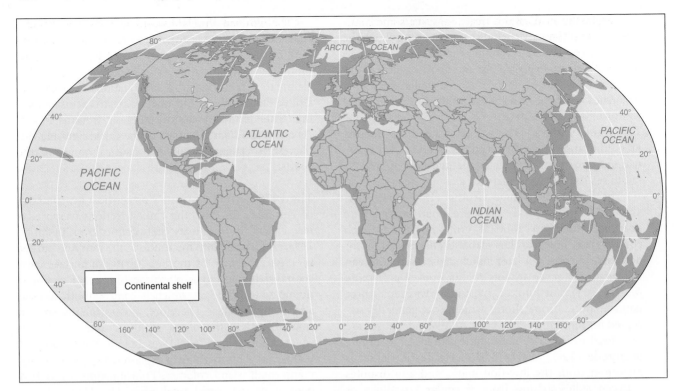

Figure 11-32 Continental shelves. The various continental shelves extend out from as little as a few hundred meters or yards to nearly 1,000 kilometers (600 miles).

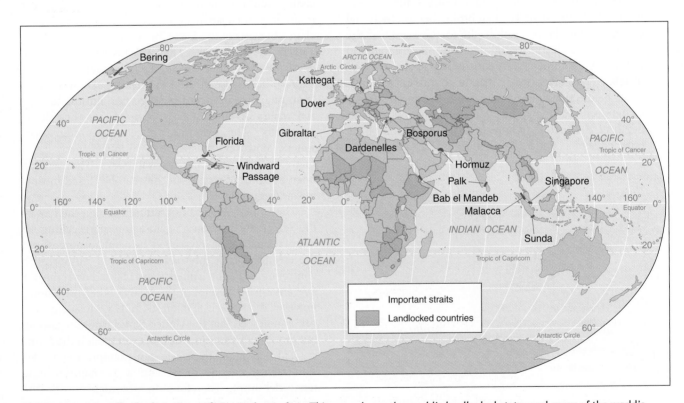

Figure 11-33 Landlocked states and strategic straits. This map shows the world's landlocked states and some of the world's most important narrow sea lanes. As states have extended their seaward claims, many of these narrow waterways have been claimed by the adjacent states. A conflict in one of these narrow straits could disrupt the world economy. The Straits of Malacca, for example, is at certain points only 2.5 kilometers (1.5 miles) wide, but it carries 40 percent of world trade.

Compromises on disputed provisions were eventually reached, and the United States signed and ratified the treaty in 1996.

Another clause of the 1982 treaty allows countries to claim some powers in a "contiguous zone" of 12 nautical miles further out beyond the 12 nautical miles of territorial waters. Therefore, on September 2, 1999, the United States announced that it would start enforcing U.S. law and would be boarding ships up to 24 nautical miles off the coast. The U.S. proclamation applies to the territorial waters around all U.S. possessions.

Landlocked states The independence of Montenegro from Serbia in 2006 added Serbia to the list of the world's now 43 landlocked states without seacoasts. Several other states are not totally landlocked, but their own coasts are unsuitable for port development, so they must rely on neighbors' ports. The landlocked states shown in Figure 11-33 must secure the right to use the high seas, the right of innocent passage through the territorial waters of coastal states, port facilities along suitable coasts, and transit facilities from the port to their own territory.

Landlocked or partially landlocked states may gain access to the sea in one of three ways. First, any navigable river that reaches the sea may be declared open to the navigation of all states. International commissions regulate navigation on many international rivers, and often these same commissions guard against pollution and regulate the drawing of irrigation waters from the river (Figure 11-34). Second, a landlocked state may obtain a corridor of land reaching either to the sea or to a navigable river. Several countries have long, thin extensions of land out to seaports. Some of these, such as the Congo's corridor to the Atlantic Ocean, are important transport routes, but others, such as Namibia's Caprivi Strip to the Zambezi River, serve no significant traffic function. The third way a landlocked state can gain access to the sea is to obtain facilities at a

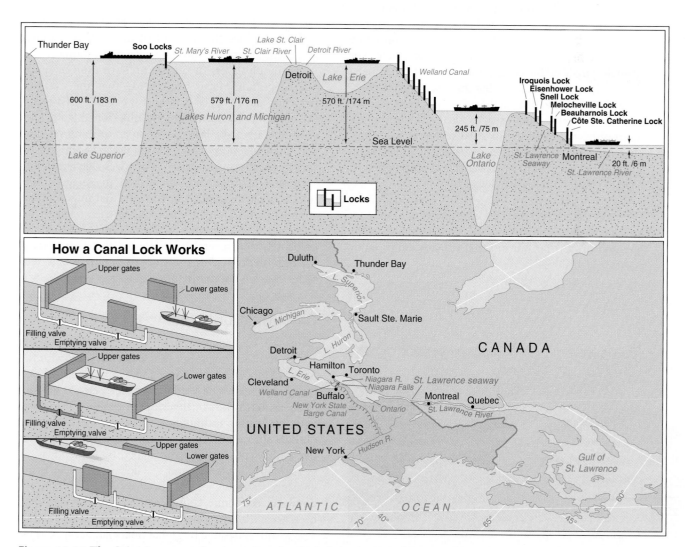

Figure 11-34 The Saint Lawrence Seaway. Completed jointly by the United States and Canada in 1959, the Saint Lawrence Seaway is an example of international cooperation on an international waterway. It allows ocean vessels to reach from the Atlantic Ocean into the Great Lakes. The joint project included the construction of hydroelectric power plants.

specific port plus freedom of transit along a route to that port. Coastal states have signed international conventions promising to assist the movement of goods across their territories from landlocked states without levying discriminatory tolls, taxes, or freight charges. Bolivia, for example, has been granted access to ports in Chile, Argentina, and Peru.

The high seas   All nations agree that beyond territorial waters are high seas, where all nations should enjoy equal rights. Several problems, however, threaten any area where rights are not specifically assigned. These include depletion of resources (such as overfishing) and pollution. International conventions regulate each of these potential abuses of the high seas. The 1972 London Dumping Convention, for example, signed by 70 countries, originally banned the dumping of radioactive and other "highly dangerous" wastes at sea, but it was extended to ban dumping of all forms of industrial wastes in 1995. The enforcement of international agreements such as these is uncertain, and their effect is limited. Problems also arise from the collective responsibility to prevent piracy on the high seas. In the absence of any one sovereign authority, all countries have the responsibility to stop piracy but the result is patchy enforcement. For example, piracy off the coast of the failed state of Somalia has worsened in recent years, but this has not led to coordinated measures to combat the problem (Figure 11-35).

Freedom on the high seas is also only meaningful when respected by states. In 2010, a ship bearing supplies for Gaza was attacked by Israel 80 miles from shore. Israel claimed its boycott of Gaza outweighed its need to respect international waters.

Airspace and outer space   The question of how far up a nation's sovereignty extends is as complex as the question of how far out to sea it extends. Ships enjoy innocent passage through territorial waters, but airplanes have never been granted innocent passage to fly over countries. Airlines must negotiate that right, and states commonly prescribe narrow air corridors, the altitudes at which aircraft must fly, and even the hours of the day the passages are open. U.S. President Dwight Eisenhower first proposed an "Open Skies" treaty in 1955, but this treaty was not realized until 1992. The treaty establishes conditions for unarmed observation flights over the entire territory of participants. In accordance with the treaty, a Russian Air Force jet spent five days in August 1997 flying over the United States, taking photographs of U.S. military installations. This astonished anyone who grew up during the Cold War.

Diplomats cannot agree on the altitude at which national airspace ends. One possible definition of national airspace would limit it to the lowest altitude at which artificial unpowered satellites can be put into orbit at least once around Earth. That ranges between 113 and 161 kilometers (70 and 100 miles). The U.N. Office for Outer Space Affairs does try to formulate recommendations for designing and flying space vehicles to reduce the amount of debris they produce and their chances of colliding with one another.

Many observers dread the militarization of space, but suggestions have been made that the eventual parceling out of space is almost inevitable, as a measure of national security. In 1997, the United States successfully targeted a laser beam on an orbiting satellite, and it spends more than a billion dollars each year on space weapons research. The U.S. government's "National Space Policy," published in August 2006, rejects any arms control agreements that might hinder "freedom of action in space." Countries have so far refrained from extending territorial claims to the Moon or planets, but several have insisted that they reserve the right to do so. Our study of geography, therefore, is so far restricted to Earth.

Globalization's Challenge

States face challenges from groups inside their borders and from other states, but they must also deal with the forces of globalization. These come in many forms. Air travel that made the world a smaller place is also an efficient network for delivering terrorism and disease. Communication technology and media that transcend borders spread new ideas around the world. Countries like China and Iran feel threatened by those ideas, as we saw in Chapters 6 and 7. Global trade can bring enormous advantages but also risks for each country's domestic economy. Global commodity markets may even wreck local economies, as we saw in Chapter 8. The wealth generated by global financial markets, as we will see in Chapter 12, also spreads devastating economic risk. These and other forces of globalization pass right through the borders of most states, beyond the reach of governments.

Figure 11-35 Fighting piracy off the Somali coast. These meagre vessels of the Somali Coast Guard are patrolling the busy sea lanes where Somali pirates made numerous attempts to hijack cargo ships in 2009.

States have been somewhat reluctant to embrace the forms of cooperation necessary to fully contain or manage these threats. Instead, governments tend to "go it alone" in solving problems until they discover they need to work with other states. Most governments tend to view the problems of globalization in terms of **national security,** meaning as risks to the state itself, not to mention whether it could affect citizens in a way that would weaken the authority of the state. In other words, states view the challenges coming from globalization as challenges to state sovereignty. This kind of thinking often means that states view problems first and foremost as issues that may require a military or police response. For example, the Mexican government responded to the H1N1 flu by imposing a curfew on large cities, using the military to keep public order and, later, to distribute facemasks. Likewise, the U.S. government, in the wake of the 9/11 terrorist attacks, has given itself new powers to monitor everything that crosses the border, such as shipping containers and also phone calls and e-mails. The enormity of the task makes for clumsy enforcement, and the government sometimes illegally crosses the line between protecting national security and violating the rights of citizens. This raises questions of how to preserve democracy in the face of global challenges.

On some issues, governments have learned that a shared response is better suited to the challenge at hand. Diplomatic relations today are swamped with shared concerns that the dangers of globalization will outweigh the benefits. Counterterrorism efforts have become an area in which even those countries that lack diplomatic ties are more willing to share information with one another. At the same time, national governments are most concerned about their own country. The challenge presented by global warming, for example, is regarded as a security issue by powerful governments rather than an environmental problem. As we describe in Chapter 13, the possibility of mass displacement, hunger, and instability are not enough for major polluting countries to cut carbon emissions because it might hurt their economies. The failure to respond to these issues has also changed the nature of diplomacy and cooperation itself. For example, many people who see free trade policies as worsening international poverty have protested against the World Trade Organization as a non-democratic institution. Other private citizens around the world donate money and volunteer their time to **non-governmental organizations (NGOs),** seeking to promote solutions to global problems that governments fail to achieve. Many NGOs address local issues, but many are also concerned with global problems, including issues such as war, violations of human rights, and environmental degradation (Figure 11-36).

Figure 11-36 Anti-globalization protests. Hundreds march to protest against unfair trade policies in Geneva, where World Trade Organization representatives met in early December 2009.

Chapter Review

Summary

Earth's division into sovereign states is the most important territorial organizing principle of human activities. The idea that the whole world should be divided into countries originated in Europe and diffused worldwide with European conquest. The neatness of the units on today's world political map nevertheless exaggerates the degree to which all people accept the current pattern.

A nation is a group of people who want to have their own government and rule themselves. The ideal state is theoretically a nation-state. Several European nations existed before they achieved their own states. The political patterns of the colonies, however, were the conquerors' creations. When independence came, only large units were at first granted independence, but eventually small units were recognized.

Many states suffer from cultural subnationalism. Governments might be able to forge a singular nation for their state. Six national institutions—a country's church or churches, armed forces, schools, media, political parties, and labor unions—become key centrifugal or centripetal forces in any state. In other cases, states cannot overcome centrifugal forces. States may be divided and their borders redrawn to accommodate national differences. Subnationalism can also lead to violence against civilians as states are ethnically cleansed of unwelcome persons, possibly resulting in genocide.

Each state demarcates its territory's borders, but states differ as to the type of political regime found within each one. Countries also subdivide their territory for political representation or administration. Each country has a unique balance of powers between its local governments and its central government. Special districts can undermine the powers of local governments, and most federal states contain territories that are administered directly from the federal center. Capitals are symbolically important.

Electoral geographers note that in any system of representative government, voting-district boundaries can be gerrymandered to determine the outcomes of the ensuing elections. The kinds of rights and freedoms enjoyed by citizens vary widely among states.

States must deal with each other and most states find it advantageous to cooperate. Cooperation is often conducted through diplomatic representatives. Treaties are another important way to formalize relationships. Sometimes, states clash over disagreements, leading to conflict that might become violent. There are large parts of the Earth's surface that are not part of any state, and governments must try to agree on how they are shared. States and citizens must also contend with challenges coming from new technologies and geographic relationships of globalization, which pass through borders.

Key Terms

autocratic p. 443
centrifugal force p. 433
centripetal force p. 433
collapsed state p. 440
continental shelf p. 459
core area p. 427
cultural boundary p. 446
electoral geography p. 452
enculturation p. 437
ethnic cleansing p. 440
failed state p. 440
federal government p. 447
genocide p. 440
geometric boundary p. 445

geopolitics p. 441
gerrymandering p. 452
iconography p. 437
indirect rule p. 431
innocent passage p. 459
irredenta p. 429
kleptocracy p. 444
meritocracy p. 456
nation p. 426
nation-state p. 427
national security p. 463
national self-determination
    p. 428
non-governmental organization
    (NGO) p. 463
partition p. 438

permanent mission p. 457
physical boundary p. 445
political community p. 426
political culture p. 427
recognition p. 457
regionalism p. 438
separatist movement p. 438
social Darwinism p. 430
socialization p. 437
sovereignty p. 426
state p. 424
subnationalism p. 438
superimposed boundaries
    p. 430
unification p. 438
unitary government p. 447

Questions for Review and Discussion

1. Differentiate a nation, a state, and a nation-state. Give examples of countries that are not nation-states.

2. What important decision was taken at the 1964 conference of the Organization of African Unity (OAU)?

3. How do governments try to forge their populations into nations?

4. What are the two principal methods of gerrymandering?

5. What are the places associated with cooperation among states?

Thinking Geographically

1. When and how was the pattern of local governments in your state drawn? What responsibilities do the various local governments have? Could you improve the pattern, reassigning responsibilities to new districts or different levels of government?

2. Read the constitutions of a few countries that you believe to be totalitarian and centralized. Do the constitutions reveal that? Does the U.S. Constitution accurately reveal the division of power between state and federal governments that exists today?

3. How much land in your state is owned by the federal government? What is its legal status?

4. Is your state overrepresented or underrepresented in the electoral college?

5. Do you think regions of a country ought to be able to secede? If you favor a vote, do you think that majority approval should be required throughout the entire country or only in the areas seeking to secede?

6. Make a list of several mountain chains or rivers around the world that serve as international borders. Which act as effective barriers? Which do not? Why?

7. Investigate the symbolism of the flags of a few countries. Look at Australia, the Comoros, Liberia, Mongolia, Namibia, Venezuela, and South Africa.

8. Why might a unitary government be more successful in Japan than in the United States?

Log in to www.mygeoscienceplace.com for videos, animations, **MapMaster**™ interactive maps, RSS feeds, case studies, and self-study quizzes to enhance your study of A World of States.

   MapMaster™

IBM was once a wholly American company. Now IBM also works around the world, including from large offices in India, one of which is in the city of Bengalooru (Bangalore).

12

Paths to Economic Growth

On June 6, 2006, Samuel Palmisano, chair of IBM, the world's largest computer services company, stood with Indian President Abdul Kalam in Bengalooru, India (formerly Bangalore), and announced a three-year plan to invest over $6 billion in India. Three years earlier, IBM had faced a public relations disaster when a disgruntled employee had publicized corporate plans to accelerate the relocation of high-paying white-collar jobs out of the United States. The relocations had proceeded despite the publicized complaints. In 2003, IBM had 9,000 employees in India. By mid-2009, it had approximately 74,000, and its Indian employees earned, on average, 12 percent of what U.S. employees doing similar work had earned. IBM had cut 14,500 U.S. jobs. IBM's Bengalooru center today monitors tens of thousands of computer servers and applications and hosts thousands of consultants offering services throughout the world. Other technology multinationals, such as Microsoft, Intel, and Cisco Systems, have also announced multi-billion-dollar investments in India.

The movement of jobs overseas is the inevitable result of not only corporations' efforts to lower costs but also of the technology that knits the world economy—the phenomenon of globalization. The first jobs to leave the United States and other rich countries were the blue-collar factory jobs, which went to new factories in low-wage countries; China has captured the lion's share of these. The relocation lowered costs to manufacturers and prices to consumers, but at the cost of job opportunity in the rich countries.

Now service-sector jobs are following the factory jobs. Everything from relatively low-paying call center jobs to software development, accounting, income tax preparation, and even legal and medical analysis can be performed by lower-paid workers in China, India, the Philippines, Russia, or elsewhere. Three million service jobs are expected to leave the United States alone by 2015, including half of the jobs in packaged software and information technology services. An IBM executive said, "Our competitors are doing it, and we have to do it."

How will economic globalization redistribute good jobs, or wealth, and how will it affect development in the countries where living standards are low today? The answers will determine future job opportunity in today's rich countries.

A Look Ahead

Analyzing and Comparing Countries' Economies

Gross domestic product and gross national income are the principal measures of nations' wealth. Analyzing the relative importance of different economic activities across various countries reveals how they may be described as preindustrial, industrial, or postindustrial. Natural resources are additional factors that help to determine national wealth. Adding value to resources or providing valuable services builds wealth. Technology, communications, and government policies constantly change the locational determinants of economic activity.

The Geography of Manufacturing

Locational determinants, such as raw materials or consumers, influence the world geography of manufacturing. As these change, they cause some manufacturing to relocate continuously.

National Economic–Geographic Policies

Countries have different systems to organize and regulate domestic economic activity in order to increase wealth and distribute that wealth around the territory.

National Trade Policies

Some countries try to build their domestic economies without trade, while others participate in international trade to attract capital and win markets.

The Formation of the Global Economy

International economic links have reduced the barriers to exchange and communication, and many activities have expanded their scale of organization to cover the whole globe. Economic globalization has far outpaced cultural or political integration.

An economy is any system for organizing the production and distribution of goods in a specific area. Typically, we think of national economies, because most people agree that one of the principal tasks of any government is to promote the welfare of its people, including their economic welfare. As we saw in previous chapters, governments play a considerable role in economic activities related to food and other resources in rural areas as well as in the formation of urban economies. Indeed, the sovereignty and internal organization of states has produced a map of local economies and economic policies. Laws and regulations on land use and trade, as well as government spending itself, have always been crucial for shaping national economies.

This chapter is divided into five sections. The first section examines several ways of measuring countries' economies. Then it maps the rich and the poor countries and examines reasons for their wealth or poverty. The second section of the chapter examines the world distribution of manufacturing and reasons for that distribution. The third section of this chapter investigates national economic–geographic policies. Each country has a unique way of organizing its economy and managing the distribution of economic activities within its territory. The fourth section looks at how countries regulate their participation in world trade. The closing section examines the formation of a global economy, as opposed to a national economy. World trade has accelerated so fast and multiplied to such a degree that international management consultant Peter Drucker (1909–2005) wrote, "The world economy has become a reality, and one largely separate from national economies. The world economy strongly affects national economies; in extreme circumstances it controls them."

welfare, and these measures do not always rank countries in the same order. We will return to the broader concept of human development in Chapter 13. In this chapter, we will emphasize economic development.

Measures of Gross Product and Their Limitations

The two most commonly used measures of a country's wealth are its gross domestic product and its gross national income. **Gross domestic product (GDP)** is the total value of all goods and services produced within a country. **Gross national income (GNI)** is the GDP plus any income that residents receive from foreign investments, minus any money paid out of the country to foreign investors. These are two different measures. Cars made by Toyota in the United States, for example, would be added to the U.S. GDP. But in calculating GNI, the net profits made by Toyota in the U.S. would be added to the Japanese total because that is where the company is headquartered (Figure 12-1).

Either GDP or GNI can be measured *per capita*— per person in the country. GDP provides a better guide to analyzing a country's domestic economy, because it measures exactly what activities are occurring and what wealth is being created inside the country. GNI is best used when we want to compare and contrast different countries' total incomes as a guide to the total wealth available to the people in that country.

For most countries, there is little difference between the two statistics, but the differences that exist are important to geographers. These differences may reveal the impact that foreign investment has on a country's economy. Some countries either receive large payments from

Analyzing and Comparing Countries' Economies

Some countries have achieved high incomes and standards of living for their people. Others have not. Throughout this book, we have generally referred to the rich countries and the poor countries. The terms *rich* and *poor* are fairly straightforward and descriptive. Rich countries are those that receive higher incomes from their economic activities and have accumulated more wealth than poorer countries. The gains made by rich countries accumulate over time and usually improve living standards for people in that country. Countries that have achieved high levels of income from manufacturing are often described as **industrialized economies.** Higher incomes also allow for improved living conditions, health, and education. These countries are often described as *developed countries* in contrast with poorer *undeveloped, underdeveloped,* or *developing countries.* There are many ways to measure economic status and

Figure 12-1 Lukoil service station. The Russian oil company Lukoil has bought thousands of Getty and Sonoco Phillips service stations throughout the United States, so profits from this station in Manhattan are part of Russia's GNI. This station's profits are not, however, part of Russia's GDP, because they are not generated inside Russia. Many oil-producing states have purchased refineries and distribution systems in order to control supplies from the ground to the ultimate consumer. Thus, they capture profit from value added at each step.

abroad or send money abroad, so these countries will record a significant difference between their GNI and their GDP. For example, as Kuwait sells oil from domestic oil wells, the income is counted as part of Kuwait's GDP. Through the years, the Kuwaitis have profited from oil sales and invested profits abroad. Today the profits from Kuwaitis' foreign investments that are returned to Kuwait are part of Kuwait's gross national income, but they are not part of Kuwait's gross domestic product. Kuwait's GNI regularly amounts to 110 to 125 percent of its GDP. This reveals that Kuwaitis have enormous investments abroad.

The ratio between a country's GDP and GNI can change over just a few years because the growth of its domestic economy may be quite different than what is happening globally. In 2005, the GNI of the United States was 104 percent of GDP. Other countries, by contrast, pay out profits to owners in other countries, so their GNIs are smaller than their GDPs. In 2004, Canada's GNI was 8 percent smaller than its GDP, Argentina's GNI was about 6 percent smaller than its GDP, and Ireland's GNI was fully 25 percent smaller than its GDP. These statistics suggest that significant shares of these countries' economies are owned by foreigners. Foreign investments were affected by the sort of economic downturns the world has experienced in recent years, affecting the ratio of GNI to GDP. In 2007, the U.S. GNI was less than 1 percent larger than GDP, Canada's GNI was only 2 percent lower than GDP, Argentina's was 8 percent lower, and Ireland's GNI had risen 5 percent in comparison to GDP.

In 2008, world GDP totaled $61 trillion. The United States generated $14.2 trillion, Japan $4.9 trillion, Germany $3.7 trillion, the United Kingdom $2.6 trillion, France $2.9 trillion, Italy $2.3 trillion, and China $4.3 trillion. Along with Brazil, Russia, and Spain, which each had a GDP of about $1.6 trillion, these countries accounted for two-thirds of the measured economic activity on Earth.

Limitations of the GNI and the GDP statistics Both GNI and GDP are deceptive. Both begin with the idea that most goods and services are of little or no value if they cannot be sold for a price. Therefore, these measuring techniques underestimate the activities of hundreds of millions of people who provide entirely for their own needs or who exchange very little except through barter. For example, many of the world's farmers buy little food because they eat the food they raise themselves. Therefore, farmers' total output and income are probably both undercounted. A peasant farmer in China whose total annual income is recorded as $250 may eat food throughout the year that would cost a resident of Chicago well over $1,000 in the local supermarket. Farmers usually make up a greater percentage of the population in the poor countries than in the rich countries, so undercounting farmers' production underestimates the material welfare of many people in the poor countries. Statistics particularly overlook the work of women (Figure 12-2).

Figure 12-2 Subsistence farming. Chapter 8 noted that women raise much of Africa's food. Yet subsistence production does not count in a country's GDP. Therefore, only if this woman sold her produce to her neighbors would its value or her work appear in the country's GDP.

By noting these points, however, we do not mean to exaggerate the quality of life those people may have or to suggest that they live well. They are still poor by the standards in most developed countries.

Because statistics undercount subsistence areas, they exaggerate the degree to which modern areas, especially cities, dominate national economies. In extreme cases, a single city can provide most of a country's measurable output. Abidjan, with 15 percent of the Ivory Coast's population, accounts for 70 percent of all economic and commercial transactions in the country. São Paulo, with 10 percent of Brazil's population, contributes a quarter of that country's measurable economic activity.

Government statistics cannot measure activity that is illegal and therefore hidden. Colombia's profitable drug exports, for example, disappear in most economic measures. The same is true for illegal drugs in the United States, yet the federal government admits that the drug trade is one of the nation's largest industries. Nor does gross product include the activities in a nation's informal sector (discussed in Chapter 10).

Statistics examined per capita over time reveal a relationship between population growth and changes in the statistic measured. A country's GNI can grow, but if the population grows faster than the GNI, GNI per capita will fall. For many countries of sub-Saharan Africa, per capita incomes rose in the 1960s, flattened out in the 1970s, and fell from the mid-1980s through the rest of the century. The per capita GNI of Zambia fell at an annual rate of 1.9 percent from 1975 to 2004. That of the Democratic Republic of the Congo fell at an annual rate of 4.9 percent. Falling per capita incomes in many African countries were due in part to the general deterioration of African economies, to civil wars, and to rapid population growth. African countries also suffer from falling prices for basic commodities, as we saw in Chapters 8 and 9. Growing global competition for cotton, for example, has meant that African farmers earn less each year for the same crop.

The geography of exchange rates   Attempts to compare economies are further complicated by the fact that there are no common measures of value. Gross products are measured in local currencies. This complicates comparisons among them because the exchange rates among currencies are changeable and may be manipulated. This is of special concern to geographers not only because it hinders comparison of one place with another but also because variations in exchange rates affect the flows of goods, investment, and people that geographers study. Between 1985 and 1990, for example, the U.S. government deliberately lowered the value of the U.S. dollar by 50 percent relative to the currencies of Japan and the Western European countries. This had extensive geographic results. American goods were cheaper for foreigners to buy, so exports of U.S. manufactured goods rose 80 percent; new exports boosted the output of the nation's manufacturing belt (the Northeast–Midwest industrial region; look ahead to Figure 12-18). Foreigners invested billions of dollars in the United States, and millions of foreign tourists came to visit. In 1989, for the first time ever, foreign visitors spent more money in the United States than U.S. citizens spent abroad. As the dollar lost its value compared to other currencies, goods flowed out and tourists and investment flowed in.

At the same time, the exports to the United States from foreign trading partners sagged as Americans bought fewer foreign made goods. These exporting economies suffered, and the internal economic geography of their countries changed as their export-dependent regions suffered most. Billions of dollars that U.S. manufacturers had planned to invest in factories abroad were invested in the United States instead, and communities from the Philippines to Brazil went without new vocational training programs, roads, and schools. Places dependent on U.S. tourists suffered recessions.

From 1990 to 2001, the exchange value of the dollar generally rose, and all these flows reversed. From 2001 through 2008, the dollar fell again, and the flows reversed once more. Geographers monitor shifting exchange rates such as we have described to understand changing patterns of travel and trade and of local prosperity.

Furthermore, exchange rates differ from what economists call *purchasing power parity (PPP)*. The currency values established in foreign exchange markets often do not accurately reflect purchasing power, but PPP attempts to measure what goods or services a certain amount of money will buy at different places. When you read that annual incomes in a certain country are a certain amount of money, do not assume that the people there live at the same level of material welfare that that amount of money would buy where you live. For example, in 2008, at the international exchange rate of Indian rupees to U.S. dollars, India's per capita GNI was equivalent to $1,070. In India, however, food, clothing, housing, and other necessities were much less expensive than in the United States. Therefore, when the Indian income was adjusted for purchasing power, India's GNI per capita was actually able to purchase $2,960 worth of goods. China's per capita GNI in 2008 was only $2,940, but, adjusted for purchasing power, that figure more than doubled, to $6,020. Those two countries alone contain about 37 percent of the total human population, so adjusting income for PPP tremendously raises our measures of humanity's standard of living.

Gross Domestic Product and the Environment

Measures of gross domestic product fail to assess the environmental damage that may result from growth. In standard techniques of bookkeeping, machines and buildings are counted as capital assets. When they are no longer useful, the cost of their replacement is deducted from income. When a country's natural resources are exploited, however, their replacement cost is not calculated. Many countries sell off their timber and minerals, destroy their fisheries, mine their soils, and deplete their water resources, and the proceeds are counted as income. The loss of the natural resources, however, is not deducted as a reduction of national assets.

If natural resources were treated as assets, then statistics might demonstrate that protecting the environment is sensible economic policy. As described in Chapter 8, sustainability means using resources so they last as long as possible and minimize harm to the environment. This concept also applies to economic development, and the United Nations in 1988 held the world's first conference on **sustainable development,** which describes economic growth that minimizes environmental degradation, especially the use of non-renewable resources. Since then many economists have been working to calculate statistics that would show the value of natural resources that are destroyed or depleted. The World Bank's 2003 *World Development Report* analyzes several variations. Economists have estimated that when resource loss or depletion is considered, many countries' wealth has been growing more slowly than their GDPs would suggest. Many of the countries that are exporting their raw materials, such as some oil-exporting states, may not be growing at all, but actually depleting their national wealth.

The Pacific Ocean island-nation of Nauru illustrates the extreme of unsustainable development (Figure 12-3). The people of this 21-square-kilometer (8-square-mile) island allowed strip-mining of phosphate for fertilizer exports to ravage their homeland. This was immensely profitable, and the government trust fund once held

Figure 12-3 Phosphate mining on Nauru. After years of phosphate mining, the island of Nauru may have to be abandoned.

over $1 billion for the national population of just 13,000 people. By 2010, however, the phosphate was depleted and the government raised money by renting out fishing rights to its exclusive economic zone.

Many economists doubt the value of trying to measure sustainable development. They point out that the countries that are rich today had exploited their environments in the past at unsustainable rates, but after these countries had achieved wealth, they were able to repair much environmental damage (Figure 12-4). Perhaps sustainability is irrelevant to a poor population struggling to raise its standard of living to some minimum level, but once the population attains that level of welfare, it can afford to maintain or even repair the environment. Many poor countries resent rich countries' suggestions that they slow their economic activity to a sustainable rate. Imposing conservation measures on poor countries has been derided as *green imperialism.*

Nevertheless, when the world's banks are financing infrastructure projects such as dams, power plants, and pipelines, they are increasingly accepting the *Equator Principles,* a set of environmental and social-impact standards drawn up by The World Bank. These principles were designed to prevent large construction projects from poisoning air and water, denuding forests, and destroying the livelihoods of local residents in poor countries, where government regulators are often ill equipped or unwilling to mitigate potential side effects of foreign-financed initiatives. The principles are strict but voluntary, and there is no enforcement mechanism except negative publicity.

The Gross National Product and the Quality of Life

Figure 12-5 is a world cartogram, with each country drawn in proportion to its GDP. It is strikingly different from a territorial map. Some of the smallest

(a)

(b)

Figure 12-4 Replacing a mountain. A mining process known as strip-mining or mountain-top removal wreaks havoc on the environment. (a) A small crew of three or four workers with heavy equipment can flatten a mountain and remove coal or other minerals in a matter of days or weeks. While the workers are digging for resources, the unwanted material is put in local streams, where it releases chemicals that kill fish. Laws require the companies to restore the mountains, but the landscape is permanently altered (b).

countries—Belgium, for instance—have large GDPs. Figure 12-5 also differs from the population cartogram pictured in Figure 5-2. Some countries with small populations have enormous economies. Clearly, neither population nor territorial size explains wealth.

Figure 12-6 establishes categories of per capita GNI on a regular world map. What we really want to

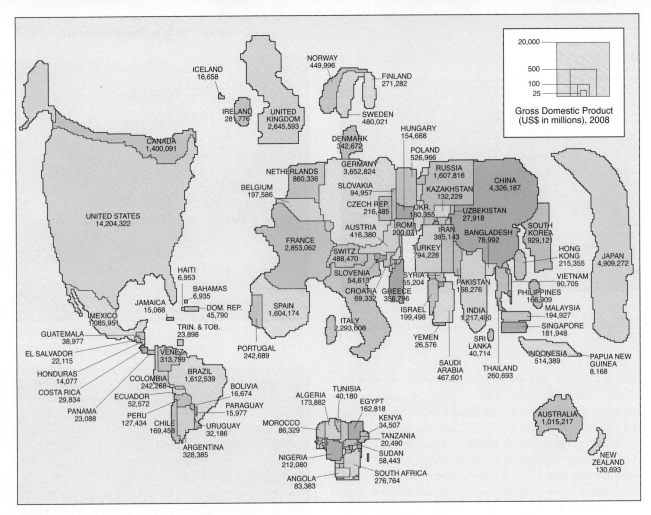

Figure 12-5 The relative sizes of nations' gross domestic products. The size of the countries on this cartogram corresponds to their total gross domestic product. The United States, Europe, and Japan grow considerably beyond the sizes of their relative territories. Africa, in contrast, shrinks.

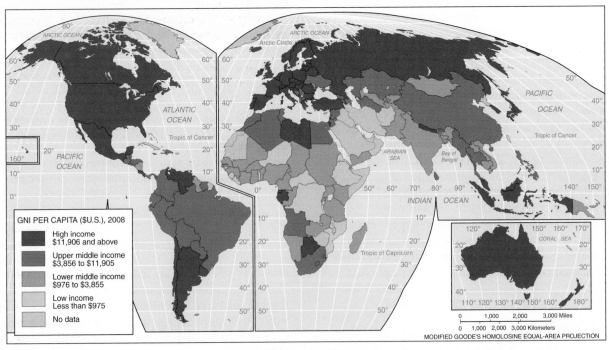

GNI PER CAPITA ($U.S.), 2008

High income
$11,906 and above

Upper middle income
$3,856 to $11,905

Lower middle income
$976 to $3,855

Low income
Less than $975

No data

MODIFIED GOODE'S HOMOLOSINE EQUAL-AREA PROJECTION

MapMaster
Layered Themati
World: Economi

Figure 12-6 Per capita gross national income. Per capita national incomes in many countries in Asia and Africa are a fraction of those in Europe and North America.

know, however, is what percentage of *people* live at various levels of income. Therefore, Figure 12-7 colors the various levels of income on the cartogram of world population. By this device, we can see that about 15 percent of the world's population lives in countries in which per capita GNI is less than $975, and another 55 percent live in countries in which per capita GNI is between $976 and $3,855. However, the distribution of incomes is not equal within these countries. There are rich people in poor countries and poor people in rich countries. In the United States, the wealthiest 10 percent of the population receive about 30 percent of all income in the country. The poorest 10 percent of the population shares just 2 percent of all income. Interestingly, other industrialized countries, including the United Kingdom and China, have about the same distribution at the top and bottom.

Some of the world's poorest countries have more extreme inequality in income. In countries such as Bolivia, Haiti, and Angola, the wealthiest 10 percent take in more than 40 percent of all income. The poorest 10 percent share less than 1 percent. Remember that 1 percent of income in these countries is much less than in wealthier countries. In Namibia, the wealthiest 10 percent of earners take an astonishing 65 percent of their country's income. The poorest 60 percent of the population share just over 10 percent. The poorest populations of the poorest countries are also most likely to suffer from higher mortality, disease, and hunger, as examined in Chapters 5 and 8. Rates of illiteracy are also high in most developing countries, not just among the poor but also among women in many traditional cultures. Some have even speculated that such inequalities are what make it possible for the world's autocrats, described in Chapter 11, to attain and keep power.

These patterns are not entirely permanent and, in many countries, incomes are rising. The proportion of people living on $1.25 per day has improved from 52 percent of the world's population in 1981 to 25 percent in 2005, while the percentage living on only $2 per day has improved from 70 percent to 47 percent. People are earning more. Most of these improvements are attributable to China's rapid economic development, which has considerably reduced the poverty rate in that country from 84 percent in 1981 to 16 percent in 2005. Excluding China, the world poverty rate has fallen very slowly and the actual number of poor has increased in South Asia and sub-Saharan Africa. Rising incomes are not lowering poverty rates and these rates, in turn, slow a country's economic development. The international effort to improve the lives of the world's poorest population is known as human development, which we address in Chapter 13. The rest of this chapter will focus on economic development.

Preindustrial, Industrial, and Postindustrial Societies

Geographers want to know what people are doing for a living and which activities are producing the wealth in each country. To do this, they must analyze more than just the total GDP. Chapter 10 noted that, over a long period, most countries that are rich today experienced shifts in the distribution of jobs between their economies' primary, secondary, and tertiary sectors. These shifts were defined as a **sectoral evolution.** The best way to examine any country's economy is to study the relative importance of each sector in its GDP.

Sectoral evolution changes national employment

Before industrialization, most of a country's labor force is occupied in the primary sector (Figure 12-8). Societies with the bulk of their employment in the primary sector are called **preindustrial societies.** Many societies still today are preindustrial and largely engaged in agriculture. As we saw in Chapter 8, however, these agricultural economies cannot always feed their own populations (Figure 12-9).

As some countries industrialized, many workers found employment in factories, and the proportion of the labor force employed in the primary sector declined. This does not mean that the primary sector became less productive, however. Instead, the use of labor-saving technologies, such as tractors or mining equipment, allows fewer people to do this work more efficiently. The primary, extractive activities remain crucial to many economies, but they provide a diminishing share of jobs.

Many jobs lost in agriculture were initially replaced by new opportunities in industry. The proportion of workers in the secondary sector increased until, at no precisely defined point, certain societies came to be called **industrial societies,** in which most jobs and a large share of the GDP were produced by manufacturing. In the United States, for example, manufacturing employment overtook farm employment during World War I, and by 1925, manufacturing workers had become the largest single occupational group. This also caused a change in population geography from small, rural family-centered farms to crowded factory-towns.

Continuing evolution of some countries' economies has drawn a higher percentage of workers into the tertiary sector, producing services instead of goods. Services accounted for almost 25 percent of all jobs in the United States by 1929. Sometime in the 1940s, the proportion first exceeded 50 percent, and the United States became the world's first **postindustrial society.** By 2000, tertiary employment represented three-quarters of the nation's jobs. The leading job categories in the United States in 2005, by numbers, were teachers, retail salespersons, secretaries and typists, truck drivers, farmers and farm laborers, janitors and cleaners, waiters and waitresses,

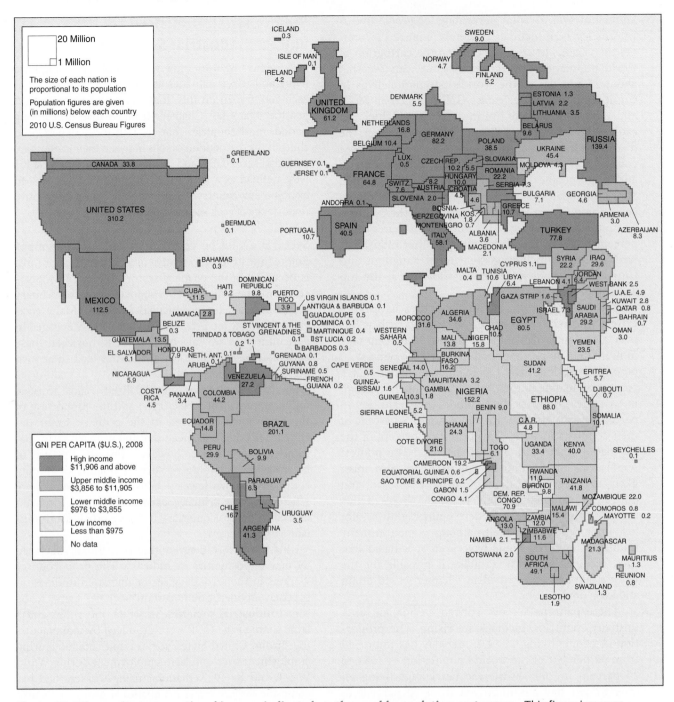

Figure 12-7 Per capita gross national income indicated on the world population cartogram. This figure improves on Figure 12-6 by placing the categories of income on the population cartogram that we first saw as Figure 5-2. Therefore, this figure clarifies how many people live at each income level. The colors correspond to those in Figure 12-6, so we can see that 40 percent of Earth's population lives in low-income countries, 44 percent in middle-income countries, and 16 percent in high-income countries.

nurses and home health aides, and freight and stock handlers. All advanced nations have seen the percentage of jobs in the tertiary sector increase. This trend has also been noted in a few poor nations, whose tertiary sectors are well integrated into the international economy—tourist destinations, for example.

At first in the United States, both the secondary and the tertiary sectors increased their shares of total

employment as the primary sector's share dropped, but in recent years, the proportion of workers in the secondary sector has begun to fall as well. The nation began losing manufacturing jobs as a percentage of all employment in about 1960, and even the absolute number of manufacturing jobs has been declining since at least 1970.

The evolution of industrial and postindustrial economies also produces greater gender differences in

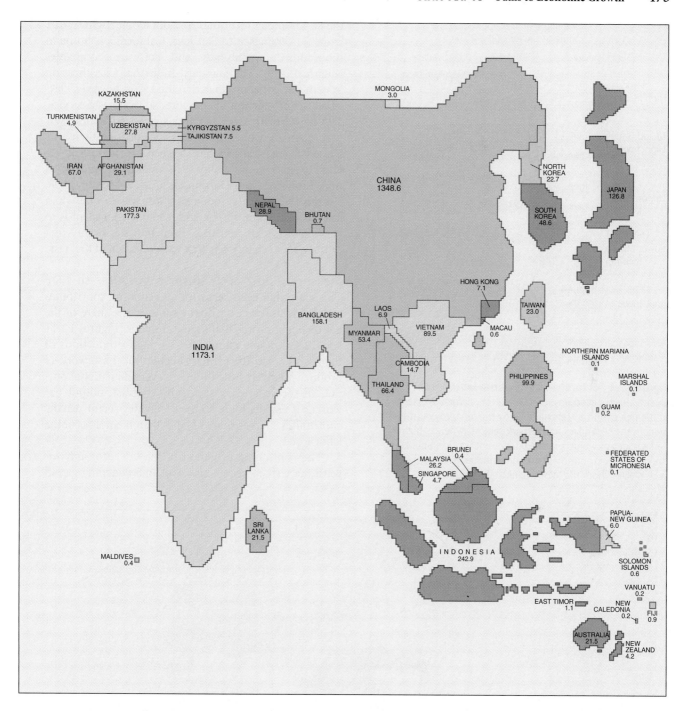

employment. In preindustrial societies, all available help—men and women as well as children—is usually put to work raising crops and tending animals. As countries become more industrialized, however, gender differences in labor become noticeable. It is often the case that men first leave rural areas to seek industrial employment, leaving women behind to work the farm. In traditional rural areas, such as in Turkey, women are often restricted by social conventions from working outside the home. Note that in Turkey, Norway, and the United States, industrial employment remains predominantly male dominated, but men dominate the service sector only in Turkey.

The "new economy" The production of computer hardware, software and services, telecommunications, and products on and for the Internet are all collectively referred to as the *new economy*. This term, however, like the term *quaternary sector* discussed in Chapter 10, is not specific or agreed upon. Sectoral evolution and technological evolution proceed so rapidly that economic statisticians cannot always agree on how to measure developments. For example, some economists believe that the production of computer software should be counted as a manufacturing activity, and they have ranked it as the third-largest manufacturing industry in the United States (after automobiles and electronics). Silicon Valley,

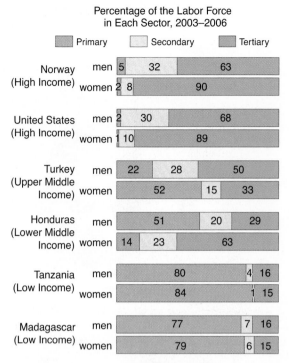

Percentage of the Labor Force
in Each Sector, 2003–2006

| | Primary | Secondary | Tertiary |

Norway (High Income)
men 5 32 63
women 2 8 90

United States (High Income)
men 2 30 68
women 1 10 89

Turkey (Upper Middle Income)
men 22 28 50
women 52 15 33

Honduras (Lower Middle Income)
men 51 20 29
women 14 23 63

Tanzania (Low Income)
men 80 4 16
women 84 1 15

Madagascar (Low Income)
men 77 7 16
women 79 6 15

Figure 12-8 Sectoral analysis of the labor force. For each country's economy, there are three sectors. Workers in the rich countries (exemplified by the top two countries) work in the secondary and tertiary sectors, whereas higher percentages of workers in the poor countries (exemplified by the bottom two countries) are concentrated in the primary sector. (Data from the World Bank)

the 80-kilometer (50-mile) corridor along Highway 101 from San Francisco to San Jose, California, is home to thousands of computer and software companies (Figure 12-10). The Standard Industrial Classification System (SIC) that originated in the 1930s, however, classified the production of computer software as a service. The new North American Industry Classification System (NAICS), which has been developed jointly by the United States, Canada, and Mexico, is continually revised to sharpen our ability to monitor economic evolution. Professor Rudi Dornbusch of MIT wrote in 1998, "In the 19th century, the great effort was to draw up a map of each country. The 20th century equivalent is to get a good set of statistics." Even into the twenty-first century, rapid change frustrates achievement of that task.

Measuring how each sector contributes to GDP

The share that each sector of an economy contributes to GDP is not necessarily the same as the share of the national labor force that sector employs (Figures 12-11 and 12-12). Some countries have high proportions of their workers in the primary sector, but those workers produce relatively little of the measurable national output. In the secondary sector, by contrast, output per worker is usually high, so the secondary sector usually contributes a larger share of GDP than it employs of the labor force.

The continuing evolution of the U.S. economy toward the tertiary sector may explain a widening gap in

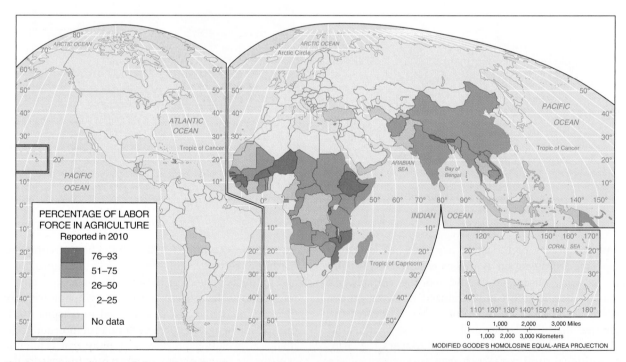

Figure 12-9 Percentage of the active labor force working in agriculture, 2010. If you compare this map with Figure 8-30, which illustrates the prevalence of undernourishment, you will see that many countries with high percentages of their labor forces in agriculture are barely able to feed themselves, whereas several countries that have only small percentages of their labor forces working in agriculture enjoy food surpluses. What can explain this paradox? The explanation is that economic and technological development have allowed some countries to invest in agriculture, raise agricultural yields, and at the same time release workers for manufacturing and service activities. Thus, a high percentage of workers in agriculture suggests simpler tools and less investment for farming.

Figure 12-10 Silicon Valley, California. By some measures, California's Silicon Valley is the manufacturing center of the United States, yet no smokestacks are visible. Clearly, technology has progressed from the "dark Satanic mills" of industrialization described by poet William Blake (1757–1827).

Percentage of GDP Contributed
by Each Sector, 2007

	Primary	Secondary	Tertiary

	Primary	Secondary	Tertiary
High Income Countries	2	26	72
Norway	1	43	56
United States	1	22	77
Upper Middle Income Countries	6	33	61
Turkey	9	28	63
Lower Middle Income Countries	13	41	46
Honduras	13	28	59
Low Income Countries	25	30	45
Tanzania	46	17	37
Madagascar	27	17	56

Figure 12-11 Sectors' contributions to GDP. This figure shows the origin of the GDP, by sector, for the same six countries that were shown in Figure 12-8, as well as the average for each income group. The percentage of a country's GDP that is produced in the secondary sector almost always exceeds the percentage of the country's labor force working in that sector. This means that value added per worker is usually highest in the secondary sector, so average workers' incomes are usually highest in that sector. The large percentages of the labor forces that work in the primary sectors of the poor countries produce relatively small shares of GDP.

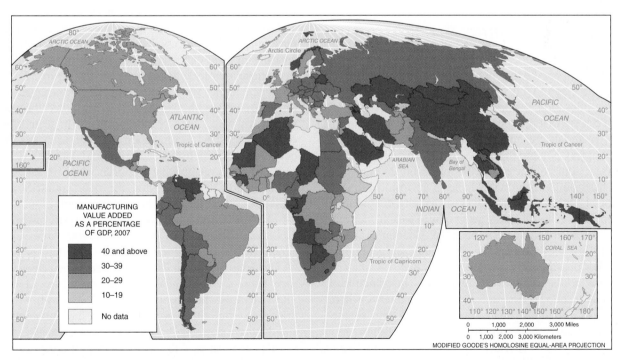

Figure 12-12 Industrial countries. The term *industrial society* has no exact formal definition, but this map reveals the countries in which industry—the secondary sector—contributes the highest share of GDP. The industrial sector is relatively small in some of the richest postindustrial countries and high in a few relatively poor countries that do process some mineral ores.

incomes. Many of the occupations that pay the highest incomes are usually in the tertiary sector (sports star, heart surgeon, lawyer, etc.), but most of the tertiary-sector jobs do not pay as well as factory jobs. The jobs listed above as most numerous in the United States—teacher, retail salesperson, secretaries and typists, and so forth—are not generally the highest-paying occupations. The inflation-adjusted median income of U.S. workers has been virtually flat since 1973. Median *family* and *household* incomes have risen only as more members of each household have gone to work—first the women, and then the youngsters—and in the twenty-first century, median incomes of households headed by someone under age 66 have actually been falling. Those people at the very top of the income scale, however, have enjoyed substantial increases. At the 95th percentile of income, inflation-adjusted median wage and salary income rose 67 percent between 1966 and 2008. Income for the top 1 percent more than doubled, and income for the top 0.1 percent increased more than 600 percent. Generally speaking, those people with more education are doing better, while the incomes of those with less education are falling. Educational requirements, however, are rising for almost all occupations. Today many college graduates hold jobs that once required only a high school diploma.

We do not know whether the widening wage gap in incomes is inevitable in a postindustrial society or whether this phenomenon is the result of uniquely U.S. tax laws or other policies. International comparative studies are not conclusive.

Why Some Countries Are Rich and Some Countries Are Poor

If you compare the map of per capita GNI in Figure 12-6 and the map of HDI rankings in Figure 13-26 with the data on location of world natural resources in Table 9-1, you can see that the countries most richly endowed with raw materials are not necessarily enjoying the highest per capita GNIs and the highest standards of living. Conversely, some of the richest peoples who do enjoy the highest standards of living do so in environments with meager natural resources. The geography of resources alone does not explain the geography of wealth.

Environmental determinists (see Chapter 6) would find this inexplicable, but the study of geography teaches that the key to wealth is not *having* raw materials, but *adding value* to raw materials. The principle of adding value was first introduced in Chapter 8, where we read of dairy farmers converting milk into cheese. Some resource-poor countries are able to import raw materials, process them, manufacture items from them, and enjoy their ultimate use.

The value added to raw materials downstream in the secondary and tertiary sectors surpasses the value of the original raw materials, whether the raw materials are mineral or agricultural. The value added by refining copper ore and manufacturing things out of it, for example, quickly surpasses the value of copper ore. The value added by grinding wheat into flour is greater than the value of the wheat. This principle holds true no matter how valuable the original raw material may be. Diamonds are valuable, but the value added to them by diamond cutters and polishers is greater than the value of the original uncut diamonds. The greater the value added, the greater the potential for profit.

The places that add the greatest value to raw materials prosper, while the places that export raw materials without adding any value to them do not enjoy the same economic growth. They may even have to buy goods manufactured out of raw materials that they originally exported themselves. For example, Jamaica exports bauxite to the United States, where that bauxite is made into aluminum and then into consumer goods such as foil wrap and cans, which Jamaica then imports back. As long as this continues, the United States will grow richer than Jamaica.

Even within the secondary sector, there is a hierarchy of value added. The manufacture of textiles, for example, adds less value than does the manufacture of clothing, so places that manufacture textiles try to develop a clothing industry, too. Other generally low-value-added manufactured goods include shoes, toys, sporting goods, simple television sets, and inexpensive cars. Countries try to progress from production of these items to production of higher-value luxury cars, computers, chemicals, and ever-more-complex electronic items. An agricultural example is to note how Chile began exporting grapes, then evolved to exporting moderately priced jug wine, and today is exporting high-value fine wines. **Economic development** is a process. It consists of progressively increasing the value of goods and services that a place is able to produce in order to enjoy or to export (Figure 12-13). Producing higher value goods normally allows a place to capture more wealth, which is then distributed through wages to workers who spend and invest in their local communities. This accumulation of wealth improves the quality of health, housing, education, and public services. This virtuous cycle is what we mean when we say rich countries are "developed" and the poor countries are "underdeveloped," or, optimistically, "developing."

Many of today's poor countries still export only unprocessed raw materials or materials with little value added, such as agricultural goods, basic metals, or textiles, for example (Table 12-1). Only a small percentage of their exports are manufactured goods (Figure 12-14). The welfare of their people rises and falls with world prices of commodities over which they have no control and that have slowly declined (Figure 12-15). In 2002, the world market prices of the 33 leading non-fuel commodities in world trade were only 55 percent of what they had been in 1970. Thus, the earnings of countries exporting these items had dropped by as much as 45 percent.

Exporters of some raw materials have tried to form cartels, such as OPEC (see Chapter 9), that

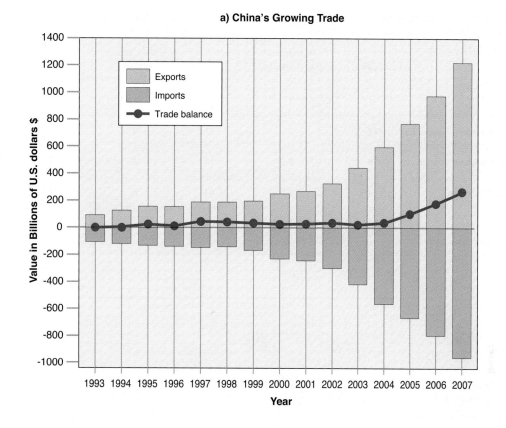

a) China's Growing Trade

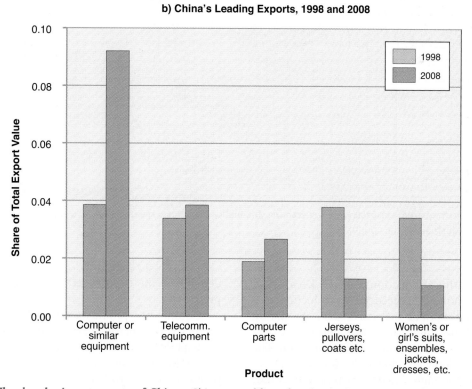

b) China's Leading Exports, 1998 and 2008

Figure 12-13 The developing economy of China. China exemplifies a fast-developing economy making enormous gains from growing export value (a). Over the past decade, it has shifted its exports (b) from low-value-added items (notably toys and miscellaneous textiles, clothing, and leather products) to progressively increase the share of its exports that are higher-value-added items, such as electronic products and machinery.

TABLE 12-1 A Few Countries' Merchandise Exports by Type, 2007 (rows equal 100 percent)

	Food	Agricultural Raw Materials	Fuels	Ores and Metals	Manufactured Goods	Other
High-Income Countries	**6**	**2**	**8**	**4**	**75**	**5**
Norway	5	1	64	8	18	4
United States	8	2	4	4	77	5
Upper-Middle-Income Countries	**11**	**2**	**21**	**8**	**55**	**3**
Turkey	8	0	5	3	81	3
Lower-Middle-Income Countries	**8**	**2**	**15**	**4**	**68**	**3**
Honduras	52	2	5	7	29	5
Low-Income Countries	**14**	**3**	**44**	**2**	**35**	**2**
Tanzania	35	7	1	13	17	27
Madagascar	31	3	5	3	57	1

Data from The World Bank.

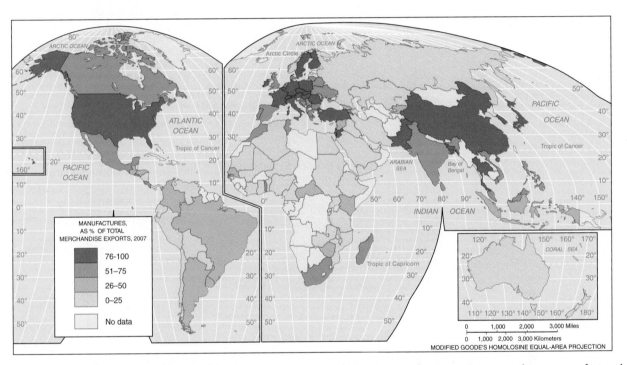

Figure 12-14 Manufactured exports. As a general rule, the higher the percentage of a country's exports that are manufactured goods, the more industrialized the country.

Video
The Coffee-Go-Round

cooperate in setting production levels to keep prices high. Few cartels have succeeded. If production of a commodity can easily be increased, then any nation can gain a short-term advantage by dumping stockpiles on world markets, causing prices to collapse. The Association of Coffee-Producing Countries tries to keep coffee prices high, but the members have not been able to cooperate as a united group. The price of coffee on world markets periodically collapses. In the countries producing coffee and depending on coffee exports, suddenly there is no money for children's schoolbooks, national roads go unpaved, and hospitals run out of medicines. Erratic swings in commodity prices can harm the economies of countries dependent on exports as much as steadily falling prices do.

Commodity production usually employs only a small fraction of the national labor force. This can cause problems in distributing the national wealth. Oil export, for example, can be very profitable, but the task requires so few workers that having oil to export is often referred to as the "petrocurse." About 90 percent of global oil and gas resources are owned by wholly or partially government-owned entities, as in Saudi Arabia, Iran, Russia, Qatar, and Kuwait. Economists agree that the wisest course is for a government to "sow the oil" (in the words of Venezuelan writer Arturo Pietri) by investing in education, infrastructure, and developing

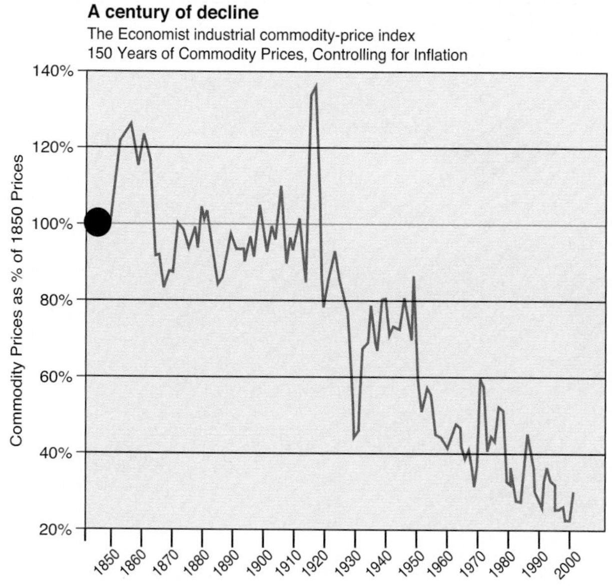

A century of decline
The Economist industrial commodity-price index
150 Years of Commodity Prices, Controlling for Inflation

Figure 12-15 Declining values of commodities.
Economic development and increasing demands on resources may yet cause commodity prices to rise on world markets, but they have been declining for over 150 years. Countries that export commodities are earning less than ever before.

industries. The temptation is great, however, to spend the income on current consumption. About 80 percent of Libya's national budget, for example, comes from oil exports, but few workers are needed, so the government "makes work" for others. The National Oil Company employs probably twice the number it needs, and another 20 percent of the Libyan working population is employed by the civil service, where most are admittedly unnecessary. Government subsidies for food and housing allow citizens to refuse menial jobs, so heavy labor is done by sub-Saharan Africans and more skilled labor is handled by Egyptians. Thus, paradoxically, the official unemployment rate is 30 percent, yet over two million foreigners are working. Libya's minister of Finance, Abdulgader Elkhair, has said, "If we hadn't had oil, we would have developed. Frankly, I'd rather we had water."

Norway is practically the only oil-exporting nation to have achieved balanced economic growth. When Norway first pumped crude oil from the North Sea in 1975, it had the advantages of being a stable democracy with honest civil servants, a well-established legal system, and a large middle class. Nevertheless, oil wealth at first seduced Norwegians: The national welfare system was expanded, and the workweek declined. In the 1990s, however, Norway segregated the oil business and diversified the rest of the economy. The government set up the Petroleum Fund to finance Norwegian retirements.

As Figure 12-1 illustrates, some countries that export commodities capture the value added downstream by buying processing plants, fleets, refineries, or even retail networks to reach ultimate consumers in rich countries. Russia's President Vladimir Putin himself cut the ribbon

to open the first U.S. Lukoil service station in Manhattan. The Venezuelan National Oil Company (PDVSA) owns Citgo, with stations across the United States, and which even, in a gesture of "foreign aid" to America's poor, has supplied home heating oil to low-income households in Boston and New York at discount prices. Mideast oil-producing states have built facilities for manufacturing petrochemicals and pharmaceuticals. Diamond-producing countries in Africa are requiring that raw stones be cut and polished before export, so local citizens are encouraged to study diamond cutting and jewelry design.

Debt-burdens as the cause of poverty Many poorer countries borrowed heavily from wealthy countries after World War II in an effort to build infrastructure and improve living conditions that would spur industrialization. These approaches to economic development were not entirely unsuccessful, but they failed to bring these countries much wealth. When investments were successful, industries in developing countries had trouble gaining access to markets in the wealthier countries. The debtor countries could not earn enough to pay back their loans. Public debt has weakened their currencies, wrecked their economies, and fueled corruption in government. Some countries owe enormous portions of their annual income just to pay back the interest on their loans (Figure 12-16). A few countries have been able to successfully industrialize, but their success has been dependent upon the advantages they enjoy from the geography of manufacturing, their national policies, and international trade patterns. The rest of the chapter examines these issues in turn.

The Geography of Manufacturing

Most countries strive to industrialize themselves so they can capture the high value added by manufacturing. One of the earliest scholars to identify the locational determinants of manufacturing was Alfred Weber (1868–1958). Weber's analysis focused on the role of transport costs, and it included models that differentiate material-oriented manufacturing from market-oriented manufacturing.

Material-oriented manufacturing is located close to the source of the raw material for one of two reasons. The first reason is that the raw material is heavy or bulky, and manufacturing reduces that weight or bulk. The steel industry is an example. The value added in manufacturing steel is low relative to the cost and difficulty of transporting the raw materials (iron ore, coal, and water), so steel mills generally are located where the raw materials can be found or cheaply assembled. The second reason why material-oriented manufacturing may be located near the raw material is that some raw materials are perishable and need immediate processing.

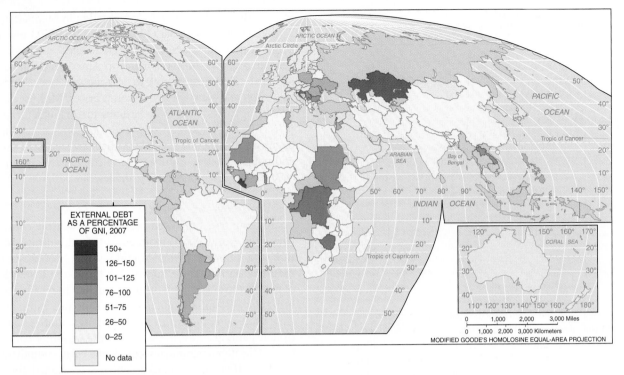

Figure 12-16 Debt burden. Many countries remain poor because their governments are struggling to pay back loans that did not raise export values.

Examples include canning and freezing foods, and manufacturing cheese from milk.

Market-oriented manufacturing, Weber's second category, is located close to the market, either because the processing increases the perishability of the product (baking bread, for instance) or because the processing adds bulk or weight to the product. A soft drink, for example, consists of a tiny amount of syrup, plus water, in a can. The syrup can be transported easily and cheaply, but canned soda is heavy and expensive to move. Therefore, the water should be added and the beverage canned close to the market.

Weber elaborated his models of the location of manufacturing by adding considerations, such as the availability of a labor force. In that three-factor model, manufacturers locate factories to minimize the cost distance from three points: the location of the raw materials, the labor force, and the market. The optimum location for any specific manufacturer depends on the balance of these costs in his or her business.

Several examples demonstrate that consideration of Weber's factors is still important in many industrial location decisions. Wood furniture making typifies an industry that may locate near its source of raw materials. In the United States, Grand Rapids, Michigan, became a center for wood furniture manufacturing in the nineteenth century, drawing on Michigan's extensive forests. When those forests were depleted, furniture makers relocated to their new sources of wood: Georgia and North Carolina. The copper industry exemplifies another bulk-reducing industry. In the United States,

most copper ore is smelted (separated into its metallic ingredients) at the mines in New Mexico and Arizona, because smelting removes a high percentage of the original ore as waste, saving the cost of transporting it before processing. The largest copper refineries and manufacturing facilities, by contrast, are found in Baltimore, Maryland; Norristown, Pennsylvania; and Perth Amboy, New Jersey. These locations are closer to markets and labor forces, and they are also situated to draw on imported copper supplies.

Many industries are a balance of these factors. The automobile industry is an example of an industry that brings together a number of previously manufactured parts and assembles them into a more complex product. In the United States, several new automobile assembly factories are located in the center of the country (Figure 12-17). In addition to its geographic convenience, this region is attractive to manufacturers because labor conditions and terms are more favorable to manufacturers in this region than they are in the regions of older, more heavily unionized industry. This region is also closer to more large population areas, which are the main sales markets.

Locational Determinants for Manufacturing Today

Weber's studies still help us understand many industrial location decisions, but several things have changed since Weber's time. Transportation costs have steadily fallen (recall the acceleration of diffusion described in

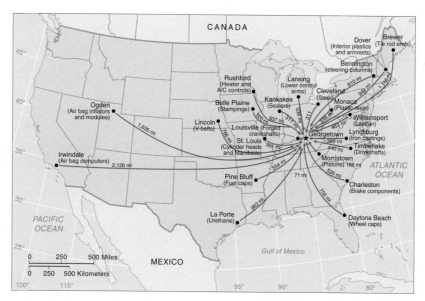

Figure 12-17 Supplying parts to an auto assembly plant. Georgetown, Kentucky, was chosen as the location of a Toyota Camry assembly plant partly in order to draw conveniently on supplies of parts from around the United States.

TABLE 12-2 Locational Determinants for Manufacturing

Factors Considered by Alfred Weber	Additional Factors that are Important Today
1. Raw materials	1. Capital
2. Labor force	2. Technology
3. Market	3. Governmental regulations, including environmental legislation, labor rights, and labor union power
4. Transport costs	4. Political stability
	5. Inertia

Chapter 6), and at the same time the value added in manufacturing has increased as manufactured products have become more and more complex. The value of the iron, copper, and other raw materials that go into a mobile telephone is only a minuscule percentage of the value of the completed unit. As transport costs fall and the value added in manufacturing rises, high-value-added manufacturing can more easily relocate; it is, as we learned in Chapter 6, increasingly mobile. Furthermore, today trade binds world regions, so locational determinants are no longer tied to local, regional, or even national scales. The entire globe must be examined as one theater of interconnected operations.

Weber's models considered the locations of the raw materials, the labor force, the market, and the transportation costs in determining the best site for a manufacturing operation. Since Weber's time the balance among locational determinants for many industrial processes has tipped away from what Weber considered. For example, the value of the labor input is generally a shrinking percentage of the value of manufactured goods. In the consumer electronics industry, for example, it is no more than 5 to 10 percent. At least five other considerations, however, must be added to Weber's

(Table 12-2). Each of these considerations is an *input*, or ingredient, for manufacturing. The first is *capital*: because manufacturing is increasingly capital intensive, it must have access to investors. The second is *technology*, including communications, which are not readily or reliably available in all places. Some technology is also protected by national laws that limit its diffusion to other countries. Third, industries seek places with *hospitable governmental regulations*; that usually means low taxes, little environmental regulation, and restraints on labor rights and unionization. *Political stability* is the fourth consideration. Manufacturers hesitate to invest in volatile political environments. The *inertia* of facilities and suppliers that are already in place is still another consideration in the geography of manufacturing. Factories are major investments, and networks of suppliers and trained labor forces develop around them; therefore, they are not quickly abandoned.

Figure 12-18 is a map showing the world's most important manufacturing regions. A map of manufacturing measured by value added, however, would show that the greatest value added is in those countries that dominate as shown in Figure 12-5, the cartogram of the nations' gross domestic products. Much of the world's

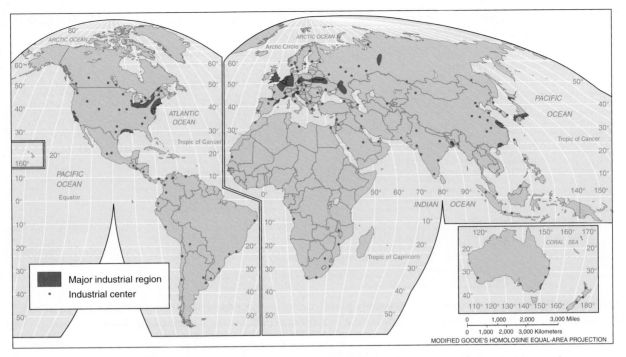

Figure 12-18 World manufacturing regions. This figure shows the greatest concentrations of manufacturing today.

manufacturing takes place in Europe, for historical reasons. Europeans brought about the Industrial Revolution first, giving Europe a technological lead which enabled it to overwhelm most of the rest of the world politically and economically. Europeans generally located the manufacturing in their own countries in order to retain the value added. It was not their intention to enrich their colonies. In some cases, they actually intended to hold back the development of manufacturing in their colonies. For example, from 1651 until 1776, at the very beginning of the Industrial Revolution, Britain's Navigation Acts prohibited Americans from manufacturing products from local raw materials. Raw materials had to be shipped to Britain, where value was added by manufacturing, and then the products were shipped back to the colonies—at higher prices. Even iron from New Jersey had to be shipped to Britain, where it was made into nails, and the nails were then shipped back to New Jersey, by which time they were very expensive. The colonists' resentment of this exploitation helped trigger the U.S. War of Independence. Many colonies in Africa and Asia later suffered the same deliberate restraint on their industrialization.

When manufacturers located factories in countries other than their own, they generally chose other rich countries where they could find markets and labor forces. Today, the rich countries still have the technological lead and the best-trained labor forces. They usually are politically stable, and they provide the eventual market for most goods. This geographic system has prevailed for so long that it has a tremendous inertia. Raw materials continue to flow from the poor countries

to the rich countries, where they are transformed into more valuable goods. Still today, for example, Ghanaian cocoa beans may enter Europe freely, but Ghanaian cocoa powder faces a stiff tax. This pattern has been labeled *economic colonialism* or *economic imperialism*. It is noteworthy, however, that neither Japan nor Switzerland, two countries with wealth but without substantial natural resources, ever had a major political empire.

In their efforts to industrialize, today's poor countries suffer from a number of problems that reinforce one another. Compared to those in rich countries, markets in poor countries are small because people can afford very little. In addition, cheap manufactured goods from the rich countries can flood the poor countries' domestic markets and discourage competitive domestic manufacturing. The countries' poverty increases their political instability. They need capital, and their people need education.

Poor countries can, however, offer three locational determinants: raw materials, inexpensive labor, and hospitable regulatory environments. Workers in poor countries receive relatively low wages, and they will often work in conditions worse than will workers in the rich countries. Poor countries' governments seldom complain about pollution coming from a new factory that provides jobs. Today's poor countries can exploit these advantages to attract new industry, and many corporations from the rich countries are placing manufacturing operations abroad or buying from suppliers abroad to take advantage of these factors. **Outsourcing** is the name given to the delegation of any of a firm's operations to an external subcontractor who specializes in that activity. If the outsourcing is to a foreign country, it is often called **offshore outsourcing.** Thus, some poor countries are industrializing by attracting the offshore outsourcing from rich countries.

Locational Determinants Migrate

The geography of locational determinants is not fixed. The locational considerations for manufacturing make it possible for manufacturing to migrate around the world as the mixes of attributes among places change and attract it.

There are, in fact, reasons the geography of locational determinants for industry—and thus the geography of industry—will continually change. Chapter 6 already suggested relevant factors: new products, new technologies, new raw materials, new sources of traditional raw materials, new methods of manufacture, new governments with new policies, growing and shrinking labor supplies, and the opening and development of new markets.

If a place first attracts manufacturing because it offers a low-wage labor force, its local standard of living and local wages will rise. As local wages rise, the industries that developed there to take advantage of the low wages will be driven away. They will migrate to still poorer places, where wages are lower. Thus, ideally, as each location develops, it surrenders lower-value-added activities to poorer regions. These regions in turn begin to climb the ladder of economic development. For example, WestPoint Stevens, Inc., America's largest textile maker, outsourced production offshore and reduced its U.S. payroll by half between 1988 and 2000. The U.S. company is concentrating on marketing and on adding psychological value by signing licensing agreements with popular and well-established designers Ralph Lauren and Martha Stewart.

The manufacture of men's dress shirts for the U.S. market offers another example. Shirt making is labor intensive, so manufacturers are continuously searching for cheaper labor. In the 1950s, U.S. shirt manufacturers located in Japan to use the low-paid workers there. When the costs of labor and real estate rose in Japan, the companies moved to Hong Kong. As Hong Kong's factories gave way to offices, the shirt makers moved again, first to Taiwan and Korea and then, in the 1970s and 1980s, to China, Thailand, Singapore, Indonesia, Malaysia, and Bangladesh. Toward the end of the 1980s and into the early 1990s, Costa Rica, the Dominican Republic, Guatemala, Honduras, and Puerto Rico became shirt-manufacturing centers. The Nien Hsing Textile Company, for example, which was founded in Taiwan in the 1970s, moved its jeans sewing factory to Managua, Nicaragua, in the 1990s. Its 1,800 Nicaraguan workers earn from $65 to $124 per month; the labor input in a pair of inexpensive jeans that retails in the United States for $25 (under several different brand names) is 50 cents. The most expensive jeans manufacturers maintain high

prices for their garments largely by adding psychological value through status advertising.

The Caribbean basin countries offer cheap labor and U.S. tax advantages. These advantages were granted under a U.S. government program, the Caribbean Basin Initiative, that was designed to help these countries industrialize. These countries' proximity to the United States offers several additional advantages. For one, U.S. managers can oversee operations more easily. Proximity also enables speedier delivery, and fashion demands the fast introduction of new styles. Also, renting a ship for an additional week's sailing time from the Far East is more expensive than the shorter trip from the Caribbean. Another advantage of great concern to business executives depends on interest rates, or what we call the *cost of money.* Shirts in the hold of a ship in transit are a form of capital, and that capital is not earning any profit. One shipload of shirts may be worth $15 million. The interest on that capital each day that the ship is on its way to the United States adds a few cents to the retail price of each shirt. "Time is money," and distance is money, too.

Video
Geraldo's
Brazil

Manufacturing in the United States

No other country yet matches America's total manufacturing output of $2.28 trillion (value added in 2006). In the United States, however, the share of the labor force in manufacturing and the sector's contribution to GDP have been falling—from 25 percent of jobs and 25 percent of GDP in 1960 to 9.6 percent of jobs and 11.5 percent of GDP in 2008. American corporations are building new factories abroad, and Americans are increasingly buying imported goods. By 2003, about half of the manufactured goods that Americans bought were made abroad, up from 31 percent in 1987. Economists debate whether this economic evolution away from the secondary sector is good for the United States.

The importation of manufactured goods, notably cars, apparel, electronic goods, computers, and furniture, has lowered their cost to American consumers. Falling prices is called *deflation.* The deflation of prices of manufactured goods comes at the cost of American jobs, and, as we have seen, the substitution of service jobs for manufacturing jobs contributes to the spreading gap in incomes in the United States. Manufacturing jobs pay an average of 20 percent higher wages than tertiary-sector jobs. In addition, manufacturing jobs have a higher multiplier effect than service jobs: The average production-sector job produces three times as many additional employment opportunities as does the average service job. Manufacturing also disproportionately employs non-college-educated workers, and as of 2008, more than 64 percent of U.S. workers lacked college degrees. We may applaud that the incomes of female workers, traditionally associated with service occupations, have risen relative to men's wages. Full-time female workers' wages held steady at about 60 percent of men's for decades after World War II, but by 2008 they

Recession and the Crisis for U.S. Automakers

The weak position of many large U.S. manufacturing firms turned catastrophic during the financial meltdown and recession that began in late 2007. Since 1931, General Motors had been the largest automaker in the world. In 2007, it was the fifth-largest corporation in the world. There were three main reasons GM was eventually forced to seek government assistance before declaring bankruptcy in 2009. First, GM was unable to cut labor costs, and unions negotiated for pay even when assembly lines were not running. As we noted in Chapter 5, large companies like GM were also burdened by their go-it-alone policy of managing their retirees' pensions rather than participating in government-backed plans or multi-industry programs. Second, GM was losing sales to other automakers whose products were considered more reliable. The automaker introduced too many new models; they were expensive to design but did not sell well. Third, sales of all vehicles, especially the best-selling SUVs, began to drop quickly in 2008, as gas prices reached $4 per gallon. The recession deepened.

During 2009, GM lost billions of dollars and could not find investors to keep it running. The U.S. government loaned GM $9.4 billion, but the company still declared bankruptcy in 2009. The new GM was owned by the U.S. government (60 percent), Canadian and provincial governments (about 12 percent), a union insurance and pension plan (about 18 percent), and the old GM (about 10 percent).

In the process, GM had to sell or close some of its once-famous brands, including Pontiac, Saab, Hummer, and Saturn. GM was not alone in its struggles; the U.S. government also provided Chrysler with a $10 billion loan. The Italian company Fiat now manages Chrysler, and the autoworkers' union is its largest investor. It is too soon to tell how the new smaller GM and Chrysler will fare and whether Ford, a long-time automobile giant, will be able to survive on its own.

The U.S. automobile industry employs more than 875,000 people, many of whom have been laid off or are hoping production will resume. The loss of jobs in U.S. auto manufacturing will be disastrous for local communities that once made cars or car parts. Auto parts manufacturer Delphi filed for bankruptcy in 2005 and announced it would sell or close 21 of its 29 plants in the United States. The last 8 plants would reduce their workforces, and remaining workers would suffer cuts in benefits and wages. Chief Executive Officer Robert S. Miller said, "We've got 60,000 workers in Mexico. Labor there is paid $7,000 a year. Before long, it's going to be very difficult for someone with a manual-labor job in the U.S. to have the kind of lifestyle they've enjoyed up to now. It ain't going to happen." Meanwhile, new competition is appearing from Chinese firms. Geely Automobile from China has an interest in the U.S. auto market and began its bid to acquire Volvo from Ford in 2009.

---

had risen to 77 percent of their male counterparts', yet wages overall were barely rising during this period.

The United States has experienced a decline in some key industries that were once a major part of its economy. In recent decades, the U.S. steel industry has steadily lost market share to less expensive steel producers from other parts of the world. To salvage what was left of the industry, the United Steelworkers Union agreed to cut wages, benefits, and jobs. The steel companies ended retiree health plans and transferred the costs of pensions to the federal Pension Benefit Guaranty Corporation (which offers lower pensions). The result was to clear billions of dollars off the firms' books and out of steelworker retiree's pockets. Today, America's airlines and auto firms are headed down that same path.

Can this evolution away from manufacturing continue? Americans will be able to import manufactured goods only as long as Americans can offer the rest of the world services plus a few sufficiently high-value goods. In 2003, David Heuther, chief economist at the U.S. National Association of Manufacturers, stated flatly that "you have to assume that manufacturing will continue to disappear," but he argued that America's high-tech advantage and its ingenuity will sustain a

small high-value manufacturing base. Nobel prize winner George A. Akerlof, by contrast, argued that the value of the U.S. dollar will continue to fall against other currencies. This will make manufacturing operations relatively less expensive in the United States, and that event will entice manufacturing operations back to the United States. How much would the dollar have to fall, and what would be the additional ramifications of that event? We do not know.

While prices of imported items are deflating, prices of goods produced locally and of services such as health care, education, and such day-to-day activities as trash collection and auto repair rise (*inflate*). In the early years of the twenty-first century, the deflation in the costs of manufactured goods has slowed the inflation of the overall cost of living.

The Economy of Japan

Japan is the most astounding and paradoxical success story in economic geography today. The Japanese islands have few natural resources, but on this meager natural geographic endowment, the Japanese have developed a great economy. How?

The answer lies in Japan's cultural resources and in its trading patterns. Japan's economic success demonstrates how a place can prosper by importing raw materials and adding value in manufacturing, or by providing other downstream services. The key cultural resources include the people's education and labor skills, technology, and the country's style and degree of cooperative organization. All these factors are carefully managed by the government. Japan's participation in international exchange is also carefully regulated. The Japanese import raw materials, transform them into manufactured goods, and export these finished goods, plus an array of valuable services, around the world. In 2008, Japan's imports were 43.2 percent raw materials by value and its exports were 75 percent manufactured products. Today, Japan's main products are automobiles, which account for 19 percent of its exports' value.

In the 1950s, Japan was a relatively poor country, but it began to manufacture products of relatively low value added. These products included cargo ships, toys and games, sporting goods, inexpensive clothing and cars, and simple electronic appliances. Japan built its prosperity by steadily adding value to its exports.

As Japan developed, it surrendered the manufacture of the items that are low in the value-added scale. Japan invested in factories to manufacture these products in less-developed countries that are dominated economically by Japan. Through the 1990s, Japan's production of lower-technology consumer electronic equipment began to fall as the production of these items was relocated by Japanese firms to Vietnam, Thailand, and Indonesia. Japan's leading position in the production of electronics and telecommunications equipment is now challenged by firms in China and other countries that are able to export products of similar quality as Japanese firms but at much lower prices. For example, what was once a Japanese monopoly on liquid crystal displays (LCD), the screens for laptops and other equipment, has become a major Chinese export.

Entering the twenty-first century, the Japanese economy continued to evolve. By 2002, with about $2 trillion in assets invested overseas around the world and $500 billion in foreign reserves in its vaults, Japan for the first time earned more money from its investments abroad than it did from exporting manufactured goods. Some observers have described this as stagnation. Japan, a nation with a fertility rate below the replacement rate, an aging population, and now beginning to live off its savings, will likely become a "post-industrial" society like other wealthy countries.

Technology and the Future Geography of Manufacturing

Because the balance of manufacturing inputs is continuously changing, maps of the world geography of wealth, of manufacturing, and of economic development will continue to change. Each nation must tailor its economic policies in order to compete. For example, the importance of

technology is increasing. When Japan was industrializing, it bought technology. By one estimate, the Japanese paid foreign manufacturers a total of only $10 billion for patents and licenses between 1950 and 1980. That must rank as the shrewdest investment any nation ever made. Today, Japan holds almost 50 percent of all U.S. patents granted to non-U.S. residents, and it has joined the United States and the other most advanced countries among recognized holders of key patents. Japan has continued to achieve advances in industrial productivity. In 2004, Japan's 356,000 installed robots outnumbered those in the United States by more than three-to-one.

Technology and capital contribute the most rapidly increasing shares of the final value of most manufactured goods, and technology and capital are exceptionally mobile. In the manufacture of computer disk drives, for example, the raw materials are an insignificant fraction of the drives' final value, so computer disk drives can be easily manufactured and distributed worldwide from almost anywhere. Why then are more than half manufactured in Singapore? They were not invented there. The decisive locational determinants for this high-value product are the availability of inexpensive skilled labor, technology, capital, and political stability in Singapore. None of these factors, however, is unique to Singapore. How long will Singapore continue to dominate world manufacture of disk drives?

Tomorrow's world geography of manufacturing may be guessed by mapping investment in research today. Some analysts feel that U.S. investment in research may be falling behind that of Japan and other international economic competitors, measured both as a percentage of GDP and in actual money terms. Singapore recognizes that its current dominance in the manufacture of electronics is uncertain, and the government has bet that pharmaceuticals and biotechnology can replace that industry. It has opened a research center called Biopolis, which has attracted some leading U.S. scientists. Dr. Neal Copeland and Dr. Nancy Jenkins left the U.S. National Cancer Institute to take positions in Singapore, saying, "We wanted to be in a place where they are excited by science and things are moving upward."

National Economic–Geographic Policies

Each country organizes its domestic economy, manages the distribution of economic activities within its territory, and regulates its participation in world trade and investment.

Political Economy

Each country defines a set of principles to organize its economic life. The study of these principles is **political economy.** In all economic and political systems, the government usually provides those services that are unprofitable to private providers but that diffuse benefits throughout the economy, called *positive externalities* (see Chapter 9). These services include national security, education, transportation, water and sewerage, and other aspects of the infrastructure.

Each country also manipulates its national budget and expenditures, interest and exchange rates, and money supply in order to promote growth and high employment. The tools for manipulating these are imperfect, but national governments' economic planning and management, their revenue and expenditure, and their role in redistributing income have made national economic policies more central to most people's lives than ever before.

Direct government participation in the economy ranges theoretically from a *communist system*, in which both the natural resources and the productive enterprises are **nationalized**—that is, owned by the government in the name of the people—to a **capitalist system,** in which the state defers to private enterprise and a stock market raises and allocates investment capital. The economy of each country lies between these two extremes: Most countries have mixed economies in which governments are involved in some activities, while private markets do the rest.

Even among those countries generally considered capitalist, there is a great range of government involvement in the economy. Each country's political economy usually reflects other characteristics of that country's culture. The United States, for example, generally favors a system that minimizes the government's role. This is called **laissez-faire capitalism,** from the French for "Leave us alone," supposedly said by the French economist Vincent de Gournay (1712–1759) to a government bureaucrat. The U.S. government has never developed an explicit industrial policy, although its regulations, research funding, and defense budget have greatly influenced the national economy. The United States has also designed its tax system to encourage investment in certain activities.

The decade from 2001 to 2010 caused some to reevaluate America's laissez-faire stance. Many people came to see the federal government as uniquely capable of performing some urgent tasks: fighting terrorism abroad; providing security at home; rescuing bankrupt airlines, banks, and automakers; and restarting a stalled economy. The *liberal* political philosophy generally favors government action to solve problems that private interests cannot or should not tackle. These sentiments caused a move toward greater governmental regulation and even governmental assumption of some responsibilities that had been privatized. For example, in 2001, Congress created a new federal force responsible for airport security, a responsibility that previously had been surrendered to private security forces.

The *conservative* political philosophy, however, favors privatization, arguing that the private sector can deliver many public services at higher quality for lower cost than can government bureaucracies. Therefore, the conservative presidential administration of George W. Bush privatized many government services. By 2006, some 40 percent of federal discretionary funding went to private companies doing everything from cleaning federal parks to feeding U.S. troops in Iraq. Some privatized programs revealed corruption, price inflation, and incompetence, but others defeated bureaucratic inaction and inspired new approaches to problem solving. Case-by-case analysis is required to determine which needs are best fulfilled by governments and which are best dealt with by private companies.

The governments of several Asian countries plan and regulate their economies, although they do not own many companies outright. This is called *state-directed capitalism*, and it conforms with these countries' Confucian bureaucratic cultural traditions (discussed in Chapter 7). South Korea, for example, is considered a capitalist free-market economy, yet the government proposes national economic plans and subsidizes new industries. Until the late 1990s, industry was dominated by combinations of corporations, called *chaebols*, which were affiliated by interlocking directorships and ownership. Daewoo, Hyundai, and Samsung were examples familiar around the world. The economies of Japan and Taiwan are similarly organized, but in the United States, such combinations would be broken up as illegal trusts.

Western European countries have been privatizing their economies through recent decades, although nationalized companies still account for about 25 percent of GDP in France, 11 percent in Germany, and 3.5 percent in Britain, which began the privatizing trend in the early 1980s. Eastern European countries are rapidly privatizing as they abandon Communism, and the special case of China will be discussed shortly. Some African countries long have insisted on national ownership of natural resources as a point of national pride, but even they have begun to privatize and invite foreign capital. In 1997, Zambia's then-President Frederick Chiluba said, "We don't care who buys the mines in Zambia so long as the mines make money and contribute to the exchequer [national treasury]." Many countries are today creating new national stock markets (Figure 12-19). The combination of privatization, reduction of government regulation, opening to international investment, and increasing participation in international trade is often called a *neoliberal* approach to political economy.

Most Latin American countries have traditions of outright government ownership of the nation's assets and industries, but in the 1980s and 1990s, these governments began privatizing their assets. Purchasers included domestic investors and corporations, but most purchasers were foreign. Early in the twenty-first century, however, nationalist politicians fanned public resentment of foreign ownership of resources, and, encouraged by rising energy prices, several Latin America governments began to assert greater control over their energy resources (Figure 12-20).

Figure 12-19 An emerging stock market. The new Ugandan Stock Market lists stock shares of only 12 companies, and it is open only a few hours per week. Nevertheless, the government hopes that its existence will stimulate Ugandans to save and invest in Uganda's growth and to incorporate new business ventures.

Figure 12-20 Nationalizing resources. The banner declares "Nationalized: Property of the Bolivians." In May 2006, Bolivia's newly elected President Evo Morales ordered the military to occupy his country's oil and gas fields. "The time has come," he declared, "the awaited day, a historic day, in which Bolivia retakes absolute control of our natural resources. The looting by the foreign companies has ended." Bolivia's resources had in fact been privatized unconstitutionally in the mid-1990s, without the approval of the Bolivian Congress.

National political-economic policies also dictate the conditions of employment. The United States favors a flexible labor market. Employers are quite free to hire workers as needed, but also to release workers they no longer need, which means that Americans change jobs at a high rate. Europeans' jobs, by contrast, have traditionally been more protected. Employers cannot so freely release unneeded workers, and therefore, it has been argued, economic growth is hampered because

employers are reluctant to hire workers in the first place. For example, through most of 2005, the unemployment rate of people 15–24 years of age in France was 22 percent; in the United States it was 12 percent. The next year, the French government attempted to increase the flexibility of the labor market and encourage the hiring of young people by allowing employers more easily to release workers under 26 years of age. The initiative triggered destructive riots across the country; the young felt that they were being asked to sacrifice job protections that older workers had traditionally enjoyed. Some individuals in any country may be described as *opportunity seeking*, whereas others are more *security seeking*. Whole cultures may tend toward one or the other of these two types, and some economists argue that rigid job protections for workers seeking security hamper an economy's ability to compete.

For private enterprises to succeed, the government must in any case perform certain basic functions. Economists regularly attempt to measure and compare countries' "business environments," asking, for example, whether a country's government secures property rights, whether the judiciary is independent and reliable, how long it takes to obtain licenses necessary to start a new enterprise, and whether corruption and bribery are endemic. For example, The World Bank has reported that it requires 2 procedures and 2 days to start a business in Australia, but 7 procedures and 32 days in Ethiopia. It takes 15 procedures to register property in Brazil, but only 3 in Denmark. An average sub-Saharan business executive takes 50 days to get the 21 signatures on 9 forms necessary to export an item, and those procedures may require several bribes. An Indian exporter needs to collect 22 signatures on 10 documents. Such studies indicate that the business environment in some countries significantly thwarts economic growth.

The results of privatization   If a country privatizes some assets, it must devise a policy for investing the windfall of money that it receives. This challenge faces many countries that are privatizing assets today. If a country is so poor that it spends the money on immediate needs, such as food, then it will soon be worse off than it was before privatization. The assets will have been sold and the money spent. In other cases, some countries' corrupt rulers have virtually stolen the money from privatization and deposited it in personal accounts in foreign banks. Our present state of economic theory and experience suggests that if a government invests the income from privatization in education, infrastructure, or health care, however, then the country might achieve economic development.

The privatization of national assets or industries affects a country's entire economy and politics. This is because nationalized activities fulfill responsibilities beyond the merely economic ones. Nationalized enterprises might be subsidized in order to maintain high employment, to sustain the economies of poor regions, or just to ensure justice and protect rights. These purposes may provide positive externalities across the society, but they may be expensive. The principal goal of a private enterprise, by contrast, is to maximize profits for shareholders. This usually requires the enterprise to be cost efficient in order to compete in world markets. Today, as more countries embrace privatization and capitalism as the surest route to economic development, thousands of workers are being laid off from economically inefficient nationalized industries or from enterprises that are being privatized. Schools, prisons, and even police duties are being given to private firms that are not elected or sworn to uphold the laws. In many countries, from Argentina to China, these workers are heading for the ballot box (or taking to the streets) in protest. National leaders are gambling that privatization will release economic dynamism and stimulate economic growth fast enough to create new job opportunities for these workers.

Crony capitalism   *Crony capitalism* is the name given to situations in which financial markets are not free and fair but are manipulated by politicians or insiders. Crony capitalism sometimes occurs in countries that practice state-directed capitalism but are run by oligarchies. In this case, a political crisis can immediately become an economic crisis. In the autumn of 1997, for example, an international loss of confidence in the stability of the government of Indonesia triggered a reevaluation of that country's financial strength, and the ensuing turmoil endangered the financial structures of several other East Asian countries. Political and economic restructuring required adopting regulatory infrastructures and legal safeguards for markets, democracy, and the rule of law.

Events early in the twenty-first century revealed that the United States is not immune to crony capitalism. Corporations were found to have been creating hundreds of millions of dollars in fake profits (Enron), disguising operating expenditures as capital investments (WorldCom), booking nonexistent revenues to inflate the stock price (Global Crossing, Lucent, and many others), and overstating earnings and backdating accounts to inflate executives' rewards (United Health Group, Fannie Mae, and many banks). *Fortune* magazine called these latter activities "stealing, pure and simple." Numerous large corporations have been virtually pillaged by their boards of directors and executives, while one of Wall Street's most trusted stock analysts urged his clients to buy stocks that he privately called "crap." America's most prestigious accounting firms and banks worked secretly together.

Multi-billion-dollar corporations went bankrupt, and hundreds of millions of dollars were levied in fines against corporations, accountants, and banks. Several corporate executives have been jailed, and new regulatory legislation was passed to rectify these faults in America's economic system. Stockholder and legal suits and criminal proceedings continue. These events,

however, have significantly damaged the prestige that the American capitalist model has held for 50 years, and the ramifications of this damage will continue in world trade negotiations for decades. The fact that other countries felt freed by the situation to develop alternative models may, however, in the long run, produce innovations in economic development theory that will benefit all countries.

Wealth Variations Within States

Countries try to boost their economic growth by organizing their populations, territory, and resources for production. A strong sense of nationalism can encourage the population to work together, but the population also needs education and training. The territory must be at peace, and the country should constitute a single market for raw materials, goods, and labor.

Internal mobility of the population is another factor that may induce growth. Mobility assists each individual in realizing his or her own potential, and it also allows employers to draw on the potential of the entire national population. We would expect workers to move from areas of high unemployment to areas of low unemployment, so variations in regional unemployment can provide a measure of labor mobility and national cultural homogeneity. Great variations in unemployment suggest that something is discouraging people from moving. Factors preventing mobility may include people's attachment to place or family; racial or ethnic animosity among different localized groups; lack of nationwide availability of certain cultural products and services (religious services, foods, or other consumer goods); regional variations in unionization and difficulty in joining unions; variations in workers' pay, rights, or benefits; nationwide acceptance of degrees or certification for professionals; and a lack of knowledge of distant opportunities. Italy exemplifies a country in which unemployment rates in one region, the south, are usually two or more times as high as in the north. This reflects continuing cultural differences between the regions.

States with internal frontiers   Not all states occupy and utilize their full territory. The world population map (on the rear endpaper) reveals that many countries have core areas of dense population and development but also **internal frontier** areas, which are undeveloped regions that may offer potential for settlement. Environmental limitations may hinder exploitation of frontiers. Brazil, for example, contains vast sparsely settled areas in which the government has encouraged settlement, but the results have often been economically disappointing and ecologically disastrous. Similarly, the island of Java is only 7 percent of Indonesia's national territory, but it is home to 60 percent of the population. The Indonesian government has tried to relocate settlers to other islands, but many new settlements have had difficulty supporting themselves.

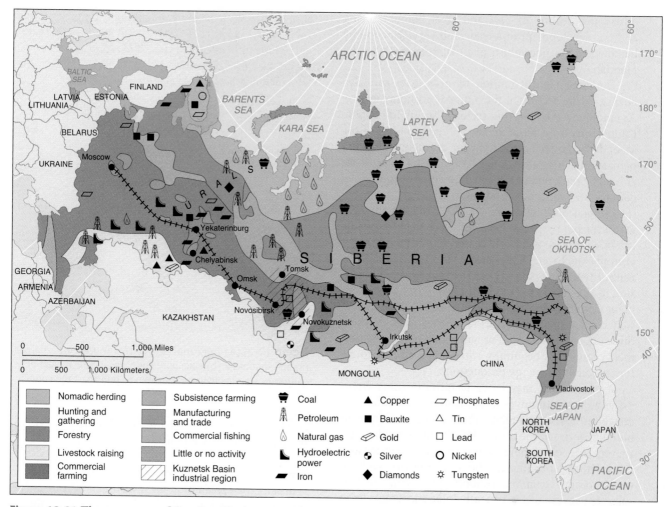

Figure 12-21 The resources of Russia. Siberia remains almost uninhabited, and enormous amounts of capital would be needed to overcome its inhospitable climate and great distances to exploit its raw materials.

Some African states seem overpopulated relative to their current ability to feed themselves, yet they are only sparsely populated. With an end to civil wars and with greater investment in agriculture, some of these territories could support greater populations. Canada's northern regions are poorly suited to agriculture (compare Figure 4-22 with Figure 11-23), but they are exploited for mineral, timber, and hydropower resources. Expanses of the Siberian region of Russia have never been densely settled or even fully explored for resources, yet most of Russia's population remains west of the Urals (Figure 12-21). Japan might become a partner with Russia in developing Siberia's potential.

How Do Governments Distribute Economic Activities?

Most countries try to maintain a fairly equal standard of living throughout their territory, although many have biases that favor some regions over others. Great disparities of wealth from region to region are a centrifugal force that may pull the state apart (Figure 12-22). Economic competition among regions may dominate national politics. Politicians weigh the domestic regional impact of every program and devise new ones to reduce disruptive imbalances. We have already seen the cultural differences between Italy's northern and southern provinces, which may be either a cause or an effect of regional variations in income (compare Figure 11-3 with Figure 12-22). Norway builds bridges and tunnels to reach even the most isolated parts of its territory. This keeps rural populations from leaving these outlying farms and villages and enables goods to flow more freely to remote areas. The apportionment of the Norwegian Parliament even deliberately underrepresents the main cities in favor of rural areas.

The geography of any country's resource endowment may favor some regions over others. The relative fortunes of regions may change, depending on the discovery of new resources, shifting trade patterns, or patterns of innovation. In the area of today's Belgium, for example, the lowland coastal Flemish grew rich from trade and commerce during the medieval and Renaissance periods, and they dominated the highland Walloons. Later the discovery of coal and the industrialization of Wallonia brought the Walloons prosperity.

Today the decline of coal and steel manufacturing and the development of a trade and service economy has swung the pendulum of prosperity back to the Flemish.

The national distribution of prosperity can also change as the sectors of the national economy evolve. As employment shifts from sector to sector, job opportunity shifts from place to place. The expansion of the secondary and tertiary sectors, for example, brings urbanization. These shifts might occur faster than workers can be retrained or relocated, so certain regions of a country may suffer while others thrive.

Countries often devise special development programs for poor regions, just as cities or states designate urban enterprise zones (discussed in Chapter 10). For example, the United States created the Tennessee Valley Authority in 1933, the Appalachian Regional Commission in 1964, and the Lower Mississippi Delta Development Commission in 1988 (Figure 12-23). Income and welfare were significantly below national levels in these three regions, and federal programs attempted to boost their economies, beginning with improvements to infrastructure and energy production.

A government may locate factories or other enterprises in poor regions, or it may distribute its offices, research institutes, and military installations. Governments may lure private enterprises by offering subsidies, tax waivers, free development sites, or loans. The Spanish government, for example, subsidizes international businesses that invest in Spain's poorer provinces, and Italy subsidizes investment in its south. India encourages firms to locate in officially designated less developed areas by offering low-interest loans, tax reductions, and low freight rates on government railroads for products from the designated areas.

No matter what economic policies a government devises, some regions may dominate a national economy while others suffer population losses. As a whole, the central regions of the United States and Canada have been losing population since World War II. Kansas alone has more than 2,000 officially registered ghost towns (Figure 12-24). While U.S. Plains states have suffered population declines, the South and Southwest, known as the Sunbelt, have absorbed most U.S. population growth.

National Transportation Infrastructures

Political considerations Some regions in Southeast Asia and in parts of South America and Africa are in open rebellion against their national governments, and the central governments invest in roads specifically to "occupy" the territory. Colombia, for example, has recently discovered oil deposits beneath its eastern plains, but exploiting these deposits will require the central government to seize control of this region from drug cartels and independence movements. China has been

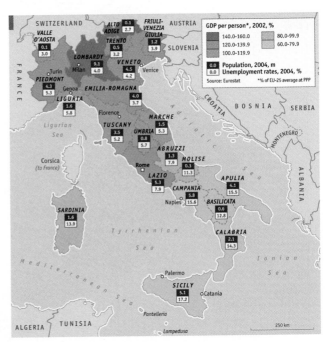

Figure 12-22(a) Regional disparities in income. Italy (a) and Indonesia (b) exemplify countries in which great disparities in regional incomes threaten to pull apart the country.

accused of building railroads to strengthen its grip on peripheral minority regions (Figure 12-25).

Transportation and economics Economic growth is usually the explanation for investment in transportation. Transportation and communication allow different territories to specialize their production and to trade. This territorial division of labor is comparable to

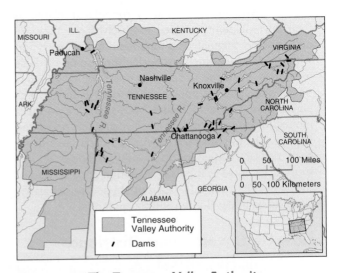

Figure 12-23 The Tennessee Valley Authority region. The Tennessee Valley Authority was created to bring electric power and jobs to this economically deprived seven-state region in 1933. Despite the authority's many dams, 60 percent of the electricity it produces today is generated in its coal-fired plants. The modernization of these plants is an issue of current political debate.

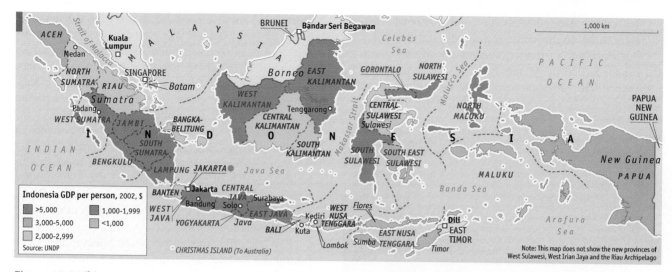

Figure 12-22(b)

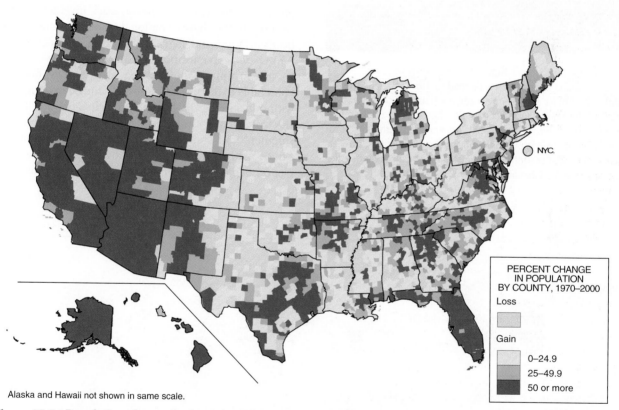

Alaska and Hawaii not shown in same scale.

Figure 12-24 Population change in the United States, by county. Some counties in the central U.S. farming region and some old cotton-farming centers in the South counted their maximum populations before 1900. The populations of many other areas across the Great Plains and in Appalachia peaked by 1940, when the prosperity that had come from farming and mining was past. The national population today seems to be concentrating at the margins of the country.

the division of labor among workers. The design and regulation of a country's transport infrastructure are two important influences on internal geography. In some countries transportation is slow and difficult. Some regions are practically inaccessible. In other countries, virtually every place is easily and cheaply accessible from every other place (Figure 12-26).

Whenever anything is in transit, the value it represents is idle. The owner is therefore losing an opportunity to invest that value in some other activity that might pay a return every minute. A financial return sacrificed by leaving capital invested in one form or activity rather than another is called an **opportunity cost.** Goods in transit cost their owner opportunity costs as well as transportation charges, and that is why people want things to move fast. If a country has a poor transportation infrastructure, then much of the country's capital is tied up in raw materials or in materials in transit. This principle applies on a small, local scale:

Figure 12-25 China's high-altitude train to Tibet. China opened a new railroad to Tibet in 2006. The difficulties of its engineering made the train one of the wonders of the world. High-tech systems stabilize the tracks over permafrost, and passenger cabins are enriched with oxygen to help passengers cope with the high elevations. Critics, however, argue that the influx of tourists and of Chinese influence will destroy the Tibetan culture.

Your local stores do not need to carry large inventories if they can get whatever you want quickly. Furthermore, if factories can receive regular supplies of the materials they need, they can reduce their inventories of parts. Minimizing inventories is the key to the modern manufacturing method called *just-in-time manufacturing.* In the early 1990s, for example, Dell Computer Company stored 30 days of inventory and components in warehouses. By 2004, however, the company had no warehouses. Instead, it required suppliers to stock 8 to 10 days' worth of goods no farther than 90 minutes from its assembly plants. Its de facto warehouses, therefore, were the rows of semitrailers lined up in the truck bays of its plants.

Fast and efficient transportation releases capital for productive investment, so developing a country's transportation system is key to economic growth. In the United States, deregulation lowered both truck and rail freight rates between 1981 and 2002, and new computer systems helped businesses track inventory better. As a result, the share of national GDP accounted for by transportation and logistics fell from 16.2 percent to 8.7 percent—an enormous national savings.

Just-in-time manufacturing saves money, but it is vulnerable to disruption, especially considered together with outsourcing. Parts come to Dell Computer factories from all over the world, yet on any day, one missing shipment can stop the entire operation. Disruptions can be natural or can be caused by humans. For example, on September 21, 1999, an earthquake measuring

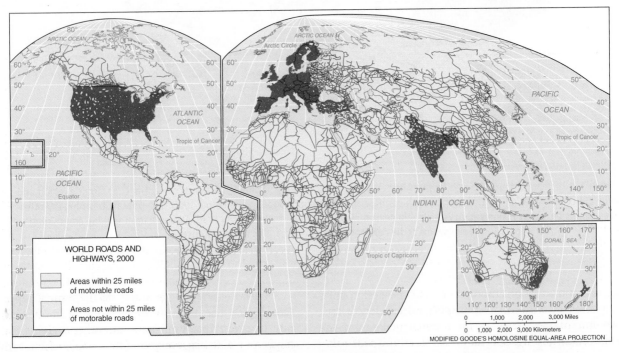

Figure 12-26 World roads and highways. Compare this map with the world population distribution shown in the map on the rear endpaper. Many densely settled regions, such as Eastern China, would benefit from improved road networks. The African continent is three times the size of the United States, yet it contains only one-quarter of the mileage of paved roads.

GLOBAL AND LOCAL

India's New Roads

In 2005, the Indian government announced a 15-year project to widen and pave approximately 65,000 kilometers (40,000 miles) of narrow, decrepit national highways. This is the most ambitious infrastructure project in India since the country gained independence in 1947. The new roads will save billions of dollars per year in transportation costs and open more parts of the country to world trade by connecting them to shipping ports.

The first stage of the National Highway Development Project has been dubbed the "Golden Quadrilateral." This rough quadrilateral's 5,840 kilometers (3,625 miles) run through 13 states and India's four largest cities: New Delhi, Kolkata (formerly Calcutta), Chennai (formerly Madras), and Mumbai (formerly Bombay). It is mostly complete. The second stage of the project is improving the north–south and east–west main corridors, totaling about 7,300 kilometers (4,536 miles). As of early 2010, this stage was nearly half complete. Importantly, these corridors will connect major ports, aiding India's trade with the rest of the world. Later stages will connect other industrial centers and improve urban roads.

The project has experienced delays in the process of acquiring the land—8,326 hectares (20,574 acres)—along the highway. The government has the power of eminent domain, but it must compensate for land taken and must rely on cumbersome regulations and local officials. Some observers contrast the Indian process with that of China, an absolutist country in which land is simply taken. China has built more roads more quickly.

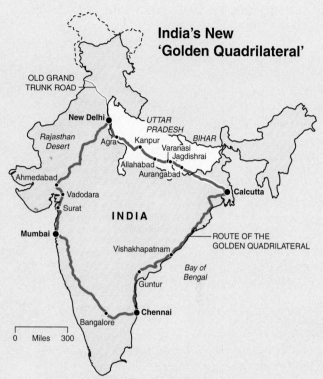

India's expanding roads network Government initiatives to improve existing roads and build new ones seek to decrease travel time across the subcontinent.

7.6 on the Richter scale struck Taiwan, where 90 percent of the world's scanners and over 50 percent of computer motherboards (over 50 percent of those in one industrial park in Hsinchu) were manufactured. Production was restored at most factories within a week, but another such disruption could do more lasting damage to global supply lines. On September 12, 2001, the day after the attacks on the World Trade Center, much of U.S. manufacturing activity came to a sudden halt when the United States closed its border and grounded all flights. Terrorist threats anywhere in the international transportation system threaten the development of global just-in-time manufacturing.

India exemplifies a country whose economic growth is held back by an inadequate road system. Traffic in passengers and goods on India's roads has increased thirtyfold since the country gained independence, but the total road mileage in India has increased only fivefold. Roads throughout India are overcrowded, and thousands of villages lack all-weather roads. The government hopes to have every village of more than 500 people connected to an all-weather road in the next few

years. Accessibility would allow regional specialization of production and cash cropping, especially of high-value perishable foods, about 40 percent of which spoils during transit delays today. Better transportation could also improve the Indian diet.

China is investing heavily in new roads and railroads. The government has created an extensive network of multiple-lane highways, but its plans include the construction of an additional 85,000 kilometers (52,817 miles) of intercity highways and urban ring roads within 30 years. The government is constructing 5,400 kilometers (3,355 miles) of high-speed rail track, and plans to build or refurbish an additional 40,000 kilometers (24,855 miles) of its railways. Plans include a 1,320-kilometer (820-mile) link from Beijing to Shanghai capable of speeds up to 354 km per hour (220 mph), and a second 177-kilometer (110-mile) magnetic levitation line from Shanghai to Hangzhou capable of 450 km per hour (280 mph). China will import the technology, but at least 70 percent of all parts will be built in China, thus greatly enhancing China's manufacturing technology.

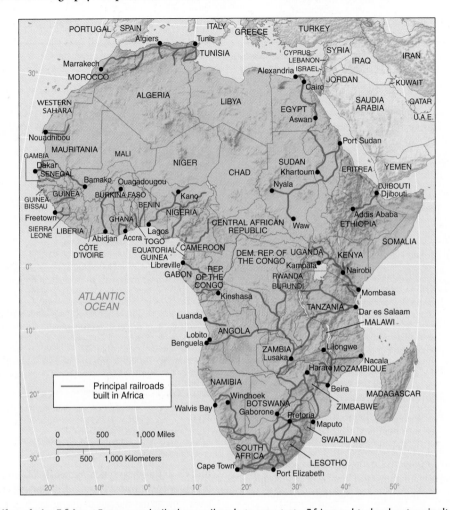

Figure 12-27 Railroads in Africa. Europeans built these railroads to penetrate Africa and to haul out agricultural products and minerals, and still today railway lines provide the only freight service in many roadless areas. Africa's coastline offers few natural harbors, and many of its rivers drop off the continental interior through rapids and waterfalls only a few miles inland from the coast (compare this figure with the map on the front endpaper). Thus, the continent is relatively impenetrable by water, which presents a major impediment to African economic development. Many lines shown on this map, however, are disrupted by civil strife or else they are inadequately maintained, so they offer only intermittent service. Deterioration of the rail network increases transport costs and lowers income for many African states. Lines in Sudan, for example, and the important line reaching into central Africa from Benguela, Angola, have not operated for 30 years.

In many countries, the pattern of the transport network is not integrated, but rather consists of lines penetrating into the country from the ports to bring out the exploitable wealth. These routes are often the legacy of colonialism or the product of neocolonialism, and they facilitate getting into or out of a country, but they do not facilitate getting around inside it (Figure 12-27). Australia, too, exhibits such a pattern of railroads, and Australia's rail routes were built at many different *gauges* (widths between the tracks). The Australian government has tried to knit one national rail system, but Sydney and Melbourne, Australia's two largest cities and only 700 kilometers (435 miles) apart, were not linked until 1995, and the first transcontinental north–south line, promised in 1911, did not open until 2004.

Regulating transportation Mapping a national transport network is only a first step toward understanding what moves, where, and why. Lines on a map are static, but movement on those lines can be manipulated by conditions and charges that are "hidden" behind the map. Most countries blend public and private transportation services, and they manage freight rates on government-owned railroads, for example, as India does, to distribute industry or other activities.

In all countries, trucks, railroads, and airlines compete for traffic. They may also compete for government subsidies, and these subsidies can in turn affect domestic economic geography. In Kenya, for example, the trucking industry has successfully prevented national investment in railroads. In most European countries, in contrast, government-owned railroads receive generous subsidies. In the United States the trucking industry, the oil industry, the highway construction industry, and the car-driving public make up a lobby powerful enough to overwhelm the railroad interests. As a result, trucks pay only a fraction of their real highway-use costs, and they do not pay for the many negative externalities that they impose, such as air pollution, traffic accidents, and costs due to congestion. This subsidy to the trucking industry enables it to capture much long-haul traffic that is actually best suited to railroads. Trucks carry about 58 percent of commercial freight by weight and railroads carry only about 12 percent. Railroads have been unable to extend their coverage while trucks benefit from taxpayer-supported road construction in every state. Thus, in the United States as in Kenya, political competition distorts economic competition in the transport system and locks inefficiency into the economy.

U.S. freight railroads are nevertheless enjoying a revival. Computers improve tracking of shipments and timing of schedules, and highway congestion (and the associated pollution and accidents) encourages the rail alternative. State and local governments are helping finance improvements in the network. For example, the six major North American railroads announced in 2003 a multi-billion-dollar project to untangle their lines around Chicago, speeding freight across the country.

Countries may also manipulate domestic freight rates to direct traffic to favored seaports, and the port facilities themselves may be subsidized to attract business and create jobs. The Netherlands, for example, organizes its national transport network and fixes prices on it to capture additional international income on freight arriving at its port in Rotterdam.

These examples provide only a hint of the manipulation of charges built into each country's transportation infrastructure. When geographers investigate economic activities, they often discover these "hidden" policies.

National communications infrastructures   The dissemination of information can be as important as the dissemination of goods and materials. The degree to which citizens of a country enjoy access to sources of information can reveal its economic development and political liberty, and it can also promote these two values.

The availability of telephone lines is a good measure of access to information (Figure 12-28). Today, fax machines, personal computers with Internet access, and other modern communications devices are diffusing to many developing countries. The connections are already transforming the geography of the twenty-first century for many once-remote places.

All aspects of infrastructure—pipelines, electricity generators, highways, telephone facilities, and more—require continuous maintenance and updating. As societies become increasingly dependent on these facilities, and as the facilities become more complex, both maintenance costs and repair costs in the event of failure rise dramatically. In 2009, however, the American Society of Civil Engineers released a "report card" on the state of America's infrastructure. The overall grade was D, with the highest mark (C+) for solid-waste handling. The most dramatic underinvestment noted involved the nation's roads. Simply maintaining existing roads would require increasing annual spending by more than 50 percent; improvements would require doubling spending. The engineers described

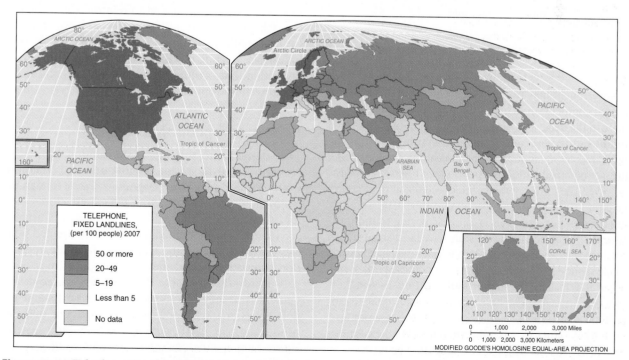

Figure 12-28 Telephone fixed landlines.   A few places that have never had telephone service are skipping landline technology and jumping right to mobile telephone technology. The existence of telephone landlines, however, still provides a good measure of the availability of information and communications in a society.

CONNECTIONS

Socks and Politics

In 2005, two cities called themselves "The Sock Capital of the World": Fort Payne, Alabama, and Datang, China. In 2004, Fort Payne produced 728 million pairs of socks; Datang produced 9 billion pairs. Workers in Fort Payne averaged $10 per hour; workers in Datang averaged 70 cents per hour. When the U.S. Congress considered a new trade pact with Central America in 2005, Fort Payne's representative, Robert B. Alderhol, expressed concern that such an agreement could unleash a new flood of imported goods into the United States. Nevertheless, to the surprise of many observers, Representative Alderhol voted for the trade pact. Five days after the vote, the U.S. government announced that it would extend limits on imports of socks from China. This action raised the prices of socks for all Americans and perhaps temporarily saved some jobs in Fort Payne.

America's infrastructure as, quite simply, "falling apart." Their conclusions are punctuated by the collapse in 2007 of a major bridge over the Mississippi River in Minneapolis and repeated failure of part of the San Francisco–Oakland Bay Bridge. These events are warnings about infrastructure maintenance.

National Trade Policies

Countries differ in the degree to which they participate in world trade and circulation. Each country wants to develop its own economy, to ensure its national security, and to some degree, to protect its national culture or identity. Chapter 8 noted how some countries subsidize domestic food production as a national security measure. Countries also protect national security industries—weapons, for example.

The Import-Substitution Method of Growth

A government may protect a national industry from imported competition as only a temporary measure. This is done in the belief that newly developing industries, called **infant industries,** cannot compete with imports. Infant industries manufacture only small quantities of the product, so manufacturing costs per unit are high. Eventually, however, the industry might build a national market for its product and produce larger quantities of it, thus reducing the cost per unit. This is called achieving **economies of scale.** When economies of scale have been achieved in protected national markets, then imports will be allowed to compete. This policy of protecting domestic infant industries is called the **import-substitution method** of economic growth. The basic premise of this policy is to promote value-added production by capturing a domestic market, rather than sending money out of the country to benefit value-added production in other countries.

This form of economic protectionism helped some countries industrialize in the past, including, to some degree, the United States, but it involves economic risks.

If a country raises tariffs to protect a particular industry, all national consumers are subsidizing that industry. Some protected industries may never be able to survive competition with imported goods, so the national public may go on subsidizing them. This may be an inefficient investment of national resources. For example, the United States long imposed high tariffs on imported apparel; the U.S. Department of Commerce estimated that this protection cost U.S. consumers $145,000 per job saved in the U.S. apparel industry, which was several times what those workers actually earned. The wages apparel workers earned, however, multiplied through the economy, and if the workers had had no jobs, some might have had to go on welfare. The United States agreed gradually to reduce barriers to trade in textiles after 2005, but China's share in the U.S. market for textile and apparel products jumped, so the United States threatened to reintroduce tariffs unless China imposed restraints on its own exports. China eventually imposed export tariffs on its own industries. Overall, it is difficult to balance the costs of protection, but manufacturers and workers still lobby for protection from foreign competition.

Export-Led Economic Growth

The alternative to the import-substitution method of economic growth is the **export-oriented method.** Countries that adopt this program welcome foreign investment to build factories that will manufacture goods for international markets. In that way, they can achieve economies of scale immediately. Export-oriented policies rely on global capital markets to facilitate international investment, and they rely on global marketing networks to distribute the products. Export-led development is usually accompanied by low tariffs and export duties, market-based domestic economic policies, privatization of most economic activity, and general laissez-faire openness to economic innovation.

The countries that have grown the fastest in recent decades have generally followed export-oriented programs. The success of these policies demonstrates again how, increasingly, what happens *at* places is the result of what happens *among* places.

For a country's economy to grow by exporting, other countries must be willing to accept the developing nation's products. Through recent decades, the United States has been the major importer of products from countries choosing this method of development, and, in doing so, the United States has accumulated substantial trade deficits. At the end of the 1970s, the United States was a creditor country, with a net stock of foreign assets worth about 10 percent of GDP. Persistent current accounts deficits, however, turned the country into a net debtor by 1985, and since then the United States has been getting more and more deeply into debt. The *current accounts balance* summarizes a country's current transactions with the rest of the world, which include trade, income from international investments, and transfers. At the end of 2008, the U.S. current accounts deficit was over $706 billion, which is about 5 percent of GDP. This debt is in addition to the national debt of about $10.7 trillion. Some economists fear that this debt threatens America's prosperity and even independence of action in foreign affairs.

China: How trade policies can affect the national distribution of wealth   International trade and investment sometimes multiply regional economic inequalities within a country. If a country adopts protectionism, all regions of the country may not profit equally, which can intensify regionalism. If a government adopts the alternative export-led growth policies, outside investment might concentrate in favored regions.

China exemplifies on a colossal scale how foreign investment can intensify regional imbalances in economic growth and how the centrifugal force of these imbalances can threaten national stability. China has always had an economic and cultural dichotomy between an outwardly focused southeast coastal region and an inwardly focused central region. Chinese capitals were in the north or center, and the government limited the trading activities of the merchants in the southern coastal cities. In the nineteenth century, however, European powers forcibly "opened" China's southeastern ports to trade, and industrial development, inspired by international trade, clustered along the coast. The Chinese government keenly felt this humiliation.

The Communist government that Mao Zedong brought into power in 1949 set out to boost the inland regions out of their relative economic backwardness by redistributing industry inland. From 1949 to 1979, foreign trade plummeted; China's coastal areas and cities stagnated.

Mao died in 1976, and in 1979, the Chinese government reversed its policies. It designated zones along the coastline where relatively free capitalistic practices were allowed. These areas were soon expanded as international corporations built factories in the zones to exploit the cheap land and labor, and the designated areas leaped to create new wealth (Figure 12-29). China had

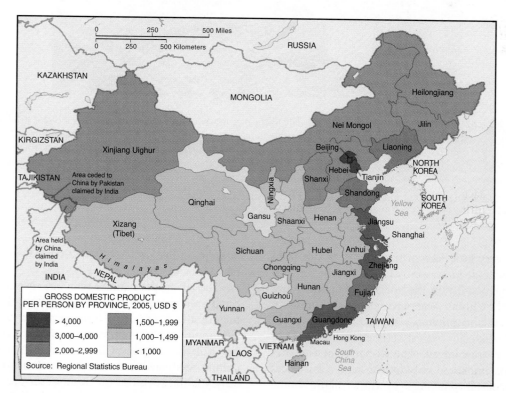

Figure 12-29 China's booming coastal provinces. The great variations in per capita gross product among China's provinces result largely from foreign investment and astounding economic development in the coastal regions. China has invited foreign investment since 1979, and capitalist-style economic growth has boomed.

Tourism

Tourism could be described as the ultimate import strategy, bringing consumers to places that are commodities unto themselves. Tourism is by some accounts the world's largest industry. Every year, a greater proportion of the world population travels, and in most countries, tourism is one of the fastest-growing sectors of the economy. International tourist arrivals numbered over 922 million in 2008, and international tourist receipts totaled approximately $944 billion, excluding airfares, which is equivalent to 6 percent of global trade in merchandise. Tourism is the top earner of foreign exchange for many countries, and the World Tourism Organization

(a U.N. organization headquartered in Madrid) believes that tourism will continue to expand despite the threat of terrorism.

Citizens of rich countries account for the largest tourist expenditures, and rich countries are also the most popular destinations. Measured by tourist arrivals, the leading destinations in 2008 were France, Spain, the United States, China, and Italy. When tourism income is measured as a share of total national income, however, the importance of tourist income to the poor countries becomes clear. Tourism made up a significant amount of the income for many poor countries. Tourism receipts equal about 14 percent of Cambodia's GNI and about 7.5 percent of Egypt's GNI.

Cancún, Mexico. This resort was carefully planned. The Mexican government selected it for its many relevant locational determinants, including annual days of sunshine, temperature range, attractiveness of local beaches, temperature of the water, currents, variety of sea life, scenery, and nearby archaeological ruins. In the 1970s, the government began investing millions of dollars to develop Cancún as a vacation destination. The problems of overdevelopment caused the government to begin limiting growth in 1994.

engaged in very little foreign trade in 1979, but from 1980 to 2006, trade increased by 12–15 percent per year.

Economic activity in the special economic zones boomed. As China's seaboard provinces grew in wealth, the inland provinces slipped behind noticeably, and the historic contrast between the interior and the periphery reemerged. The disparity of incomes across China improved only by workers' remittances from the coastal provinces to their inland home provinces. Today the government is investing some of the country's new wealth in education, health care, and public works in the interior, yet about one-third of the

population live on less than $2 per day and many are peasant farmers.

The rift in the nation's economic geography duplicated a widening rift between the state and private sectors of the economy. China remained dedicated to Communism in theory, but in practice, it continued to privatize the economy. Chapter 8 noted that when agriculture was privatized, food production soared. Similarly, the industries of the state sector declined as the private industries flourished (Figure 12-30). Closures, mergers, and privatizations of state-owned enterprises lowered the state-owned sector of the economy from

The tourist potential of any spot depends upon its "three A's":

- *Accessibility*, usually via convenient air routes
- *Accommodations*, at all levels from simple pensiones to grand resort hotels
- *Attractions*, which can include a pleasant climate, as well as beaches, museums, architecture, historic sites, festivals and performances, shopping opportunities, and more

Areas spectacularly endowed with wildlife may feature **ecotourism**, which is travel to see distinctive examples of scenery, unusual natural environments, or wildlife. The rich bird life on the islands of Lake Nicaragua, the lions of South Africa, and the tropical vegetation in many countries are important tourist destinations. Ecotourism today provides the money needed to save many countries' natural environments as well as many species of rare animals. Overall, the countries best endowed for tourism have both natural and cultural attractions, pleasant climates, good beaches, and local staff who can communicate with and understand the expectations of wealthy foreigners. Political stability is a necessity. Therefore, despite Africa's wealth of ecological and cultural attractions, political instability, poor infrastructure, and lack of accommodations have restricted its income to only 5 percent of international tourists and just over 3 percent of global tourist dollars, and those were divided almost exclusively between the countries of Mediterranean North Africa and the country of South Africa.

Tourism creates many jobs and offers many multiplier effects. For example, a new hotel creates demand for a host of related goods and services. Tourism may also stimulate the local agricultural economy, and it is generally a high-value-added and labor-intensive industry that trains local labor. Tourism may, however, corrupt local culture or provoke inflation, and too many tourists may overwhelm the natural environment.

Ecotourism. Tourists on an elephant-back safari in Zambia.

almost 100 percent in 1980 to less than one-third by 2006. China's 1954 Constitution said that business executives must be "used, restricted, and taught." The 1988 Constitution softened those words to "guided, supervised, and managed," but amendments added in 1999 declare private business "an important component of the socialist market economy."

In August 1997, *The People's Daily*, China's principal Communist Party newspaper, announced, "The words 'market economy' have been writ large on the flag of socialism. . . . We ran a planned economy for more than 20 years and created the foundations of industrialization, but . . . ahead is a new world. There is no way back." Within weeks, the Fifteenth Communist Party Congress declared that the government would continue to insist on ownership of only a few key industries, and the definition of ownership will include holding only a minority stake in enterprises. This is privatization in all but name. In 2006, the Chinese government approved the 11th Five Year Plan for future economic development, but for the first time, that document retreated from the phrase *wu nian ji hua* ("plan") and substituted the less-controlling phrase *wu nian gui hua* ("blueprint").

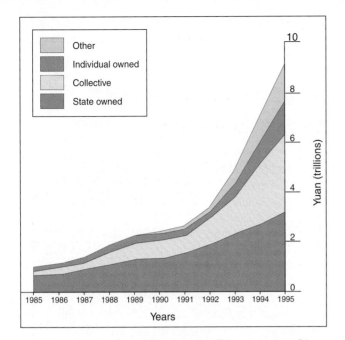

Figure 12-30 Industrial output in China, measured in trillions of yuan. The private sector of the Chinese economy soared after economic liberalization. The share of the total economy created by state-owned enterprises fell from about 75 percent in 1975 to about 28 percent by 2000. If we accept the economic definition that communism is state ownership of industry, then China ceased being a communist country about 1993.

China's economic evolution exemplifies the ladder of economic development: Wages are rising, and so is the value added to Chinese exports. In 2002, its exports were still 70 percent garments, toys, shoes, and furniture—items with low value added—but other products of increasing sophistication are following: from plastic to better-quality shoes, clothing, ceramic tiles, and even automobiles. In 2004, China surpassed the United States as the world's leading exporter of technology and telecommunications goods—$180 billion against $149 billion. China had earlier pulled ahead of Europe and Japan. China is already experiencing a shortage of skilled workers—both industrial and managerial—which is inflating salaries. Multinational corporations, including General Motors, Intel, Honda, and Motorola, are relocating factories deeper into the Chinese interior, where wages and land costs are lower, or to lower-cost countries such as Vietnam and Indonesia.

Some observers feared that an avalanche of cheaply made goods from China might inundate world markets and virtually bankrupt every other country, but that will not necessarily happen. Chinese exports have indeed been responsible for deflating prices of manufactured goods in the world's rich countries, but, as China develops, the Chinese themselves are absorbing more of their own production and importing more of other countries'

products. By the early twenty-first century, eastern China already boasted a lower-middle-income market of 500 million people. The domestic market for cell phones, cars, cameras and film, and DVD players was already number one on Earth. Toshiba, the Japanese electronics and industrial equipment giant, once used China as an export base, but by 2002, Toshiba was selling over 60 percent of what it made at its 34 Chinese factories to the Chinese themselves: televisions, refrigerators, computers, and automation equipment. On April 20, 2006, while visiting the United States, China's President Hu Jintao insisted, "China is pursuing a policy of boosting domestic demand, which means that we'll mainly rely on domestic demand to further promote economic growth." Within a week of his return to China, the government changed taxes and interest rates in ways designed to encourage Chinese to spend more and to save less: Thus, Chinese development may lift both the country's own and the world's standard of living.

Taiwan Taiwan is a principal source of investment funds for China. The 23 million people of this island enjoy a per capita GDP of almost $31,100 per year (PPP), contrasted to about $6,000 in China. This contrast underlines a bitter political division.

In the 1930s, a civil war raged in China between the Communists, led by Mao, and the Kuomintang, a political party led by General Chiang Kai-shek (1887–1975). In 1949, the Communists drove the Kuomintang offshore to Taiwan, where it established a separate government. The ethnic Chinese, although a minority among the Taiwanese, continued to control the government until the elections of 2001. Meanwhile, state-directed capitalism achieved phenomenal economic growth. In the 1990s, Taiwanese began to invest and transfer low-wage, low-technology manufacturing to China, but the sophistication of the technology being transferred has steadily risen. In 2003, for example, Taiwan Semiconductor Manufacturing Company, the world's largest contract chipmaker, built a huge semiconductor factory in Shanghai. The Chinese government welcomes such investment from its "rebel" province.

Both the government in Beijing and that on Taiwan claim that there is only one China and that they represent it, but at present China and Taiwan are in fact two separate countries. The U.N. General Assembly ousted the Taiwan government and seated the People's Republic in its place in 1971, and the United States recognized the People's Republic as the sole government of China in 1978. China says it will attack Taiwan if—but only if—Taiwan explicitly declares independence; Taiwan says it will declare independence only if China attacks. Some political accommodation will probably eventually be reached between the two governments—perhaps following the slogan explaining China's reincorporation of Hong Kong: "One country, Two systems." China and

Taiwan might agree on joint economic ventures and joint representation in foreign affairs, for instance, without completely merging their domestic governments. The economic integration of Hong Kong, Taiwan, China, and the considerable economic vitality of the overseas Chinese is proceeding rapidly.

As a result of China's slowing rate of population growth (see Chapter 5), its surge in food production (see Chapter 8), and foreign investment, China has achieved a rate of economic growth of about 10 percent per year for the last two decades. Overall, the Chinese economy quadrupled in size between 1980 and 2000, and since 2000, China's contribution to global GDP growth has been greater than America's and more than half again as big as the combined contribution of India, Brazil, and Russia, the three next-largest emerging economies. This has come about with repressive government, and any number of factors might slow or halt this development, including political disruption, environmental disruption, an energy supply crisis, or inadequacies in China's infrastructure. If the growth continues, however, China will overtake the United States as the world's largest economy before 2015. The population of China, a full 22 percent of the human population long living in poverty, might achieve considerable material comfort within 50 years. China's dramatic economic and industrial growth will likely transform many aspects of global political, economic, and environmental patterns in the twenty-first century. The growth is taking place over a period of years rather than suddenly, so it does not make the headlines, but in

terms of its impact on the geography of wealth and power in the world, it is one of the most important "news items" of our time.

The Formation of the Global Economy

International trade has been carried on for millennia. Trade always triggers regional specialization and increases in productivity, as well as new cultural possibilities and combinations (see Chapter 6). As new means of transportation and communication have linked world regions at lower cost, more goods are produced for trade, so trade multiplies faster than total production. From 1950 to 2000, world GDP sextupled, but world exports multiplied many times over, and in 2008, world trade reached a total of over $15.7 trillion in merchandise and an additional $3.8 trillion in commercial services (Figure 12-31).

International economic and communication links have "shrunk" Earth in terms of time or cost distance, and economic globalization has far outpaced cultural or political integration. Tensions among these economic, cultural, and political organizing activities generate much of today's international news. The globalization of the economy is already redistributing economic activities in every economic sector, and countries are continuously renegotiating conditions of production, trade, and exchange.

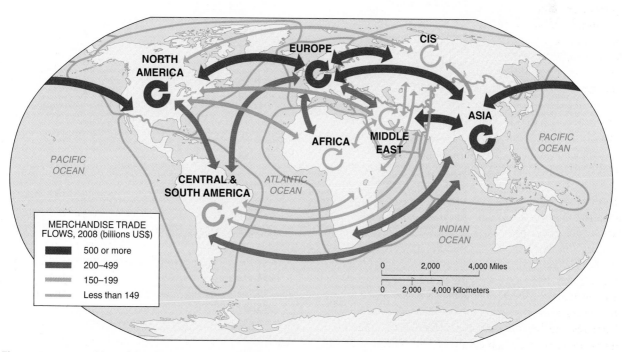

Figure 12-31 World merchandise trade by region. Most trade is among developed regions, but some developing regions, such as Latin America, are capturing an increasing share.

Transnational Investment and Production

Two developments in economic globalization accelerated in the late nineteenth century. One was the evolution of a world market for certain primary products, foods, and minerals. Chapters 8 and 9 traced the results of that development. The second development was an increase in international investments. At first these investments were limited to shares of stocks and bonds representing minority holdings in foreign companies, called *portfolio investments*, but foreign direct investment soon followed. **Foreign direct investment (FDI)** means investment in enterprises that are actually operated by a foreign owner.

Modern transnational manufacturing was launched in 1865, when the German chemical magnate Alfred Bayer built a factory in Albany, New York. Today, technology allows the integration of manufacturing processes worldwide. For example, South Korea's Kia Motors Corporation manufactures a popular sport utility vehicle, the Kia Sorrento, named for a town in Italy. The Sorrento is assembled in South Korea, but it is full of components from around the world. The optical pickup units of the vehicle's CD player, which read the CDs, are made in China, for example. Their mechanical

structure and electronic components are added in Thailand, and then the units are shipped to Reynosa, Mexico. From there the units are trucked to Matamoros, Mexico, where they are assembled into an audio system. The audio systems are trucked to California and shipped to South Korea, where the Sorrentos are then shipped to the United States.

Enormous enterprises have grown up that own and coordinate production and marketing facilities in several countries. These are called **multinational corporations** or **transnational corporations** (Table 12-3). Transnational companies (as defined by the United Nations) that have international production facilities now number 60,000 and boast 500,000 foreign affiliates, accounting for an estimated 25 percent of total global production.

The evolution of transnational corporations historically followed four stages. In the first stage, the corporation exported products from its home country to meet demand abroad. In the second stage, the corporation established production facilities abroad to supply specific markets. Exports from the home country dropped. In the third stage, the foreign production facilities began to supply foreign markets other than their own local markets. In the fourth stage, the foreign production facilities began to export back into the home country. These stages appeared first in trade in primary materials such as copper

TABLE 12-3 Sales of the World's Largest Corporations Compared to National GDPs, 2008 (in millions of dollars)

Countries Ranked by GDP	Countries or Corporations	2008 GDP or Revenue	Corporations Ranked by Revenue
23	Saudi Arabia	467,601	
	Royal Dutch Shell	**458,361**	**1**
24	Norway	449,996	
	Exxon Mobil	**442,851**	**2**
25	Austria	416,380	
	Wal-Mart Stores	**405,607**	**3**
26	Iran	385,143	
	BP	**367,053**	**4**
27	Greece	356,796	
28	Denmark	342,672	
29	Argentina	328,385	
30	Venezuela	313,799	
31	Ireland	281,776	
32	South Africa	276,764	
33	Finland	271,282	
	Chevron	**263,159**	**5**
34	Thailand	260,693	
35	Portugal	242,689	
36	Colombia	242,268	
	Total (oil)	**234,674**	**6**
	ConocoPhillips	**230,764**	**7**
	ING Group	**226,577**	**8**
37	Czech Republic	216,485	
38	Hong Kong, China	215,355	
39	Nigeria	212,080	
	Sinopec	**207,814**	**9**
	Toyota Motor	**204,352**	**10**
40	Romania	200,071	

and oil. Later the stages were repeated during the evolution of international manufacturing.

Today, the evolution of transnational activity can skip the historic first stage of export from the transnational's home country. Transnationals can invest in countries that have chosen an export-led growth strategy, and they can export globally from those countries (called *export platforms*) before those host countries are themselves markets for the goods they produce. Transnational corporations' ownership and activities are so widespread that it is difficult to tell which corporations are corporate citizens of which countries. Nestlé, the world's largest food company, with corporate facilities around the globe, is generally regarded as a Swiss corporation, yet its shares are traded on the London Stock Exchange. The company's board of directors includes citizens of several countries, but no records reveal Nestlé's actual stockholders. One economist estimates that U.S. citizens own about 14 percent of Nestlé. Gilbert Williamson, president of another global concern, NCR Corporation, has said, "I was asked the other day about United States competitiveness, and I replied that I don't think about it at all. We at NCR think of ourselves as a globally competitive company that happens to be headquartered in the United States."

Multinational corporations challenge individual nations to regulate them or to tax them. Imagine, for example, that a multinational corporation manufactures sneakers for $1 per pair at a subsidiary in Malaysia and sells them to a U.S.-based subsidiary for $40 per pair. The multinational company has kept enormous profits in Malaysia, where tax rates may be lower than in the United States. Trade among different divisions of individual corporations today accounts for an estimated one-third of all international trade, and the corporations avoid any country's taxes by allocating investment and profits wherever they wish. In 2007, 28 percent of all U.S. multinationals' trade in goods occurred among arms of the same companies.

The U.S. government takes the position that any company that employs and trains U.S. workers and that adds value in the United States is a U.S. corporation, no matter which flag it flies at its international headquarters or where those headquarters are. Accordingly, the U.S. government, acting on behalf of a Japanese-owned office machine factory in Tennessee, has protested the trading practices of a Singapore corporation that is a subsidiary of the U.S.-based Smith-Corona Corporation, which is owned by a British corporation.

The International Tertiary Sector

When most people hear the words *foreign trade*, they think of oil and iron, bananas and coffee, cars and clothes being shipped around the world. It is easy to imagine such "things." The international economy, however, has transcended trade in primary products and manufactured goods. Trade in tertiary sector services is expanding, and so is international investment in the tertiary sector.

The United States is the world's leading exporter of services (14 percent of the total of global service exports in 2008), followed by the United Kingdom, Germany, and France. In the United States, the net surplus in exporting commercial services amounted to $153 billion, which partly offsets the country's annual deficit in merchandise trade. For example, in education (which students come from abroad to receive, but which is counted as an export), the United States has enjoyed an export surplus of almost $5 billion per year. However, new restrictions on visas to the United States instituted since 9/11 have deflected the global stream of students to Canada and to Europe, so U.S. education exports have declined. This has harmed America's ability not only to educate these students but to retain them as skilled immigrants. Other U.S. tertiary exports include real estate development and management, accounting, medical care, business consulting, computer software development, legal services, advertising, security and commodity brokerage, and architectural design (Figure 12-32). U.S. entertainment productions dominate the world's airwaves and movie screens. Movies, music, television programs, and home videos together account for a significant annual trade surplus.

In many cases, corporations that provide services to manufacturing corporations in their homelands have followed these out into the world arena. This partly explains the internationalization of legal counsel, business consulting, accounting, and advertising (Figure 12-33). Many professionals, such as architects and physicians, market their skills around the world. Financial services, entertainment, and still other services are produced for global markets. For many nations, welcoming tourists

Figure 12-32 Design is a U.S. export. U.S. architect Frank Gehry (born in Canada) designed this spectacular Guggenheim Museum that opened in Spain in 1997. Architectural design is a significant U.S. export. This building carries a political message, for it was entirely planned, built, and paid for by Basques, a minority group within Spain, in Bilbao, their largest city, to assert their presence in the world (see Chapter 11). On the façade of the building, huge letters proclaim "Museoa," which is not Spanish but the Basque language. The shimmering metal used to coat the petal-like steel panels is titanium.

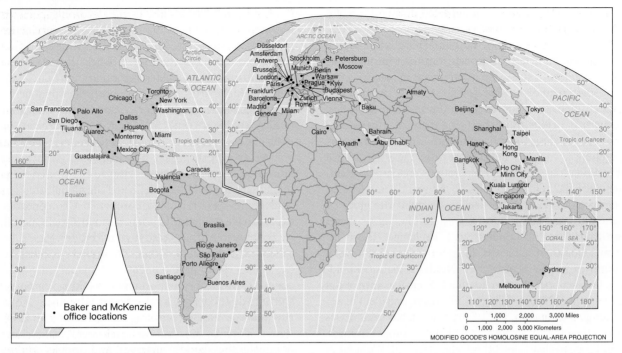

Figure 12-33 One of the largest international law firms. The Chicago-based law firm Baker & McKenzie can provide legal advice on every area of business law in 67 offices in more than 39 countries.

and foreign students to their lands have become major elements of their economies. These activities count as exports of services.

International trade in services demonstrates the same trends as trade in goods. Countries are losing their autonomy to an increasingly global market. The costs of global electronic transmissions are virtually negligible, so transnational corporations treat the world as "one place" and exploit resources, locate facilities, and market their products accordingly. The global market for tertiary labor skills was originally restricted to such back-office work as key punching and standardized data processing. Today, global clients can draw on providers of even the most specialized skills and services anywhere, and in many places, these services cost much less than they would if they were provided by providers based in the rich countries. For example, local accountants in India and the Philippines working for the firm Ernst & Young calculate taxes for the firm's U.S. and global clients. U.S. and Canadian law firms send legal work to be done in India, with almost 1 million English-speaking lawyers trained in common law who can research cases and draft contracts, patent applications, wills, and even divorce papers. X-rays taken in hospitals in the United States are today routinely transmitted to India, where they are analyzed by capable doctors at a fraction of the price the same service would command at home. A California-based company offers tutoring services to U.S. students by Indian tutors. American parents pay $20 per hour, and the Indian tutors earn $230 per month, which is twice local teachers' salaries. Even several U.S. states' welfare benefits and food stamps toll-free "help lines" are located in Mumbai, India (Figure 12-34). White-collar workers in the rich

Figure 12-34 An Indian telephone "help line" for U.S. welfare recipients. Manisha Martin in Mumbai, India, is answering questions telephoned in by welfare recipients in New Jersey. Manisha's monthly $200 salary is five times the per capita income in India but less than that of the welfare recipients whose questions she answers.

countries enjoyed the deflating prices of manufactured goods when domestic blue-collar jobs migrated to the newly industrializing countries. Today, however, many of even the most skilled white-collar workers see their own jobs departing.

In the short run, the transfer of jobs abroad puts deflationary pressure on wages for people doing these jobs in the rich countries; in the long run, many jobs may disappear from the rich countries. For the populations of the rich countries, the crucial question is: Will the white-collar workers in the rich countries be able to add ever-higher value to services, and thus advance into jobs paying ever-higher incomes than they now enjoy? Or will they fall into lower-paying service-sector jobs, as most former manufacturing workers have? In other words, how many current or would-be data processors or computer programmers in the rich countries will capture positions as pharmaceutical chemists, lawyers, bank executives, or surgeons? How many, by contrast, will find jobs only flipping fast-food hamburgers or greeting customers at discount stores at minimum wages? We do not know the answers to these questions. As more and more jobs become mobile and the world becomes one labor market, each worker must compete with every other worker with his or her skills in the world. The only workers protected against the globalization of the labor market are those whose work requires personal contact, but much personal contact is already being replaced in many contexts by videoconferencing and other innovations in communications. As *Fortune* magazine, one of America's leading business periodicals, editorialized colloquially in May 2006, "The hard truth is, we ain't seen nothin' yet."

The early-nineteenth-century English economist David Ricardo wrote that "Nothing contributes so much to the prosperity and happiness of a country as high profits." Globalization, however, has broken down the historic relationship between corporate profits and national prosperity. Today the profits of multinational companies based in the developed economies can soar even as the home-country economies stagnate and workers' real incomes stay flat or fall. This is because the firms make profits in countries other than their homelands. The world's 40 biggest multinationals now employ on average 55 percent of their workforces in countries other than their homelands, and they earn 59 percent of their revenues abroad. If profits and executive pay soar while ordinary employees' wages stagnate and pensions and health benefits erode, workers might demand higher taxes on profits, restrictions on overseas investment, import barriers, or guaranteed employment. These actions, however, might be suicidal, because the companies would just move their headquarters to friendlier jurisdictions. Many have already done so, as some large companies and financial institutions have established their headquarters offshore.

The Geography of Foreign Direct Investment

The amount of FDI today is several times the amount of foreign aid being given around the world, and FDI has been much more successful in triggering economic growth. Developing countries used to seek aid from foreign governments and international lending institutions, but today they are more likely to court foreign companies to invest in their countries. Since the 1980s, global FDI has grown three times faster than world trade and four times faster than total world output, reaching

$1.7 trillion in 2008. This flow of FDI has greatly increased and significantly redistributed the world's productive capacity. No world agency tracks portfolio investment, but including portfolio investment would multiply the total amount of capital invested internationally.

Companies from an increasing number of countries have become active investors in other countries. Transnational corporations from developing countries, including Brazil, Thailand, South Korea, and Mexico, have reached around the world. Today, developing countries account for 19 percent of foreign direct investment in other countries. The United States is the most active recipient and sender of FDI, totaling $316 billion and $312 billion, respectively, but three sources—the United States, the combined European Union countries (see Chapter 13), and Japan—account for about 56 percent of FDI.

U.S. firms account for about 18.3 percent of global FDI, and the United States also has attracted the highest accumulation of FDI from other countries. By mid-2008, FDI in the United States totaled about $2.3 trillion, and U.S. FDI abroad totaled over $3.2 trillion. In addition, foreign investors owned $1.4 trillion worth of U.S. stocks and $900 billion worth of corporate bonds. These foreign investments and purchases testified to foreign confidence in the American economy and political stability, and they supported U.S. prosperity. Foreign firms carry on research, introduce new technology and management techniques, and change Americans' daily lives in the workplace as well as their consumption habits. Nevertheless, since 9/11, the United States has occasionally vetoed foreign investment in "sensitive" or "strategic" industries. There was a public uproar when a Dubai investment company tried to purchase a company that manages U.S. ports, and protests in 2005 stopped the Chinese oil company CNOOC from buying U.S.-based Unocal.

FDI in developing countries Three trends characterize the geography of FDI over the past several decades. First, the greatest share of global FDI—about 75 percent—is investment from one rich country to another. For example, U.S. firms have their greatest investments in Europe.

Second, FDI in developing countries has been geographically selective. The countries that have attracted the most investment are countries that have chosen the export-led method of economic growth. Their economies have boomed. Four Asian nations—Singapore, South Korea, Taiwan, and Hong Kong (reunited with China in 1997, but still often considered separately in international trade statistics)—attracted FDI and grew so fast during the 1980s that they earned the nickname The Four Tigers. These areas reached European levels of prosperity within a decade and quickly became sources of capital and technology for the next tier of Asian developing lands: Thailand, China, Malaysia, and Indonesia.

The areas that have not attracted FDI have been slower to develop. Africa has attracted only about 2 to 3 percent of total FDI, and almost all of that investment has been in extractive activities—oil and mining. Political instability has been a major factor discouraging international investors there.

A third characteristic of the pattern of investment is that the largest source countries of FDI provide a majority of the FDI for a cluster of countries that have become their economic satellites. Japanese investment dominates in East and Southeast Asia, U.S. investment in Latin America, and Western European investment in Eastern Europe and Central Asia. These clusters help explain world patterns of trade. For example, in recent years, the United States has had a trade deficit in electronic consumer goods with Thailand. Thai electronics factories are mostly owned by Japanese corporations. Thus, the consumer electronics that the United States imported from Thailand profited the Japanese corporations as well as the Thai economy. In a similar fashion, many of Europe's imports from Latin America are the products of U.S. corporations. This pattern of global investment carries political ramifications: Trade sanctions against any individual country with which the United States runs a large trade deficit (such as China) may not necessarily succeed in reducing total U.S. imports because the transnational corporations simply move the production of the lower-cost goods from the sanctioned country to another low-cost country and ship the goods to the United States from there. Thus, the U.S. trade deficit with the sanctioned country falls, but the deficit with another country rises.

Geographers watch capital flows carefully, because the pace of development within individual countries is increasingly determined by their ability to attract foreign capital. The world map of wealth continues to change kaleidoscopically.

Some observers have hypothesized broad political ramifications of the rise in global FDI. To attract private investment, governments must adopt policies that are very painful, such as cutting public funding of important services like education. Furthermore, no country can attract investment unless it protects the right to own private property. Arbitrary government actions—often typical of autocratic regimes—will isolate a country from international capital flows. For example, several major international corporations have decided against investing in Malaysia out of concern that Malaysia might try to regulate Internet access. International capital cannot be attracted without acceding to international rules, and local political leaders realize that the authority they exercise over their own people cannot be exercised over international investors. The process of compromising and accommodating investor's concerns is often called the *discipline of the market*. Therefore, the substitution of private investment for foreign aid may have a liberalizing and democratizing influence worldwide.

Critics contend that measures intended to attract FDI often hurt local populations. Many have been made landless and poor because wealthy international firms purchased land and resources once shared by local communities struggling for survival. In Uganda, for example, land that was once shared within traditional social groups is being sold out from under them, leaving farmers without farms. More generally, FDI is helpful for countries with a diversified economy, but those countries that are primary commodity exporters are hurt by FDI, where it extracts wealth without expanding economies. Since 1996, wealth has been transferring out of the developing world at a rapid pace. By 2009, economic trade was extracting $900 billion per year from poor countries in Africa, Asia, and Latin America and transferring it to developed economies, which captured that value.

The Globalization of Finance and Risk

The FDI that is redrawing the world economic map is only a fraction of the vast amounts of capital moving around the world. Multinational banking and telecommunications make it possible to monitor and trade instantaneously in national currencies, stocks, and bonds listed anywhere in the world. In 2008 over $2 trillion was traded daily on the world's foreign exchange markets—over 100 times the average daily merchandise trade. Over $45.5 trillion was tied up in stocks on the world's major stock exchanges at the end of 2008. These statistics are provided by the World Federation of Exchanges, but as noted previously, no international agency records ownership of all global assets, even by national source.

The revolution in global networked technology has brought with it electronic funds transfers (EFT). The same basic computer and telecommunications technology that allows consumers to purchase goods online allows investors and banks to transfer billions of dollars between markets almost instantly. These technologies allow investors to avoid risks to their portfolios by rapidly buying and selling investment on markets around the world. Investors may transfer funds several times a day, often moving money between different types of investment. In some ways, EFT detaches investors from the geography of banks and stock exchanges, allowing an investor to be remote and his or her money highly mobile. In other ways, these changes have created new geographies. Today, there is always a bank or stock exchange open somewhere around the world. Profit chasing moves with the sun. As banks and exchanges close in North America, investors may remain active in East Asian trading, followed by Middle East exchanges, on to Europe, and then back to New York.

Nations are expected to regulate their own stock exchanges, and the home countries of global banks are responsible for supervising the banks' operations, but countries do not agree on how to do these things. Many giant banks have located their international headquarters in places renowned for lenient banking regulations, a practice known as **offshore banking.** Prior to the financial crisis in 2008, tiny Monaco (population 30,000) counted more than 340,000 bank accounts holding almost $50 billion. The 260-square-kilometer (100-square-mile) British Caribbean colony of the Cayman

Islands (population 36,000) is one of the world's greatest financial centers, with assets of $700 billion held in over 600 chartered banks. By comparison, the state of New York had only 154 chartered banks. The Cayman Islands, however, has almost no vaults, tellers, security guards, or even bank buildings. Assets are held electronically in computers. Many banks consist of one-room offices, and the confidentiality of their transactions is protected by local laws. For example, international agencies cannot trace the profits of the international drug trade or terrorist funding. Renewed scrutiny resulting from the financial crisis may change the geography of what is known as offshore banking. In 2009, the Obama administration made some progress in opening the long-secret Swiss banking system, where the world's wealthiest have hidden enormous sums of money to avoid paying taxes or to hide stolen profits.

Electronic funds transfers and hidden bank accounts are a sign of investors' fears that financial risks will wipe out their assets. Financial trading is notoriously nervous about geopolitical events that might sap the value from capital investments. But when investors quickly flee a market, it often makes things worse. War in the Middle East raises worries of oil disruptions and rising fuel costs that eat into profits. Political instability in developing countries raises worries that currency values will fall. In 1994, investors caused the Mexican peso to crash by rapidly selling their holdings in the wake of

RAPID CHANGE

The Financial Crisis

The 2000s was not a great financial decade in the United States. In 2000, the U.S. economy was just coming back from the collapse of technology stocks in the late 1990s. Then the terrorist attacks of 9/11 slowed the economy and raised the specter of a war that would increase fuel prices. Yet, as the economy emerged from the first half of the decade, its fundamentals were not terrible. Inflation was under control, unemployment was moderate, and the stock market was booming again. The Dow Jones Industrial Average, an index of the value of many Americans' savings, peaked in October 2007 at 14,100. Then, a financial crisis that was the worst since the Great Depression of 1929 began. The Dow plummeted, losing 54 percent of its value during the 17 months before stocks started rising in value again. A dollar invested in October 2007 was worth about 54 cents in March 2009. Unemployment increased, reaching over 10 percent in 2009, as companies laid off employees they could not afford to pay.

The financial crisis that began in 2007 had a lot to do with how people and especially banks borrowed and invested money. As we saw in Chapter 10, few in the United States could foresee that the rapid growth of subprime mortgages during the 2000s would lead to loans that would go sour. It was even more surprising that so many banks had made enormous risky loans on the gamble that those mortgages were safe. The problem was that the large banks had invested in complex products called *mortgaged-backed securities* that bundled together thousands of individual mortgages, including risky mortgages. Normally, mortgage borrowers reliably pay back their loans with interest; banks were depending on those loan payments so they could loan out more money to businesses. As an unexpected increase in mortgage defaults occurred in 2007 and 2008, investors dumped any assets that included mortgage-backed securities, driving their value to zero because no one would buy them. Banks were stuck with bad mortgage loans and, in turn, stopped lending money, leading to a credit crunch.

Without access to loans, businesses could not pay employees or order materials to keep working. The credit crunch also ended the housing sales boom, and house prices dropped quickly. The homes that reverted to the banks by default were now worth less than the original amount loaned, a further loss for the banks.

The implications of the financial crisis were different on Wall Street and Main Street, the nicknames given to the banks and the rest of America. On Wall Street, the market panic began in June 2007, when Bear Stearns suffered losses from subprime mortgages totaling $1.9 billion. The government stepped in to save some banks, and Congress approved a controversial $700 billion bailout program to keep banks from collapsing. Taxpayers were ultimately Wall Street's lifeline. By 2010, however, the banks were reporting profits and paying executives large bonuses, yet they were still reluctant to loan money and fought against new regulations meant to prevent a repeat of the crisis.

The rage against Wall Street's gross incompetence was felt on Main Street, where the financial crisis had triggered a full-scale recession, meaning the economy shrank. Many small businesses were hurt by the lack of credit, and the recession dashed sales. Rising unemployment and stagnant wages combined with rising fuel and health care costs, effectively decreasing household incomes for average Americans. Individuals and institutions lost money that they had invested in banks, as well as in stocks, forcing some retirees to go back to work. The recession ultimately led to the bankruptcy of GM and the sell-off of Chrysler as consumers looked for ways to save money. Many smaller companies had to close permanently, and most struggled to survive. The effect of the financial crisis and recession was also felt by local governments that had to cut services in the face of falling tax revenues and institutions such as universities, whose students struggled to pay for school. But it was felt most sharply by the more than 8 million Americans who lost their jobs during this period.

violence in Chiapas and a political assassination. Similar risk-aversion appears to have triggered the collapse of Southeast Asian economies in 1997, which almost led to a global financial meltdown. Risk increases with the complexity of investment schemes, as witnessed in the U.S. financial meltdown in 2008.

Today, Japan is one of the world's principal sources of capital, so the regulations governing the Tokyo stock market affect investments and jobs throughout the world. The Arab oil states are another source of capital, so a costly Mideast war raises the interest rate on U.S. college tuition loans. Bank policies in Frankfurt, Germany, and Johannesburg, South Africa, affect the checking account charges of U.S. banks. Financial manipulations in Rio de Janeiro, Brazil; Bangkok, Thailand; and Mumbai, India, affect the security of pensions of many U.S. elderly people, which have been invested abroad. And the interest rates set at the Federal Reserve Bank in Minneapolis affect business taxes in Indonesia. The explanations of the "where" and "why there" of local economic affairs lie in global finance.

International Regulation of the Global Economy

The rules regulating business vary so widely from nation to nation that international trade requires cooperation or even political regulation. Under the auspices of the United Nations, the General Agreement on Tariffs and Trade (GATT) was signed in 1947, but it was superseded by the creation of the World Trade Organization (WTO) in 1995. The WTO establishes panels to investigate trade disputes, and countries that are found to have broken WTO rules must change their own policies or laws or else offer compensation; if they do not, they can face trade sanctions.

In mid-2008, 153 countries belonged to the WTO. Russia and Iran still stood outside, accused of avoiding the WTO's free-market principles. Membership in the WTO brings nations a significant advantage. Each member receives from every other member the terms of trade that that nation grants its most favored partners—called *most favored nation* status.

Negotiations aimed at continuing reductions in global trade barriers were optimistically begun in Doha, Qatar, in November 2001. These negotiations, however, collapsed in failure in 2006 and again in 2008. Poorer countries would not further open their markets to goods and services from rich countries unless the rich reduced the subsidies to their own agriculture, which the rich refused to do. The U.S. farm bill, described in Chapter 8, is a major source of the problem because it economically closes one of the world's largest markets to food grown from developing countries. The collapse of the "Doha Round" may cripple the usefulness of the WTO and lead to a proliferation of bilateral agreements.

The financial crisis that began in 2007 brought attention to the need to better regulate financial services. But investment banking is one of the most thoroughly globalized forms of trade and also one of the largest markets in the world. Who would manage such a titan? The advantages that each country sees in its own laws would need to be aligned through mutual agreement. Prior to 2008, there was little hope that such cooperation was possible. As governments in Europe and Asia realized their own exposure to the U.S. financial crisis, it became clear that coordination was necessary to at least stop the crisis from expanding (Figure 12-35).

In October 2008, the United States, the United Kingdom, the European Union, Japan, and several other major economies made a coordinated reduction of their basic lending rate, a move meant to improve the circulation of money in the world's markets. They also moved together in bailing out their banks. The moves were more than an effort to restore confidence; they were also an attempt to reduce the chance that nervous investors would flee one country for another and create new problems. A month later, the leaders of the G-20 met in Washington, D.C., to coordinate further policies towards resolving the crisis. The G-20 has replaced the G-8 as the primary forum for coordinating the national policies of the world's largest economies. The leaders have agreed to coordinated action to regulate the sorts of investing practices and weak oversight that led to the crisis, including possible changes to the International Monetary Fund.

Other international organizations oversee specific aspects of international commerce, such as the International Organization of Securities Commissions mentioned earlier. The London-based International Accounting Standards Board works to standardize corporate accounting procedures, and the International Labour Organization (ILO) defines labor standards with the goal of improving working conditions. The Geneva-based International Standards Organization (ISO) determines and

Figure 12-35 Northern Rock. This British bank was another casualty of exposure to the U.S. subprime mortgage crisis. Here, depositors lined-up to withdraw their money from the bank. It was nationalized in 2008 by the United Kingdom to keep it from collapsing until it could be privatized.

The Rise of BRIC and the G-20

Once upon a time, the leaders of the world's largest economies could fit around a small table. The seven leaders of France, the United States, the United Kingdom, Germany, Italy, Canada, and Japan began meeting annually in the mid-1970s to discuss the oil crisis and recession at that time. The G-7 continued to meet annually to discuss economic growth, trade, and finance, as well as supporting democracy and resolving conflicts. In 1998, they included Russia in their annual summit meetings, becoming the G-8.

But the organization has had to change to adjust to the realities of a world in which they are no longer the only economic powers. New and vibrant economies have emerged, and by 1999, the G-20 formed to discuss the global financial crisis of the late 1990s. At first the G-20 was just a meeting of finance ministers from Argentina, Australia, Brazil, Canada, China, France, Germany, India, Indonesia, Italy, Japan, Mexico, Russia, Saudi Arabia, South Africa, South Korea, Turkey, the United Kingdom, the United States, and the European Union.

The global financial crisis that began in 2007 brought together the heads of government from the G-20 countries at a 2008 summit in Washington, D.C. They met again in Pittsburgh in September 2009, and at that point, they also decided that the G-20 would replace the G-8 as the primary meeting of international economic cooperation. Together, the G-20 represents about 90 percent of world economic output, 80 percent of world trade, and about two-thirds of world population.

The G-8 will continue to serve as the primary forum on international security and political issues. But it is worth considering that their power may be eclipsed by the world's fastest-growing economies that are also among the world's largest countries. Brazil, Russia, India, and China—the so-called BRIC—are already a part of the G-20, and Russia also sits on the G-8. Although the BRIC are not organized in any formal way, they share certain concerns that they will bring to bear on discussion in both the G-20 economic meetings and the G-8 political summits. As rapidly growing economies, they have compelling interest in gaining access to world markets. They also form the largest challengers to the traditional leaders in world trade, especially the European Union, the United States, and Japan. Being among the largest economies in the world, they are also aware that they are a desirable market for their partners in the G-20.

Perhaps the emergence of the G-20 and the BRIC are a reminder that the world's historical powers in Europe and the United States are being challenged for influence in the world of the twenty-first century. The geopolitical position of the United States as a world leader was obtained through economic dominance and insured by its military might. The world of the current century looks more and more like one in which the United States will be joined by other countries with more vibrant economies. As the editor of Newsweek International, Fareed Zakaria, described it, this is "not about the decline of America but rather the rise of everyone else."

Leaders of the world's largest economies. The leaders of the world's largest economies met at the G-20 summit in Washington, D.C., in 2008. The leaders met to deal with the financial crisis.

publishes standards for manufacture of items in world trade. Virtually all items in international trade must conform to ISO standards, so items manufactured even for nations' domestic markets do, too. For example, the specifications of your tennis racket are covered by eight pages in ISO 11415:1995. A new series, ISO 14000, establishes environmental standards for manufacturing processes and products. The hope is that these standards will reduce global pollution and environmental degradation.

Left to follow their own interests, economic players have promoted economic development that has raised the standard of living in some once very poor countries. But the benefits of economic development have been geographically quite limited. Continued growth of developed economies currently exploits resources in poor countries. Investment has worsened conditions in many countries where value is being extracted but not returned. If globalization is to be truly successful—a global economy that produces universal benefits—it will require better coordination and sufficient regulation. Regulations are needed to end the most reckless investment practices that destroy poor economies and environments. Better international coordination might also lessen the risks that cause investors to flee countries. Yet many of the problems facing economic development in rich and poor countries are not wholly economic. In the next chapter we describe how humanity is seeking to address the global problems, including the problem of poverty.

Chapter Review

Summary

Every country devises a program to build and regulate its national economy and to distribute economic well-being across the national territory. Gross national income, and gross domestic product are two ways of measuring wealth, although each undervalues subsistence production and unpaid labor. Maps of these values reveal great differences in global incomes. Welfare may vary greatly within countries, and statistics examined per capita and through time may reveal a relationship between population growth and economic growth.

Sectoral evolution reveals that some countries are preindustrial, some industrial, and some postindustrial. Various sectors' contribution to national product is not the same as their share of employment.

The world distribution of wealth is not the same as the distribution of raw material resources. Downstream secondary- and tertiary-sector activities add value and produce wealth. Today's map of global manufacturing is a result of historical developments modified by new balances among relevant locational determinants, including capital and technology, as well as labor and raw materials. Several locational determinants are themselves mobile, so manufacturing will probably be redistributed continually.

Each nation's political economy usually reflects other aspects of geographic situation. Nations usually try to knit and develop the national territory and to maintain an equal standard of living throughout it. Transportation networks allow regional specialization and trade, and their improvement releases national resources for production. Communication infrastructures both reflect and boost economic development as well as, potentially, political liberty.

Import-substitution policies protect national growth behind tariff walls, whereas export-led policies encourage participation in international trade as a source of capital and as markets. Both methods entail both economic risks and political risks.

International trade has multiplied faster than total world production. Capital has been invested internationally, and world markets for primary products, foods, and minerals have evolved. Today manufacturing industries are globalizing, and flows of FDI are redistributing world productive capacity. A global capital market has formed, and international trade in tertiary-sector services is growing. The World Trade Organization attempts to regulate international commerce, which generally means lowering barriers to free trade. The global financial crisis has triggered the need for greater coordination and regulation of banking practices.

Key Terms

capitalist system p. 488
economic development p. 478
economies of scale p. 498
ecotourism p. 501
export-oriented method p. 498
foreign direct investment (FDI)
    p. 504
gross domestic product (GDP) p. 468
gross national income (GNI) p. 468

import-substitution method p. 498
industrialized economy p. 468
industrial society p. 473
infant industry p. 498
internal frontier p. 490
laissez-faire capitalism p. 488
market-oriented manufacturing
    p. 482
material-oriented manufacturing
    p. 481
multinational corporation p. 504

nationalized p. 488
offshore banking p. 508
offshore outsourcing p. 485
opportunity cost p. 493
outsourcing p. 484
political economy p. 487
postindustrial society p. 473
preindustrial society p. 473
sectoral evolution p. 473
sustainable development p. 470
transnational corporation p. 504

Questions for Review and Discussion

1. Define and explain the three sectors of an economy. What is sectoral evolution?

2. What locational determinants for manufacturing did Alfred Weber identify? What new determinants are important today?

3. Differentiate laissez-faire capitalism from state-directed capitalism.

4. Why is transportation important to economic development?

5. What are the geographies of women's and men's economic activity? How do they vary between richer and poorer countries?

6. Aside from GNI or GDP, what are some alternative noneconomic measures of welfare in different places?

Thinking Geographically

1. Americans have difficulty understanding privatization because Americans live in a country in which most activities are and always have been private. Can you, however, imagine selling the national parks to the Disney Corporation? What would be the advantages and disadvantages of doing so? Would the corporation run the parks better?

2. What physical and cultural attributes are most conducive to a nation's economic development?

3. What industries in your town have relocated offshore in recent years?

4. Latin America is rich in natural resources, and yet it has long suffered economic and political problems. Many of these problems stem from class and ethnic structures and from political history. Why?

5. At the opening of international currency markets on June 7, 2010, the following values were quoted as the value of each currency, in U.S. dollars:

Australian dollar	0.8153
Brazilian real	0.5316
Euro	1.1942
Indonesian rupiah	0.00011
Japanese yen	0.0109
South African rand	0.1279

Check these values in today's newspaper. Calculate the percentage each has risen or fallen against the dollar or against one another. What flows of goods, of investment, or of tourists should these fluctuations have triggered?

6. Did the United States develop in a sustainable way, or at a sustainable rate, between 1870 and 1910? Does it develop in a sustainable way, or at a sustainable rate, today?

7. Could foreign ownership of businesses hurt the economy of a developing country?

8. What countries have frontiers? What barriers exist that are slowing development of those areas? How might these barriers be overcome? Investigate and compare the real development potential of the frontiers of Canada, Russia, and a South American or African country.

9. How does the quotation from Peter Drucker at the opening of this chapter (p. 470) echo that of Clifford Geertz in Chapter 6 (p. 210)?

10. Find the latest edition of the *Human Development Report* in your library and read the authors' justification for what they have done. If you disagree with their methods, find any one of the other statistical yearbooks published by the United Nations or The World Bank, browse among the quantified descriptions of life in each country, and devise your own overall "ranking of the quality of life" among the countries.

MapMaster™

Protestors clash with police in Copenhagen at world climate talks. More than 100 people protesting for greater controls on greenhouse gas emissions were arrested by Copenhagen police. Police also used tear gas to control the crowd, which was generally peaceful.

13

Global Challenges and the Scale of Response

For two weeks in December 2009, representatives of the world's countries met in Copenhagen to discuss the problem of climate change. Outside, tens of thousands of protestors took to the streets, reminding delegates that the whole world was watching. The protestors represented different groups, most of them environmental activists, others advocates of the poor, many from Europe and a few from farther away. Their banners and slogans, as well as speeches by nongovernmental organization leaders, expressed frustration over the inability of the world's leaders to come to agreement on ways to deal with a problem of enormous global significance that has been recognized for decades but continues to evade effective action.

The demonstrations in Copenhagen were only the most immediate in a long list of attempts to increase awareness of the problem and urge action. In October 2009, the government ministers of The Maldives put on diving gear and held an underwater cabinet meeting as a way of dramatizing the plight faced by that low-lying island nation and others threatened by rising sea level. In New York City, a group painted a line around parts of the city, marking an elevation of 3 meters (10 feet) above sea level to symbolize the degree to which future flooding might inundate the city.

Global-scale problems may demand global-scale solutions, but the diversity of economic, cultural, and physical conditions around the world is a continuing and daunting challenge for all. Increasing economic and physical linkages among diverse parts of the world generates a need for concerted approaches, but at the same time these linkages reveal our differences and so they make agreement difficult if not impossible. Cooperation among governments is one response, but meetings of world leaders tend to be undemocratic and overly cautious. Protests that raise awareness are another response by a burgeoning global civil society made up of networks of people united by their common concerns but with little control over public policy. Are these the best ways to deal with global problems? In this chapter, we explore some of the challenges facing the global community today and some opportunities for addressing them.

515

The world today is challenged by environmental, political, and economic change. Pollution is threatening public health. War endangers innocent civilians. Financial crises have left countries bankrupt and families destitute. The causes of the problems are not all new, but the rapid pace at which some challenges have become truly global in their reach is nothing short of breathtaking. Each of these challenges produces novel geographic problems; some are unevenly distributed around the world and some have a continental or even global impact. The previous chapters in this book have noted some of the rapid changes in our world today, as well as the connections between global trends and local places. This chapter goes further, examining how humanity is organizing itself to confront four rapidly changing global challenges.

First, we examine the challenge of protecting the global environment. Responding to the problems of environmental change requires that problems be addressed on several scales at once. Locally, the problems of soil erosion, pollution, and rising sea levels demand energetic responses from local communities and governments, sometimes requiring changes with painful economic outcomes. Globally, problems such as greenhouse gases and their effects do not respect national boundaries, so building consensus on how to manage them calls for an international effort.

The second section of this chapter discusses whether state sovereignty, a government's right to rule over its territory and people, is compatible with international political cooperation, especially in pursuit of peace and human rights. The United Nations is one organization that attempts to bridge the gap between the interests of individual states and the need for cooperation on complex political, economic, and social issues. Yet cooperation among states has not ensured that the rights of individuals are respected. Protecting human rights necessarily means limiting the absolute sovereignty of states.

Third, this chapter examines instances in which neighboring countries are joining together at the regional scale to tackle common problems. The European Union is particularly significant because the participants have great economic and military power and because in each case the members have surrendered considerable autonomy to a central body and common policies. Other international organizations coordinate policies among countries, although their goals may be limited.

The fourth section examines the difficult task of improving the quality of life of all people, commonly known as **human development.** As we have seen throughout this book, the world's many regions and localities must respond to the challenges presented by a changing environment, the legacies of colonialism, and the uneven distribution of economic opportunity. Improving human development around the globe requires an understanding of forces operating at different scales.

The chapter concludes by emphasizing the importance of a geographic approach to understanding contemporary global problems.

The scale and mode of response   Humans are creative problem solvers. One of the most powerful tools we have to solve problems is our capacity to find new forms of cooperation. Like our early ancestors, neighbors are still held together by the fact that living in the same place means sharing the same problems. On another level, states emerged in the seventeenth century as an effective way to promote security and economic development. The exploration of space, the most technologically advanced activity ever, is today an international effort. Each of these modes of organization has required human cooperation at different scales, whether local, national, regional, or global.

In this chapter, we consider some of the ways that humans currently cooperate in responding to global challenges. First and foremost, people must trust each other in order to cooperate. The many geographic differences in cultural, political, and economic systems described in previous chapters suggest that building trust is not a simple task. We will focus primarily on the geographic reasons why the scale of response is not truly global in each case.

In many cases, the mode of response is stuck at the level of the nation-state. As described in Chapter 11, states possess sovereign power to control their own affairs, and they have long guarded those powers in the face of new challenges. States have found reason to cooperate on some issues, however. Most forms of political cooperation are based on formal agreements, such as treaties or conventions, between countries. **Bilateral agreements** can be made between two countries on any issue of shared concern. **Multilateral agreements** are made between any number of countries, often on the basis of regional proximity, economic advantage, or mutual security. In some cases, these agreements create new organizations that allow countries to coordinate their efforts. Most organizations are based on the idea of **intergovernmentalism,** by which the participating states must agree to the policy. A different form of organizational cooperation is **supranationalism,** which means member states agree to give up some of their individual authority to the larger organizational body.

Economic cooperation, including trade, is one area where states have found it advantageous to agree on limiting sovereign power. Lowering barriers to trade between countries has been a long-desired goal because it is thought that free trade produces greater advantages for more people. Yet some countries became wealthy on trade relations that advantage them and disadvantage other countries. Some developing countries, meanwhile, fear free trade because of this. As we explore later in this chapter, these different national policies have produced regions with different forms of economic cooperation.

A **free-trade area** has no internal tariffs, but its members are free to set their own tariffs on trade with the rest of the world. A **customs union** enforces a common external tariff. A **common market** is a still closer association of states; it is a customs union, and in addition, common laws create similar conditions of production within all members.

States are not the only entities that can cooperate to confront global challenges. In fact, states are responsible for many of the problems in the world today. The loudest demands for greater human rights, whether in China, Iran, or the Western democracies, are not from other states but from citizens themselves. Networks of individuals who are not controlled by states or businesses are termed **civil society**. In recent years, networks of activists have employed the technologies of global communications to promote a **global civil society** to debate complex issues from climate change to human rights to safe drinking water. Some successful **nongovernmental organizations** have emerged from these networks and now press governments to respond to the interests of citizens.

Protecting the Global Environment

Just as the world has become more integrated and interdependent in recent decades, so has it become apparent that we share common concerns about environmental quality. While local environmental quality issues such as urban air pollution remain important—and are growing in importance in industrializing regions—global environmental issues have risen in importance. International governmental and nongovernmental organizations are increasingly involved in finding ways to manage concerns about the environment in a changing world.

Changing patterns of economic activities at the global scale mean changing patterns of resource consumption and pollution outputs. Whereas most of the world's industrial activity and most output of industrial pollutants are still concentrated in the rich countries, a significant shift in pollution output to the poorer countries is occurring (Figure 13-1). Two factors are driving this shift. First, heavy manufacturing, and the pollution that accompanies it, is shifting from rich countries to poor ones. Second, populations are growing faster in poor countries than in rich ones, so consumer-generated pollution such as sewage, solid waste, and automobile emissions are growing rapidly in poor countries. Unfortunately, the poor countries are less able to afford pollution-control systems, so their populations increasingly suffer the negative effects of living with pollution (Figure 13-2).

Energy Consumption

In general, poor countries are using rapidly rising amounts of natural resources, whereas the use of natural resources in the rich countries is growing slowly. Rich countries were provoked to decrease their natural resource use and to increase the energy efficiency of their

(a)

(b)

Figure 13-1 Changing locations of heavy industry.
(a) An abandoned steel mill in Volkingen, Germany, and
(b) a steel mill in Benxi, China. The locus of heavy industry in the 1950s and 1960s was in Europe, North America, and Japan, but today much of this industry has shifted to developing nations, and along with it, much energy consumption and pollutant emissions.

Figure 13-2 Air pollution in China. Air pollution in Beijing, China. Rapid economic growth has increased the use of automobiles, contributing to both air pollution and traffic congestion.

economies by the rise in energy prices in the 1970s. One obvious example of this increase in energy efficiency is the change in automobile design between the early 1970s and the 1990s. Modern autos are much more fuel efficient: A typical mid-1990s five-passenger sedan traveled at least 50 percent farther on a liter of gasoline than a comparable-sized car built 20 years earlier. At the same time, in many parts of the world, rising incomes have led people to buy larger vehicles such as minivans and sport-utility vehicles, which are less efficient than the sedans and station wagons that were more common earlier. This has caused a reduction in our overall vehicle fuel efficiency that persists, even though rising gasoline prices have slowed sales of inefficient vehicles.

Air-pollution-control regulations are another factor in this increased efficiency. Using the auto as an example again, one of the easiest ways to reduce emissions is to improve the efficiency of fuel use. The replacement of carburetors and mechanical ignition points with computer-controlled fuel injection and electronic ignition both reduces pollution output and improves fuel efficiency. Hybrid power systems that combine internal combustion engines with battery-stored electric power further improve energy efficiency. Similar changes that have occurred throughout all industries have reduced both the amount of resources used and the pollution generated.

At the same time that the rich countries increased their efficiency of fossil-fuel use, many also experienced a rapid economic transition away from energy- and pollution-intensive heavy industries toward light-industry and service-based economies. Steel manufacturing and related heavy-equipment industries were especially hard hit by the oil price increases of the 1970s. The steel industry in the United States, Canada, and much of Europe was aging, and

many mills used less efficient, older technologies. These mills also faced increased competition from other regions. Developing countries built their own steel industries, and global production capacity was greater than demand. The domestic automobile industry in the United States and Canada faced increasing competition from Japan. When fuel prices rose, consumers demanded smaller foreign-made cars, not the large gas-guzzlers that the North American plants were producing. Increases in energy costs caused a general slowdown in manufacturing. The combined effects of these factors contributed to the decline of heavy industry in the northeastern United States and adjacent Canadian cities, in northern Britain, and to a lesser extent, in Germany. Along with this came a significant drop in fossil-fuel use, which helped to reduce pollution emissions in these regions (Figure 13-3).

The decline of the traditional heavy industrial regions of the developed world was accompanied by the growth of manufacturing in rapidly industrializing regions of East Asia and Latin America. Much of this new manufacturing, however, is labor and capital intensive rather than energy intensive, so it has made a smaller contribution to increased energy consumption in these regions than would otherwise have occurred. The increase in wealth that this manufacturing brings, especially in urban regions, is much more significant. However, this added wealth raises demand for energy-consuming machines, especially automobiles. The number of autos in use in China, Malaysia, Brazil, and other nations is rising with rapid economic growth rates, so consumer-driven petroleum use and air pollution also rise.

Energy Efficiency Trends

It takes energy to generate income. Some ways of generating income take more energy than others, however. Making steel uses large amounts of fuel per dollar earned, whereas making televisions requires much less. Typically, manufacturing takes much more energy than providing services. Also, some ways of making a given product are more efficient than others. Modern factories or office buildings usually use much less energy per unit of output or office space than old ones. We can see changes in energy efficiency by examining the relationship between how much economic activity changes in relation to changes in energy use.

The link between economic and environmental approaches over recent decades can be seen in trends of economic efficiency of energy use. **Energy intensity,** the amount of energy used to produce a dollar of gross domestic product (GDP), is shown for a few representative countries for the period 1980–2007 in Table 13-1. In this period, the purchasing power parity (PPP)-based GDP of the United States grew at an average rate of 3.0 percent per year (adjusted for inflation), yet energy use grew at only 1.0 percent per year. In 2007, a unit of

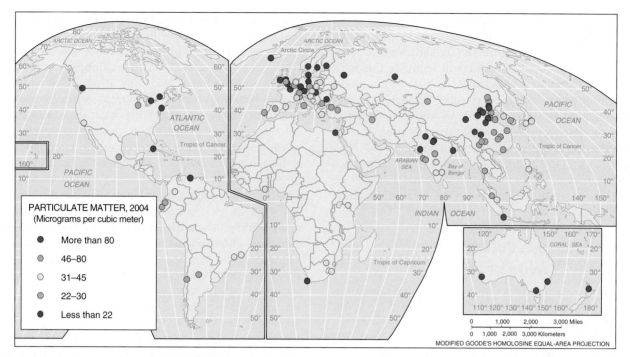

Figure 13-3 Particulate matter concentrations in urban areas. This map shows concentrations of particulate matter in select urban areas. Particulate matter is a major cause of health problems such as asthma, chronic bronchitis, and acute respiratory problems.

TABLE 13-1 Change in energy intensity for selected countries, 1980–2007 (PPP; constant dollars)

Country	GDP Growth Rate (PPP)	Energy Growth Rate	Percentage Change in Energy Intensity
China	10.0	5.7	−65.5
United States	3.0	1.0	−41.8
Canada	2.8	1.3	−32.6
France	2.1	1.1	−23.8
Kenya	4.0	3.0	−23.6
Japan	2.3	1.5	−19.9
Tanzania	4.2	3.8	−9.0
South Korea	6.7	6.5	−4.4
India	6.1	5.9	−3.4
Mexico	2.6	2.7	0.6
Indonesia	5.0	5.5	11.1
Malaysia	6.1	6.7	15.7
Turkey	4.6	5.4	24.7
Thailand	5.9	7.8	63.4
Vietnam	5.6	7.9	79.6

Source: Energy Information Administration.

energy use produced about 72 percent more income in the United States than it did in 1980. Most of the wealthy industrial powers experienced a similar pattern. Income grew faster than energy use. In contrast, in many poorer countries, energy use grew faster than income. Thailand, for example, experienced impressive economic growth, averaging 5.9 percent per year (adjusted for inflation), but energy use grew at an average of 7.8 percent per year. In China, economic growth averaging 10.0 percent per

year dramatically exceeded growth in energy use. Individual countries vary considerably, but the general trend over that time period is for faster growth of energy use in the poorer world than in the wealthy world. However, because the poorer world uses much less energy overall than does the wealthy world, its impact on global patterns is less significant. On average, the world is using significantly less energy to generate a dollar of GDP than it did a few decades ago.

Energy use is an imperfect measure of pollution, but it is a useful one. In general, the more fossil fuels we use, the more pollution we produce. Certainly this is true for carbon dioxide emissions; we will return to this issue later in the context of attempts to negotiate a global warming treaty. For emissions other than CO_2, the correspondence between economic growth and pollution is not so close, because pollution depends a lot on what sources the energy comes from and how it is used. Nevertheless, to the extent that energy use can be equated with pollution output, both energy use and pollution are increasing rapidly in the poor world, especially in those countries with high economic growth rates.

Video
Warming Up
in Mongolia

Development, Pollution, and the Quality of Life

Both monetary wealth and environmental conditions contribute to our quality of life. They contribute in different ways, however. Monetary income provides us with the ability to purchase sufficient quantities of good-quality food and electric or gas stoves for cooking, routine medical checkups, and safe water delivered to our homes. It also makes available automobiles, telephones, computers, and large, warm houses and vacations at the beach. Good environmental quality gives us clean air to breathe, fish that are free from disease-carrying bacteria and toxins, and soil that produces good crops without excessive inputs of fertilizer. It also gives us diversity of wildlife to observe and enjoy, parks for relaxing, and the satisfaction of knowing that future generations will have at least some of the same opportunities to enjoy these things as we do.

Economic activity and environmental quality both contribute to make life materially and observably better in ways that virtually everyone agrees are important, especially health-related factors such as occurrence of disease, infant mortality, or longevity. Wealth helps prevent disease by paying for educational expenses and salaries of medical personnel, as well as the offices, laboratories, medicines, and supplies they need to do their jobs well. A clean environment also helps to prevent diseases such as chronic lung problems that occur in areas with bad air pollution. Wealth helps prevent infant mortality and improves longevity by making possible balanced diets that include sufficient protein and fresh vegetables, regardless of season. A clean environment that yields drinking water that is free from chemical and biological contaminants also helps reduce infant mortality and lengthen lifespans.

While everyone needs good health and sufficient food supplies, the value of other things may vary considerably from one individual to another, or one culture to another. Clean water in a lake may be especially important to someone who likes to swim in lakes or who likes to eat fish. But for someone who spends every day in a city and enjoys indoor urban activities for entertainment, clean lakes may not be very important.

Wealth and environmental quality provide amenities that are valued differently by different cultures. Wealth brings us televisions, on which we can watch comedy or sports or the news. Such entertainment is a central part of some cultures today, but it would be hard to argue objectively that this makes us better people or that it makes our lives materially better than alternative activities such as conversation, participant sports, or reading. A wilderness park can be an exhilarating place to hike, learn, or simply contemplate the beauty of nature, but who is to say that this is a more meaningful experience than attending an opera? The value of many of the goods we produce through extracting resources and polluting the environment depends solely on individual or cultural attitudes.

Balancing development and environmental preservation If there were simple and effective ways to increase both wealth and environmental quality simultaneously, then the process of development would be a straightforward one. Many of the paths to greater monetary wealth, however, actually reduce environmental quality, and improving environmental quality often has significant costs.

For example, in the wealthy countries of North America and Western Europe, industrialization generated a standard of living and level of technology that were high enough so that people could give cleaner air a high priority. These countries could afford to make the needed investments in pollution control. Industrial production created the pollution, as consumers bought the things that were made in the smoke-belching factories, but production also provided the wealth for eliminating the pollution.

In many poorer countries, the situation is quite different. The contrast in wealth between rich and poor countries is enormous. People in poor countries are anxious to raise their standards of living, and rapid industrial development is seen as the quickest way to gain that wealth. Governments thus channel any available funds toward industrial development and actively seek foreign investment in new plants. Environmental protection may be a concern, but it is certainly a secondary one, so any environmental regulations that might slow development are few and weak.

The industrial growth that is taking place in poorer countries is heavily concentrated in urban areas, where the physical, political, and economic infrastructure necessary to support it is most available. Seeking employment opportunities, people migrate to these cities from rural areas, thus swelling urban populations. The resulting urban growth is rapid and uncontrolled, and governments find it difficult to provide the basic services that urban dwellers in rich countries expect, such as safe drinking water, sewage collection and treatment, and solid-waste removal.

The result is that some of the worst air and water pollution in the world occurs not in the rich countries,

Water Privatization

As the population increases around the world, and as economic growth stimulates new demands for water, availability problems are increasing. Water supply problems are also worsened by pollution, which can make otherwise useful water unacceptable. Water supply problems—whether issues of quality or quantity or both—do have technological solutions. In most of the world, water availability is strongly seasonal: there is a time of year when rivers have high flow and a time when flow is low. But human water use is less variable, so we often need to store water from the wet season in order to make it available in the dry season. We accomplish this by building reservoirs and the pipe or canal systems that deliver water from the reservoirs to where it is used.

Fortunately, water, unlike oil or coal, doesn't lose its fundamental properties when it is used. The addition of pollutants that takes place when water goes through industrial or other uses can be reversed. We can increase water supply by treating polluted water to remove unhealthful substances so that it can be used again. Reuse of treated water in irrigation is common in some areas, and in the future, we are likely to see increasing use of treated wastewater for drinking purposes. In a sense, we already do this, because downstream cities use water from rivers that carry treated wastewater from upstream. In some areas, total water use exceeds total renewable water supply (see Figure 9-24).

Increasing water supplies by building reservoirs or treatment systems requires money. Typically, the construction of water supply or treatment systems is undertaken by government agencies because water is something upon which everyone depends, and governments usually have the necessary financial resources. In addition, in most societies, water is commonly owned, unlike minerals or manufactured goods, which are usually the property of miners or manufacturers before they are sold in a market.

Many of the world's governments, however, do not have sufficient capital to build new water systems. As a result, they are turning to private corporations to make the investments needed. This is especially true in poor countries, but governments in many wealthy regions are also finding it desirable to outsource, or **privatize**, construction and operation of water systems to private companies. These companies are motivated by the prospect of profits earned by selling water or water-related services (such as pipeline operation and maintenance) to water users. In other words, a resource that many consider to be publicly owned, that should be provided to all at minimal cost, is being privatized and sold for a profit.

Water privatization is going on all around the world. The key players are a handful of large multinational corporations, mainly owned in Europe or North America. They bid for, and win, contracts with government agencies to construct water systems and operate them, charging water users for the service. In many cases, the fees these companies collect are substantial—and greater than those that water users paid before privatization. It is inevitable that wealthier people will find it easier to pay for water than poor people, and so privatization leads to a situation where rich people's water needs are met while those of the poor are not.

In Bolivia, water privatization has been especially controversial, and the dispute over water contributed to a major political upheaval. In 1999, under pressure from the World Bank, Bolivia granted a 40-year contract to a subsidiary of London-based multinational International Water Ltd. to operate water systems serving a half-million people in Cochabamba. Water prices quickly rose, and widespread protests ensued. At first, the protests were relatively peaceful, but they became increasingly violent as the government responded with troops and tear gas. After two months of demonstrations, during which hundreds were injured and six were killed, then Bolivian President Hugo Banzer Suárez finally cancelled the contract. The Cochabamba fiasco awakened many citizens in Bolivia to the privatization controversy, and Evo Morales rose to presidency of Bolivia in part through opposition to privatization. Similar controversies over privatization of resources are playing out in many parts of the world.

Wastewater treatment in Israel. Israel is a world leader in wastewater reuse, with much of the reclaimed water being used for crop irrigation.

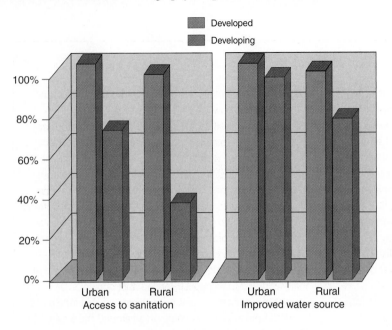

Developed

Developing

Figure 13-4 Access to improved sanitation and safe drinking water in rural and urban areas of developed and developing countries, 2006. Significant progress has been made in recent years, particularly on safe drinking water, but much remains to be done.

which consume most of the world's resources, but in the poor ones. Many newly industrialized cities, such as Mexico City, Beijing, Cairo, Yangon, and São Paolo, have severe air-pollution problems. The situation for water pollution is no better (Figure 13-4).

In rural areas of poor countries, air quality is usually better than in urban areas, but the problem persists. This is because indoor air pollution is perhaps even more a problem in rural areas than in urban ones. Clean energy sources such as electricity or gas are rarely available in rural areas, and most people use wood, dung, or other biomass fuels for cooking and heating. These fuels are burned in simple stoves or fireplaces, often poorly ventilated. Therefore, chronic lung diseases are common, especially among women and children because they are the ones most exposed to the polluted air inside. In rural areas, sewage collection and improved water supplies are generally less available than in urban areas.

Can we trade wealth for environmental quality? It is estimated that at least 5 million people, mostly children, die each year in Asia, Africa, and Latin America from diarrhea caused by contaminated drinking water. A similar number, in the same regions, is estimated to die from chronic respiratory problems caused by poor air quality. These numbers are very crude estimates, but they certainly highlight some key issues regarding wealth, environmental quality, and the quality of life. Why do these deaths occur? Are they caused by poverty or pollution? How can they be prevented?

One way to eliminate these deaths would be to increase incomes. A small amount of money in the hands of a mother in Mumbai would allow her to buy kerosene to boil drinking water for her children, or it might buy simple rehydration therapies that would allow her children to survive the diarrhea. The availability of cheap electricity or gas in Beijing might allow people to cook on electric stoves or heat their homes with gas instead of burning coal and thus reduce the severe particulate air pollution of that city. If, however, that increase in income is achieved through an intensification of coal-fueled industry with few emission controls, as it is in China and India, aren't we just trading one problem for another, with no net gain?

There are no simple answers. Clearly, there is a need to increase incomes, especially in the poorer countries, while avoiding the environmental destruction that all too often accompanies that development. As discussed in Chapters 8 and 12, such an approach is usually called *sustainable development*. It has become the stated goal of virtually every governmental, nongovernmental, and international agency involved in promoting economic development.

Sustainable development as a concept makes sense, but making it a reality is difficult. Many people who are involved in promoting sustainability agree that the key is to assign realistic economic values to environmental attributes and to make sure these values are considered in the development process. This is similar to calculating national income in terms of gross sustainable product, as discussed in Chapter 12, but doing so at the project level. For example, when a natural resource such as land covered in forest is converted to another use—say, agriculture—we normally think only of the positive value of the agricultural land created, and perhaps subtract the value of the continuing timber production it replaced. But what of the value of the clean water that came from the forest and that has been replaced by streams polluted with agricultural runoff? What of the value of the biodiversity that may have been lost in the process? Planning projects and measuring their achievements in these terms may help to ensure that the values in the present environment are given due consideration when we plan modifications in the name of economic development.

Malaria and DDT

The case of malaria and the approaches we take to controlling it are an interesting example of global contrasts in managing the environment and the trade-offs between immediate human needs and protecting the natural environment. While AIDS is indeed the most serious health issue in Africa, malaria remains a serious problem. Malaria kills about 1 million people per year, mostly children, in Africa. Some children who survive malaria suffer from learning impairments or brain damage. The disease is also a leading cause of perinatal death, low birth weight, and maternal anemia.

Although several treatments for malaria are available after an individual has been infected, the best way to prevent infection is through preventing mosquito bites. The most effective agent for controlling mosquitoes is DDT, a pesticide invented in 1937 (its inventor won the Nobel Prize for his work) and used with great success ever since. While DDT has been banned in most developed nations, it is still widely used for mosquito control in many relatively poor tropical nations, where it is considered an essential tool in the battle against malaria.

As discussed in Chapter 4, DDT is a persistent pesticide—one that remains in the environment for a relatively long period of time. DDT is harmless to humans in the low concentrations in which it is normally found, but it is passed up the food chain, reaching lethal concentrations in some carnivores. DDT is one of a group of similar toxic chemicals known as persistent organic pollutants (POPs). POPs can evaporate in relatively warm conditions, such as are common in the tropics, and can be carried on the wind to distant locations where they can be redeposited on Earth's surface. Through this mechanism, such chemicals have been redistributed throughout the globe. They are found in even the most remote places, including Arctic regions where they can accumulate because it is too cold for them to evaporate again.

People who live in Arctic regions depend largely on meat for their food, and they are traditionally hunters of seals, whales, and other carnivorous animals that, by their position at the top of the food chain, tend to have high concentrations of POPs. The problem is worsened by the fact that these people have diets that are high in fat, and POPs are most concentrated in animal fat. POPs have been found in mothers' breast milk among the Inuit of northern Canada at concentrations nine times higher than among women in southern Canada. Inuit people are now being warned to reduce their consumption of foods that are staples in their culture, particularly seal.

Health officials thus face a dilemma. The most effective available tool for fighting malaria in the tropics is also one that is poisoning the food upon which arctic people depend. The environmental rules banning DDT that are viewed by North Americans as essential may be wholly unacceptable to many tropical nations. Donor agencies in wealthy nations are reluctant to fund malaria-control programs that include the use of DDT. In 1995, under international pressure, South Africa eliminated DDT from its malaria programs. Mosquitoes quickly became resistant to the substitute chemical, and the number of malaria cases rose dramatically. In 2001, 90 countries signed a treaty to eliminate POPs from the environment, and in that treaty, an exemption was made for the use of DDT in poorer tropical nations. South Africa resumed the use of DDT in 2001, and the number of cases quickly dropped to near their 1995 level. The topic remains highly controversial For example, the U.S.-based Congress of Racial Equality supports the use of DDT to reduce malaria in Africa, and many see the debate over DDT as having racial dimensions. In 2006 the World Health Organization reversed long-standing policy and declared support for indoor use of DDT in malaria-infested regions. It remains to be seen whether the new U.S. initiative to improve health in Africa and elsewhere will be effective at controlling malaria in ways that do not cause DDT to continue to diffuse around the globe.

Applying DDT. These workers are spraying DDT in a village in Madagascar, to eradicate fleas responsible for bubonic plague.

The central tensions over globalization of economic activity and the increased free trade under World Trade Organization (WTO) rules focus on the equitable distribution of costs and benefits of production. Those who argue that industrial production should not be shifted from wealthy nations to poor ones without adequate environmental protections feel that the benefits of income generated by manufacturing jobs are less than the costs endured by those living near the factories. From their perspective, the costs of production are being transferred from rich countries to poor ones while the benefits, in the form of profits and consumer goods, remain in rich countries. Those who argue that free trade is good for poor countries obviously see the benefits of accepting those factories as outweighing their environmental costs.

These tensions were prominent at the World Summit on Sustainable Development, held at Johannesburg, South Africa, in 2002. The summit focused on improving people's lives and conserving natural resources in a world that is experiencing increasing demands for food, water, shelter, sanitation, health services, and economic security. The conference was attended by leaders from around the world; the United States was represented by Secretary of State Colin Powell. When Powell spoke, he stated that "the United States is taking action to meet environmental challenges, including global climate change," but a skeptical audience responded with jeers and heckling. Clearly, many outside the United States feel that the most powerful and influential nation in the world bears a responsibility to address these issues more effectively.

International Equity in Environmental Management

Measuring the value of environmental quality and natural resources is difficult enough in itself, but it is even more difficult when we recognize that people in different countries value resources differently. Many North Americans and Europeans place a high value on tropical rain forest biodiversity and argue that deforestation should be halted in order to protect that diversity, even if it means curtailing some economic benefits of deforestation. The governments of most tropical nations, however, are much less concerned about biodiversity and more concerned with the economic gains to be made by exploiting the forest. For environmental issues of only regional or national concern, such issues are usually resolved at that level. But when we consider resources of global importance, international negotiations must resolve conflicts derived from differences in valuing resources. In this section, we will discuss some examples of major environmental issues that have been the subjects of global-scale negotiations or that are likely to become so in the near future.

Ozone depletion and the Montreal Protocol As we learned in Chapter 2, several trace gases in the atmosphere are important in absorbing radiation. One of these is ozone, a form of oxygen with three atoms per molecule instead of the usual two. Ozone is relatively abundant in the stratosphere, a layer of the atmosphere located between about 15,000 and 55,000 meters (50,000 and 180,000 feet) above sea level. Stratospheric ozone absorbs significant amounts of incoming solar radiation, especially in the ultraviolet wavelengths. Ultraviolet radiation is harmful to many things at Earth's surface, but our greatest concern about it is that it is a cause of skin cancer.

As early as the 1960s, scientists raised concerns about the potential impact of air pollution on ozone concentrations in the stratosphere. Among the particularly worrisome substances was a class of chemicals called chlorofluorocarbons (CFCs). CFCs are human-made chemicals that have been used in a wide variety of applications, including refrigeration and air-conditioning and many industrial processes.

The usefulness of CFCs stems from the fact that they are chemically very stable. But this very stability allows them to remain in the atmosphere for a long time after they are released—long enough to reach the stratosphere. There they are broken down by the intense solar radiation. When the CFCs break down the chlorine atoms that are released enter reactions that contribute to a loss of ozone in the stratosphere.

By the 1970s, there were strong theoretical reasons to believe that CFCs would harm stratospheric ozone, and they were sufficient to cause most nations to ban the use of CFCs as propellants in aerosol spray cans. It was not until the mid-1980s, however, that we accumulated clear evidence that damage to the ozone shield was in fact occurring, especially in high latitudes, and that this damage was attributable to CFCs.

When this became known, international action was obviously necessary. CFCs were being manufactured and used in many, mostly wealthy industrialized, countries. In 1987, a multilateral treaty was negotiated that has since been ratified by 196 countries. The agreement, called the Montreal Protocol, called for signatory nations to freeze CFC production at 1989 levels and then cut production 50 percent by 1999. Developing nations were allowed to increase CFC production for 10 more years, and they were not required to phase out CFCs until 2010. As additional evidence confirming damage to the ozone layer accumulated, the agreement was strengthened. Amendments in 1990 and 1992 accelerated the phaseout, resulting in a complete cessation of CFC production in rich countries (which accounted for about two-thirds of all production) in 1996. CFC emissions by the rich countries dropped faster than expected, and by 1997 levels of chlorine in the stratosphere became stabilized and began to decline (Figure 13-5). CFCs are no longer produced, although production of a less harmful class of substitute compounds called HCFCs continues.

The Montreal Protocol is generally regarded as a success story in international environmental management.

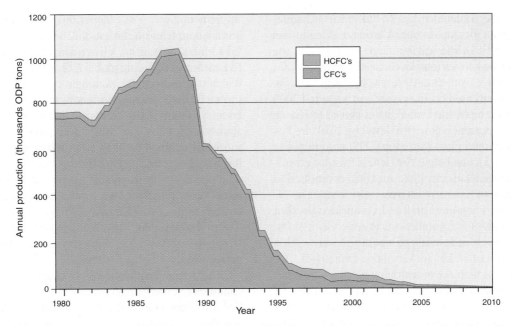

Figure 13-5 Chlorofluorocarbon production. Production of chlorofluorocarbons peaked in 1988 and declined dramatically thereafter as a result of an international agreement to phase out these ozone-destroying chemicals.

One of the factors in this success was that the principal nations involved had similar stakes in the outcome. (1) The countries for which skin cancer represents the greatest health problem are generally those in which (a) other health problems, such as those associated with basic sanitation, are less critical, and (b) populations are mainly in high latitudes. It is also noteworthy that skin cancer is most common among fair-skinned people. Thus, the wealthy high-latitude nations of Europe and North America are among those with the greatest concern about the negative impacts of ozone destruction. (2) The industrial powers of North America were the major producers and consumers of CFCs. These same wealthy nations were the ones with the greatest chemical research and development capabilities and were thus most likely to be able to develop substitute chemicals quickly. (3) Although CFCs had a wide range of important uses, the challenge of replacing them with other substances was not as onerous as some other environmental problems and thus could be accomplished without undue economic disruption. (4) Finally, the agreement established different provisions for poor nations than for rich ones. The two-tiered agreement recognized the greater need of poor countries to continue producing these useful chemicals, and perhaps, these countries' lower level of concern about their environmental impacts. Thus the countries that bore the greatest burden of complying with the terms of the treaty were also the ones that benefited the most from its enactment.

Global warming and international efforts to control CO_2 emissions Among the many international meetings on environmental issues that have been sponsored by the United Nations, two major international

conferences stand out: the Conference on the Human Environment held in Stockholm in 1972 and the World Conference on Environment and Development (Earth Summit) held in 1992, in Rio de Janeiro. In both cases, the contrasting agendas of rich and poor nations have been major barriers to international agreements and action to protect the global environment, although international friction was less in 1992 than in 1972.

At the 1992 Earth Summit in Rio de Janeiro, an international treaty known as the United Nations Framework Convention on Climate Change (UNFCCC) was created, committing most countries to undertake serious efforts to reduce global emissions of carbon dioxide and other greenhouse gases. That agreement, which took force in 1994, did not require any country to actually reduce emissions. Rather, it was a promise to talk about reducing emissions, with the expectation that significant action would be agreed to within five years. Nonetheless, it marked the beginning of an ongoing effort to tackle this difficult international environmental issue. As of June 2010, 192 countries were parties to the UNFCCC. It is governed by a "Conference of the Parties" (COP) that makes decisions concerning the UNFCCC and its implementation.

The COP has met many times since 1994, with the most significant meeting to date being the 1997 conference in Kyoto, Japan, that resulted in the Kyoto Protocol. Under the pressure of the five-year deadline for agreement set in 1992, delegates struggled to hammer out an agreement that would bridge the enormous gap between countries that wanted major reductions in CO_2 output and countries that wanted no restrictions at all.

With such differences among the countries represented, compromise on many issues was difficult or impossible. The conference finally did reach an agreement, though an incomplete one. A group of 37 countries comprising most of the industrialized countries of the world committed to various levels of reduction in CO_2 emissions relative to their 1990 levels, by 2010, with the final figure being based on an average for the period 2008 to 2012. For example, the United States agreed to reduce emissions to 7 percent below 1990 levels by 2010; the European Community (as it existed in 1997) committed to an 8 percent reduction below 1990 levels, Russia committed to no change relative to 1990, Australia committed to not more than an 8 percent increase, and so on. The remaining, mainly less-industrialized countries were not required to make any commitments to reduce emissions. Among these less-industrialized countries were China and India, both of which are very large emitters of CO_2 (though still less than the industrialized nations on a per capita basis) with high growth rates in their emissions.

The Kyoto Protocol also includes mechanisms for international cooperation for emissions reduction. One such mechanism is global-scale emissions trading, in which countries that exceeded their emissions quotas could purchase emissions allowances from countries whose emissions were below the target levels. For example, between 1990 and 2007, France reduced its emissions by almost 12 percent, compared to a targeted 8 percent reduction. On the other hand, Spain's emissions grew by 55 percent in this time period. Under a carbon trading scheme, France could sell to Spain permission to emit additional CO_2 and thus compensate for some of Spain's excess emissions. Another allows for "Joint Implementation" of emissions-reduction projects among countries on the list of 37 industrialized nations. Like emission trading, this encourages the most effective means of reducing global CO_2 emissions, regardless of where they occur, by making it possible for one country to effectively pay for emissions reductions in another country (Figure 13-6). A country that can reduce emissions easily will do so while a country that finds this more difficult will be likely to purchase emissions credits instead of reducing emissions. Thus the emissions reductions occur in the places where they are less costly, rather than requiring every place to reduce emissions regardless of cost. The Kyoto Protocol also includes a "clean development" mechanism whereby industrialized nations could be credited for emission-reduction projects in developing countries.

The Kyoto Protocol was to enter into force when countries representing 55 percent of global emissions ratified the treaty. This level was finally achieved in late 2004, and the treaty entered into force in February 2005. As of June 2010, 186 countries and the European Union had ratified the treaty. The protocol effectively ceases to function when the 2008–2012 averaging period ends.

Although the United States signed the protocol when it was drafted in 1997, it has not ratified the treaty and thus is not bound by its provisions. Treaty ratification in the United States is the authority of the Senate, and there never was sufficient support in the Senate for the Kyoto Protocol to even bring ratification to a vote. The reasons for this lack of support are many, but concerns about the costs of reducing carbon emissions were most important.

Figure 13-6 Klinki Forestry Project, Costa Rica. These Klinki Pine trees were planted using funding from a U.S. Corporation as a Joint Implementation project under the Kyoto Protocol. They will both sequester carbon, and produce nuts that support wildlife.

Much of the U.S. economy depends on cheap fossil energy, from petroleum that fuels our cars, to coal that produces our electricity, to natural gas that heats our homes and powers industry. Opponents of the protocol successfully argued that costs of these fuels and the products that depend on them would rise, harming the U.S. economy. In addition, opponents of the treaty often cited the fact that China, India, and other growing industrial powers with which the United States competes in world markets would not be subject to emissions limitations.

Has the Kyoto Protocol been a success or a failure? On the positive side, it did represent a major step forward toward global cooperation in CO_2 emissions reduction—the first binding commitment of its nature. Under the protocol, many countries have begun meaningful programs that have resulted in emissions reductions in both industrialized and developing countries. For example, carbon emissions trading systems are in place in the European Union, and the EU system allows crediting of emissions reductions achieved through joint implementation and clean development projects. A few countries (mainly in Europe) have actually met or exceeded their emissions reductions targets, and even those that have not met their targets have likely reduced emissions below what they would have been had the protocol not been negotiated.

On the other hand, global emissions have continued to rise, in many cases much more than the Kyoto Protocol allows. The protocol does not require any emissions reductions in China, and that country's emissions grew more than 150 percent between 1992 and 2009. China surpassed the U.S. in 2007 to become the world's largest CO_2 emitter. U.S. emissions have grown also; they were about 26 percent greater in 2010 than in 1990 (as compared to the 7 percent reduction negotiated in the original Kyoto agreement).

The parties to the UNFCCC have met many times since the Rio conference of 1992, but the most significant meeting since Kyoto took place in Copenhagen in late 2009. The scientific evidence that human-caused global warming is underway grew dramatically in the years after Kyoto, and the Intergovernmental Panel on Climate Change 4th Assessment report released in 2007 was much stronger than any previous report in its predictions of significant impacts of climate change (see Chapter 2).

By 2009, as the Kyoto Protocol was approaching expiration, there was much stronger sentiment in support of an international climate treaty than there had been ever before, including strong support from U.S. President Barack Obama. The nations of the world gathered in Copenhagen in the hopes of negotiating a new treaty that would strengthen the commitments made in Kyoto and result in meaningful reductions in CO_2 emissions. In advance of the meeting, many nations made announcements pledging substantial emissions cuts, hoping to spur others to action. China and the United States, representing 40 percent of global CO_2 emissions,

were key. China promised a 40 to 45 percent decrease in energy intensity by 2020, and President Obama pledged that the U.S. would reduce emissions by about 17 percent below 2005 levels by 2010. But as the negotiations proceeded to consider the all-important details of a treaty, optimism was replaced with pessimism, accusations, and discord. The Copenhagen summit ended without substantial agreements.

Challenges to International Cooperation on Climate

With nearly 200 nations participating in the discussions on a global climate treaty, consensus is difficult to impossible. In order for diverse sovereign nations to join in a treaty, each nation must individually see the benefits of treaty membership outweighing the costs. We have not yet found a formula that would meet that criterion.

What costs and benefits do different countries see? Who would be the winners, and who would be the losers if carbon emissions were effectively controlled? In general, Europe is strongly in favor of emissions controls. Environmental protection has strong support there. Europe has an added advantage in that it has already made substantial progress in reducing emissions, both through policies specifically promoting clean energy and as a result of the transition from coal-based industries to a service-based economy since the 1970s. The collapse of heavy industry in Eastern Europe resulting from the breakup of the Soviet bloc caused a dramatic reduction in emissions after 1990, and the Kyoto Protocol's use of a 1990 base gave Europe a head start in meeting the goals.

Arguments in favor of emissions controls also come from small island nations like The Maldives, Jamaica, Trinidad, Samoa, and the Seychelles that are most threatened by sea-level rise. For many of these nations, tourism is a vital industry, and beaches and adjoining development are the first to be damaged by rising sea level. The U.S. insurance industry sometimes joins in the argument because of its vulnerability to major losses caused by hurricanes and other storms. Such disasters are most likely in nations that have long coastlines and valuable investments such as resorts along those coasts. Hence small island nations think that preventing global warming is very much in their interest.

Developing nations in general favor emission controls but insist that the developed nations shoulder the burden because they are the source of most of the increased CO_2 in the atmosphere. A large group of about 130 developing countries known as the G-77 took a very strong stance at the Copenhagen conference, opposing controls on emissions in developing countries unless major emissions cuts were made by the United States, Europe, Japan, and other industrial powers. The developing nations also generally insist that it is the obligation of the developed world to finance clean economic development in the developing world that will improve their economies without contributing to global warming. One outcome of the Copenhagen conference was a pledge of a fund providing $100 billion per year, contributed by developed nations, to support clean development in developing countries as well as measures to help those countries adapt to global warming.

China is a member of the G-77 group, as is India, but China is in a unique position because its economy is growing very rapidly, and it has become the largest emitter of CO_2 in the world. India's economy is also growing rapidly, and as of 2009, it was the fourth largest emitter of CO_2, just behind Russia. China recognizes its need to reduce carbon emissions, not only to slow global warming but also because it is suffering from its own pollution from coal combustion. However, China strongly resists international monitoring of its emissions that would be necessary under any binding treaty. India also recognizes the need to control emissions and has said it will reduce its carbon intensity by 20 to 25 percent relative to 2005 levels by 2020. India still feels that the primary responsibility lies with developed nations.

The most vocal groups opposing carbon controls are those that benefit most from fossil-fuel use—especially the major oil-producing nations. Within the United States there is considerable diversity of opinion, with strong interests in the fossil fuel industries (especially coal-burning electric utilities and oil producers) opposing emissions controls, and equally strong environmental and clean-energy interests that favor controls.

The former Soviet republics and some of their former allies in Eastern Europe are in an interesting position relative to CO_2 emissions controls. Most of them were large CO_2 emitters prior to their political and economic collapse in the early 1990s. Their industrial economies have not recovered, and their emissions today are typically much lower than they were in 1990. For this reason, it is easy for them to support emissions cuts that have 1990 levels as their starting point, as in the Kyoto Protocol. These countries have also benefitted considerably from the joint implementation provisions of the Kyoto Protocol, under which Western European countries have made substantial investments in former Soviet countries. At the same time, Russia is a major oil producer and exporter, and so is not keen to constrain oil use.

The failure of the Copenhagen conference came as a surprise to some, given the strong evidence that CO_2 is affecting climate, widespread calls for action, and pledges of action from most of the world's largest carbon-emitting nations. The disappointing result is mainly attributable to disagreements over the details of a treaty, rather than the broad principles.

For example, the U.S. has strongly favored a cap-and-trade system for controlling CO_2 emissions. Cap-and-trade systems depend on a market for emissions permits (or credits) in which the permits are in short supply and thus high priced. Only then will emissions be reduced, because if emissions permits are cheap, it is easier to buy permits than to reduce emissions. Such a system has been successfully used in the United States for sulfur oxide pollution control, and a CO_2 trading system has already been implemented by Europe and other nations under the Kyoto Protocol. The structure of the Kyoto Protocol, with 1990 as the base year for emissions, allowed Russia to accumulate substantial amounts of emissions credits, which it could then use to attract investment under the joint implementation program. Russia's possession of a substantial supply of credits threatens the cap-and-trade system because of a potential oversupply of credits on the market, and so the details of how credits are to be managed and priced can be critical to the success or failure of a treaty.

Many believe that because each country has different and often unique needs and goals relative to global warming and CO_2 emissions controls, the most effective approach will be individual actions not coordinated or mandated under any international treaty. This approach recognizes that governments of sovereign nations act primarily on the basis of domestic political considerations rather than in response to international pressures. The United States is an excellent example: If internal politics won't allow a treaty to be ratified by the Senate, then no amount of negotiation by representatives at international conferences will have any significance. But there is strong sentiment in the United States and many other nations for reduced fossil fuel use, and this sentiment itself may be the most powerful motivator for emissions controls, at least in the short run.

Global Security and Human Rights

The search for peace echoes through human history. Yet war, violence, and oppression have been common in most places around the world. Traditional rulers usually believed they had a divine right to rule and saw little reason to cooperate with others outside a local population. In fact, violence was easier to justify when it was directed against other peoples. Political rulers easily took human lives, whether among their own subjects or others. Things began to change with the rise of the modern nation-state ideal in the seventeenth century. Political authority was invested in independent, sovereign states. As we saw in Chapter 11, sovereign states had their own territories and were made responsible for taking care of their own populations. Citizens in some European states demanded that their governments recognize the rights of individuals. The revolutionary governments set up in the United States and France detailed many of the basic rights that became the basis for **human rights,** including the right to one's own life and liberty; freedom to own property; freedom of religion, thought, and expression; and the right to a fair trial.

The system of independent sovereign states also created a new international order that was supposed to keep peaceful relations between states. Sovereignty is basically a form of cooperation in which states agree not to interfere in each other's internal affairs—the boundary between territories is to be respected. As the number of independent states grew, however, the number of possible conflicts among states increased and states sometimes violated one another's sovereignty. The causes of international warfare among sovereign states has not changed much in the last three hundred years. Causes include aggressive ambitions, as with Iraq in the early 1990s; border disputes or rival claims on territory, as between Ethiopia and Eritrea; and domestic crises or policies that have effects abroad, causing other states to threaten to intervene, as with the former Yugoslavia.

Achieving world peace has long been the ultimate goal of international cooperation. Doing so means convincing states that national security, reducing threats to the state, can only be achieved by cooperating with other states in pursuit of **global security,** the reduction of threats to all states. In a world made up of independent sovereign states, there is no global government, judicial system, or police that rules over all states. Can greater security be achieved through state cooperation, or does it require global government? And how do we measure security? Does global security mean a reduction of threats against states or against people? This debate is a global one, engaging national leaders, activists, and international organizations, and answers rest on different scales of response. International organizations have set high standards and seek to promote dialogue on these issues. Human rights advocates argue that increasing security means furthering human rights at the individual and community level, and that securing these rights necessarily limits state sovereignty. States' rights advocates argue that sovereignty remains the best system to promote peace.

Interests Versus Principles

States are regularly confronted with challenges and threats that cause them to privilege national interests above all others. This makes cooperation difficult, as countries try to build trust and universal goals while also pursuing their own priorities. As we saw in Chapter 11, creating treaties is one way that states have learned to cooperate by writing out agreements about the obligations and promises that each state makes to another. Multilateral treaties, such as those addressing global environmental problems described earlier in this chapter, often attract participation by keeping the terms of the agreement very specific.

Tackling more principled ideals, such as peace and development, often challenges national interests. States often have to make concessions or change the way they govern. National leaders are often unsure whether such changes are worth what they would gain from international cooperation. Complex political issues, such as preventing war, have fostered the creation of special organizations with rules, routines, and meeting spaces that provide a situation for states to discuss and hopefully agree on positive actions that promote greater peace.

The first lasting international political structure was the League of Nations. It was meant to appeal to the national security interests of individual states by promising **collective security,** whereby states commit to one another's security, lowering the risk of war. The League of Nations was founded after World War I, to help enforce the peace treaties and avoid a return to war. Several powerful states, including the United States, were not members, and the Soviets did not join until later. The League of Nations provided a forum for discussion of world problems, and it carried out many humanitarian projects. It exercised little real power, however, and was unable to resolve the disputes that eventually led to World War II. The lesson provided by the League of Nations is that national interests all too easily trump global interests. Indeed, the League of Nations was championed by U.S. President Woodrow Wilson, for which he received the Nobel Peace Prize, but Congressional Republicans kept the United States out of the League of Nations, afraid that it would limit American sovereignty.

The United Nations The advantages of collective security were made clear by World War II because aggression by the Axis powers—Germany, Italy, Hungary, Romania, Bulgaria, and Japan—could be stopped only with political unity. In 1942, the 26 countries allied against the Axis powers adopted a "Declaration by the United Nations." The victorious Allies, led by the United States, the United Kingdom, the Soviet Union, and France, met in San Francisco after the war to adopt the charter for a U.N. organization that would ensure the peace; by the end of 1945, the new organization had 51 members. As more colonies received independence and joined the United Nations, membership rose to 192 by 2006—virtually all Earth's sovereign states. The United Nations Charter is a multilateral treaty that commits all countries to cooperation on matters of peace and security. The United Nations, however, is an intergovernmental organization, meaning it only has the powers and authority given to it by its member states. Once again, this produces tension between national interests and international principles.

The U.N. General Assembly serves as a global forum on issues of peace and security, but it has no power of enforcement (Figure 13-7). The power that the United Nations does exercise is vested in the 15-member Security Council. Five members of the council (the United States, China, France, the United Kingdom, and Russia—originally the U.S.S.R.) hold their seats permanently, and the remaining 10 members are elected by the General Assembly for two-year terms. This allocation of seats reflects the major world powers in 1945, but recent discussions have suggested that granting additional seats to, perhaps, Germany, Japan, India, and Brazil might better reflect the present world balance of population and power. The U.N. Secretariat, with the

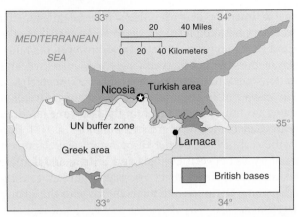

Figure 13-8 Cyprus. Since 1974, Cyprus has been divided into Greek and Turkish portions, with little mingling between the two groups. The Turkish sector has declared itself to be the Turkish Republic of northern Cyprus, but only Turkey recognizes it as an independent country. The Greek portion joined the EU in 2004. The two sides have been discussing reconciliation.

Figure 13-7 United Nations headquarters in New York City. The United Nations found a permanent home on the east side of Manhattan Island, where its headquarters buildings were designed by an international committee of architects, including Le Corbusier. The high slab of the Secretariat dominates the group, with the domed General Assembly to the north.

Secretary General at its head, handles all administrative functions. On January 1, 2007, Kofi Annan ended two five-year terms as Secretary General and was replaced by the South Korean diplomat Ban Ki-moon.

The Security Council has occasionally voted to authorize U.N. member states to use military force to stop threats to world peace. U.N. forces fought in Korea from 1950 to 1953. In 1990, a Security Council resolution gave the United States and its allies sanction to force Iraq to retreat from Kuwait, and on September 28, 2001, the United States obtained a Security Council resolution obliging U.N. members to cooperate in combating terrorism. In a greater number of cases, the United Nations has sent observer troops to patrol world trouble spots, and these incidents have been increasing. Examples

include Kashmir; the Golan Heights; Lebanon; Cambodia; the former Yugoslavia; the Mediterranean island of Cyprus, split by feuding Greeks and Turks; and Haiti (Figure 13-8). U.N. troops have attempted both *peacekeeping* and *peace enforcement* efforts. Peacekeeping involves patrolling a situation in which combatants have agreed to a peace treaty. Peace enforcement, however, is trying to impose a cease-fire or peace on combatants, and United Nations forces have had less success in these efforts. In some cases, the United Nations has assumed responsibility for monitoring the freedom of national elections, helping to ensure fair elections that will be acceptable to voters and legitimize national governments.

The International Court of Justice (ICJ) in The Hague is another branch of the United Nations that seeks cooperation among states. The ICJ hears disputes between states, not individuals, and bases its decisions on international law drawn from treaties and the normal actions of states. The states involved in disputes must choose to come before the ICJ, however, and the court has no power to enforce its decisions.

Perhaps the greatest accomplishment of the United Nations is when nothing happens; in other words, when countries are at peace. States that are at least talking to each other are seldom making war on each other. In September 2005, the United Nations hosted the greatest meeting of heads of state ever: More than 170 came together for the 2005 World Summit, which also recognized the sixtieth session of the U.N. General Assembly. Media coverage of the event focused on the sometimes off-kilter rhetoric of leaders such as Hugo Chavez. More substantially, the meeting reaffirmed the global pledge to cooperate through the United Nations on peace and collective security, strengthening the United Nations, upholding human rights and the rule of law, and supporting human development, which we address later in this chapter. That so many of the countries represented at that summit were nonetheless engaged in war and oppression is a stark reminder that we live in a world of states driven by national interests. But there is no greater forum than the United Nations to seek international cooperation on problems that are truly global.

U.N. special agencies The United Nations has more than 40 specialized agencies created to address specific problems. These include the Intergovernmental Panel on Climate Change, described above. Some agencies facilitate public and technical exchange among member states: the Universal Postal Union, the International Civil Aviation Organization, the World Meteorological Organization, and the Intergovernmental Maritime Consultative Organization. The U.N. International Telecommunications Union, for example, has allocated frequencies on the radio spectrum, thus allowing international expansion of mobile telephones, satellite computer transmissions, and digital radio broadcasting.

The U.N. Conference on Trade and Development (UNCTAD) works to assist the economic development and trade of poor countries. UNCTAD has, for example, developed a standardized computer customs collection system that is in use in over 50 countries. Importantly, the international financial institutions described in Chapter 12, were created under U.N. auspices. The World Bank, International Monetary Fund, and related organizations were created at a U.N. conference in 1944.

Other U.N. agencies encourage international cooperation toward specific scientific and cultural goals, including the World Tourist Organization and the United Nations Educational, Scientific, and Cultural Organization (UNESCO; Figure 13-9). In 2000, the United Nations sponsored the Millennium World Peace Summit of Religious and Spiritual Leaders; about 800 attended from around the world.

Some of the most important U.N. agencies are those that address humanitarian needs. The fight against world hunger is led by the Food and Agriculture Organization, which encourages countries to adopt better policies, as described in Chapter 8. The U.N. High Commissioner for Refugees was created after World War II to help refugees and displaced persons. The World Health Organization was created in 1948 to deal with combating disease around the world.

Nuclear disarmament One of the most pressing issues of global security in the twentieth century was the development of nuclear weapons. Encouraged by U.S. President Eisenhower, the International Atomic Energy Agency was created in 1957 to promote the peaceful use of nuclear energy and to prevent the diffusion of nuclear weapons. There was a real fear that the nuclear weapons created by the United States would cause other countries

Figure 13-9 Yellowstone National Park. Yellowstone Park was the first World Heritage Site. UNESCO supervises the Convention Concerning the Protection of the World Cultural and Natural Heritage. Signatories to the convention—virtually all of the world's countries today—have agreed that certain areas have such unique worldwide value that they should be treated as part of the heritage of all humanity. Designated sites range from Sinharaja Forest Reserve in Sri Lanka to the ancient city of Hatra, Iraq; Canterbury Cathedral in the United Kingdom; Dinosaur Provincial Park in Alberta, Canada; and Bolivia's historic capital of Sucre. Yellowstone National Park was the world's first national park, and its 100th anniversary in 1972 inspired the global convention.

to do the same in fear of being dominated by the United States. This is exactly what the Soviet Union and other countries began to do after the first atomic weapons were dropped on Japanese cities. The arms race during the twentieth century raised the prospect that humans would destroy themselves through nuclear war. These weapons programs also drained money away from other priorities and poisoned the environment.

As politicians faced the domestic and international impact of their weapons programs, they recognized that they shared basic principles beyond simple national interest. Beginning in the 1960s, governments agreed to put limits on their weapons programs, culminating in the U.S.–Russian arms reduction treaties in the 1990s. Nonetheless, other countries have since tried to start nuclear weapons programs, a violation of the nonproliferation treaty in effect since 1970. Countries have also

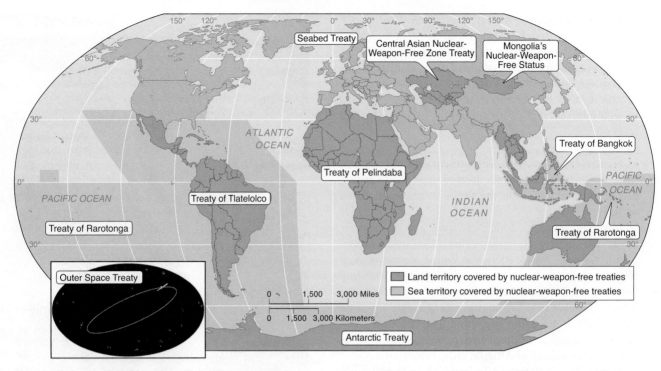

Figure 13-10 Nuclear-Weapons-Free Zones. Different treaties have established parts of the Earth, and outer space, to be areas that should remain free of nuclear weapon development, deployment, or use. The countries within the South America, Africa, Central Asia, Southeast Asia, and Oceania NWFZs signed and ratified their respective agreements. (Data from United Nations)

created Nuclear-Weapons-Free Zones that ban the development or deployment of such technologies in those areas (Figure 13-10).

Since the revelation of secret facilities and missile tests, Iran has received close international scrutiny by Western governments and the International Atomic Energy Agency (IAEA). International cooperation will be required to further reduce the stocks of existing nuclear weapons and to prevent the creation of new weapons programs in countries that do not have them. President Barack Obama has set a goal of ridding the world of nuclear weapons, although his administration will "maintain a strong deterrent" as long as other countries have weapons. In other words, national interests prevail even on the path to principled global security.

Human Rights and National Sovereignty

The United Nations works on two different scales. On the one hand, it seeks to promote global cooperation among sovereign states. On the other hand, the United Nations promotes the human rights of individuals and communities who are being oppressed by sovereign and independent states that are also U.N. members. This requires delicate diplomacy that is often criticized for catering to state interests at the expense of human rights. The United Nations is not the only entity concerned with human rights. The U.S. Department of State publishes regular reports on human rights around the world, as do nongovernmental organizations, such

as Amnesty International and Human Rights Watch. International scrutiny of government treatment of their citizens has diffused human rights standards that are increasingly considered universal. But this presumes global acceptance of fundamentals of a political culture, which rests in turn on fundamentals of human culture. Despite the fact that he himself owned slaves, Thomas Jefferson wrote confidently that certain truths regarding human rights were "self-evident." The same cannot be said today because the diffusion of human rights has collided with different ideas of what those rights should be.

The U.N. General Assembly adopted a Universal Declaration of Human Rights very early, in 1948. At that time, however, most U.N. members were Western countries, and they were still holding most African and Asian societies as colonies. As these societies have achieved political independence, they have begun to challenge Western ideas of human rights. They deny that any human rights are in fact universal and "self-evident." Human rights, they insist, may vary from culture to culture.

Two questions have arisen in international organizations. First, what is the balance between the rights of individuals and those of the state? Governments argue that their responsibility to provide stability and security often require curtailing individual rights. The failure to achieve gender equality has been defended in some countries as necessary to maintain the social traditions that provide political stability. The government of Iran unleashed violent militias to stop peaceful protests in Tehran during 2009 on the basis that the protesters were endangering public security. Even the United States after September 11, 2001, argued that civil liberties were sometimes less important than national security.

The second question is the balance between political rights and what some people call "economic rights." In 1997, China's President Jiang Zemin insisted, "As a developing country of 1.2 billion people, China's very reality determines that the right of subsistence and development is the most fundamental and most important human right in China. Before adequate food and clothing is insured for the people, the enjoyment of other rights would be out of the question."

At the last major U.N. Conference on Human Rights, in 1993, many countries were concerned that Western countries would use their definitions of human rights as excuses for withholding aid or other economic benefits from developing nations that may not accept the same definitions. Singapore's Foreign Secretary Kishore Mahbubani insisted, "Too much stress on individual rights over the rights of the community will retard progress. In the future, these agreements will assert the rights of society over the rights of individuals." Indonesia's foreign minister stated flatly that "Human rights exist as a function of history, culture, value systems, geography, and phases of development." Several Asian countries jointly offered a statement that fairness and justice should be measured against "regional particularities and various historical, cultural, and religious backgrounds." An African statement demanded that observers must take account of "the historical and cultural realities of each nation." Not all cultures recognize the equality of women with men, and the assembled nations could not even agree on a condemnation of torture. They insisted, however, that "the right to development is inalienable." The United States hesitates to accept a statement that development is a "right," because that could lead to demands that the rich countries redistribute their wealth among the poor countries.

After many days of debate, a final document was adopted, stating that while differences of "historical, cultural and religious background must be borne in mind, it is the duty of states, regardless of their political, economic and cultural systems, to promote and protect all human rights and fundamental freedoms." In other words, the states agreed to disagree.

The United Nations long had a Commission on Human Rights, the chairmanship of which was awarded not by merit but by rotation among members. Thus, the chairmanship periodically fell to U.N. members whose own records on human rights were not good. Therefore, in 2006, the U.N. replaced the Commission with a new Human Rights Council, with restrictions on membership. When the new Council first met, however, Iran, China, Cuba, Pakistan, Russia, and Saudi Arabia were counted among its members, thus continuing to illustrate the difficulty of reaching international agreement on definitions of human rights and how to promote them.

Humanitarian Intervention

National sovereignty means that international borders are unchallengeable. No matter how monstrous any regime may be, no matter how much it persecutes its own people or oppresses minorities, no outside state or international agency has the right to interfere. This is true even when internal persecution triggers international flows of refugees. The U.N. Charter insists, "Nothing contained in this present Charter shall authorize the United Nations to intervene in matters which are essentially within the domestic jurisdiction of any state." This is the legacy of the Treaty of Westphalia (see Chapter 11).

Disarray in several states, however, and the formation of powerful nongovernmental organizations may have signaled the beginnings of change in international law. The first war against Iraq in 1991 did not overthrow Iraq's government, but the United Nations did subsequently protect some Iraqis from their own government. In 1992, the Security Council voted to send a U.S.-led force to Somalia, which was wracked by civil war, to guarantee the distribution of food-relief supplies. The purpose of that mission was defined as "humanitarian," but the mission expanded to restoring central civil government to Somalia. Several states contributed troops (notably Canada and Pakistan), but Somalia still has not achieved peace. U.N. authorized interventions in the former Yugoslavia, Rwanda, and East Timor violated state sovereignty, strictly defined, but they did so in the interest of people facing imminent harm at the hands of their own governments.

In 1999, then-U.N. Secretary General Kofi Annan, reviewing cases of civil war and slaughter in Africa, argued, "Nothing in the Charter precludes a recognition that there are rights beyond borders," and referred to a "developing international norm in favor of intervention to protect civilians from wholesale slaughter." The secretary general pressed the United Nations formally to establish the principle that massive and systematic violations of human rights must be stopped by international intervention. Many world leaders, however, agree with President Abdelaziz Bouteflika of Algeria, who has argued that "interference can only occur with the consent of the state concerned." Many countries fear that U.N. intervention would serve as a cloak for Western interference. Russia's Foreign Minister Igor Ivanov has said bluntly, "Human rights are no reason to interfere in the internal affairs of a state." China's President Jiang Zemin has emphasized, "Dialogue and cooperation in the field of human rights must be conducted on the basis of respect for state sovereignty. So long as there are boundaries between states, and people live in their respective countries, to maintain national independence and safeguard sovereignty will be the supreme interests of each government and people." To date, no state has invaded another on strictly humanitarian grounds. Invading states have instead argued that their own national security is at risk if humanitarian crises are not stopped.

The globalization of justice After World War II, the victorious Allies tried the leading German and Japanese officials for crimes against humanity and for genocide. Since then, international law has affirmed that these crimes are subject to universal jurisdiction. Any state may prosecute such crimes, regardless where the crime was committed. Few states, however, have been willing to do so. Therefore, in 1993, the United Nations created a War Crimes Tribunal to try local political and military leaders for "crimes against humanity" that had occurred during the fighting that accompanied the breakup of Yugoslavia. During the rule of Slobodan Milosevic (1987–2000), Serbia fought and lost five wars, bringing upon itself terrible destruction, charges of international crimes, and international disgrace. Milosevic was eventually overthrown by nationwide rioting and brought to trial for crimes against humanity, but he died of heart ailments before the end of his trial. The wartime leader of the Bosnian Serbs, Radovan Karadzic, was arrested in 2008 and now faces a list of charges, including genocide. Other U.N. tribunals have investigated African wars, and another is trying the leaders of the Khmer Rouge that killed millions in Cambodia during its rule (1975–1979).

In 1998, the United Nations voted to create a permanent International Criminal Court (ICC), and two years later, delegates from more than 100 nations agreed on a catalog of acts that constitute international crimes. The required 60 signatory nations ratified the treaty by 2002, so 21 justices were sworn in, and the ICC began operating in The Hague, the Netherlands. The ICC must defer to national courts; its prosecutors can issue indictments only when national courts are unwilling or unable to deal with atrocities such as war crimes, crimes against humanity, and genocide. The ICC has investigated atrocities in Africa and issued indictments against those responsible.

The United States originally signed the treaty to establish the ICC, but it never ratified the treaty, and in 2002, the U.S. government repudiated even having signed the treaty. American critics of the ICC believe that groundless charges would frequently be brought against U.S. military personnel and officials. The United States has more troops engaged in more actions in more countries all around the world than any other country, and such trials might be used to embarrass the U.S. government. These fears are baseless, however, because the ICC would defer to the U.S. military and civilian justice system.

Serbia and Iraq: Signs of the Times?

The wars in Serbia (1999) and Iraq (beginning in 2003) provide two stark examples of why states violate the sovereignty of other states. In both cases, the United States led military interventions against countries that had not directly harmed U.S. territory or civilians. Yet in both cases, first under President Bill Clinton and then under President George W. Bush, the United States argued that pressing U.S. security interests were at stake. In both cases, the United States initiated the war, although its targets were already embroiled in domestic conflict.

The war in Serbia came about in 1999, when the Serbian nationalist regime led by Slobodan Milosevic began to ethnically cleanse the Albanians who had long lived in the province of Kosovo in southern Serbia. The campaign to expel civilians looked much like the violence that Milosevic and others unleashed against civilians in Croatia and Bosnia-Herzegovina between 1991 and 1995 (see Chapter 11). The major Western powers, including the United States, Britain, and France, were too slow in trying to stop the violence in the early 1990s and only put peacekeeping troops in the region after a peace deal was concluded. By 1999, the U.S. and allied governments were determined to stop Milosevic from committing further crimes against civilians. The ethnic cleansing of Albanians from Kosovo produced an enormous humanitarian crisis as more than 1 million refugees poured over the borders into neighboring countries. The refugee crisis jeopardized the stability of neighboring countries and the delicate peace in Bosnia. Clinton and his war planners did not want to put combat troops on the ground in Kosovo, knowing that the American public would not tolerate U.S. casualties for a humanitarian mission.

As the violence in Kosovo escalated, NATO initiated a bombing campaign against Serbian troops in Kosovo and against strategic sites in Serbia itself. It was the first time that the NATO alliance declared war on another country. The NATO allies did not have the permission of the U.N. Security Council, which includes Russia, an ally of Serbia. From March 24 to June 10, NATO aircraft attacked Serbia, eventually forcing Milosevic to concede to NATO's demands (Figure 13-11).

Figure 13-11 NATO's humanitarian intervention.
U.S. warplanes were not alone in the war against Serbia. Here, a German Tornado fighter departs a NATO airfield in Northern Italy to attack targets in Serbia.

NATO's primary goal was to stop the ethnic cleansing of civilians, allowing them to return to Kosovo under protection of NATO peacekeepers and under U.N. administration. The U.N. Security Council approved this postwar plan but declared that Kosovo remained a part of Serbia's sovereign territory. Further negotiations between Kosovar Albanian and Serbian politicians failed to find common ground. On February 17, 2008, the parliament of Kosovo declared independence from Serbia, which was recognized by the United States, Britain, France, and most other European countries. By 2010, some 64 countries had accepted Kosovo as an independent state. Serbia challenged the decision in the International Court of Justice, and Russia has blocked U.N. efforts to normalize Kosovo's status. The case of Kosovo suggests that sovereignty is not an acceptable defense for countries that commit massive violations of human rights. Yet NATO's war against Serbia on behalf of civilians was widely criticized as an unacceptable violation of the principle of sovereignty, as was the U.S. invasion of Iraq four years later.

Iraq was a weak state by the time the U.S. invasion began in 2003. During most of the 1980s, the Iraqi government under the rule of Saddam Hussein's totalitarian regime had fought a bloody war with Iran. In the late 1980s, Iraq conducted a genocidal campaign against its own Kurdish civilian population living in the north. In 1990, Iraq invaded Kuwait, claiming that it was Iraqi territory. The U.N. approved the use of force to expel Iraq from Kuwait, after which Iraq's government once again attacked its own civilians, whom Saddam Hussein considered a threat to national security. The United States and Britain set-up no-fly zones over northern Iraq to protect Iraqi Kurds and southern Iraq to protect Shiite communities from further violence by government forces. These interventions to protect Iraqi civilians from their own government continued from the war until 2003, when the U.S. invaded.

The 2003 invasion by the U.S.-led coalition was prompted by fears that Iraq might possess weapons of mass destruction that could be passed to terrorist groups. Although there was little evidence of such dangers, the U.S. government argued that it could invade a country preemptively, meaning attacking Iraq before it attacked the United States (Figure 13-12). Although humanitarian issues were also cited by the U.S. government and its allies, the primary purpose of the mission was to remove Saddam Hussein from power. The United States, Britain, and Spain tried to obtain an endorsement from the U.N. Security Council but withdrew the proposal when it was clear that France and other NATO allies opposed the use of military force that so clearly violated Iraq's sovereignty. The subsequent civil war in Iraq was not foreseen by the U.S. war planners or their allies, although

Figure 13-12 Powell's presentation to the United Nations. Secretary of State Colin Powell holds up an empty vial to warn that if Iraq were to give even a small amount of anthrax to terrorists, the results could be catastrophic.

experts had warned of that very scenario. The United States had instead expected that regime change in an already weak country would be easy. Together with U.S. operations in Afghanistan, the United States would then have had troops on either side of Iran, which many think was the ultimate target of the Bush administration.

Opinion on these wars was mixed. Some thought a "humanitarian war" for Kosovo had a perverse logic because it used violence to stop violence. Others thought the relative accuracy of modern air power, though far from perfect, made it justifiable to use a bombing campaign to stop massive violations of human rights and prevent regional instability. The war in Iraq brought much louder condemnation against the United States, even among its long-time European allies. British Prime Minister Tony Blair committed to the invasion, despite strong public opinion in Britain against the war. Spain's participation stood in contrast to the opposition of 90 percent of Spaniards. Critics contend that these wars are a sign that the world's most powerful military, possessed by the United States and its NATO allies, has added to global insecurity without really improving anyone's national security. Some argue that these interventions weakened the sovereignty of all countries. Others think that these controversies are part of a bumpy historical transition out of which will emerge a new international norm on humanitarian intervention and the limits of state sovereignty.

Living with Landmines

Landmines have been around for hundreds of years. They can be simple explosives attached to a trigger or complex mechanisms that cannot be defused. Landmines explode when someone or something moves the trigger. They are usually hidden to deter an enemy from passing through an area. Militarily, they are very effective. But landmines routinely harm innocent civilians, sometimes years or decades after a war. In ideal circumstances, militaries map their landmines so they can be removed after the war. Wars rarely go as planned, however, and recent conflicts in places like Bosnia and Afghanistan left many thousands of landmines in place during the chaos. These landmines remain hidden dangers in the fields, on the roads, even around schoolhouses, where civilians return after the war to restart their lives. These weapons are meant to kill or maim, often leaving those who survive without limbs or sight. Cluster bombs are another instrument of war that leave behind small unexploded bomblets, which last long after the battle.

In war-ravaged countries where landmines and other remnant weapons exist, civilians face an added challenge. Public funds to clear landmines are scarce. The war in Bosnia left thousands of communities in close proximity to landmines. More than 12 years since the end of the war, there are an estimated 1,683 square kilometers (650 square miles) of contaminated land. Some farmers remove the mines themselves. In other cases, the mines lie undetected until accidentally detonated. Demining operations are slow: just a few square kilometers a year. As one international expert put it, Bosnia will never get rid of all the mines. Yet Bosnia is a relative success compared to parts of Afghanistan and Iraq, where several layers of landmines have accumulated in some strategic areas, put down during one conflict after another.

Since 1999, these weapons are estimated to have caused nearly 18,000 deaths and almost 52,000 injuries in more than 100 countries. More than 60 percent of these victims are thought to be civilians. Children are more than one-quarter of all victims. In Afghanistan, deaths and injuries caused by landmines exceeded more than 12,000 persons in the past decade.

In 1996, the government of Canada organized a conference to create a treaty that would end the use of landmines. The International Campaign to Ban Landmines (ICBL) was launched to promote the treaty and the treaty has been signed by 156 countries since it was created in 1997. The treaty obligates signatory states to stop the production and transfer of landmines, to destroy their stockpiles, and to clear landmines from their territories. The ICBL and its campaigner Jody Williams received the Nobel Prize for Peace in 1997. The use of landmines appears to be dropping as more states sign the treaty and stockpiles are destroyed. The United States has not signed the treaty, but it appears to be limiting its use of these weapons. The other 36 states that have not signed the treaty include China, Russia, North and South Korea, Iran, India, and Israel. Most of these countries argue that their national security interests cannot preclude the use of landmines.

Landmine decontamination in Cambodia.

Regional Cooperation

States that share the same region often face similar problems. New forms of cooperation among neighboring states have created regional organizations that allow states to work together. **Regional cooperation** allows these states to combat their common problems while sharing the advantages of improved cross-border security, trade, and travel. European integration, particularly the European Union, is the most ambitious example of regional cooperation in the world. What started as economic cooperation has become a form of political unification. Other regional groupings, such as the North American Free Trade Agreement and the Commonwealth of Independent States, have more limited goals but reflect the interests of major powers.

European Integration

World War II ended in Europe in 1945, with the surrender of the Axis powers. The victory of the Allied powers helped boost the postwar status of its principle countries, namely the United States, Canada, Great Britain, France, the U.S.S.R., and China. In Europe, the Allies had squeezed the Axis powers from West and East, and their victorious armies met at the River Elbe. For the moment, Europe was at peace, but its bloody history of war and division suggested it might not last.

The Allies themselves were already divided by mistrust, and Europe soon split into two competitive blocs of states. British Prime Minister Winston Churchill provided the enduring image of Europe's new division: "From Stettin in the Baltic to Trieste in the Adriatic, an iron curtain has descended across the continent." The states of Western Europe, aided by the United States, rebuilt their countries as market democracies with programs meant to ensure basic social services. East of the Iron Curtain, however, the territories seized by Soviet armies during and after World War II were absorbed into the Soviet Union, as you recall from Figure 11-5. The Soviets also installed Communist regimes in Poland, Hungary, Romania, Czechoslovakia, and Bulgaria. Their constitutions were amended to guarantee Communist Party dominance in national politics, and their economies were largely centralized and run according to state interests. Local Communist leaders seized power in Yugoslavia and Albania without Russian help.

The Iron Curtain was at first just a term to describe a divided Europe. Only later did barbed wire and soldiers physically divide Europe, most memorably in Berlin. The Allies occupied the capital city of Berlin and, like they did to the rest of Germany, divided it into four separate military zones. Already by 1949, however, the U.S., French, and British zones were united functionally, and a new Federal Republic of Germany (West Germany), with Bonn as its capital, gained sovereignty in 1955. The Soviet Union reacted by granting its occupation zone independence as the German Democratic

Republic (East Germany). Like Europe as a whole, Germany and Berlin were now divided between West and East. West Berlin, however, lay about 200 kilometers (125 miles) inside East Germany. West Berlin showcased Western prosperity and provided a means of escape from oppression in East Germany. So many East Berliners fled into West Berlin that in August 1961, East Germany built a wall across the city to seal in its citizens (Figure 13-13).

The Berlin Wall was also an iconic landscape that highlighted the very different postwar politics in Eastern and Western Europe. The Communist regimes to the East were satellites of the Soviet Union, dominated by the centralized power of governing officials in Moscow. The independent countries of Western Europe worked to improve postwar relations with each other. These countries pursued military, economic, and political cooperation in hopes of shared peace and prosperity. They were helped by the United States, which wanted strong allies in Western Europe to counter the threat of Soviet expansion. Their efforts outlasted the Soviet bloc, which collapsed in 1989.

Military pacts The Allies' postwar occupation of Europe had divided countries into two opposing military alliances. In the West the **North Atlantic Treaty Organization (NATO),** established in 1949, united Belgium, Denmark, France, Great Britain, Iceland, Italy, Luxembourg, the Netherlands, Norway, and Portugal with Canada and the United States. They were later joined by Greece, Turkey, West Germany, and Spain. In 1955, the *Warsaw Pact* linked the U.S.S.R. and its satellites. The U.S.S.R. used this treaty to justify the continued occupation of Eastern European countries

Figure 13-13 The Berlin Wall. Erected in 1961, the Berlin Wall was the physical manifestation of Cold War division. It was erected quickly and brutally enforced from the East German side as a means of keeping out Western "imperialists." In fact, it was meant to keep East Germans from leaving for the West.

by Soviet troops, which suppressed occasional anti-communist uprisings. The Western and Eastern blocs faced one another in a protracted **Cold War**: Although the two blocs never engaged directly in combat, they competed for economic growth and for political influence around the world.

The military standoff in Europe dissolved in 1989. The U.S.S.R. directed its attention to its own internal problems and released its grip on the countries of Eastern Europe. These countries, in turn, rejected the privileged role of the Communist Party in their own countries and within a few months scheduled free elections. The Warsaw Pact formally dissolved in 1991. NATO eventually expanded to include most of its former adversaries: Poland, the Czech Republic, Hungary, Latvia, Lithuania, Estonia, Bulgaria, Slovakia, Slovenia, Romania, Albania, and Croatia. In 2002, NATO signed an agreement with Russia on terrorism, arms control, and international crisis management, so today even Russia attends NATO conferences. Macedonia, Ukraine, and Georgia have been told they will receive invitations to join.

Despite the end of the Cold War, NATO has become increasingly active in international affairs. In 1993, NATO used force for the first time in Bosnia and remains on the ground in Kosovo. After September 11, 2001, NATO invoked, for the first time, a clause that binds the members to recognize an attack directed from abroad on any member as an attack on all members. This clause originally was designed to protect Western Europeans from the U.S.S.R.; few observers would have ever suspected that it first would be invoked to protect the United States from terrorists. European NATO forces helped topple the Taliban regime in Afghanistan and have remained there. The European NATO allies are a part of the largest and most powerful military alliance in world history. At the same time, some European politicians have expressed a desire to set military priorities independently of the NATO framework long dominated by the United States.

Economic integration   The end of World War II required enormous reconstruction and economic recovery, which the countries of Europe could not achieve on their own. In 1947, U.S. Secretary of State George C. Marshall announced a plan of financial aid to war-shattered Europe (Figure 13-14). Aid was offered to the Soviet Union and to the countries of Eastern Europe, but Soviet dictator Joseph Stalin (1879–1953) rejected it. The European Recovery Program, or **Marshall Plan,** thus became a program for the economic rehabilitation of Western Europe. It helped to rebuild infrastructure and restore industrial production. It also encouraged economic cooperation among the European countries in hopes that promoting free trade might lessen the tensions that had produced past wars.

Encouraged by the success of the Marshall Plan, six Western European countries began to cooperate economically by forming the European Coal and Steel

Figure 13-14 George C. Marshall.   George C. Marshall (1880–1959) is shown here (at right) with Mrs. Marshall and President Eisenhower in 1957, at ceremonies commemorating the tenth anniversary of the Marshall Plan. As secretary of state (1947–1949), he devised the Marshall Plan for the recovery of Europe, for which he won the Nobel Peace Prize in 1953.

Community in 1952: Belgium, the Netherlands, Luxembourg, West Germany, France, and Italy. Steel was a strategic material for both military and industry. The European economies, especially France, Germany, and Italy, required huge amounts of iron and steel for their reconstruction. The coal, iron ore, and production facilities were unevenly distributed across France, Germany, and their smaller neighbors. In fact, controlling these resources was part of why they had gone to war. Producing a common market for the coal and steel sector eliminated trade barriers and increased the output of steel in all European Coal and Steel Community (ECSC) countries. More importantly, cooperation in this one sector improved relations between former enemies. The success of this common market approach encouraged further cooperation, and the European Economic Community (EEC) and European Atomic Energy Community (Euratom) came into existence in 1957.

The Soviet Union and its Eastern European satellites, meanwhile, formed the Council for Mutual Economic Assistance (Comecon) that achieved a partial integration of their economies. Some of the Eastern European economies grew under their Communist regimes, enough to surpass the Soviet Union itself in per capita measures, but they did not keep up with Western Europe. Comecon formally disbanded in 1991, and the Eastern European countries began to privatize their economies. At the same time, firms in Eastern Europe that had been protected from outside competition were now exposed to global markets that could provide higher quality goods at lower prices. Many companies collapsed, laying off millions of workers and causing a protracted and painful recession during the 1990s.

Political cooperation The devastating effects of World War II left many political leaders suspicious of one another. In Eastern Europe, states under Soviet influence were more concerned with conforming to Moscow's expectations than cooperating with each other. In Western Europe, however, a number of visionary diplomats were busy at work building new organizations to serve their common interest in growth and stability. The **Council of Europe** was formed in 1949 to promote European unity on democratic principles, human rights, and the rule of law. It has grown from its original 10 members and today includes almost every European country, including Russia and Turkey. The Council has played a major role in establishing and protecting human rights through the European Convention on Human Rights. This treaty is enforced by the European Court of Human Rights, which protects individual citizens by, for example, outlawing the death penalty and guaranteeing the freedoms of expression and religion. Another organization, the **Organization for Security and Cooperation in Europe**, was created by a 1975 meeting between leaders of the United States, Soviet Union and 33 European countries. Created to lower Cold War tensions in Europe, this intergovernmental organization now focuses on conflict resolution, election reform, and arms reduction.

The most comprehensive form of cooperation in Europe, however, emerged as common markets evolved into political agreements. The European Economic Community grew during the 1970s in part because political leaders thought that cooperation among the countries of Europe would bring greater peace and stability to their countries. They also hoped a stronger, more unified Europe would be more influential in world affairs, rather than dominated by the United States and Soviet Union. In 1993, the 12 countries of the EEC changed the name of their organization to the **European Union (EU)** to reflect broad cooperation in areas such as monetary union, common citizenship, freedom of movement, common foreign policy and security, and cooperation on police and justice matters.

The formation of the European Union also came on the heels of the collapse of the Eastern bloc. Less than one year after the dramatic fall of the Berlin Wall in 1989, East Germany was absorbed by West Germany (Figure 13-15). The EU then expanded its membership by adding Sweden, Finland, and Austria in 1995. A far greater challenge, however, was adding formerly communist countries whose citizens were still adjusting to democratic politics and capitalist economies. The EU supported these countries with enormous aid packages and technical assistance. When the EU enlarged in 2004

Figure 13-15 Germans celebrate the opening of the Berlin Wall. At 7 o'clock on the evening of November 9, 1989, an East German official mistakenly handed a television speaker a draft of a new regulation suggesting that East Germans could go to the West without a visa at once. By 10 o'clock, hundreds of thousands stood at the wall, shouting, "Open the Gate!" Confused guards did, and East Germans surged through, to be met by West Germans and doused with beer and champagne. The Berlin Wall had fallen, and East Germany was soon swept away by history.

to include the Czech Republic, Estonia, Hungary, Latvia, Lithuania, Poland, Slovakia, Slovenia, Malta, and Cyprus, the event was celebrated as the end of Europe's Cold War division. Bulgaria and Romania joined on January 1, 2007, and other Eastern European countries are in line for eventual membership (Figure 13-16).

The European Union today The European Union largely achieved the major goals for cooperation set by European leaders at the end of World War II: to end conflict and division, to share in economic growth and prosperity, and to play a larger role in world affairs.

As a political entity, the European Union is unlike the governments of individual states you read about in Chapter 11 because it lacks a constitution and exists only as long as its member states want it. The EU is a unique political creature that handles some matters, such as business law, on a supranational basis while other matters, such as new members, are strictly intergovernmental issues. Most issues are still decided and enforced by individual states, although EU policies may shape or set minimal expectations. It is thus more accurate to talk about EU governance as shared between national governments and institutions of the European Union. The three main EU institutions are the European Commission, the Council of the European Union, and the European Parliament. The Commission, with one member from each country, proposes legislation, ensures compliance with

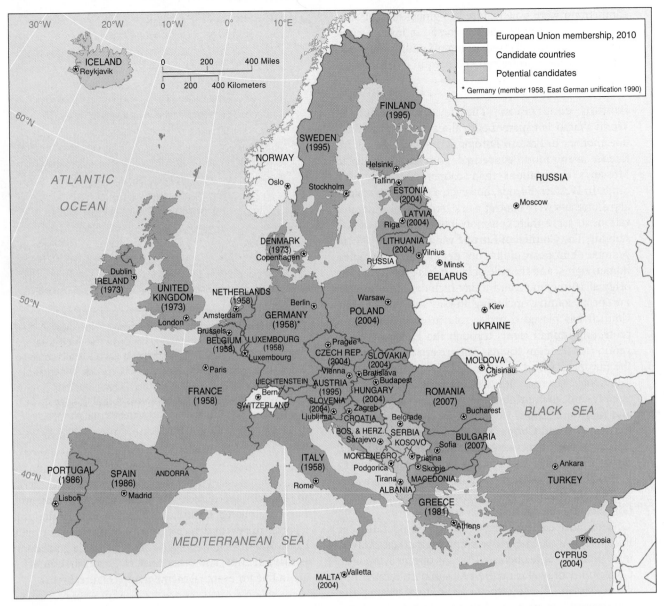

Figure 13-16 The European Union. What started as a common coal and steel market among six Western European countries after World War II became the European Economic Community in 1958. It outlived the Cold War and grew to become the European Union, now encompassing most of Europe. The expansion of the European Union in 2004 and 2007 added enormously to the total size of the EU but far less to its GDP.

the treaties, and is responsible for administration. It sits in Brussels, Belgium, which has enjoyed considerable growth and prosperity as de facto capital of the Union. Commissioners are expected to operate independently of the member states, including their own, and to represent the interests of the citizens of the EU as a whole.

The main legislative body is the Council of the European Union, which is made up of ministers from each member states' elected government. Votes in the Council of the European Union are weighted roughly according to the populations of the EU members. Depending on the issue, decisions may be approved by simple majority, qualified majority, or unanimous agreement. A qualified majority currently requires support by at least 55 percent of the majority of the member states on behalf of at least

65 percent of the EU population. The Treaty of Lisbon, which came into force in December 2009, established that the president of the European Council would represent the European Union to the rest of the world.

The European Parliament meets in Strasbourg, France, and in Brussels, with additional offices in Luxembourg. The apportionment of its seats tries to balance population-based representation, which benefits large countries, and the equality of member states, which benefits small countries. Its members are elected directly by the citizens of the member countries, and the members who represent affiliated political parties from different countries join together as international voting blocs. Therefore, the Parliament, and to a degree the Commission, are genuinely supranational. When the Treaty of Lisbon came into force in 2009, the Parliament was given additional legislative powers, on par with the Council of the European Union. These two bodies now look like a two-chamber legislature, similar to the U.S. House of Representatives and U.S. Senate. The European Parliament must approve proposed EU laws and can offer amendments. It must also approve the EU budget and can dismiss the European Commission. Parliament is considered the most democratic of Union institutions because its members are directly elected by EU citizens every five years.

European law is applied by the Court of Justice of the European Communities. In cases of conflict between EU law and national laws, the supremacy of EU law is being steadily affirmed. Union institutions have become so important to the member states that about 60 percent of each member's new domestic legislation is drafted in Brussels and simply translated into national law. Final adoption of the European Convention on Human Rights in 2000 profoundly changed government in several European states that, unlike the United States, had no Constitution or independent judiciary to enforce guarantees against official abuse. After 2000, for the first time, national judges were bound to measure acts of their national legislatures against the EU Convention. This is truly supranational law.

European Union integration   The EU largely achieved the creation of one economic community by the end of 1992. The following six criteria, however, demonstrate that the EU is much more than an economic union. It is uniting its members and binding their territories in many ways.

1. *The creation of a single market for goods, services, capital, and labor.*   The common market approach for coal and steel grew to include agriculture and eventually all goods. EU laws have standardized business regulations across countries and lowered barriers to free trade, allowing European manufacturers to achieve the economies of scale in one European market that U.S. manufacturers have long enjoyed in the huge U.S. domestic market. Companies in the EU can provide services to clients anywhere in the union and hire workers from other EU countries. Financial firms and private investors can invest in activities throughout the union.

2. *The creation of a single currency and central bank.*   Sixteen EU members have adopted one currency, the euro (Belgium, Germany, Greece, Spain, France, Ireland, Italy, Luxembourg, the Netherlands, Austria, Portugal, Finland, Slovenia, Cyprus, Malta, and Slovakia). Coordination among the countries that use the euro has been strained because European Central Bank rules require individual countries that use the euro to keep down national budget deficits. This rule may force cuts in social welfare programs and other governmental

spending. Ending large deficits in Greece led to deep cuts in public spending that led to rioting in 2010. Nevertheless, the euro has been one of the most popular EU initiatives.

3. *Infrastructure systems.*   The members of the European Union are redesigning and adding to Europe's transport, energy, and communication systems to better connect what had been separate national networks (Figure 13-17). All railroads, for example, must use the same gauge and electricity systems, and all highways, bridges, and tunnels must have the same construction specifications. One air-traffic control system has evolved and was extended in 2006 to 10 European countries outside the EU.

4. *Regional policy.*   The Single European Act commits the countries to the idea of shared prosperity by lessening the discrepancies between wealthy and poor areas within the EU (Figure 13-18). The EU tailors development policies for poor agricultural regions and declining urban areas with the intent that these programs will produce new economic growth. It pays for these programs with funds paid by the wealthier areas and countries in the union. Ireland, once a relatively poor European country, benefitted enormously from this program and is no longer receiving aid.

5. *Social policy.*   EU policies equalize worker health and safety standards, the international transfer of pension rights, the guarantee of rights to migrant workers, and mutual recognition of technical qualifications. The Workers' Charter regulates the workweek, overtime pay, holidays, and other terms of employment. The EU Court of Justice has enforced standard civil rights and antidiscrimination legislation. Union citizens may travel, settle, and work anywhere within the union, whereas non-EU citizens face common immigration and visa requirements. Regardless where they reside within the union, EU citizens have the right to vote in, and stand for, election in local and European elections. The concept of "European citizenship" is being made real.

6. *Environmental policy.*   The EU has become a leader in international efforts to protect the global environment. The EU has established policies of protecting the quality of the environment, protecting human health, ensuring prudent use of natural resources, and promoting international measures to deal with regional or global environmental problems. These policies have led to new standards for air and water quality, restrictions on pollutant discharges, and procedures for environmental impact assessment. The EU still faces environmental challenges, however. Some of the worst environmental problems are found in the new member states of Eastern Europe where Communist era industrial pollution continues to threaten public health and the environment.

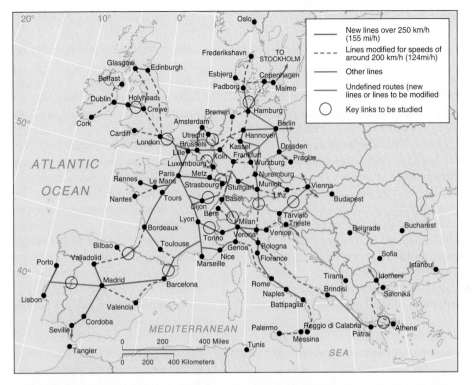

Figure 13-17 Europe's integrated rail network. An integrated network of high-speed trains has been superimposed on the existing national networks. It required laying thousands of kilometers of new track and improving much of the rest. In an area as compact and densely populated as Europe, city-to-city train travel is a sensible alternative to air travel at little or no additional time cost.

Through the years, statistical descriptions of life in the EU member countries have converged. These include statistics as diverse as median educational attainment of the population, degree of urbanization, percentage of females in the labor force, crude birth rates, and total fertility rates. This convergence demonstrates that life in the member countries is growing measurably more alike. At the same time, the various European national cultures are growing less distinct than they were just decades ago. Styles of dress, cuisine, architecture, and other artifacts are merging. Almost two-thirds of the Union's citizens report speaking at least one other European language in addition to their native tongue.

The challenges of expansion and international relations The EU has struggled to "digest" the many new members that joined in 2004 and 2007. Several of these new member states have large populations that are economically far behind the earlier members. Eastern European countries generally have lower per capita incomes, shorter life expectancies, higher percentages of their populations in agriculture, and lower measures of human development (discussed later in this chapter). Some Eastern Europeans are migrating west in search of jobs and education. So far, citizens in the new member states have not yet received the right to move freely within the EU, although the United Kingdom and Ireland have allowed some to immigrate. Hundreds of thousands have already entered the United Kingdom, while immigration from Eastern Europe boosted the number of foreign-born living in Ireland to 400,000 people by 2010, double the figure from 2002.

The accession of new members has also presented political and even cultural challenges. For example, some have called for greater democratization of EU governance by shifting power toward the European Parliament, where larger countries wield substantial political power. This has caused frictions with smaller states. It also highlights differences in social and cultural attitudes. Polish politicians, for example, have balked at granting equal rights to women and gay people and at protecting natural habitats and endangered species.

Croatia, Macedonia, and Turkey remain candidates for membership, and Switzerland enjoys close links without membership. The EU has also extended an opportunity for Albania, Bosnia and Herzegovina, Kosovo, Montenegro, and Serbia to join. Some in the EU are opposed to admitting these countries because they are too poor and conflict-ridden. Others in the EU are opposed to adding countries with cultures that they consider too different from Western Europe. Both reasons, in fact, have been expressed in public reaction to the prospect of Turkish membership. This raises questions about what regional integration seeks to achieve. It also raises questions about what "Europe" means and where its boundaries lie. The EU has partially addressed

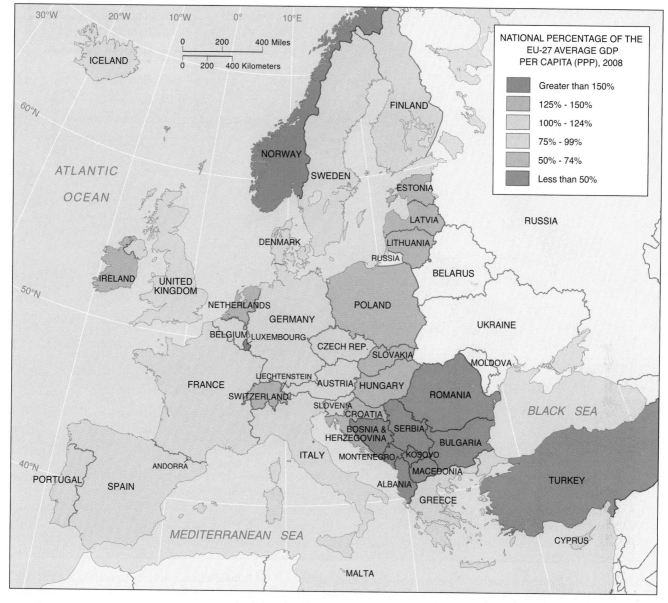

Figure 13-18 Economic disparities. This map reveals that enormous disparities in GDP per capita exist across the Europe. The average GDP per capita of the 27 EU member states in 2008 was €25,100 per person, and the countries are shown here as a percentage of that average. The EU hopes to achieve high living standards for all, including prospective countries in southeastern Europe.

this through its European Neighborhood Policy that identifies an outer ring of countries it will cooperate with but not offer membership (Figure 13-19).

The European Union negotiates trade pacts with outside countries and other blocs, but it cannot set foreign policy independently of its member states. This reveals how far the countries remain from true union. Nevertheless, the Treaty of Lisbon provided for a High Representative for Foreign and Security Policy who will hold meetings of foreign ministers of the EU member states and serve on the European commission. The EU has also formed peacekeeping forces, separate from NATO, which have deployed several times in the past decade.

European Union policies also affect foreign corporations doing business there. Many U.S. corporations do business in Europe, and EU regulations increasingly

affect U.S. economic affairs. The European Commission has intervened in everything from U.S. antitrust policy to food safety. European governmental review of corporate mergers is also much stricter than in the United States, so it was Europeans who in 2001 blocked the proposed merger of the American giant corporations General Electric and Honeywell. The European Commission also imposed huge fines against Microsoft for anticompetitive practices related to its Windows operating system, hoping to change the way the company does business.

NAFTA and the Americas

On January 1, 1994, the **North American Free Trade Agreement (NAFTA)** came into effect, linking the United States, Canada, and Mexico. The agreement

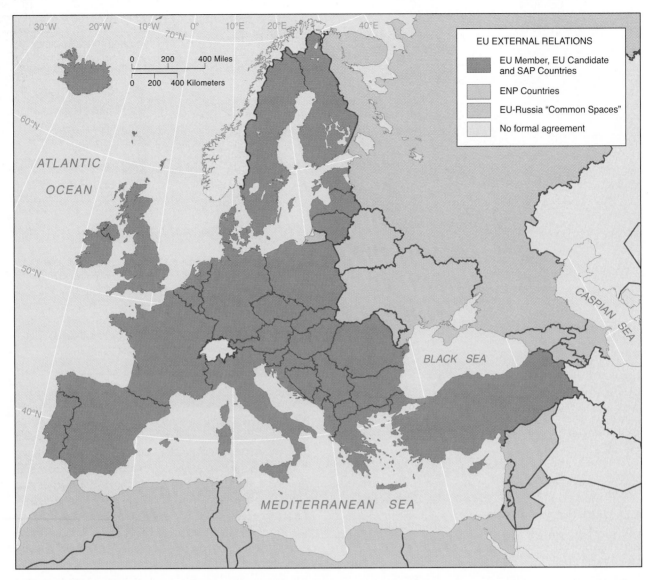

Figure 13-19 European neighborhood policy. The future geography of the European Union will look something like this. The EU will likely incorporate the countries that are currently candidates. The countries along the southern shore of the Mediterranean and in far Eastern Europe will be part of the European Neighborhood, a framework for cooperation but not membership with the EU. Russia has negotiated its own framework for cooperation with the EU.

does not propose a full customs union or common market, nor are there any plans for a supranational governing body. NAFTA concerns only trade and investment. At the beginning of 2009, the North American partners outstripped the 27-member European Union in total area (20 million square kilometers to 4.5 million), but not in total population (456 million to 500 million). The NAFTA bloc economies had a GDP in 2008 of over $17 trillion while the EU produced about $15 trillion.

The Canada–U.S. free trade agreement Canada and the United States signed the Free Trade Pact in 1988; the terms of this pact eliminated all barriers to trade in goods and services between the partners by 1999. The treaty also removed or reduced the hurdles

to cross-border investment, government procurement, agricultural sales, and the movements of employees.

Canada and the United States have long been by far the world's two leading trading partners, as well as major investors in each other's economies (Figure 13-20). About two-thirds of foreign investment in Canada comes from the United States, representing one-third of all U.S. investment abroad, and Canada is the fourth largest investor in the United States. About 10 percent of the chief executives of the largest U.S. corporations are Canadian born.

U.S. corporations were generally strong supporters of the pact, because Canadian tariffs were two to three times higher than U.S. tariffs, but Canada's labor unions fought the pact. The unions pointed out that employment fringe benefits in Canada are generally

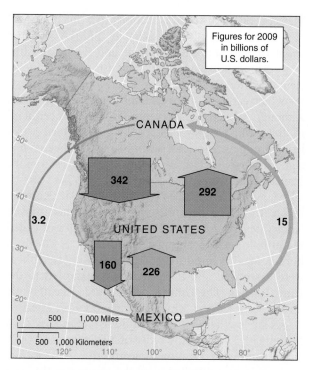

Figure 13-20 North American trade flows. The United States buys about 76 percent of Canada's and 80 percent of Mexico's exports each year, whereas Canada takes about 20 percent of U.S. exports and Mexico just over 12 percent. Trade between Canada and Mexico is limited but increasing rapidly.

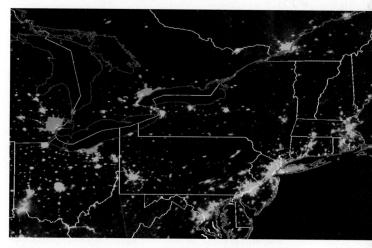

Figure 13-21 The blackout of August 14, 2003. The power grids of the United States and Canada are so interconnected that the border itself could not be identified on this satellite image of the international blackout of August 14, 2003. This image was created from the difference between two other images, one collected before the blackout and the other during it. The areas with power outages are either dark or red.

much more generous than they are in the United States, and they feared a drop to U.S. standards. In Canada, labor unions represent 40 percent of the labor force, whereas in the United States, unions represent only 19 percent. Canada's farmers also fought the pact. Canadian agriculture is at a competitive disadvantage because of the harsh climate, vast distances, and small scale of regional markets. Canadian small producers suffer competition from big food-processing firms that can buy agricultural goods more cheaply in the United States.

Some Canadians feared that strengthening economic ties with the United States would weaken Canadian nationalism (Figure 13-21). If you look back to Figure 11-22, you will notice a thin line of Canadian settlement along the border and an almost empty empire stretching away to the north. About 75 percent of Canadians live within 241 kilometers (150 miles) of the U.S. border, and they absorb U.S. news and culture. Canadian means of transportation and communication have always been designed to foster east–west linkages. Nevertheless, the international border to the south is punctured by more than 75 million border crossings per year, as well as by common telephone lines, pipelines, computer links, power lines, contracts, cross-ownership, and even by common professional sports allegiances. Canadians worry that they might suffer by tying themselves too closely to a country that knows very little about its northern neighbor.

America's security concerns after 9/11, however, have caused a tightening of this border. For 189 years, the U.S.–Canadian border was famously "the world's longest undefended border," but in 2006, the United States mounted armaments on Coast Guard ships in the Great Lakes. The militarization of the border had required an addendum to the 1817 Rush-Bagot Treaty. The addendum allowed the weaponry but restricted its use to the U.S. side of the lakes. Another U.S. initiative requires all travelers between the United States and Canada to have passports. Because only 32 percent of Americans and 55 percent of Canadians held passports in 2006, it is expected that these new regulations will strangle border crossings. The United States has suggested erecting a joint security cordon around the two countries, but Canada resists the appropriation of Canadian sovereignty that step would entail.

The trade pact provides protection for Canadian broadcasting, publishing, and related cultural industries, but it may be difficult even today to distinguish Canadian cultural products from those of the United States. The majority of Canadians do not watch Canadian television, read Canadian books, listen to Canadian music, or go to Canadian movies and plays. Moreover, many U.S. citizens are probably unaware of the Canadian nationality of entertainers Nelly Furtado, k.d. lang, Celine Dion, Shania Twain, Avril Lavigne, the Barenaked Ladies, Mike Myers, James Cameron, Michael Bublé, and Jim Carrey. The Canadian company Alias/Wavefront developed the animation software prominent in virtually every special-effects-based feature film created in recent years (*Spiderman, Ice Age, Hollow Man, The Matrix, The Perfect Storm*, and the *Lord of the Rings* and *Star Wars* series of films).

The degree of cooperation between the United States and Canada may be represented symbolically by the fact that Canada is the only country that the United States has allowed to build an embassy on Pennsylvania Avenue in Washington, D.C., America's grand ceremonial boulevard; and in October 1999, President Clinton traveled to open a new U.S. embassy in Ottawa. This was the first time that a U.S. president ever traveled abroad for this purpose.

Mexico NAFTA, among the United States, Canada, and Mexico, went into effect on January 1, 1994, to bring down tariffs and other barriers to trade in four five-year steps.

The integration of Mexico into a North American trade pact might be more difficult than the integration of the U.S. and Canadian economies. The United States and Canada are both rich, postindustrial societies with populations that are growing slowly. Mexico is a less-developed nation with a population still growing rapidly—from 25 million in 1950 to 111 million by 2009. To provide work for this rapidly growing labor force, Mexico needs economic growth, which might reduce the push for Mexicans to migrate to the United States. Mexico's Hispanic culture and language contrast with the British traditions and English language held in common in the two countries to the north (except in Quebec), but that cultural contrast may not be as troublesome in building a free trade pact as it would be if the three countries were building a full common market.

Signing NAFTA was part of an evolution of Mexican policies toward export-led economic development. Mexico's economy had been state directed and centralized through most of the twentieth century, but in 1986, Mexico began privatizing its state sector, selling off factories, shopping centers, mines, and public utilities. Only PEMEX, the state oil monopoly created when

Mexico's oil industry was nationalized in 1938, was viewed as politically untouchable. PEMEX supplies about 10 percent of Mexico's GDP and about two-fifths of government revenue, but economists view it as inefficient, with too many employees and too little reinvestment of earnings in exploration or adding value. Although Mexico is the seventh-largest producer of oil in the world, it imports 20 percent of its refined petroleum products. Mexico's second-largest source of foreign exchange, after oil, is workers' remittances from the United States, which total about $25 billion per year.

Many U.S. companies were active in Mexico even before the trade pact was signed. They had moved labor-intensive operations to Mexican factories called **maquiladoras** concentrated along the U.S.–Mexican border. Mexico imported components duty free; these were assembled and the products re-exported to the United States. The United States charged tariffs only on the value added, which was low because Mexican wages are low. The existence of these factories by 1994 had already significantly redistributed Mexican population and urbanization toward the border, but *maquiladoras* multiplied after the signing of the pact (Figure 13-22). New foreign direct investment (FDI) in Mexico, about $2 billion per year in the 1980s, rose to $25 billion per year by 2007. Exports of manufactured goods rose from 43 percent of Mexico's merchandise exports in 1990 to 80 percent by 2004. In 2000, about 800,000 *maquiladora* jobs existed in the border region.

Already early in the twenty-first century, however, Mexico's *maquiladoras* began to suffer competition from China, where wages were lower. In the 1990s, Mexico had risen to the rank of number two among nations exporting to the United States (after Canada), but China seized that position by 2003. The number of *maquiladoras* began to fall, at a cost of hundreds of thousands of jobs, and Mexico realized that continuing

Figure 13-22 Assembling steering wheels in a *maquiladora*. A Delphi Delco employee stitches the cover on a steering wheel in Matamoros, Mexico. She is one of about 11,000 employees working for the company in seven factories located just south of the border from Brownsville, Texas.

economic growth depended on lowering the cost of energy and improving the transportation infrastructure and education. The government encouraged the study of engineering, and the number of students rose to 451,000 by 2006 (compared to 370,000 in the United States). Multinational corporations responded by moving higher value-added activities to Mexico. For example, General Electric established facilities to design and test jet engines and turbines in the city of Querétaro, where engineers earn about one-third of U.S. salaries. Similarly, Tijuana is the world capital of television manufacture, and many companies that started manufacturing television sets are now assembling computers and other higher-value electronics.

Many Americans feared that the pact with Mexico would trigger a migration of U.S. manufacturing jobs to Mexico, where wages, working conditions, and environmental protection laws are all below U.S. standards. A principal reason that Mexican wages are lower than U.S. wages, however, remains that Mexican productivity is lower than U.S. productivity; that is, each Mexican worker's output is much lower than the output of each U.S. worker. A worker's productivity is usually determined by the amount of capital investment that has been made per worker, and that investment is much higher in the United States than in Mexico.

Some of the new FDI in Mexico comes from European and Asian corporations that re-export goods from Mexico into the United States. This occasionally causes trade disputes. Negotiators of trade pacts must agree what percentage of the total value of a good should have been added in a country for that product to qualify as being a product of that country. This is called a **local content requirement.** Arguments over local content requirements typically arise whenever two countries negotiate free-trade pacts that are not customs union agreements. Disagreements have arisen between the United States and Mexico about whether goods assembled in Mexico out of components made in other countries should enjoy free access to the U.S. market. Such disputes have also arisen with Canada.

The *maquiladoras* are not the only border phenomenon that geographers have noticed. Wherever two cities—one Mexican, one U.S.—face each other across the border, these so-called "twin cities" are being compelled to integrate and often internationalize their infrastructures. In 1998, for example, Mexican and U.S. officials opened the International Wastewater Treatment Plant, which treats Tijuana's excess sewage on the San Diego side of the border, the first international facility of its kind in the world. A unified air-quality district covers El Paso, Texas, and Ciudad Juarez.

Mexican politics From 1929 until 1997, Mexico was ruled by one political party, the Institutional Revolutionary Party (PRI), whose political apparatus made little distinction between the party and the state. Accusations of corruption and repression abounded. Some Americans

hesitated to join in a pact with Mexico because they doubted Mexico's democracy, but others hoped that closer ties to the United States would strengthen it. In 1997, the PRI lost its majority in the lower house of Mexico's Congress, and by 2006, the PRI had slipped in status to the third largest party in Congress. In 2000, the PRI lost the presidency to Vicente Fox, of the National Action Party (PAN). Many observers attribute this political change to the economic changes that have come with NAFTA. President Fox himself later regretted that he had not been able to achieve all of the "historic transformations our times demand," but he insisted that democracy and the rule of law had made progress under his presidency. Felipe Calderón, of PAN, won the presidency in 2006, to succeed President Fox.

President Fox suggested that the United States and Mexico seal the two countries' borders against outsiders but open the borders to freer movements between the two countries. About 11.6 million Mexican-born persons live in the United States, and of this number, about 31 percent do not have U.S. citizenship. These Mexican citizens in the United States account for about 8.5 percent of Mexico's population. Whether these people will eventually retire in the United States or return to Mexico is an important question in both countries' future. Regardless of what happens with U.S. immigration rules, however, there are 20 million legal immigrants from Mexico now living in the United States, many of whom maintain strong family and economic ties with that country. There are probably more than 1 million Americans living in Mexico, which is becoming an attractive retirement location for some.

Agriculture remains a sore point among NAFTA nations, and each country continued to protect its domestic markets for years after the agreement was passed. U.S. subsidized products can overwhelm the farm economies of either neighbor, bankrupting Canadian farmers and impoverishing millions of Mexican peasants. Agriculture produces only about 5 percent of Mexico's GDP, but roughly 20 percent of the labor force is directly involved in it. Mexico has been flooded with imported corn. In 2006, one-third of the corn used in Mexico was imported from the United States, and the price of corn in Mexico had fallen over 70 percent since 1994, thus reducing the incomes of the 15 million Mexicans who depend on producing corn for their livelihood. Canada, meanwhile, imposed a tariff on subsidized corn from the United States. Mexico's exports of fruits and vegetables to the United States, however, had increased so much as to double Mexican agricultural exports to the United States overall between 1994 and 2004. Therefore, in 2007, U.S. fruit and vegetable growers launched a lobbying campaign in Congress to add their products to the list of agricultural products already subsidized in the United States. Tariffs were removed on all agricultural products traded between the United States and Mexico on January 1, 2003, except beans and white corn, which Mexico stopped protecting in 2008, despite protests by angry Mexican farmers.

Just as the U.S.–Mexican pact took effect, a guerrilla organization calling itself the Zapatista National Liberation Army, in honor of Emiliano Zapata (discussed in Chapter 11), rose up in Mexico's southern state of Chiapas. The rebels were motivated in part by fears that imports of U.S. corn and other market pressures would wipe out their traditional agricultural economy. "The free trade agreement," said the leader of the group, "is a death certificate for the Indian peoples of Mexico." Chiapas is one of the poorest states in Mexico, and largely Indian in population. The insurrection has received support from local Roman Catholic priests, who are followers of liberation theology as discussed in Chapter 7. The Mexican government negotiated land tenure and other issues with the rebels, but the uprising serves as a reminder of how far Mexico has to go to achieve the goals of economic welfare and political equity for all of its peoples. Some Mexicans have demanded the renegotiation of agricultural trade rules.

The geography both of manufacturing and of agriculture in Mexico highlights a growing split between the country's north and south. New manufacturing plants have concentrated in northern Mexico, most of the new productive commercial agriculture is in the north, and the north has rapidly urbanized, resulting in a regional economic growth rate over 4.5 percent. The south, by contrast, retains the traditional *ejido* agriculture system (see Chapter 8) and has enjoyed little industrialization. Growth there has occurred at less than 1 percent per year. PAN generally represents the capitalist, globalized economic interests of the north, where it won its first governorship in the late 1980s, and whence it has grown to national power. President Calderón, for example, holds a master's degree in public administration from Harvard University.

Results of the NAFTA pact   Since the formation of NAFTA, the individual economies of the three signatories have experienced many changes, and it is difficult to isolate the impact of NAFTA. Trade and investment among the three partners have risen substantially. Imports into the United States from both Mexico and Canada have surged even more strongly than U.S. exports, so that net value of trade between the United States and the two partners has swung from a net surplus to a net deficit.

U.S. labor unions argue that the United States has lost jobs, that wages and living standards are down, and that workers' rights and bargaining power weakened in all three signatory countries. Both U.S. and Canadian labor unions have launched strong recruitment and organizational training drives in Mexico. Some factories are undoubtedly being closed and moved to Mexico. In February 1997, for example, the RCA Corporation announced its intention to close the world's largest TV assembly site, a plant in Bloomington, Indiana, and to move 1,100 jobs to Mexico. The company cited wage costs as the principal reason for the move. RCA is not a U.S. corporation but is owned by Thomson, a French multinational conglomerate.

International manufacturing and marketing have begun to reorganize in scale from national- to North American–scale planning. The U.S.-based Ford Motor Company, for example, makes its Courier, Fiesta, and EcoSport models in Cuautitlan Izcalli, Mexico, and most of its larger models in the United States. NAFTA's effects on Canada are difficult to ascertain. Total U.S. exports to Canada have risen, and some Canadian industries seem unable to compete, but overall, Canada's trade surplus with the United States has doubled, creating many new jobs in Canada.

The transport infrastructures of the three countries are being knit (Figure 13-23). In 1998, the Canadian National railroad, with a main line stretching across Canada from Halifax, Nova Scotia, to British Columbia, bought the Illinois Central Railroad, giving the Canadian line a route south to the Gulf of Mexico ports of New Orleans, Louisiana, and Mobile, Alabama. Canada's deepwater port at Halifax was already winning trade from New York, Boston, and other U.S. East Coast ports. The Canadian National cooperates with America's Kansas City Southern Railroad, whose line to Mexico City carries 40 percent of Mexico's rail traffic. The United States has promised to allow Mexican trucks to haul goods throughout the United States, but the question of harmonizing truck inspections and driver qualifications has slowed implementation of this agreement.

Trade between Canada and Mexico grew at an average annual rate of 26 percent from 1993 to 2007, but it is still a small fraction of either country's trade with the United States.

Many environmentalists remain critical of NAFTA. Although the agreement includes provisions for environmental protection, critics argue that Mexico's environmental policies allow U.S. and Canadian firms to pollute more by moving factories south of the border. NAFTA's Commission on Environmental Cooperation issues an annual report called *Taking Stock*, in which it has criticized the worst polluters in North America, but the Commission has no powers to enforce cleanups.

Expanding Western Hemispheric Free Trade

The United States continues to sign free-trade pacts with individual countries. Between 2001 and 2006, the United States signed free-trade agreements with 15 countries, with a total population of 230 million and combined GDP of $2.2 trillion.

The United States has suggested the creation of a Free Trade Area of the Americas, excluding only Cuba, and negotiations have continued on that initiative. A number of trade blocs formed in the Western Hemisphere in the

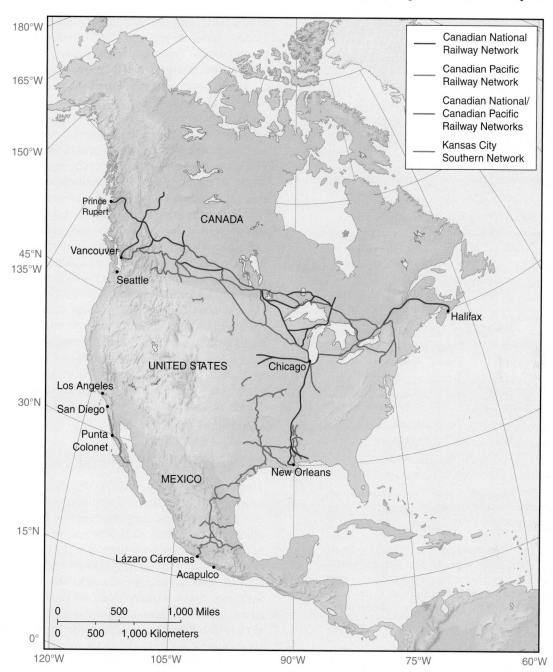

Figure 13-23 Western North American ports. Asian–North American trade is rising so fast that west coast ports in both Mexico and Canada are expanding and integrating with the U.S. transport network. In 2005, the Los Angeles–Long Beach port still handled 80 percent of U.S. imports from Asia, but in Mexico, Lázaro Cárdenas, which is served by the Kansas City Southern Railroad, is being upgraded and expanded. Developers want to build a huge new port facility in Punta Colonet, a bay on the Baja Peninsula 250 kilometers (155 miles) south of the U.S. border. Canadian National Railways is expanding port facilities at Prince Rupert on Canada's west coast. Lower port fees and fear of terrorist threats on U.S. soil increase these ports' attractiveness to U.S. importers.

early 1990s. Venezuela, Colombia, Peru, Bolivia, and Ecuador formed the Andean Pact. The English-speaking countries of the Caribbean united in a Caribbean Union and Common Market (CARICOM). In 2006, Venezuela joined Argentina, Brazil, Paraguay, and Uruguay in a customs union called Mercosur; Chile and Bolivia are associate members and may join. Citizens of member countries may live and work in any

country and be granted the same rights as citizens of those nations. Perhaps a genuine United States of South America is in the works.

The unification of transport infrastructures again reveals the expanding territory being organized (Figure 13-24). A new railroad tunnel and new gas and oil pipelines have pierced the Andes Mountains between Chile and Argentina and now link Bolivia and

Figure 13-24 New transport routes knitting South America. These roads, railroads, pipelines, and electrical grids are pulling together the population of South America, which the map on the rear endpaper shows to be, for the most part, scattered around the fringes of the vast continent.

Brazil. New roads join Brazil and Venezuela, and a new power line brings electricity from Venezuela's Guri Dam to Brazil's Amazon cities. Argentina uses Chilean ports for its exports to Asia. Brazil and Argentina have reconciled the gauges of their railroad systems, and new highways link several countries. A new inland waterway allows barges to travel 3,424 kilometers (2,140 miles) down the Paraguay and Paraná rivers from eastern Bolivia to the sea.

Trade within Mercosur has soared, but in 2005, only 23 percent of total Mercosur trade was with other Mercosur countries. This is a small amount compared with the level of trade typical between industrialized neighbors, and it suggests that the various Latin American states did not produce items they could exchange to mutual profit. The new trade agreements, however, encourage specialization of production, increases in

productivity, and intraregional trade and growth. The United States is the principal supplier of imported goods to most Latin American countries, so growth in Latin America boosts the U.S. economy.

While supporters for expanded free trade in the Western Hemisphere argue that it will encourage economic development in the regions' poorer countries, vocal opposition has emerged. Many Central Americans and South Americans are suspicious of the United States, given its domineering history in the region during the course of the nineteenth and twentieth centuries. President Monroe's doctrine that the Western Hemisphere should remain under the guidance of the United States was often used to justify U.S. invasions and meddling in Latin American affairs. Puerto Rico, the U.S. Virgin Islands, and the naval base at Guantánamo Bay are among the territories taken by the United States during this period. Most Americans are unaware of the repeated U.S. military interventions in these countries and the role that U.S. corporations and government had in supporting dictatorships there. Expanding trade is seen as simply a new form of U.S. domination that will do little to benefit local populations. These suspicions have played into the hands of autocratic rulers like Hugo Chavez of Venezuela, and many in the region are leery of U.S. plans for expanded trade.

Other Forms of Regional Cooperation

The European Union, NAFTA, and the Western Hemisphere trade areas are among the most important regional international organizations today, but many others exist (Figure 13-25). Some of these are primarily military-defensive. Since the end of the Cold War, however, they have been turning into general economic and cultural associations. For example, the Association of Southeast Asian Nations (ASEAN) began as a military alliance, but it became a trading organization that has already significantly reduced trade tariffs among its 10 members. By 2015, it hopes to have completed a series of additional agreements on goals of common defense, law enforcement and police cooperation, and other forms of cooperation, as well as establishing good relations with its many neighbors, including Australia, Japan, China, and New Zealand. Other organizations, such as the Southern African Development Coordination Conference (SADCC), are economic. The purpose of each international organization is simply to provide a framework for consultation so as to reach agreement whenever possible in areas of mutual concern. In some cases, however, the members of these organizations can agree on very little and have even gone to war with one another. The African Union, for example (called the Organization of African Unity before 2002), and the Arab League serve few functions, although both are negotiating trade agreements. Some of today's patterns of cooperation are legacies of older relationships between imperial powers and their colonies. As we saw in

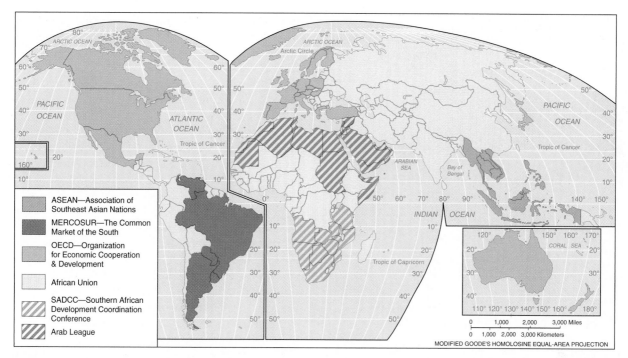

Figure 13-25 International organizations. These are just a few of the many international organizations that provide frameworks for cooperation.

Chapter 11, many former British colonies joined the British Commonwealth, a loose collection of far-flung states that seek to foster democratization and free trade among its members.

Russia and its neighbors   Russia today has extensive relationships with many of the surrounding countries as a result of its imperial and Communist past. A look back at Figure 11-2 helps to remind us that at the same time Western European countries were building overseas empires, Russia was building an empire across Asia. During World War I, the empire suffered internal rebellion as well as external attack. The Communists, led by Vladimir Lenin (1870–1924), seized control in 1917 and offered a new propagandistic ideal of a federal union among the former subject nations. This union, the Union of Soviet Socialist Republics (U.S.S.R., or Soviet Union), came into being in 1922.

The configuration and internal organization of the U.S.S.R. changed through the years, but in 1990, the Soviet Union covered 17 percent of Earth's land area (excluding Antarctica), and its population of about 290 million made it Earth's third most populous country, after China and India.

The Soviet Union was subdivided into 15 union republics, each of which was theoretically the homeland of a national group and therefore was named for that group (Figure 13-26). Several republics contained within them homelands of smaller ethnic groups. Russia was the largest republic, occupying fully 76 percent of the total territory of the U.S.S.R., and Moscow was the capital of both the Russian Republic and also of the

U.S.S.R. Within Russia itself, 88 constituent units, 20 of which were ethnically based republics, enjoyed varying degrees of autonomy. Russia is, therefore, still officially referred to as the Russian Federation.

The individual Soviet republics ostensibly enjoyed considerable autonomy, but in fact all power was centralized in Moscow. Russian Communists dominated the Union government and even the non-Russian republics. Russians attempted to acculturate the other peoples. The Russian language, for example, was preferred for central government affairs, and it was one of two official languages in all non-Russian republics. Migration of Russians into the other homelands was yet another form of Russian influence over its neighbors.

The Soviet Union contained 55 million Muslims in 1990—concentrated in Kazakhstan, Turkmenistan, Uzbekistan, Tajikistan, and Kyrgyzstan in Central Asia plus Azerbaijan in the Caucasus. Birth rates were high among these peoples, and many of them were politically dissatisfied. Their dissatisfaction threatened the Union's political stability.

The economic geography of the U.S.S.R.   The two principal characteristics of Soviet economic development were economic self-sufficiency for the Union as a whole and centralized state control. Natural resources were the property of the state, and no one was allowed to own "means of production" in such a way as to profit from the labor of others. Only a little private property and a few private-sector services were allowed.

Russia's treasury and supplies of raw materials subsidized inefficient industries in other republics.

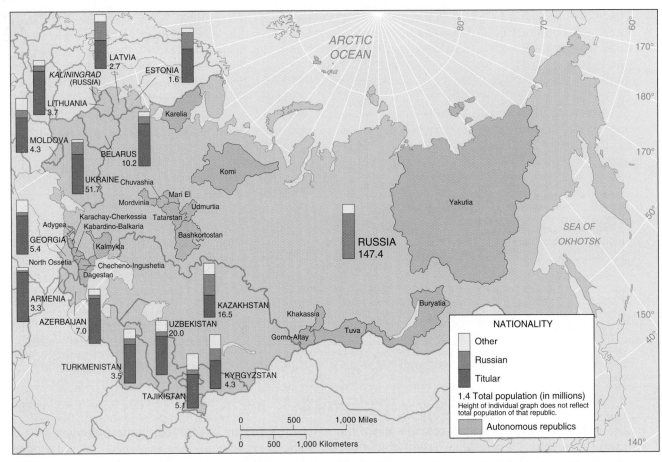

Figure 13-26 The political and population geography of the U.S.S.R. in 1990. Most of the states of the former Soviet Union are dominated by nationalities for which the republic was named, here shown as "titular" nationalities. The map illustrates that Kazakhs were a minority in "their own" republic, and in several republics immigrant peoples—often Russians—formed substantial minorities.

Agriculture was brutally collectivized. New lands were opened to agriculture, notably the steppes of northern Kazakhstan, but conditions across much of the U.S.S.R. territory—particularly northern Russia and the desert regions of central Asia—remain unfavorable for agriculture. Agriculture also suffered from mismanagement, lack of worker incentive, lack of investment, poor transport, and government policies that kept farm prices low. Heavy industry, guided by planning bureaucracies, made immense strides, but the Soviet Union lagged behind the West in light industry, in the production of consumer goods, and in high-technology products. Both production and internal trading were highly centralized, and extremely inflexible. These three factors—centralization, rigidity, and geographic concentration of production—bound the republics together in trade.

Environmental protection was sacrificed for industrialization. In 1990, Alexei Yablokov, director of the Institute of Biology at the Academy of Sciences, estimated that 20 percent of Soviet citizens lived in "ecological disaster zones" and 35 to 40 percent more in "ecologically unfavorable conditions."

The breakup of the U.S.S.R. Mikhail Gorbachev assumed power as the leader of the Soviet Union in 1985, and he tried to boost the Union's economy by launching two liberalizing initiatives: *glasnost*, a loosening of restraints against freedom of speech and the press, and *perestroika*, a restructuring of the economy and of politics. Gorbachev gambled that his policies could unleash individual initiative and productivity while retaining the centralized control of the Communist Party and of the central Union government. Gorbachev lost. The tensions unleashed brought down the old governmental system before Gorbachev could bring a new one into place.

During 1991, much was done to reform or overthrow the government of the U.S.S.R., but by the end of the year the individual republics had declared their independence. On December 26, 1991, the Soviet Parliament formally voted the U.S.S.R. out of existence. Representatives of the three Slavic republics (Belarus, Russia, and Ukraine) created a new Commonwealth of Independent States (CIS), which was eventually joined by nine more former republics. Not included in the CIS are Estonia, Latvia, and Lithuania (called the Baltic

Republics), which were independent democracies after World War I but forcibly incorporated into the Soviet Union in 1939. They achieved relative prosperity under Soviet rule, but joined the European Union in 2004.

Developments in Russia

After the fall of communism and the dissolution of the Soviet Union, Russia suffered a virtual implosion of calamitous proportions from which it seems to be emerging only now, early in the twenty-first century. Every measure of income, health, or welfare still reveals a society facing enormous challenges. The economy has shrunk, the birthrate and the total fertility rate have dropped, and the death rate has risen. Life expectancy has shortened, and Russia's population, which was 147 million in 1999, fell to 140 million in 2009 and is projected to fall to 109 million in 2050. Neither attempts to achieve democracy nor capitalism seem to have succeeded.

Vladimir Putin was elected president in 2000, in elections that would not have been considered open and fair in Western democracies. State-owned media, for example, relentlessly attacked Putin's opposition. Press freedoms have been limited and some 300 Russian journalists have been murdered since 1993. Many suspect the government had a role in their deaths. The central authority of the Kremlin has been maintained over the rest of the Russian Federation, but one ethnic unit, Chechnya (formally, Checheno-Ingushetia, as seen in Figure 13-26) launched a long and bloody struggle to achieve full independence. The 20 million Muslims in Russia constitute the majority in 7 of the autonomous republics, and fertility rates among them are high, so in 2003, Russia joined the Organization of the Islamic Conference. Russia's Muslims, said President Putin, "have every right to feel part of the Muslim world."

Meanwhile, the privatization of the state-owned economy meant seizure of many assets by the members of the old Communist government bureaucracies, by organized crime, or by a few ruthless oligarchs. *Privatizatsia* ("privatization") turned into *prikhvatizatsia* ("to grab"). Hundreds of billions of dollars were siphoned out of Russia and deposited in foreign banks during unscrupulous privatizations of state-owned companies between 1990 and 2000. Only history will decide how many leaders or profiteers in Russia in the 1990s were naïve, incompetent, corrupt, or all three. One oligarch, Mikhail Khodorkovsky, made hundreds of millions of dollars during this period, much of it from his oil company, Yukos. He was also critical of Putin's government. In 2003, he was arrested and charged with fraud. His oil company was sold off to interests close to Putin and he remains in jail.

Russia still has immense natural resources, a large skilled and educated population, and excellent scientific institutes. Soaring energy prices early in the twenty-first century returned enormous profits to Russia and it has been exercising enormous influence over world energy supplies, especially affecting the European Union, one of its biggest markets.

The other formerly Soviet republics

Experience in most of the other formerly Soviet republics (aside from the Baltics) has echoed the Russian example. Politically, they have renegotiated their relationships with the former autonomous regions within them, although several newly independent republics, including Moldova, Tajikistan, and Georgia, have experienced civil wars. In 2008, Russia and Georgia went to war over the breakaway region of South Ossetia in Georgia. As a result, Georgia withdrew from the CIS.

Almost two decades after winning independence, most of these republics guarantee few freedoms and continue to be ruled by autocrats or oligarchs. The highly centralized command economy of the U.S.S.R. left a legacy ensuring that none of the successor states has a balanced economy. Most are dependent on other former republics for essential supplies and rely on enterprises in other former republics as their sole markets. Russia continues to supply most of them with natural resources at prices below those on world markets. It is proving very difficult to unravel this economic interdependence. Most of the republics have become involved with the Eurasian Economic Community, a customs union that grew out of the CIS. Most of the new countries are privatizing their economies, but not all are privatizing at the same rate. As in Russia itself, the process in several of these countries has been corrupt, with significant meddling by Russian interests. Several of the states have skidded toward worsening poverty, instability, and international isolation.

Ukraine was the second most powerful of the Union republics, but since the breakup, its industrial might has proved largely illusory. Many factories were obsolete and represented more of a liability than an asset to a new economy. The government has been slow to reform the economy and reluctant to privatize, so inefficient state-owned industries still pile up losses. Income in Ukraine fell approximately 50 percent between 1990 and 2000, and the country remains dependent upon Russia for fuel and raw materials. Ukraine, known as the "breadbasket" of the former Union and potentially one of the richest agricultural regions on Earth, has even had to import food. The president, Viktor Yushchenko, would like to move Ukraine's alliances to the West by joining the European Union and NATO. Russia continually interferes in Ukrainian affairs.

Many former republics want to diversify away from economic dependence on Russia by giving priority to trade and investment with other partners in the West, Asia, and the Middle East. The newly independent Central Asian countries retain economic links to Russia, but their Muslim religion and other cultural ties pull them away. Turkey offers a democratic and secular model, and it is strengthening its links with Turkic peoples that dominate much of Central Asia. The Iranians, who are related to the Tajiks in language and culture and who share Shiite Islam with the Azerbaijani, have subsidized new mosques and schools in Azerbaijan. Iran is

Figure 13-27 Central Asian transport routes. The Iranian government completed a rail link from Mashad to Tedzhen in 1996, so today Iran's port of Bandar Abbas is central Asia's principal outlet to the sea. Completion of a railroad from Mashad to Bafq will shorten that route by 900 kilometers (560 miles). When the line is completed from Bafq to Zahedan, goods from central Asia will be able to bypass war-torn Afghanistan to reach Pakistan and India. A pipeline is also under construction from oil-rich Turkmenistan around the southern shore of the Caspian Sea through Iran to Turkey.

also becoming an important transit route for Central Asian countries that have long been cut off from the rest of the world (Figure 13-27). To the east, China has cultivated economic and political contacts, and even Japan and South Korea have invested in Kazakhstan and Kyrgyzstan.

Some of the Central Asian Republics—particularly Kazakhstan—may prove to be rich in oil, so the questions of who will take the lead in developing that resource and by what routes the oil will reach world markets have launched diplomatic initiatives among many concerned countries. Each possible route carries political and economic complications. China has made huge investments in Central Asia. In 2005, the China National Petroleum Corporation bought Kazakhstan's largest oil company and built a pipeline to the Chinese border. A natural gas pipeline from Turkmenistan was announced in 2009.

Even the United States has assumed a role in central Asia. The United States has used air bases in Kyrgyzstan and Uzbekistan in exchange for long-term commitment to friendship and assistance. This was partly to protect these governments from their own Islamic fundamentalist organizations. These agreements triggered concern among many regional governments that the United States intends to establish a permanent military presence in central Asia—perhaps to obtain the oil. Russia, meanwhile, formed in 2002 a Collective Security

Treaty Organization to create military cooperation with Armenia, Belarus, Kazakhstan, Kyrgyzstan, Russia, Tajikistan, and Uzbekistan. Azerbaijan and Georgia have withdrawn from the pact. The military alliance with Russia is clearly meant to exclude further U.S. military cooperation with the former Soviet republics.

One of the most volatile issues among the new central Asian republics is their competitive demands for the limited regional water supplies. Problems associated with the unequal water resources of mountainous Tajikistan and Kyrgyzstan and their low-lying neighbors Uzbekistan, Kazakhstan, and Turkmenistan broke up a U.N.-sponsored conference on regional environmental issues in 2003 and have even raised fears of armed conflict.

The dissolution of the Union of Soviet Socialist Republics created new units on the world political map, but these countries have largely remained tied together. Russia remains the political and economic leader of most of the former Soviet republics. It has gained enormous wealth from its energy exports. After the chaotic 1990s and experiments with reforms, the Kremlin once again wields enormous power in the region and has regained much of its influence on world events. Russia is now one of the major world powers, along with the United States and Europe, which has transformed regional leadership into global power. Many wonder if China or India will take the same approach as their surging economic power gives them greater influence over neighboring countries within their regions.

Human Development

In previous chapters, we have explored the geographic aspects of many problems that negatively affect the lives and livelihoods of people around the world. We have seen that different patterns of health, diet, wealth, housing, and even political relations are related to environmental and cultural factors in complex ways. A geographic approach is necessary to appreciate the enormity of the challenges facing people in different areas. These insights are also fundamental to understanding human development, which is a broad concept that describes any effort at improving the human condition. Most development efforts concentrate on improving the material conditions of health and wealth, in hopes that they will form the basis for greater contentment, social stability, and productive creativity.

Comparing the United States today to just 100 years ago reveals much about what we idealize as development. In 1900, the average life expectancy was about 48 years, and most deaths were due to infectious disease. Americans born today can expect to live until 80 on average and will most likely die from age- and lifestyle-related diseases. In 1900, about one-quarter of boys between the ages of 10 and 15 worked jobs, as did one-tenth of girls. Only half of U.S. children were enrolled in schools. Child labor laws and public education have turned that

Figure 13-28 Human development in America. Children as young as four were shucking oysters in this Louisiana cannery in 1911. Better nutrition, education, and levels of income helped improve life for most Americans during the last century. Laws banning childhood labor helped to ensure that more children received schooling.

around so that today, child labor is illegal, and school enrollment rates are over 95 percent (Figure 13-28). Average incomes are today eight times greater than in 1900, and there have been major advances for women and minorities. The improvement in human development in the United States owes much to economic growth spurred on by vast natural resources and continuous immigration. But the United States has also gained enormously from large government projects, such as highways, dams, housing, and public services. It has also gained from the geopolitical power gained in World War II, which has translated into advantageous conditions for U.S. trade and finance.

As in the United States, European levels of development grew tremendously over the nineteenth and twentieth centuries, despite several wars. Western European countries grew wealthy at the expense of their colonies, while producing an impoverished working class at home. It was the effort to address the difficulties of the working poor in Western Europe that began the idea of development as the improvement of society. In time, it became a hallmark of democracies that they should enable individuals to improve their lives. After World War II, the only societies that experienced significant improvement in development were those that industrialized, including the Soviet Union, Japan, and a few other countries.

It is essential to recognize that today's developed countries have had natural and historical advantages, often at the expense of poorer countries. The geographic and historical conditions of poor countries

means that they cannot wholly emulate the developed countries. Bangladesh does not have the farmland and mineral wealth of the United States. Vietnam will not become an imperial power, prospering from the misery of others. They must find other paths to development.

Most of the world's poor are on their own, left to struggle against disease, environmental degradation, ignorance, and poverty. Governments are too weak or unwilling to help and international programs cannot reach everyone. Amid the growing wealth of India's cities, for example, migrants from rural areas set up camp between railroad lines. They come in search of work at the bottom end of the economic ladder. Malnourished and uneducated, they live desperate lives just inches from bustling crowds of relatively healthy well-paid workers. These people and most people living in the squatter settlements described in Chapter 10 are seeking to improve their conditions, and they've moved to the city as a kind of do-it-yourself development program. Some may eventually gain better wages, improve their housing and nutrition, and send their children to school. Others will fall prey to disease, debt, and violence. It's quite a gamble, considering they hold none of the cards.

On the scale of the nation-state, improving the quality of life for citizens is a basic expectation of modern governments, although many fail in this regard. Successful states have many ways of promoting development. Public services can directly target health problems or provide minimal assistance to those in need. Compared to the social costs of epidemics and crushing poverty, these programs are relatively affordable. But because the counties with the greatest needs for development are among the poorest, government assistance is at best a "safety net" rather than a ladder out of poverty. Public education, however, has been one of the most successful government interventions in the lives of citizens in the modern world (Figure 13-29). Education makes people better able to provide for themselves and better able to participate in public life, strengthening their communities and society.

On a global scale, development since the 1940s has been understood as an international effort to reduce disparities in living conditions. This effort was given added importance because of the numerous countries that gained their independence during this time. The Green Revolution and advances in food production were hoped to end hunger, as we saw in Chapter 8. International bodies such as the World Health Organization, United Nations Development Program, and UN-Habitat were created to improve the health and living conditions of persons living in poorer countries. International financial institutions, such as the World Bank described in Chapter 12, are supposed to promote development by improving economic conditions, although their economic policies have actually worsened conditions in many countries.

Although global agencies are important, most international development programs are run by national

Figure 13-29 Girls' school in Afghanistan. The importance of education in improving the lives of Afghani girls since the U.S. invasion in 2001 has been complicated by the ongoing fighting and traditional attitudes about women's roles. Some girls' schools have been targeted by Islamic fundamentalists.

governments working in other countries. There are many motivations for wealthier governments to promote development around the world. Some are driven by a sense of social justice to reduce suffering and perform acts of charity. Norway and Sweden, for example, spend the largest share of their national income (about 1 percent) on international development work. They are joined by Luxembourg, the Netherlands, and Denmark as the only five countries to give more than the 0.7 percent pledged in 1970 by the countries of the world. On average, the wealthier, industrialized countries direct about 0.45 percent of their national incomes toward international development. On this measure, the United States ranks lowest among other industrialized countries, spending only 0.16 percent of its national income. However, because the U.S. economy is so large, this small share of national income yields the highest level of spending in actual dollars by any government, totaling some $27 billion in 2008. The U.S. Agency for International Development is the main conduit for these funds, but aid is also spent through contributions to U.N. programs such as the United Nations High Commissioner for Refugees (UNHCR).

International aid comes in many forms, from bags of food to complex interventions in a country's currency. The $121 billion spent by developed countries in 2008 targeted all regions of the world, with about 20 percent going to sub-Saharan Africa and about 13 percent to the Middle East and North Africa. Iraq was the largest single recipient of international aid in 2008, totaling $9.5 billion, followed by Afghanistan, which received some $3.5 billion. It is somewhat surprising that in spite of its rapid economic development, China was the third largest recipient of international aid. Aid to China sparked a controversy in the United Kingdom, where the Conservative party raised objections to sending $61 million on poverty alleviation to a country that hosted an expensive Olympic games and a space exploration program. One politician said, "British taxpayers need to know that their aid money is helping the poorest people in the world, not going to countries which have enough money to tackle poverty themselves." Such controversies in donor countries highlight the political dimension to international aid. Some countries receive aid at the expense of other countries with greater needs, often because donor governments are pursuing national security interests. China, for example, gives aid to countries where it sees future energy supplies, much like other wealthy donor governments.

Measuring Development

A concept like human development is too broad to capture in one statistic. In its fullest sense, development means helping people to improve their own lives and it is not always possible to measure this improvement. There are also differences about whether development means meeting basic needs or achieving personal satisfaction with one's condition. Is it enough to have good health, enough to eat, and basic shelter? Or does development also entail a sense of contentment and happiness, which might reflect various cultural and religious traditions, as we saw in Chapters 6 and 7. Similarly, some people might define development as an endless process of accumulating money and consumer goods. And how do we compare levels of development?

The previous chapters in this book demonstrate that the conditions of human existence are interrelated in complex ways. Understanding and improving development, therefore, has produced piles of statistics and reports of all the ways we might try to measure it. The World Bank Development Report for 2009 is 456 pages long and contains 90 data tables, each with multiple indicators related to development. These range from measures of economic output, such as gross domestic product (GDP) and income, to rates of HIV infections and urbanization rates. The United Nations Development Program reports are similarly complex and national development programs sometimes report data at the village level.

Many economists and politicians focus on measures such as GDP or GNI, which were introduced in Chapter 12 to represent the sum total of a country's annual economic activity. Looking back at Figures 12-6 and 12-7, we see the familiar pattern of wealthy Western countries, a much poorer Africa and south Asia, and the rest of the world somewhere in between. Of course, levels of economic productivity and levels of poverty are closely related. But statistics such as GDP or GNI per capita assume that wealth is shared equally in a country, which is never the case. In developing countries, the more typical level of income is a fraction of per capita statistics. It is impressive that India has a GDP of $1.2 trillion but far less so that it has a GNI per person of only $1,070. It is even more revealing to know that 75.6 percent of India's population live on less than $2 per day ($730 per year).

At the same time, psychologists have suggested that the focus on material consumption is misguided and that we should consider how content people are with their situation. Some intrepid researchers have tried to measure Gross National Happiness, which would increase with improvements in economic conditions, the quality of the natural environment, cultural preservation, and democratic governance. Bhutan, a remote country in the Himalaya Mountains, whose king proposed this measure, is one of few countries that ranks highly on happiness despite its relative poverty in terms of GDP, presumably because its cultural values place greater emphasis on spiritual and community matters. Rapid cultural change and globalization, as described in Chapter 6, suggest that more objective measures are more reliable, even if they do not fully capture happiness.

The Human Development Index (HDI) The United Nations Development Program introduced the **Human Development Index (HDI)** in 1990 to capture three of the most widely accepted criteria of human development: health, knowledge and a decent standard of living. This measure views development as something rooted in having a long and healthy life, an education, and a reasonable level of material wealth. Health is measured as life expectancy at birth, as described in Chapter 5. Knowledge is a combination of adult literacy rates and the rate of school enrollment at all levels. The standard of living is measured as GDP per capita, and these amounts are converted to U.S. dollars and adjusted for purchasing power parity (PPP), as described in Chapter 12. These figures are combined in a rather complex calculation to yield the Human Development Index score, which can range between 0.000 (very low) and 1.000 (very high), depending on the factors. It is therefore possible that countries with a long life expectancy and high levels of education but a weak economy would score as "more developed" than a country with a strong economy whose citizens have less education and lower life expectancies.

The 2009 Human Development Report listed Norway (HDI=.971) as the most developed country in the world (Figure 13-30). Norwegian life expectancy is 80.5 years, literacy rates and school enrollments are about 99 percent, and GDP per capita is $53,433. Close behind Norway are Australia, Iceland, Canada, and Ireland. The top 25 countries are predominantly European and include Japan, the United States, New Zealand, Singapore, and Hong Kong. The 25 countries on the low end of the HDI are all in sub-Saharan Africa except for East Timor and Afghanistan, which are both countries that have recently experienced conflicts. The lowest-ranked country is Niger (HDI = .340). Life expectancy is 50.8 years, adult literacy is 28.7 percent, only about 27 percent of school-aged persons are enrolled, and the GDP per capita is a mere $627.

It is interesting to note that if we use the HDI to measure the level of U.S. development over a century ago, it would today rank about 135 out of today's 182 countries, roughly on par with the level of development today in India or Cambodia.

Gender and development As we saw in previous chapters, the status of women in many cultures is far below that of men and undervalues their fundamental importance in society. Gender discrimination limits the ability of women as individuals to fully participate and contribute to society. Returning to our discussion of population dynamics in Chapter 5, women not only bear the biological burdens of reproduction but, as we saw in Chapter 6, culturally defined gender roles often assign women the primary responsibility for raising children. Recall that female education has had a striking effect on lowering total fertility rates, primarily because an educated woman is less likely to be limited by cultural biases regarding the value of large families. In turn, they are more likely to earn wages outside the home and participate in community life, raising their status and giving them more control over their lives. It is easy to see that development is meaningless if it does not improve the lives of women, both as individuals with rights to participate in society and as mothers and caregivers.

The United Nations Development Program collects data on women to produce the Gender-Related Development Index, using the same measurements as the

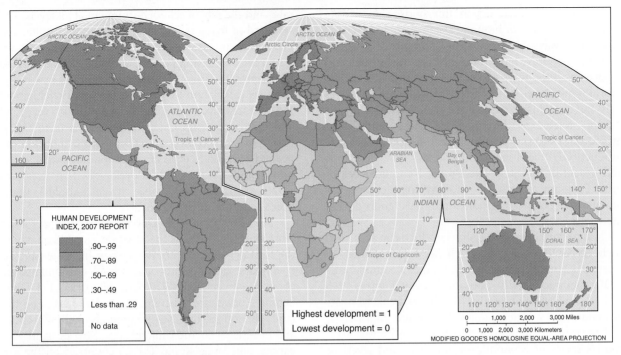

Figure 13-30 Human Development Index. The United Nations Development Program created the HDI to provide a comparable measure of three basic components of human development: wealth, health, and education.

HDI. Most countries' rank on the GDI is the same as or very close to their rank on the HDI. This does not mean that men and women's development is equal, however. Norway, ranks first in HDI and second in GDI. Norwegian women live longer and are better educated than Norwegian men; women in Norway, however, have an average estimated income that is only 77 percent that of men. Similar differences characterize the other highly developed countries. Among countries that rank lower on HDI, however, women tend to fare much worse on all measures except life expectancy.

One of the most striking differences between women in countries with high and low development is access to education. This may be visualized by comparing female to male enrollments in primary and secondary school (Figure 13-31). There are many obstacles to female education. In many countries, basic education is not free. A family may be too poor to send all their children to school and so one boy is sent while the other boys and girls work around the home. In other cultures, women are not permitted outside the home or are considered unable to learn. Often, it is simply sexism. Whatever the reason, women without an education are less equipped to work outside the home or participate in public life. As we saw in Chapter 5, improving women's education significantly lowers total fertility rates in some developing countries. Women's participation in the labor force and levels of pay discrimination, as we saw in Chapter 12, are useful ways of measuring the fulfillment of women's individual rights.

Likewise, women's participation in politics and civil society serves as a proxy for their role in public life, especially at the local level.

In 2007, the World Economic Forum ranked 58 countries according to a "gender gap" measure of economic opportunity, economic participation, political empowerment, educational attainment, and health and well-being (Table 13-2). The gap measured narrowest in four Nordic countries, where men and women experience very similar levels of development. The gaps between men and women were widest in many Muslim Middle Eastern countries, as well as much of South Asia and sub-Saharan Africa.

The Role of the Environment

A good share of what enables development is wealth, and, in a basic sense, wealth is derived from the natural environment. As we saw in Chapters 8 and 9, humanity's basic material needs are fulfilled by transforming our environment. Forests are cut down for wood and the land is cleared for planting crops. Mines are dug to access metal ores that are melted by burning coal or oil. We noted above that the more developed countries in the world today industrialized early, often by extracting raw materials from countries that are today less developed. Rubber tappers, fur trappers, and tea and cotton plantations from Africa, Asia, and Latin America provided much of the raw goods over the past 300 years that have made Western European countries wealthy.

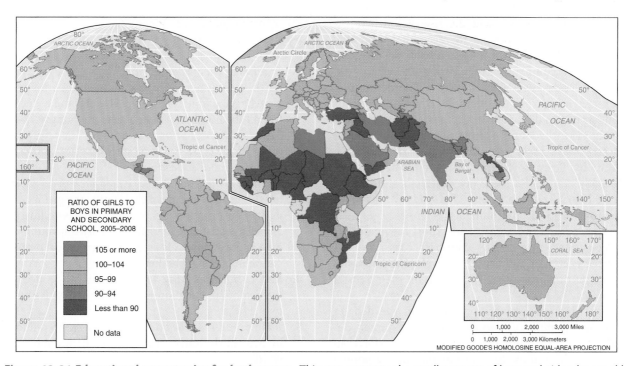

Figure 13-31 Educational opportunity for both sexes. This map compares the enrollment rate of boys and girls who are old enough to be in primary or secondary school. Countries scoring below 100 have higher proportions of boys than girls in school and those above 100 have higher proportions of girls than boys. The differences suggest important cultural differences or the effect of national policies on educational equality.

TABLE 13-2 Women's Empowerment, Measuring the Global Gender Gap, 2009

25 Highest-Ranked Countries			25 Lowest-Ranked Countries		
Country	Rank Out of 134 Countries	2009 Score	Country	Rank Out of 134 Countries	2009 Score
Iceland	1	0.8276	Nepal	110	0.6213
Finland	2	0.8252	Guatemala	111	0.6209
Norway	3	0.8227	United Arab Emirates	112	0.6198
Sweden	4	0.8139	Jordan	113	0.6182
New Zealand	5	0.7880	India	114	0.6151
South Africa	6	0.7709	Korea, Rep.	115	0.6146
Denmark	7	0.7628	Bahrain	116	0.6136
Ireland	8	0.7597	Algeria	117	0.6119
Philippines	9	0.7579	Cameroon	118	0.6108
Lesotho	10	0.7495	Mauritania	119	0.6103
Netherlands	11	0.7490	Burkina Faso	120	0.6081
Germany	12	0.7449	Syria	121	0.6072
Switzerland	13	0.7426	Ethiopia	122	0.5948
Latvia	14	0.7416	Oman	123	0.5938
United Kingdom	15	0.7402	Morocco	124	0.5926
Sri Lanka	16	0.7402	Qatar	125	0.5907
Spain	17	0.7345	Egypt	126	0.5862
France	18	0.7331	Mali	127	0.5860
Trinidad and Tobago	19	0.7298	Iran	128	0.5839
Australia	20	0.7282	Turkey	129	0.5828
Barbados	21	0.7236	Saudi Arabia	130	0.5651
Mongolia	22	0.7221	Benin	131	0.5643
Ecuador	23	0.7220	Pakistan	132	0.5458
Argentina	24	0.7211	Chad	133	0.5417
Canada	25	0.7196	Yemen	134	0.4609

Scores can range from 1.00 (highest) to 0.00 (lowest).

Source: World Economic Forum, Global Gender Gap Report 2009.

The end of empires and the rise of independent nation-states were supposed to mean that economic benefits would no longer flow to already wealthy, industrialized countries. But existing industries did move to lesser-developed countries. The developing countries have tried the various economic development strategies described in Chapter 12. Few of these strategies worked, and populations in less-developed countries still make their livelihoods directly from nature by farming, fishing, foresting, and mining. Meanwhile, much of these raw materials still flow to the industrialized countries. This leads to increased pressure on local environments as growing populations try to feed themselves and export raw materials for consumers in wealthier countries. Trying to improve the human condition in developing countries means putting even greater demands on their natural resources.

This dilemma should remind us of the debates about overpopulation and the burden of human demands on environmental resources described in Chapter 5. Certainly, food production in developing countries could be made more responsive to local needs rather than grown for export at prices too low to improve many lives. Chapter 8 also introduced *sustainability* to describe the use of resources that can continue without a drop in productivity and without degrading the environment. **Sustainable development**, in fact, has become a basic idea about how to improve the lives of impoverished populations while also preserving the Earth's capacity to support future generations. Because so many of our basic human needs are satisfied by using natural resources, this idea will continue to dominate development programs and goals. Slowing population growth is thus a necessary component of achieving sustainable development, especially with increasing life expectancies that add to population size. But sustainably utilizing Earth's resources also requires a conversation about what constitutes enough development and for whom. We already know that the consumption patterns in industrialized countries are unsustainable if everyone in the world were to live that way. Yet some in the poorer countries see sustainable development as limiting their opportunities for economic growth so that wealthier countries can continue to enjoy high levels of resource use.

Millennium Development Goals

Video
The Millennium Goals—Dream or Reality?

One of the most ambitious international programs to promote sustainable development was sponsored by the United Nations at the start of the second millennium. The countries of the world and the major international organizations involved in development met in 2000 and agreed to a set of shared priorities. The U.N. Millennium Conference set eight development goals that the world could achieve by 2015. Some of the Millennium Development Goals use 1990 as a baseline for comparison:

1. *Eradicate extreme poverty and hunger.* The target for this goal is to reduce by half the number of people whose income is less than $1 a day and reduce by half the number of people who are hungry.
2. *Achieve universal primary education.* The target for this goal is that all boys and girls have access to a primary school education.
3. *Promote gender equality and empower women.* Meeting this goal would mean eliminating the gender disparity in primary and secondary school enrollments, as well as equalizing literacy rates between young men and women. It also seeks women's equal participation in national parliaments.
4. *Reduce child mortality.* This goal seeks to reduce the death rate of children under the age of five by two-thirds.
5. *Improve maternal health.* This goal seeks to reduce by two-thirds deaths caused by pregnancy and childbirth.
6. *Combat HIV/AIDS, malaria, and other diseases.* The target for this goal is stopping and beginning to reverse the spread of these diseases.
7. *Ensure environmental sustainability.* This goal seeks to promote adoption of environmentally sustainable practices by national development programs and to reverse the loss of forests. It also seeks to halve the number of people without access to improved drinking water and sanitation, while also improving the lives of slum dwellers.
8. *Develop a global partnership for development.* This goal includes a wide range of targets, among which the most important is the creation of fair international trade and finance systems committed to promoting development, good governance, and poverty reduction. Another important target is eliminating the national debt that burdens the least developed countries.

Reaching any of these goals would be a major accomplishment, yet they are achievable with concerted effort by the world's major development agencies, development programs among the wealthier countries, and developing countries themselves. Much of the burden of implementing and sustaining the eight Millennium Development Goals falls to individual countries and most countries can afford to cover most of the expenses. The least developed countries, however, are also the poorest and have few domestic resources to confront some of these challenges. The United Nations Development Program estimates that international assistance by wealthier countries will amount to 0.5 percent of their GNI until 2015. This is substantially less than the 0.7 percent of annual GNI pledged by countries four decades ago but amounts to twice their actual level of spending. Other goals, such as women's empowerment, require something far more

The Importance of Water for Human Development

When the countries of the world agreed to the Millennium Development Goals, they specifically made access to safe drinking water and sanitation part of ensuring environmental sustainability. This basic commodity is perhaps the perfect symbol of humanity's reliance on the natural environment. And the lack of access to safe water for basic needs is a cornerstone for all other development.

People in developed countries have few problems getting the water they need and perhaps take for granted the fundamental role it plays in our daily lives. Count the number of times you use a sink, toilet, or shower during a day. Now imagine if you had to walk 15 minutes to find that water, wait in line to fill your containers, and then walk back with more than 18 kilograms (40 pounds) of water! This is a typical situation for those in sub-Saharan Africa and millions more around the world. Drinking, cooking, and keeping clean suddenly become tiring tasks that take up much of the day.

Water is a basic need for keeping clean and healthy. The great strides made in public health across the industrialized world in the nineteenth century required increased water use. Moving human waste away from residential areas, cleaning food, and reducing exposure to germs through hand washing helped to lower the mortality rate and improve the quality of life. The 2.6 billion people in the world today who cannot sanitarily remove their waste from their surroundings live amid open sewage. In such conditions, as are common in many developing countries, it is even more important to keep oneself and one's food clean but, in fact, it becomes harder without water.

The World Health Organization and the U.N. Children's Fund note that the lack of access to clean water is especially hard on children who suffer from water shortage– related diseases while also bearing the burden of water collection. Some worry that population increase will strain water

resources, leading to water shortages in drier areas. Many areas are already using their fresh water supplies unsustainably. Others argue that there is enough freshwater, if properly used. Steps such as fixing leaking water mains, managing reservoirs, and charging nominal fees for water use would conserve enough water. Ensuring that water reaches the most vulnerable people in any country is more difficult.

A child in Gulu, Uganda, carries water home from a well.

contentious than money because they challenge cultural norms regarding women's role in society.

Geography and Thinking Globally

The most pressing issues in the world today are not limited to any one place or country. Rapid environmental degradation, war and humanitarian crises, political and economic fragmentation, and the hardships experienced by much of the world's population are all part of our connected and globalized world. Geography is the study of the Earth in its whole and its local variety. It is fundamentally about the connections between places, peoples, and their environments. It has been a very long time since humans in one part of the world were untouched by events in other places. Today, our lives and our future are more interconnected than ever before, and we are more aware of how global trends shape the world around us. Global warming, H1N1, financial crises, and wars are obvious to everyone as issues that link far-flung parts of the world. But so too are rising grain prices, regional integration, and efforts toward human development. Geography provides a way of understanding how these issues unfold at different scales and connect places around the globe. Geography also explains why these changes have different effects on different places.

As we consider the many consequences of global integration and greater cultural and economic interaction between rich and poor nations, the greatest challenges will be those arising from differences in the value systems, both economic and cultural, of the people involved. When European countries open the borders they once fought over and share the same economic destiny, then they become part of the same political–economic system. That joint membership requires that they share certain values and agree to operate by the same set of rules. The political gulf between the United States and China, on the other hand, is much larger. These two nations are the two largest emitters of carbon dioxide in the world, thus making it more difficult for them to reach agreement on controlling global warming.

The enormous economic gaps between rich and poor, both between and increasingly within countries, only multiply the challenges posed by increased global interaction. Will human development be the key to unlocking economic development in the poorer nations of the world? Can economic development be rapid enough to reduce the disparities in national income between these countries and the rich countries? Or will it perpetuate and strengthen the inequities that result from the greater ability of rich countries to capture value added? As we grapple with the many social and political tensions created by global change and integration, will we have the opportunity to protect what remains of the natural environment? Or will we instead be so focused on economic development alone that the idea of sustainable development will remain more an idea than a reality?

These questions cannot be answered today. In searching for the answers, however, it is clear that we need to consider a wide range of topics simultaneously: climate, resources, culture, religion, politics, and economics. Such a wide-ranging viewpoint is open to students of any subject, but geography especially encourages such broad-minded analysis.

Chapter Review

Summary

Globalization, population growth, and resource and environmental stresses have created and necessitated increased interaction among the peoples and governments of the world. The division of the world into sovereign states that hold the greatest power poses problems for resolution of global-scale problems.

In recent decades, the location of manufacturing activities that generate a large share of the world's atmospheric pollution has changed, with the postindustrial economies transitioning to service-based activities, and rapid industrialization in less developed parts of the world such as south and east Asia and Latin America. Control of ozone-depleting chemicals such as CFCs could be achieved because replacing CFCs with less harmful substances could be accomplished with relatively minor economic hardship and because the countries most concerned about their effects were also the ones responsible for the greatest production. Control of CO_2 emissions is much more difficult. While past emissions of carbon dioxide and other atmospheric pollutants were almost entirely generated by the wealthier countries, the greatest growth in emissions is occurring in rapidly industrializing countries. This unevenness is a major barrier to international efforts to limit the emissions that are causing global warming.

Although state sovereignty remains the primary form of political organization in the world, states often find it useful to cooperate. The United Nations was created after World War II to foster greater cooperation and preserve global peace and security through collective security. Today, it serves as a forum for international negotiation and coordination of peacekeeping activities, and as a means of organizing social programs, economic development programs, and other international cooperative undertakings. The diffusion of universal human rights has not been adopted in all countries and cultures, but it is an important set of political rights that applies to individuals. Some states think that preserving human rights and averting humanitarian conflicts is reason to violate another state's sovereign territory. Other states disagree.

Some of the strongest forms of economic and political cooperation among states and societies occur at a regional level. New regional blocs are appearing, most notably the European Union and the North American Free Trade Area, but also Mercosur in South America and the Association of Southeast Asian Nations. The European Union is unique because its treaties bind the countries together, and each gives a central supranational government certain power, but it is not a federal government. The EU's forms of cooperation are broad, while other regional organizations are limited to economic or defense agreements.

Economic growth over the past five centuries has produced enormous wealth in developed countries, but many parts of the world remain poor. Human development seeks to improve the quality of life among the world's poor not only by encouraging economic growth but also ensuring access to sufficient housing, education, and health care. Developed countries have pledged many times to help pay for poverty reduction programs, which would cost a small portion of their incomes, but have never fully delivered on their promise. The most common measure of human development is the Human Development Index, which provides a measure of wealth, health, and education. Factors such as women's equality and the quality of one's environment are other important measures of development. The Millennium Development Goals represent the most concerted international effort to date to significantly reduce poverty and improve access to basic needs.

Key Terms

bilateral agreement p. 516
civil society p. 517
Cold War p. 538
collective security p. 529
common market p. 517
Council of Europe p. 539
customs union p. 517
energy intensity p. 518
European Union (EU) p. 539

free-trade area p. 517
global civil society p. 517
global security p. 529
human development p. 516
Human Development Index (HDI)
  p. 557
human rights p. 528
intergovernmentalism p. 516
local content requirement p. 547
maquiladora p. 546
Marshall Plan p. 538

multilateral agreement p. 516
nongovernmental organization p. 517
North American Free Trade
  Agreement (NAFTA) p. 543
North Atlantic Treaty Organization
  (NATO) p. 537
Privatization p. 521
Organization for Security and
  Cooperation in Europe p. 539
regional cooperation p. 536
supranationalism p. 516
sustainable development p. 560

Questions for Review and Discussion

1. What is energy intensity? How is energy intensity changing? What are the causes and significance of these changes?

2. The Montreal Protocol is generally regarded to have been successful in controlling problems of ozone depletion, while the Kyoto Protocol has not been successful in controlling carbon emissions. What factors contribute to the success or failure of these treaties?

3. What arguments have governments presented to explain why they refuse to sign global environmental treaties?

4. How have governments tried to control the diffusion of nuclear weapons?

5. Name the members of the European Union. Describe the government of the European Union.

6. What activities and results of the European Union prove that it is more than a free-trade area or customs union, but rather is a full common market?

7. Is NAFTA a free-trade area, customs union, or common market? What kind of market exists among South American countries?

8. What is human development? How does the United Nations measure it?

9. Why isn't greater wealth sufficient to increase human development?

Thinking Geographically

1. In what ways does economic development promote improved environmental quality, and in what ways does it reduce environmental quality?

2. International diplomacy generally observes the principle of nonintervention in sovereign states, and this principle often preserves misgovernment and unnecessary suffering. On what grounds could the principle of nonintervention be overturned?

3. Investigate a few of the projects of the specialized agencies of the United Nations, such as UNICEF, the FAO, or WHO.

4. List factors of European cultural unity and disunity. As the European Union expands, how will various factors boost or retard cooperation among the members and economic development?

5. How has physical geography encouraged or discouraged the formation of economic relationships in Europe, in the CIS states, or in North America? How have transportation or settlement patterns altered these relationships?

6. What measures other than those in the human development index do you think are also important? Why do you think your measures are not used in the HDI?

Log in to www.mygeoscienceplace.com for videos, animations, Map**Master**™ interactive maps, RSS feeds, case studies, and self-study quizzes to enhance your study of Global Challenges and the Scale of Response.

MapMaster™

The Köppen Climate Classification System

Climate classification is a challenging intellectual and scientific enterprise. Climates vary in many different ways, the most significant of which are temperature, precipitation, and seasonal variations in these variables. Each of these dimensions of variation should be portrayed by a useful classification system.

In addition to the multiple criteria that must be considered, none of these variables falls into simple categories, like male and female. Temperature and precipitation vary through wide ranges, with gradual transitions from place to place. Climate classification is a little like naming colors. Most people would agree on what is meant by "blue" and what is meant by "green." Finding agreement on the difference between "bluish-green" and "greenish-blue" (let alone names like "turquoise" or "teal") is not so simple.

Wladimir Köppen (1846–1940) was a German climatologist who undertook this difficult task. Köppen recognized that there is a strong correlation between climate and vegetation types, and he reasoned that if climate types corresponded to significant vegetation types, the classification would be more useful and meaningful. A correlation between vegetation and climate is also useful in mapping climates in areas where few weather data are available but where vegetation can be mapped. In this book, we use a modified version of Köppen's system, summarized in this appendix.

Köppen's approach was to identify five basic climate types and name them A, B, C, D, and E, in rough order of their latitudinal location on Earth (A in the tropics; E at the poles). Four of these are humid climate types and one, B, is an arid type. Each is divided into subtypes, with the second letter providing information about rainfall, and the third indicating temperature, except for polar climates in which only two letters are used and the second letter pertains to temperature. All climates are distinguished on the basis of mean monthly data.

For the humid climates, the second letter of the classification (f, s, m, or w) is used to describe the seasonal distribution of rainfall. In midlatitude climates,

f indicates evenly distributed rainfall, s indicates summer dry, and w indicates winter dry. In tropical (A) climates, the letter m is used to indicate a monsoonal precipitation regime, and the summer (or high-sun) dry type, s, does not exist. The third letter (a, b, c, or d) is used to distinguish warmer and cooler variations of the subtropical and midlatitude (C and D) climates, with a being the warmest and d the coolest.

For dry climates (B), the second letter (W or S) distinguishes desert from semiarid climates. The specific criteria for distinguishing these two from humid climates and from each other are based on mathematical formulas comparing temperature and rainfall. This is because plant water needs are temperature dependent. In a cool climate, less rainfall is needed to support vegetation than in a warm climate. This is also the reason for differences in rules for identifying "winter dry" and "summer dry" climates.

Some of the climates distinguished by Köppen do not relate clearly to vegetation differences. In eastern North America, for example, the transition from Dfa to Dfb climates passes through the middle of the broadleaf deciduous forest region; neither does the line distinguish any major agricultural regions. However, the boundary between Dfb and Dfc corresponds relatively closely to the northern limit of commercial agriculture in this area.

In this book, we have simplified Köppen's scheme somewhat, especially in Figure 2-33 and the section describing climate types. We chose to identify 11 different major climate types that correspond reasonably well to significant vegetation differences. These types are named humid tropical, seasonally humid tropical, desert, semiarid, humid subtropical, marine west coast, Mediterranean, humid continental, subarctic, tundra, and ice cap and ice sheets. Each of these corresponds to one to four subtypes of a Köppen major type. Our simplifications are designed to retain distinctions that relate clearly to natural vegetation and agricultural regions, while eliminating those that divide major vegetation regions.

Letter			Derivation	Description	Definition	Climate Types as Used in This Book
First	Second	Third				
A			Alphabetical	Tropical humid climates	Average temperature of each month above 18°C (64°F)	
	f		German: *feucht* ("moist")	No dry season	Average rainfall of each month at least 6 cm (2.5 in.)	Humid tropical (Af)
	m		Monsoon	Monsoonal; short dry season compensated by heavy rains in other months	1 to 3 months with average rainfall less than 6 cm (2.5 in.)	Humid tropical (Am)
	w		Winter dry	Dry season in low-sun season	3 to 6 months with average rainfall less than 6 cm (2.5 in.)	Seasonally humid tropical (Aw)
B			Alphabetical	Dry climates; potential evaporation exceeds precipitation	Dry climates are determined by a set of formulas relating temperature and precipitation. Different formulas are used to determine whether precipitation is concentrated in the summer, winter, or neither. If the wettest winter month has three or more times the rainfall of the driest summer month, the climate is considered to have a winter concentration of rainfall. If the wettest summer month has 10 or more times the rainfall of the driest winter month, the climate is considered to have a summer concentration of rainfall. If neither of these is true, rainfall is considered evenly distributed.	
	W		German: *wuste* ("desert")	Arid climates	If precipitation is concentrated in the high-sun season, average annual precipitation (cm) less than mean annual temperature (°C). If precipitation is concentrated in the low-sun season, average annual precipitation (cm) less than mean annual temperature (°C) plus 14. If precipitation is evenly distributed through the year, average annual precipitation (cm) less than mean annual temperature (°C) plus 7.	Desert
	S		Steppe, or semiarid	Semiarid climates	If precipitation is concentrated in the high-sun season, average annual precipitation (cm) less than 2 times mean annual temperature (°C). If precipitation is concentrated in low-sun season, average annual precipitation (cm) less than 2 times mean annual temperature (°C) plus 14. If precipitation is evenly distributed through the year, average annual precipitation (cm) less than 2 times mean annual temperature (°C) plus 7.	Semiarid
		h	German: *heiss* ("hot")	Low-latitude dry climate	Average annual temperature above 18°C (64°F)	Hot desert (BWb) or hot semiarid (steppe) climate (BSk)
		k	German: *kalt* ("cold")	Midlatitude dry climate	Average annual temperature below 18°C (64°F)	Cool desert (BWk) or cool semiarid (steppe) climate (BSk)
C			Alphabetical	Midlatitude humid climates	Average temperature of coldest month between 18°C (64°F) and –3°C (27°F)	
	s		Summer dry	Dry summer	Driest summer month has less than one-third the precipitation of the wettest winter month	Mediterranean (Csa, Csb)
	w		Winter dry	Dry winter	Driest winter month has less than one-tenth the precipitation of the wettest summer month	Humid subtropical (Cwa)
	f		German: *feucht* "moist"	No dry season	Does not fit either s or w above	Marine west coast (Cfb, Cfc)
		a	Alphabetical	Hot summers	Average temperature of warmest month above 22°C (72°F)	Humid subtropical (Cfa)
		b	Alphabetical	Warm summers	Average temperature of warmest month below 22°C (72°F); at least 4 months with average temperature above 10°C (50°F)	Marine west coast (Cfb, Cfc)

(Continued)

Letter						
First	**Second**	**Third**	**Derivation**	**Description**	**Definition**	**Climate Types as Used in This Book**
		c	Alphabetical	Cool summers	Average temperature of warmest month below 22°C (72°F); fewer than 4 months with average temperature above 10°C (50°F)	
D			Alphabetical	Humid midlatitude climates with cold winters	4 to 8 months with average temperatures above 10°C (50°F)	Humid continental (Dfa, Dfb, Dwa, Dwb); Subarctic (Dfc, Dfd, Dwc, Dwd)
Second and third letters same as in C climates						
		d	Alphabetical	Very cold winters	Average temperature of coldest month below −38°C (−36°F)	
E			Alphabetical	Polar climates	No month with average temperature above 10°C (50°F)	
	T		Tundra	Tundra climates	At least one month with average temperature above 0°C (32°F)	Tundra (ET)
	F		Frost	Ice cap climates	No month with average temperature above 0°C (32°F)	Ice cap and ice sheets (EF)

GLOSSARY

A

absolute location The location of a place as pinpointed in terms of the global geographic grid

accent A dialect difference that involves pronunciation only

acculturation The process of adopting some aspect of another culture

acid deposition The deposition of acidic substances on the ground, primarily as a result of sulfur and nitrogen oxide pollution of the atmosphere

actual evapotranspiration (ACTET) The amount of water evaporated and/or transpired in a given environment

adiabatic cooling The cooling of air as a result of expansion of rising air; *adiabatic* means "without heat being involved"

advection The horizontal movements of air or substances by wind or ocean currents

African diaspora The migration of black peoples out of Africa

agglomeration The bringing of people and activities together in one place for greater convenience

agricultural inputs All the inputs (materials, labor, capital) that go into growing and harvesting a crop

Agricultural Revolution The application of science and technology to agriculture, resulting in greatly increased yields and releasing workers for other occupations

alluvial fan A fan-shaped deposit of sediment formed where a stream emerges from a narrow canyon onto a wider valley floor

alpine glacier A glacier occupying a valley in a mountainous area. The movement of an alpine glacier is primarily governed by the underlying topography

angle of incidence The angle at which solar radiation strikes a particular place at a point in time

apartheid A policy of racial segregation enforced in South Africa 1948–1993

aquaculture Herding or domestication of aquatic animals and farming of aquatic plants

arithmetic density The number of people per unit of area

artifact A material object of culture; literally, "a thing made by skill"

asylum Safety that a country grants to refugees

atmosphere A thin layer of gases surrounding Earth to an altitude of less than 480 kilometers (300 miles)

autocratic A form of government that is run according to the interests of the ruler or ruling elite rather than the people

autumnal (fall) equinox September 22 or 23 in the Northern Hemisphere, or March 20 or 21 in the Southern Hemisphere, when at noon the perpendicular rays of the Sun strike the equator (meaning that the Sun is directly overhead along the equator)

B

basic sector The part of a city's economy that is producing exports

beach A deposit of wave-carried sediment along a shoreline, on which waves break

behavioral geography The study of how people perceive their environment and of how their thoughts and perception influence their behavior

bilateral agreement A formal agreement made between two countries

biodiversity The amount of variety of living things in a given environment

biogeochemical cycle The environmental recycling process that passes essential substances such as carbon, nitrogen, and other nutrients among the biosphere, atmosphere, hydrosphere, and/or lithosphere

biogeochemical oxygen demand The amount of dissolved oxygen in a water body that is consumed by decay of organic pollutants added to the water

biomagnification The tendency for substances that accumulate in body tissues to increase in concentration as they are passed to higher levels in a food chain

biomass The dry mass of living or formerly living matter in a given environment

biome A large grouping of ecosystems characterized by particular plant and animal types

biosphere All living organisms on Earth

biotechnology New techniques for modifying biological organisms and their physiological processes for applied purposes

blog A "Web log," an online diary or newsletter on the Internet

boreal forest An evergreen needleleaf forest characteristic of cold continental climates

brain drain The emigration of a country's best-educated people and most skilled workers

broadleaf deciduous forest A forest with broadleaved trees that lose their leaves in the winter; characteristic of humid midlatitude environments

brownfields Abandoned polluted industrial sites in central cities, many of which are today being cleaned and redeveloped

C

capital-intensive activity An activity in which a large amount of capital is invested per worker

capitalist economic system An economic system in which the state defers to private enterprise and a stock market raises and allocates capital

carbon cycle The movement of carbon among the atmosphere, hydrosphere, biosphere, and lithosphere as a result of processes such as photosynthesis and respiration, sedimentation, weathering, and fossil-fuel combustion

carbon dioxide A trace gas found in the atmosphere with chemical formula CO_2; a major contributor to the greenhouse effect

carbon monoxide A pollutant with chemical formula CO formed by the incomplete combustion of fossil fuels

carnivore An animal whose primary food supply is other animals

carrying capacity The population that can be supported by a given resource, as in the number of people that can be supported by agricultural land.

cartel An organization formed to control the market for a particular commodity, usually by restricting supply

cartogram A maplike image designed to convey the magnitude of something rather than exact spatial locations

cartography Map making

caste A group in the rigid social hierarchy of Hinduism

central business district (CBD) The traditional core of a city, where office buildings and retail shops tend to be concentrated

central place theory A model of the distribution of cities across an isotropic plain

centrifugal forces Forces that tend to pull states apart

centripetal forces Forces that bind a state together

chain migration A migration pattern resulting when migrants follow the paths of previous migrants from the same area of origin to the same destination

chemical weathering The breakdown of rocks or minerals through chemical reactions at Earth's surface

city A concentrated nonagricultural human settlement

civil society Networks of individuals who are not controlled by states or businesses

climate The totality of weather conditions over a period of several decades or more

climax community The end point of community succession

cloning The production of identical organisms by asexual reproduction from a single cell of a preexisting organism

cognate A word that clearly looks or sounds like one in another language to which it is related historically

cognitive behavioralism The theory that people react to their environment as they perceive it

cold front The boundary formed when a cold air mass advances against a warmer one

Cold War The competition between the United States and the Soviet Union for global influence during the period between World War II and the collapse of the Soviet Union

collapsed state A state that has proven incapable of providing its citizens with either economic development or even peace and security

commercial agriculture The raising of food to sell

commercial revolution The tremendous expansion of global trade between about 1650 and 1750

commodification The marketing of cultural practices and material artifacts for sale to outsiders

common market A customs union within which common laws create similar conditions of production

community succession or succession A process of ecosystem change in which organisms modify their immediate environments in ways that allow other species to establish themselves and dominate

composite cone volcano A volcano formed by a mixture of lava eruptions and more explosive ash eruptions

concentration The distribution of a phenomenon within a given area

condensation Water changing from a gas state (vapor) to a liquid or solid state

conformal map A map that distorts size but preserves shapes

congregation Residential clustering by choice

consolidated metropolitan statistical area (CMSA) Two or more contiguous MSAs

contiguous diffusion or contagious diffusion Diffusion that occurs from one place directly to a neighboring place

continental glacier A thick glacier hundreds to thousands of kilometers across, large enough to be only partly guided by underlying topography

continental shelf An area of relatively shallow water that surrounds most continents between the shore and the point where the continental slope drops more sharply to the deep sea floor

convection Circulation in a fluid caused by temperature-induced density differences, such as the rising of warm air in the atmosphere

convergent plate boundary A boundary between tectonic plates in which the two plates move toward one another, destroying or thickening the crust

core area The historic homeland of a nation or area of greatest settlement

Coriolis effect The tendency of an object moving across Earth's surface to be deflected from its apparent path as a result of Earth's rotation

Council of Europe A multilateral treaty organization of European countries formed in 1949 to promote cooperation on democratic principles, human rights, and the rule of law

creole A pidgin language that has become a mother tongue

crude birth rate The annual number of live births per thousand people

crude death rate The annual number of deaths per thousand people

cultural boundaries Political boundaries drawn with respect to existing patterns of human activity

cultural diffusion The spreading of cultural attributes

cultural ecology The study of the ways societies adapt to environments

cultural geography The study of the geography of human cultures

cultural imperialism The substitution of one set of cultural traditions for another, either by force or by degrading those who fail to acculturate and rewarding those who do

cultural landscape A landscape that reveals the many ways people modify their local environment

cultural mosaic A phrase describing Canada and suggesting that various cultural groups retain individuality there

cultural preservation The effort to document, popularize, and rejuvenate traditional cultures

cultural realm The region throughout which a culture prevails

culture A bundle of attributes of shared behavior or belief, including virtually anything about the way a people live

culture area or region The entire region in which a cultural trait occurs

culture core area The region in which a cultural trait is predominant

culture domain The area in which a cultural trait is common but not predominant

culture realm The area in which a cultural trait occurs but may be common than other comparable traits

customs union A free-trade area that enforces a common external tariff

cyberspace The extension of reality through global electronic means of communication

cyclone Large low-pressure area in which winds converge in a counter-clockwise swirl in the Northern Hemisphere (or clockwise in the Southern Hemisphere)

D

delta A deposit of sediment formed where a river enters a lake or an ocean

demographic equation or balancing equation A population's rate of natural increase (or decrease) plus the net migration rate (which may be negative)

demographic transition model A model that describes the historical experience of population growth in the countries that are today rich

demography The analysis of a population in terms of specific characteristics, such as age or income levels

density The frequency of occurrence of a phenomenon in relation to its geographic area

dependency ratio The ratio of the combined population less than 15 years old and adult population over 65 years old to the population of those between 15 and 64 years of age

desert A vegetation type with sparsely distributed plants, specifically adapted for moisture gathering and moisture retention

desert climate A climate with low precipitation and temperatures warm enough to cause potential evapotranspiration to be substantially higher than precipitation for most or all of the year

desert pavement The stony surface of a desert soil formed by selective removal of fine particles by surface erosion

desertification The process of a region's soil and vegetation cover becoming more desertlike as a result of human land use, usually by overgrazing or cultivation

dialect A variation within a language

diffusion The process of an item or a feature spreading through time

diffusionism The theory that aspects of civilization were developed in very few places and then diffused from those places to the rest of the world

digital divide The gap between regions with high and low levels of digital technology use, especially electronic communications

diminishing returns A condition that exists when, upon adding equal amounts of one factor of production, such as fertilizer or labor, each successive application yields a smaller increase in production than the application just preceding

discharge The quantity of water flowing past a point on a stream per unit time

dissolved oxygen Oxygen found in dissolved form in water; it is essential for aquatic animals, and depleted by pollution

distance The extent of space between two objects or places; it can be measured absolutely, in terms of miles or kilometers, or in terms of other units, such as time or cost to cross

distance decay The diminution of the presence or impact of any cultural attribute away from its hearth area

distribution The position, placement, or arrangement of a phenomena

divergent plate boundary A boundary between tectonic plates in which the two plates move away from each other, and new crust is created between them

domestication The process of adapting plants and animals to obtain their intimate association with humankind, to the advantage of humankind

double cropping Harvesting two crops from each field per year

doubling time The number of years it would take any country's population to double at the present rate of increase

downstream activities Economic activities that are second, third, or even fourth steps in the transformation of a raw material into goods for ultimate consumers

drainage basin The geographic area that contributes runoff to a particular stream, defined with respect to a specific location along that stream

drainage density The total length of streams in a drainage basin divided by the drainage area

drift Changes and errors in how a language is used that accumulate over time

dune An accumulation of windblown sand, shaped by the wind

E

earthquake A sudden release of energy within Earth, producing a shaking of the crust

ecology The scientific study of ecosystems

economic development The process of progressively increasing the value of goods and services that a place is able to produce in order to enjoy or export

economic geography The study of how various peoples make their living and what they trade

economies of scale Economic factors which determine that as the number of units of a good produced increases, the production cost per unit generally falls

ecosystem An interrelated collection of plants and animals and the physical environment with which they interact

ecotourism Travel to see distinctive examples of scenery, unusual natural environments, or wildlife

edge cities New larger urban areas located outside old central cities

ejido A Mexican form of land tenure in which a peasant community collectively owns a piece of land along with the natural resources and houses on it

El Niño A circulation change in the eastern tropical Pacific Ocean, from westward flow to eastward flow, that occurs every few years

electoral geography The study of voting districts and voting patterns

electronic networks Interconnected communication systems that allow users to exchange information

emigration Movement away from a place

enculturation or socialization Teaching youngsters a society's values and traditions, its political and social culture

endogenic processes Forces within Earth that affect its surface, such as plate tectonics, volcanic eruptions, and earthquakes

energy intensity The amount of energy required to produce a unit of economic output

environmental determinism The simplistic belief that human events can be explained entirely as the result of the effects of the physical environment

epicenter The location on Earth's surface immediately above the focus of an earthquake

epidemiological transition The shift within a country of the principal causes of death from infectious to degenerative diseases

epidemiology The study of the incidence, distribution, and control of disease

equal-area map A map projection that preserves size but distorts shape

equator Earth's imaginary midline perpendicular to the axis and midway between the poles

ethnic cleansing The forcible and violent expulsion from a political territory of persons of a particular culture

ethnic enclave An area that contains a high concentration of one cultural group in contrast to the surrounding area

ethnic group A dubious term suggesting that a particular cultural group is in the minority or "not normal" for a particular place and time

ethnocentrism A tendency to judge foreign cultures by the standards and practices of one's own; and usually to judge them unfavorably

ethnonationalism The claim that a particular culture group has special political rights over other groups in a given area

etymology The study of the origin and history of words

European Union (EU) A bloc of European countries enjoying free trade and committed to some degree of political union

eutrophication An increase in growth of aquatic plants such as algae and associated increase in nutrient levels in a water body

evapotranspiration The sum of water converting from liquid to vapor state via evaporation or transpiration

evolutionism The theory that a culture's sources of change were embedded in the culture from the beginning, so the course of development was internally determined

exclusive economic zone (EEZ) A 200-nautical-mile zone within which a coastal state controls both mining and fishing rights from its shores

exogenic processes Forces originating in the atmosphere that, aided by gravity, shape Earth's surface; erosion by running water, glaciers, wind, and waves are examples

export-orientation or export-oriented growth A national economic policy of welcoming foreign investment to build factories that will manufacture goods for international markets

external economies The range of goods and services in a city that can be rented or hired temporarily

externality An exchange of a good or service that takes place without agreement of the parties in a market; pollution is generally considered an externality

exurbs The settlements that make up the outermost ring of expanding metropolitan areas

F

failed state A state that has proven incapable of providing its citizens with either economic development or even peace and security

famine Food shortages that lead to extensive starvation

fault A fracture in Earth's crust along which displacement of rocks has occurred

federal government A form of government in which a central government shares power with subunits

fertility transition A historical decrease in births in which populations move from high to low birth rates

fishery A concentration of aquatic species suitable for commercial harvesting

floodplain A low-lying surface adjacent to a stream channel and formed by materials deposited by the stream

focus (of an earthquake) The location in Earth where motion originates in an earthquake

folk culture A culture that is handed down and preserves traditions

food chain The sequential consumption of food in an ecosystem, beginning with green plants, followed by herbivores and carnivores, and ending with decomposers

food security Having access to sufficient and appropriate food to be healthy

foreign direct investment (FDI) Investment by foreigners in wholly owned enterprises that are operated by the foreigner

formal region A region defined by essential uniformity in one or more physical or cultural features

fossil fuel A source of chemical energy stored in formerly living plant and animal tissue. Coal, oil, and natural gas are fossil fuels

free-trade area An international territory having no internal tariffs, but that give its members the freedom to set their own tariffs on trade with the rest of the world

friction of distance The effort, time, or cost necessary to move or transport items

front A boundary between warm air and cold air

functional region A region defined by interaction among places, such as trade and communication

fundamentalism The strictest adherence to traditional religious beliefs

G

Gaia hypothesis A holistic view that likens Earth to a living organism with the ability to regulate critical functions, such as climate, through interactions between the atmosphere, hydrosphere, biosphere, and lithosphere

gateway city An urban area that emerges at the intersection of two different physical or cultural areas

gender roles The culturally defined duties and behaviors associated with being a man or a woman

gene splicing or recombinant DNA The joining of the genes of two or more organisms to produce recombinant (recombined) genetic material

genetic engineering The manipulation of species' genetic material through selective breeding or recombinant DNA

genocide The practice of intentionally trying to eliminate a national, ethnic, racial, or religious group

gentrification The occupation and restoration of select urban residential neighborhoods by wealthy urban white-collar workers

geographic information system (GIS) A computer system used to organize, store, analyze, and display geographic information

geography The study of the interaction of all physical and human phenomena at individual places and of how interactions among places form patterns and organize space

geometric boundaries Political boundaries drawn with respect to lines of latitude, longitude or specific coordinates

geomorphology The study of the shape of Earth's surface and the processes that modify it

geopolitics The influence of physical or human geography on international affairs

gerrymandering The drawing of voting district lines in ways that include or exclude specific groups of voters, so that one group gains an unfair advantage

glacier A large mass of flowing, perennial ice

global civil society International networks of activists organized around specific issues

global positioning system (GPS) A navigational tool consisting of a fleet of satellites orbiting Earth, broadcasting digital codes, and a portable receiver that can receive those codes and determine its location

global security The reduction of risks that present common threats to all states

global warming A general increase in temperatures over a period of at least several decades believed to be caused primarily by increased levels of carbon dioxide in Earth's atmosphere

globalization The organization of any activity treating the entire globe as one place

grade A condition in which a stream's ability to transport sediment is balanced by the amount of sediment delivered to it

green revolution An intensive effort, starting about 1950, to develop new grain varieties and associated agronomic systems and to establish them in developing countries. It focused on certain crops (wheat, rice) and certain techniques (breeding for responses to fertilizer inputs) and was driven largely by private foundations

greenhouse effect Atmospheric warming that results from the passage of incoming shortwave energy and the capture of outgoing longwave energy

greenhouse gases Trace substances in the atmosphere that contribute to the greenhouse effect; water vapor, carbon dioxide, ozone, methane, and chlorofluorocarbons are important examples

Greenwich Mean Time (GMT) The time at the prime meridian or 0° longitude at Greenwich, England

gross domestic product (GDP) The total value of all goods and services produced within a country

gross national income (GNI) A country's GDP plus any income that residents receive from foreign investments, minus any money paid out of the country to foreign investors

groundwater The water beneath Earth's surface at a depth where rocks and/or soils are saturated with water

growth rate The value of a population's annual demographic equation divided by the population size at the beginning of that year

gyre A circular ocean current beneath a subtropical high-pressure cell

H

hearth The place where a distinctive culture originated

herbivore An animal whose primary food supply is plants

heritage site A place where a dead or dying folk culture is preserved or commemorated

hierarchical diffusion Diffusion that occurs downward or upward through a organizational hierarchy; when mapped, it shows up as a network of spots

hinterland The region to which any city provides services and upon which it draws for its needs

historical consciousness A people's consciousness of past events insofar as that consciousness influences their present behavior

historical geography The study of the geography of the past and how geographic distributions have changed

historical materialism The belief that technology has historically increased humankind's control over the environment and improved material welfare, and that this improvement prompts other historical events and movements

horizon A layer in the soil with distinctive characteristics derived from soil-forming processes

Human Development Index (HDI) An index that combines statistics of life expectancy, school enrollment, literacy, and income to compare the quality of life around the world

human geography The study of the geography of human groups and activities

human rights The set of rights, such as the freedom of speech and the right to own property, that are thought by some to apply to all people

humid continental climate A climate characterized by cold winters and warm summers, with moderate levels of precipitation

humid subtropical climate A climate with cool winters, hot summers, and moderately high levels of precipitation

humid tropical climate A climate with high temperatures and high rainfall amounts all the year

hunger The deficiency of a person's or population's diet in calories, protein, vitamins, or micronutrients

hunter-gatherer A person who live on what he or she can hunt or harvest from Earth

hurricane An intense tropical cyclone that develops over warm ocean areas in the tropics and subtropics, primarily during the warm season. Hurricanes in the Pacific Ocean are called typhoons; in the Indian Ocean they are called cyclones

hybridity The idea that two different cultures can be combined

hydrocarbon A chemical substance composed of carbon and hydrogen; hydrocarbons in the atmosphere contribute to the formation of photochemical smog

hydroelectric power Electricity generated by water passing through turbines at a dam

hydrologic cycle The movement of water from the atmosphere to Earth's surface, across that surface, and back to the atmosphere

hydrosphere The water realm of Earth's surface, including the oceans, surface waters on land (lakes, streams, rivers), groundwater in soil and rock, water vapor in the atmosphere, and ice in glaciers

I

ice-cap climate A climate with very cold temperatures all year, including summer temperatures that are rarely above freezing

iconography A state's set of symbols, including a flag and anthem

identity The characteristics that are used to describe the unique qualities of an individual or that are thought to be shared by a group of people

igneous rock Rock formed by crystallization of magma

immigration Movement into a place

import-substitution A national economic policy of protecting domestic infant industries

incorporation The process of defining a city territory and establishing a government

indigenous peoples As defined by the United Nations, "descendants of the original inhabitants of a land who were subjugated by another people coming after them"

indirect rule The imperialist use of native rulers as intermediaries between the imperialists and the people

Industrial Revolution The evolution, which first occurred in Europe between about 1750 and 1850, from agricultural and commercial society to industrial society relying on inanimate power and complex machinery

industrial society A society with a significant share of its output from the secondary sector

industrialized economy A country that has achieved a high level of income from manufacturing

inertia The force that keeps things stable or fixed in place

infant industry A newly developing industry that probably cannot compete with imports

infant mortality rate The number of infants per thousand who die before reaching 1 year of age

infiltration capacity The maximum amount of water that can soak into a soil per unit time

informal sector or underground sector Economic activities that do not appear in official accounts

infrastructure Fixed assets in place, such as buildings, dams, and roads

innocent passage The internationally guaranteed right of the ships of one state to pass through the territorial waters of another on their way to a third

insolation The amount of solar energy intercepted by a particular area of Earth

intergovernmentalism The creation of organizations through the cooperation of many individual states

internal economies Goods and services that a large company can provide for itself

internal frontier A sparsely populated and underdeveloped region within a country that may have potential for settlement and development

internally displaced person (IDP) A person who has been forced to flee his or her home but has not left his or her country of origin

International Date Line An imaginary line on Earth's surface where, by international agreement, travelers traveling eastward subtract one calendar day and travelers traveling westward add one calendar day. The line generally follows the 180° meridian, but it deviates for political and economic convenience

intertropical convergence zone (ITCZ) A low-pressure zone between the Tropic of Cancer and the Tropic of Capricorn where surface winds converge

irredenta Territory that one state claims from another

isogloss A line around places where speakers use a linguistic feature in the same way

isolate A language unrelated to its neighbors

isostatic adjustment A vertical movement of Earth's crust, caused by the loading or unloading of the buoyant crust

isotropic plain A theoretical perfectly flat surface with absolutely no variations across it

K

karst An assemblage of landforms found in areas of intense subsurface chemical weathering that often include features such as caves and underground drainage

kleptocracy Government by thieves or theft

L

labor-intensive activity An activity that employs a high ratio of workers to invested capital

laissez-faire capitalism A capitalist system that minimizes the government's role in the economy

land reform The redistribution of large land holdings such as plantations to poor would-be farmers

landform A characteristic shape of the land surface, such as a hill, valley, or floodplain

language A set of words, plus their pronunciation and methods of combining them, that is used and understood as communication within a group of people

language family Languages that are related by descent from a common protolanguage

large-scale map A map that shows a given area in a large space

latent heat Heat stored in water and water vapor, not detectable by a thermometer; *latent* means "hidden"

latent heat exchange The exchange of energy necessary to change water from one of its states to another—solid, liquid, or gaseous

latitude The location of a place measured as angular distance north and south of the equator

lava Magma that reaches Earth's surface and erupts

liberation theology The belief in putting the problems of overcoming poverty at the heart of Christian theology

life expectancy The average number of years that a newborn baby within a given population can expect to live

lingua franca A second language held in common for international discourse

liquidity The quality of being readily convertible into cash

lithosphere The solid Earth, composed of rocks and sediments overlying them

Little Ice Age The period between about 1500 and 1750, when climates on Earth were especially cool

local content requirement A definition of the percentage of the total value of a good entering one country that must have been added in the second country for that product to qualify as a product of the second country

location The place where a thing is; it can be defined absolutely or relatively

loess An accumulation of windblown silt

longitude The location of a place measured as angular distance east and west from the prime meridian

longshore current A current in the surf zone along a shoreline, parallel to the shore

longshore transport Sediment transport by a longshore current

longwave energy Energy radiated by Earth in wavelengths of about 5.0 to 30.0 microns

M

magma Molten rock beneath Earth's surface

malnourished A person's having not enough nutrients

Malthusian theory The pessimistic argument that population increases will always outpace increases in food production, causing cycles of war, famine, and disease; articulated by Thomas Malthus (1766–1834)

mantle The portion of Earth above the core and below the crust

map A two-dimensional (flat) representation of some portion of Earth's surface

maquiladora A factory in Mexico specializing in assembling items for export to the U.S. market

marine terrace A nearly level surface along a shoreline, elevated above present sea level, formed by coastal erosion at a time when sea level at the location was higher than at present

marine west coast climate A climate with moderately cool winters, moderately warm summers, and moderate to high rainfall all year

market-oriented manufacturing Manufacturing that locates close to the market either because the processing increases the perishability of the product or because the processing adds bulk or weight to the product

Marshall Plan A plan of financial aid for the economic rehabilitation of Europe after World War II; named for U.S. Secretary of State George C. Marshall

mass movement Downslope movement of rock and soil at Earth's surface, driven mainly by the force of gravity acting on those materials

material-oriented manufacturing Manufacturing that locates close to the source of the raw material either because the raw material is heavy or bulky or because it is perishable

meandering The tendency of flowing water to follow a sinuous course with alternating right- and left-hand bends

mechanical weathering The breakdown of rocks into smaller particles caused by application of physical or mechanical forces

Mediterranean climate A climate with warm, dry summers and cool, moist winters

megacity An urban area with 10 million or more inhabitants

melting pot Name given to the United States in a 1914 novel of that title, suggesting that ethnic and racial differences among immigrants melt together to form one culture

meltwater channel A river channel carved by water from a melting glacier

mental map The ideas that people have about places, regardless of whether those ideas are true or false

meridians Imaginary lines extending from pole to pole and crossing all parallels at right angles

meritocracy A society in which the most capable people can rise to the top based on merit alone

metamorphic rock Rock formed by modification of other rock types, usually by heat and/or pressure

methane A trace gas found in the atmosphere with chemical formula CH_4; a major contributor to the greenhouse effect

metropolitan statistical area (MSA) According to the U.S. Census Bureau, "an integrated economic and social unit with a recognized large population nucleus"

microclimate An area with local climatic conditions, as in a city, that differ from those of surrounding areas

midlatitude cyclone A storm characterized by a center of low pressure in the midlatitudes usually associated with a warm front and a cold front

midlatitude low-pressure zones Regions of low pressure with air converging from the subtropical and polar high-pressure zones

migration chain A network of social and communication linkages that attracts migrants to follow others who have previously migrated

model An idealized, simplified representation of reality

monoculture The specialized production of one crop

monotheism Belief in the existence of only one god

monsoon circulation Seasonal reversal of pressure and wind in Asia, in which winter winds from the Asian interior produce dry winters, and summer winds blowing inland from the Indian and Pacific oceans produce wet summers

moraine An accumulation of rock and sediment deposited by a glacier, usually in or near the melting area

multilateral agreement A formal agreement made between more than two states

multinational corporation An enterprise that produces and markets goods in several countries

multiplier effect The fact that jobs in a city's basic sector multiply jobs in the nonbasic sector

N

nation A group of people who want to have their own government and rule themselves

national security A government's assessment of the risks that threaten its rule

national self-determination The idea of the nation-state as defended by U.S. President Woodrow Wilson after World War I

nation-state A state ruling over a territory containing all the people of a nation and no others

natural landscape A landscape without evidence of human activity

natural population increase or decrease The difference between the number of births and the number of deaths

natural resource Something that is useful and that exists independent of human activity

net migration rate The number of emigrants subtracted from the number of immigrants for an area

network hypothesis The theory that central city unemployment is caused by a lack of social networks

nitrogen oxide A compound of nitrogen and oxygen with chemical formula NO_x; a component of air pollution

nonbasic sector The part of a city's economy serving the needs of the city itself

nongovernmental organization (NGO) A group of private individuals committed to promoting particular issues or providing material assistance, often across international boundaries

nonrenewable resource A resource that is either not being produced by nature or is produced much more slowly than it is used by humans

North American Free Trade Agreement (NAFTA) An agreement between the United States, Canada, and Mexico to reduce barriers to trade and investment between their countries

North Atlantic Treaty Organization (NATO) A military bloc founded in 1949. Its membership and activities have expanded significantly, bringing its very purpose into question

O

official language The language in which legal documents are kept in a country

offshore banking Banks locating their headquarters in places renowned for lenient banking regulations

offshore outsourcing A company's relocation of some economic activities to another country where costs are lower

omnivore An animal that feeds on both plants and other animals

opportunity cost A capital return sacrificed by leaving capital invested in one form or activity rather than another

Organization for Security and Cooperation in Europe An intergovernmental organization focused on conflict resolution, election reform, and arms reduction

orthography The study of writing, or a system of writing

outsourcing A company's decision to hire another company to do part of its activities

outwash plain An accumulation of sand and gravel carried by meltwater streams from a glacier, usually deposited immediately beyond the terminal moraine from the glacier

overland flow Water flowing across the soil surface on a hillslope, usually resulting from precipitation falling faster than the ground can absorb it

Overseas Chinese Chinese migrants who often remain linked across international borders by bonds of family, clan, and common home province

ozone A gas composed of molecules with three oxygen atoms; it is a highly corrosive gas at ground level, but in the upper atmosphere essential to protecting life on Earth by absorbing ultraviolet radiation

P

parallels Lines connecting all points of the same latitude

parent material Mineral matter such as rocks or transported sediments from which soil is formed

particulate A small solid particle in the air; a component of air pollution

partition A boundary drawn to create new states from the territory of an existing state

pastoral nomadism A group's style of life that does not have fixed residences; the group drives flocks from place to place to find grazing lands and water

pathogen A disease-causing organism

pattern The arrangement of objects within an area

permafrost Soil or rock with a temperature below 0°C (32°F) all year

permanent mission One country's diplomatic officers and staff who reside and work in another country

photochemical smog A mixture of air pollutants including oxidants such as ozone, formed by interaction of sunlight and pollutants such as nitrogen oxides and hydrocarbons

photosynthesis A chemical reaction that occurs in green plants in which carbon dioxide and water are converted to carbohydrates and oxygen

photovoltaic cell A device that converts light to electricity

physical boundaries Political boundaries drawn with respect to features of the physical environment

physical geography The study of the characteristics of the physical environment

physiological density The density of population per unit of arable land

plate tectonics theory A theory describing and explaining the movement of large, continent-sized slabs of Earth's crust relative to one another

Pleistocene Epoch A period of geologic time consisting of the first part of the Quaternary Period beginning about 3 million years ago and ending about 12,000 years ago

polar front A boundary between cold polar air and warm subtropical air that circles the globe in the midlatitudes

polar high-pressure zones Regions of high pressure and descending air near the North and South Poles

political community Governments that join together to form a government to solve common problems

political culture A set of unwritten ways in which written rules are interpreted and actually enforced

political economy The study of individual countries' organization and regulation of their economies

political geography The study of the interaction between political processes and the distributions of all other activities and transformations of the landscape

pollution A human-caused increase in the amount of a substance in the environment

pollution prevention A strategy for reducing pollution that focuses on reducing the amount of pollutants created rather than on removing them from waste streams

polyculture The raising of a variety of crops

polyglot state A country that grants legal equality to two or more languages

polytheism The worship of many gods

popular culture The culture of people who embrace innovation and conform to changing norms

population geography The study of the distribution of human-kind across Earth

population momentum The lag between falling birth rates and continued population growth that persists until the larger cohorts complete their reproductive years

population projection A forecast of the future population, assuming that current trends remain the same or else change in defined ways

population pyramid A graphic device that shows the shares of a nation's population by age groups

possibilism The theory that the physical environment itself will neither suggest nor determine what people will attempt, but it may limit what people can profitably achieve

postindustrial society A society with the bulk of its economic activity in the tertiary sector

potential evapotranspiration (POTET) The maximum amount of water that could be evaporated from a moist surface and/or transpired by plants if it were available

potential resource Something that is not useful today but may become so in the foreseeable future

prairie A vegetation type characterized by dense grass up to 2 meters high, found in midlatitude semiarid climates

preindustrial society A society with the bulk of its economic activity in the primary sector

primary sector The part of the economy that extracts resources directly from Earth, including agriculture, fishing, forestry, and mining

primate city A large city concentrating a national population or national political, intellectual, or economic life

prime meridian The meridian passing through the Royal Observatory in Greenwich, England, from which longitude is measured

privatize or privatization To give or sell government assets to private individuals or investors

projection A method of portraying Earth or any portion of it on a flat map

proselytize To try to convert others to your religious beliefs

protolanguage See root language

proxemics The study of how people perceive and use space

psychological value added An increase in the cost of an item due not to an increase in its actual functionality or usefulness, but in its design, packaging, or "status" advertising

pull factors Considerations that attract people to new destinations

push factors Considerations that make a person want to leave a place and seek a better life elsewhere

Q

Quaternary Period The period of geologic time encompassing approximately the past 3 million years

R

race A group of people variously defined by relatively minor biological differences within the human species

racism The false belief in the inherent superiority of one race over another and the linking of human ability, potential, and behavior to racial inheritance

radiation Energy in the form of electromagnetic waves that radiate in all directions

recognition The formal acknowledgement by one state of another state's existence

recombinant DNA See gene splicing

refugee As defined by the 1951 Geneva Convention, someone outside his or her country with "a well-founded fear of being persecuted in his country of origin for reasons of race, religion, nationality, membership of a particular social group or political opinion"

region A territory that exhibits a certain uniformity

regional geography An inventory analysis of all characteristics of any individual place

regionalism Political identities based on areas within a state that are culturally or economically different from the rest of the state

relative humidity The actual water content of the air, expressed as a percentage of how much water the air could hold at a given temperature

relative location The location of a place relative to other places

relocation diffusion Diffusion from one widely separated point to another

remote sensing The acquisition of data about Earth's surface from a satellite orbiting the planet or from high-flying aircraft

renewable resource Something that is produced by nature at rates similar to those at which it is consumed by humans

replacement rate A total fertility rate of about 2.1, which stabilizes a population

respiration A chemical reaction that occurs in plants and animals in which carbohydrates and oxygen are combined, releasing water, carbon dioxide, and heat

return migration The return of migrants to their homes or areas of origin

root language or protolanguage The common ancestor language to any group of several of today's languages

runoff Flow of water from the land, either on the soil surface or in streams

S

sanitary landfill A site at which solid waste is deposited and covered with layers of earth

saturation vapor pressure The maximum amount of water vapor that air can hold, expressed as a pressure

savanna A vegetation type characterized by grasses and scattered trees, characteristic of seasonally dry tropical climates

scale A quantitative statement of the relative sizes of an object on a map and in reality

scientific revolution in agriculture The continuing application of science to agriculture

sea level The general elevation of the sea surface, averaging out variations caused by waves, storms, and tides

seafloor spreading The creation of new oceanic crust where two tectonic plates are diverging on the seafloor

seasonally humid tropical climate A climate with warm temperatures all the year, a season with high rainfall, and a pronounced dry season

secondary sector The part of the economy that transforms raw materials into manufactured goods

sectoral evolution A shift in the concentration of activity from an economy's primary sector to its secondary and tertiary sectors

secularism A lifestyle or policy that deliberately ignores or excludes religious considerations

sediment transport The movement of rock particles by surface erosion

sedimentary rock Rock formed through accumulation of many small rock fragments at Earth's surface

segregation Residential clustering as a result of discrimination

seismic waves Vibrations or shock waves originating at the focus of an earthquake and transmitted through Earth

seismograph A device for recording movements of Earth's crust, such as earthquakes

semiarid climate A climate with precipitation less than potential evapotranspiration for much of the year, but not as dry as a desert.

sensible heat Heat detectable by sense of touch, or with a thermometer

separatism A subnational group's seeking to establish its own state on territory taken from an existing state

service sector That part of the economy that services the primary and secondary sectors

sexuality A person's sexual orientation and behavior

shamanism A belief in the power of mediums (shamans) who characteristically go into autohypnotic trances, during which they are thought to be in communion with the spirit world

Sharia Islamic teachings that are often incorporated into civil law in Islamic countries

shield The ancient core of a continent

shield volcano A volcano with relatively gentle slopes formed by eruption of relatively fluid lavas

shortwave energy Radiant energy emitted by the Sun in wavelengths of about 0.2 to 5.0 microns

SIAL Crust formed of relatively less dense minerals, dominated by silicon and aluminum (an acronym for silicon-aluminum)

SIMA A crust formed of relatively dense minerals, dominated by silicon and magnesium (an acronym for silicon-magnesium)

site The characteristics of the absolute location of a place

situation The characteristics of the relative location of a place

small-scale map A map that shows the land in a very small space

social Darwinism The theory of British sociologist Herbert Spencer that "Nature's law" calls for "the survival of the fittest," even among cultures and entire peoples

socialization *See* enculturation

soil A dynamic, porous layer of mineral and organic matter at Earth's surface

soil creep The slow downslope movement of soil caused by many individual, near-random particle movements such as those caused by burrowing animals or freeze and thaw

soil fertility The ability of a soil to support plant growth through the storing and supplying water, air, and nutrients

soil order A major category in the U.S. soil classification system

sojourner A migrant who intends to stay in a new location only long enough to save capital to return home to a higher standard of living

solar energy Radiant energy from the Sun

sovereignty The exclusive right to rule over a demarcated space and all the people and resources within it

spatial identity The attachment of a certain identity to a particular place or region

spatial mismatch hypothesis The hypothesis that central city unemployment is caused by the removal of job opportunity to the suburbs and the concentration of the poor in the central city

speech community A group of people who speak together

state An independent political unit that claims exclusive jurisdiction over a defined territory and over all the people and activities within it

steppe A vegetation type characterized by relatively short, sparse grasses, found in midlatitude semiarid climates

storm surge An area of elevated sea level in the center of a hurricane that may be several meters high; storm surge does much of the damage when a hurricane comes ashore

stratus clouds Flat layers of clouds formed along a warm front

structural landform A landform whose major characteristics are derived from endogenic processes or by erosional exposure of rock structures

subarctic climate A high-latitude climate characterized by brief, cool summers and long, cold winters

subculture A group that shares a smaller bundle of attributes with a larger, more diverse society

subnationalism Regional and other alliances within a state that rival the state's dominant nationalism

subsistence agriculture The raising of food only for oneself, not to sell

substitutability The degree to which one commodity can be substituted for another in various uses

subtropical high-pressure (STH) zones Regions of high pressure and descending air at about 25° north and south latitudes

succession *See* community succession

sulfur oxide An air pollutant consisting of compounds of sulfur and oxygen, derived mainly from combustion of coal and oil

summer solstice For places in the Northern Hemisphere, June 20 or 21, when at noon the Sun is directly overhead along the parallel of 23.5° north latitude; for places in the Southern Hemisphere, December 21 or 22, when at noon the Sun is directly overhead at places along the parallel of 23.5° south latitude

superimposed boundaries Boundaries drawn over existing territorial demarcations

supply and demand The interplay of buyers and sellers of a commodity in the marketplace

supranational organization An organization that exercises power over countries

surface erosion The downslope movement of rock and soil at Earth's surface, driven mainly by air, water, or ice moving across the surface

sustainability Human use of Earth's limited resources in ways that do not constrain future resource use

sustainable agriculture Food production that can be continued indefinitely and that limits or even reverses environmental degradation

sustainable development Economic development that can be continued indefinitely and that limits or even reverses environmental degradation

sustained yield A way of managing a renewable natural resource such that harvest can continue indefinitely

swidden Slash-and-burn clearing and cultivation

syncretic religion A religion that combines two or more traditional religious practices

system An interdependent group of items that interact in a regular way to form a unified whole

systematic geography The study of universal laws or principles that apply to all places; topics may be as diverse as the geography of soils (pedology), of life forms (biogeography), of politics (political geography), of economic activities (economic geography), and of cities (urban geography)

T

tectonic plate A large, continent-sized piece of Earth's crust that moves in relation to other pieces

telecommuting Working at home at a computer terminal connected to an office

temperature inversion A layer in the atmosphere in which relatively warm air lies above cooler air

terminal moraine An accumulation of rock and sediment at the toe of a glacier

territoriality Organizing social and material practices by controlling a space

terrorism Violent acts intended to frighten and to intimidate for political ends a civilian population beyond the immediate victims

tertiary sector The part of the economy that offer services, such as retail, consulting, and education

thematic map A map designed to show a particular aspect of the region portrayed

theocracy A form of government where a church rules directly

threshold The minimum number of potential customers that is needed for a product or service to be offered

topical geography See systematic geography.

topographic map A map that shows variation in elevation

topography The shape of earth's surface; surface relief.

toponymy The study of place names

tornado A rapidly rotating column of air usually associated with a thunderstorm, often having winds in excess of 300 kilometers per hour (185 miles per hour)

total fertility rate The average number of children that would be born to each woman in a given society if, during her childbearing years (15–49), she bore children at the current year's rate for women of that age

toxic substance A pollutant that can be harmful even at very low concentrations

trade wind The prevailing wind in subtropical and tropical latitudes that blows toward the Intertropical Convergence Zone, typically from the northeast in the Northern Hemisphere and from the southeast in the Southern Hemisphere

transform plate boundary A boundary between tectonic plates in which the two plates pass one another in a direction parallel to the plate boundary

transnational corporation An enterprise that produces and markets goods in several countries

transpiration The use of water by plants, normally drawing it from the soil via their roots, evaporating it in their leaves and releasing it to the atmosphere

trophic level A position in the food chain relative to other organisms, such as producer, herbivore, or carnivore

Tropic of Cancer The parallel of 23.5° north latitude

Tropic of Capricorn The parallel of 23.5° south latitude

tropical rain forest Broadleaf evergreen vegetation characteristic of humid tropical environments

tsunami An extremely long sea wave created by an underwater earthquake; the wave may travel hundreds of kilometers per hour

tundra A low, slow-growing vegetation type found in high-latitude and high-altitude conditions in which snow covers the ground most of the year

tundra climate A climate characterized by long, very cold winters and short, cool summers.

typhoon A hurricane in the Pacific Ocean

U

underground sector See informal sector

undernourished Having one's bodily functions physically degraded by hunger

undocumented immigrant A person who crosses a border without completing legal papers to do so

unification The erasure or movement of boundaries to unite cross-border groups

unitary government A form of government in which the balance of power lies with the central government

untouchable One of a group of people considered so low that their status is below the formal structure of the Hindu caste system

urban enterprise zones Areas within which governments create generous conditions for enterprises to encourage the creation of jobs

urban form The distribution pattern of land use and activities within cities

urban geography The geographic study of cities

urban heat island Warmer temperatures in a city compared to the surrounding rural area created by urban activities and conditions

urbanization The process of concentrating people in cities

V

value added by manufacturing The difference between the value of a raw material and the value of a product manufactured from that raw material

vernacular region A region defined by widespread popular conception of its existence

vernal (spring) equinox March 20 or 21, in the Northern Hemisphere, or Sept. 22 or 23 in the Southern Hemisphere, when at noon the perpendicular rays of the Sun strike the equator (meaning that the Sun is directly overhead along the equator)

virtual reality A "place" created by intense involvement either with interactive electronic devices or else with distant people through electronic devices

volcano A vent in Earth's surface where lava emerges

W

warm front A boundary formed when a warm air mass advances against a cooler one

water budget An accounting of the amounts of precipitation, evapotranspiration, soil moisture storage and runoff at a given place

wavelength The distance between successive waves of radiant energy or of successive waves on a water body

weather Patterns of atmospheric circulation, temperature, and precipitation over short time periods such as hours to days

weathering The chemical and/or mechanical breakdown of rocks into smaller particles at Earth's surface

Web log *See* blog

winter solstice For places in the Southern Hemisphere, June 20 or 21, when at noon the Sun is directly overhead at places along the parallel of 23.5° north latitude; for places in the Northern Hemisphere, December 21 or 22, when at noon the Sun is directly overhead at places along the parallel of 23.5° south latitude

workers' remittances Money that migrant workers send home from elsewhere

X

xerophyte A plant adapted to living in arid conditions

Z

zero population growth A stabilized world population

Zionism The belief that the Jews should have a homeland of their own

zoning Restricting or prescribing the use to which parcels of land may be put

CREDITS

toc-1 vi: © First Light / Alamy; **toc-2 vii:** NASA; **toc-3 viii:** TORSTEN BLACKWOOD/AFP/Getty Images; **toc-4 ix:** © Nexus7 | Dreamstime.com; **toc-5 x:** (cc) oledoe - http://www.flickr.com/photos/oledoe/264341670/; **toc-6 xi:** Ron Gilling/Peter Arnold; **toc-7 xii:** ADRIAN DENNIS/AFP/Getty Images

1-CO: featurechinapix/Newscom; **1-01:** Courtesty of Ryukoku University Library; **1-02:** EFB Photo; **1-03:** USDA NRCS; **1-05:** © Jeremy Nicholl / Alamy; **01-08:** After Wilbur Zelinsky, "North America's Vernacular Regions," *Annals, Association of American Geographers,* 70 (March 1980): 14. By permission of the Association of American Geographers; **01-09:** After James Rubenstein, *The Cultural Landscape: An Introduction to Human Geography,* updated 9th ed. Upper Saddle River, NJ: Prentice Hall, 2008; **1-12:** Indian Buddhist Sculpture "Bodhisattva", c. 2nd-3rd century. Schist, chlorite, 45 in. h x 15 in. w x 7 in. d. Eugene Fuller Memorial Collection, Seattle Art Museum. Photo by Paul Macapia.; **1-14:** Reprinted with permission from Brian J. L. Berry, "The Geography of the United States in the Year 2000," *Transactions of the Institute of British Geographers LI,* November 1970, 21–53; **1-21:** Courtesy of ACSM; **1-23:** Image by Scientific Visualization Studio, NASA Goddard Space Flight Center; data courtesy Landsat Project; **1-24 a& b:** USGS; **1-25a:** NASA Goddard Space Flight Center, Scientific Visualization Studio; **1-25b:** NASA Goddard Space Flight Center, Scientific Visualization Studio; **1-26:** © TheProductGuy / Alamy; **1-28:** After Keith Clarke, *Getting Started with Geographic Information Systems,* 4th ed. Upper Saddle River, NJ: Pearson Education, 2003; **1-29:** WHR map; **1-30:** NASA/JPL-Caltech; **1-31:** WHR map

2-CO: © First Light / Alamy; **2-1:** Data from Rutgers University, Global Snow Lab, http://climate.rutgers.edu/snowcover/, and Alaska Department of Natural Resources; **2-4:** From Christopherson, R. W., *Geosystems,* 3rd ed. Upper Saddle River, NJ: Prentice Hall, 1997; **2-05:** Getty Images Inc. - Stone Allstock; **2-7:** Adapted from National Climatic Data Center, *Monthly Climatic Data for the World* 47, no. 1, January 1994. Prepared in cooperation with the World Meteorological Organization. Washington, D.C.: National Oceanic and Atmospheric Administration. From Christopherson, R. W., *Geosystems,* 5th ed. Upper Saddle River, NJ: Prentice Hall, 2003; **2-09:** ballycroy/iStochphoto; **2-10:** © Svetik | Dreamstime.com; **2-13:** © Corbis RF; **2-14:** Credit Image by Jesse Allen, Robert Simmon, and the MODIS science team.; **2-15:** © Surfotter | Dreamstime.com; **2-16a:** Atlas of Oregon, 2nd edition, Copyright 2001, University of Oregon Press.; **2-16b:** Atlas of Oregon, 2nd edition, Copyright 2001, University of Oregon Press.; **2-17:** From Christopherson, R. W., *Geosystems,* 5th ed. Upper Saddle River, NJ: Prentice Hall, 2003; **2-19:** From Christopherson, R. W., *Geosystems,* 5th ed. Upper Saddle River, NJ: Prentice Hall, 2003; **2-20:** From Christopherson, R. W., *Geosystems,* 5th ed. Upper Saddle River, NJ: Prentice Hall, 2003; **2-21:** From Tarbuck, E. J., and Lutgens, F. K., *Earth Science,* 7th ed. New York: Macmillan, 1994; **2-22:** From Christopherson, R. W., *Geosystems,* 5th ed. Upper Saddle River, NJ: Prentice Hall, 2003; **2-24:** From Remko Scharroo, Walter H.F. Smith and John L. Lillibridge, "Satellite Altimetry and the Intensification of Hurricane Katrina," EOS, Vol. 86, No. 40, 4 October 2005. Copyright 2005 American Geophysical Union. Modified by permission of American Geophysical Union; **2-25:** NOAA; **2-26:** courtesy NASA/NOAA & OceanRemote Sensing Group, John Hopkins Univ. Applied Physics Laboratory; **2-27:** From Christopherson, R. W., *Geosystems,* 5th ed. Upper Saddle River, NJ: Prentice Hall, 2003; **2-28:** © Jim Mahoney/Dallas Morning News/Corbis; **2-29:** NASA, Goddard Space Flight Center.; **2-32:** From Christopherson, R. W., *Geosystems,* 5th ed. Upper Saddle River, NJ: Prentice Hall, 2003; **2-35:** © Reuters/CORBIS; **2.37:** Mac99/iStockphoto; **2-39:** © Bob Gibbons / Alamy; **2-41:** © Andrew Shlykoff / Alamy; **2-43:** © Lynne Harty Photography / Alamy; **2-45:** (c) Neil Rabinowitz / CORBIS All Rights Reserved; **2-47a:** © Ellen Isaacs / Alamy; **2-47b:** © Ellen Isaacs / Alamy; **2-49:** Gregory K. Scott/Photo Researchers, Inc.; **2-51:** Andrew N. Ilyasov/Shutterstock; **2-53:** Geray Sweeney/Alamy **2-55:** Armin Rose/Shutterstock; **2-56:** From D.M. Sigman and E.A. Boyle, "Glacial/interglacial variations in atmospheric carbon dioxide." Adapted

by permission from Macmillan Publishers Ltd: *Nature,* 407 (2000): 859, copyright 2000; **2-60:** Adapted from Figure SPM-4 of Intergovernmental Panel on Climate Change (IPCC), *Climate Change 2007: Synthesis Report.* Geneva, Switzerland: IPCC, 2007

3-CO: Logan Abassi/MINUSTAH via Getty Images; **3-01:** LIU JIN/AFP/Getty Images; **3-04:** From E. J. Tarbuck and F. K. Lutgens, *The Earth,* 4th ed. New York: Macmillan, 1993; **3-07:** From E. J. Tarbuck and F. K. Lutgens, *Earth Science,* 7th ed. New York: Macmillan, 1994; **3-08:** From E. J. Tarbuck and F. K. Lutgens, *Earth Science,* 7th ed. New York: Macmillan, 1994; **3-09:** Dana Stephenson//Getty Images; **3-11:** From E. J. Tarbuck and F. K. Lutgens, *The Earth,* 4th ed. New York: Macmillan, 1993; **3-12:** Tom McHugh/Photo Rsearchers Inc; **3-13:** From E. J. Tarbuck and F. K. Lutgens, *The Earth,* 4th ed. New York: Macmillan, 1993; **3-14:** Courtesy: NASA/US Dept. of the Interior/ USGS/EROS; **3-15:** Norm R. Catto; **3-16:** From E. J. Tarbuck and F. K. Lutgens, *Earth,* 7th ed. Upper Saddle River, NJ: Prentice Hall, 2002; **3-17:** William Renwick; **3-18:** Xico Putini/Shutterstock; **3-20:** Courtesy USGS; **3-22:** Image courtesy of NASA/USGS/USDA Farm Services Agency; **3-23:** From E. J. Tarbuck and F. K. Lutgens, *Earth Science,* 7th ed. New York: Macmillan, 1994; **3-25:** NASA; **3-26:** Tom Bean/Corbis/Stock Market; **3-27:** Marli Bryant Miller; **3-28:** courtesy of Bruce F. Molnia/ USGS; **3-29:** USGS; **3-30:** From R. W. Christopherson, *Geosystems,* 5th ed. Upper Saddle River, NJ: Prentice Hall, 2003; **3-32:** From E. J. Tarbuck and F. K. Lutgens, *Earth,* 4th ed. New York: Macmillan, 1993; **3-35:** Jeff Schmaltz, MODIS Rapid Response Team/NASA; **3-36:** NOAA; **3-37:** From E. J. Tarbuck and F. K. Lutgens, *Earth,* 7th ed. Upper Saddle River, NJ: Prentice Hall, 2002; **3-38:** From E. J. Tarbuck and F. K. Lutgens, *Earth,* 7th ed. Upper Saddle River, NJ: Prentice Hall, 2002; **3-40:** William Renwick; **3-42:** Lynn Davis/Corbis; **3-43:** Adapted from R. W. Christopherson, *Geosystems,* 5th ed. Upper Saddle River, NJ: Prentice Hall, 2003

4-CO: NASA; **4-02:** J. A. A. Jones, *Global Hydrology.* London: Addison Wesley Longman, 1997; **4-04:** Modified from R. W. Christopherson, *Geosystems,* 5th ed. Upper Saddle River, NJ: Prentice Hall, 2003; **4-05:** NASA; **4-08:** WHR Photo/NASA; **4-09:** From R. W. Christopherson, *Geosystems,* 3rd ed. Upper Saddle River, NJ: Prentice Hall, 1997; **4-10a:** NOAA; **4-10b:** NOAA; **4-10c:** NOAA; **4-10d:** NOAA; **4-11:** Data courtesy of Carbon Dioxide Information and Analysis Center, Oak Ridge National Laboratory; **4-15a:** Loyal A. Quandt/National Soil Survey Center, Natural Resources Conservation Service, U.S.D.A.; **4-15b:** Loyal A. Quandt/National Soil Survey Center, Natural Resources Conservation Service, U.S.D.A.; **4-15c:** Loyal A. Quandt/National Soil Survey Center, Natural Resources Conservation Service, U.S.D.A.; **4-15d:** Loyal A. Quandt/National Soil Survey Center, Natural Resources Conservation Service, U.S.D.A.; **4-15e:** Loyal A. Quandt/National Soil Survey Center, Natural Resources Conservation Service, U.S.D.A.; **4-15f:** Loyal A. Quandt/National Soil Survey Center, Natural Resources Conservation Service, U.S.D.A.; **4-16:** USDA; **4-17:** WHR Photo; **B-X4-1a:** NASA; **B-X4-2b:** NASA; **4-18:** Bernard Boutrit/Woodfin Camp; **4-20:** From T. L. McKnight, *Physical Geography: A Landscape Appreciation,* 4th ed. Englewood Cliffs, NJ: Prentice Hall, 1993; **4-21:** From R. W. Christopherson, *Geosystems,* 3rd ed. Upper Saddle River, NJ: Prentice Hall, 1997. Adapted from E. p. Odum, *Fundamentals of Ecology,* 3rd ed., Figure 9-4, p. 261, © 1971 Saunders College Publishing. Adapted by permission; **4-22:** From R. W. Christopherson, *Geosystems,* 5th ed. Upper Saddle River, NJ: Prentice Hall, 2003; **4-24:** © Gary Braasch/CORBIS; **4-25:** From R. W. Christopherson, *Geosystems,* 5th ed. Upper Saddle River, NJ: Prentice Hall, 2003; **4-26:** © Andre Jenny / Alamy; **4-27:** ricardoreitmeyer/iStockphoto; **4-28:** Russell Munson; **4-29:** NASA

5-CO: TORSTEN BLACKWOOD/AFP/Getty Images; **5-01:** © Nouk | Dreamstime.com; **5-03:** © Bob Krist/CORBIS; **5-04:** Jeff Schmaltz, MODIS Rapid Response Team/NASA; **5-05:** © Steve Vidler / SuperStock; **5-08:** Data courtesy of the Population Reference Bureau; **5-09:** U.S. Census Bureau; **5-10:** U.S. Census Bureau; **5-12:** Population Reference Bureau; **5-13:** © Jose Luis Pelaez, Inc./CORBIS; **5-14:** Population Reference Bureau;

INDEX

Note: Page numbers followed by f refer to Figures

POPULATION DENSITY
(PERSONS PER
SQUARE KILOMETER)

1,000 and above
250–999
25–249
5–24
Below 5

ARCTIC OCEAN

ATLANTIC
OCEAN

PACIFIC
OCEAN

Tropic of Cancer

Equator

Tropic of Capricorn

Antarctic Circle

80°

60°

40°

20°

0°

20°

40°

60°

80°

160° 140° 120° 100° 80° 60° 40°

0   1,000   2,000 MILES
0   1,000   2,000 KILOMETERS

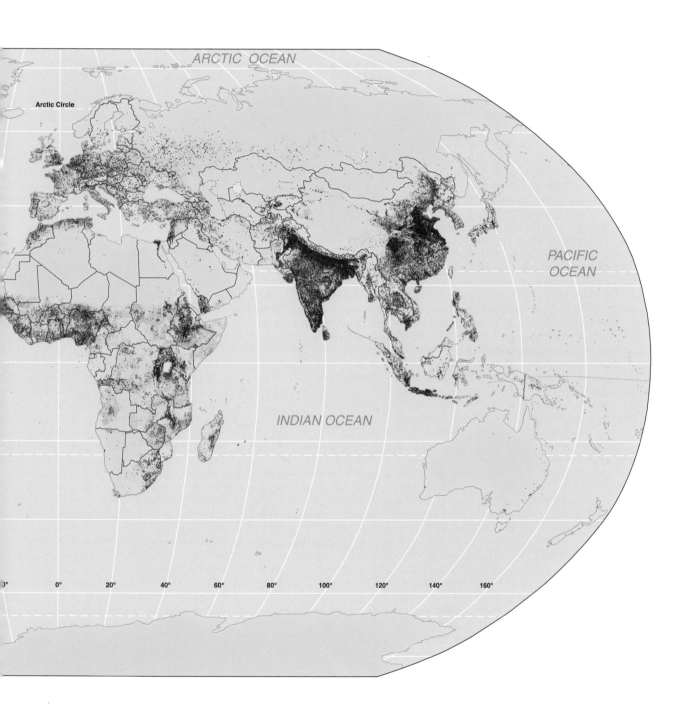

ARCTIC OCEAN

Arctic Circle

PACIFIC
OCEAN

INDIAN OCEAN

0°   20°   40°   60°   80°   100°   120°   140°   160°